TABLE 6

AREAS UNDER THE NORMAL CURVE
VALUES OF $A(z)$ BETWEEN ORDINATE AT MEAN (Y_o) AND ORDINATE AT z

Y_0

.19847

0 0.52 z

Example:
$z = 0.52$ (or -0.52),
$A(z) = 0.19847$ or 19.847%

$z\left(=\dfrac{x}{\sigma}\right)$	.00	.01	.02	.03	.04	.05	.06	.07	.08	.09
0.0	.00000	.00399	.00798	.01197	.01595	.01994	.02392	.02790	.03188	.03586
0.1	.03983	.04380	.04776	.05172	.05567	.05962	.06356	.06749	.07142	.07535
0.2	.07926	.08317	.08706	.09095	.09483	.09871	.10257	.10642	.11026	.11409
0.3	.11791	.12172	.12552	.12930	.13307	.13683	.14058	.14431	.14803	.15173
0.4	.15542	.15910	.16276	.16640	.17003	.17364	.17724	.18082	.18439	.18793
0.5	.19146	.19497	.19847	.20194	.20540	.20884	.21226	.21566	.21904	.22240
0.6	.22575	.22907	.23237	.23565	.23891	.24215	.24537	.24857	.25175	.25490
0.7	.25804	.26115	.26424	.26730	.27035	.27337	.27637	.27935	.28230	.28524
0.8	.28814	.29103	.29389	.29673	.29955	.30234	.30511	.30785	.31057	.31327
0.9	.31594	.31859	.32121	.32381	.32639	.32894	.33147	.33398	.33646	.33891
1.0	.34134	.34375	.34614	.34850	.35083	.35314	.35543	.35769	.35993	.36214
1.1	.36433	.36650	.36864	.37076	.37286	.37493	.37698	.37900	.38100	.38298
1.2	.38493	.38686	.38877	.39065	.39251	.39435	.39617	.39796	.39973	.40147
1.3	.40320	.40490	.40658	.40824	.40988	.41149	.41309	.41466	.41621	.41774
1.4	.41924	.42073	.42220	.42364	.42507	.42647	.42786	.42922	.43056	.43189
1.5	.43319	.43448	.43574	.43699	.43822	.43943	.44062	.44179	.44295	.44408
1.6	.44520	.44630	.44738	.44845	.44950	.45053	.45154	.45254	.45352	.45449
1.7	.45543	.45637	.45728	.45818	.45907	.45994	.46080	.46164	.46246	.46327
1.8	.46407	.46485	.46562	.46638	.46712	.46784	.46856	.46926	.46995	.47062
1.9	.47128	.47193	.47257	.47320	.47381	.47441	.47500	.47558	.47615	.47670
2.0	.47725	.47778	.47831	.47882	.47932	.47982	.48030	.48077	.48124	.48169
2.1	.48214	.48257	.48300	.48341	.48382	.48422	.48461	.48500	.48537	.48574
2.2	.48610	.48645	.48679	.48713	.48745	.48778	.48809	.48840	.48870	.48899
2.3	.48928	.48956	.48983	.49010	.49036	.49061	.49086	.49111	.49134	.49158
2.4	.49180	.49202	.49224	.49245	.49266	.49286	.49305	.49324	.49343	.49361
2.5	.49379	.49396	.49413	.49430	.49446	.49461	.49477	.49492	.49506	.49520
2.6	.49534	.49547	.49560	.49573	.49585	.49598	.49609	.49621	.49632	.49643
2.7	.49653	.49664	.49674	.49683	.49693	.49702	.49711	.49720	.49728	.49736
2.8	.49744	.49752	.49760	.49767	.49774	.49781	.49788	.49795	.49801	.49807
2.9	.49813	.49819	.49825	.49831	.49386	.49841	.49846	.49851	.49856	.49861
3.0	.49865	.49869	.49874	.49878	.49882	.49886	.49889	.49893	.49897	.49900
3.1	.49903	.49906	.49910	.49913	.49916	.49918	.49921	.49924	.49926	.49929
3.2	.49931	.49934	.49936	.49938	.49940	.49942	.49944	.49946	.49948	.49950
3.3	.49952	.49953	.49955	.49957	.49958	.49960	.49961	.49962	.49964	.49965
3.4	.49966	.49968	.49969	.49970	.49971	.49972	.49973	.49974	.49975	.49976
3.5	.49977	.49978	.49978	.49979	.49980	.49981	.49981	.49982	.4	
3.6	.49984	.49985	.49985	.49986	.49986	.49987	.49987	.49988	.4	
3.7	.49989	.49990	.49990	.49990	.49991	.49991	.49992	.49992	.4	
3.8	.49993	.49993	.49993	.49994	.49994	.49994	.49994	.49995	.	
3.9	.49995	.49995	.49996	.49996	.49996	.49996	.49996	.49996		
4.0	.49997									

Statistics for Business and Economics

Third Edition

Stephen P. Shao
Old Dominion University

Charles E. Merrill Publishing Company
A Bell & Howell Company
Columbus, Ohio

16.50
gift

To my brothers:
    Peter Pinkang
    Gordon Pinsan
    Pin Tsung

This book was set in Times New Roman and Kabel Shaded. The production editors were Marilyn Neyman and Laura Harder. The cover was designed by Will Chenoweth.

Library of Congress Catalog Card Number:
76-294

International Standard Book Number:
0-675-08640-X

Printed in the United States of America

1 2 3 4 5 6 7 8 —80 79 78 77 76

Preface

The third edition of *Statistics for Business and Economics,* like the two previous editions, is designed primarily for the first course in statistics for students of business and economics. However, the new edition has up-dated and reorganized many illustrations and problems appearing in the previous editions.

The application of statistical methods to problems in business and economics has increased continuously during recent years. Therefore, general topics in traditional and modern statistics have been carefully selected to deal with these problems. In addition, some advanced topics are also included as optional material to make this book more comprehensive. The scope of *Statistics for Business and Economics* will make it flexible enough to be used for a basic course, a second course in statistics at undergraduate level, or an accelerated course for graduate students. Note that the optional chapters and sections are marked by a star (★) for easy recognition. The omission of the starred material will not interrupt the continuity of the text organization.

Experience shows that students of business and economics have varying backgrounds in mathematics; some are weak and some are strong. In this book, discussions of complicated topics, such as those covered in probability distributions, sampling studies, and correlation analysis, begin with very elementary concepts. Whenever possible, simple sentences are used to present text discussions; formulas are derived intuitively and are presented in their simplest forms. Mathematical proofs of some complicated formulas are placed in footnotes for those students who wish to have further understanding of the formula derivations.

Special terms, such as "unbiased estimate" and "number of degrees of freedom" given in Chapter 13, are clearly defined and explained before they are used. After basic principles are presented, detailed examples are used to support the discussion of the principles. The examples are distinctly separated from discussion material in print so that students can easily follow the procedures of solving problems. In addition, diagrams and charts are frequently employed as aids to illustrate the problems. By this approach, it is hoped that students who have weak mathematical backgrounds will find the book suitable to their capabilities; while on the other hand, those who have stronger backgrounds will find challenging material throughout the book, since it includes many topics which require analytical thinking.

Text discussions in each chapter are followed by exercises including questions and problems. Computed answers to odd-numbered problems are placed in the Appendix. Detailed solutions to all problems are provided in the instructor's manual.

To review the material in the book, a summary of the discussion and formulas is placed at the end of each chapter. Reference pages for the formulas in each summary are provided for convenience in locating the definition of each symbol in the formulas. Detailed tables for general purposes are located in the Appendix. Simplified tables, which are condensed from the tables in the Appendix (such as the table for areas under the normal curve and the t-distribution table), are used for illustrative purposes and are placed in the text discussion for easy reference.

This book is divided into seven parts of four chapters each. The material covered in the first two parts is basic to all topics in the later parts. Part 1 presents the material for beginning statistical study. It includes the basic aspects of statistics, the methods of collecting, organizing, and presenting statistical data for analysis, and reviews of basic arithmetic and algebraic operations. If the instructor finds that his students are well prepared in mathematics, he may, of course, omit the fourth chapter or simply assign it to the students as outside reading material for review. Part 2 covers methods for simple statistical analysis. The important topics discussed in this part are average, dispersion, skewness, and kurtosis.

Parts 3, 4, and 5 deal with the increasingly important topics of statistical induction and decision theory. Part 3 provides basic tools—probability and various types of probability distributions. Part 4 illustrates the process and application of statistical induction by the use of normal (z) and student's (t) distributions. It also introduces the application of statistical induction in quality control. Part 5 explains various types of nonparametric tests, including chi-square (X_2) tests, variance ratio (F) distributions, and decision theories.

Statistical methods included in the last two parts are closely related, especially in the applications of the method of least squares. Part 6 presents time series analysis, including the construction of index numbers, secular trend, seasonal variation, cyclical fluctuation, and irregular movement. Part 7 discusses mainly the methods used in regression and in correlation analysis.

Details concerning the assignments for various types of courses are given in the instructor's manual. The following diagram shows the chapters and the starred sections in the text organization which may be used as a basis for making brief course outlines. (Chapter numbers are circled in the diagram. The starred chapters and sections, as mentioned before, are optional material for a basic course, whether a one-semester or one-year course.)

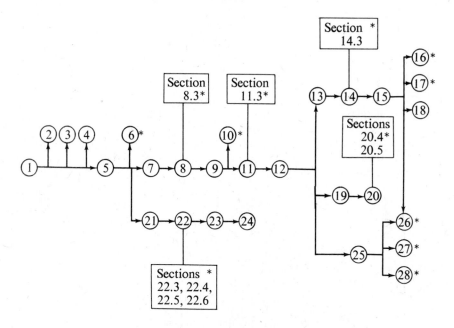

The following suggested course outlines are based on the diagram above.

One-year (or two-semester) basic course—Chapters 1, 2, 3, 4, 5, 7, 8, 9, 11, 12, 13, 14, 15, 16, 19, 20, 21, 22, 23, 24, and 25. (Total: 21 chapters. Chapters 2 and 3 may be assigned as outside reading material. Chapters 4 and 9 may be omitted if students have adequate mathematical backgrounds.)

One-semester basic course—Chapters 1, 3, 5, 7, 11, 12, 13, 14, 21, 22, and 25. (Total: 11 chapters.)

One-semester advanced course—Chapter 6, Section 8.3, Chapter 10, Sections 11.3 and 14.3, Chapters 17 and 18, Sections 20.4, 20.5, 22.3, 22.4, 22.5 and 22.6, and Chapters 26, 27, and 28. (Total: 7 chapters and 9 sections.) Additional reviewing material may be selected from the one-year course plan as outlined above.

I am indebted to the literary executor of the late Sir Ronald A. Fisher, F. R. S., Cambridge, to Dr. Frank Yates, F. R. S., Rothamsted, and to Messrs. Oliver & Boyd, Ltd., Edinburgh, for permission to reprint Tables III and IV from their book *Statistical Tables for Biological, Agricultural and Medical Research*.

I am also indebted to my colleagues of the School of Business Administration of Old Dominion University for their encouragement and suggestions made from reading and teaching the first edition of this book. Above all, I am deeply grateful to my late wife, Betty Outen Shao, for her expert services in editing the manuscript of the first edition of this book.

Stephen Pinyee Shao
Virginia Beach, Va.

Contents

*Each chapter (or section) marked by a large boldface star (★) is optional material. It may be omitted without interrupting the continuity of the text organization.

Contents

Contents

part 1

Beginning Statistical Study

Most of this text is devoted to the methods of statistical analysis. The first part in this text has four chapters. These chapters will provide the groundwork for statistical analysis. The first chapter introduces the basic concept of statistics and the first step in statistical study—collection of statistical data. Chapter 2 explains the second step—organization of statistical data. Chapter 3 illustrates the third step—presentation of statistical data. Chapter 4 gives a review of the basic mathematics. All topics included in these four chapters are essential for statistical analysis.

1 Introduction

The basic concept of statistics and the first step in statistical study—collection of statistical data—are introduced in this chapter. The nature of statistics is explained in the first section, Section 1.1. Statistical data provide the necessary information for statistical study. If the data are inadequate, the conclusion drawn from them obviously is not valid. The concept of statistical data is discussed in Section 1.2. Statistical data obtained in business and economic activities are presented in Section 1.3. The methods used in a statistical study and the areas covered in this text are outlined in Section 1.4. The procedures of collecting data from published sources and from surveys are included in Sections 1.5 and 1.6, respectively.

1.1 The Nature of Statistics

The word *statistics* has double meanings. It has been referred to as numerical or quantitative information. It also has been referred to as the methods of dealing with that information. However, statisticians prefer to call the information *statistical data* and the methods *statistical methods*.

When a reader has few numerical facts, he may utilize numerical information to the maximum extent without spending much time or thought in analyzing the facts. Examine the statement:

John is 22 years old, and Mary is 18 years old.

A reader may easily interpret the information in many different ways. For example, John is a young man 22 years of age, but he is four years older than Mary. However, when the reader has a large volume of numerical facts, he may find that the information is of little value to him, since he cannot interpret it all at one time. Note the following statement:

> John is 22 years old, Mary is 18 years old, Jack is 25 years old, Jean is 16 years old, and so on for 1,000 selected students in the Swan College on October 1, 1976.

A reader will certainly have a hard time in interpreting the distribution of ages intelligently.

The large volume of numerical information gives rise to a need for systematic methods which can be used to organize, present, analyze, and interpret the information effectively. Thus valid conclusions can be drawn and reasonable decisions can be made from the use of these methods. Statistical methods are primarily developed to meet this need.

1.2 Statistical Data

Abundant quantitative or numerical information may be found in almost every type of activity in our daily living. For example, the price tag on a hat is shown in a *number* of dollars, the employment condition in a nation is expressed in a *number* of persons, the enrollment of a college is recorded in a *number* of students, the distance traveled by a salesman is reported in a *number* of miles, and the age of a person is represented by a *number* of years. However, not all quantitative information is regarded as statistical data. Quantitative information suitable for statistical analysis must be a set (or sets) of numbers showing *significant relationships*. In other words, statistical data are numbers that can be compared, analyzed, and interpreted. A single number which will not compare with or which shows no significant relationship to another number is not statistical.

In the example above, John's age alone is not statistical if there is no other age available for comparison. However, the ages of 1,000 students are statistical data, since the ages can be compared and analyzed, and the findings of the analysis can be interpreted. Also, the so-called "statistics" of a patient as measured by a doctor are not statistical data, since each measurement, such as the patient's height, shows no significant relationship to other measurements, such as the number of heart beats per minute or the eyesight measurement of the patient. However, information regarding the heights of all patients seen within a certain period of time is statistical, since the heights can be compared, analyzed, and interpreted according to their relationships.

The area from which statistical data are collected is generally referred to as the *population* or *universe*. A population can be *finite* or *infinite*. A finite population has a limited number of individuals or objects, whereas an infinite population has an unlimited number. For example, an English class of 25 students is a finite population. The number of college students in the United States during the past, present, and future is unlimited; therefore, such students form an infinite population.

The task of collecting a complete set of data from a small finite population is relatively simple. If we wish to obtain the ages of the 25 students in the English class, we may simply ask each student his age; thus we have a complete set of data. However, collecting such data from a large finite population is sometimes impossible or impractical. To collect a complete set of data concerning the ages of all students from the United States schools on October 1, 1976, for example, would be impractical, although it is possible, because of the time and the cost involved. Collection of the complete data from an infinite population is definitely impossible.

In order to avoid the impossible or impractical task, a *sample* consisting of a group of representative items is usually drawn from the population. The sample is then used for statistical study and the findings from the sample are used as the basis to describe, estimate, or predict the characteristics of the population. Assume that the 1,000 students presented above are representative students on the Swan College campus and are selected from the entire student body (population) of 30,000 students in 1976. The set of collected data concerning the ages of the 1,000 students is a sample, and a researcher may use these findings to estimate or predict the ages of all of the students in the college. The most common methods of collecting sample data are discussed in Section 1.6.

1.3   Statistical Data in Business and Economic Activities

Statistical study is important in business and economic areas since vast amounts of statistical data can be obtained in the two related areas.

The growing complexity of business activities in recent years has definitely increased the use of statistics for decision-making on every level of management. In earlier days, when business firms were small, owners of the firms were directly engaged in almost all of the areas of business activities. An owner of a small firm then might act as the store manager, producer, salesman, and sometimes as the janitor. He would make personal contacts with his customers and would know exactly what they wanted from him. As the size of business firms grew, owners were often separated from management because of special skills required for management. A manager may spend his entire time in planning, organizing, supervising, and controlling business operations. Personal contact with each of the thousands of customers has become impossible. Thus the manager of a modern business firm faces a much greater degree of uncertainty concerning future operations than he did when the size of the business was small. If he wishes to be successful in his decision-making, the manager must be able to deal systematically with the uncertainty itself by careful evaluation and application of statistical methods.

Business activities can be grouped into areas of sale, purchase, production, finance, personnel, accounting, and marketing research. Abundant quantitative information can be obtained for each of the major areas; applications of statistical methods are numerous. For example, statistical tables and charts are frequently used by sales managers to present numerical facts of sales; sampling methods are used by marketing researchers in making surveys of consumer preferences for certain brands of competitive merchandise; quality control methods such as those discussed in Part 4 of this text are increasingly applied in production activity.

Economics deals with production, distribution, and consumption of goods and services in a much broader realm than business activities. Economists are more concerned with economic activity as a whole, whereas business managers are primarily concerned with activities concerning individual business firms.

In economics, production is defined as the creation of utility or usefulness. For example, a factory manufacturing automobiles creates "form utility," a railroad transporting goods from one place to another creates "place utility," a bank lending money for a period of time creates "time utility," and a salesman transferring the ownership of goods or services from one person to another creates "possession utility." All of the activities, either providing goods or providing services, are examples of production. Distribution in economics usually refers to the distribution of incomes among the factors of production. The factors of production are (1) labor, (2) land, (3) capital, and (4) entrepreneurship. The income received by a laborer is called *wage*; by a land-owner, *rent*; by a capital investor, *interest*; and by an entrepreneur, or enterpriser who combines the other three factors of production to produce goods or services, *profit*. Consumption in economics refers to the final use of goods or services for the satisfaction of human beings' needs.

Everyone is directly concerned with the economy of his time. There is no doubt that statistical methods are indispensable tools in analyzing the numerous quantitative facts concerning economic activity.

1.4 Statistical Methods

Statistical methods can be divided into five basic steps according to the order of application in a statistical study: (*1*) *collection*, (*2*) *organization*, (*3*) *presentation*, (*4*) *analysis, and* (*5*) *interpretation*.

In the example above, if the Dean of Students of Swan College wishes to know the typical age group of the students in the college, he may first *collect* statistical data concerning the ages of a group of representative students in the college, say 1,000 students from the population of 30,000 students. (Proper size of a sample is discussed in detail in Chapter 13.) Second, he may *organize* the collected ages by classifying them into different age groups. Third, he may *present* the organized data in a tabular form such as that shown in Table 1–1. Fourth, he may *analyze* the ages presented in the table to get the desired information. For example, he may find that the typical age group of the students in the college is the age group "18 and under 20" since it has the highest number of students (600 students as shown in the table). Finally, the dean may *interpret* the finding from his analysis of the sample by pointing out that the typical ages of all of the students in the college are from 18 to under 20 years.

Strictly speaking, there should be no definite dividing lines separating the five basic steps. Some of the methods may be used in more than one step. In this example, the method of classifying the age groups used in organization is closely

TABLE 1-1

AGES OF 1000 SELECTED STUDENTS IN SWAN COLLEGE
OCTOBER 1, 1976

Ages	Number of male students	Number of female students	Total
Under 18	120	50	170
18 and under 20	500	100	600
20 and under 22	100	80	180
22 and above	30	20	50
Total	750	250	1,000

Source: Hypothetical data.

related to the methods used in the analysis. Actually, the classifications of the age groups are determined by the dean's intention of finding the type of information from the data in the analysis. However, the division does give us a logical order for studying the statistical methods.

Statistical methods form the major part of this text. The scope of the basic steps and the plan for presenting them in the text are further discussed below.

Collection—Collecting Statistical Data

Quantitative information supplies facts for solving business and economic problems. After a problem has been clearly defined and understood, facts that can be expressed quantitatively, if any are relevant, should be collected. According to locations of information, statistical data can be classified into two types: (1) internal data and (2) external data.

When the quantitative information is obtained within the organization that makes a statistical study, the information is called *internal data*. A business firm may maintain many kinds of quantitative information in various departments for a long period. The sales department may keep sales invoices. The personnel department may keep the individual records of employees. The production department may keep the daily records concerning the amount of raw material, direct labor, and manufacturing expenses used and the units of finished product produced. However, the far more common types of records kept in most business firms which can be used for a statistical study are the accounting records, such as employee earnings from a payroll, sales amounts from a sales journal, and cash receipts from a cash book.

When the information is obtained from the outside of the organization, it is called *external data*. External data are usually classified in two types by method of obtaining them: (1) published data and (2) survey of original data.

The most convenient and economic way of obtaining external statistical data is to find the information from material that is published by outsiders, such as federal, state, and city governments, corporations, trade and professional associations, banks, newspapers, magazines, research institutions, colleges, and other private publishers. When published data are not available for a particular study, a survey of original data, or first-hand collection, may become necessary. Collecting external data by survey is usually a costly, tedious, and time-consuming process. However, effective methods of survey have been developed to save money, energy, and time. The methods of collecting statistical data from publications and surveys are presented in the next two sections in this chapter.

Organization—Organizing Collected Data

Data collected from published sources are usually organized already. However, a large mass of figures that are collected from a survey frequently needs organization. The first step in organizing a group of data is *editing*. The collected data must be edited very carefully so that the omissions, inconsistencies, irrelevant answers, and wrong computations in the returns from a survey may be corrected or adjusted. The next step is *classifying*. The purpose of this step is to decide the proper classifications in which the edited items will be grouped. Proper classifications are important to statistical studies, since the successive steps—namely, presentation, analysis, and interpretation of the statistical data—are affected by given classifications. The last step is *tabulating*. Similar items are counted and recorded according to the proper classifications in this step. A detailed discussion concerning the organization of collected data is presented in Chapter 2.

Presentation—Presenting the Organized Data in an Easy-to-Read Form

After collected data are organized according to proper classifications, the data are ready for presentation. Data presented in an easy-to-read form can facilitate statistical analysis. There are three ways to present organized data: (1) word statement, (2) statistical table, and (3) statistical chart.

Word statement is convenient for presenting data including only a few items. When a large group of numbers is included in a set of data, word statement presentation becomes inefficient and burdensome, since the classifications, units of measurements, and other detailed explanations might have to be repeated many times in the statement. Statistical tables, such as Table 1–1, are usually appreciated by readers if they can be effectively constructed. A statistical chart or a graph is a pictorial device for presenting statistical data. A well-organized and vivid graphical presentation can help readers acquire much knowledge in a short period of time. However, a graph usually gives a reader only an approximate value of the facts. If an exact amount is desired, a statistical table is preferred. Detailed discussions concerning the construction of statistical tables and charts are presented in Chapter 3.

Analysis—Analyzing the Presented Data

The majority of the material in this text is devoted to the methods used in analyzing the presented data, mostly in a tabular form. Methods used in analyzing statistical data are numerous, ranging from simple observation of the data to complicated, sophisticated, and highly mathematical investigation. This text is designed to introduce the basic statistical methods applied in business and economics. Only the most generally used methods of statistical analysis are included in this text. They are grouped into five major areas: simple statistical analysis (Part 2 of the text), statistical induction (Parts 3, 4, and 5), time series analysis (Part 6), and relationship analysis (Part 7).

Part 2 provides the basic background for statistical analysis. It includes the methods of finding averages and dispersions which are necessary for anyone studying the material included in the later chapters of the text.

Parts 3, 4, and 5 of this text present the background and the applications of statistical induction. Statistical induction is also frequently called *inductive statistics* or *statistical inference*. It includes the methods of drawing inferences from sample data. To be specific, statistical induction includes the methods of generalizing, estimating, or predicting the characteristics of a population or universe based on a sample. Many statisticians frequently classify statistical methods as either *descriptive statistics* or *inductive statistics*. The term "descriptive statistics" is broadly applied to any statistical methods except the methods used in inductive statistics. In other words, the purpose of descriptive statistics is not to draw any conclusions or inferences about a population from the sample if the statistical study deals with sample data. Descriptive statistics includes the methods of collection, organization, presentation, analysis, and interpretation of a group of data, either sample data or complete information, without an attempt to make a *prediction* based on the data. The facts shown in Table 1–1 could be descriptive statistics. However, if the information shown in the table were used as a sample of 1,000 students drawn from a population of 30,000 students in Swan College *and* the findings from the sample were used to estimate or predict the ages of all students in the college, the process of estimation, usually based on probability theory (as presented in Chapter 9), is within the scope of inductive statistics.

Part 5 also introduces decision-making methods for problems under conditions of uncertainty. These methods are included in this part since they also deal with sample information.

Part 6 presents the analysis of time series. The subject of time series is important in studying the changes in business and economic activities within various time periods, such as variations in a specified number of years or in a period of less than a year. The federal government frequently publishes and analyzes quantitative information concerning such changes for policy-making or as a public service. Individual business firms also keep records chronologically for decision-making involving future operations such as in budgeting and forecasting.

Part 7 of this text presents techniques for analyzing the relationship between two or more sets of statistical data. Many business and economic activities are related to each other. The degree of relationship between them may be found through analysis. If a person investigates the information concerning one activity, it is possible for him to predict or estimate the future movements of another related activity.

Interpretation—Interpreting the Findings from the Analysis

After the analysis of the statistical data is complete, the findings from the analysis must be interpreted. Correct interpretation will lead to a valid conclusion of the study and thus can aid one in making decisions. In this respect, care should be directed to avoiding misuse of statistical data. One common way of misusing statistical data is the injection of personal bias. *Bias* in this case means that a user of data emphasizes the facts which support his predetermined position or views. Because bias is so common, statisticians are frequently degraded by slogans such as "There are three kinds of lies: lies, damn lies, and statistics," and "Figures don't lie, but liars figure."

There are two kinds of bias: conscious and unconscious. Both are prevalent in statistical analysis. Examples of conscious bias are numerous. An advertiser frequently uses statistics to prove that his product is much superior to the product of his competitor. A politician prefers to use statistics to support his point of view. Management and labor leaders may simultaneously place their respective statistical figures on the same bargaining table to show that their rejections or requests are justified.

It is almost impossible for unconscious bias to be completely absent in a statistical work. As long as human beings are involved, a completely objective attitude in attacking a problem is difficult to obtain, although ideally a scientist should have an open mind. A statistician should be aware of the fact that his interpretation of the findings from statistical analysis is influenced by his own experience, knowledge, and background concerning the given problem.

For convenience, the work of interpretation will follow each analysis in this text. No separate chapter is provided for interpreting findings.

1.5 Collecting Published Data

The quality of data collected from a published source can be evaluated in a more meaningful way if the kind of source is known. One can further obtain published data readily if he knows the published guides for locating the sources and the names of sources of published data.

Primary and Secondary Sources

The sources of published data can be classified into two kinds: primary and secondary. A source is *primary* when the data are obtained from the publication which is published by the original data collector. A source is *secondary* when the data are obtained from a publication which is published by an organization other than the original collector. For example, the United States Office of Education collects and publishes quantitative information concerning public schools each year. When the information is obtained from a publication of the Office of Education, it is from a primary source. The Newspaper Enterprise Association, Inc. republishes the information in its *World Almanac* annually. When the information is obtained from the republication, it is from a secondary source.

In general, data of primary sources are preferred over the data of secondary sources, since the former are more meaningful and reliable than the latter. Republications often omit some detailed information such as the explanation of certain terms and the methods of collecting data. In addition, it is possible for the data from a secondary source to contain typographical errors and other errors because of reclassification, conversion of units, and other revisions used for the purpose of republication. When the secondary sources are used, several republications should be checked to see if any errors are involved.

However, it is often more convenient to obtain data from secondary sources than from primary sources. Republications are more abundant than the original publications. Many organizations collect data from various primary sources and assemble them in a comprehensive volume for easy reference. The user thus may find a large quantity of desired information in one volume instead of checking many primary sources.

Guides for Locating Published Data

The volume of published data in the areas of business and economics is vast and comprehensive. It is almost impossible for any statistician to know all sources of the vast amount of published data. A reader may find that it is frequently profitable to consult the published *guides* in locating the desired information. Some of the well-known guides to published data are available in public and college libraries. Examples of some up-to-date guides to published data in the areas of business and economics are:

Readers' Guide to Periodical Literature, published by the H. W. Wilson Company, New York. This publication is the major current guide to general periodicals. Its indexing service began in 1900, with the first permanent cumulative volume appearing in 1905. It is issued semimonthly from September to June and monthly in July and August, with an annual cumulation appearing in one alphabet. At present, the guide indexes approximately 160 periodicals. Each article indexed is referred to by author and by subject.

Business Periodicals Index, published by the H. W. Wilson Company. This publication is a subject index to periodicals in various fields of business. It is published monthly except August and cumulated to annual volumes. It now indexes more than 150 periodicals. This index and the *Applied Science and Technology Index* succeeded the *Industrial Arts Index* in January, 1958. The *Industrial Arts Index* began its service in 1913 and indexed more than 200 periodicals through 1957.

Bureau of the Census Catalog, published quarterly and cumulative-to-annual, with monthly supplement, by the Bureau of the Census, Department of Commerce. This catalog contains publications by the bureau's divisions of Agriculture, Construction, Housing, Foreign Trade, Geography, Governments, Manufactures, Mineral Industries, Population, Retail Trade, Wholesale Trade, Selected Service Industries, and Population. It also contains statistical compendia and special publications, selected publications of other agencies, and selected technical papers by staff members.

Sources of Published Data

Instead of looking up the guides first, a student may wish to be familiar with the major sources of data so that he can find the desired information directly from the sources and thus save time. The following seven groups are the most important sources of published data concerning business and economic activities, although they are by no means a complete list of sources.

Government Agencies

Government agencies provide the majority of published statistical information available to readers. Nearly every level of local, state, and federal government agencies publishes some sort of statistical data concerning business and economic activities. However, the most important sources of the government published data are federal government agencies. The following list of the federal government publications will furnish a reader data concerning almost any area of interest related to the United States. Most of the sources, however, are secondary. If more detailed information is needed, footnotes or sources of the data are placed under individual tables in the publication; thus the reader can investigate the primary sources. The federal governmental publications can ordinarily be purchased from the issuing agency. Many popular government publications are also available through the Division of Public Documents, U. S. Government Printing Office, Washington, D. C. 20402.

Business Conditions Digest. This digest is published monthly by the Bureau of Economic Analysis, Department of Commerce. It brings together many of the economic time series found most useful by business analysts and forecasters. Its predecessor was *Business Cycle Developments*, which emphasized the cyclical indicators approach to the analysis of business conditions. (Cyclical movements are discussed in Chapter 24 in this text.)

Economic Indicators. This volume is prepared by the Council of Economic Advisers, but is published monthly by the Joint Economic Committee of the Congress. It is an important compilation of current economic statistics.

Economic Report of the President. This report is published annually by the Council of Economic Advisers, Executive Office of the President. It contains statistical data on price, employment, wages, production, purchasing power, finance, and other subjects related to economic conditions.

Federal Reserve Bulletin. This bulletin is published monthly by the Board of Governors of the Federal Reserve System. It contains detailed statistical data on member banks, bank debits, money supply, business loans, interest rates, security prices, consumer credit, and other United States financial and business statistics. It also shows international financial statistics and a list of Federal Reserve Board publications.

Monthly Labor Review. This review is published by the Bureau of Labor Statistics, Department of Labor. It contains current labor statistical data on employment, labor turnover, earnings and hours, consumer and wholesale prices, work stoppages, and work injuries.

Statistical Abstract of the United States. This volume has been published annually since 1878 by the Bureau of the Census, Department of Commerce. It summarizes statistics on the industrial, social, political, educational, and economic organization of the United States and comparative international statistics. It includes a comprehensive selection of data from most of the important statistical publications, both governmental and private, and an extensive guide to sources of statistics.

Survey of Current Business. This survey is published monthly with weekly supplement by the Office of Business Economics, Department of Commerce. It gives statistical data for the most recent 13 months on gross national product, national income, industrial production, commodity prices, domestic trade, employment and population, finance, international transactions of the United States, transportation and communication, and other economic and business figures.

Treasury Bulletin. This bulletin is published monthly by the Office of the Secretary, Treasury Department. It contains analytical material on fiscal operations and related Treasury activities, including financial statements of government corporations and business-type enterprises.

Trade Associations

Trade associations are organized by businessmen engaged in similar industries or occupations. These associations flourish in almost every field of business activity. The total number of national and local trade associations now is more than 12,000 associations as reported in the *Directory of National Trade Associations*, published every five years or so since the first in 1913 by the Department of Commerce. Many of the national associations have thousands of members and maintain central headquarters with large numbers of staff members. They frequently publish quantitative information concerning production, sales, prices, wages, employment, and other activities of the trade represented. The following annual statistical publications are among the most commonly used sources of data:

Air Transport Facts and Figures, Air Transport Association of America, Washington, D. C.

Gas Facts, American Gas Association, New York.

Railroad Car Facts, American Railway Car Institute, New York.

Bituminous Coal Annual, Bituminous Coal Institute, Washington, D. C.

Life Insurance Fact Book, Institute of Life Insurance, New York.

Magazines and Trade Periodicals

The publishers in this category are not the headquarters of trade associations but are either independent publishers or some of the major firms in a particular industry or profession. Magazines included here are only those whose chief concerns are in general business and economic activities. Examples of such magazines are:

Barron's, Barron's Publishing Company, Inc., New York (weekly).

Business Week, McGraw-Hill Publishing Company, Inc., New York (weekly).

Commercial and Financial Chronicle, William B. Dana Company, New York (weekly for statistical issue).

Dun's Review and Modern Industry, Dun & Bradstreet, Inc., New York (monthly).

Trade periodicals are those concerning particular trades. Examples of the trade periodicals are:

Advertising Age, Advertising Publishing Company, Chicago (weekly).

Iron Age, Chilton Company, Philadelphia (weekly).

Printers' Ink, Printers' Ink Publishing Corporation, New York (weekly).

Railway Age, Simmons-Boardman Publishing Corporation, New York (weekly).

Steel, Penton Publishing Company, Cleveland (weekly).

Newspapers and Almanacs

Many newspapers give extensive daily reports on prices of stocks and bonds, sales and productions of certain commodities, and other useful statistical information. Almanacs, which are published mostly by large newspapers, also include important business and economic statistical data. Examples of such newspapers and almanacs are:

Wall Street Journal, Dow Jones and Company, Inc., New York (daily except Saturdays, Sundays, and holidays).

New York Times, New York Times Company, New York (daily).

The World Almanac, Newspaper Enterprise Association, Inc., New York (annual).

The Economic Almanac, National Industrial Conference Board, New York (annual).

Private Statistical Services Organizations

Organizations of this type publish statistical data for the purpose of serving their subscribers. The organizations not only supply statistical data but also frequently give analyses, interpretations of the data, forecasts, and recommendations concerning financial or investment problems. The "Moody's bond ratings," for example, indicate investment risks on bonds from the least investment risk (i.e., highest investment quality) to greatest investment risk (i.e., lowest investment quality) in the *Moody's Manuals of Investment* (see below). Examples of the publications by such services are:

Trade and Securities Statistics, published by the Standard and Poor's Corporation, New York. This publication is issued monthly in a loose-leaf form and the monthly issues are combined in an annual volume. It contains detailed information concerning all important corporations, industries, and general business conditions. This organization also publishes the weekly issue *Outlook* to supply additional information concerning investments.

Moody's Manuals of Investment, published annually by the Moody's Investors Service, Inc., New York. There are five manuals each year: (1) *Industrial*, (2) *Transportation*, (3) *Public Utility*, (4) *Bank and Finance*, and (5) *Municipal and Government*. Current material concerning the manuals is issued semiweekly in loose-leaf binders. In addition, the Service publishes weekly *Bond Survey* and *Stock Survey*.

United Business Service, published weekly by United Business Service Company, Boston. It contains analyses of investments, industries, and general business conditions, with forecasts and recommendations to investment problems.

The Value Line Investment Survey, published weekly by Arnold Bernhard and Company, Inc., New York. It reviews approximately 1,100 stocks and gives a summary of advices and an index of the stocks. The qualities of the stocks are graded in the survey.

International Organizations

Many international organizations make foreign and international statistical data available to readers. Examples are:

Monthly Bulletin of Statistics, Statistical Yearbook, Yearbook of International Trade Statistics, and *Yearbook of National Accounts*, published by Statistical Office, United Nations, New York.

International Financial Statistics (monthly), *Annual Report, Balance of Payments Yearbook*, published by International Monetary Fund, Washington, D. C.

Other Business and Educational Organizations

There are many other business and educational organizations which publish useful statistical information. Examples of such organizations are chambers of commerce, individual corporations, banks, and business research bureaus of colleges and universities. Frequently unpublished reports are available from this type of organization.

1.6 Collecting Survey Data

When the internal records and the available published data are not suitable for a particular study, a survey of original data may have to be conducted to meet the need. The work of making a survey is usually limited by time, money, and manpower available for the study. Instead of collecting complete information concerning the study, a sample consisting of a group of representative items is usually drawn from the source of information (population) in a survey. Details concerning the types of samples, sampling designs, and sampling theories are presented in Parts 3, 4, and 5 of this text. In this section only the most common methods of collecting sample data are discussed. The methods are *direct observation* and *asking questions*.

Direct Observation

Statistical data may be obtained by direct observation. For example, in order to decide the hours during which a street will be used for one-way traffic, the traffic officer may actually observe the traffic movement on the street. He may go to the street to obtain a sample of traffic movement by counting the number of cars, trucks, buses, and other vehicles moving on the street during different hours. If a researcher wishes to know the retail food prices in a city, he may go to a group of selected food stores to observe the price tags on the food counters. In recent years, many different types of mechanical devices have been developed for observations. For example, the A. C. Nielsen Company installed *audimeters* on television sets as a recording device to record on tape the time that the sets are turned on and off and the station to which they are tuned.

The observation method can give accurate information and is usually preferred when it can be used effectively and economically. However, it is limited to a few types of studies and is often too inconvenient in actual observation of certain operations. For example, if we wish to know the income received in a week by a group of taxicab drivers, it would be very inconvenient to observe them. It would be more practical and easier to find the result by asking them certain questions.

Asking Questions

There are three ways to ask questions in order to collect original data: (1) personal interview, (2) telephone, (3) mail questionnaire. Each of the ways has its advantages and disadvantages. The decision of which one of the three ways should be used in a particular study must be made according to the amount of time, money, and manpower allowed and the degree of accuracy needed for this study.

Personal interviews usually result in immediate responses from the people being questioned. This method also results in more accurate answers than other methods, since personal contact during an interview provides an opportunity for explaining some points which are not clearly stated. Frequently, a recipient of a mail questionnaire does not answer the questions until it is convenient for him. When he answers the questions, he answers them according to his own interpretation. Recipients may completely ignore the questionnaires if they can not give favorable answers or if the questions are not clear to them.

The method of asking questions by mail has the chief advantages of saving money and manpower and of covering a larger geographical area. The cost of mailing a group of letters is much cheaper than that of sending persons to make personal interviews. When the sources of data are spread over a large area, the method of questioning by mail may be the best method in terms of convenience as well as economy.

The method of questioning by telephone has been used increasingly in recent years for some simple studies within a single locality. The chief advantage of this method is that a researcher may hire a small number of telephone operators at a low cost and can reach a large group of people within a short period of time. However, the telephone subscribers may not be a representative group of the entire population being sampled. Thus the value of the sample obtained may be doubtful. In addition, only a small amount of information can be obtained through a telephone conversation. A long list of questions or questions involving technical terms usually can not be answered effectively over a telephone.

Before a large-scale survey is begun, a good practice is to test the drafted questions with a small group of persons who are representative of the population being surveyed. The *pretest* will give opportunities to discover any oversight in the draft, to improve the final form of the questionnaires, and to gain some valuable experience in interview techniques; thus a better result can be obtained.

Exercises 1

1. What are the meanings of the word *statistics*? State the reason for developing statistical methods.

2. Give one of your own examples to illustrate that a group of numbers is statistical data. Find a group of numbers which are not suitable for statistical analysis.

3. What is a finite population? An infinite population? Why would you use a sample instead of a complete set of data including all of the cases in a population?

4. Why are statistical methods useful in business? In economics? Can statistical methods be applied in areas other than business and economics? If so, give an example to illustrate.

5. What are statistical methods? Describe briefly the basic steps which are used to group such methods. Give your own example to illustrate.

6. What information, other than that illustrated in this chapter, can you find from your analysis of Table 1-1? Interpret your findings. Do not refer to future chapters.

7. What are internal data and external data? Give examples to illustrate.

8. What are the basic steps in organizing data collected from a survey? Why are these steps required?

9. What are the three ways to present organized data? State the advantages and disadvantages in using each of the ways.

10. State the difference between inductive statistics and descriptive statistics. Give your own example to illustrate the difference.

11. State the difference between primary and secondary sources of published data. Data of primary sources are preferred over the data of secondary sources because the former are more meaningful and reliable than the latter. Why, then, should we ever use secondary sources?

12.   Why are guides for published data important to a statistical study? If a study concerning the future of the steel industry is assigned to you, how would you collect information to support your study?

13.   Name the major publications published by the following governmental agencies: (a) Department of Commerce, (b) Board of Governors of the Federal Reserve System, (c) Treasury Department, (d) Department of Labor, (e) Executive Office of the President, and (f) the Joint Economic Committee of the Congress.

14.   State the similarity and the difference between trade association publications and trade periodicals. Give an example to illustrate.

15.   What is the difference between magazines and the publications by private statistical services organizations? If you were making a study concerning investment problems, which one of the two sources do you believe would give greater assistance?

16.   If you are interested in statistical data concerning international or foreign financial problems, what sources would you check for such information?

17.   How would you make a survey by the direct observation method? Give an example to illustrate.

18.   Which one of the three ways is the most effective way to ask questions in making a survey? Why?

2 Organizing Statistical Data

Organizing statistical data is the second step in a statistical study. The problems of editing are discussed in Section 2.1. The bases of classifying statistical data are presented in Section 2.2. There are various methods of tabulating statistical data. The methods by manual devices are explained in Section 2.3. The methods by punched cards and computer systems are given in Sections 2.4 and 2.5 respectively.

2.1 Editing Collected Data

Editing is frequently needed in data organization. The work of editing data collected from internal records and published sources is relatively simple. However, data collected from a survey may need extensive editing. Answers from a survey recorded by observers, interviewers, or respondents frequently are not presented in a manner so that they can be readily classified or tabulated. An editor may find one or several of the following things that should be corrected or otherwise dealt with:

1. *Answers are inconsistent.* When one answer conflicts with another answer, the reason for the inconsistency should be discovered. Requestioning the respondent may be necessary. For example, a respondent stated in 1976 that he had been living continuously in a certain house in City X for 15 years but in another question stated that he had moved in the city in 1966. Since there are only 10 years from 1966 to 1976, the discrepancy should be corrected after the respondent is requestioned.

2. *Writings are not determinable.* The meaning of an answer cannot be determined if the writing is too poor to read or the check mark is not placed properly. When an answer is not determinable, it should be verified or may simply be discarded.

3. *Answers are incomplete.* When an unanswered question is discovered, an effort should be made to obtain an answer. If it is inconvenient or not worth a requestioning, the words "not reported" should be filled in the space for the answer.

4. *Computations are required.* If the figures are reported in detail, computations should be completed by the editor. For example, if respondents reported cigarette consumptions in various units of time, such as five packages a week, two packages a day, and ten cartons a month, the consumptions should be computed to the same basis, such as the unit being a year. If the computation has already been completed by a respondent, the computation should be verified.

2.2 Classifying Edited Data

The work of classifying statistical data collected from a survey and that from internal records is basically the same. Collections of published data are usually in classified forms, although in some cases reclassifications are necessary in order to suit a particular study. Generally speaking, when a group of collected data consists of only a few items, there may not be a need for classification. However, when a large number of figures is collected, the figures must be classified if they are to be very meaningful to a reader.

There are many ways to classify statistical data. In general, classifications may be determined according to four bases: time, place, quantity, and quality. The classified data should be arranged in an orderly manner so that they can be used more effectively by a reader in making analysis and comparison. The most common methods of arranging the order of the classifications according to the four bases are given below.

Time

Classifications of statistical data are frequently based on time intervals, such as years, months, weeks, and days. Data classified by time intervals are usually arranged in chronological order, either beginning with the earliest period or with the latest period. In general, the former arrangement, beginning with the earliest period, is preferred. However, if the recent events are to be emphasized, the latter arrangement is used more frequently. For example, sales amounts of a department store may be classified by years as follows:

Year (class)	Sales Amounts
1974	$1,350,000
1975	1,600,000
1976	1,840,000
1977	2,190,000
1978	2,450,000

Place

Geographical areas such as continents, countries, states, cities, and other regions, are logical classifications for many types of data. The areas may be arranged in alphabetical order or according to the importance of certain areas. In some cases, the classifications are listed in a traditional established order. For example, the sixteen branches of a company may be classified by location in traditional order as follows:

Location of Branches (class)	Number of Branches
Eastern region	3
Southern region	4
Western region	7
Northern region	2

Quantity

All types of quantitative information can be classified according to the magnitude of the figures included in the information. Classifications based on quantity are usually arranged in either ascending order (from small to large) or descending order (from large to small). The grade points of twenty students in an English class, for example, may be classified as follows:

Grade Points (class)	Number of Students
0 and under 20	2
20 and under 40	1
40 and under 60	5
60 and under 80	8
80 and under 100	4

More detailed discussion concerning classifications by quantity is presented in Section 5.1, "Frequency Distribution."

Quality

Statistical data may be classified by any dominant characteristic. Classifications based on quality are usually arranged in order of importance. However, if the importance of individual classes is not to be emphasized, alphabetical order is preferred for easy reference. In some cases, the order may have been established by tradition or custom. For example, shareowners of a company may be classified by amount of education as follows (order of importance):

Education (class)	Number of Shareowners
Attended high school or less	5,000
Graduated from high school	6,000
Attended college	7,500
Graduated from college	8,500

Average hourly earnings in the United States during a given period may be classified by kinds of industries (alphabetical order):

Industry (class)	Average Hourly Earnings
Contract construction	$7.13
Manufacturing	4.64
Mining	5.63
Transportation and public utilities	5.72
Wholesale and retail trade	3.65

2.3 Tabulation
by Manual Devices

After the proper or desired classifications have been chosen, the next step in organizing is to arrange the mass of quantitative facts in summary form based on the classifications. This process is called *tabulation*. The principal methods of tabulation can be grouped into three categories: (1) manual devices, (2) punched cards, and (3) computer systems. The common manual devices are (a) manual written cards, (b) tally sheet, and (c) rod-sorting cards. They are presented in this section. The methods in other categories are discussed in the next two sections.

Manual Written Cards

This method is usually used in tabulating simple facts that can be written on one card for each case or respondent. The cards in a deck representing all facts in the collection are sorted into piles which are placed separately according to different classes. The number of cards in each pile is then recorded and summarized.

For example, if an English professor wishes to find out the number of his freshman students by major fields and sexes during the period from 1974 to 1976, he gathers the following information from his record of each student on a card:

Name of the student _____
Major field _____
Sex _____

The cards are first sorted into different piles representing individual fields. The cards in each pile are then sorted into two subpiles representing male and female students, respectively. The cards in each pile or subpile are counted and recorded. The results of the counting are summarized in Table 2–1.

TABLE 2-1

NUMBER OF STUDENTS IN FRESHMAN ENGLISH CLASSES, 1974–1976
BY MAJOR FIELDS AND SEXES

Major Fields	Number of Students (cards in first sorting)	Number of Students (cards in second sorting)	
		Men	Women
Business	20	12	8
Education	16	6	10
Engineering	17	15	2
English	10	7	3
Social studies	15	10	5
Science	14	8	6
Others	4	2	2
Total	96	60	36

Tally Sheet

A tally sheet usually provides spaces for classifications, tally marks, and totaling the number of tallies. Each of the facts in the collection is recorded on the tally sheet by entering a tally mark (/) in the proper space. After all the facts are recorded, the tallies are counted and the total number of tallies is then entered in the total column immediately following the tallies.

For example, in the illustration above, instead of sorting the cards into different piles or subpiles, the information on the professor's record may be transferred directly to the tally sheet by using tally marks as in Table 2–2. The tally marks are separated into groups of five to facilitate counting.

TABLE 2-2

NUMBER OF STUDENTS IN FRESHMAN ENGLISH CLASSES, 1974–1976
BY MAJOR FIELDS AND SEXES

Major Field	Number of Male Students		Number of Female Students		Total Students
	Tally	Total	Tally	Total	
Business	卌卌//	12	卌///	8	20
Education	卌/	6	卌卌	10	16
Engineering	卌卌卌	15	//	2	17
English	卌//	7	///	3	10
Social studies	卌卌	10	卌	5	15
Science	卌///	8	卌/	6	14
Others	//	2	//	2	4
Total		60		36	96

Rod-sorting Cards

Rod-sorting cards are useful for small scale studies. The cards, such as the Keysort card manufactured by Royal McBee Corporation shown in Figure 2–1, have holes around the edges. The holes are numbered. The numbers against individual holes are used to code the collected facts, which are usually written on the center of the card. Notches are cut on each card through the holes above the numbers representing the fact to be sorted from other facts. When a rod is inserted in the holes at a chosen location of the cards in a group filed in a cabinet or a box and is then lifted, all cards with a notch in the location will drop down. The dropped card are counted and the results of counting are then summarized in a tabular form.

For example, a survey concerning the smoking habits among students of different classes is conducted on a college campus. When a freshman is interviewed, a notch is cut through a hole above number 4, which is the code number representing the freshman class (see Figure 2–2). Later, if we wish to find out how many freshmen have been interviewed in the survey, a rod is simply inserted through the hole above 4. When the rod is lifted, the cards in the group with no notches in the location above 4 will hang on the rod, while the cards with notches will drop down. The dropped cards, which represent individual freshmen, are counted and the results of counting are recorded.

The notches represent the license tag number 46280.

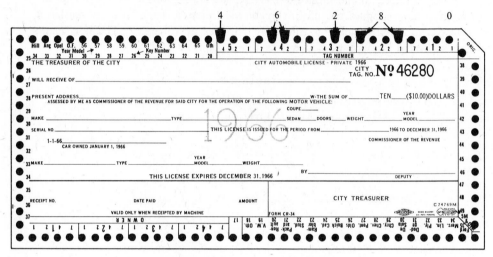

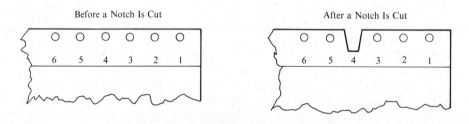

Punch   Sort

FIGURE 2-1

ROD-SORTING CARD AND OPERATING EQUIPMENT

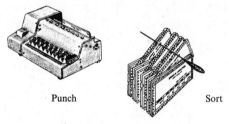

Before a Notch Is Cut   After a Notch Is Cut

FIGURE 2-2

DETAIL OF A ROD-SORTING CARD

2.4   Tabulation
by Punched Cards

Cards with punched holes have been used popularly in many different capacities by various types of organizations, such as industrial units, government agencies, and educational institutions. Sales invoices, paychecks, coupons for advertising, wage statements, inventory records, and permits to attend classes are frequently issued by the organizations in the form of punched cards.

Punched cards can be used in tabulation by both the method of punched cards and the method of computer systems. However, the facilities used in tabulating data by the two methods are different. The method of punched cards for tabulating statistical data basically involves two types of facilities: the cards and the machines used to operate the cards.

The Punched Card

The punched card is a piece of hard-surfaced paper on which information is recorded by punching holes in proper positions arranged on the card. Once the punching is completed, the punched cards become a permanent record which may be processed at machine speeds to obtain desired results.

The International Business Machines Corporation (IBM) is one of the major manufacturers of punched card equipment in the United States. An IBM card has 80 columns and the holes in the card are punched in rectangular form. There are other types of cards, such as the 96-column cards produced by IBM and Burroughs with holes punched in round form. Since all types of cards serve the same functions, only the uses of the 80-column IBM cards, which are widely used on the market today, are discussed here.

An IBM card is $7\frac{3}{8}$ inches long, $3\frac{1}{4}$ inches wide, and 0.007 inches thick. (See Figure 2–3.) The eighty vertical columns are numbered one to eighty from the left side of the card to the right. Each column is divided into twelve punching positions: 12, 11 (or X), 0, 1, 2, 3, 4, 5, 6, 7, 8, and 9, designated from the top to the bottom of the card. Each column is able to accommodate a digit, a letter, or a special character by punching a hole or holes in the column. The punching position for any one of digits 0 to 9 in each column corresponds to the number printed on the card. For example, in Figure 2–3, there is a 0 punched in column 15, a 1 punched in column 16, and a 9 punched in column 24.

The three punching positions on the top of the card, 12, 11 (or X), and 0, are known as the zone punching area. Thus, the 0 punch may be either a zone punch or a digit punch. A combination of a zone punch and a digit punch is used to accommodate any of the 26 alphabetic letters in one column. The various combinations of punches representing the 26 letters are based on a special arrangement. The first nine letters (A through I) are represented by the combination of a 12 punch and the digit punches 1 through 9. The second nine letters (J through R) are the combination of an 11 (or X) punch and the digit punches 1 through 9. The last eight letters (S through Z) are the combination of a zero punch and the digits 2 through 9. The combinations representing the letters are shown in columns 31 to 56 in Figure 2–3. Special characters are recorded by one, two, or three punches. Examples of eleven special characters are shown in columns 63 to 73 in the figure.

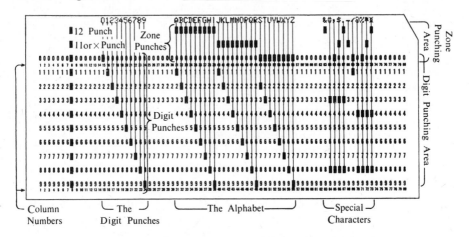

FIGURE 2-3

Illustration of a Punched Card

When the collected data are recorded on the cards, each card is used to record only the facts about an individual case or a respondent. The columns on the card are divided into segments called *fields*. A field may consist of one or eighty columns, depending on the length of the unit of information. Since there is a limited number of columns in the card, a code system is usually applied in recording certain longer information. For example, if a company has 2,000 to 3,000 employees in a year, a four-column field would be proper for recording employee numbers. Thus, instead of recording the name of an employee, Frederick Lawrence Montgomery, which takes up 29 columns, including two spaces between the names, a coded employee number such as 1568 may be assigned to represent his name on the card. Figure 2–4 illustrates the fields assigned for recording information concerning accounts receivable on the card. The date field, for example, is coded and assigned to occupy six columns. The first two columns are assigned to represent the number of a month, the second two columns represent a day in a month, and the last two columns represent a year. Thus code 05 13 76 represents May 13, 1976.

The cards in Figures 2–3 and 2–4 have upper right corner cuts. Corner cuts are frequently used to identify card types or to insure that the cards in a group are all facing the same direction and are all right side up.

The Punched Card Machines

A basic installation of punched card machines normally consists of a card punch (Figure 2–5), a sorter (Figure 2–6), and an accounting machine (Figure 2–7). In addition to the three basic types, there are many other punched card machines developed to meet various needs. To discuss all types of punched card machines is beyond the scope of this text. Only the basic types of punched card machines are discussed.

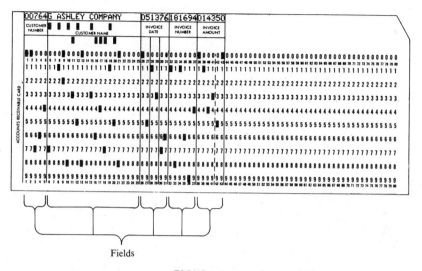

Fields

FIGURE 2-4

AN ILLUSTRATION OF FIELDS ON A PUNCHED CARD

FIGURE 2-5

A PRINTING CARD PUNCH MACHINE

Card punch machine. Statistical information is first recorded on the form of punched holes by means of a card punch machine. The machine has a keyboard similar to that of a typewriter. When the information is being recorded, the operator reads the information and presses the keys of the keyboard to punch the proper letters, digits, or special characters on a card. The punching process is operated in an orderly manner; one column is punched at a time. After one column is punched, the card is automatically positioned for the next punching. Data being punched may be printed at the top of the column, depending upon the type of machine used. The Printing Card Punch machine shown in Figure 2–5 prints the data as they are punched on the card.

Sorter machine. Second, the punched cards are arranged to meet a specific requirement by the use of a sorter machine. There are many different types of sorters, ranging in speed from 450 cards per minute to 2,000 cards per minute.

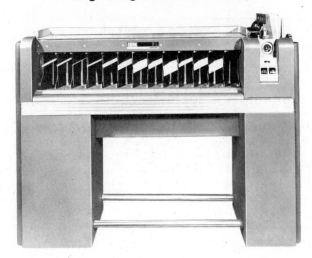

FIGURE 2-6

A SORTER MACHINE

FIGURE 2-7

AN ALPHABETICAL ACCOUNTING MACHINE

The sorter shown in Figure 2–6 has 13 pockets to receive sorted cards: one pocket for each of the 12 punching positions in a column and one pocket for rejecting cards with no hole in the column being sorted.

The sorter can perform three basic types of arrangement: sequencing, grouping, and selecting. Sequencing is the process of arranging information represented by a given field of cards in alphabetic or numerical order, in either ascending or descending order.

Grouping is the process of arranging like items into individual groups. For example, if we wish to have a report showing sales by individual salesmen, the cards representing sales during a given period are sorted on the salesman number field.

Selecting is the process of extracting a desired card or cards from a larger file of data. This process can be done by the sorter without disturbing the sequence of the remainder of the file. For example, if all students majoring in English in a college are needed to prepare a special report, the cards representing the students can be removed from the file of all students in the college by the sorter.

Accounting machine.   The last step in tabulating data by machine punched cards is to print the information on a tabular form. The Alphabetical Accounting Machine shown in Figure 2-7 can be used to print alphabetic or numerical characters and to total data by proper classification from punched cards. The machine prints in two different manners: detail printing and grouping printing. When data are printed from cards with one line printed per card, it is called detail printing or listing printing. Figure 2-8 illustrates the monthly payroll report

The Sutton Machine Company
MONTHLY PAYROLL REPORT

Date: 4/30/72

Employee Number	Name of Employee	Hours Worked	Hour Rate	Total Wages
1123	BAKER JOHN	40	2 50	100 00
1123	BAKER JOHN	40	2 50	100 00
1123	BAKER JOHN	35	2 50	87 50
1123	BAKER JOHN	30	2 50	75 00
		145*		362 50*
1245	DAVIS ROBERT	40	2 20	88 00
1245	DAVIS ROBERT	38	2 20	83 60
1245	DAVIS ROBERT	40	2 20	88 00
		118*		259 60*
1268	EDWARDS GEORGE	36	2 00	72 00
1268	EDWARDS GEORGE	40	2 00	80 00
1268	EDWARDS GEORGE	40	2 00	80 00
1268	EDWARDS GEORGE	32	2 00	64 00
		148*		296 00*

FIGURE 2-8

AN EXAMPLE OF DETAIL PRINTING BY ACCOUNTING MACHINE

in detail printing. When data from cards are summarized by individual classifications and only the totals of the classes are printed, it is called grouping printing. Figure 2-9 illustrates the grouping printing of the same data listed in Figure 2-8.

The Sutton Machine Company MONTHLY PAYROLL REPORT Date: 4/30/72				
Employee Number	Name of Employee	Hours Worked	Hour Rate	Total Wages
1123	BAKER JOHN	145	2 50	362 50
1245	DAVIS ROBERT	118	2 20	259 60
1268	EDWARDS GEORGE	148	2 00	296 00

FIGURE 2-9

AN EXAMPLE OF GROUPING PRINTING BY ACCOUNTING MACHINE

2.5 Tabulation by Computer Systems

An electronic data processing system, or simply called *computer system,* can do many things in many ways. It can tabulate statistical data at a speed far greater than the speeds of the methods previously mentioned. It can also perform highly complicated mathematical operations that were often impractical, or even impossible, only a few short years ago. This section will introduce the basic concept and the general operations of a computer system. The topics included below are (1) data representation within a computer system, (2) the basic functional units of a computer system, and (3) the procedure of using a computer system to solve a given problem.

Data Representation

Data within the computer system are represented by electronic signals called *binary indications.* A binary indication is the result of the presence or the absence of electric current in a circuit. This method of representing data corresponds to the presence or absence of holes in a punched card. Specific values are assigned to the binary indications and are used as the language representing data in the computer system.

For example, in some computers, the values associated with the binary indications are related directly to the binary number system. Although this numbering system is not used in all computers, the method of assigning the values by using the system is useful in learning the general concept of data representation. In the decimal number system, we use ten digits to represent all numbers. The place value of the ten digits signifies units, tens, hundreds, thousands, etc. The binary number system uses only two digits: 0 and 1. The place value of 1 is based on the progression of powers of 2. The units position has the value of 1; the next position to the left, a value of 2; the next, 4; the next 8, and so on, as shown in Table 2–3A. The numbers which are not powers of 2 can be obtained by addition (or its reverse operation—subtraction) and multiplication (or its reverse operation —division), as in the following illustrations:

Addition		*Multiplication*	
Decimal Values	*Binary Values*	*Decimal Values*	*Binary Values*
1	1	1	1
+ 2	+ 10	× 2	× 10
3	11	2	10
4	100	5	101
+ 1	+ 1	× 3	× 11
5	101	15	101
8	1000		101
4	100		1111
+ 1	+ 1		
13	1101		

The binary values 0 to 16 are listed in Table 2–3B. Any number written in the decimal number system can be expressed in the binary number system. We thus may assign 0 to the absence and 1 to the presence of electric current in a circuit for representing data in a computer system.

TABLE 2-3

ILLUSTRATION OF BINARY NUMBERING SYSTEM AS COMPARED
TO DECIMAL NUMBERING SYSTEM

A. Place Value of Binary Numbers 0 and 1		B. Binary Values as Compared to Decimal Values 0 to 16	
Decimal Value	*Binary Value*	*Decimal Value*	*Binary Value*
0	0	0	0
1	1	1	1
$2^1 = 2$	10	2	10
$2^2 = 4$	100	3	11
$2^3 = 8$	1000	4	100
$2^4 = 16$	10000	5	101
$2^5 = 32$	100000	6	110
$2^6 = 64$	1000000	7	111
		8	1000
		9	1001
		10	1010
		11	1011
		12	1100
		13	1101
		14	1110
		15	1111
		16	10000

Basic Functional Units of a Computer System

A computer system basically consists of a combination of four functional units: *input*, *storage*, *central processing*, and *output*. Among them, the most important unit is the central processing unit, a high-speed electronic computer. Figure 2–10 shows a partial view of the IBM 370 data processing system, a large and powerful system capable of doing millions of calculations in a few moments.

FIGURE 2-10

IBM 360 Data Processing System, partial view

Source: International Business Machines Corporation. Reprinted with permission.

The ordinary order of processing simple data through the four basic units of a computer system is as follows:

a. The input unit enters data via electronic language into the arithmetic-logic section in the central processing unit for calculation or organization.

b. The central processing unit sends the result of the calculation or the organized data to the storage unit until needed.

c. The storage unit sends the result to output unit for recording or printing. This is diagrammed in Figure 2–11. The functions of the four basic units in a computer system are further discussed in the following paragraphs.

Input Unit

The function of the input unit is to enter data represented by an appropriate medium into the computer system by the use of an input device. An input device is a machine linked directly to the computer system and operated under the control section of the central processing unit. It senses or *reads* data from a certain input medium when the medium physically moves through the machine. The information is then transmitted to the arithmetic-logical section for processing or directly to a storage unit until needed.

The input medium may be punched cards, punched paper tape, magnetic tape with magnetized spots, paper documents with magnetic ink characters inscribed, typed words, or handwriting. Data represented by the various media are read by different types of input devices. For example, punched cards are read by a card reader and punched paper tapes are read by a paper tape reader.

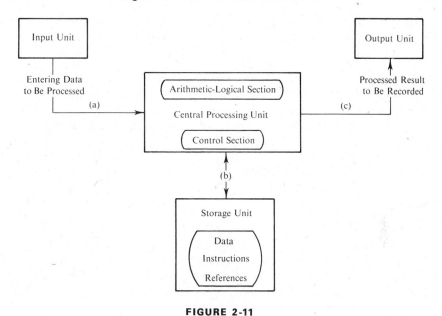

FIGURE 2-11

BASIC UNITS OF A COMPUTER SYSTEM

Central Processing Unit

The central processing unit serves as the controlling center of the entire data processing system. It can be divided into two sections.

1. The control section.
2. The arithmetic-logical section.

The control section directs and co-ordinates the entire computer system according to the instructions received from the storage but originated by its human operators. The instructions involve controlling the operations performed by the input unit, the arithmetic-logical section and the output unit. They also include the work of transferring data to and from storage for the operations.

The arithmetic-logical section can perform many operations, including adding, subtracting, multiplying, dividing, shifting, transferring, comparing, and storing data. Its logical ability can test various conditions encountered during processing and can take proper action called for by the result.

Storage (Memory)

The function of the storage unit is to provide an electronic filing cabinet. Information placed in storage can be data, instructions, or references. Stored data can be the data received from an input device, data rearranged by the computer, or calculated new information for output. Stored instructions, also called *stored programs*, are used to direct the control section of the central processing unit in operating the entire computer system. References are the information associated with processing, such as a table of square roots and code charts. The various types of information may be stored in any one of the three common types of storage devices: magnetic core, magnetic drum, and magnetic disk. The location of individual stored information in a storage device is numbered. When needed, the stored information can be readily located by the computer according to the numbers of location.

Output Unit

The function of an output unit is to record or *write* information received from the storage unit on cards, paper tape, magnetic tape, or as printed information on paper. Different output media require different types of output devices or machines.

Procedure of Solving Problems

The procedure of solving a given problem by the use of a computer system usually includes three steps.

1. Construct a Flowchart

A *flowchart*, also called a *block diagram*, is used to show a sequence of steps to be performed by a computer for solving a given problem. It uses graphic symbols, such as boxes, circles, and lines with arrowheads, as shown in Figure 2–12, to represent the steps in a definite order. An example of a flowchart which shows the steps of averaging a group of numbers is also presented in Figure 2–12. The numbers are punched into cards. The computer reads the first number card, adds the number to the previous total, and adds one to the number of cards read. The steps are repeated in reading the second and all subsequent cards until the last card. The remaining steps in the flowchart are self-explanatory.

2. Write a Program

A *program* is a complete set of written instructions that enables the computer to process the information for solving a given problem. The instructions are usually prepared according to the steps outlined in the flowchart and must be written in a language that the computer can accept. There are many types of programming languages available. The selection of a particular language depends on the type of hardware and software available in a computer center, the nature of the problem, and the knowledge possessed by the programmer.

The term *hardware* generally refers to the physical equipment that makes up a computer system. *Software* includes computer programs, procedures, manuals, and other computer operation instructions. Thus a statistician, who does not know the techniques of programming, may also use computer facilities to solve a complicated problem if the computer center maintains adequate prewritten programs.

The common programming languages are described below. All are designed for certain purposes. Among them, however, FORTRAN and COBOL are more widely used languages.

FORTRAN (*FOR*mula *TRAN*slator). FORTRAN is often referred to as a *scientific language* and is designed primarily to simplify the work of solving mathematical problems. It was introduced in 1957 by IBM. Since its initial introduction, it has progressed through several versions. The current version is FORTRAN IV which has become an extremely popular language.

COBOL (*CO*mmon *B*usiness *O*riented *L*anguage). COBOL specifications were jointly designed by a group of computer users, computer manufacturers,

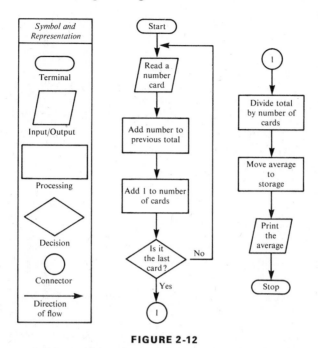

FIGURE 2-12

ILLUSTRATION OF A FLOWCHART TO AVERAGE NUMBERS

and government agencies, and were initially published in 1960 by the U.S. Government Printing Office. COBOL is particularly applicable to business data processing problems. It is written in Englishlike form. This feature offers readability and understandability of program details.

PL/1 (*Programming Language One*). PL/1 can be used to solve all types of business and scientific problems efficiently. Thus the PL/1 could replace both FORTRAN and COBOL programs. However, computer users now are anxious to protect their vast investment in the two types of programs.

ALGOL (*ALGO*rithmic *Language*). Like FORTRAN, ALGOL is designed as a scientific language. It was developed by a group of international mathematicians in 1957.

BASIC (*B*eginner's *A*ll-purpose *S*ymbolic *I*nstruction *C*ode). BASIC is a widely used language in time-sharing. Time-sharing refers to a computer system with many users; each user has direct access to the central processing unit by a remote terminal. A student can learn to write BASIC programs at the terminal in a short period of time.

RPG (*R*eport *P*rogram *G*enerator). RPG is generally used for small computers. The small computers are primarily designed for business processing applications.

3. Execute the Program

The prepared program is now transformed into an input medium, such as punched cards. The medium is then read into the computer for processing and writing the final result as the program directed.

Electronic computers are designed to multiply man's ability to do mental work. They have been developed rapidly and used to a great extent in the past two decades. Many industrial, governmental, and educational organizations are using electronic computers to do highly complicated or complex work, such as processing huge volumes of data, making managerial decisions, and doing scientific research. Electronic computers are still being explored and improved drastically in both capacities and speeds. The extent of their applications in the future is expected to be even greater. A knowledge of the basic units of a computer system and its uses thus is highly desirable for a modern statistician.

Exercises 2

1. What are the reasons that data collected from a survey may need extensive editing?

2. What are the four bases of classifying statistical data? How would you classify and arrange data according to the different bases?

3. What are the methods of manually written cards and tally sheet in tabulating collected statistical data? Give your own examples to illustrate each method.

4. Explain the use of rod-sorting cards. Give your own example in your explanation.

5. What are zone punches, digit punches, columns, and fields on an IBM card? How can we record a large amount of information on the limited number of columns? How many columns and fields are on an IBM card?

6. What are the basic punched card machines? How would you use the machines in tabulating a group of data?

7. Convert each of the following decimal values to its equivalent binary value:

(a) $1 + 3 = 4$ (b) $2 + 4 + 5 = 11$ (c) $6 + 7 + 8 = 21$
(d) $2 \times 3 = 6$ (e) $4 \times 5 = 20$ (f) $6 \times 7 = 42$

8. State briefly the basic functional units of a computer system. What is the ordinary order of processing simple data through the basic units?

9. What is the procedure of solving a problem by the use of a computer system?

10. What is a program? Explain the names FORTRAN and COBOL.

3 Presenting Statistical Data

Presenting statistical data is the third step in a statistical study. There are three ways to present organized data: (1) word statement, (2) statistical table, and (3) statistical chart. When a series of data includes only a few items, a word statement may be properly used to present the simple facts. However, when a large group of data is being presented, word statement presentation becomes inefficient and burdensome. In this case, statistical tables and charts are preferred.

When statistical data are presented in tabulated form, the data are systematically arranged in columns and rows. Details concerning statistical tables are discussed in Sections 3.1 through 3.3. A statistical chart or a graph is a pictorial device for presenting statistical data. It is usually constructed according to the information provided in a table. Details relating to charts and illustrations concerning the construction of charts from tables are presented in Sections 3.4 through 3.6.

3.1 Statistical Tables Classified by Purposes

Statistical tables can be grouped into two types according to the purposes served by the tables: (1) *general purpose table* (also called the *reference table* or *repository table*) and (2) *special purpose table* (also called *summary table*, *text table*, or *analytical table*).

General purpose tables provide information for general use or reference. They are not constructed for specific discussion. In other words, these tables serve as repositories of information. Thus general purpose tables frequently include detailed information. They are arranged for easy reference. Government agency tables, such as the *Statistical Abstract of the United States*, *Survey of Current Business*, and *Federal Reserve Bulletin*, are mostly of this kind. For example, the *Survey of Current Business* has a table entitled "Employment and Population" showing the number of employees in manufacturing, mining, contract construction, transportation, wholesale and retail trade, government, and other U.S. vocational areas. This table is a general purpose table, since it merely tells facts which are not there for a particular discussion. When the general purpose tables are used by a researcher, they are usually placed in the appendix of the report for easy reference.

Special purpose tables provide information for particular discussion. A special purpose table should be designed so that a reader may easily refer to the table for comparison, analysis, or emphasis concerning the particular discussion. Thus the table should be constructed in a short and simple manner and should be placed near the pertinent textual discussion. The tables used in this chapter are of this kind. For example, Table 3-1 is designed to show the principal parts of a statistical table and is placed near the discussion of the parts.

3.2 Principal Parts of a Table

The number of parts in statistical tables may vary. In general, a complete table, such as Table 3-1, may include seven principal parts. They are (1) title, (2) caption, (3) stub, (4) body, (5) headnote, (6) footnote, and (7) source of data. The first four parts are basic and must be included in any statistical table. The remaining three parts are additional and may not be present in some tables, depending on the given information. However, whenever they are applicable, the additional parts must also be present in the table.

All parts should be presented in a clear and simple, yet attractive and complete manner, so that a reader may spend the smallest amount of time and get the most information from the table presentation. Statisticians frequently use ruling, spacing, and special type styles such as boldface, italics, roman capitals, and lowercase letters to improve the appearance and clarity of a table.

Title

The title is a description of the contents of the table. It should be compact and complete. A complete title usually indicates (1) What are the data included in the body of the table? (2) Where is the area represented by the data? (3) How are the data classified? (4) When do the data apply? The title of Table 3-1 illustrates the four complete descriptions: what—civilian noninstitutional population, where—United States, how—by employment status and by sex, and when —1973 and 1975.

When more than one table is presented in a discussion, each table should be numbered. The table number is especially important, as it is easier to refer to a table number than to the entire title of the table.

TABLE 3–1

TITLE⟶ UNITED STATES CIVILIAN NONINSTITUTIONAL POPULATION*, 20 YEARS AND OVER, BY EMPLOYMENT STATUS AND SEX, 1973 AND 1975

HEADNOTE   ⟶ (Numbers in Millions†)

Employment Status	1973			1975		
	Total	Male	Female	Total	Male	Female
All population	130.2	60.9	69.2	133.9	62.8	71.1
Civilian labor force	80.3	49.5	30.7	83.1	50.5	32.6
Employed	77.2	47.9	29.2	77.4	47.5	29.9
Agriculture	3.1	2.5	.6	2.9	2.4	.5
Nonagriculture industries	74.1	45.4	28.7	74.5	45.1	29.4
Unemployed	3.1	1.6	1.5	5.6	3.0	2.6
Unemployment rate	3.8%	3.2%	4.8%	6.8%	6.0%	8.1%
Not in labor force	49.9	11.4	38.5	50.8	12.3	38.5

CAPTIONS applies to the *Employment Status* / 1973 / 1975 heading block; STUBS ⟶ applies to the left-hand row labels.

FOOTNOTE *The noninstitutional population comprises all persons who are not inmates of penal or mental institutions, sanitariums, or homes for the aged, infirm, or needy.

†Details shown may not add to totals shown because of rounding.

SOURCE ⟶ Source: U. S. Department of Labor, *Monthly Labor Review*, March, 1975, p. 88.

Captions

The caption, also called boxhead, is the heading at the top of a column or columns. The simplest table may consist of only two columns and two captions: one for stubs and one for data. However, many tables have more than two captions and sometimes have main captions and subcaptions. For example, in Table 3-1, in addition to the caption of stubs, there are two main captions, 1973 and 1975. Each of the two main captions has three subcaptions: total, male, and female.

Stub

The descriptions of rows of the table are called stubs. Stubs are placed at the left side of the table. They usually represent the classifications of the figures included in the body of the table. The nature of classifications is indicated by the caption of the column including the stubs. For example, the classifications in the stub column of Table 3–1 are based on the employment status.

As in our example, tabular data are frequently cross-classified. In this case both stubs and captions of the table are used to represent classifications of data. This type of table is called a *cross-classified table*. The two main captions—1973 and 1975—represent the classifications based on time, whereas the stubs represent the employment status; both stubs and captions are used to show the population data. While it is possible for each classification subgroup to be further divided into many small groups, the purpose of classifying data in a table is to place similar items in a group so that the details are reduced and the data can be effectively presented for analysis. Too many small groups may defeat the purpose of a table presentation. If the data included in a table must be divided and subdivided into many small groups for analysis, several simple tables are preferred to a crossclassified table.

Body

The body is the content of the statistical data. Data presented in the body are arranged according to descriptions or classifications of the captions and stubs. Thus, effective presentation of the data in a table depends on the arrangement of the columns and rows. Statistical data can be classified according to four bases: time, place, quantity, and quality. For the most common methods of arranging the order of the classifications according to the four bases, see Section 2.2.

If comparisons are to be made, the figures included in a table should be so arranged that they can easily be compared. In general, figures can be compared more easily when they are placed in a column rather than in a row. When two or more sets of figures are to be compared, they should be placed in adjacent columns or as close as possible. For example, the arrangement of the captions in Table 3–1 is easy for making comparisons between male forces and female forces for individual years. If comparisons between male forces and between female forces for different years are desired or emphasized, the captions may be rearranged as follows:

Employment Status	*Total*		*Male*		*Female*	
	1973	1975	1973	1975	1973	1975

Important figures should be placed in the most noticeable positions in a table. Since readers' habits are generally such that they read from left to right and from top to bottom, the column nearest to the stubs and the row immediately below the captions are considered to be the most noticeable positions. In Table 3–1, the column showing the total in 1973 and the row showing the total for all population are thus considered the most noticeable positions and are used to emphasize the important figures.

Important figures may also be emphasized by the use of different colors or styles of type. In many cases, the figures representing totals are printed in bold-face.

Headnote

Headnotes are usually written just above the captions and below the title. They are used to explain certain points relating to the whole table that have not been included in the title, captions, or stubs. For example, the unit of the data is frequently written as a headnote, such as "Numbers in Millions" shown in Table 3–1.

Footnote

Footnotes are usually placed below the stubs. They are used to clarify some parts included in the table that are not explained in other parts.

Source

The source of data, or simply *source*, is usually written below the footnotes. If the data are collected and presented by the same person, it is customary not to state the source in the table. Details concerning the collection are mentioned in the discussion along with the tabular presentation. However, if the data are taken from other sources, such as primary or secondary sources of published data, these sources should be stated in the table. The statement will enable the reader to check or evaluate the data, or to obtain additional information from the original source, if needed, and will give proper credit or responsibility to the original collector of the data.

3.3 Rounded Numbers in a Table

Many statistical tables are not designed to show the exact figures, especially if each figure consists of many digits. Approximate figures are sufficient for many uses. In rounding the exact figures to approximations, it is customary to round the digits after one of the commas which are used to separate the digits into groups of three. That is, numbers should be rounded off to the units of thousands, millions, or billions, rather than other units. For example, 2,346,783 may be rounded to 2,347 thousands or to 2.347 millions, but not to 234.7 ten-thousands.

The most common method of rounding a number to its nearest value in statistical studies may be summarized in three main rules:

1. If the portion of digits to be dropped begins with 4 or less, leave the preceding digit unchanged.

478.49987 is rounded to 478 if the decimal places (49987) are dropped. 45,356,589 is rounded to 45 millions if the portion 356,589 is dropped.

2. If the exact digit 5, or 5 followed by zeros only, is to be dropped, mark the preceding digit *even* or follow the two rules below.

a. If the preceding digit is odd, add 1 to the digit.

37.5 or 37.5000 is rounded to 38, since 7 is an odd digit. 683,500 is rounded to 684,000, since 3 is an odd digit.

b. If the preceding digit is even, leave the digit unchanged.

48.5 is rounded to 48.0, since 8 is an even digit. 34,500 is rounded to 34,000, since 4 is an even digit.

This method is intended to reduce the cumulative rounding error to a minimum. When this intention can not be accomplished, the need of rounding must be carefully examined. For example, if the figures in a group all end in exactly 5, such as the receipts of a store in 25¢, 45¢, 65¢, and 85¢, the rounding process will eliminate all 5's. In such cases, it might be better to retain the 5's instead of rounding 25¢ to 20¢, 45¢ to 40¢, and so on.

3. If the portion of digits to be dropped begins with 5 followed by nonzero digits, add 1 to the preceding digits. Thus, 13,467.5008 is rounded to 13,468 if

the decimal places (5,008) are dropped. 456,982 is rounded to 457,000 if the portion 982 is rounded.

Frequently the total of the rounded figures in a table does not agree exactly with the rounded total of the original figures. For example, the original total population figure in 1973 as shown in Table 3–1 is rounded to 130.2 million. The total male population and the total female population are also rounded to millions. However, the total of the rounded totals of individual types is not 130.2 but is 130.1, or

$$60.9 + 69.2 = 130.1 \text{ (million)}$$

In this case, one of the two commonly used methods may be adapted:

1. Add a statement in the footnote, such as the second footnote in Table 3–1, "Details shown may not add to totals shown because of rounding."

2. Adjust one of the rounded numbers in the addition so that the total of the rounded numbers agrees with the rounded total. The number to be adjusted should give the *smallest amount of change*. For example, the rounded numbers shown below would not add to the rounded total 89. (The rounded total 89 is obtained by rounding the original total 88.52). In this case, 18.40 is adjusted to 19, since this adjustment gives the smallest amount of change, or 0.60 ($= 19 - 18.40$). If 7.20 had been adjusted to 8, the change would be larger, or 0.80 ($= 8 - 7.20$). The adjusted rounded numbers shown in the third column now add to the total 89.

Original Number	Rounded Number (round individually)	Adjusted Rounded Number
18.40	18 ⟶	19
7.20	7	7
62.92	63	63
	88 (sum)	
88.52 ⟶	89	89

3.4 Fundamentals of Chart Construction

A chart or graph is a pictorial expression of given information. Nearly all types of quantitative information can be expressed in the form of charts. There is no single rule based on which an interesting and effective chart may be drawn. However, an understanding of the fundamentals of chart construction, principal parts of a chart, common types of charts, and methods of constructing different types of charts should be of great value to both the reader and the maker of charts. The fundamentals are discussed in this section; other topics are presented in the next two sections of this chapter.

A person who understands how to construct good charts can present quantitative information to his readers much more quickly than if he had to arrange the figures in tabulated form or write the same information in sentence form. We frequently hear the expression: "One good picture is worth a thousand words." However, a chart gives a reader only an approximate value of the information. If an exact amount is desired, the tabulated figures or the original source of the chart should be consulted.

CHART 3-1

RECTANGULAR CO-ORDINATES

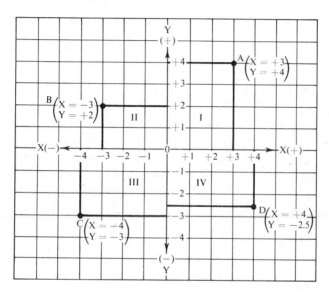

Basically, charts are drawn according to the system of *rectangular co-ordinates*. Rectangular co-ordinates are based on two straight reference lines perpendicular to each other in a plane, also called a *grid*, as shown in Chart 3–1. The horizontal line is usually referred to as the *X-axis*, or the *abscissa*, whereas the vertical line is referred to as the *Y-axis*, or the *ordinate*. The two lines divide the plane into four parts called *quadrants*, which are indicated in the chart by numbers I, II, III, and IV. The point of intersection of the two lines is called the *origin*, or zero point. Scales are labeled along the two axes, beginning at the point of origin. The abscissas (on the *X*-axis) to the right of the origin are conventionally designated as positive, whereas those to the left of the origin are negative. The ordinates (on the *Y*-axis) above the origin are positive, and those below the origin are negative. Any point on the plane may refer to two values according to the two scales. For example, in quadrant I the abscissa of point *A* is +3, and the ordinate of the same point is +4. The values of +3 and +4 constitute the co-ordinates of *A*.

In statistical charts, the co-ordinates are assigned to represent two corresponding items, such as one representing a class and the other representing the quantitative information of the class. When numbers included are all positive, only the area of quadrant I, or the upper right-hand portion of the plane, is needed for

CHART 3-2

ILLUSTRATION OF DATA REPRESENTATION BY POINTS ON A STATISTICAL CHART

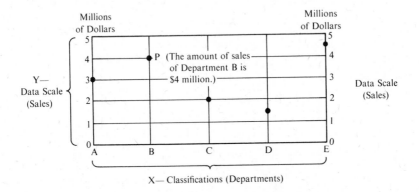

X— Classifications (Departments)

showing the numbers. The other three quadrants are omitted for the sake of simplicity and saving space. Chart 3–2 shows that the letters on the X-axis represent various departments of a department store and the numbers on the Y-axis represent the sales in millions of dollars during a given period. Points representing the sales of the various departments in the store are plotted on the chart. Point P, for example, indicates that the sales of department B in the store during a given period are $4 million.

3.5 Principal Parts of a Chart

Since charts are pictorial devices, the details included in charts may vary greatly, ranging from a few points to very complicated graphical presentation. The various complications depend not only on the amount of data to be presented but also on the artistic designs of graphs to be included in the charts. However, the principal parts as shown in Chart 3–3 may be found frequently in many charts. In most cases, the functions of the parts in the chart resemble those of the parts in a table. Some exceptions are discussed below.

Title

As in a table, the title is a description of the contents of the chart. The guides for making a good title are similar to those concerning a table. However, the title of a chart may be placed either on the top of the chart as in a table or at the bottom of the chart.

Graphs

Graphs, like the body of a statistical table, are used to represent data shown in a

CHART 3–3

PRINCIPAL PARTS OF A STATISTICAL CHART

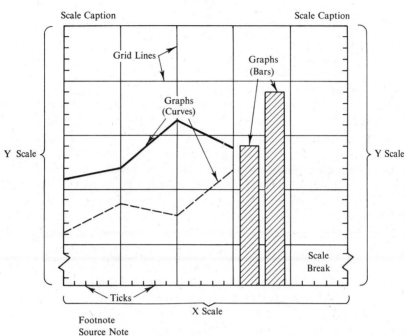

chart. There are many different types of graphs used in statistical charts, including lines, bars, dimensions, symbols, maps, and combinations of the various graphs. The graphs should be printed in much heavier print than the grid lines in order to show the importance of the data represented.

Scales

The Y and X scales are basically labeled according to the system of rectangular co-ordinates as discussed in the above section. However, whereas the Y-scale is used to measure the magnitudes of the graphs representing data, the X-scale is frequently used to designate the classifications of the data.

Source Note

The *source of data* from which the chart was constructed should be placed at the bottom part of the chart. If the chart was taken from another publication, the *source of the chart* should also be indicated in the source note.

3.6 Common Types
of Charts

The most common types of charts are (1) line chart, (2) bar chart, (3) component part chart, (4) dimensions chart, (5) pictogram, and (6) statistical map. In practice, charts are frequently constructed from the data already presented in statistical tables. The method of constructing each type of chart from a table is illustrated below.

Line Chart

A chart that consists of lines or broken lines (also called *curves*) for representing data is referred to as a line chart. To construct a line chart, first plot the data by points according to the scales on the two reference lines. Then connect the points by straight lines. The scales used on the two reference lines are usually quantitative and are continuously labeled. They may or may not be equal, and may be arithmetic or logarithmic. On an arithmetic scale, equal distances represent equal amounts. Details concerning logarithmic scale are discussed in the next chapter.

Line charts are used mostly to show data classified by quantity or time. The application of line charts in showing data classified by quantity is presented in detail in Chapter 5. The presentation of data classified by time is discussed in this section to illustrate the method of constructing line charts with arithmetic scales.

Data classified on the basis of time intervals are referred to as *time series*. In a time series line chart, the scale representing time intervals is usually placed on the horizontal line (*X*-axis), and the scale representing data is on the vertical line (*Y*-axis). The time designation is labeled *continuously* from left to right on the horizontal line, beginning with the earliest time of the entire period. Vertical grid lines or ticks are used to guide the time designation on the time scale. Each time designation may represent a *period of time* or a *specified time*. In either case, it is written in the middle of the space representing the time, as shown in Charts 3–4 to 3–7. The charts show that there are two ways to write time designations in relation to vertical guide lines:

1. The guide lines can be used as the boundaries of the periods labeled, such as the illustrations in Charts 3–4 and 3–6. Each of the time designations is placed in the center of the two guide lines. For example, in Chart 3–4 the line at the left of the time designation 1975 indicates the beginning of the year, January 1, 1975, whereas the line at the right indicates the end of the year, December 31, 1975, which may also be considered January 1, 1976, since the time scale is labeled continuously.

2. Each guide line can be considered as the middle of the period labeled, such as the illustrations in Charts 3–5 and 3–7. In Chart 3–5, for example, the line above 1975 indicates the middle of the year, or July 1, 1975. The beginning of the year, January 1, 1975, thus falls in the middle of the lines labeled 1974 and 1975, and the end of the year, December 31, 1975, or January 1, 1976, is located in the middle of the lines labeled 1975 and 1976.

Time designation representing a *specified time* can be indicated in the two ways:

1. The specified time, such as "January 1 of each year," represented by each guide line is indicated in the title of the chart, as shown in Chart 3-7.

2. The specified time, such as "January 1 of each year," is written directly under the guide line as shown below:

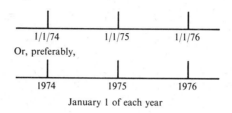

January 1 of each year

Time series are of two kinds: *period data* and *point data*. Period data are the figures representing either the information accumulated during some *period of time*, such as sales made during a week and units produced during a month, or the average of individual figures representing certain information in a given period of time, such as the average price of sugar per pound based on the daily prices during the month of June. Point data are the figures representing the information at a *specific point of time*, such as the price of sugar per pound on a certain date in June, number of employees in a company at the end of each month, and cash on hand at the end of each year.

When period data are plotted, each of the figures of the given data is represented by a point. The point is customarily plotted directly above the *middle* of the space designated for the period. If there are two guide lines indicating the beginning and the end of the period, such as in Chart 3-4, the point is plotted in the middle of the two lines. If the guide line is erected in the middle of the period designated, such as in Chart 3-5, the point is plotted on the line. Example 1 is used to illustrate the methods of constructing a line chart of period data.

Example 1

The amount of the annual sales of the Fulton Drug Store from 1969 to 1976 is listed below. Construct a line chart showing the given data.

TABLE 3-2

ANNUAL SALES OF FULTON DRUG STORE
1969 TO 1976

Year	Sales	Year	Sales
1969	$ 500	1973	$ 8,000
1970	1,000	1974	10,000
1971	2,000	1975	12,000
1972	4,000	1976	14,000

Source: Hypothetical data for illustration purpose.

Solution.   These are period data. The data are plotted on Charts 3–4 and 3–5 on different time scales. In any scale, however, the time designation is written in the middle of the space representing the year. Each of the points representing data is also plotted in the middle of each year.

CHART 3–4

A LINE CHART (PERIOD DATA)
ANNUAL SALES OF FULTON DRUG STORE, 1969 TO 1976

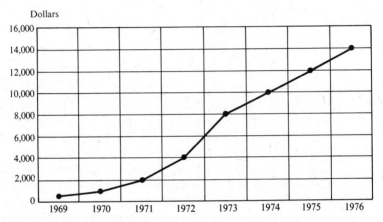

Source of data: Table 3-2, Example 1.

CHART 3–5

A LINE CHART (PERIOD DATA)
ANNUAL SALES OF FULTON DRUG STORE, 1969 TO 1976

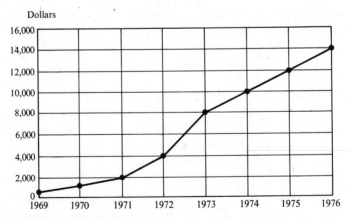

Source of data: Table 3-2, Example 1.

When point data are plotted, each figure should be plotted on the point directly above the time designation to which it refers. This can be done in two ways: (a) Point data are plotted above the time scale with time designation for period of time. (b) Point data are plotted above the time scale with time designation for the specified time. Example 2 is used to illustrate the methods of constructing a line chart of point data.

Example 2

The amount of cash on hand in the Gontor Motor Company at the beginning of each year from 1968 to 1976 is given in Table 3–3. Construct a line chart showing the given data.

TABLE 3-3

CASH ON HAND IN THE GONTOR MOTOR COMPANY
JANUARY 1 OF EACH YEAR, 1968–1976

(Thousands of Dollars)

Year	Cash	Year	Cash
1968	20	1972	50
1969	40	1973	45
1970	30	1974	70
1971	25	1975	65
		1976	75

Source: Hypothetical data for illustration purpose.

Solution. The amounts shown in the table are point data. The data are plotted on different time scales:

(a) Above the time scale with time designation for period of time. This is shown on Chart 3–6. Each point is plotted directly above the beginning of the space designated for each year, or January 1.

(b) Above the time scale with time designation for the specified time. This is shown on Chart 3–7. The specified time of each guide line, or January 1 of each year, is stated in the title of the chart. An alternative method to indicate the specified time of each guide line is to label January 1 of each year directly under the years on the time scale.

In any scale, however, each of the points representing the data is plotted above the point of time to which it refers.

Bar Chart

A bar chart has a number of rectangular bars. The width of each bar is usually equal to others. The length of each bar shows the data represented. In comparison to line charts, bar charts are effective for emphasizing a few items of one or two series of data, whereas line charts are preferable in presenting many items of one or several series of data. Bar charts emphasize the differences among individual items, but line charts emphasize the continuous changes or general trend among the items. Bar charts are used frequently in presenting data classified by any basis (time, place, quantity, or quality). However, line charts are used mostly in presenting data classified by time or quantity. Furthermore, it takes more time to draw bars in a bar chart than to plot points and connect the points by straight

CHART 3–6

A LINE CHART (POINT DATA)
CASH ON HAND, GONTOR MOTOR COMPANY, 1968–1976

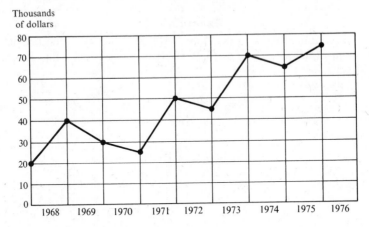

Source of data: Table 3-3, Example 2.

CHART 3–7

A LINE CHART (POINT DATA)
CASH ON HAND, GONTOR MOTOR COMPANY
JANUARY 1 OF EACH YEAR, 1968–1976

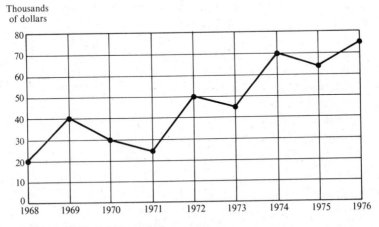

Source of data: Table 3-3, Example 2.

lines in a line chart for the same data. Nevertheless, a vivid, well-balanced, and attractive bar chart is appreciated by most readers.

The bars in a bar chart may be arranged in vertical or horizontal manner, depending on the artistic preference of the chart maker. In general, as in a line chart, vertical bars are used in presenting data classified by time and quantity, whereas horizontal bars are preferred in presenting data classified by place and quality. Examples 3 and 4 are used to illustrate the methods of constructing bar charts from statistical tables.

Example 3

The number of trucks manufactured by H. K. Noble Co. during each year from 1967 to 1976 is given in Table 3–4. Use the information to draw a vertical bar chart.

TABLE 3–4

Number of Trucks Manufactured By H. K. Noble Company
1967–1976

(Thousands of Trucks)

Year	Trucks	Year	Trucks
1967	100	1972	200
1968	140	1973	225
1969	130	1974	250
1970	150	1975	275
1971	175	1976	285

Source: Hypothetical data for illustration purpose.

Solution (see Chart 3–8). The production of trucks for each year is represented by a vertical bar. Each bar is placed in the middle of the space representing the year to which the production refers. There are spaces of equal width between individual bars.

CHART 3–8

A Vertical Bar Chart

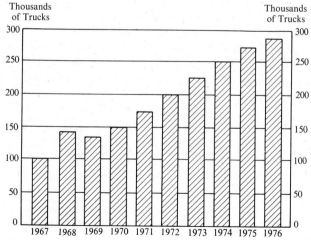

Source of data: Table 3-4, Example 3.

Example 4

Use the numbers of students in the total column of Table 2–2, page 23, to draw a horizontal bar chart.

CHART 3–9

A Horizontal Bar Chart

Number of Students in Freshman English Classes
by Major Fields, 1974–1976

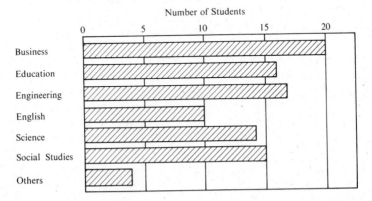

Source of data: Table 2–2, Example 4.

Solution (see Chart 3–9).   The number of students for each major field is represented by a horizontal bar according to the scale above the bars. Each bar is placed at the right-hand side of the major field to which the number of students refers.

Component Part Chart

A component part chart shows the relationships among the individual parts as well as the total or totals of the parts of one or several series of data. The relationships can be expressed in either the actual amounts of the data or the relative values in percentage rates of the data. The most common devices for showing component part charts are bars, lines, or segments of a circle or a pie. The methods of constructing a component part line chart, a component part bar chart, and a pie chart are illustrated in Example 5.

Example 5

The annual production in the three departments of the Loren Manufacturing Company from 1966 to 1976 is shown in Table 3–5. Construct:

(a) Two component part line charts showing the complete information in the table by

(1) The actual units produced by individual departments.

(2) The relative values in percent (%) of units produced by individual departments compared to the total for each year.

(b) Two component part bar charts showing the facts of years 1966 and 1976 by

TABLE 3-5

ANNUAL PRODUCTION, LOREN MANUFACTURING COMPANY
BY DEPARTMENTS, 1966 TO 1976

(Thousands of Units)

Year	Department A	Department B	Subtotal	Department C	Total Units Produced
1966	150	190	340	160	500
1967	170	230	400	170	570
1968	200	150	350	200	550
1969	240	210	450	150	600
1970	200	280	480	220	700
1971	250	300	550	100	650
1972	270	230	500	200	700
1973	300	220	520	260	780
1974	280	320	600	200	800
1975	350	280	630	270	900
1976	400	250	650	150	800

Source: Hypothetical data for illustration purpose.

(1) The actual units produced by individual departments.
(2) The relative values in percent (%) of units produced by individual departments compared to the total for each of the two years.
(c) A pie chart showing the facts in 1976.

Solution. (a) (1) Chart 3–10
(2) Chart 3–11
(b) (1) Chart 3–12
(2) Chart 3–13
(c) Chart 3–14

In Chart 3–10, the individual parts representing the actual units produced by the three departments from 1966 to 1976 are shaded in different ways. The area in each part for each year is determined according to the units produced by each department. For example, the area representing the number of units produced by Department B in 1966 is located on the data scale between 150 and 340 (see subtotal column in Table 3–5), since the difference between the two points on the scale is 190 (thousands of units).

Before Chart 3-11 is constructed, the percentages of the units produced by individual departments based on the total for each year must be computed. The percentages computed from the data given in Table 3–5 are shown in Table 3–6. Each percentage is obtained by dividing the units produced in the department by the total units produced in the year. The percentage of the total for each year is always 100% ($=1$).

CHART 3–10

ILLUSTRATION OF COMPONENT PART LINE CHART SHOWING
ACTUAL AMOUNTS OF DATA

*Annual Production, Loren Manufacturing Company
by Departments, 1966 to 1976*

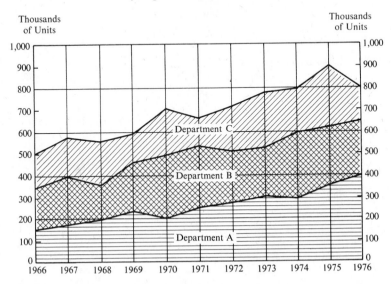

Source of data: Table 3–5, Example 5.

In Table 3–6, the method of rounding as stated on p. 41 is used in computing the percentages to two decimal places for some years. When the total of the individual percentages after the rounding process is not equal to 100.00%, the larg-

TABLE 3-6

ANNUAL PRODUCTION, LOREN MANUFACTURING COMPANY
BY DEPARTMENTS, 1966 TO 1976

(In Percent of Total Production)

Year	Department A	Department B	Subtotal	Department C	Total
1966	30.00	38.00	68.00	32.00	100.00
1967	29.82	40.36	70.18	29.82	100.00
1968	36.36	27.28	63.64	36.36	100.00
1969	40.00	35.00	75.00	25.00	100.00
1970	28.57	40.00	68.57	31.43	100.00
1971	38.46	46.16	84.62	15.38	100.00
1972	38.57	32.86	71.43	28.57	100.00
1973	38.46	28.21	66.67	33.33	100.00
1974	35.00	40.00	75.00	25.00	100.00
1975	38.89	31.11	70.00	30.00	100.00
1976	50.00	31.25	81.25	18.75	100.00

Source: Computed from Table 3-5.

est percentage among the group is increased or decreased to bring the total to 100%. For example, the percentages in 1971 are computed and rounded to two decimal places as follows:

Department A 250 ÷ 650 = 0.384615 or 38.4615%, rounded to 38.46%
Department B 300 ÷ 650 = 0.461538 or 46.1538%, rounded to 46.15%
Department C 100 ÷ 650 = 0.153846 or 15.3846%, rounded to 15.38%
                                                    Total   99.99%

However, 38.46% + 46.15% + 15.38% = 99.99%. Thus, the largest percentage rate, 46.15%, is increased to 46.16% to bring the sum of the three rates to 100%.

The above method is simple in application and can give the *smallest relative error based on individual percentage*. It is, therefore, followed throughout this text.

Note that there are various methods of forcing a total to 100%. For example, it is also possible to increase or decrease a percentage which will give the *smallest amount of change*. However, this method requires detailed analysis on the computed relatives. The work of the analysis is too tedious for practical applications.

CHART 3–11

ILLUSTRATION OF COMPONENT PART LINE CHART SHOWING
PERCENTAGES OF DATA

*Annual Production, Loren Manufacturing Company
by Departments, 1966 to 1976*

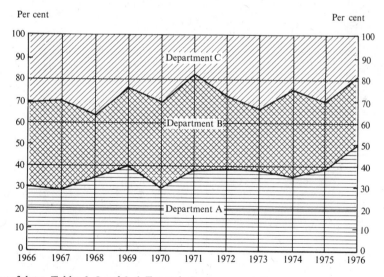

Source of data: Tables 3–5 and 3–6, Example 5.

In Chart 3–12, the height of each part of a bar is made according to the number of units produced by each department. The location of each part in individual bars is determined in the same manner as the method used in locating each part in Chart 3–10. Again, the individual parts are shaded in different ways to show distinctively the data represented.

Before Chart 3–13 is constructed, the percentages of the units produced by individual departments based on the total of each year for 1966 and 1976 must be computed. The percentages can be obtained from Table 3–6 for the two years. The percentage chart shows the relative changes of the production by departments.

In Chart 3–14, the pie chart, a circle is divided proportionally into component parts according to the units produced by the individual departments. The circle, which represents the total of the parts, may be conveniently divided into either 360 degrees or 100 equal parts by means of a printed form or a protractor. However, if a 360 degree protractor is used, each of the percentage rates should be multiplied by 3.6 before the data are plotted. The size of each part in the pie chart is computed in Table 3–7.

CHART 3–12	CHART 3–13
ILLUSTRATION OF COMPONENT PART BAR CHART, SHOWING ACTUAL AMOUNTS OF DATA	ILLUSTRATION OF COMPONENT PART BAR CHART, SHOWING PERCENTAGES OF DATA
Annual Production *Loren Manufacturing Company* *by Departments, 1966 and 1976*	*Annual Production* *Loren Manufacturing Company* *by Departments, 1966 and 1976*

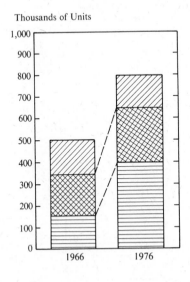 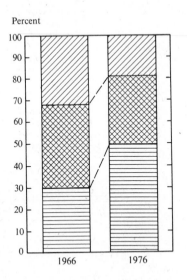

Department C   Department B   Department A

Source of data: Table 3–5, Example 5.

Source of data: Tables 3–5 and 3–6, Example 5.

CHART 3–14

ILLUSTRATION OF PIE CHART

Production, Loren Manufacturing Company,
by Departments, 1976

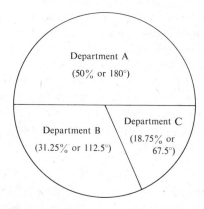

Source of data: Table 3-7, Example 5.

TABLE 3-7

COMPUTATION FOR CONSTRUCTING THE PIE CHART
PRODUCTION, LOREN MANUFACTURING COMPANY, BY DEPARTMENTS, 1976

(1) *Department*	(2) *Units Produced* *(thousands)*	(3) *Percent (%)* *(2) ÷ 800*	(4) *Degree (°)* *(3) × 3.6*
A	400	50.00	180°
B	250	31.25	112.5°
C	150	18.75	67.5°
Total	800	100.00	360.0°

Source: Table 3-5.

The number of degrees may also be computed directly from the original data. The total units produced (800) correspond to the total number of degrees in the circular arc (360°). Thus, 1 unit corresponds to 360°/800. The production of Department A, 400 units, now corresponds to an arc of 400(360°/800) = 180°. In the same manner, the productions of Departments B and C can be computed as 250(360°/800) = 112.5° and 150(360°/800) = 67.5°, respectively.

Area and Volume Charts (Dimensions Charts)

Instead of heights (one dimension) of the bars with the same width being used to represent data, areas (two dimensions) or volumes (three dimensions) may be used in a chart. However, the work of comparing one area with another area is more difficult than that of comparing bars. It is even more difficult to compare volumes. For this reason, many statisticians prefer not to use area or volume charts except for comparisons in special cases.

The common types of area graphs are squares, rectangles, and circles. Three-dimensional graphs may be in forms of cubes, spheres, and cylinders. In the construction of area or volume charts, certain mathematical knowledge is necessary. For example, the size of one side of a square area is the square root of the area. If the area is 9 square feet, each of the two equal sides (length and width) is 3 feet, since $3 \times 3 = 9$. The size of one side of a cube is the cubic root of the volume. Thus, if the volume of the cube is 8 cubic inches, each of the three equal sides (length, width, and height) is 2 inches, since $2 \times 2 \times 2 = 8$. The methods of finding a square root and a cubic root are further discussed in Chapter 4, because the methods are useful in statistical analysis.

Example 6

Assume that the number of employees in the Sooner Machine Company is 2,000 in 1970 and is 16,000 in 1976. Construct the following types of charts to represent the given data:

    (a) An area chart using square areas
    (b) An area chart using circles
    (c) A volume chart using cubes

Solution. (a) The square areas are drawn in Chart 3–15. The number of employees in 1976 is 8 times the number in 1970, or $16,000/2,000 = 8$. Thus, if the area representing the number of employees in 1970 is $\frac{1}{4}$ square inch (each side being $\frac{1}{2}$ inch, since $\frac{1}{2} \times \frac{1}{2} = \frac{1}{4}$), the area representing the number in 1976 must be

CHART 3–15

AN AREA (SQUARE) CHART

Number of Employees, Sooner Machine Company, 1970 and 1976

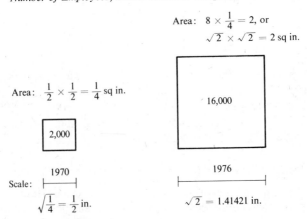

Source of data: Example 6(a).

2 square inches, or $\frac{1}{4} \times 8 = 2$. Each side of a 2 square-inch area is the square root of 2, or 1.4 inches. (See table of squares, Table 1 in the Appendix. $\sqrt{2} = 1.41421$.)

(b) The circle areas are drawn in Chart 3-16. In general, the area of a circle is the product of the squared radius (r^2) and the constant value 3.1416 (usually represented by the symbol π), or

$$\text{Area} = 3.1416(r^2)$$

Let A represent the number of employees in 1970 and A equal the area of a circle with the radius $\frac{1}{2}$ inch, or

$$A = 3.1416 \times (\tfrac{1}{2})^2 = .7854 \text{ sq in.}$$

Let B represent the number of employees in 1976 and B equal the area of a circle with the radius r inches, or

$$B = 3.1416(r^2)$$

Also,

$$B = 8A = 8 \times 3.1416 \times (\tfrac{1}{2})^2$$

since area B (representing 16,000 employees) is 8 times area A (representing 2,000 employees).

Equate B's and solve for r:

$$3.1416(r^2) = 8 \times 3.1416 \times (\tfrac{1}{2})^2$$
$$r^2 = 8 \times (\tfrac{1}{2})^2$$
$$r = \sqrt{8(\tfrac{1}{2})^2} = \sqrt{2} = 1.41421 \text{ inches}$$

The answer indicates that if we wish to have a circle with an area B which is 8 times area A, the radius of B must be the square root of 8 times the squared radius of A, or $\sqrt{8(\tfrac{1}{2})^2}$.

CHART 3–16

AN AREA (CIRCLE) CHART

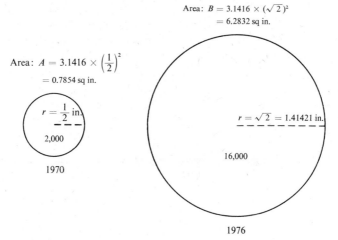

Area: $A = 3.1416 \times \left(\dfrac{1}{2}\right)^2$
= 0.7854 sq in.

$r = \dfrac{1}{2}$ in.

2,000

1970

Area: $B = 3.1416 \times (\sqrt{2})^2$
= 6.2832 sq in.

$r = \sqrt{2} = 1.41421$ in.

16,000

1976

Source of data: Example 6(b).

Check: $B = 3.1416(\sqrt{2})^2 = 3.1416 \times 2 = 6.2832$ sq. in. which is equal to $8A = 8 \times .7854 = 6.2832$.

(c) The volumes are drawn in Chart 3–17. If the volume representing the number of employees in 1970 is $\frac{1}{8}$ cubic inch (each side being $\frac{1}{2}$ inch, since

CHART 3–17

A VOLUME (CUBE) CHART

Number of Employees, Sooner Machine Company, 1970 and 1976

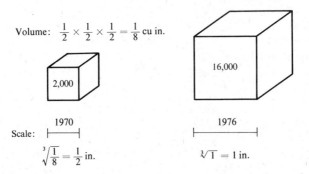

Scale: 1970   1976

$\sqrt[3]{\frac{1}{8}} = \frac{1}{2}$ in. $\sqrt[3]{1} = 1$ in.

Source of data: Example 6(c).

$\frac{1}{2} \times \frac{1}{2} \times \frac{1}{2} = \frac{1}{8}$), the volume representing the number in 1976 must be 1 cubic inch ($\frac{1}{8} \times 8 = 1$). Each side of a 1 cubic inch is the cubic root of 1, or $\sqrt[3]{1} = 1$.

Pictogram

A chart that consists of a number of picture symbols is called a pictogram or pictograph. The symbols are of the same size, and each of them represents the same kind of information with a fixed value. A pictogram is essentially a modified type of bar chart. Whereas the length of each bar represents the magnitude of a given item in a bar chart, the number of picture symbols shows the magnitude in a pictogram. Statistical presentation by pictogram is particularly useful in stimulating the reader's interest or in showing the data to a layman, because it is readily self-explanatory and usually is presented in an interesting and pleasant manner. Example 7 is used to illustrate the application of a pictogram.

Example 7

Assume that the number of television sets produced by the Northern Electric Company in 1974 was 6,000 sets and in 1976 was 8,500 sets. Draw a pictogram to show the above data.

Solution. The pictogram is shown in Chart 3–18. Each symbol in the chart represents 1,000 sets, and the symbols are of the same size.

CHART 3–18

A PICTOGRAM

Annual Production of Television Sets
Northern Electric Company, 1974 and 1976

Each symbol represents 1000 television sets.

Source of data: Example 7.

Statistical Maps

Statistical maps show the quantitative information on a geographical basis. Comparison by geographical areas is greatly facilitated when the information is plotted on the geographical units being compared. The data may be shown in a map in one or a combination of the following common forms: dots, picture symbols, crosshatchings, colors, bars, and numbers. Chart 3-19 shows the use of statistical maps in presenting data.

CHART 3–19

A STATISTICAL MAP

Sales of Early Department Stores
in the United States, by States,
1976

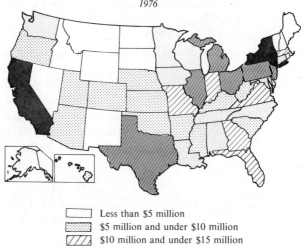

☐ Less than $5 million
▦ $5 million and under $10 million
▨ $10 million and under $15 million
▤ $15 million and under $20 million
■ $20 million and over

SOURCE: Hypothetical data.

Exercises 3

1. Explain the three ways of presenting organized data. Indicate the differences.

2. What are the differences between general purpose tables and special purpose tables? How would you place them in a research project report?

3. State the principal parts of a table. Does every statistical table have all of the principal parts mentioned in this chapter?

4. What is a cross-classified table? How would you use it effectively?

5. Round the following numbers.
   a. To whole numbers (drop decimal places): 95.624; 43.5001; 3.48923; 621.500; 218.5000.
   b. To thousands: 52,863.46; 67,490; 317,500; 2,674,005.467; 7,986,500.

6. The *Federal Reserve Bulletins,* issued monthly by the Board of Governors of the Federal Reserve System, have the following information concerning gross national product:

Gross national product amounted to $1,397.3 billion in 1974, $102.4 billion more than the amount in 1973. The changes of the four components of the gross national product from 1973 to 1974 are as follows:

(1) Personal consumption expenditures rose to $876.7 billion in 1974, including durable goods, $127.5 billion; nondurable goods, $380.2 billion; and services, $369.0 billion. The expenditures in 1973 were $805.2 billion, including durable goods, $130.3 billion; nondurable goods, $338.0 billion; and services, $336.9 billion.

(2) Gross private domestic investment showed $209.4 billion in 1974. The breakdowns of the investment in 1974 were fixed investment, $195.2 billion and business inventories increased, $14.2 billion. The total investment in 1973 was $209.4 billion. The breakdowns in 1973 were fixed investment $194.0 billion and business inventories increased, $15.4 billion.

(3) Net exports of goods and services were $2.0 billion in 1974, $1.9 billion less than the net amount of 1973.

(4) Government purchases of goods and services totaled $309.2 billion in 1974, whereas the purchases in the previous year totaled $276.4 billion. The 1974 purchases included $116.9 billion for federal and $192.3 billion for state and local governments. The federal government purchased $106.6 billion, and the state and local governments purchased $169.8 billion in 1973.

All the above figures are rounded to billions of dollars from the original figures. The details stated may not add to the totals because of rounding. Construct a table showing the above information and the changes (indicating increase or decrease) from 1973 to 1974 for each item.

7. The number of units shipped by Donax Company for each year from 1967 to 1976 is given in Table 3–8.

a. State if these are period data or point data.

b. Construct a line chart.

c. Construct a vertical bar chart.

TABLE 3-8

UNITS SHIPPED BY DONAX COMPANY, 1967–1976

(Thousands of Units)

Year	Number of Units	Year	Number of Units
1967	30	1972	100
1968	50	1973	70
1969	40	1974	110
1970	80	1975	90
1971	60	1976	130

Source: Hypothetical data.

8. The assets of United Lindy Company at the end of each year from 1969 to 1978 is given in Table 3–9.

a. State if these are period data or point data.

b. Construct a line chart.

c. Construct a vertical bar chart.

d. Give your findings from the charts constructed.

TABLE 3-9

ASSETS OF UNITED LINDY COMPANY
1969 TO 1978

(Millions of Dollars)

End of Year	Assets	End of Year	Assets
1969	105	1974	145
1970	112	1975	157
1971	120	1976	168
1972	128	1977	176
1973	136	1978	195

Source: *Life Insurance Fact Book,* 1969, p. 66.

9. Construct a line chart showing the number of commercial transits through the Panama Canal and the amount of tolls received for each year from Table 3–10.

TABLE 3-10

TABLE 3-10

NUMBER OF COMMERCIAL TRANSITS THROUGH PANAMA CANAL
AND TOLLS COLLECTED, 1962 TO 1973

Year	Number of Transits	Tolls ($1,000)
1957	8,579	38,444
1958	9,187	41,796
1959	9,718	45,529
1960	10,795	50,939
1961	10,866	54,128
1962	11,149	57,290
1963	11,017	56,368
1964	11,808	61,098
1965	11,834	65,443
1966	11,925	69,095
1967	12,412	76,769
1968	13,199	83,907
1969	13,146	87,423
1970	13,658	94,654
1971	14,020	97,380
1972	13,766	98,765
1973	13,841	111,032

Source: Panama Canal Company, *Annual Report.*

10. From the information in Table 3–11 construct a line chart showing the comparison between reserves and borrowings of member banks. Are they period data or point data?

TABLE 3-11

RESERVES AND BORROWINGS OF MEMBER BANKS
DECEMBER, 1973 TO DECEMBER, 1974

(Averages of Daily Figures, In Millions of Dollars)

Period	Total Reserves Held	Borrowings at Federal Reserve Banks
1973		
December	35,430	1,039
1974		
January	36,655	1,044
February	35,242	1,186
March	34,966	1,352
April	35,929	1,714
May	36,519	2,580
June	36,390	3,000
July	37,338	3,308
August	37,029	3,351
September	37,076	3,287
October	36,796	1,793
November	36,837	1,285
December	36,941	703

Source: *Federal Reserve Bulletin.*

11. The short-term claims on foreigners at the end of September, 1974, reported by nonbanking concerns in the United States are given in Table 3–12. Construct a horizontal bar chart.

TABLE 3-12

SHORT-TERM CLAIMS ON FOREIGNERS REPORTED BY NONBANKING CONCERNS
END OF SEPTEMBER, 1974

Area or Country	Claims (millions of dollars)
Europe	4,438
Canada	1,567
Latin America	2,236
Asia	2,030
Africa	321
All other	179
Total	10,771

Source: *Federal Reserve Bulletin*

12. From the information in Table 3–13, construct a bar chart showing the comparison between the production by Division *A* and that by Division *B* in Stevenson Factory for each year from 1972 to 1978.

TABLE 3–13

PRODUCTION IN STEVENSON FACTORY
BY DIVISIONS, 1972 TO 1978

Year	Units Produced	
	Division A	Division B
1972	2,400	8,500
1973	3,200	8,700
1974	4,300	9,200
1975	5,700	9,700
1976	6,900	10,200
1977	8,200	12,600
1978	9,500	13,800

13. The data in Table 3–14 give the civilian labor force in the United States from 1970 to 1974. Construct component part line charts showing the complete data in (1) numbers of persons and (2) percentage rates.

14. Refer to the information given in Problem 13. Construct:
   a. Two component part bar charts showing the facts of years 1970 and 1974 in (1) numbers of persons and (2) percentage rates.
   b. A pie chart showing the facts in 1974.

15. The following information was obtained from Bureau of Labor Statistics:
   The average hourly earning for the workers in contract construction companies in January, 1975 was $7.14 per hour, and that for the workers in leather

TABLE 3-14

CIVILIAN LABOR FORCE IN THE UNITED STATES
1970 TO 1974

Employment Status	1970	1971	1972	1973	1974
	(Millions of persons)				
Civilian labor force	82.7	84.1	86.5	88.7	91.0
Employed in nonagricultural industries	75.2	75.7	78.2	81.0	82.4
Employed in agriculture	3.5	3.4	3.5	3.4	3.5
Unemployed	4.0	5.0	4.8	4.3	5.1
	(Percentage distribution %)				
Civilian labor force	100	100	100	100	100
Employed in nonagricultural industries	90.9	90.0	90.4	91.3	90.6
Employed in agriculture	4.2	4.1	4.0	3.8	3.8
Unemployed	4.9	5.9	5.6	4.9	5.6

Source: *Monthly Labor Review.*

products industries was $3.00 per hour during the same period. Using the above data, construct:

    a. An area chart with squares.

    b. An area chart with circles.

16. Refer to the information given in Problem 15. Construct a pictogram.

17. The total annual sales of Campbell Company was $500,000 in 1974 and $2,000,000 in 1976. Construct a volume chart to show the annual sales for each year. Let each side of the cube be $\frac{1}{2}$ inch long for 1974 data. (Hint: $\sqrt[3]{\frac{1}{2}} = .79$)

18. From the data in Table 3–15 construct a crosshatched map showing the employees on nonagricultural payrolls by states. Use the following classifications (in thousands of persons):

2,000 and over

1,000 to 1,999

500 to 999

Under 500

TABLE 3-15

EMPLOYEES ON NONAGRICULTURAL PAYROLLS, BY STATES
DECEMBER, 1974

(Thousands of Persons)

State	Number	State	Number	State	Number
Alabama	1,131	Louisiana	1,192	Ohio	4,199
Alaska	119	Maine	347	Oklahoma	883
Arizona	741	Maryland	1,458	Oregon	830
Arkansas	623	Massachusetts	2,386	Pennsylvania	4,467
California	7,881	Michigan	3,170	Rhode Island	353
Colorado	910	Minnesota	1,502	South Carolina	1,029
Connecticut	1,279	Mississippi	676	South Dakota	220
Delaware	234	Missouri	1,777	Tennessee	1,557
Florida	2,759	Montana	238	Texas	4,406
Georgia	1,767	Nebraska	557	Utah	447
Hawaii	337	Nevada	262	Vermont	159
Idaho	269	New Hampshire	297	Virginia	1,799
Illinois	4,462	New Jersey	2,751	Washington	1,211
Indiana	1,986	New Mexico	365	West Virginia	557
Iowa	1,012	New York	7,082	Wisconsin	1,702
Kansas	794	North Carolina	2,015	Wyoming	136
Kentucky	1,063	North Dakota	192		

Source: United States Department of Labor, Bureau of Labor Statistics.

4 Essential Mathematics for Statistical Analysis

This chapter provides selected basic topics in mathematics that are essential in computing many problems in this text. After statistical data are collected, organized, and presented in an easy-to-read form, mostly in statistical tables, the data are ready to be analyzed. Since statistical analysis deals with quantitative information, mathematical operations are frequently employed during the process of analysis.

4.1 Linear Equation in One Unknown

The basic principles of the fundamental operations in algebra are the same as those in arithmetic. Algebraic operations are essentially extensions of arithmetic operations. However, an understanding of certain terminology in algebra should be helpful in solving algebraic equations.

An algebraic expression, or simply an *expression*, is any symbol (such as the letters a, b, and so on) or combination of symbols that represents a number. When an expression consists of several parts that are connected by plus and minus signs, each of the parts, together with the sign preceding it, is called a *term*. An expression consisting of one term is called a *monomial*, whereas an expression having more than one term is called a *multinomial*, or a *polynomial*. An expression of two terms is also called a *binomial*, and one of three terms is a *trinomial*. For example, expression $+5ax$ or $5ax$ is a monomial; $3x + y$ is a binomial; and the expression $ax^2 + 4x + 7$ is a trinomial.

If two or more numbers are multiplied together, such as $5ax$, each number $(5, a, x)$ or the product of any of the numbers $(5a, 5x, ax)$ is called a *factor*. Any factor of a term is called the *coefficient* of the remaining factors. When a factor is an explicit number, it is called the *numerical coefficient* of the term; other factors are called *literal coefficients*. In the term $5ax$, the number 5 is the numerical coefficient of ax, and the factor ax is the literal coefficient of 5. When no numerical coefficient is indicated in the term, it is understood that the numerical coefficient is one, such as $y = 1y$.

An equation shows that two algebraic expressions are equal. For example, $3x + 4 = 10$ is an equation. The equation shows that the left *side* of the equation $(3x + 4)$ is equal to the right side of the equation (10). The letter x, whose value is desired, is called the *unknown*. When the desired value is found, the value is called the *solution* or the *root*. Thus, the value 2, which satisfies the equation $[3(2) + 4 = 10]$ is the solution.

The number of powers of the single unknown in an equation indicates the degree of an equation. Equations of the first power are called *linear equations*. Thus, $3x + 4 = 10$ is a linear equation because the unknown value $x = x^1$.

There are two types of equations. The equation discussed above is called a *conditional equation* because only when x represents 2 are the sides equal. The other type of equation is called *identity*. In an identity, x may represent any value, and the two sides are equal. For example, $2x + 3x = 5x$ is an identity, because the two sides are equal when x represents any value. For example, let $x = 1$. The left side becomes $2(1) + 3(1) = 5$, and the right side becomes $5(1) = 5$. Our interest, however, is concentrated on conditional equations.

If both sides of an equation have the same number added to or subtracted from them, or are multiplied by or divided by the same number, the two sides remain equal. This property is the basis for solving linear equations.

Example 1

Solve $5x + 2 = 8x - 19$.

Solution.   Subtract $8x$ from both sides to remove $8x$ from the right.

$$5x + 2 - 8x = 8x - 19 - 8x$$

Combine like terms.

$$-3x + 2 = -19$$

Subtract 2 from both sides to remove 2 from the left.

$$-3x + 2 - 2 = -19 - 2$$

Combine like terms.

$$-3x = -21$$

Divide both sides by (-3), the coefficient of x.

$$\frac{-3x}{-3} = \frac{-21}{-3}$$
$$x = 7$$

The solution is checked by substituting $x = 7$ in the original equation:

$$5(7) + 2 = 8(7) - 19$$
$$35 + 2 = 56 - 19$$
$$37 = 37$$

The operations presented above may be simplified by moving the terms from one side of the equation to the other side after changing their signs; that is, from $+$ to $-$, $-$ to $+$, $\times$ to $\div$, and $\div$ to $\times$. The order of the operations is usually to move all the terms containing the unknown to the left side and all other terms to the right side until the unknown remains alone on the left side.

Example 1 is simplified as follows:

$$5x + 2 = 8x - 19$$
$$5x - 8x = -19 - 2$$
$$-3x = -21$$
$$x = \frac{-21}{-3} = 7$$

Example 2

Solve for x:

$$\frac{9}{x + 4} = \frac{5}{x - 8}$$

Solution.

$$\frac{9(x - 8)}{x + 4} = 5$$
$$9(x - 8) = 5(x + 4)$$
$$9x - 72 = 5x + 20$$
$$9x - 5x = 20 + 72$$
$$4x = 92$$
$$x = \frac{92}{4} = 23$$

Check:

$$\frac{9}{23 + 4} = \frac{5}{23 - 8}$$
$$\frac{9}{27} = \frac{5}{15}$$
$$\frac{1}{3} = \frac{1}{3}$$

Example 3

Solve for x:

$$3ax + b = 6c$$

Solution.

$$3ax = 6c - b$$
$$x = \frac{6c - b}{3a}$$

Check:

$$3a\left(\frac{6c - b}{3a}\right) + b = 6c$$
$$6c - b + b = 6c$$
$$6c = 6c$$

4.2 Systems of Linear Equations

A *system* of equations is a group of two or more related equations. If there are two linear equations in two unknowns and if there is only one solution for each unknown that satisfies both equations, the two equations of the system are called *independent simultaneous equations*, or simply *independent equations*. For example, $2x + y = 5$ and $3x - 2y = 11$ are two linear equations in two unknowns (x and y), and they are independent equations because only $x = 3$ and $y = -1$ will satisfy both equations (see Example 4 below).

If two equations can be changed to the same linear equation in two unknowns, they are called *dependent equations*. A system of dependent equations has an un-limited number of solutions. For example, $x + y = 5$ and $2x + 2y = 10$ are dependent equations, because the latter can be changed to the form of the previous one by dividing each term by 2. The equation $x + y = 5$ is satisfied by unlimited pairs of numbers such as $x = 1, y = 4$; $x = 2, y = 3$; and $x = 3, y = 2$. If x is equal to any value, there is a solution for y in the equation.

If two linear equations in two unknowns have no common solution, they are *inconsistent equations*. For example, $x + y = 3$ and $x + y = 4$ are inconsistent equations, since there is no common solution for x or y in the two equations.

Independent equations are of particular interest to us. There are four common methods of solving independent linear equations simultaneously: (1) elimination by addition or subtraction, (2) elimination by substitution, (3) graph, and (4) matrix algebra. The use of matrix algebra to solve equations is presented in Chapter 27. Example 4 is used to illustrate the first three methods.

Example 4

Solve the two equations simultaneously:

Equation (**1**) $2x + y = 5$
Equation (**2**) $3x - 2y = 11$

Solution. (a) *Addition or Subtraction.*

Eliminate y by Addition	*Eliminate x by Subtraction*

Eliminate y by Addition

Multiply Eq. (1) by 2.

$$4x + 2y = 10 \qquad (3)$$

Rewrite Eq. (2)

$$3x - 2y = 11 \qquad (2)$$

Add Eqs. (3) + (2)

$$7x = 21$$

$$x = \frac{21}{7} = 3$$

Substitute $x = 3$ in Eq. (1).

$$2(3) + y = 5$$

$$y = 5 - 6 = -1$$

Eliminate x by Subtraction

Multiply Eq. (1) by 3:

$$6x + 3y = 15 \qquad (4)$$

Multiply Eq. (2) by 2:

$$6x - 4y = 22 \qquad (5)$$

Subtract Eq. (4) − (5)

$$+7y = -7$$

$$y = \frac{-7}{7} = -1$$

Substitute $y = -1$ in Eq. (1).

$$2x + (-1) = 5$$

$$2x = 5 + 1 = 6$$

$$x = \frac{6}{2} = 3$$

Check: Substitute $x = 3$ and $y = -1$ in Eqs. (1) and (2).

In Eq. (1): $2(3) + (-1) = 5$ $6 - 1 = 5$ $5 = 5$
In Eq. (2): $3(3) - 2(-1) = 11$ $9 + 2 = 11$ $11 = 11$

(b) *Substitution.* Solve Eq. (1) for y in terms of x,

$$y = 5 - 2x \qquad (6)$$

Substitute Eq. (6) in Eq. (2). [Note: Do not substitute Eq. (6) in Eq. (1) since (6) is derived from (1).]

$$3x - 2(5 - 2x) = 11$$
$$3x - 10 + 4x = 11$$
$$7x = 11 + 10 = 21$$
$$x = \frac{21}{7} = 3$$

Substitute $x = 3$ in Eq. (1).

$$2(3) + y = 5$$
$$y = -1$$

(c) *Graph.* A linear equation in one unknown has only one solution, but a linear equation in two unknowns has an unlimited number of solutions. Here, only three pairs of the answers for x and y are selected for each of the two given equations and are arranged as follows:

Equations	x	y
	0	5
(1) $2x + y = 5$	2	1
	5	-5
	1	-4
(2) $3x - 2y = 11$	5	2
	-1	-7

The selected answers are plotted in Chart 4–1 according to the system of rectangular co-ordinates presented on p. 43. When the three points plotted for each equation are connected, straight lines (1) and (2) are formed. Any point in the straight line representing each equation will satisfy the equation. Thus, equations of the first power are called linear equations. Lines (1) and (2) in Chart 4–1 intersect at the point where $x = 3$ and $y = -1$. The values thus satisfy both Eqs. (1) and (2).

CHART 4-1

GRAPHIC METHOD (SOLUTION TO EXAMPLE 4)

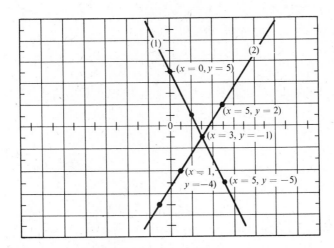

In a similar manner, a system of three linear equations in three unknowns may be solved by the methods of elimination and substitution.

Example 5

Solve the three equations simultaneously:

$$x + y + z = 9 \tag{1}$$
$$3x - y + 2z = 11 \tag{2}$$
$$2x + 3y - 5z = -7 \tag{3}$$

Solution.   Eliminate y from Eqs. (1) and (2).

(1) + (2) $\qquad\qquad 4x + 3z = 20$ (4)

Eliminate y from Eqs. (2) and (3).

(2) × 3 $\qquad\qquad 9x - 3y + 6z = 33$ (5)

(3) + (5) $\qquad\qquad 11x + z = 26$ (6)

Now a new system of two linear equations, (4) and (6), in two unknowns, x and z, is obtained.

Eliminate z from Eqs. (4) and (6).

(6) × 3 $\qquad\qquad 33x + 3z = 78$ (7)

(7) − (4) $\qquad\qquad 29x = 58$

$$x = \frac{58}{29} = 2$$

Substitute $x = 2$ in Eq. (6).

$$11(2) + z = 26$$
$$z = 26 - 22 = 4$$

Substitute $x = 2$, $z = 4$, in Eq. (1).

$$2 + y + 4 = 9$$
$$y = 9 - 6 = 3$$

Check: Substitute $x = 2$, $y = 3$, and $z = 4$ in Eq. (2).

$$3(2) - 3 + 2(4) = 6 - 3 + 8 = 11$$

4.3 Ratio and Proportion

Quantitative information of many things may be expressed in either actual values or relative values for the purpose of comparison. For example, if there are 5,000 students in a college, including 1,000 girls and 4,000 boys, the comparison between the number of girls and the number of boys may be carried out in two ways: (1) 1,000 girls to 4,000 boys (actual values), and (2) 1 girl to 4 boys (relative values) or

$$\frac{1,000 \text{ girls}}{4,000 \text{ boys}} = \frac{1 \text{ girl}}{4 \text{ boys}}$$

The expression of the relative values of various things is called *ratio*. Thus, the latter comparison may be stated: "The ratio of the number of girls to the number of boys in the college is 1 to 4," which may be written $1 : 4$.

In general, the ratio of the first term (the number mentioned first in a statement) to the second term is the quotient of the first term divided by the second term.

$$\text{Ratio of the first term to the second term} = \frac{\text{The first term}}{\text{The second term}}$$

$$= \text{The first term} : \text{the second term}$$

Thus, the ratio of a to b is expressed as a/b, or $a \div b$, or $a : b$. The ratio of b to a is expressed as b/a, or $b \div a$, or $b : a$. The ratio of the number of boys to the number of girls in the above example is $4000/1000 = \frac{4}{1}$, or $4 : 1$, or simply 4. When the division indicated by a fraction representing a ratio is carried out, the quotient then represents the value of the first term, and the value of the second term is always considered to be 1. Thus, $\frac{4}{1} = 4$ has the same meaning as $4 : 1$; $\frac{3}{5} = 0.6$ is equivalent to 0.6 to 1.

Because a ratio may be expressed as a fraction, the rules applying to fractions apply to ratios as well. For example, a fraction is usually changed to its lowest terms in a final answer. Thus, $10/20$ may be changed to $\frac{1}{2}$, and the ratio of 10 to 20 is equal to the ratio of 1 to 2, or $10 : 20 = 1 : 2$.

In a fraction, when each term of the fraction is multiplied or divided by the same number (other than zero), the value of the fraction is not changed. The principle may be extended in a ratio when more than two relative values are expressed in ratio form. For example, the relative values of three things, A, B, and C, are 2, 3, and 7.5, respectively. If $A = 2$, then $B = 3$, and $C = 7.5$; if one of them is doubled, then the others and the total of them are also doubled; if one of them is multiplied by 8, then the others and the total of them must also be multiplied by 8; and so on. This property is illustrated in Table 4–1.

TABLE 4-1

ILLUSTRATION OF EQUIVALENT RATIOS

Terms (things)	Ratio in Lowest Terms	Ratio After Each Term Is Doubled	Ratio After Each Term Is Multiplied by 8
A	2	$4 (= 2 \times 2)$	$16 (= 2 \times 8)$
B	3	$6 (= 3 \times 2)$	$24 (= 3 \times 8)$
C	7.5	$15 (= 7.5 \times 2)$	$60 (= 7.5 \times 8)$
Total	12.5	$25 (= 12.5 \times 2)$	$100 (= 12.5 \times 8)$

$$A : B : C = 2 : 3 : 7.5 = 4 : 6 : 15 = 16 : 24 : 60$$

The illustration indicates that when a group of relative numbers needs to be changed to a group with the same ratio and a desired total, each term in the given group must be multiplied by the quotient of the desired total divided by the given total. Observe the illustration. If the group of the three relative numbers, 2, 3, and 7.5, needs to be changed to a group with the same ratio and a desired total of 100, each term must be multiplied by 100/12.5, or 8. The method is useful in changing an obtained total to a desired total, such as in some precentage problems.

A *proportion* is a statement which indicates that two ratios are equal. Thus, $a/b = c/d$, or $a : b = c : d$ is a proportion. The proportion is read, "the ratio of a to b is equal to the ratio of c to d," or "a is to b as c is to d." The letters a, b, c, and d are the *terms* of the proportion; a and d are the extremes; b and c are the *means*. When any three of the four terms in a proportion are given, the other unknown term can always be found by either one of the following two methods:

1. Cross multiplication. This method is used when the proportion is written as an equation of two fractions, or

$$\frac{a}{b} = \frac{c}{d}$$

The result of cross multiplication is written $ad = bc$. If both sides of the proportion are multiplied by bd, the common denominator, the proportion becomes

$$\frac{a}{\not b}(\not b d) = \frac{c}{\not d}(b\not d) \quad \text{or} \quad ad = bc$$

2. Expression that the product of the extremes (ad) equals the product of the means (bc). This method is used when the proportion is written as an equation of two ratios, each ratio being connected by a colon, or

$$a : b = c : d$$

The result can also be written as $ad = bc$.

Example 6

Solve for x: $7 : 9 = x : 8$.

Solution. Because the product of the extremes equals the product of the means,

$$9x = 7(8) = 56$$
$$x = \frac{56}{9} = 6\tfrac{2}{9}$$

$$\text{Check: Right side} = x : 8 = \frac{6\tfrac{2}{9}}{8} = \frac{\frac{56}{9}}{8} = \frac{56}{9} \times \frac{1}{8} = \frac{7}{9}$$
$$= 7 : 9 = \text{left side}$$

Example 7

Solve for x: $\dfrac{x}{24} = \dfrac{5}{16}$

Solution.   By cross multiplication,

$$16x = 24(5) = 120$$

$$x = \frac{120}{16} = 7\tfrac{8}{16} = 7\tfrac{1}{2}$$

$$\text{Check: Left side} = \frac{x}{24} = \frac{7\tfrac{1}{2}}{24} = \frac{\tfrac{15}{2}}{24} = \frac{15}{2} \times \frac{1}{24} = \frac{5}{16} = \text{Right side}$$

4.4   Percentage

When the second term in a ratio is 100, the ratio may be written in a special form called *per cent*. The symbol for per cent is %. The unit value of a per cent, or 1 %, is one-hundredth (1/100 or 0.01). For example, the ratio of 25 to 100 may be written as 25%, which is equivalent to the common fraction 25/100 or the decimal fraction 0.25. The word percentage refers to problems in which hundredths or per cents are used as a basis of calculation or comparison.

Basic Operations of Per Cent

Because per cent (%) may be written as a fraction (1/100) or a decimal (0.01), the following basic operations should be regarded as essential in solving percentage problems.

Convert a Per Cent to a Decimal or a Whole Number
   Rule. Move the decimal point in the per cent two places to the *left* and drop the per cent sign %.

Example 8

$$100\% = 1\,00\,\% = 1 \qquad\qquad 3580\% = 35\,80\,\% = 35.80$$

$$74\tfrac{1}{2}\% = 0\,74\,\tfrac{1}{2}\% = 0.74\tfrac{1}{2} \qquad\qquad 64.7\% = 0\,64\,7\% = 0.647$$

$$2.35\% = 0\,02\,35\% = 0.0235 \qquad\qquad 25\% = 0\,25\,\% = 0.25$$

Convert a Per Cent to a Common Fraction
   Rule. First drop the per cent sign %. Second, use the number as the numerator and 100 as the denominator. Third, reduce the fraction to its lowest terms.

Example 9

$$100\% = \frac{100}{100} = 1$$

$$3580\% = \frac{3580}{100} = 35\frac{80}{100} = 35\frac{4}{5}$$

$$74\tfrac{1}{2}\% = \frac{74\tfrac{1}{2}}{100} = \frac{\dfrac{149}{2}}{100} = \frac{149}{2} \times \frac{1}{100} = \frac{149}{200}$$

$$64.7\% = \frac{64.7}{100} = \frac{647}{1,000}$$

$$2.35\% = \frac{2.35}{100} = \frac{235}{10,000} = \frac{47}{2,000}$$

$$25\% = \frac{25}{100} = \frac{1}{4}$$

Convert a Decimal or a Whole Number to a Percentage

Rule. Move the decimal point two places to the *right* and annex a per cent sign %.

Example 10

$$45 = 45.00.\% = 4500\% \qquad 0.56 = 0.56.\% = 56\%$$

$$2.6 = 2.60.\% = 260\% \qquad 0.038 = 0.03.8\% = 3.8\%$$

$$0.0027 = 0.00.27\% = 0.27\% \qquad 1 = 1.00.\% = 100\%$$

Convert a Common Fraction to a Percentage

Rule. First convert a common fraction to a decimal. Next, convert the decimal to a percentage as above.

Example 11

Carry the following numbers to two decimal places before converting them to percentages.

$$\frac{3}{5} = 3 \div 5 = 0.60 = 60\%$$

$$\frac{6}{25} = 0.24 = 24\%$$

$$\frac{21}{4} = 5\tfrac{1}{4} = 5.25 = 525\%$$

$\dfrac{1}{3} = 0.33\tfrac{1}{3} = 33\tfrac{1}{3}\%$, or round to 33% since $\tfrac{1}{3}\%$ is less than $\tfrac{1}{2}$ of the unit to be retained (% unit)

$6\tfrac{5}{9} = 6.55\tfrac{5}{9} = 655\tfrac{5}{9}\%$, or round to 656%

$\dfrac{2}{31} = 0.06\tfrac{14}{31} = 6\tfrac{14}{31}\%$, or round to 6%

Computation

In a percentage problem, the given number of per cents is usually expressed on a *base*. The base is the number which is regarded as a whole or 100% and from which a certain number of per cents is expressed or taken. The number of per cents (%) is called the *percentage rate* or simply *rate*. For example, we state that 10% of the students in a college are girls. The number of students in the college is the base (100%) from which we take 10% as the answer to the number of girls. If there are 2,000 students in the college, 1% of the whole is

$$\frac{2,000}{100} = 20 \text{ students}$$

and 10% of the whole is

$$20 \text{ students} \times 10 = 200 \text{ students}$$

The two stages of computation may be combined as follows:

$$\frac{2,000}{100} \times 10 = 200 \text{ students}$$

or

$$2,000 \times \frac{10}{100} = 2,000 \times 10\% = 200 \text{ students}$$

or the problem may be stated in proportional language as follows:

2,000 students are to 100% as x students are to 10%

Thus,

$$2,000 : 100\% = x : 10\%$$
$$100\% x = 2,000(10\%)$$
$$x = 2,000(10\%) = 200 \text{ students}$$

In general, the computation may be expressed as follows:

Base × rate = product

or simply

$$BR = P$$

If any two of the three items are given, the other unknown item can be solved from the expression.

Example 12

What is 15% of $600?

Solution.   15% is the rate, and $600 is the base from which 15% is taken. Thus,

$$\text{Product} = \$600\% \times 15\% = \$90$$

Check: Since $600 is 100% (base), then 1% should be $600/100 = $6, and 15% should be $6 × 15 = $90.

Example 13

What percentage of $580 is $139.20?

Solution. $580 is the base from which the unknown percentage rate is taken, and the part taken (product) is $139.20. Thus,

$$580 \times \text{rate} = 139.20$$

Solve the above equation.

$$\text{Rate} = \frac{139.20}{580} = 0.24 = 24\%$$

Check: 1% of $580 is $5.80. 24% of $580 is $5.80 × 24 = $139.20.

Example 14

In 1975 the total sales of a company were $35,000; in 1976 sales were $41,300. What was the percentage rate of increase?

Solution.   (a) The amount of increase based on the sales of 1975 is

$$\$41,300 - \$35,000 = \$6,300$$

The percentage rate of increase must also be based on the sales of 1975, or

$$35,000 \text{ (base)} \times \text{rate} = 6,300$$

$$\text{Rate} = \frac{6,300}{35,000} = 0.18 = 18\%$$

The sales of 1976 were 18% more than the sales of 1975.

(b) This problem may be solved in a different way. The percentage rate of 1976 sales based on 1975 sales is 118% and is computed as follows:

$$35,000 \text{ (base)} \times \text{rate} = 41,300$$

$$\text{Rate} = \frac{41,300}{35,000} = 1.18 = 118\%$$

Since the base is always equal to 100%, the precentage rate of increase of 1976 sales over 1975 sales is 118% − 100% = 18%

Check: 35,000 + (35,000 × 18%) = 35,000 + 6,300 = $41,300 (1976 sales)

Example 15

If 45% of a number is 162, what is the number?

Solution.   The number is the base from which 45% is taken, and the product is 162.

$$\text{Base (the number)} \times 45\% = 162$$

$$\text{Base} = \frac{162}{45\%} = \frac{162}{0.45} = 360$$

Check: 1% of the number is 360/100 = 3.6. 45% of the number is 3.6 × 45 = 162.

Example 16

In 1976 the total enrollment in a college was 4,770, which shows a 6% increase over the enrollment in the previous year. What was the enrollment in 1975?

Solution.   The enrollment in 1975 is the base from which the 6% increase was computed.

$$\text{Base} + (\text{base} \times 6\%) = 4{,}770$$

or

$$B + 6\%B = 4{,}770$$
$$B(1 + 6\%) = 4{,}770$$
$$B = \frac{4{,}770}{1 + 0.06} = \frac{4{,}770}{1.06} = 4{,}500$$

The enrollment in 1975 was 4,500 students.

Check: $4{,}500 + (4{,}500 \times 6\%) = 4{,}500 + 270 = 4{,}770$ students enrolled in 1976.

4.5 Exponents and Radicals

Multiplication of Exponentials

When two or more *equal* factors are multiplied together, the product is called a *power* of the factor and can be written in a special form called *exponential*. For example,

$$a \times a \times a = a^3$$

The symbol a is called the *base*. The number 3, which indicates the number of equal fractors of a, is called the *exponent*. The expression a^3 is called the exponential. The exponential a^3 is also called the third *power* of the factor a or *cube a*. The exponent for the first power of a base is 1 (which is usually not written) such as $a^1 = a$. A second power is called a *square*, such as a^2.

When multiplication involves exponentials, the following laws and definitions may be applied to simplify computation.

Law 1.   Bases Are the Same.

$$a^m \cdot a^n = a^{m+n}$$

Example 17
$$2^2 \cdot 2^3 = (2 \cdot 2)(2 \cdot 2 \cdot 2) = 2^5; \quad \text{or} \quad 2^2 \cdot 2^3 = 2^{2+3} = 2^5$$

Example 18
$$x^3 \cdot x^4 = (x \cdot x \cdot x)(x \cdot x \cdot x \cdot x) = x^7; \quad \text{or} \quad x^3 \cdot x^4 = x^{3+4} = x^7$$

The law gives the following two definitions:

1. $$a^0 = 1 \ (a \neq 0)$$

The sign $\neq$ means "not equal to."

Example 19

$$2^3 \cdot 2^0 = 2^{3+0} = 2^3. \text{ Thus,}$$

$$2^0 = \frac{2^3}{2^3} = \frac{8}{8} = 1$$

or

$$2^0 = 1$$

It also can be proved that $5^0 = 1$, $37^0 = 1$, and so on.

2.

$$a^{-m} = \frac{1}{a^m}$$

Example 20

$$2^3 \cdot 2^{-3} = 2^{3+(-3)} = 2^0 = 1$$

Divide both sides of $2^3 \cdot 2^{-3} = 1$ by 2^3.

$$2^{-3} = \frac{1}{2^3}$$

Law 2.   Bases Are Different but Exponents Are the Same.

$$a^m \cdot b^m = (ab)^m$$

Example 21

$$3^2 \cdot 4^2 = 3 \cdot 3 \cdot 4 \cdot 4 = (3 \cdot 4)(3 \cdot 4) = (3 \cdot 4)^2$$

or

$$3^2 \cdot 4^2 = (3 \cdot 4)^2$$

Example 22

$$x^3 \cdot y^3 = (x \cdot x \cdot x)(y \cdot y \cdot y) = (xy)(xy)(xy) = (xy)^3$$

or

$$x^3 \cdot y^3 = (xy)^3$$

Law 3.   Base Is an Exponential.

$$(a^m)^n = a^{mn} \quad \text{or} \quad (a^{1/m})^n = a^{n/m}$$

Example 23

$$(2^3)^2 = (2 \cdot 2 \cdot 2)(2 \cdot 2 \cdot 2) = 2^6, \quad \text{or} \quad (2^3)^2 = 2^{3 \cdot 2} = 2^6$$

Example 24

$$(x^2)^3 = (x \cdot x)(x \cdot x)(x \cdot x) = x^6, \quad \text{or} \quad (x^2)^3 = x^{2 \cdot 3} = x^6$$

Example 25

$$(5^{1/2})^4 = (5^{1/2})(5^{1/2})(5^{1/2})(5^{1/2}) = 5^{(1/2)+(1/2)+(1/2)+(1/2)} = 5^{4/2}$$
$$= 5^2 = 25$$

or

$$(5^{1/2})^4 = 5^{(1/2) \cdot 4} = 5^{4/2} = 5^2 = 25$$

Example 26

$$(x^{1/4})^3 = (x^{1/4})(x^{1/4})(x^{1/4}) = x^{(1/4)+(1/4)+(1/4)} = x^{3/4}$$

or

$$(x^{1/4})^3 = x^{(1/4)3} = x^{3/4}$$

Division of Exponentials

When division involves exponentials, the following laws can be applied to simplify computation.

Law 1.   Bases Are the Same.

$$a^m \div a^n = \frac{a^m}{a^n} = a^{m-n}$$

Example 27

$$3^5 \div 3^3 = \frac{3 \cdot 3 \cdot \cancel{3} \cdot \cancel{3} \cdot \cancel{3}}{\cancel{3} \cdot \cancel{3} \cdot \cancel{3}} = 3^2 = 9$$

or

$$3^5 \div 3^3 = \frac{3^5}{3^3} = 3^{5-3} = 3^2 = 9$$

Example 28

$$2^3 \div 2^7 = \frac{2^3}{2^7} = 2^{3-7} = 2^{-4} = \frac{1}{2^4} = \frac{1}{16}$$

Example 29

$$x^6 \div x^4 = \frac{x^6}{x^4} = x^{6-4} = x^2$$

Example 30

$$y^2 \div y^8 = \frac{y^2}{y^8} = y^{2-8} = y^{-6} = \frac{1}{y^6}$$

Law 2.   Bases Are Different but Exponents Are the Same.

$$\frac{a^m}{b^m} = \left(\frac{a}{b}\right)^m$$

Example 31

$$3^2 \div 6^2 = \frac{3 \cdot 3}{6 \cdot 6} = \frac{3}{6} \cdot \frac{3}{6} = \left(\frac{3}{6}\right)^2 = \left(\frac{1}{2}\right)^2 = \frac{1}{4}$$

or

$$\frac{3^2}{6^2} = \left(\frac{3}{6}\right)^2 = \left(\frac{1}{2}\right)^2 = \frac{1}{4}$$

Example 32

$$\frac{x^5}{y^5} = \left(\frac{x}{y}\right)^5$$

Basic Aspects of Radicals

If $x^4 = A$ and x is unknown, the unknown value may be expressed in a *radical* form $\sqrt[4]{A} = x$. The number 4 written at the upper left of the *radical sign* $\sqrt{}$ is the *index* of the radical, and the quantity represented by the letter A under the sign is called the *radicand*. The radical $\sqrt[4]{A}$ indicates the *4th root* of A. Here, the root is represented by x, which is one of the *4* equal factors of quantity A; that is, $(x)(x)(x)(x) = A$.

When the index is 2, the number 2 is usually omitted and the radical is called the square root. Thus, $\sqrt[2]{25}$ is understood to be the same as $\sqrt{25}$.

When the index is an *even* number, a *positive radicand* has two numerically equal roots; one is positive and the other is negative. However, the radical sign $\sqrt{}$ is used to represent only the *positive root*, called the *principal root*.

Example 33

25 has two square roots:

$$+5 \text{ and } -5 \text{ because } (+5)(+5) = 25 \text{ and } (-5)(-5) = 25$$

However, in keeping with the definition of the radical sign, $\sqrt{25}$ is a positive number. $\sqrt{25} = +5$, not -5. But $-\sqrt{25} = -(+5) = -5$, which is a negative number. Likewise, 16 has two fourth roots: $+2$ and -2, because $(+2)(+2)(+2)(+2) = 16$ and $(-2)(-2)(-2)(-2) = 16$. However, $\sqrt[4]{16} = +2$, not -2.

When the index is an *odd* number, a *positive radicand* has only a positive root, and a negative radicand has only a negative root.

Example 34

$$\sqrt[3]{27} = +3 \text{ because } (+3)(+3)(+3) = 27$$
$$\sqrt[3]{-27} = -3 \text{ because } (-3)(-3)(-3) = -27$$
$$\sqrt[5]{-32} = -2 \text{ because } (-2)(-2)(-2)(-2)(-2) = -32$$

When the index is an *even* number, a radical with a *negative radicand* represents an *imaginary number*, such as $\sqrt{-9}$, since neither $+3$ nor -3 qualifies as a square root of the radicand -9. $(+3)^2 = +9$, and $(-3)^2 = +9$.

When an exponential has a fractional exponent, it can be changed to a radical. The change should be based on the laws of exponents as presented above.

Example 35

Express the exponential $a^{2/3}$ in its radical form.

Solution.   When $a^{2/3}$ is raised to third power,

$$(a^{2/3})^3 = a^2 \quad \text{or} \quad a^{2/3} \cdot a^{2/3} \cdot a^{2/3} = a^2$$

Since $a^{2/3}$ is one of the three equal factors whose product is a^2, the third root of a^2 should be $a^{2/3}$, or, as it is written,

$$a^{2/3} = \sqrt[3]{a^2}$$

In general, when an exponential has a fractional exponent, it can be changed to a radical form according to the definition below:

$$a^{n/m} = \sqrt[m]{a^n}$$

Examples 36 and 37 are used to illustrate the applications of the above definition.

Example 36

Express the exponentials in their radical forms.

$$15^{2/2} = \sqrt[2]{15^2} = \sqrt{15^2} \qquad x^{1/2} = \sqrt{x}$$
$$28^{3/4} = \sqrt[4]{28^3} \qquad\qquad y^{1/3} = \sqrt[3]{y}$$

Example 37

Express the radicals in their exponential forms.

$$\sqrt{16} = 16^{1/2} \qquad \sqrt[5]{x} = x^{1/5}$$
$$\sqrt{45^3} = 45^{3/2} \qquad \sqrt[7]{y^2} = y^{2/7}$$

The coefficient of a radical may be changed to a factor of the radicand. The change should also be made according to the laws of exponents.

Example 38

Express the coefficient of the radical as part of the radicand.

Solution.

$$a\sqrt{b} = a^{2/2} \cdot b^{1/2} = (a^2 b)^{1/2} = \sqrt{a^2 b}$$
$$x \cdot \sqrt[3]{y} = x^{3/3} \cdot y^{1/3} = (x^3 y)^{1/3} = \sqrt[3]{x^3 y}$$

Observe the answers in Example 38.
1. The coefficient is raised to the power of the index of the radical.
2. The result obtained in (1) is a factor of the original radicand.

Example 39

Express the coefficient of the radical as part of the radicand based on the observations from Example 38.

Solution.

$$6 \cdot \sqrt{2} = \sqrt{6^2 \cdot 2} = \sqrt{72}$$
$$3 \cdot \sqrt[4]{2} = \sqrt[4]{3^4 \cdot 2} = \sqrt[4]{162}$$
$$2^3 \cdot \sqrt{5} = \sqrt{(2^3)^2 \cdot 5} = \sqrt{64 \cdot 5} = \sqrt{320}$$

Computation of Radicals

Since the fundamental operations involving radicals in this text are limited to square roots, only the radicals with index 2 are presented below.

Addition and Subtraction
The like radicals (radicals with the same indexes and radicands) can be added or subtracted. The sum or remainder of the coefficients of the like radicals is the coefficient of the common radical in the answer.

Example 40

Combine $15\sqrt{7} + 4\sqrt{7} - 8\sqrt{7} + \sqrt{7}$

Solution.

$$15\sqrt{7} + 4\sqrt{7} - 8\sqrt{7} + \sqrt{7} = (15 + 4 - 8 + 1)\sqrt{7}$$
$$= 12\sqrt{7}$$

Example 41

Combine $\sqrt{72} - \sqrt{8} + \sqrt{50}$

Solution.   Here the three terms have unlike radicals. They are first changed to like radicals. The changes are made by reversing the procedure employed in Examples 38 and 39.

$$\sqrt{72} = \sqrt{6^2 \cdot 2} = 6\sqrt{2}$$
$$\sqrt{8} = \sqrt{2^2 \cdot 2} = 2\sqrt{2}$$
$$\sqrt{50} = \sqrt{5^2 \cdot 2} = 5\sqrt{2}$$

Thus,

$$\sqrt{72} - \sqrt{8} + \sqrt{50} = 6\sqrt{2} - 2\sqrt{2} + 5\sqrt{2}$$
$$= (6 - 2 + 5)\sqrt{2} = 9\sqrt{2}$$

Multiplication
In multiplying radicals, apply the following rule:

$$\sqrt{a} \cdot \sqrt{b} = \sqrt{ab}$$

where a and b are greater than zero.

Proof:

$$\sqrt{a} \cdot \sqrt{b} = a^{1/2} \cdot b^{1/2} = (ab)^{1/2} = \sqrt{ab}$$

Example 42

Multiply $(5\sqrt{4})$ by $(7\sqrt{9})$

Solution.

$$(5\sqrt{4})(7\sqrt{9}) = (5)(7)(\sqrt{4})(\sqrt{9}) = 35\sqrt{(4)(9)} = 35\sqrt{36} = 35(6) = 210$$

or

$$5\sqrt{4} \cdot 7\sqrt{9} = 5(2) \cdot 7(3) = 10 \cdot 21 = 210$$

Example 43

$$8\sqrt{2} \cdot 6\sqrt{7} = (8)(6)\sqrt{(2)(7)} = 48\sqrt{14}$$

Division

In dividing radicals, apply the following rule:

$$\frac{\sqrt{a}}{\sqrt{b}} = \sqrt{\frac{a}{b}}$$

where a and b are greater than zero.

Proof:

$$\frac{\sqrt{a}}{\sqrt{b}} = \frac{a^{1/2}}{b^{1/2}} = \left(\frac{a}{b}\right)^{1/2} = \sqrt{\frac{a}{b}}$$

Example 44

Divide $\sqrt{75}$ by $\sqrt{3}$

Solution.

$$\frac{\sqrt{75}}{\sqrt{3}} = \sqrt{\frac{75}{3}} = \sqrt{25} = 5$$

The quotient obtained from the division of radicals may be a fraction. A fraction is considered to be in its simplest form when the denominator includes no radical.

Example 45

$$\frac{\sqrt{5}}{\sqrt{7}} = \frac{\sqrt{5} \cdot \sqrt{7}}{\sqrt{7} \cdot \sqrt{7}} = \frac{\sqrt{35}}{7} \text{ (simplest form)}$$

Finding the Square Root

The value of the square root of a number may be directly read from Table 1 provided in the appendix. However, an understanding of finding the square root by computation will aid one's efficiency in locating the answer in the table. Furthermore, it will enable a reader to find the square root when a table is not available or the answer is not provided in the table.

Example 46

Find the square root of 54.2831 to two decimal places.

Solution.   The steps of finding the square root are:

(a) Separate the given number (54.2831) into groups of two digits, starting at the decimal point in both directions. Each group will give one digit in the final answer. This problem requires three decimal places before rounding to the final number (two places.) Thus three groups ($\overline{28}\ \overline{31}\ \overline{00}$) after the decimal point are supplied for the process of finding the square root.

The first digit of the quotient (7) is the largest integral square root of the leftmost group of the given number (54). Write the square of the quotient ($7^2 = 49$) under the first group and subtract as in long division, or

$$\begin{array}{r} 7. \\ \sqrt{54.\ \overline{28}\ \overline{31}\ \overline{00}} \\ 7^2 = 49\,(-) \\ \hline 5 \end{array}$$

(b) Bring down the next group of two digits (28) to make the first remainder (528). Multiply the quotient obtained in (a) by 20 to get the partial divisor ($7 \times 20 = 140$). Divide the remainder by the partial divisor to get the next digit in the quotient (3, since $528 \div 140 = 3+$). Add this quotient digit to the partial divisor, giving the complete divisor ($140 + 3 = 143$). Multiply the complete divisor by the quotient digit and subtract, again as in long division, or

$$
\begin{array}{r}
7.\quad 3 \\
\sqrt{54.\ \overline{28}\ \overline{31}\ \overline{00}} \\
7^2 = 49 \\
\hline
5\ \ 28
\end{array}
$$

$$7 \times 20 = 140$$
$$\underline{+\quad 3}$$
$$143 \times 3 = \underline{4\ \ 29\,(-)}$$
$$99$$

Approximating quotient digit
$$528 \div 140 = 3 +$$

(c) In each successive step for each group of two digits, repeat step (b). The accumulated digits in the quotient are multiplied by 20 for the partial divisor. The complete divisor is the sum of the partial divisor and the digit used as the quotient in the step. Each digit in the quotient after the first digit may be approximated by dividing the remainder by the partial divisor, as done in (b). The complete computation is shown below.

Step (a)

$$\begin{array}{r} 7.\ \ 3\ \ 6\ \ 7 \\ \sqrt{54,\ 28\ 31\ 00} \\ 7^2 = 49 \end{array}$$ (Answer)

Approximating quotient digit

Step (b) $\begin{cases} 7 \times 20 = 140 \\ \quad + \ \ 3 \\ \overline{143 \times 3 =} \end{cases}$

Step (c) $\begin{cases} 73 \times 20 = 1,460 \\ \quad + \quad 6 \\ \overline{1,466 \times 6 =} \\ 736 \times 20 = 14,720 \\ \quad + \qquad 7 \\ \overline{14,727 \times 7 =} \end{cases}$

5	28
4	29
	99 31
	87 96
	11 35 00
	10 30 89
0	01 04 11

$528 \div 140 = 3 +$

$9,931 \div 1,460 = 6 +$

$113,500 \div 14,720 = 7 +$

(Remainder)

(d) *Answer.* $\sqrt{54.2831} = 7.367$, rounded to 7.37.

Check: $7.37^2 = 54.3169$, which is slightly above the given value.

$\qquad$ $7.36^2 = 54.1696$, which is slightly below the given value.

Also, the square of the answer plus the final remainder equals the given number, or

$$7.367^2 + 0.010411 = 54.272689 + 0.010411$$
$$= 54.2831 \text{ (the given number)}$$

Example 47

Find the square root of 542.831 to two decimal places.

Solution.

$$\begin{array}{r} 2\ \ 3.\ \ 2\ \ 9\ \ 8 \\ \sqrt{5\ \ 42,\ 83\ \ 10\ \ 00} \\ 2^2 = 4 \end{array}$$ (answer)

$2 \times 20 = 40$

$\quad + \ 3$

$\overline{43 \times 3 =}$

$23 \times 20 = 460$

$\quad + \ \ 2$

$\overline{462 \times 2 =}$

$232 \times 20 = 4,640$

$\quad + \quad 9$

$\overline{4,649 \times 9 =}$

$2,329 \times 20 = 46,580$

$\quad + \qquad 8$

$\overline{46,588 \times 8 =}$

1	42
1	29
13	83
9	24
4	59 10
4	18 41
	40 69 00
	37 27 04
0	03 41 96

$142 \div 40 = 3 +$

$1,383 \div 460 = 3 +$
(The actual quotient is 2, since 3 is too high.)

$45,910 \div 4,640 = 9 +$

$406,900 \div 46,580 = 8 +$

Answer: $\sqrt{542.831} = 23.298$, rounded to 23.30.

Check: $23.3^2 = 542.89$. Also, $23.298^2 = 542.796804$

$$+ \quad 0.034196$$
$$\overline{ 542.831000}$$

4.6 Logarithms

Finding the Logarithm of a Number

An exponential and its expanded result may be expressed in a different way as follows:

$$7^2 = 49 \text{ may be written as } \log_7 49 = 2$$
$$5^3 = 125 \text{ may be written as } \log_5 125 = 3$$
$$10^3 = 1,000 \text{ may be written as } \log_{10} 1,000 = 3$$

The expression $\log_{10} 1,000 = 3$ reads: "The logarithm of the number 1,000 to the base 10 is 3." From the example, we can see that *a logarithm is an exponent.*

When the base 10 is used, the logarithms are called *common logarithms.* Our discussion in this section will be limited to the use of common logarithms. The subscript, which indicates the base 10 in a common logarithm, is omitted for the sake of simplification. Thus, $\log_{10} 1,000$ has the same indication as $\log 1,000$.

When a number is an exact power of 10, the logarithm of the number is a whole number and can be determined in the following manner:

Exponentials	*Logarithms*
$10^4 = 10,000$	$\log 10,000 = 4$
$10^3 = 1,000$	$\log 1,000 = 3$
$10^2 = 100$	$\log 100 = 2$
$10^1 = 10$	$\log 10 = 1$
$10^0 = 1$	$\log 1 = 0$
$10^{-1} = \dfrac{1}{10} = 0.1$	$\log 0.1 = -1$
$10^{-2} = \dfrac{1}{10^2} = \dfrac{1}{100} = 0.01$	$\log 0.01 = -2$
$10^{-3} = \dfrac{1}{10^3} = \dfrac{1}{1,000} = 0.001$	$\log 0.001 = -3$

When a number is not an exact power of 10, the logarithm of the number is the sum of a whole number (called *characteristic*) and a positive decimal (called *mantissa*), or

log number = characteristic + mantissa

The value of the characteristic can be determined in the same manner as shown above. The mantissa, determined by a complex formula, can be found in a table of mantissas such as Table 4–2. The mantissas of most numbers are unending decimal fractions and can be rounded to any desired number of decimal places. The mantissas in Table 4–2 are rounded to four places and those in Table 2 in the Appendix are rounded to six places. For illustration purposes, only the simple four-place table (Table 4–2) is used in this section.

Example 48

Find log 854.

Solution. The number 854 is larger than 100 but is smaller than 1000. Thus, the logarithm of 854 is larger than the logarithm of 100 but is smaller than the logarithm of 1,000, or

$$\log 100 < \log 854 < \log 1{,}000$$

Since

$$\log 100 = 2 \quad \text{and} \quad \log 1{,}000 = 3$$

then

$$2 < \log 854 < 3$$

or

$$\log 854 = 2 + \text{a positive decimal}$$
$$= 2 + 0.9315 \text{ (from the mantissa table—Table 4–2)}$$
$$= 2.9315$$

The equation can be expressed in exponentials as follows:

$$(10^2 = 100) < (10^{2.9315} = 854) < (10^3 < 1{,}000)$$

Example 49

Find log 0.0854.

Solution. The number 0.084 is larger than 0.01 but is smaller than 0.1. Thus, the logarithm of 0.0854 is larger than the logarithm of 0.01 but is smaller than the logarithm of 0.1, or

$$\log 0.01 < \log 0.0854 < \log 0.1$$

Since

$$\log 0.01 = -2 \quad \text{and} \quad \log 0.1 = -1$$

then

$$-2 < \log 0.0854 < -1$$

or

$$\log 0.0854 = -2 + \text{a positive decimal}$$
$$= -2 + 0.9315 \text{ (from the mantissa table)}$$
$$= 8.9315 - 10$$

or

$$= -1.0685$$

This can be expressed in exponentials as follows:

$$(10^{-2} = 0.01) < (10^{-1.0685} = 0.0854) < (10^{-1} = 0.1)$$

TABLE 4–2

LOGARITHMS OF NUMBERS (N) 100 TO 999

FOUR-PLACE MANTISSAS

(WITH THE HIGHEST DIFFERENCE [D] BETWEEN ADJACENT MANTISSAS ON EACH LINE)

N	0	1	2	3	4	5	6	7	8	9	D
10	0000	0043	0086	0128	0170	0212	0253	0294	0334	0374	43
11	0414	0453	0492	0531	0569	0607	0645	0682	0719	0755	39
12	0792	0828	0864	0899	0934	0969	1004	1038	1072	1106	36
13	1139	1173	1206	1239	1271	1303	1335	1367	1399	1430	34
14	1461	1492	1523	1553	1584	1614	1644	1673	1703	1732	31
15	1761	1790	1818	1847	1875	1903	1931	1959	1987	2014	29
16	2041	2068	2095	2122	2148	2175	2201	2227	2253	2279	27
17	2304	2330	2355	2380	2405	2430	2455	2480	2504	2529	26
18	2553	2577	2601	2625	2648	2672	2695	2718	2742	2765	24
19	2788	2810	2833	2856	2878	2900	2923	2945	2967	2989	23
20	3010	3032	3054	3075	3096	3118	3139	3160	3181	3201	22
21	3222	3243	3263	3284	3304	3324	3345	3365	3385	3404	21
22	3424	3444	3464	3483	3502	3522	3541	3560	3579	3598	20
23	3617	3636	3655	3674	3692	3711	3729	3747	3766	3784	19
24	3802	3820	3838	3856	3874	3892	3909	3927	3945	3962	18
25	3979	3997	4014	4031	4048	4065	4082	4099	4116	4133	18
26	4150	4166	4183	4200	4216	4232	4249	4265	4281	4298	17
27	4314	4330	4346	4362	4378	4393	4409	4425	4440	4456	16
28	4472	4487	4502	4518	4533	4548	4564	4579	4594	4609	16
29	4624	4639	4654	4669	4683	4698	4713	4728	4742	4757	15
30	4771	4786	4800	4814	4829	4843	4857	4871	4886	4900	15
31	4914	4928	4942	4955	4969	4983	4997	5011	5024	5038	14
32	5051	5065	5079	5092	5105	5119	5132	5145	5159	5172	14
33	5185	5198	5211	5224	5237	5250	5263	5276	5289	5302	13
34	5315	5328	5340	5353	5366	5378	5391	5403	5416	5428	13
35	5441	5453	5465	5478	5490	5502	5514	5527	5539	5551	13
36	5563	5575	5587	5599	5611	5623	5635	5647	5658	5670	12
37	5682	5694	5705	5717	5729	5740	5752	5763	5775	5786	12
38	5798	5809	5821	5832	5843	5855	5866	5877	5888	5899	12
39	5911	5922	5933	5944	5955	5966	5977	5988	5999	6010	11
40	6021	6031	6042	6053	6064	6075	6085	6096	6107	6117	11
41	6128	6138	6149	6160	6170	6180	6191	6201	6212	6222	11
42	6232	6243	6253	6263	6274	6284	6294	6304	6314	6325	11
43	6335	6345	6355	6365	6375	6385	6395	6405	6415	6425	10
44	6435	6444	6454	6464	6474	6484	6493	6503	6513	6522	10
45	6532	6542	6551	6561	6571	6580	6590	6599	6609	6618	10
46	6628	6637	6646	6656	6665	6675	6684	6693	6702	6712	10
47	6721	6730	6739	6749	6758	6767	6776	6785	6794	6803	10
48	6812	6821	6830	6839	6848	6857	6866	6875	6884	6893	9
49	6902	6911	6920	6928	6937	6946	6955	6964	6972	6981	9
50	6990	6998	7007	7016	7024	7033	7042	7050	7059	7067	9
51	7076	7084	7093	7101	7110	7118	7126	7135	7143	7152	9
52	7160	7168	7177	7185	7193	7202	7210	7218	7226	7235	9
53	7243	7251	7259	7267	7275	7284	7292	7300	7308	7316	9
54	7324	7332	7340	7348	7356	7364	7372	7380	7388	7396	8
N	0	1	2	3	4	5	6	7	8	9	D

TABLE 4-2

(CONTINUED)

N	0	1	2	3	4	5	6	7	8	9	D
55	7404	7412	7419	7427	7435	7443	7451	7459	7466	7474	8
56	7482	7490	7497	7505	7513	7520	7528	7536	7543	7551	8
57	7559	7566	7574	7582	7589	7597	7604	7612	7619	7627	8
58	7634	7642	7649	7657	7664	7672	7679	7686	7694	7701	8
59	7709	7716	7723	7731	7738	7745	7752	7760	7767	7774	8
60	7782	7789	7796	7803	7810	7818	7825	7832	7839	7846	8
61	7853	7860	7868	7875	7882	7889	7896	7903	7910	7917	8
62	7924	7931	7938	7945	7952	7959	7966	7973	7980	7987	7
63	7993	8000	8007	8014	8021	8028	8035	8041	8048	8055	7
64	8062	8069	8075	8082	8089	8096	8102	8109	8116	8122	7
65	8129	8136	8142	8149	8156	8162	8169	8176	8182	8189	7
66	8195	8202	8209	8215	8222	8228	8235	8241	8248	8254	7
67	8261	8267	8274	8280	8287	8293	8299	8306	8312	8319	7
68	8325	8331	8338	8344	8351	8357	8363	8370	8376	8382	7
69	8388	8395	8401	8407	8414	8420	8426	8432	8439	8445	7
70	8451	8457	8463	8470	8476	8482	8488	8494	8500	8506	7
71	8513	8519	8525	8531	8537	8543	8549	8555	8561	8567	6
72	8573	8579	8585	8591	8597	8603	8609	8615	8621	8627	6
73	8633	8639	8645	8651	8657	8663	8669	8675	8681	8686	6
74	8692	8698	8704	8710	8716	8722	8727	8733	8739	8745	6
75	8751	8756	8762	8768	8774	8779	8785	8791	8797	8802	6
76	8808	8814	8820	8825	8831	8837	8842	8848	8854	8859	6
77	8865	8871	8876	8882	8887	8893	8899	8904	8910	8915	6
78	8921	8927	8932	8938	8943	8949	8954	8960	8965	8971	6
79	8976	8982	8987	8993	8998	9004	9009	9015	9020	9025	6
80	9031	9036	9042	9047	9053	9058	9063	9069	9074	9079	6
81	9085	9090	9096	9101	9106	9112	9117	9122	9128	9133	6
82	9138	9143	9149	9154	9159	9165	9170	9175	9180	9186	6
83	9191	9196	9201	9206	9212	9217	9222	9227	9232	9238	6
84	9243	9248	9253	9258	9263	9269	9274	9279	9284	9289	6
85	9294	9299	9304	9309	9315	9320	9325	9330	9335	9340	6
86	9345	9350	9355	9360	9365	9370	9375	9380	9385	9390	5
87	9395	9400	9405	9410	9415	9420	9425	9430	9435	9440	5
88	9445	9450	9455	9460	9465	9469	9474	9479	9484	9489	5
89	9494	9499	9504	9409	9513	9518	9523	9528	9533	9538	5
90	9542	9547	9552	9557	9562	9566	9571	9576	9581	9586	5
91	9590	9595	9600	9605	9609	9614	9619	9624	9628	9633	5
92	9638	9643	9647	9652	9657	9661	9666	9671	9675	9680	5
93	9685	9689	9694	9699	9703	9708	9713	9717	9722	9727	5
94	9731	9736	9741	9745	9750	9754	9759	9763	9768	9773	5
95	9777	9782	9786	9791	9795	9800	9805	9809	9814	9818	5
96	9823	9827	9832	9836	9841	9845	9850	9854	9859	9863	5
97	9868	9872	9877	9881	9886	9890	9894	9899	9903	9908	5
98	9912	9917	9921	9926	9930	9934	9939	9943	9948	9952	5
99	9956	9961	9965	9969	9974	9978	9983	9987	9991	9996	5
N	0	1	2	3	4	5	6	7	8	9	D

Note: For more detailed mantissas, see Table 2 (six-place mantissas) in the Appendix

In general, the logarithm of a number may be found according to these rules and lists.

Rule 1.   The number is 1 or above. The characteristic is positive and is 1 less than the number of digits in the whole-number part.

Number		Logarithm of the Number	
Value Range	*Digits in Whole-Number Part*	*Charact- eristic*	*Mantissa*
1 to 9.999999 but less than 10	1	0	
10 to 99.99999 but less than 100	2	1	Four-
100 to 999.9999 but less than 1,000	3	2	place
1,000 to 9,999.99 but less than 10,000	4	3	mantissa
10,000 to 99,999.9 but less than 100,000	5	4	table
etc.	etc.	etc.	

Rule 2.   The number is less than 1. The characteristic is negative and is 1 more than the number of zeros between the decimal point and the first nonzero digit.

Number		Logarithm of the number	
Value Range	*Zeros between Decimal Point and the First Nonzero Digit*	*Charac- teristic*	*Mantissa*
0.1 to 0.999999 but less than 1	none	-1	Four-
0.01 to 0.099999 but less than 0.1	1	-2	place
0.001 to 0.009999 but less than 0.01	2	-3	mantissa
0.0001 to 0.000999 but less than 0.001	3	-4	table
etc.	etc.	etc.	

The above lists show that the characteristic of a number depends only on the position of the decimal point, regardless of the value of the individual digits in the number. However, the mantissa of a number is independent of the position of the decimal point in the number. The mantissa is always a positive decimal fraction and is determined by the *significant digits* of the number. Here the significant digits of a number are the digits excluding zeros either on the right or left side of the first nonzero digit. Thus, the significant digits of the numbers 60,100, 6.01, and 0.00601 are 6, 0, and 1. The logarithms of the numbers which have the same significant digits arranged in the same order thus have the same mantissa, but not the same characteristic. Example 50 gives additional illustrations of finding the logarithms based on the above lists.

Example 50

$$\log 601 = 2.7789$$
$$\log 60{,}100 = 4.7789$$
$$\log 6.01 = 0.7789$$
$$\log 0.00601 = -3 + 0.7789 = 7.7789 - 10, \text{ or } -2.2211$$
$$\log 0.601 = -1 + 0.7789 = 9.7789 - 10, \text{ or } -0.2211$$

Finding the Antilogarithm

When the logarithm of a number is known but the number is unknown, the unknown number is called *antilogarithm*, abbreviated *antilog*. The process of finding the unknown number from a given logarithm is the inverse of the process used in finding the logarithm of a number.

Example 51

If $\log N = 2.9315$, find N.

Solution.   First, find the mantissa 9315 from the table of mantissas.
  Second, locate the digits representing the mantissa 9315. The digits are 854.
  Third, place a decimal point in the digits according to the value of characteristic.
  Consult *Rule 1* on p. 94. When the characteristic is 2, the value range of the number is from 100 to 999.99$\cdots$ but less than 1000, or a three-digit number. Thus,

$$N = \text{antilog } 2.9315 = 854$$

Check the answer in Example 48.

Example 52

If $\log N = 7.6928 - 10$, find N.

Solution. $N = \text{antilog } (-3 + 0.6928) = 0.00493.$

Interpolation

The interpolation method may be employed to extend the use of the table of mantissas in finding an approximate mantissa (Example 53) or antilog (Example 54). Under this method, it is *assumed* that the differences between numbers in the table and the differences between their mantissas are proportional.

Example 53

Find $\log 27.36$.

Solution.   Since the N columns and rows in the table of mantissas provide only numbers with three digits, the mantissa for the four-digit number 27.36 can not be found directly in the table. Thus, the interpolation method is employed.

The number 27.36 is between numbers 27.3 and 27.4. The mantissas of numbers 27.3 and 27.4 can be determined directly from the table. The numbers and their mantissas are arranged correspondingly in the following manner: (The decimal point is disregarded at this stage.)

Number	Mantissa	
2740	4378	(1)
2736	x	(2)
2730	4362	(3)

$$\frac{(2)-(3)}{(1)-(3)} \quad \frac{6}{10} = \frac{x-4362}{16}$$

$$x = 4{,}362 + 16(\tfrac{6}{10}) = 4{,}362 + 9.6 = 4{,}371.6$$

In interpolation, the larger number and mantissa are placed on the top, line (1). The arrangement will facilitate the subtraction. The differences between the numbers are assumed to be proportional to the differences between the corresponding mantissas; that is, 6 is to $(x - 4362)$ as 10 is to 16. Thus, the value of x is $\tfrac{6}{10}$ of 16 above 4362, the smaller mantissa.

The characteristic of log 27.36 is 1. Thus,

$$\log 27.36 = 1.4371.6 \quad \text{or} \quad 1.43716$$

since the second decimal point is not needed in a number.

Note: If Table 2 in the Appendix is used, the interpolation process is not needed in this case. The mantissa of the number 2736 obtained directly from Table 2 is 437116, or log 27.36 = 1.437116.

Example 54

If log $N = 2.5709$, find N to five significant digits.

Solution. The mantissa 5709 is between mantissas 5717 and 5705, which correspond to numbers 373 and 372 respectively, in the table. Since five significant digits are required, two zeros are added to numbers 373 and 372. The decimal point is disregarded in computing the digits of N in the interpolation. The numbers and their corresponding mantissas may be arranged as follows:

Number	Mantissa	
37,300	5717	(1)
x	5709	(2)
37,200	5705	(3)

$$\frac{(2)-(3)}{(1)-(3)} \quad \frac{x-37{,}200}{100} = \frac{4}{12}$$

$$x = 37{,}200 + 100\left(\frac{4}{12}\right)$$

$$= 37{,}200 + 33.3 = 37{,}233.3, \text{ rounded to } 37{,}233$$

Since the characteristic is 2, the value of N must have three digits in the whole-number part. Thus,

$$N = \text{antilog } 2.5709 = 372.33$$

Computation

Logarithms may be used to simplify computations in multiplication, division, raising to powers, and extracting roots. The disadvantage of using logarithms in computations is that the results of the computations are *approximations* in most cases, although more accurate results can be obtained through the use of more places in mantissas. The interpolation process is omitted in the following illustrations for the sake of simplification. The antilog of a computed logarithm is thus found according to the nearest value in the table of mantissas.

Multiplication

In multiplying two or more numbers by the use of logarithms, employ the law for multiplication: *The logarithm of the product equals the sum of the logarithms of the factors.* Thus, if A is multiplied by B, then

$$\log (AB) = \log A + \log B$$

Example 55

Multiply 5.26 by 21.3.

Solution.   Log $(5.26 \times 21.3) = \log 5.26 + \log 21.3 = 2.0494$ or

$$\begin{aligned} \log 5.26 &= 0.7210 \\ (+) \log 21.3 &= 1.3284 \\ \hline &2.0494 \end{aligned}$$

The mantissa 0494 is close to 0492, which corresponds to digits 112 in the table. Thus,

$$\text{antilog } 2.0494 = 112 \text{ approximately (answer)}$$

Check by the conventional method: $5.26 \times 21.3 = 112.038$.

Example 56

Multiply $138 \times 4.92 \times 0.0264$.

Solution.   Log $(138 \times 4.92 \times 0.0264) = \log 138 + \log 4.92 + \log 0.0264$
$$= 1.2535 \text{ or}$$

$$\begin{aligned} \log 138 &= 2.1399 \\ \log 4.92 &= 0.6920 \\ (+) \log 0.0264 &= 8.4216 - 10 \\ \hline &11.2535 - 10 \end{aligned}$$

The mantissa 2535 is close to 2529, which corresponds to digits 179 in the table. The characteristic is $11 - 10 = 1$. Thus,

$$\text{antilog } (11.2535 - 10) = 17.9 \text{ approximately (answer)}$$

Check by the conventional method: $138 \times 4.92 \times 0.0264 = 17.924544$.

Division

In dividing a number by another, use the law for division: *The logarithm of a quotient equals the logarithm of the dividend minus the logarithm of the divisor.*

Thus, if A is divided by B, then

$$\log (A \div B) = \log A - \log B$$

Example 57

Divide 0.437 by 256.

Solution.   Log $(0.437 \div 256) = \log 0.437 - \log 256 = 7.2323 - 10$

$$\log 0.437 = 9.6405 - 10$$
$$(-) \log \quad 256 = 2.4082$$
$$\overline{7.2323 - 10}$$

The mantissa 2323 is close to 2330, which corresponds to digits 171 in the table. The characteristic is $7 - 10 = -3$. Thus,

$$\text{antilog } (7.2323 - 10) = 0.00171 \text{ (answer)}$$

Check by the conventional method: $0.437 \div 256 = 0.001707$

Powers and Roots

In raising to powers and extracting roots by using logarithms, apply the law for powers and roots: *The logarithm of an exponential equals the exponent times the logarithm of its base.* Thus, if A is raised to the pth power, then

$$\log A^p = p (\log A)$$

If A is raised to a power of $(1/q)$, then

$$\log A^{1/q} = \frac{1}{q} (\log A)$$

or

$$\log \sqrt[q]{A} = \frac{1}{q} (\log A)$$

Example 58

Compute 1.06^{74}

Solution.

$$\log 1.06^{74} = 74 (\log 1.06) = 74 (0.0253) = 1.8722$$

The mantissa 8722 corresponds to digits 745. Thus,

$$\text{antilog } 1.8722 = 74.5$$

or

$$1.06^{74} = 74.5$$

Example 59

Compute $\sqrt[15]{26.8}$

Solution.

$$\sqrt[15]{26.8} = 26.8^{1/15}$$

$$\log 26.8^{1/15} = \left(\frac{1}{15}\right)(\log 26.8) = \left(\frac{1}{15}\right)(1.4281) = 0.0952$$

The mantissa 0952 is close to 0969, which corresponds to digits 125 in the table. The characteristic is 0. Thus,

$$\text{antilog } 0.0952 = 1.25$$

or

$$\sqrt[15]{26.8} = 1.25$$

Semilogarithmic or Ratio Chart

The line charts presented in Chapter 3 are of two arithmetic scales: that is, equal distances on each reference line represent equal amounts. A *semilogarithmic* (abbreviated *semilog*) *chart* or a *ratio chart* has an arithmetic scale on one line and a logarithmic scale on the other. The logarithmic scale is usually placed on the vertical line in a semilog chart, and the arithmetic scale is placed on the horizontal line.

The logarithmic scale is divided according to the logarithms of the numbers labeled on the scale. The division is illustrated in Chart 4–2. The chart indicates that the logarithms of a group of numbers, such as numbers from 1 to 10, can be shown graphically on two different scales: (1) The numbers are plotted on a logarithmic scale. (2) The logarithms of the numbers are plotted on an arithmetic scale. Both scales give the same locations of the points plotted. In practice, the logarithmic scale is preferred because of the following reasons: (a) Printed semilog papers are available in most office supply or book stores. (b) The work of finding the logarithms of the numbers in a series of data is not required. (c) The process of plotting the numbers on a printed paper with logarithmic scale is simpler than that of plotting the logarithms of the numbers on an arithmetic scale. The logarithm of a number usually includes many decimal places.

It is important to note that on a logarithmic scale the distance between two numbers does not represent the difference between the actual numbers. Equal distances represent equal ratios of the numbers labeled on the scale. For example, in Chart 4–2, the distance between log 2 and log 1 is equal to the distance between log 8 and log 4. The equal distances indicate that the ratio of 2 to 1 is the same as the ratio of 8 to 4.

A printed semilog paper may have one *cycle*, two cycles, or sometimes three cycles. A cycle covers the area from 1 to 10. Thus, a two-cycle paper may have two sets of labels of 1 to 10, such as the scale shown in Chart 4–3. The value at the top of a cycle on the logarithmic scale is always ten times the value at the bottom of the cycle. The value at the bottom of a cycle is the base value and may be assigned arbitrarily for any number *other than zero*. Since the ratio of any number to zero is meaningless, a logarithmic scale cannot have zero as the base value. Once the base value is assigned, other numbers on the scale can be determined.

The method of plotting data on a ratio chart is similar to that on the line chart. After the points are plotted, the points are connected by straight lines. Thus, a ratio chart is essentially a special type of line chart. Example 60 is used to illustrate the method of plotting the data on a ratio chart with a logarithmic scale (vertical) and an arithmetic scale (horizontal).

CHART 4-2

COMPARISON OF LOGARITHMIC SCALE AND ARITHMETIC SCALE

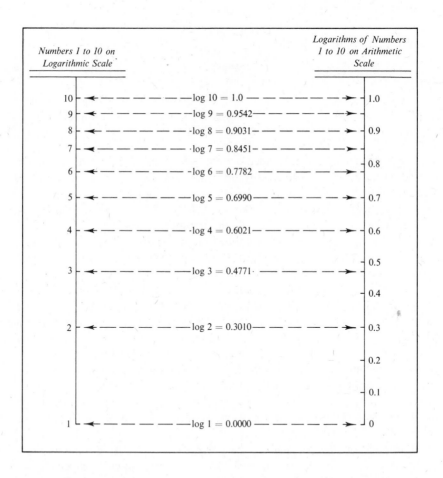

Example 60

Use the data shown in Table 4–3 to construct a ratio chart on a two-cycle semi-log paper.

TABLE 4-3

ANNUAL SALES OF FULTON DRUG STORE
1969 THROUGH 1976

Year	Sales	Year	Sales
1969	$ 500	1973	$ 8,000
1970	1,000	1974	10,000
1971	2,000	1975	12,000
1972	4,000	1976	14,000

Source: Table 3-2. p. 47.

CHART 4-3

A RATIO CHART

*Annual Sales of Fulton Drug Store 1969
Through 1976*

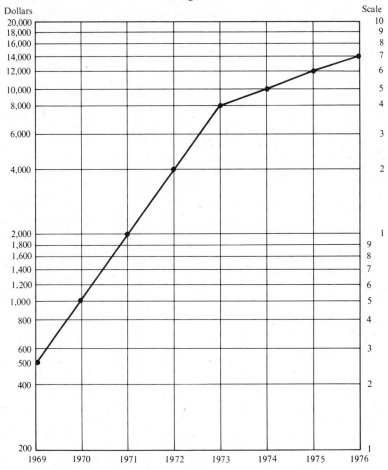

Source of data: Example 60.

Solution.　See Chart 4–3. The table shows that the smallest value is 500 and the largest value is 14,000. The largest value is 28 times (or $14{,}000 \div 500 = 28$) as large as the smallest value. Since the value at the top of a one-cycle paper is only ten times the value at the bottom of the cycle, a two-cycle paper is needed to cover the given data.

The appearance of a chart can be pleasant if the curve which connects the plotted points is placed in a central position of the printed paper. Thus, the base value is arbitrarily assigned as 200. The value on the top of the first cycle is $2{,}000(= 200 \times 10)$, and that of the second cycle is $20{,}000(= 2{,}000 \times 10)$. Other numbers on the scale also may be computed from the base value. For example, $400 = 200 \times 2$, $600 = 200 \times 3$, and so on for the first cycle; $4{,}000 = 2{,}000 \times 2$, $6{,}000 = 2{,}000 \times 3$, and so on for the second cycle.

Observe the plotted Chart 4–3. When the ratio of two numbers is equal to the ratio of two other numbers, the distances between the points representing the pairs of numbers are the same. For example, the ratio of 1,000 to 500 ($= 2$) is the same as the ratio of 2,000 to 1,000($= 2$). Thus, the distance between 1,000 and 500 has the same length as the distance between 2,000 and 1,000. Further, when

CHART 4-4

ILLUSTRATION OF EXPANDING LOGARITHMIC SCALES

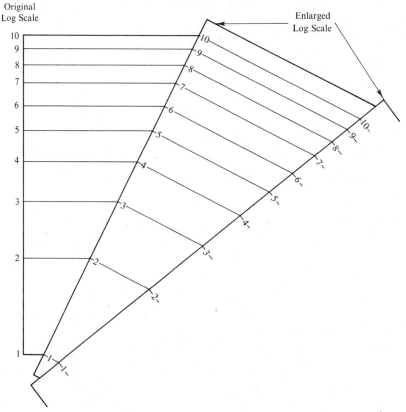

the ratios during a period are constant, the points in the period are on a straight line. For example, all the ratios of 8,000 to 4,000, 4,000 to 2,000, 2,000 to 1,000, and 1,000 to 500 are 2. Thus, the points representing the sales for the years from 1969 to 1973 are on a straight line when the logarithmic scale is used. This is not true in Chart 3–5, where the arithmetic scale is used. However, the points for the years 1973, 1974, 1975, and 1976 are not on a straight line in Chart 4–3, but are on a straight line in Chart 3–5. When a series of numbers increases by a constant amount, the points representing the numbers are on a straight line on an arithmetic scale. The constant amount of increase for each year is $2,000 from 1973 to 1976.

If a large ratio chart is needed, a blank sheet may be placed diagonally on the printed logarithmic paper. An enlarged logarithmic scale may then be marked on the blank sheet according to the scale on the printed paper. If several blank sheets are used in the same manner as shown in Chart 4–4, any desired size of enlargement may be obtained. If a smaller ratio chart is needed, the above procedure is reversed; that is, the printed logarithmic paper is placed diagonally on a blank sheet before the logarithmic scale is marked.

Exercises 4-1

Reference: Sections 4.1 and 4.2

A. Solve the following equations for x.

1. $4x + 3 = 5x - 1$

2. $9x - 15 = 7x + 17$

3. $\dfrac{3}{x + 2} = \dfrac{4}{x + 1}$

4. $\dfrac{x}{14} + \dfrac{2x}{5} = 3\dfrac{3}{10}$

5. $y = a + bx$

6. $5xy + 2 = 15a + 7$

7. $28 + x = \dfrac{8x - 16}{3}$

8. $x - 7 = \dfrac{17x + 1}{5}$

9. $4dx - 5y + 7z = 0$

10. $3ax + 2bx = 5c$

B. Solve for x and y (elimination by addition or subtraction).

11. $3x - 2y = 5$
    $4x + 3y = 18$

12. $2x + y = 16$
    $3x - y = 29$

13. $4x + 7y = 11$
    $3x - 2y = -28$

14. $5x - 28y = -7$
    $2x - 13y = -10$

C. Solve for x and y (elimination by substitution).

15. $2x - y = -5$
    $x + 3y = 29$

16. $x + 2y = -19$
    $2x - 3y = 11$

D. Solve for x and y by the graphic method.

17. $5x + 7y = 11$
 $2x - 4y = -16$

18. $x + 3y = 1$
 $3x - 2y = 14$

E. Solve for $x, y,$ and $z.$

19. $x + y + 2z = -3$
 $6x - y + z = 1$
 $2x + 3y - 3z = 17$

20. $x - y + z = 15$
 $3x - 2y - 3z = -2$
 $2x + y - 2z = -13$

Exercises 4-2

Reference: Sections 4.3 and 4.4

A. Change each of the following groups of relative values to a group with the same ratio and a total of 100.

1. $2 : 8 : 7 : 3$

2. $2 : 7 : 8 : 15 : 3$

B. Solve for x in each of the following proportions.

3. $x : 9 = 8 : 13$

4. $5 : x = 12 : 17$

5. $\dfrac{3}{11} = \dfrac{2x}{7}$

6. $\dfrac{7}{22} = \dfrac{14}{3x}$

C. Convert each of the following to a decimal or a whole number.

7. 0.52%

8. 7.6%

9. 43%

10. 283%

11. $8,600\%$

12. 126.75%

D. Convert each of the following to a common fraction in its lowest terms.

13. 6%

14. 235%

15. $3,186\%$

16. 5.2%

17. 0.028%

18. 0.0025%

E. Convert each of the following to a percentage.

19. 0.45

20. 0.052

21. 0.0032

22. 76

23. 5.18

24. 128.4

F. Convert each of the following to a percentage (carry to two decimal places before converting the decimals to percentages; round the fractional percentages, if any).

25. $\dfrac{3}{25}$

26. $\dfrac{18}{23}$

27. $4\dfrac{3}{4}$

28. $3\dfrac{14}{15}$

29. $12\dfrac{5}{7}$

30. $\dfrac{7.4}{13.28}$

G. Statement problems.

31. What is 32% of $\$460$?

32. A store sold an article at 40% more than its cost. What was the selling price if the cost of the article was $85?

33. John, Jack, and Joe are co-owners of a retail store. They have agreed that profits should be divided as follows: 45% to John, 30% to Jack, and the remaining part to Joe. If the profit amounts to $12,500 this year, how much will each receive?

34. What per cent of $470 is $164.50?

35. $45 is what per cent less than $150?

36. A television bought for $230 is sold for $326.60. What per cent of the cost is the profit?

37. If 12% of a number is 9, what is the number?

38. What number decreased by 5% of itself is 133?

39. Bob received a dividend of $200, which is 8% of his investment. What is the value of the investment?

40. A man sold a car for $3290, a loss of 6% on his purchase price. What was his purchase price?

Exercises 4-3

Reference: Section 4.5

A. Multiplication and division of exponentials.

1. $3^2 \cdot 3^4$ **2.** $5^3 \cdot 5^0$ **3.** $4^2 \cdot 4^{-3}$ **4.** $x^2 \cdot x^5$

5. $a^4 \cdot a^{-4}$ **6.** $b^3 \cdot b^{-7}$ **7.** $3^2 \cdot 5^2$ **8.** $4^3 \cdot 2x^3$

9. $a^4 \cdot b^4$ **10.** $(xy)^5 \cdot a^5$ **11.** $(c^{1/5})^2$ **12.** $(xy^{2/3})^6$

13. $3^7 \div 3^4$ **14.** $5^2 \div 5^4$ **15.** $a^5 \div a^2$ **16.** $(xy)^3 \div (xy)^5$

17. $(+5xy)(-4x^2yz^3)$ **18.** $(-6abc)(-3ab^2w)$

19. $(-45xyz^3) \div (5xzw^2)$ **20.** $(8ab^2 - 15a^3bc^5 + 20a^2b^3) \div (5a^2b)$

B. Express the following in radical form.

21. $a^{1/2}$ **22.** $d^{2/3}$ **23.** $x^{4/7}$

24. $26^{1/2}$ **25.** $185^{2/5}$ **26.** $58^{3/4}$

C. Express the following in exponential form.

27. $\sqrt[3]{b^2}$ **28.** $\sqrt{g^3}$ **29.** $\sqrt[5]{y^4}$

30. $\sqrt{410}$ **31.** $\sqrt[6]{258^2}$ **32.** $\sqrt[4]{68.2^3}$

D. Express the coefficient of the following radicals as part of the radicand.

33. $a\sqrt{ab}$ **34.** $x \cdot \sqrt[3]{xy}$ **35.** $k \cdot \sqrt[4]{2m}$

36. $5 \cdot \sqrt[3]{4}$ **37.** $3 \cdot \sqrt[4]{2}$ **38.** $12\sqrt{3}$

E. Compute the following.

39. $3\sqrt{2a} + 7\sqrt{2a} - 4\sqrt{2a}$ **40.** $5\sqrt{6d} - 8\sqrt{6d} + 10\sqrt{6d}$

41. $\sqrt{27} + \sqrt{48} - \sqrt{12}$ **42.** $\sqrt{18} - \sqrt{50} + \sqrt{72}$

43. $(4\sqrt{5})(3\sqrt{2})$ **44.** $(7\sqrt{6})(10\sqrt{3})$

45. $(6\sqrt{5})(-15\sqrt{11})$ **46.** $(-5\sqrt{12})(2\sqrt{7})$

47. $(3\sqrt{5})(2\sqrt{3})(5\sqrt{14})$ **48.** $(4\sqrt{3})(6\sqrt{3})(2\sqrt{3})$

49. $\sqrt{100x} \div \sqrt{25x}$ **50.** $\sqrt{117y} \div \sqrt{13y}$

51. $\sqrt{7} \div \sqrt{3}$ **52.** $\sqrt{15} \div \sqrt{41}$

F. Compute the square root.

53. $\sqrt{0.0576}$ **54.** $\sqrt{5.76}$ **55.** $\sqrt{225}$

56. $\sqrt{2.25}$ **57.** $\sqrt{18,769}$ **58.** $\sqrt{973.44}$

59. $\sqrt{179.8}$ (to one decimal place)

60. $\sqrt{2778}$ (to one decimal place)

Exercises 4-4

Reference: Section 4.6

A. Find the logarithm of each number.

1. 341   **2.** 23.6   **3.** 3.12   **4.** 560

5. 7,620   **6.** 0.00345   **7.** 0.82   **8.** 0.0732

B. Find N if log N is

9. 0.8704   **10.** 1.8014   **11.** 2.5465   **12.** 3.1461

13. $9.6551 - 10$ **14.** $8.4082 - 10$ **15.** $7.4639 - 10$ **16.** 0.0969

C. Using the interpolation method, find the answers.

17. Find log 1,453.

18. Find log 42.36.

19. If log $N = 2.9045$, find N to four significant digits.

20. If log $N = 0.4688$, find N to five significant digits.

D. Compute by using logarithms (omit interpolation).

21. 38×45 **22.** 123×2.74 **23.** $56 \div 4.22$ **24.** $427 \div 15.2$

25. 123^{14} **26.** 2.45^{28} **27.** $\sqrt[4]{23,600}$ **28.** $\sqrt[12]{56.3}$

E. Ratio charts.

29. The number of units produced by A. C. Fisher Company from 1969 to 1976 is given in Table 4–4. Construct a semilogarithmic chart showing the facts.

TABLE 4-4

DATA FOR PROBLEM 29

Year	Number of Units	Year	Number of Units
1969	750	1973	2,400
1970	600	1974	4,800
1971	1,200	1975	9,600
1972	1,800	1976	19,200

30. The annual advertising expense and net income of Coxes Department Store are given in Table 4–5. Construct a semilogarithmic chart showing the information and state your findings from the chart.

TABLE 4-5

DATA FOR PROBLEM 30
(Millions of Dollars)

Year	Advertising Expense	Net Income
1970	0.4	0.5
1971	0.8	1.0
1972	1.6	2.0
1973	1.8	1.7
1974	1.6	1.1
1975	1.2	0.8
1976	0.7	0.6

part 2

Simple
Statistical
Analysis

This part presents the basic tools for statistical analysis. It includes averages (Chapters 5 and 6), dispersion (Chapter 7), skewness and kurtosis (Chapter 8). The material included in this part will be referred to frequently in future chapters. A clear understanding of the material is thus essential to the study of the subject of statistics.

5 Averages: Arithmetic Mean, Median, and Mode

An *average* is a single value that is typical or representative for a group of numbers. The values included in a group of data usually vary in magnitude; some of them are small and some are large. Obviously, a representative value for a group of numbers is normally neither the smallest nor the largest value, but is a number whose value is somewhere in the middle of the group. Thus, an average is frequently referred to as *a measure of central tendency*.

Averages have many uses. An average is frequently used as a device in summarizing a group of numbers, especially a large group of numbers, for describing the statistical data. Examples are the average age of students in a college, the average weekly wage of manufacturing workers in a state, and the average family income in a nation.

Averages are also frequently used for comparison of one group of data with another. Examples are the average years of education of the employees in one company compared with the average in another company, the average units produced in one plant compared with the average produced in another plant, and the average miles traveled by a group of salesmen compared with the average miles traveled by another group.

The most commonly known averages in statistics are (1) the arithmetic mean, (2) the median, (3) the mode, (4) the geometric mean, and (5) the harmonic mean. Each average has its particular characteristics. The determination of which one of the different types of averages should be used under various circumstances largely depends on the characteristics of the averages. In general, the first three averages are used more frequently, and they are discussed in this chapter. The geometric mean and the harmonic mean are used only in special cases; they are presented in the next chapter.

There are more details involved in computing the averages for grouped data than for ungrouped data, although the methods are basically the same for the two types of data. The numbers included in the ungrouped data are single values and are not classified into groups. The grouped data, also called *frequency distribution*, are organized data and are classsified according to quantity, as discussed in Section 2.2. Details of the grouped data are further discussed in Section 5.1. The three common types of averages are presented individually in Sections 5.2 through 5.4. The chief characteristics of the averages are compared in Section 5.5. This last section also summarizes the discussion presented in this chapter.

5.1   Frequency Distribution (Grouped Data)

Generally speaking, when a group of collected data consists of only a few items, there may not be a need for organization. The collected data which have not been organized numerically are frequently called *raw data*. However, when a large group of items is collected, the values of the items should be organized in order to facilitate statistical analysis. The values may first be arranged according to ascending or descending order of magnitude. The data so arranged are called an *array*. Thus, the values 4, 6, 2, 9, 8, 4, 8, and 8 are raw data which can be arranged as an array: 2, 4, 4, 6, 8, 8, 8, 9. There are repeating values in the array. When repeating values are indicated, the arrangement is then called a *frequency array* and the number indicating the times a value is repeated is called the *frequency*. The frequency array may be constructed by using tally marks as shown in Table 5-1.

TABLE 5-1

FREQUENCY ARRAY

Value	Tally	Frequency
2	/	1
4	//	2
6	/	1
8	///	3
9	/	1
Total		8 values

Further, when the values are grouped into several classes based on quantity and the number of values within each class is indicated, a more compact tabular presentation of the data can be obtained. The table showing grouped data based on quantity is called a *frequency distribution*, such as Table 5-2.

The arrangement of a frequency distribution has a great effect in computing the various averages as well as in other phases of statistical analysis. However, there are no definite rules that can be used to construct a perfect frequency distribution table. The following discussions concerning number of classes, class limits, and sizes of class intervals, nevertheless, should be carefully examined in constructing such a table from given raw data.

TABLE 5-2

FREQUENCY DISTRIBUTION

Class Interval	Frequency
1–3	1
4–6	3
7–9	4
Total	8 values

Source of data: Table 5–1.

Number of Classes

The number of classes depends upon the number of values to be grouped and the type of information that the investigator wishes to have. In general, the number of classes should not be too large or too small. If the number of classes is too large, many classes may have too few items or no frequencies at all. Too many classes may also defeat the purposes of classifying the data. On the other hand, if the number of classes is too small, very little information concerning the original data may be given. The number of classes in a published frequency distribution table is usually between five and twenty.

Class Limits

Class Limits and Mid-Point Value

The lower and upper class limits stated in a frequency distribution indicate the boundaries of each class in the distribution. However, in many cases, the *stated class limits* are not the *real class limits*. There are gaps between classes. In such cases, the *mid-point of each gap* is considered as the real limit between the two classes forming the gap.

The *mid-point of each class* is usually used to represent each original value grouped into the class for purposes of further mathematical analysis. The class mid-point may be computed from either the stated class limits or the real class limits of the class.

$$\text{Class mid-point} = \frac{\text{stated lower class limit} + \text{stated upper class limit}}{2}$$

or

$$= \frac{\text{real lower class limit} + \text{real upper class limit}}{2}$$

Example 1

Find the real class limits and the mid-point value for each of classes 1–3, 4–6, 7–9, etc. stated in the first column of Table 5–3. Assume that all classes are the same size.

Solution. There is a gap between each stated upper limit and the stated lower limit of the next class, such as between 3 (the upper limit of the first class) and 4 (the lower limit of the second class). The mid-point of the gap is the real limit.

TABLE 5-3

COMPUTATION FOR EXAMPLE 1

Stated Class Interval		Real Class Interval		Class mid-point
Lower limit	Upper limit	Lower limit	Upper limit	
1	3	0.5 —— 3.5		$2\left(=\dfrac{1+3}{2} \text{ or } \dfrac{0.5+3.5}{2}\right)$
4	6	3.5 —— 6.5		$5\left(=\dfrac{4+6}{2} \text{ or } \dfrac{3.5+6.5}{2}\right)$
7	9	6.5 —— 9.5		$8\left(=\dfrac{7+9}{2} \text{ or } \dfrac{6.5+9.5}{2}\right)$
etc.				

Thus, $(3 + 4)/2 = 3.5$ is the real upper limit of the first class and is also the real lower limit of the second class.

The size of each class is equal to the difference between two successive lower class limits or two successive upper class limits, either based on the stated or real class intervals. The size of the class interval based on the first and the second stated class intervals is computed to be 3 ($=4 - 1$ or $=6 - 3$). Thus, the real lower limit for the first class is 0.5 ($=3.5 - 3$), the real upper limit for the second class is 6.5 ($=3.5 + 3$), and so on. There is no gap between each real upper limit and the real lower limit of the next class. The difference between two successive mid-point values is also equal to the size of class interval.

Example 2

Find the real class limits and the mid-point value for each class stated in the first column of Table 5–4: 0 and under 2, 2 and under 4, 4 and under 6, etc. Assume that all classes are the same size.

TABLE 5-4

COMPUTATION FOR EXAMPLE 2

Stated Class Interval		Real Class Interval		Mid-Point Value
Lower limit	Upper limit	Lower limit	Upper limit	
0 and under 2		0 —— 2		$1 = (0 + 2)/2$
2 and under 4		2 —— 4		$3 = (2 + 4)/2$
4 and under 6		4 —— 6		$5 = (4 + 6)/2$
etc.				

Solution.   In this type of class designation, each stated upper limit may become very close to the stated lower limit of the next class. For example, the stated upper limit in the first class "under 2" may become very close to 2. The gap between "under 2" and 2 is very small. Thus, the real upper limit in the first class is *con-*

sidered to be 2 for purposes of mathematical analysis. The mid-point values are computed from the real class limits.

In selecting class limits for a frequency distribution, two important points should be carefully considered: (1) The mid-point value of each class should be selected, if possible, at the point which is the concentration of the original values grouped into the class. (2) The mid-point value of each class should be a simple or a whole number so that the further computation and analysis can be simplified. For example, if most of the salaries of a group of employees during a period are multiples of $5, the salaries may be classified as in Table 5–5.

TABLE 5-5

SELECTING CLASS LIMITS AND MID-POINT VALUES

Salary (class interval)	Mid-Point	Number of Employees (frequency)
$203–$207	$205	3
208–212	210	5
213–217	215	6
etc.		

The class limits in Table 5–5 give: (1) The mid-points are selected at the points of concentration, since they are multiples of $5. (2) The mid-point of each class is a whole number, which may simplify further computation and analysis of the data.

Class Limits and Variables

A *variable* is a set of values and is usually represented by a symbol, such as X variable consisting of 1, 2, 5, 7, 8, and so on. If a symbol representing a number has a fixed value, the symbol is called a *constant*, such as a if $a = 4$.

There are two types of variables: *continuous variables* and *discrete variables*. A continuous variable can theoretically assume any value between two given values. For example, there are unlimited values between numbers 70 and 71, such as 70.1, 70.6, 70.80047, ... Data which can be described by a continuous variable are called *continuous data*, such as the measurement of the height of a person which can be 70.1 inches, 70.6 inches, or 70.80047 inches. If the variable can not assume any value between two given values, it is called a discrete variable. Data represented by a discrete variable are called *discrete data*, such as the number of students in a class, which can assume any of the values 0, 1, 2, 3, 4, . . . , but cannot be 1.2, 2.6, or 3.4. In general, values representing measurements are continuous data, whereas countings or enumerations are discrete data.

In selecting class limits for the two types of data, care should be taken to avoid ambiguous class limits. The problem of avoiding ambiguous class limits for discrete data is relatively simple. Gaps between class limits of successive classes may be used to avoid ambiguity. For example, the number of clubs in a college according to the size of each club may be stated as in Table 5–6.

TABLE 5-6

SELECTING CLASS LIMITS FOR DISCRETE DATA

Number of Students (class interval)	Mid-Point	Number of Clubs (frequency)
5–9	7	12
10–14	12	35
15–19	17	23
20–24	22	14
etc.		etc.

In Table 5–6, the class limits give no doubt about the class into which a given size of club is grouped. However, ambiguous class limits are often employed in continuous data. For example, the heights of a group of students may be ambiguously stated as in Table 5–7.

TABLE 5-7

AMBIGUOUS CLASS LIMITS

Height in Inches (class interval)	Mid-Point	Number of Students (frequency)
60–62	61	4
62–64	63	7
64–66	65	10
etc.		

In the absence of any additional information, a reader may not be able to tell from Table 5–7 whether a student who is 64 inches tall is grouped into the second or third class. This type of ambiguous class limits should be avoided. Clear class limits for the above example may be stated as in Table 5–8. A reader now can tell exactly that the student who is 64 inches tall must be grouped into the third class, not the second class, since 64 is not under 64.

TABLE 5-8

SELECTING CLASS LIMITS FOR CONTINUOUS DATA

Height in Inches (class interval)	Mid-Point	Number of Students (frequency)
60 and under 62	61	4
62 and under 64	63	7
64 and under 66	65	10
etc.		

In many cases, the values in continuous data are rounded numbers. If the heights of the students are reported in the nearest inches, it is proper to state the class limits as in Table 5–9.

TABLE 5-9

SELECTING ALTERNATIVE CLASS LIMITS FOR
CONTINUOUS DATA

Height in Inches (class interval)	Mid-point	Number of Students (frequency)
60–62	61	Depends on the
63–65	64	roundings of the
66–68	67	original measurements
etc.		

In the class intervals in Table 5–9, although the real class limits are overlapping between individual classes, such as 62.5 inches being the real upper limit of the first class and the real lower limit of the second class, there can be no doubt about the class into which a student is grouped. For example, students A, B, C, D are grouped according to the classes as shown in Table 5–10.

TABLE 5-10

GROUPING ITEMS TO CLASSES

Students	Actual Height in Inches	Rounded Height in Inches	Class to Be Grouped
A	60.4	60	60–62
B	62.5	62	60–62
C	62.7	63	63–65
D	65.5	66	66–68

Sizes of Class Intervals

In general, there are three types of class intervals according to the sizes of the classes in a frequency distribution: (1) classes of equal size, (2) classes of unequal sizes, and (3) open-end classes. The size of a class interval is the difference between the real lower and upper class limits, and is also referred to as the *class width* or *class size*. The determination of the sizes of class intervals in a frequency distribution depends on the number of classes, the types of information desired, and the degree of variation of the original values.

Classes of Equal Size

This type of class designation is usually preferred and has been used in the illustrations above. When all classes are the same size, computations concerning the frequency distribution are greatly simplified. For example, the number of classes in a frequency distribution may be computed by dividing the range of the raw data (the difference between the largest value and the smallest value) by the size of the class interval.

Classes of Unequal Sizes

Unequal class intervals are undesirable in most cases, but are sometimes used to serve particular purposes, such as to cover values varying in a wide range. When the unequal class intervals must be used, the class intervals should be increased in an orderly manner if possible.

The types of unequal class intervals in Table 5–11, for example, are sometimes used in reporting data concerning assets valuation or personal income. The lower limit in each class, except the first class, is a multiple of $500. Each mid-point is computed from the real limits of each class.

TABLE 5-11

CLASSES OF UNEQUAL SIZES AND OPEN-END CLASSES

Class Interval	Real Class Limits		Mid-Point
	Lower	Upper	
Less than $500	?	$ 500	?
$500 to less than $1,000	$ 500	1,000	$ 750
$1,000 to less than $5,000	1,000	5,000	3,000
$5,000 to less than $10,000	5,000	10,000	7,500
$10,000 to less than 25,000	10,000	25,000	17,500
$25,000 and over	25,000	?	?

Open-End Classes

An open-end class has one of its two class limits not stated numerically, such as the first class "*Less* than $500" and the last class "$25,000 and *over*" in the example above. This type of class should be avoided if possible, since we cannot tell exactly what is the mid-point value or other representative value of the class for computational purposes. However, the open-end class is sometimes conveniently used to include a few extremely large or small values. In such cases, the sum of the values in the open-end class should be indicated in order to facilitate computation of the frequency distribution.

5.2 Arithmetic
Mean

The *arithmetic mean*, or simply the *mean*, is the most commonly used type among the five types of averages. The methods of computing the mean for ungrouped data and for grouped data are presented below.

Ungrouped Data

Basic Method

The mean for ungrouped data is the quotient of the sum of the values divided by the number of values in the given set of data.

$$\text{Mean} = \frac{\text{sum of values}}{\text{number of values}}$$

or, symbolically,

$$\bar{X} = \frac{\Sigma X}{n} \qquad\qquad (5\text{–}1a)$$

In formula (5–1a),

> X represents the set of values, or X variable.
> n represents the number of values in the set.
> Σ is the Greek letter *sigma* and represents "the sum of" or "the summation of."
> $\bar{X}$ represents the mean of X variable, called "X bar" or "bar X."
> The bar on the top of a letter or letters usually represents "the arithmetic mean of."

Example 3

The miles traveled by five students in coming to the Larchmont School from their homes are given below. Compute the mean of the miles traveled by the five students.

Students	Miles traveled (X variable)
A	1
B	4
C	10
D	8
E	10
Total	33

Solution.   Use formula (5–1a). The sum of the values in the given information is

$$\Sigma X = 1 + 4 + 10 + 8 + 10 = 33$$

The number of values is

$$n = 5 \text{ (students)}$$

The mean is

$$\bar{X} = \frac{33}{5} = 6.6 \text{ miles}$$

Short Method

Example 3 shows that every value in the set of values is taken into consideration in computing the mean. When the deviations of the individual values from the mean are added, the *algebraic sum* of the deviations is equal to zero. In other words, the computed mean lies at the point of balance; that is, at the point at which the sum of the positive deviations is equal to the sum of the negative deviations (see Table 5–12). If a value other than the exact mean is selected, the algebraic sum of the deviations above and below the selected value will not be zero. However, if the algebraic sum is divided by the number of values in the data and the quotient is added to the selected value, the result will be equal to the exact mean. This fact provides a short method of computing the mean. The short method will enable us to save a considerable amount of time when a large group of data is involved in computing the mean, especially in a frequency distribution.

TABLE 5-12

ALGEBRAIC SUM OF DEVIATIONS FROM TRUE
ARITHMETIC MEAN BEING ZERO

Values X	Deviations from Mean (6.6) $X - \bar{X}$
1	$\left.\begin{array}{c}-5.6 \\ -2.6\end{array}\right\} = -8.2$
4	
10	$\left.\begin{array}{c}+3.4 \\ +1.4 \\ +3.4\end{array}\right\} = +8.2$
8	
10	
Total	0

In general, let

   $A =$ the assumed mean or the arbitrarily selected value
   $v =$ the deviation of each value from the assumed mean $= X - A$

then

$$\bar{X} = A + \frac{\Sigma(X - A)}{n}$$

or

$$\bar{X} = A + \frac{\Sigma v}{n} \qquad \text{(5-1b)*}$$

The fraction $(\Sigma v)/n$ is also called the correction factor.

Example 3 is computed according to the short method, or formula (5-1b) below.

Let 8 be the assumed mean, or $A = 8$. The deviations of the individual values from the assumed mean are shown in Table 5-13.

TABLE 5-13

COMPUTATION FOR EXAMPLE 3 (SHORT METHOD)

Values X	Deviations from Assumed Mean $v = X - 8$
1	-7
4	-4
10	2
8	0
10	2
Total	$\Sigma v = -7$

*Formula (5-1b) may be proved as follows:

$$X = X - A + A = v + A$$
$$\Sigma X = \Sigma v + nA$$
$$\frac{\Sigma X}{n} = \frac{\Sigma v}{n} + \frac{nA}{n}$$

Thus,

$$\bar{X} = A + \frac{\Sigma v}{n}$$

$$\bar{X} = A + \frac{\Sigma v}{n} = 8 + \frac{-7}{5} = 8 + (-1.4) = 6.6 \text{ (the true mean)}$$

Weighted Mean

In Example 3, the number of miles traveled by each of the five students is included only once in computing the mean. The number of trips made by each student during a period is disregarded. When each value is considered equally important, the mean is called *unweighted mean*. When each of the values in a set of data is assigned a weight according to the relative importance in the group, the computed mean is called *weighted mean*. The weighted mean is obtained as follows: first, multiply each value by the weight assigned to the value; second, add these products; and third, divide the sum of the products by the sum of the weights.

Let $w =$ the weight assigned to each value in X variable; then

$$\bar{X} = \frac{\Sigma(wX)}{\Sigma w} \tag{5-2}$$

Note that when the weight assigned to each value is 1, formula (5-2) becomes the same as formula (5-1a). The latter formula is for an unweighted mean.

Example 4

The miles traveled during each trip and the number of trips made by each of the 5 students in coming to Larchmont School from their homes in a week are given in columns (2) and (3) of Table 5-14. The number of miles traveled during each trip X is weighted by the number of trips w made by each student in column (4).

Solution. The weighted mean is computed according to formula (5-2):

$$\bar{X} = \frac{\Sigma(wX)}{\Sigma w}$$

$$\bar{X} = \frac{112}{20} = 5.6 \text{ miles}$$

Note that the divisor in the division is 20 (the sum of the weights), not 5 (the number of students).

TABLE 5-14

COMPUTATION FOR EXAMPLE 4

(1) Student	(2) Miles Traveled per Trip X	(3) Number of Trips w	(4) Total Miles Traveled wX
A	1	6	6
B	4	5	20
C	10	4	40
D	8	2	16
E	10	3	30
Total (Σ)	...	20	112

Grouped Data

In the computation of the arithmetic mean for grouped data, the mid-point of each class is used to represent the value of each item included in the class. The mean computed from a frequency distribution thus may differ from the mean computed from the original data, since not every one of the actual values in a class would be the same value as the mid-point. However, the difference is usually slight.

The method of computing the mean for grouped data is needed in many cases. The work of computing the mean from a frequency distribution is much simpler than that from the ungrouped data of a large number of values. In addition, the original data may not be given if the frequency distribution table is obtained from a published source.

Basic Method

The mean for grouped data is basically obtained as follows: first, multiply each mid-point by the frequency in the class; second, add these products; and third, divide the sum of the products by the sum of the frequencies.

Let

X = the mid-point of individual class
f = the frequency of individual class
n = the sum of the frequencies $= \sum f$

then

$$\bar{X} = \frac{\sum (fX)}{n} \tag{5-3a}$$

Example 5

The miles traveled by 20 students in coming to Swan College from their homes are given below:

0.8	1.2	2.6	2.8	3.3
3.4	3.7	4.0	4.5	5.3
5.8	6.1	6.2	6.5	7.1
7.3	7.4	7.6	7.8	9.2

(*Total: 102.6 miles*)

Solution. (a) The mean for the given raw data is 5.13 miles, or

$$\text{Mean} = \frac{102.6}{20} = 5.13 \text{ miles}$$

(b) The data are classified into five groups as shown in Table 5–15.

TABLE 5-15

GROUPING DATA FOR EXAMPLE 5

Miles Traveled	*Tally*	*Number of Students*
0 and under 2	II	2
2 and under 4	~~IIII~~	5
4 and under 6	IIII	4
6 and under 8	~~IIII~~ III	8
8 and under 10	I	1
Total		20

Table 5–16 is arranged to compute the mean for the grouped data.

TABLE 5-16

COMPUTATION FOR EXAMPLE 5 (BASIC METHOD)

Miles Traveled (class interval)	Average Miles (mid-point) X	Number of Students (class frequency) f	Total Miles Traveled fX
0 and under 2	1	2	2
2 and under 4	3	5	15
4 and under 6	5	4	20
6 and under 8	7	8	56
8 and under 10	9	1	9
Total		$\Sigma f = n = 20$	$\Sigma (fX) = 102$

$$\text{Mean} = \bar{X} = \frac{\Sigma (fX)}{n} = \frac{102}{20} = 5.1 \text{ miles} \qquad \text{(formula 5–3a)}$$

The difference between the two computed means (5.13 miles for the ungrouped data and 5.1 miles for the grouped data) is very small.

Short Method—Deviations in Original Units (v)

The short method of computing the arithmetic mean for grouped data is preferred in most cases, especially when the number of classes in the given data is large. This method is easy to apply when the classes are the same size. When the classes are not the same size, the procedure illustrated above in the basic method is simpler.

The principle used in the illustration of the short method of computing the mean for ungrouped data can also be applied to the short method for grouped data: the algebraic sum of the deviations of individual values from the *exact mean* is zero but from the *assumed mean* is not zero. The deviations from the assumed mean can be expressed in original units of data or in class intervals in a frequency distribution. When the deviations are expressed in original units, the procedure of the short method of computing the mean for grouped data is arranged as follows:

1. Select an assumed mean A. The answer is not affected by the value selected as the assumed mean. Any point, including zero, may be used as the assumed mean. However, in order to simplify the computation, the mid-point of one of the centrally located classes in the given data should be selected as the assumed mean.

2. Find the deviation of each class mid-point from the assumed mean in original units of data, such as dollars, miles, inches, and so on.

Let

$$v = \text{the deviation in original units}$$
$$X = \text{each class mid-point}$$

Then

$$v = X - A$$

3. Multiply each deviation v by the frequency in the class f to obtain the total deviation of the class, or fv.

4. Add these products to obtain the total deviation of all items included in the data, or $\Sigma (fv)$. The sum of the products is usually not equal to zero. If it is a zero, the assumed mean must be the exact mean also.

5. Divide the sum of the products $\Sigma (fv)$ by the sum of the frequencies (Σf or n) to obtain the correction factor

$$\frac{\Sigma (fv)}{\Sigma f} \quad \text{or} \quad \frac{\Sigma (fv)}{n}$$

6. Add the correction factor to the assumed mean to obtain the exact mean of the grouped data,

$$\bar{X} = A + \frac{\Sigma (fv)}{n} \tag{5–3b}$$

Example 6

Refer to Example 5. Use the short method with deviations in original units to compute the mean for the grouped data.

Solution.   Let $A = 5$ miles. Table 5–17 is arranged to compute the mean by the short method—deviations in original units.

TABLE 5-17

COMPUTATION FOR EXAMPLE 6 (SHORT METHOD—DEVIATIONS IN ORIGINAL UNITS)

Miles Traveled	Average Miles (mid-point) X	Number of Students (frequency) f	Deviation from Assumed Mean in Miles $v = X - A$	Total Deviation in Miles fv
0 and under 2	1	2	-4	-8
2 and under 4	3	5	-2	-10
4 and under 6	5	4	0	0
6 and under 8	7	8	2	16
8 and under 10	9	1	4	4
Total		$n = 20$		$\Sigma (fv) = 2$

$$\bar{X} = A + \frac{\Sigma (fv)}{n} = 5 + \frac{2}{20} = 5 + 0.1 = 5.1 \text{ miles} \qquad \text{(formula 5–3b)}$$

Short Method—Deviations in Class-Interval Units (d)

The short method illustrated above may further be simplified if the deviations of the individual values above and below the assumed mean are expressed in class-interval units instead of in original units. When all classes in a frequency distribution are the same size, the deviations (v) from the assumed mean must have the common factor—the size of class interval. Thus, the deviations from the assumed mean may be factored according to the class size.

Let

$$i = \text{the class size}$$
$$d = \text{the deviation of class mid-point from the}$$
$$\text{assumed mean in class-interval units}$$

Thus, d is also the number of classes deviated from the *assumed mean class* (the class in which the assumed mean falls), such as $-1, -2, -3, \ldots$ (that is, one, two, three, ... classes smaller than the assumed mean class) and 1 (or, $+1$), 2, 3, ... (that is, one, two, three, ... classes larger than the assumed mean, respectively).

Then

$$d = \frac{v}{i} \quad \text{and} \quad v = d(i)$$

The last two columns in Table 5–17 (v and fv) may be factored as shown in Table 5–18, where

$$i = 2 \text{ miles, the common factor}$$

and

$$d = \frac{v(\text{miles})}{2(\text{miles})}$$

TABLE 5-18

CLASS SIZE (i) AS A COMMON FACTOR OF DEVIATIONS (v)

X	f	$v = d(i)$	$fv = fd(i)$
1	2	$-4 = -2(2)$	$-8 = -4(2)$
3	5	$-2 = -1(2)$	$-10 = -5(2)$
5	4	$0 = 0(2)$	$0 = 0(2)$
7	8	$2 = 1(2)$	$16 = 8(2)$
9	1	$4 = 2(2)$	$4 = 2(2)$
Total	$n = 20$		$2 = 1(2)$ $\sum(fv) = \sum(fd)\cdot(i)$

When the term $\sum(fv)$ is replaced by $\sum(fd)\cdot(i)$, formula (5–3b) may be written

$$\bar{X} = A + \frac{\sum(fd)}{n}i \qquad (5\text{–}3c)$$

By using formula (5–3c), Example 6 may be computed in the manner shown in Table 5–19. Again, the assumed mean $A = 5$ miles (the mid-point of class "4 and under 6.")

Example 6 illustrates the short method applied to a series of continuous data in which the stated class limits are also the real class limits. The short method may also be applied in a similar manner to a series of data in which the stated class limits are not the real class limits. This is illustrated in Example 7.

TABLE 5-19

COMPUTATION OF THE ARITHMETIC MEAN BY THE SHORT
METHOD (EXAMPLE 6)—DEVIATION IN CLASS-INTERVAL UNITS

Miles Traveled (class interval)	Average Miles (mid-point) X	Number of Students (frequency) f	Deviation from Assumed Mean in Class Interval Units $d = \frac{v}{i}$	Total Deviation in Class Interval Units fd
0 and under 2	1	2	-2	-4
2 and under 4	3	5	-1	-5
4 and under 6	5	4	0	0
6 and under 8	7	8	1	8
8 and under 10	9	1	2	2
Total		20		1

$$\bar{X} = A + \frac{\Sigma\,(fd)}{n}\,i = 5 + \frac{1}{20}(2) = 5 + 0.1 = 5.1 \text{ miles}$$

Example 7

The salaries earned by a group of 25 employees in the Barnes Company in a given
period are given in the first and second columns of Table 5–20. Compute the
mean by using the short method with deviations in class-interval units.

Solution. Let the mid-point of class "7–9" (one of the two middle classes) be
the assumed mean, or $A = \$8$. The d value in the assumed mean class must be
zero. The value of d for each of the other classes is shown in the third column.
The mid-point values (X's) are not shown in Table 5–20, since they are not needed
in the computation by the short method.

TABLE 5-20

COMPUTATION OF THE ARITHMETIC MEAN (EXAMPLE 7)

Salaries	Number of Employees f	Deviation from Assumed Mean in Class-Interval Units d	Total Deviation in Class-Interval Units fd
$1–3	1	-2	-2
4–6	4	-1	-4
7–9	9	0	0
10–12	6	1	6
13–15	2	2	4
16–18	3	3	9
	25		13

$$\bar{X} = A + \frac{\Sigma\,(fd)}{n}\,i = 8 + \frac{13(3)}{25} = 8 + 1.56 = \$9.56$$

Chief Characteristics of the Mean

From the above discussion, we may now list the chief characteristics of the mean as follows:

1. The computation of the arithmetic mean is based on all values of a set of data. The value of every item in the data thus affects the value of the mean. When some extreme values are included in the data, the mean may become less representative of the entire values. For example, the mean of values 1, 2, 4, and 93 is 25. The mean is not close to any one of the four values. The mean of values 24, 25, 25, and 26 is also 25. It is obvious that the mean 25 is less representative of the group of values 1, 2, 4, and 93 than of the latter group of values.

2. Basically, the mean is computed as follows:

$$\text{Mean} = \frac{\text{sum of values}}{\text{number of values}}$$

Thus, if any two of the three terms in the expression (mean, sum of values, and number of values) are known, the third one can be determined. For example, if the mean is 5 and the number of values is 8, the sum of the values in the data can be determined, or $5 \times 8 = 40$.

3. The mean has two important mathematical properties which provide further mathematical analysis and which have made its use more popular than any other type of averages.

    a. The algebraic sum of the deviations of individual values from (above and below) the mean is zero. This property has been indicated in the discussion concerning the short method of computing the mean. In general, let $x = X - \bar{X}$, or the deviation of each value from the mean; then

$$\Sigma x = \Sigma (X - \bar{X}) = 0$$

    b. The sum of the squared deviations from the mean is the least; or, symbolically,

$$\Sigma x^2 = \Sigma (X - \bar{X})^2 \text{ is smaller than } \Sigma (X - \text{any value})^2$$

Properties a and b are further shown in Table 5–21.

TABLE 5-21

MATHEMATICAL PROPERTIES OF THE ARITHMETIC MEAN

Values X	Deviations from the Mean (5)		Deviations from an Arbitrarily Selected Value (3)	
	Deviations $x = X - \bar{X}$	Squared deviations x^2	Deviations $X - 3$	Squared deviations $(X - 3)^2$
1	-4	16	-2	4
3	-2	4	0	0
6	1	1	3	9
10	5	25	7	49
20	$\Sigma x = 0$	$\Sigma x^2 = 46$	8	62

$$\text{Mean} = \bar{X} = \frac{20}{4} = 5$$

$$\Sigma x^2 = \Sigma (X - 5)^2 = 46$$

which is smaller than

$$\Sigma (X - 3)^2 = 62$$

5.3 Median

The median of a set of values is the value of the middle item when the values are arranged in order of magnitude. To find the median, first arrange the values in an array, either from the smallest to the largest or from the largest to the smallest. Second, locate the middle value; that is, the number of values above the median is the same as the number of values below the median. The methods of locating the median for ungrouped data and for grouped data are discussed below.

Ungrouped Data

If the number of values in a set of ungrouped data is *odd*, the median is determined in the following manner:

1. Arrange the raw data in an array.
2. Locate the middle item as the median.

Example 8

Find the median of values 1, 4, 10, 8, and 10, representing the miles traveled by five students (same as in Example 3).

Solution.   The values are first arranged in an array according to their magnitude, from the smallest to the largest, as follows:

Item number	Miles traveled (values)
1	1
2	4
3	8 ←—median
4	10
5	10

The third item is the middle item. Thus, median = 8 miles. There are two items (1 and 4) below the median and the same number of items (10 and 10) above the median.

If the number of values in a set of ungrouped data is *even*, there is no true median. The value of the median is thus assumed to be halfway between the two middle items in the array.

Example 9

Find the median of the values 9, 6, 2, 5, 18, and 12 (dollars).

Solution.　The given values are first arranged in an array:

Item number	Values
1	$2
2	5
3	6
	←—median = 7.5
4	9
5	12
6	18

The median is the mid-point of values 6 and 9 (the third and the fourth items in the array), or

$$\text{Median} = \frac{6+9}{2} = \$7.5$$

Grouped Data

When the values in raw data are grouped into a frequency distribution table, each of the values loses its identity in the table. Thus, the median obtained from an array may not be the same as the median obtained from frequency distribution of the same raw data. The median for grouped data is an approximation of the true median. The approximation may be obtained by the interpolation method. When the interpolation method is applied, all types of data are treated as continuous data; that is, discrete data are interpolated as if they were continuous data. In addition, the values included in a class are assumed to be distributed evenly. For example, if there are four items in the class with the real lower limit 0 and the real upper limit 8, the value of the four items are assumed to be 2, 4, 6, and 8, respectively, and can be diagrammed as follows:

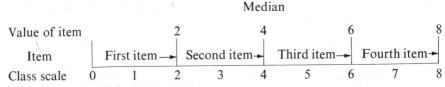

The value of each item is located at the *end* of the item on the class value scale. The median of the four items is 4 which is the value at the end of the second item.

　　The interpolation method may be carried out in two ways: (1) Compute the median from a frequency distribution table. (2) Locate the median from a frequency distribution chart. Both ways will give the same answer. They are presented below.

Compute the Median from a Frequency Distribution Table

　　1. Provide a column in the table to record the cumulative frequencies. The cumulative frequency of each class is the sum of the frequency of the class and the frequencies of the previous classes. For example, the cumulative frequency of the third class in the table is the sum of the frequencies in the first three classes.

2. Find the median class. The median class is the class that contains the median.

3. Interpolate the median from the values included in the median class.

The details of the procedure are illustrated in Example 10

Example 10

The miles traveled by 20 students in coming to Swan College from their homes are given in the first two columns below (same as the values given in Example 5). Find the median of the values.

Solution.   The median is computed in Table 5–22.

TABLE 5-22

LOCATING THE MEDIAN CLASS FOR EXAMPLE 10

Miles Traveled (class interval)	Number of Students (frequency)	Cumulative Frequency	
0 and under 2	2	2	
2 and under 4	5	7	
4 and under 6	4	11	Median class
6 and under 8	8	19	
8 and under 10	1	20	
Total	20		

There are 20 items (students) in the distribution. The median must be the value at the *end* of the tenth ($\frac{20}{2} = 10$) item in the distribution. The cumulative frequency column shows that the cumulative frequency which is just below 10 is 7. Thus, the median must fall within the class which is just above the seventh item, or within class "4 and under 6." The interpolation for the median from the values in the median class is diagrammed below:

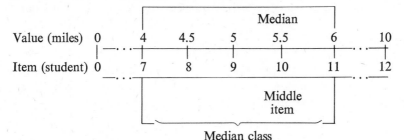

The interpolation is as follows:

Value in miles (median class limits)		Cumulative frequency (end of each item)	
6	corresponds to	11	(1)
x	corresponds to	10	(2)
4	corresponds to	7	(3)

$$\frac{(2) - (3)}{(1) - (3)} \quad \frac{x - 4}{2} \qquad \frac{10 - 7 = 3}{4}$$

The differences between the corresponding values form a proportion; thus

$$\frac{x-4}{2} = \frac{10-7}{4} \qquad x = 4 + \frac{10-7}{4}(2)$$

$$x = 4 + 2\left(\frac{3}{4}\right) = 4 + 1.5 = 5.5 \text{ miles (median)}$$

In summary, the above computation may be stated as follows:

Let

M_d = the median
L = the real lower limit of the median class
n = the number of total frequency in the given data
C = the cumulative frequency in the class just before the median class
f = frequency of the median class
i = size of the class interval in median class

Then

$$M_d = L + \frac{\frac{n}{2} - C}{f} i \qquad (5\text{--}4)$$

When formula (5–4) is applied, the median of Example 10 may be computed as follows:

$$M_d = 4 + \frac{\frac{20}{2} - 7}{4}(2) = 4 + \frac{3}{4}(2) = 5.5 \text{ miles}$$

Example 10 illustrates the method of finding the median for a series of continuous data in which the stated class limits are also the real class limits. The method may also be applied in a similar manner to a series of data in which the stated class limits are not the real class limits. This is illustrated in Example 11 below.

Example 11

The salaries earned in a given period by a group of 25 employees in the Barnes Company are given in the first and the second columns of Table 5–23 (same as the values given in Example 7). Compute the median.

Solution. The median is computed from Table 5–23. It is the 12.5th item in the distribution, or

$$\frac{n}{2} = \frac{25}{2} = 12.5$$

The median class is "7–9." The real lower limit of the median class is 6.5, or

$$L = \frac{(6+7)}{2} = 6.5$$

TABLE 5-23

COMPUTATION OF THE MEDIAN—EXAMPLE 11

Salaries (class interval)	Number of Employees (frequency)	Cumulative Frequency	
$ 1-3	1	1	
4-6	4	5	
7-9	9	14	Median class
10-12	6	20	
13-15	2	22	
16-18	3	25	
Total	$n = 25$		

The cumulative frequency in the class just before the median class is $C = 5$.

The frequency of the median class is $f = 9$. The size of the median class is $i = 3$, since the real lower and upper limits are 6.5 and 9.5, respectively.

Use formula (5-4).

$$M_d = 6.5 + \frac{\frac{25}{2} - 5}{9}(3) = 6.5 + \frac{7.5}{3} = 6.5 + 2.5 = 9$$

Locate the Median from a Frequency Distribution Chart

The median for a frequency distribution may be obtained graphically by using an *ogive* plotted on a line chart. An ogive shows the cumulative frequencies, which may be arranged on either (1) a "less than" basis or (2) an "or more" or "more than" basis. The methods of locating the median from each type of the ogives are presented in Example 12.

Example 12

From the information in Table 5-24, find the median graphically (same as the values given in Example 10 above).

TABLE 5-24

MILES TRAVELED BY 20 STUDENTS IN COMING
TO SWAN COLLEGE FROM THEIR HOMES

Miles Traveled	Number of Students (frequency)
0 and under 2	2
2 and under 4	5
4 and under 6	4
6 and under 8	8
8 and under 10	1
Total	20

Solution. *Method I. By an ogive on a "less than" basis.* First, prepare a cumulative frequency table (Table 5-25) on a "less than" basis. Second, construct Chart 5-1 showing the cumulative frequencies. Third, locate the median item on the Y (vertical) scale, which shows the cumulative frequencies. The median is the tenth item, or $20/2 = 10$. Fourth, starting from the point 10, draw a horizontal line to intersect the ogive. Then locate the median by drawing a perpendicular line from the intersection to the X (horizontal) scale. The median is 5.5 miles.

The median may simply be located from Chart 5-2. The chart shows only the median class section of Chart 5-1.

TABLE 5-25

CUMULATIVE FREQUENCIES ON "LESS THAN" BASIS

Miles Traveled (cumulative classes)	Number of Students (cumulative frequencies)
Less than 0	0
Less than 2	2
Less than 4	7
Less than 6	11
Less than 8	19
Less than 10	20

CHART 5-1

LOCATING THE MEDIAN FROM THE OGIVE ON A "LESS THAN" BASIS

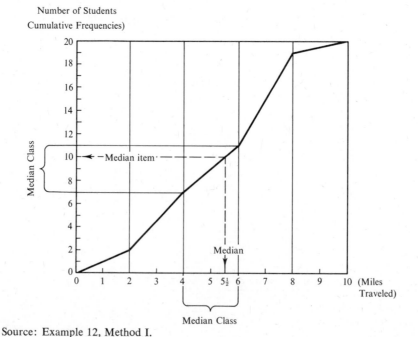

Source: Example 12, Method I.

Method II. By an ogive on an "or more" basis. First, prepare a cumulative frequency table (Table 5–26) on an "or more" basis. Second, construct Chart 5–3 (page 134) showing the cumulative frequencies. Then, locate the median as indicated in Method I above. The median is also 5.5 miles.

CHART 5-2

LOCATING THE MEDIAN FROM THE MEDIAN CLASS SECTION
OF THE OGIVE ON A "LESS THAN" BASIS

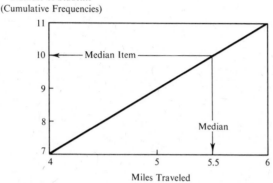

Source: Chart 5–1.

TABLE 5-26

CUMULATIVE FREQUENCIES ON "OR MORE" BASIS

Miles Traveled (cumulative classes)	Number of Students (cumulative frequencies)
0 or more	20
2 or more	18
4 or more	13
6 or more	9
8 or more	1
10 or more	0

Method III. By two ogives plotted on the same chart, one on a "less than" basis and another on an "or more" basis. The two ogives are plotted on Chart 5–4 (page 135). The median is located by drawing a perpendicular line from the intersection of the two ogives to the X-axis. The median is also 5.5 miles.

The points representing the cumulative frequencies on the ogives shown in Charts 5–1 through 5–4 are connected by straight lines. An ogive may also be drawn in the form of a smooth curve. The smooth curve is affected by the points in the median class as well as other points representing the cumulative frequencies. Thus, the median located by using the smooth ogive is considered even more meaningful than the median interpolated by the straight lines ogive. However, the median located by the smooth ogive will not be the same as the median obtained by using formula (5–4).

CHART 5-3

LOCATING THE MEDIAN FROM THE OGIVE ON AN "OR MORE" BASIS

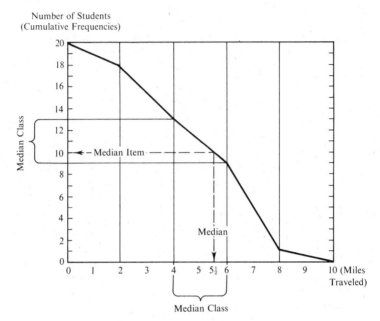

Source: Example 12, Method II.

Note that if we wish to locate the median from a frequency distribution chart for the data given in Table 5–23 of Example 11, we should label the horizontal scale continuously by using the real class limits as shown below:

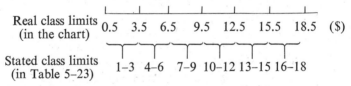

Real class limits (in the chart)	0.5	3.5	6.5	9.5	12.5	15.5	18.5	($)

| Stated class limits (in Table 5–23) | 1–3 | 4–6 | 7–9 | 10–12 | 13–15 | 16–18 |

Chief Characteristics of the Median

1. The median is a positional average. It is not affected by extreme values as is the mean, since the median is not computed from all values. For example, the median of values 4, 5, and 6 is 5 and the median of values 1, 5, and 1,000 is also 5.

2. The median is not defined algebraically as is the arithmetic mean. For example, if the median is 8 and the number of values is 5, the sum of the 5 values is not necessarily 40 (or 8×5). Note that the sum of values in Example 8 is 33 (or $1 + 4 + 10 + 8 + 10 = 33$), and the median is 8.

CHART 5-4

LOCATING THE MEDIAN FROM THE OGIVES ON
"LESS THAN" AND "OR MORE" BASES

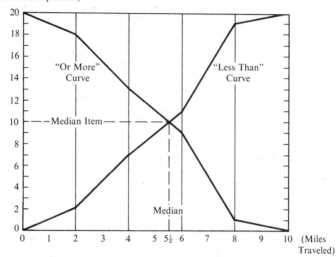

Source: Example 12, Method III.

3. The median, in some cases, cannot be computed *exactly* as can the mean. When the number of items included in a series of data is even, the median is determined *approximately* as the mid-point of the two middle items.

4. When the location of the middle item can be determined and the median class limits are known, the median for a frequency distribution with class intervals of equal or different sizes or open-ends can always be obtained by using the interpolation methods.

TABLE 5-27

COMPARISON OF THE DEVIATIONS OF INDIVIDUAL VALUES
FROM THE MEDIAN AND FROM AN ARBITRARILY SELECTED VALUE

Values (miles)	*Deviation from the Median* (8)	*Deviation from Selected Value* (6)
1	7 (= 1 − 8)	5 (= 1 − 6)
4	4	2
8	0	2
10	2	4
10	2	4
Total (disregarding signs)	15	17

Source: Example 8.

5. The median is centrally located. The absolute sum (disregarding positive and negative signs) of the deviations of the individual values from the median is *minimum*. In other words, the absolute sum of the deviations (or the sum of the distances) of the individual values from *a value other than the median* will exceed (or at least be equal to) the absolute sum of the deviations from *the median*. For example, the absolute sum of the deviations from the median 8 in Example 8 is 15, whereas the absolute sum of the deviations from the arbitrarily selected value 6 is 17 as shown in Table 5–27. The difference between the two sums is 2 miles ($=17-15$).

5.4 Mode

The mode of a set of values is the value which occurs most frequently in the set. If a value is selected at random from the given set, a modal value is the most likely value to be selected. Thus, the mode is generally regarded as the most typical value in a series of data.

Ungrouped Data

The mode for ungrouped data of a few values can be obtained by inspection, as illustrated in Example 13 below.

Example 13

Find the mode of the values 1, 4, 10, 8, and 10, representing the miles traveled by five students (same as the values given in Example 3).

Solution.   The mode of the five values is 10. The value 10 occurs twice, but each of the other values 1, 4, and 8 occurs only once.

Example 14

Find the mode of the values 1, 3, 3, 7, 7, and 8.

Solution.   There are two modes, 3 and 7, since each of the two values occurs twice.

When there are two or more modes in a set of data, the data are called *bimodal* or *multimodal*.

Example 15

Find the mode of the values 1, 2, 4, and 9.

Solution.   There is no mode, since none of the values occurs more than once.

To obtain the mode for ungrouped data of a large number of values, use the tally marks as used in the illustration of the frequency array in Table 5–1, page 113.

Grouped Data

The mode for grouped data can be obtained by various methods. Each of the methods may give a different value of mode. The four methods presented below are among the most common methods of computing the mode for a frequency distribution: the crude mode method, the interpolation by graph method, the interpolation by formula method, and the empirical mode method.

The Crude Mode Method (CM_0)

The crude mode of a frequency distribution is the value of the mid-point of the modal class. The modal class is the class with the highest frequency in the distribution.

Example 16

The miles traveled by 20 students in coming to Swan College from their homes are given in Table 5–28 (same as the values given in Example 5). Compute the crude mode.

TABLE 5-28

LOCATING THE MODAL CLASS FOR EXAMPLE 16

Miles Traveled (class interval)	Number of Students (frequency)	
0 and under 2	2	
2 and under 4	5	
4 and under 6	4	
6 and under 8	8	Modal class
8 and under 10	1	
Total	20	

Solution. The modal class is "6 and under 8" since the class has the highest frequency, 8. The mid-point of the modal class is

$$\text{The crude mode } (CM_0) = \frac{6+8}{2} = 7 \text{ (miles)}$$

The Interpolation by Graph Method (M_0)

When the interpolation method is used, the two real class limits of the modal class, the frequency in the modal class, and the frequencies in the classes immediately before and after the modal class are employed. By using the values given in Example 16, the interpolation method may be performed graphically in the following manner:

(1) Construct a bar chart based on the given data as shown in Chart 5–5. The bar chart which shows the frequency distribution continuously (that is, there are no gaps between adjacent bars) is called a *histogram.*

(2) Draw two lines diagonally in the inside of the modal class bar, starting from the upper corners of the bar to the upper corners of the adjacent bars.

(3) Draw a perpendicular line from the intersection of the two diagonal lines to the X axis (horizontal scale). The mode is located on the X axis and is 6.7 miles.

CHART 5-5

INTERPOLATING MODE IN A HISTOGRAM

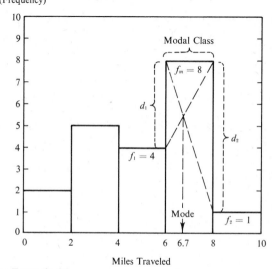

Miles Traveled by 20 Students
In Coming to Swan College from Their Homes

Source: Example 16.

The Interpolation by Formula Method (M_0)

When the interpolation is performed by the formula method, the real lower limit and the size of the modal class, the frequency in the modal class, and the frequencies in the classes immediately before and after the modal class are also employed. Thus, if the graph is drawn accurately, the mode computed from the formula should be the same value as the mode obtained from the histogram.

The formula for computing the mode for a frequency distribution is

$$M_0 = L + \frac{d_1}{d_1 + d_2} i \qquad (5\text{–}5)$$

where M_0 = the mode
$\quad\quad\quad L$ = the real lower limit of the modal class
$\quad\quad\quad i$ = size of the class interval of the modal class

f_m = frequency of the modal class

f_1 = frequency of the class before (or smaller than) the model class

f_2 = frequency of the class after (or larger than) the modal class

$d_1 = f_m - f_1$

$d_2 = f_m - f_2$

By using the values given in Example 16, the mode of the grouped data is computed according to the formula as follows:

$L = 6$, the real lower limit of the modal class "6 and under 8"

$i = 2$ miles

$d_1 = f_m - f_1 = 8 - 4 = 4$

$d_2 = f_m - f_2 = 8 - 1 = 7$

Substitute the above values in the formula (5–5).

$$M_0 = 6 + \frac{4}{4 + 7}(2) = 6 + \frac{8}{11} = 6.727, \text{ rounded to 6.7 miles}$$

The application of the two interpolation methods to a series of data in which the stated class limits are not the real class limits is similar to a series of continuous data as given in Example 16 above. Example 17 illustrates the similarity.

Example 17

The salaries earned by a group of 25 employees in the Barnes Company in a given period are given in Table 5–29 (same as the values in Example 7). Use (a) the graph method, and (b) formula (5–5) to find the mode.

TABLE 5-29

LOCATING THE MODAL CLASS FOR EXAMPLE 17

Salaries (class interval)	Number of Employees (frequency)	
$ 1–3	1	
4–6	4	
7–9	9	Modal class
10–12	6	
13–15	2	
16–18	3	
Total	25	

Solution. (a) For simplicity and meeting the requirements, only the modal class and its two adjacent classes are graphed in Chart 5–6. The chart shows that the mode is $8.38.

(b) $L = 6.5$, the real lower limit of the modal class 7–9

$$i = \$3$$
$$d_1 = 9 - 4 = 5$$
$$d_2 = 9 - 6 = 3$$

Substitute the values in formula (5–5).

$$M_0 = 6.5 + \frac{5}{5+3}(3) = 6.5 + \frac{15}{8} = 8.375, \text{ or rounded to } \$8.38$$

CHART 5-6

INTERPOLATING MODE IN A PARTIAL HISTOGRAM

Salaries Earned by Employees in Barnes Company

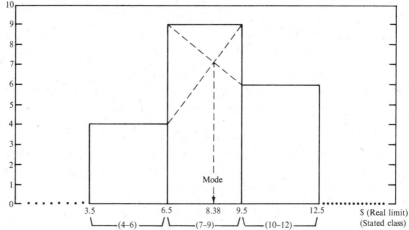

Source: Example 17.

The Empirical Mode Method (EM_0)

In a symmetrical frequency distribution, the values of mean, median, and mode are the same; that is, $\bar{X} = M_d = M_0$. A symmetrical distribution means that the middle class has the highest frequency and that the frequencies above the middle class of the distribution exactly correspond to the frequencies below the middle class. If the frequency distribution is shown by a bar chart (histogram) or a line chart (frequency polygon), a vertical line through the mid-point of the middle class may divide the graph into two perfect identical halves. The intersection of the vertical line and the X axis gives the values of the mean, the median, and the mode.

Example 18

The grades of 28 students on a statistics test are given in Table 5–30. Compute $\bar{X}$, M_d, and M_0.

TABLE 5-30

A Symmetrical Frequency Distribution (Example 18)

	Grades (class interval)	Class Mid-Point	Number of Students (frequency)
	30 and under 40	35	2
	40 and under 50	45	3
Mean class	50 and under 60	55	5
Median class	60 and under 70	65	8
Modal class	70 and under 80	75	5
	80 and under 90	85	3
	90 and under 100	95	2
	Total		28

$$\bar{X} = A + \frac{\Sigma(fd)}{n}i = 65 + \frac{0}{28}(10) = 65 + 0 = 65$$

$$M_d = L + \frac{\frac{n}{2} - C}{f}(i) = 60 + \frac{\left(\frac{28}{2}\right) - (2 + 3 + 5)}{8}(10) = 65$$

$$M_0 = L + \frac{d_1}{d_1 + d_2}(i) = 60 + \frac{8 - 5}{(8 - 5) + (8 - 5)}(10) = 65$$

Thus,

$$\bar{X} = M_d = M_0 = 65 \text{ (mid-point of the middle class)}$$

The frequency distribution in Example 18 is shown graphically in Chart 5-7. The chart shows the histogram as well as the frequency polygon. The *frequency polygon* is constructed by joining the mid-points of the top sides of the rectangles in the histogram. The two ends of the polygon may or may not be connected with the X axis. When the two ends are connected with the X axis, such as in Chart 5-7, two hypothetical classes with zero frequencies are employed; one is placed on each end of the distribution.

If the frequencies above the modal class are not the same as the frequencies below the modal class, the distribution is not symmetrical, and the values of the three averages ($\bar{X}$, M_d, and M_0) of the distribution are not the same.

When the distribution is not symmetrical, the frequency polygon, when drawn in the form of a smoothed curve, will skew either to the right side on the X scale (called *positively skewed*) or to the left side on the X scale (called *negatively skewed*). When the distribution is skewed to the right, as shown on Chart 5-8 (II), the sum of the deviations of the individual values from the mode above the modal class must be larger than that below the modal class. Thus, the value of the mean is affected by the high values in the distribution, and the position of the mean on the X scale is pushed to the right or higher value on the scale. The median is not affected as greatly by the high values as is the mean, since the

CHART 5-7

HISTOGRAM AND THE FREQUENCY POLYGON OF A SYMMETRICAL DISTRIBUTION
Grades of 28 Students on a Statistics Test

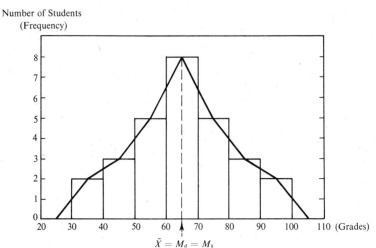

Grades of 28 Students on a Statistics Test

Source: Example 18.

median is the middle item and divides the curve into two equal areas. The median, although higher than the mode, is between the mode and the mean. The mode is always under the highest point of the curve. When the distribution is skewed to the left, the relationship between the three averages is the same, but the direction is opposite; that is, the median is still between the mode and the mean but is on the left side of the mode, as shown in Chart 5–8 (III).

In general, when the distribution is not symmetrical but is nearly symmetrical, the median locates approximately one-third of the distance from the mean toward the mode on the X scale. The relationship is expressed by the formula below:

$$\text{Empirical mode} = \text{mean} - 3(\text{mean} - \text{median})$$

or

$$EM_0 = \bar{X} - 3(\bar{X} - M_d) \tag{5-6}$$

Example 16 is computed as follows when formula (5–6) is used.

$$\bar{X} = 5.1 \qquad\qquad\qquad\qquad \text{(page 121)}$$
$$M_d = 5.5 \qquad\qquad\qquad\qquad \text{(page 133)}$$
$$EM_0 = 5.1 - 3(5.1 - 5.5) = 5.1 - 3(-0.4) = 6.3 \text{ miles}$$

This example shows a negative skewed distribution, since the mean is smaller than the median and the mode, or, expressed graphically (on page 146),

CHART 5-8

LOCATIONS OF MEAN, MEDIAN AND MODE ON X SCALE
IN SYMMETRICAL AND ASYMMETRICAL DISTRIBUTIONS

I Symmetrical Distribution

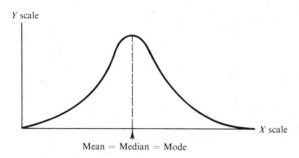

II Asymmetrical Distribution—Skewed to the Right or Higher Values (A Positively Skewed
Distribution)

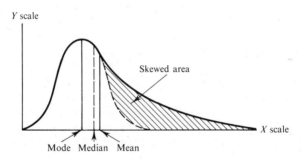

III Asymmetrical Distribution—Skewed to the Left or Lower Values (A Negatively Skewed
Distribution)

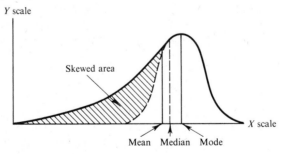

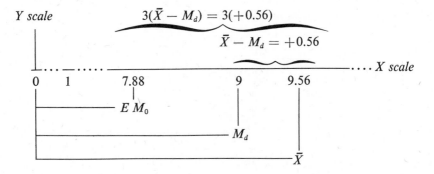

Example 17 is computed as follows when formula (5–6) is used:

$$\bar{X} = 9.56 \qquad \text{(page 125)}$$
$$M_d = 9 \qquad \text{(page 131)}$$
$$EM_0 = 9.56 - 3(9.56 - 9) = 9.56 - 3(0.56) = \$7.88$$

This example shows a positive skewed distribution, since the mean is larger than the median and the mode, or, expressed graphically,

The value of the mode obtained by formula (5–5) for Example 16 is 6.7 miles and that for Example 17 is $8.38. The above illustrations indicate that the value of the mode obtained by formula (5–5) is different from the empirical mode in each example. The different result is due to the fact that formula (5–5) is based on the values and frequencies in the modal class and its two adjacent classes only, whereas the empirical mode is based on the median and the mean, which are affected by every value in the given data.

Chief Characteristics of Mode

1. The mode is the value with the highest frequency in the set of values. It represents more items than any other value could represent in the set. The mode is not computed from all values and is not defined algebraically as is the mean. For example, if the mode is 10 and the number of values in the data is 5, the sum of the values is not necessarily 50 (or 10 × 5). Note that the sum of values 1, 4, 8, 10, and 10 is 33 (Example 13).

2. The mode, by definition, is not affected by extreme values. For example, the mode of values 1, 5, 5, and 8 is 5, and the mode of values 1, 5, 5, and 800 is also 5.

3. The mode of a set of discrete data is easy to compute. However, the true mode of a set of continuous data, strictly speaking, may never exist. The values of items included in continuous data are seldom exactly alike before rounding. For example, the heights of a group of persons may be measured as 5.60001, 5.60002, 5.60821, ... feet, and no two persons are exactly the same height. Thus, it is doubtful if there is a perfect method to compute the value of mode for continuous data.

4. The mode for a frequency distribution cannot be computed exactly, as the mean can. The four methods introduced above may give four different modal values for a given distribution. Thus, a greater exercise of judgment is needed in interpreting the significance of the computed mode.

5. The value of the mode may be greatly affected by the method of designating the class intervals. For example, there is no mode in the following group of 8 values since each value occurs only once:

$$1, 2, 3, 5, 6, 7, 8, \text{ and } 12$$

When the same values are grouped into the different types of class intervals, as given in distributions A, B, and C in Table 5–31, the value of the mode for each distribution is different from others.

TABLE 5-31

LOCATING MODES FROM THE SAME DATA BUT DIFFERENT DISTRIBUTIONS

Distribution A		Distribution B		Distribution C	
Class Intervals	Frequency	Class Intervals	Frequency	Class Intervals	Frequency
1–2	2	1–4	3	1–5	4
3–4	1	5–8	4	6–10	3
5–6	2	9–12	1	11–15	1
7–8	2				
9–10	0				
11–12	1				
Total	8		8		8

There are three modes in distribution A: classes "1–2," "5–6," and "7–8." The mode of distribution B is

$$M_0 = 4.5 + \frac{1}{1+3}(4) = 5.5$$

The mode of distribution C is

$$M_0 = 0.5 + \frac{4}{4+1}(5) = 4.5$$

5.5   Summary

An average is a single value which is considered as the most representative or typical value of a group of values. It is also referred to as a measure of central tendency. An average is frequently used as a device in summarizing a group of numbers for describing statistical data and for comparing one group with another.

The most commonly known averages in statistics are (1) the arithmetic mean, (2) the median, (3) the mode, (4) the geometric mean, and (5) the harmonic mean. The formulas and the methods of computing the first three averages for ungrouped and grouped data are summarized on page 147.

The determination of which one of the different types of averages should be used under various circumstances largely depends on the characteristics of the averages. The characteristics of the three common averages are summarized in Table 5-32.

The numbers included in the ungrouped data are single values and are not classified into groups. When the values are grouped into several classes based on quantity in a table and the number of values within each class is indicated, the table is called a frequency distribution.

The number of classes in a frequency distribution depends upon the number of values to be grouped and the type of information that the investigator wishes

TABLE 5-32

SUMMARY OF CHIEF CHARACTERISTICS OF THREE
COMMON AVERAGES

Characteristics	*Arithmetic Mean* $(\bar{X})$	*Median* (M_d)	*Mode* (M_0)
Computation based on	Every value	Middle value	Value with the highest frequency
Affected by extreme values	Greatest	No (affected by items only)	No
Algebraic manipulation	Yes: $$\bar{X} = \frac{\sum X}{n}$$	No (positional average, interpolated value in many cases)	No (concentrated average, four methods for grouped data)
Mathematical properties	$\sum (X - \bar{X}) = 0$ $\sum (X - \bar{X})^2$ is minimum	$\sum (X - M_d)$ is minimum (disregarding signs)	—
Application to open-end classes	Indeterminate	Determinate	Determinate
Comparison of answers to same data	May be larger or smaller than M_d and M_0	Between $\bar{X}$ and M_0	May be larger or smaller than M_d and $\bar{X}$
Type of data preferred	Most types	Middle value is typical, excluding extremes	Data with distinct central tendency

SUMMARY OF FORMULAS

Application	Formula	Formula number	Reference page
Arithmetic mean			
Ungrouped data			
Basic method	$\bar{X} = \dfrac{\sum X}{n}$	(5-1a)	117
Short method	$\bar{X} = A + \dfrac{\sum(X - A)}{n}$		
or	$\bar{X} = A + \dfrac{\sum v}{n}$	(5-1b)	119
Weighted mean	$\bar{X} = \dfrac{\sum(wX)}{\sum w}$	(5-2)	120
Grouped data			
Basic method	$\bar{X} = \dfrac{\sum(fX)}{n}$	(5-3a)	121
Short method (original units)	$\bar{X} = A + \dfrac{\sum(fv)}{n}$	(5-3b)	123
Short method (class-interval units)	$\bar{X} = A + \dfrac{\sum(fd)}{n} i$	(5-3c)	124
Median			
Ungrouped data	Odd number of values: the middle item		127
	Even number of values: the mid-point of the two middle items		127
Grouped data			
From a table	$M_d = L + \dfrac{\dfrac{n}{2} - C}{f} i$	(5-4)	130
From ogives	I. On a "less than" basis		132
	II. On an "or more" or "more than" basis		133
	III. One on a "less than" basis and another on an "or more" basis		133
Mode			
Ungrouped data	By inspection or tally marks to find the value that occurs most frequently		136
Grouped data			
Crude mode	$CM_0 = $ Mid-point of modal class		137
Interpolated mode	I. Use a histogram		137
	II. $M_0 = L + \dfrac{d_1}{d_1 + d_2} i$	(5-5)	138
Empirical mode	$EM_0 = \bar{X} - 3(\bar{X} - M_d)$	(5-6)	142

to have. In general, the number of classes should not be too large or too small. The lower and upper class limits stated in a frequency distribution indicate the boundaries of each class in the distribution. However, in many cases, the stated class limits are not the real class limits. There are gaps between classes. In such cases, the mid-point of each gap is considered as the real limit between the two classes forming the gap. The mid-point of each class is usually used to represent each original value grouped into the class for purposes of further mathematical analysis.

A variable is a set of values. There are two types of variables: continuous variables and discrete variables. A continuous variable can theoretically assume any value between two given values. Data which can be described by a continuous variable are called continuous data. If the variable can not assume any value between two given values, it is called a discrete variable. Data represented by a discrete variable are called discrete data. In selecting class limits for the two types of data, care should be taken to avoid ambiguous class limits.

There are three types of class intervals according to the sizes of the classes in a frequency distribution: classes of equal size, classes of unequal sizes, and open-end classes. The size of a class interval is the difference between the real lower and upper class limits and is also referred to as the class width or class size.

Exercises 5

1. What is an average? Name the most commonly known averages.

2. What is a frequency distribution? How would you determine the class limits for a variable grouped into a frequency distribution?

3. Compute (a) the arithmetic mean, (b) the median, and (c) the mode of the following values which represent numbers of visits to a city by a group of eight farmers during one month:

$$1, 2, 2, 4, 5, 11, 11, 20$$

Which one of the averages may be regarded as the most representative number of visits for the group of farmers?

4. The amounts of money saved by eleven boys are given below:

$$\$78; 76; 71; 73; 75; 3,000; 7,000; 69; 72; 78; 78$$

Find:
  (a) The arithmetic mean
  (b) The median
  (c) The mode

Which one of the averages may be regarded as the most representative amount for the given data? the next? the least representative amount?

5. Find the arithmetic mean of the numbers 20, 30, 45, 68, 72, 85, and 86 by the short method. Let $A = 50$.

6. Refer to Problem 5. Find the arithmetic mean by letting $A = 60$.

7. The final grades and the credits for the five courses received by a student are given below. Find an appropriate average grade for the student.

Courses	Grades	Credits
Accounting	86	5
English	82	3
Music	74	1
Mathematics	70	2
History	86	3

8. Find the arithmetic mean of the hourly wages of the 80 workers: $3.60 per hour (5 workers), $3.40 per hour (15 workers), $4.70 per hour (30 workers), $5.60 per hour (10 workers), and $3.50 per hour (20 workers).

9. The annual salaries of 100 men are given below.
   (a) Find the median.
   (b) Can you find the mode?
   (c) Can you find the mean?
Explain.

Salaries	Number of men
Less than $500	4
$500 to less than $1,000	10
$1,000 to less than $5,000	28
$5,000 to less than $10,000	40
$10,000 to less than $25,000	12
$25,000 and over	6

10. The hourly wages paid to a group of ten workers are distributed as follows:

Hourly Wages	Number of Workers
$4.50	1
5.10	2
6.20	3
8.40	4

Find (a) the arithmetic mean, (b) the median, and (c) the mode of the hourly wages earned by the ten workers.

11. Compute the mean from the following table:

Class interval	f	d	fd
3–7	1		
8–12	9		
13–17	12	0	
18–22	15		
23–27	13		

12. Compute the mean from the following table to 3 decimal places:

Class interval	f	d	fd
0.00–9.99	4		
10.00–19.99	12		
20.00–29.99	18		
30.00–39.99	42		
40.00–49.99	58		
50.00–59.99	79	0	
60.00–69.99	45		
70.00–79.99	33		
80.00–89.99	16		
90.00–99.99	8		

13. The following frequency distribution table shows the heights of the 18 salesmen in Bayside Company.

Height (inches)	Number of salesmen
60 and under 62	2
62 and under 64	3
64 and under 66	7
66 and under 68	5
68 and under 70	1

Find:
  (a) The arithmetic mean by using
       (1) the basic method
       (2) the short method—deviations in original units (v)
       (3) the short method—deviations in class-interval units (d)
  (b) The median from
       (1) the frequency distribution table
       (2) the ogive chart on an "or more" basis
  (c) The mode by using
       (1) the crude mode method
       (2) the interpolation by graph method
       (3) the interpolation by formula method
       (4) the empirical mode method

14. The grades of the 80 students in a statistics class are given below:
    90 85 70 75 30 55 80 43 70 40 80 99 95 55 78 70 95 80 60 30
    68 85 25 60 48 60 62 45 50 23 71 95 75 70 85 53 65 95 70 75
    63 30 90 82 80 78 65 75 90 85 75 75 25 72 60 88 73 74 75 52
    55 75 35 76 79 77 74 78 76 73 75 43 38 75 80 78 50 85 63 35
In the following computations, round each answer to two decimal places.
    (a) From the above data, set up a tally sheet showing the frequencies according to the grades; then find from the frequency array:
         (1) The arithmetic mean
         (2) The median
         (3) The mode

(b)   Use the following class intervals to group the above data in a frequency distribution:

   20–29, 30–39, 40–49, 50–59, 60–69, 70–79, 80–89, 90–99

Compute:

   (1)   The arithmetic mean by using:
      a. The basic method
      b. The short method—deviations in original units (v)
      c. The short method—deviations in class-interval units (d)
   (2)   The median from:
      a. The frequency distribution table
      b. The ogive chart on a "less than" basis
   (3)   The mode by using:
      a. The crude mode method
      b. The interpolation by graph method
      c. The interpolation by formula method
      d. The empirical mode method

(c)   Is the frequency distribution the better form for computing and interpreting the information?

(d) Discuss the significance of each of the averages computed in a and b.

6 ★ Averages: Geometric Mean and Harmonic Mean

The geometric mean and the harmonic mean are used as averages for special types of statistical data. They are presented in Sections 6.1 and 6.2, respectively. The chief characteristics of all five types of averages (the arithmetic mean, the median, the mode, the geometric mean, and the harmonic mean) are summarized in Section 6.3 for easy comparison.

6.1 Geometric Mean

The geometric mean of a set of n values is the nth root of the product of the values in the set. If there are two values, the square root of the product of the two values is the geometric mean; if three values, the cube root is the geometric mean; and so on. Note that when logarithms are employed in the computations of this section, the six-place mantissas of Table 2 in the Appendix are used.

Ungrouped Data

The geometric mean for ungrouped data can be expressed as follows:

Geometric mean $(G) = \sqrt[n]{\text{the product of the } n \text{ values}}$

Symbolically, let n values be represented by $X_1, X_2, X_3, \ldots, X_n$; then

$$G = \sqrt[n]{X_1 \cdot X_2 \cdot X_3 \cdot \cdots \cdot X_n} \tag{6-1a}$$

When logarithms are used in computing the value of G, the formula may be written

---

Each chapter (or section) marked by a large boldface asterisk (★) is optional material. It may be omitted without interrupting the continuity of the text organization.

$$\log G = \frac{1}{n} \log (X_1 \cdot X_2 \cdot X_3 \cdot \ \cdots \ \cdot X_n)$$

or

$$\log G = \frac{\log X_1 + \log X_2 + \log X_3 + \cdots + \log X_n}{n}$$

or simply

$$\log G = \frac{\Sigma \log X}{n} \qquad (6\text{-}1b)$$

The value G is the antilog of the quotient at the right side of the formula.

Example 1

Compute the geometric mean of the values 1, 4, 10, 8, and 10 (same as the values given in Example 3, Section 5.2.

Solution. (a) Use formula (6-1a).

$$G = \sqrt[5]{1 \times 4 \times 10 \times 8 \times 10} = \sqrt[5]{3{,}200} = 5.024$$

The answer is obtained by the use of logarithms as follows:

$$G = \sqrt[5]{3{,}200} = 3{,}200^{1/5}$$

Take the logarithms of the numbers on both sides.

$$\log G = \log 3{,}200^{1/5} = \left(\frac{1}{5}\right)(\log 3{,}200) = \left(\frac{1}{5}\right)(3.505150)$$

$$= 0.701030 \ (\text{Table 2.})$$

$$G = \text{antilog } 0.701030 = 5.024$$

The mantissa 701030 is close to 701050, which corresponds to digits 5024 in log table. The characteristic is 0. Thus, $G = 5.024$. (Interpolation is omitted.)

(b) Use formula (6-1b).

Values X	log X
1	0.000000
4	0.602060
10	1.000000
8	0.903090
10	1.000000
Total	$3.505150 = \Sigma \log X$

$$\log G = \frac{\Sigma \log X}{n} = \frac{3.505150}{5} = 0.701030$$
$$G = \text{antilog } 0.701030 = 5.024$$

Grouped Data

If $f_1, f_2, \ldots, f_k$ represent the frequencies of values $X_1, X_2, \ldots, X_k$, respectively, and $\Sigma f = n$, then formula (6-1a) can be written

$$G = \sqrt[n]{(X_1X_1 \cdots X_1)(X_2X_2 \cdots X_2) \cdots (X_kX_k \cdots X_k)} \qquad (6\text{–}2a)$$
$$\underbrace{\qquad}_{f_1 \text{ factors}} \underbrace{\qquad}_{f_2 \text{ factors}} \underbrace{\qquad}_{f_k \text{ factors}}$$

and formula (6–1b) can be written

$$\log G = \frac{f_1 \cdot \log X_1 + f_2 \cdot \log X_2 + \cdots + f_k \cdot \log X_k}{f_1 + f_2 + \cdots + f_k}$$

or simply

$$\log G = \frac{\Sigma \, (f \cdot \log X)}{n} \qquad (6\text{–}2b)$$

Formula (6–2a) is inconvenient in computing the value of G, since the product of the values in a frequency distribution is usually very large.

Example 2

The miles traveled by 20 students in coming to Swan College from their homes are given in the first two columns of Table 6–1 (same as the values given in Example 5, Section 5.2). The geometric mean is computed by formula (6–2b) as follows:

TABLE 6–1

Computation for Example 2

Miles Traveled (class interval)	Number of Students (frequency) f	Class Mid-point X	$\log X$	$f \log X$
0 and under 2	2	1	0.000000	0.000000
2 and under 4	5	3	0.477121	2.385605
4 and under 6	4	5	0.698970	2.795880
6 and under 8	8	7	0.845098	6.760784
8 and under 10	1	9	0.954243	0.954243
Total	$n = 20$		$\Sigma \, (f \cdot \log X) = 12.896512$	

$$\log G = \frac{\Sigma \, (f \log X)}{n} = \frac{12.896512}{20} = 0.644826$$

The mantissa 644826 is close to 644832, which corresponds to digits 4414 in log table. The characteristic is 0. Thus,

$$\text{antilog } (0.644826) = 4.414, \quad \text{or} \quad G = 4.41 \text{ miles}$$

In practice there is a rare demand for finding the geometric mean of the miles traveled by a group of persons. However, the geometric mean method is frequently used for averaging ratios, rates of change, and values showing a geometric progression. The application of the method to practical problems must rely on the understanding of the characteristics of the geometric mean.

Chief Characteristics of the Geometric Mean

(1) Basically, the geometric mean is computed as follows:

$$G = \sqrt[n]{\text{product of } n \text{ values}}$$

Thus, if any two of the three elements (G, n, and the product of n values) in the expression are known, the third one can be determined. For example, if the geometric mean is 4 and the number of values is 3, the product of the three values is 64, or

$$G^n = 4^3 = 64$$

However, the individual values cannot be ascertained, since there are many groups of three values that will give a product of 64, such as

$$4 \times 4 \times \ 4 = 64$$
$$1 \times 4 \times 16 = 64$$
$$2 \times 4 \times \ 8 = 64$$

(2) The computation of the geometric mean is based on all items in a set of data. The value of every item in the data thus affects the value of the geometric mean. If any one of the values is zero, the value of G is zero, such as

$$G = \sqrt[3]{2 \times 4 \times 0} = \sqrt[3]{0} = 0$$

If the product of the values is negative and the number of values included in the data is even, the value of G becomes an imaginary number, such as

$$G = \sqrt{(-2)(+8)} = \sqrt{-16}$$

Neither $+4$ nor -4 qualifies as a square root of the radicand -16. Further, if any of the values is negative, the value of G may not be representative of the values, such as

$$G = \sqrt[3]{(-1) \times 4 \times 16} = \sqrt[3]{-64} = -4$$

(3) The geometric mean is affected by extreme values to a smaller extent than is the arithmetic mean. For example, the geometric mean of values 1, 4, and 16 is 4 (or $G = \sqrt[3]{1 \times 4 \times 16} = \sqrt[3]{64} = 4$), whereas the arithmetic mean of the same values is 7 (or $\bar{X} = (1 + 4 + 16)/3 = 7$). The value 7 is closer to the high value 16 than the value 4 is. The value G is always smaller than the value $\bar{X}$ of the same data, except when all values in a series are equal, such as the G of values 4, 4, and 4 being 4 and the $\bar{X}$ of the three values also being 4.

(4) The geometric mean balances the ratio deviations of the individual values from the geometric mean. For example, the geometric mean of values 1, 4, 9, 12, and 18 is 6. The product of the ratios of the individual values, which are smaller than G, to G is

$$\frac{1}{6} \times \frac{4}{6} = \frac{1}{9}$$

The product of the ratios of G to the individual values, which are larger than G, is

$$\frac{6}{9} \times \frac{6}{12} \times \frac{6}{18} = \frac{1}{9}$$

Thus, the products of the ratios above and below the geometric mean are equal:

$$\frac{1}{6} \times \frac{4}{6} = \frac{6}{9} \times \frac{6}{12} \times \frac{6}{18}$$

Stated in a different way: The deviations of the logarithms of the original values above and below the logarithm of the geometric mean are equal. Table 6–2 and Chart 6–1 are used to show the relationship.

TABLE 6–2

THE RELATIONSHIP BETWEEN THE LOGARITHMS
OF INDIVIDUAL VALUES AND THE LOGARITHM OF THE GEOMETRIC MEAN

Item	Values X	log X	Deviations log X − log G	Total of Positive and Negative Deviations
A	1	0.00000	−0.77815	−0.95424
B	4	0.60206	−0.17609	
	G = 6 log G = 0.77815			
C	9	0.95424	+0.17609	+0.95424
D	12	1.07918	+0.30103	
E	18	1.25527	+0.47712	

(5) The geometric mean gives equal weight to equal rates of change. In other words, in averaging rates of change geometrically, the rate showing twice of its base is offset by another rate showing one-half of its base (Table 6–3); the rate showing five times of its base is offset by another rate showing one-fifth of its base (Table 6–4), and so on. The rates are usually expressed in percentages. Since the base of each rate expressed in per cent is always equal to 1, or 100%, the average of the two rates which are offsetting should be 100% also. The geometric mean will give the satisfactory answer, since geometric mean of a number and its reciprocal is always equal to 1, such as

$$G = \sqrt{\frac{1}{2} \times 2} = \sqrt{1} = 1$$

since $\frac{1}{2}$ is a reciprocal of 2 or $\frac{2}{1}$, and

$$G = \sqrt{\frac{1}{4} \times 4} = \sqrt{1} = 1$$

since $\frac{1}{4}$ is a reciprocal of 4 or $\frac{4}{1}$.

The illustrations in Tables 6–3 and 6–4 indicate that the geometric mean in each case provides a better answer than the arithmetic mean provides.

CHART 6–1

THE RELATIONSHIP BETWEEN THE INDIVIDUAL VALUES
AND THE GEOMETRIC MEAN ON A LOGARITHMIC SCALE

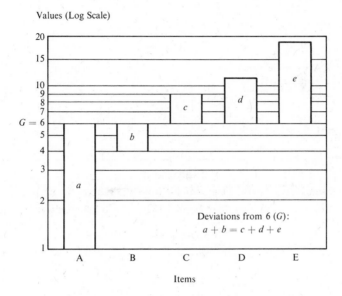

Values (Log Scale)

Deviations from 6 (G):
$$a + b = c + d + e$$

Items

Source: Table 6-2.

TABLE 6–3

COMPARISON OF THE UNITS SOLD BY COMPANY H IN 1975 AND 1976

Item	Units Sold		Rates of Change	
	1975	1976	1975 (base)	1976
A	15 lb	30 lb	100%	200%
B	40 yd	20 yd	100%	50%
Arithmetic mean			100%	$125\% = \dfrac{200\% + 50\%}{2}$
Geometric mean			100%	$100\% = \sqrt{200\%\ (50\%)}$ $= \sqrt{2(0.5)} = \sqrt{1}$ $= 1$

(6) The geometric mean of the ratios of the individual values to each immediate preceding value in a sequence of values is the only appropriate average for the ratios. The arithmetic mean of the ratios will not give the consistent result. Example 3 is used to illustrate the two different types of means in averaging the ratios.

TABLE 6–4

COMPARISON OF THE UNITS SOLD BY COMPANY K IN 1975 AND 1976

Item	Units Sold		Rates of Change	
	1975	1976	1975 (base)	1976
C	5 yd	25 yd	100%	500%
D	50 lb	10 lb	100%	20%
Arithmetic mean			100%	$260\% = \dfrac{20\% + 500\%}{2}$
Geometric mean			100%	$100\% = \sqrt{20\% \, (500\%)}$ $= \sqrt{0.2(5)} = \sqrt{1}$ $= 1$

Example 3

The annual sales of Johnson Department Store and the ratio of the annual sales to the sales in each previous year from 1972 to 1976 are given in Table 6–5.

TABLE 6–5

COMPUTATION FOR EXAMPLE 3

Year	Annual Sales	Ratio to Previous Year's Sales, X	log X
1972	$ 5,000	—	—
1973	3,600	0.72 (= 3,600/5,000)	9.857332 — 10
1974	5,760	1.60 (= 5,760/3,600)	0.204120
1975	5,184	0.90 (= 5,184/5,760)	9.954243 — 10
1976	10,368	2.00 (= 10,368/5,184)	0.301030
Total	29,912	5.22	20.316725 — 20 = 0.316725

The geometric mean of the ratios is 1.20 or 120% and is computed as follows:

$$\log G = \frac{\Sigma \log X}{n} = \frac{0.316725}{4} = 0.079181$$

The mantissa 079181 corresponds to digits 1200 in the log table. Thus,

$$G = \text{antilog } 0.079181 = 1.20$$

The arithmetic mean of the ratios is 1.305, or 130.5% and is computed as follows:

$$\bar{X} = (0.72 + 1.60 + 0.90 + 2.00)/4 = 5.22/4 = 1.305$$

The sales based on the two different average ratios (G and $\bar{X}$) are compared in Table 6–6. Only the geometric mean gives the satisfactory result, since the amount of sales computed from the geometric mean for the last year (1976) is consistent with the actual sales in the year.

TABLE 6–6

COMPARISON OF THE ANNUAL SALES COMPUTED BY THE GEOMETRIC
MEAN AND BY THE ARITHMETIC MEAN, JOHNSON DEPARTMENT
STORE, 1972 TO 1976

Year (1)	Actual Sales (2)	Sales Based on G, 120% of Previous Year's Sales (3)	Sales Based on $\bar{X}$, 130.5% of Previous Year's Sales (4)
1972	5,000	—	—
1973	3,600	6,000 (= 5,000 × 120%)	6,525 (= 5,000 × 130.5%)
1974	5,760	7,200 (= 6,000 × 120%)	8,515 (= 6,525 × 130.5%)
1975	5,184	8,640 (= 7,200 × 120%)	11,112 (= 8,515 × 130.5%)
1976	10,368	10,368 (= 8,640 × 120%)	14,501 (= 11,112 × 130.5%)
Total	29,912		

When each number following the first one is obtained by multiplying the preceding number by the average ratio, the sequence of numbers is called a *geometric progression*. Thus, the group of values 5,000, 6,000, 7,200, 8,640, and 10,368 in Table 6–6 is a geometric progression with the average ratio 120%. Generally, in a sequence of values of geometric progression, let

B = the value at the beginning of the period, or the first value
E = the value at the end of the period, or the last value
n = the number of values excluding the first value (or the actual number of years during the period)
G = the average ratio

Then the computation for the sales based on the geometric mean in Table 6–6 may be symbolically written as follows:

1972 (base year): B (= 5,000)
1973 (end of 1st year): BG (5,000 × 120% = 6,000)
1974 (end of 2nd year): $BG(G) = BG^2$ (6,000 × 120% = 7,200)
1975 (end of 3rd year): $BG^2(G) = BG^3$ (7,200 × 120% = 8,640)
1976 (end of 4th year): $BG^3(G) = BG^4$ (8,640 × 120% = 10,368)

Extending this idea further, we find that the value at the end of the nth year is

$$BG^n = E$$

Divide both sides by B.

$$G^n = \frac{E}{B}$$

Thus,

$$G = \sqrt[n]{\frac{E}{B}} \qquad (6\text{–}3)$$

The base of a ratio expressed in percent is always equal to 100%.

Let r = the average rate of increase; then

$$r = G - 100\% \qquad (6\text{-}4)$$

Formulas (6-3) and (6-4) provide a short method for computing the average rate of increase when the first value, the last value, and the number of values in a sequence are known. The application of the two formulas is illustrated in Examples 4 and 5.

Example 4

Find the average rate of increase of the annual sales from 1972 to 1976 if the Johnson Department Store had $5,000 sales in 1972 and $10,368 sales in 1976.

Solution.

$B = 5,000$

$E = 10,368$

$n = 4$ years, 1973 to 1976 inclusive. 1972 is excluded, since it is the base year. (Or $n = 1976 - 1972 = 4$.)

$G = ?$

$r = G - 100\% = ?$

Substitute the above values in formulas (6-3) and (6-4).

$$G = \sqrt[n]{\frac{E}{B}} = \sqrt[4]{\frac{10,368}{5,000}} = \sqrt[4]{2.074} = 1.20 \text{ or } 120\% \text{ (average ratio)}$$

$r = G - 100\% = 120\% - 100\% = 20\%$ (average annual rate of increase)

The value of $\sqrt[4]{2.074}$ is computed by the use of logarithms as follows:

$$\log \sqrt[4]{2.074} = \frac{1}{4} \log 2.074 = \frac{1}{4} (0.316809) = 0.079202$$

The mantissa 079202 is close to 079181, which corresponds to digits 1200 in the log table. The characteristic is 0. Thus,

$$G = \text{antilog } (0.079202) = 1.200 \text{ or } 1.20$$

The answer already has been verified in Table 6-6. Note that the sales in the intermediate years (1973 to 1975) are not used in computing the average ratio.

Example 5

The Bureau of Census reported that the population of the United States had increased from 76,094,000 in 1900 to 206,039,000 in 1970. Assume that the increase was in a form of geometric progression. Find the average rate of increase if (a) the population is computed every year, and (b) the population is computed every ten years.

Solution.   (a) $n = 70$ years (1901 to 1970, or 1970-1900 = 70)

$$G = \sqrt[n]{\frac{E}{B}} = \sqrt[70]{\frac{206,039,000}{76,094,000}} = \sqrt[70]{2.708}$$

Take logarithms of both sides.

$$\log G = \frac{1}{70} \log 2.708 = \frac{1}{70}(0.432649) = 0.006181$$

$$G = \text{antilog } 0.006181 = 1.014 \text{ or } 101.4\%$$

(The mantissa 006181 is close to 006038, which corresponds to digits 1014 in the log table.)

$$r = G - 100\% = 101.4\% - 100\% = 1.4\%$$

(average rate of increase every year)

(b) $n = 7$ decades (1900 to 1970)

$$G = \sqrt[n]{\frac{E}{B}} = \sqrt[7]{\frac{206,039,000}{76,094,000}} = \sqrt[7]{2.708}$$

Take logarithms of both sides.

$$\log G = \frac{1}{7} \log 2.708 = \frac{1}{7}(0.432649) = 0.061807$$

$$G = \text{antilog } 0.061807 = 1.153 \text{ or } 115.3\%$$

(The mantissa 061807 is close to 061829, which corresponds to digits 1153 in the log table.)

$$r = G - 100\% = 115.3\% - 100\% = 15.3\%$$

(average rate of increase every 10 years)

Check: Average ratio based on one-year period: 1.0144 (by interpolation).
Average ratio based on ten-year period: $1.0144^{10} = 1.153$

6.2 Harmonic Mean

To obtain the harmonic mean of a set of values, first find the arithmetic mean of the reciprocals of the individual values. Second, find the reciprocal of this arithmetic mean. The reciprocal of a whole number is the quotient of 1 divided by the whole number, e.g., $\frac{1}{4}$ is the reciprocal of 4; the reciprocal of a fraction is the fraction with the terms inverted, e.g., $\frac{2}{5}$ is the reciprocal of $\frac{5}{2}$. The methods of computing the harmonic mean for ungrouped and grouped data and the chief characteristics of the harmonic mean are presented below.

Ungrouped Data

The method for computing the harmonic mean for ungrouped data is illustrated in Example 6.

Example 6

Compute the harmonic mean of values 1, 4, 10, 8, and 10 (same as the values given in Example 3, Section 5.2).

Solution. The arithmetic mean of the reciprocals of the five values is

$$\frac{\frac{1}{1} + \frac{1}{4} + \frac{1}{10} + \frac{1}{8} + \frac{1}{10}}{5} = \frac{\frac{40 + 10 + 4 + 5 + 4}{40}}{5} = \frac{63}{40} \cdot \frac{1}{5} = \frac{63}{200}$$

The harmonic mean of the five values is the reciprocal of this arithmetic mean, or

$$\frac{200}{63} = 3\frac{11}{63} = 3.2 \text{ (the harmonic mean)}$$

In general, the arithmetic mean of the reciprocals of n values is

$$\frac{1/X_1 + 1/X_2 + 1/X_3 + \cdots + 1/X_n}{n}$$

The harmonic mean (H) is the reciprocal of the arithmetic mean above, or

$$H = \frac{n}{1/X_1 + 1/X_2 + 1/X_3 + \cdots + 1/X_n}$$

or

$$H = \frac{n}{\Sigma\left(\frac{1}{X}\right)} \qquad (6\text{-}5)$$

Substitute the values of Example 6 in formula (6–5); the harmonic mean can be directly computed as follows:

$$H = \frac{5}{\frac{1}{1} + \frac{1}{4} + \frac{1}{10} + \frac{1}{8} + \frac{1}{10}} = \frac{5}{\frac{63}{40}} = 5 \times \frac{40}{63} = \frac{200}{63} = 3\frac{11}{63} = 3.2$$

Grouped Data

The method of computing the harmonic mean for grouped data is similar to that for ungrouped data. However, each reciprocal of the original value must be weighted by the frequency representing the value in the computation. Thus, for grouped data formula (6–5) is adjusted as follows:

$$H = \frac{n}{\Sigma\left(f \cdot \frac{1}{X}\right)} \qquad (6\text{-}6)$$

where $n = \Sigma f$

Example 7

Find the harmonic mean of the frequency distribution in Table 6–7 (same as the values given in Example 5, Section 5.2).

TABLE 6–7

DATA FOR EXAMPLE 7

Class Interval	Mid-point X	Frequency f
0 and under 2	1	2
2 and under 4	3	5
4 and under 6	5	4
6 and under 8	7	8
8 and under 10	9	1
Total		$n = 20$

The harmonic mean is computed by using formula (6–6):

$$H = \frac{20}{2\left(\frac{1}{1}\right) + 5\left(\frac{1}{3}\right) + 4\left(\frac{1}{5}\right) + 8\left(\frac{1}{7}\right) + 1\left(\frac{1}{9}\right)}$$

$$= \frac{20}{\dfrac{630 + 525 + 252 + 360 + 35}{315}} = 20 \times \frac{315}{1,802} = 3.5$$

Chief Characteristics of the Harmonic Mean

(1) The harmonic mean, like the arithmetic mean and the geometric mean, is computed from all items in a set of values. The value of every item in the data thus affects the value of the harmonic mean. However, the harmonic mean is even less affected by extreme values than the geometric mean. The relative magnitude of the three different means for the same data may be expressed as follows:

$$\bar{X} > G > H$$

It means that the value of $\bar{X}$ is larger than the value of G, which in turn is larger than the value of H.*

Examine the three means for the values given in Example 6:

$\bar{X} = 6.6$ (page 118), $G = 5.024$ (p. 153), and $H = 3.2$ (page 162)

---

*Another type of average, called quadratic mean (Q), is also computed from all items in a set of values. It is the square root of the arithmetic mean of the squared values, or $Q = \sqrt{\sum X^2/n}$. The value of Q is even larger than $\bar{X}$ for the same data, or $Q > \bar{X} > G > H$. For the values given in Example 6,

$$Q = \sqrt{\frac{1^2 + 4^2 + 10^2 + 8^2 + 10^2}{5}} = 7.5$$

The quadratic mean of original values (X) is seldom used in business and economics studies. However, the quadratic mean of deviations ($x = X - \bar{X}$) is an important measure in this text. (See Section 7.4 for finding standard deviation, $s = \sqrt{\sum x^2/n}$.)

Also, examine the three means for the values given in Example 7:

$$\bar{X} = 5.1 \text{ (page 122)}, \quad G = 4.41 \text{ (p. 154), and } H = 3.5 \text{ (page 163)}$$

(2) The harmonic mean is not as frequently used as an average of a set of values as is the mean. However, it is useful in special cases for averaging rates. The rate usually indicates the relation between two different types of measuring units that can be expressed reciprocally. For example, if a man walked 10 miles in two hours, the rate of his walking speed can be expressed:

$$\frac{10 \text{ miles}}{2 \text{ hours}} = \frac{5 \text{ miles}}{1 \text{ hour}} = 5 \text{ miles per hour}$$

where the unit of the *first term is a mile* and the unit of the *second term is an hour*. Or, reciprocally,

$$\frac{2 \text{ hours}}{10 \text{ miles}} = \frac{\frac{1}{5} \text{ hours}}{1 \text{ mile}} = \frac{1}{5} \text{ hours per mile}$$

where the unit of the *first term is an hour* and the unit of the *second term is a mile*.

Examples 8 and 9 are used to illustrate the use of the harmonic mean in averaging rates of this type.

Example 8

Three students made a 3,600–mile relay race. Each of them drove 1,200 miles. Their driving speeds are given below. Find the average driving speed per hour of the three students during the race.

Student	Driving speed rate
A	40 miles per hour
B	50 miles per hour
C	60 miles per hour

Solution. First analyze the problem by finding the $\bar{X}$ and H of the given rates. Observe that the *first* term of each given rate is a *mile* and the *second* term is an *hour*, such as the given rate for student *A* being 40 *miles* per *hour*.

(1) Take the arithmetic mean of the driving speed rates:

$$\bar{X} = \frac{40 + 50 + 60}{3} = 50 \text{ miles per hour}$$

The answer is correct only if we can assume that the constant value is the number of hours, which is applicable to each student. That is, our assumption is: each student drove the *same number of hours* during the race. (Observe that the unit of the *second term* in the mean and in each given ratio is an hour, same as the unit of the constant value used in the assumption in our analysis.) However, this is *not* the case, since the number of hours spent during the race by each of the three students is:

$$\text{Student A } \frac{1,200}{40} = 30 \text{ hours}$$

$$\text{Student B } \frac{1,200}{50} = 24 \text{ hours}$$

$$\text{Student C } \frac{1,200}{60} = 20 \text{ hours}$$

$$\text{Total} \qquad \overline{74 \text{ hours}}$$

The total number of miles traveled by the three students at the average of 50 miles per hour would be

$$50 \times 74 = 3,700 = \text{miles}$$

which disagrees with the given distance, 3,600 miles of the relay. Thus, the arithmetic mean method *does not* give the correct answer in this case.

(2) Take the harmonic mean of the driving speed rates. First, find the reciprocals of the given rates.

Student	*Driving speed rate*
A	$\frac{1}{40}$ hours per mile
B	$\frac{1}{50}$ hours per mile
C	$\frac{1}{60}$ hours per mile

Next, take the arithmetic mean of the reciprocal rates:

$$\frac{\frac{1}{40} + \frac{1}{50} + \frac{1}{60}}{3} = \frac{\frac{15}{600} + \frac{12}{600} + \frac{10}{600}}{3} = \frac{37}{1,800} \text{ hours per mile}$$

$$= \frac{37 \text{ hours}}{1,800 \text{ miles}}$$

Then the reciprocal of the average reciprocal rate is

$$H = \frac{1,800 \text{ miles}}{37 \text{ hours}} = 48\frac{24}{37} \text{ or } 48.6 \text{ miles per hour}$$

Or simply compute the harmonic mean directly from the given rates as follows:

$$H = \frac{3}{\frac{1}{40} + \frac{1}{50} + \frac{1}{60}} = \frac{1,800}{37} = 48\frac{24}{37} \text{ miles per hour}$$

The total miles traveled in 74 hours at $48\frac{24}{37}$ miles per hour is

$$48\frac{24}{37} \times 74 = 3,600 \text{ miles}$$

which agrees with the distance given in the example. The actual number of miles traveled (1,200 miles each) is not included in the above computation.

When the harmonic mean method is used, it is assumed that the constant value is the number of miles, which is equally applicable to each student. That is, each student drove the *same number of miles*. (Note that the unit of the *first term* in the harmonic mean and in each given ratio is a mile, same as the unit of the constant value used in the assumption.) The assumption that each student drove the same number of miles is true in this case. Thus, the harmonic mean method gives the correct answer: $48\frac{24}{37}$ miles per hour.

Example 9

A toy factory has assigned a group of four workers to complete an order of 700 toys of a certain type. The productive rates of the four workers are given below.

Workers	Productive rates
H	10 minutes per toy
I	6 minutes per toy
J	15 minutes per toy
K	4 minutes per toy

Find the average minutes per toy by the group of workers.

Solution.   First, analyze the problem by finding the $\bar{X}$ and H of the given rates. Observe that the *first* term of each given rate is a *minute* and the *second* term is a *toy*.

   (1) Take the arithmetic mean of the productive rates.

$$\bar{X} = \frac{10 + 6 + 15 + 4}{4} = 8.75 \text{ minutes per toy}$$

The result is true only if we assume that each of the four workers is assigned the *same number of toys* (constant value) to meet the order, or

$$\frac{700}{4} = 175 \text{ toys per worker}$$

Note that the unit of the *second term* in the mean $\bar{X}$ is a toy, same as the unit of the constant value used in the assumption.

   Check:

Workers	Time required for producing 175 toys
H	10 × 175 = 1,750 minutes
I	6 × 175 = 1,050 minutes
J	15 × 175 = 2,625 minutes
K	4 × 175 = 700 minutes
	Total 6,125 minutes

$$\frac{6,125}{700} = 8.75 \text{ minutes per toy}$$

   (2) Take the harmonic mean of the productive rates.

$$H = \frac{4}{\frac{1}{10} + \frac{1}{6} + \frac{1}{15} + \frac{1}{4}} = \frac{4}{\frac{6 + 10 + 4 + 15}{60}} = 4 \times \frac{60}{35}$$

$$= 6\frac{6}{7} \text{ minutes per toy}$$

The result is true only if we assume that each of the four workers has worked the *same number of minutes* (constant value) or

$$6\tfrac{6}{7} \times 700 = 4,800 \text{ minutes (required for the complete order)}$$

$$\frac{4,800}{4} = 1,200 \text{ minutes per worker}$$

Note that the unit of the *first term* in the harmonic mean is a minute, same as the unit of the constant value used in the assumption.

Check:

Workers	Toys produced in 1,200 minutes
H	$\dfrac{1,200}{10} = 120$
I	$\dfrac{1,200}{6} = 200$
J	$\dfrac{1,200}{15} = 80$
K	$\dfrac{1,200}{4} = 300$
	Total = 700 toys

Next, find the correct answer based on the analysis above. In practice, it is likely for each worker in a factory to work the same amount of time but to produce different numbers of toys. The harmonic mean is more realistic in this case than the arithmetic mean.

Each of the above two examples shows that the constant value, which is equally applicable to each person (fixed number of miles for each student in Example 8 and fixed number of minutes for each worker in Example 9), has the same unit as the first term of each given rate. The analysis in the two examples indicates the following two rules:

(1) When a constant value, which has the same unit as in the *second term* of each given ratio, is equally applicable to each item in the data, the arithmetic mean should be used.

(2) When a constant value, which has the same unit as in the *first term* of each given ratio, is equally applicable to each item in the data, the harmonic mean should be used.

An experienced statistician would use the two rules directly in finding the correct answer, without going through the complete analysis as illustrated above.

6.3   Summary

This chapter continues the previous chapter in discussing the two remaining averages: the geometric mean and the harmonic mean. The methods of computing the two types of averages are summarized at the end of this section (see

TABLE 6-8

SUMMARY OF THE CHIEF CHARACTERISTICS OF THE FIVE AVERAGES

Characteristics	Arithmetic Mean ($\bar{X}$)	Median (M_d)	Mode (M_0)	Geometric Mean (G)	Harmonic Mean (H)
Computation—based on	Every value	Middle value	Value with the highest frequency	Every value	Every value
Affected by extreme values	Greatest	No (affected by items only)	No	Less than $\bar{X}$	Less than G (more weight to smaller numbers)
Algebraic manipulation	Yes: $\bar{X} = \dfrac{\sum X}{n}$	No (positional average, interpolated value in many cases)	No (concentrated average, four methods for grouped data)	Yes: $G = \sqrt[n]{X_1 \cdot X_2 \cdots X_n}$	Yes: $H = \dfrac{n}{\sum (1/X)}$
Mathematical properties	$\sum (X - \bar{X}) = 0$ $\sum (X - \bar{X})^2$ is minimum	$\sum (X - M_d)$ is minimum (disregarding sings)	—	$\sum (\log X - \log G) = 0$	—
Application to open-end classes	Indeterminate	Determinate	Determinate	Indeterminate	Indeterminate
Comparison of answers to same data	Larger than G and H	Between $\bar{X}$ and M_0	May be larger or smaller than M_d and $\bar{X}$	Larger than H, but smaller than $\bar{X}$	Smaller than G and $\bar{X}$
Type of data preferred	Most types	Middle value is typical, excluding extremes	Data with distinct central tendency	Ratios, rates, and geometric progression	Certain rates that can be expressed reciprocally

SUMMARY OF FORMULAS

Application	Formula	Formula number	Reference page
Geometric mean			
Ungrouped data	$G = \sqrt[n]{X_1 \cdot X_2 \cdot X_3 \cdots X_n}$	(6-1a)	152
or	$\log G = \dfrac{\sum \log X}{n}$	(6-1b)	153
Grouped data	$G = \sqrt[n]{(X_1 \cdot X_1 \cdots)(X_2 \cdot X_2 \cdots) \cdots}$	(6-2a)	154
or	$\log G = \dfrac{\sum (f \log X)}{n}$	(6-2b)	154
Geometric progression—Average ratio	$G = \sqrt[n]{\dfrac{E}{B}}$	(6-3)	159
Average rate of increase	$r = G - 100\%$	(6-4)	160
Harmonic mean			
Ungrouped data	$H = \dfrac{n}{\sum (1/X)}$	(6-5)	162
Grouped data	$H = \dfrac{n}{\sum (f \cdot 1/X)}$	(6-6)	162

formulas). The chief characteristics of the two types of averages, along with those of the previous chapter, are tabluated for easy comparison in Table 6–8.

Exercises 6

1. Why is the geometric mean especially useful in averaging rates and ratios?

2. Under what conditions can the harmonic mean be correctly used in averaging rates? Give an example to explain.

3. Compute (a) the arithmetic mean, (b) the geometric mean, and (c) the harmonic mean of the values 2, 4, 7, 8, and 15 to two decimal places. Which one of the three averages is the largest? next largest? the smallest?

4. Compute (a) the arithmetic mean, (b) the geometric mean, and (c) the harmonic mean of the values given in the following frequency distribution table. Which one of the three averages is the largest? next? the smallest?

Values (class intervals)	Frequencies
1 and under 3	1
3 and under 5	3
5 and under 7	4
7 and under 9	2

5. The following table shows the number of days that employees of the four production units of a factory were absent from work during 1975 and 1976.

Unit	Number of Days	
	1975	1976
A	5	1
B	3	15
C	4	2
D	8	16
Total	20	34

Find the following:

(a) The ratio of the number of days in 1976 to that in 1975 for each unit.

(b) The geometric mean of the ratios found in (a).

(c) The ratio of the total number of days absent by the four units in 1976 to the total in 1975.

(d) The arithmetic mean of the ratios found in (a).

(e) The arithmetic mean of the ratios found in (a) weighted individually by the number of days in 1975 (the base year).

(f) Discuss the significance of the findings in (b) through (e).

6. The ratios of the annual production to the production in each previous year from 1974 to 1978 in the Griffin Manufacturing Company are given below:

Year	Ratio to previous year's production
1974	50%
1975	125%
1976	80%
1977	260%
1978	140%

(a) Find (1) the arithmetic mean, and (2) the geometric mean of the given ratios. Which one of the means is appropriate for the ratios? Explain.

(b) The annual production in 1973 was 10,000 units. How many units were actually produced on 1975 by the company?

7. The Bureau of the Census reported that the population of New York City had increased from 5,620,048 in 1920 to 7,895,563 in 1970. Assume that the increase was in the form of geometric progression. What is the average rate of increase if (a) the population is computed every year, and (b) the population is computed every ten years? (Compute to two decimal places in the percentage rate.) How would you use your answer to compare with the average rate of increase of the population in the United States in each case? (See Example 5, pp. 160–61.)

8. A company hired three girls to type certain documents for a week of 40 hours. The typing speed of each girl is:

Jane: 10 minutes per page
Pat: 12 minutes per page
Donna: 15 minutes per page

(a) What is the average number of minutes per page?

(b) How many pages were typed by the girls during the week based on your answer to (a)?

9. A grocery store owner mixed four different grades of apples into one pile for resale. The costs of the different grades to him are as follows:

Grades	Costs
A	10¢ per pound
B	12¢ per pound
C	8¢ per pound
D	15¢ per pound

What is the average cost per pound of the mixed grade?

(a) Assume that the owner purchased the same quantity of apples for each grade.

(b) Assume that the owner spent the same amount of money for each grade.

10. A salesman travels from City X to City Y at an average speed of 36 miles per hour and returns by the same route at an average speed of 60 miles per hour. Find the appropriate average speed for the round trip.

7 Dispersion

The values included in a group of statistical data usually vary in magnitude, as indicated in Chapter 5. Some of the values in the group are large, but some are small. The variation of the values is called *dispersion* and can be measured by many different methods.

A measure of dispersion is important in two ways: First, it may be used to show the degree of variation among values in the given data. For example, a very low dispersion of the hourly wages of a group of workers in a factory will give the indication that the workers in the factory are paid approximately equal wages. But on the other hand, a high dispersion will give a reader the impression that the workers are paid in a wide variation of the hourly wages.

Second, the measure of dispersion may be used to supplement an average to describe a group of data or to compare one group of data with another. When the dispersion is high, the average is of little or no significance. When the dispersion is low, the value of the average becomes increasingly significant; that is, the average is a highly representative value. For example, the mean of the group of numbers 1, 2, and 12 is 5, or $(1 + 2 + 12)/3 = 5$. Since 5 is not close to any number in the group, a high dispersion is expected. The mean of a second group of numbers 4, 5, and 6 is also 5, or $(4 + 5 + 6)/3 = 5$. Since 5 is close to (or equal to) each number in the second group, a low dispersion is expected. The fact that the measure of dispersion of the first group is higher than that of the second group gives a better understanding of the comparison of the means of the two groups of data.

A measure of dispersion can be expressed either in an absolute value or in a relative value. The most common types of dispersions expressed in absolute values are: (1) range, (2) quartile deviation, (3) average deviation, and (4) standard deviation. Among them, the range is the simplest type with respect to its concept and computation. The standard deviation is by far the most important type, since it is mathematically logical and can be used in further calculations. The methods of computing the four different measures of dispersion and the characteristics of the measures are discussed in Sections 7.1 through 7.4. The relative dispersion is presented in Section 7.5.

7.1 Range

Method of Computing the Range

The range of a group of values is the difference between the lowest and the highest values, or

$$R = X_n - X_1 \qquad (7\text{-}1)$$

where R = range, X_n = the highest value, and X_1 = the lowest value.

Example 1

Find the range of the values 1, 4, 8, 10, 10.

Solution.   The highest value is 10 and the lowest value is 1.

$$R = 10 - 1 = 9$$

Chief Characteristics of the Range

1. The range is based on the lowest and highest values in a group of data. It is easy to compute and is the most convenient single value as a supplement to the mean. For example, the average weekly production of a group of workers in Factory A may be 40 units with a range from 15 to 60 units. The average in Factory B may also be 40 units, but with a range from 30 to 50 units. Considering the two different ranges, we may conclude that the average of Factory B is more representative of the units produced by the workers than the average for Factory A.

2. The range may be influenced greatly by unusual values in the given data. If there is an unusual value in the data, either very small or very large, the range may not be a proper measure of dispersion for the group of values.

3. The range is not affected by the values between the lowest and highest values. Thus, the range is only a rough estimate of the measure of dispersion.

7.2 Quartile Deviation

The quartile deviation of a group of data, like a range, is based on only two values. The two values are not extreme values, but they are the first quartile and the third quartile of the group. To find the quartiles, first divide the items of the group into four equal parts according to their values. The first quartile (Q_1) is the point on the value scale below which there are one-fourth of the items. The second quartile (Q_2) is the point below or above which there are one-half of the items. Thus, Q_2 corresponds to the median. The third quartile (Q_3) is the point below which there are three-fourths of the items. The difference between the first quartile and the third quartile is called the *interquartile range*. When this difference is divided by 2, the quotient is the *quartile deviation (Q.D.)*, or *semi-interquartile* range.

$$Q.D. = \frac{Q_3 - Q_1}{2} \qquad (7\text{-}2)$$

The methods of locating quartiles for ungrouped data and for grouped data are slightly different. They are presented below.

Quartiles for Ungrouped Data

The Number of Items is Divisible by 4.
When the number of items in the data is divisible by 4, the values of quartiles can easily be determined. In the following computation, when a quartile is located between two values, the mid-point of the two values is considered the value of the quartile. Example 2 is used to illustrate the procedure.

Example 2

Find Q_1, Q_2, Q_3, and *Q.D.* of the eight values 2, 5, 10, 3, 7, 13, 20, and 18.

Solution.   First, arrange the data in ascending order as shown below.

Q_1 is the point on the value scale below which there are one-fourth of the items. There are 8 items in the group. One-fourth of the 8 items is 2 items. The value of Q_1 must be above 3 (the second item) but below 5 (the third item). Thus, Q_1 is determined to be 4, the mid-point of 3 and 5.

Q_2 is the median, above or below which there are 50% or one-half of the items. The two middle items are 7 (the fourth item) and 10 (the fifth item). Thus, Q_2 is 8.5, the mid-point of 7 and 10.

$$
\begin{array}{l}
2 \\
3 \\
\quad \longleftarrow Q_1 = \dfrac{3+5}{2} = 4 \\
5 \\
7 \\
\quad \longleftarrow Q_2 = \dfrac{7+10}{2} = 8.5 \\
10 \\
13 \\
\quad \longleftarrow Q_3 = \dfrac{13+18}{2} = 15.5 \\
18 \\
20
\end{array}
$$

Q_3 is the point below which there are three-fourth of the items. The value of Q_3 must be above 13 (the sixth item, since $8 \times \frac{3}{4} = 6$) but below 18 (the seventh item). Thus, Q_3 is 15.5, the mid-point of 13 and 18.

$$Q.D. = \frac{Q_3 - Q_1}{2} = \frac{15.5 - 4}{2} = 5.75$$

The Number of Items is Not Divisible by 4.

When the number of items in the data is not divisible by 4, the values of quartiles can be determined in the following manner:

(1) If the number of items in the data is even, such as 6 or 10 items, Q_1 is the median obtained from the lower 50% of the values. (See Example 3.)

(2) If the number of items in the data is odd, such as 5 or 9 items, approximate Q_1 by two steps: First, disregard the middle item (Q_2). Next, locate Q_1 as done in (1) above. (See Example 4 and its footnote.)

(3) Locate Q_3 by the methods stated in (1) and (2) except that the upper 50% of the values in the data are used in the process.

Example 3

Find the $Q.D.$ of the six values 1, 6, 14, 4, 8, and 20.

Solution. First, arrange the data in ascending order according to the given values.

Lower $\begin{cases} 1 \\ 4 \leftarrow Q_1 \\ 6 \end{cases}$ Q_1 is the median of the lower 50% values,
50%   or
values $Q_1 = 4.$

Upper $\begin{cases} 8 \\ 14 \leftarrow Q_3 \\ 20 \end{cases}$ Q_3 is the median of the upper 50% values,
50%   or
values $Q_3 = 14.$

$$Q.D. = \frac{Q_3 - Q_1}{2} = \frac{14 - 4}{2} = 5.$$

Example 4

Find the $Q.D.$ of the five values 2, 3, 5, 7, and 10.

Solution.* First, arrange the data in ascending order according to the given values.

Lower $\begin{cases} 2 \\ 3 \end{cases}$ $\leftarrow Q_1 = \frac{2 + 3}{2} = 2.5$
50%
values

               $5 \leftarrow Q_2$ (the median) is disregarded in locating Q_1 and Q_3.

Upper $\begin{cases} 7 \\ 10 \end{cases}$ $\leftarrow Q_3 = \frac{7 + 10}{2} = 8.5$
50%
values

$$Q.D. = \frac{Q_3 - Q_1}{2} = \frac{8.5 - 2.5}{2} = 3$$

---

*More accurate values of Q_1 and Q_3 can be obtained by the interpolation method, which is also used for grouped data. (See Formulas 7–3 and 7–4.)

Extending this concept further, we may divide a group of values, according to their values, into 10 equal parts to obtain a *decile*; or into 100 equal parts to obtain a *percentile*.

Quartiles for Grouped Data

The quartiles for grouped data can be obtained in the same way as the median by (a) formulas or (b) interpolation. The formulas resemble the formula for the median:

$$Q_1 = L + \frac{\frac{n}{4} - C}{f}\text{---}(i) \tag{7-3}$$

In general, let

$$n = \text{number of items in the data}$$
$$Q_1 = \text{value } \textit{centered} \text{ at the } \textit{end} \text{ of the } (n/4)\text{th item}$$
$$Q_3 = \text{value } \textit{centered} \text{ at the } \textit{end} \text{ of the } (3n/4)\text{th item}$$

In Example 4:

$Q_1 = $ value centered at the end of the $\frac{5}{4}$th $= 1\frac{1}{4}$th item. The Q_1 item thus covers $\frac{1}{2}$ item above and below the center point, or from $\frac{3}{4}$th item to $\frac{7}{4}$th item in the given data (since $\frac{5}{4} \pm \frac{1}{2}$ $= \frac{3}{4}$ and $\frac{7}{4}$).

$$Q_1 = 2 \times \frac{1}{4} \text{ (value from } \tfrac{3}{4}\text{th to 1st item)} +$$

$$3 \times \frac{3}{4} \text{ (value from 1st to } \tfrac{7}{4}\text{th item)} = 2.75$$

$Q_3 = $ value centered at the end of the $\frac{15}{4}$th $= 3\frac{3}{4}$th item. The Q_3 item thus covers $\frac{1}{2}$ item above and below the center point, or from $3\frac{1}{4}$th item to $4\frac{1}{4}$th item in the given data [since $\frac{15}{4} \pm \frac{1}{2}$ $= \frac{13}{4}$ (or $3\frac{1}{4}$) and $\frac{17}{4}$ (or $4\frac{1}{4}$)].

$$Q_3 = 7 \times \frac{3}{4} \text{ (value from } 3\tfrac{1}{4}\text{th to 4th item)} +$$

$$10 \times \frac{1}{4} \text{ (value from 4th to } 4\tfrac{1}{4}\text{th item)} = 7.75$$

Now,

$$Q.D. = \frac{Q_3 - Q_1}{2} = \frac{7.75 - 2.75}{2} = 2.5$$

The interpolation is diagrammed below:

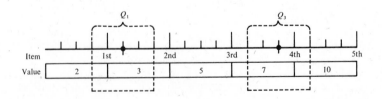

$$Q_3 = L + \frac{\frac{3n}{4} - C}{f}(i) \qquad (7\text{-}4)$$

where $L =$ the real lower limit of the applicable quartile class; that is, the first (Q_1) or the third (Q_3) quartile class, whichever applies

$n =$ the number of total frequencies in the given data

$C =$ the cumulative frequency in the class just before (or smaller) the applicable quartile class

$f =$ the frequency of the applicable quartile class

$i =$ the size of the class interval in the applicable class

Example 5

The miles traveled by 20 students in coming to Swan College from their homes are given in the first two columns of Table 7–1 (same as the values given in Example 10, p. 129). Find the Q.D.

Solution. The Q.D. is computed as follows:

TABLE 7–1

COMPUTATION FOR EXAMPLE 5

Miles Traveled	Number of Students (frequency)	Cumulative Frequency	
0 and under 2	2	2	
2 and under 4	5	7	Q_1 class
4 and under 6	4	11	
6 and under 8	8	19	Q_3 class
8 and under 10	1	20	
Total	$n = 20$		

$Q_1 =$ value at the end of $(n/4)$th or 5th item, $(20/4) = 5$

$Q_1 = L + \frac{n/4 - C}{f}(i) = 2 + \frac{20/4 - 2}{5}(2) = 2 + \frac{6}{5} = 3.2$

$Q_3 =$ value at the end of $(3n/4)$th or 15th item, $(3 \times 20)/4 = 15$

$Q_3 = L + \frac{3n/4 - C}{f}(i) = 6 + \frac{(3 \times 20)/4 - 11}{8}(2) = 6 + \frac{8}{8} = 7$

$$Q.D. = \frac{Q_3 - Q_1}{2} = \frac{7 - 3.2}{2} = 1.9$$

If the interpolation method is used, Q_1 and Q_3 may be obtained as follows:

For Q_1

Value in miles		Cumulative frequency
4	is to	7 (1)
Q_1	is to	5 (2)
2	is to	2 (3)

$$\frac{(2) - (3)}{(1) - (3)} \frac{Q_1 - 2}{2} = \frac{3}{5}$$

$$Q_1 = 2 + 2\left(\frac{3}{5}\right) = 3.2$$

For Q_3

Value in miles		Cumulative frequency
8	is to	19 (1)
Q_3	is to	15 (2)
6	is to	11 (3)

$$\frac{(2) - (3)}{(1) - (3)} \frac{Q_3 - 6}{2} = \frac{4}{8}$$

$$Q_3 = 6 + 2\left(\frac{4}{8}\right) = 7$$

Chief Characteristics of the Quartile Deviation

1. The quartile deviation is based on two values: Q_1 and Q_3. It is not affected by extreme values which are smaller than Q_1 or larger than Q_3. There are 50% of the items of the data within Q_1 and Q_3. A low quartile deviation thus indicates the small variation among the central 50% items. On the other hand, a high quartile deviation means that the variation among the central items is large.

2. The quartile deviation is one-half of the distance between Q_3 and Q_1. However, if the distribution is not symmetrical, the distance within "median or $Q_2 \pm$ Q.D." will not coincide with the interquartile range (from Q_1 to Q_3). In Example 2, for instance, the distance within "$Q_2 \pm$ Q.D." is measured from 2.75 to 14.25 (or 8.5 ± 5.75), whereas $Q_1 = 4$ and $Q_3 = 15.5$.

3. The quartile deviation is a refined measure of dispersion when it is compared to the range. However, like the range, it has the weak point that it is not based on every value included in a given distribution.

7.3 Average Deviation (A.D.)

The range and the quartile deviation are positional measures of dispersion. They are based on the positions of certain items in a distribution. The average deviation and the standard deviation are based on all items and are designed to measure the dispersion *around an average.*

The average deviation is the arithmetic mean of the deviations of the individual values from the average of the given data. The average which is frequently used in computing the average deviation is either the arithmetic mean or the median. However, only the arithmetic mean will be used here for illustration purposes. In computing the average deviation, the *absolute values* of the deviations are used; that is, the positive or negative signs of the deviations are ignored. The methods of computing the average deviation for ungrouped and for grouped data are presented below.

Ungrouped Data

There are four steps in computing the average deviation for ungrouped data.
1. Find the arithmetic mean $\bar{X}$ (or median if desired).
2. Find the deviation of each value from the arithmetic mean, or $x = X - \bar{X}$.
3. Find the sum of the absolute values of the deviations, or $\Sigma |x|$.
4. Find the average deviation by dividing this sum by the number of values (n) in the data.

$$A.D. = \frac{\Sigma |x|}{n} \tag{7-5}$$

Example 6

Find the the average deviation of values $2, 3, 5, 7, and 10.

Solution. The $A.D.$ is computed in Table 7–2.

$$\bar{X} = \frac{27}{5} = 5.4$$

$$A.D. = \frac{\Sigma |x|}{n} = \frac{12.4}{5} = \$2.48$$

Column (2) is usually omitted in practical computation.

TABLE 7-2

COMPUTATION FOR EXAMPLE 6

Values X	Deviations $x = X - \bar{X} = X - 5.4$	Absolute Values of Deviations
$ 2	−3.4	3.4
3	−2.4	2.4
5	−0.4	0.4
7	1.6	1.6
10	4.6	4.6
Total 27	$\Sigma x = 0$	$\Sigma |x| = 12.4$

Grouped Data

There are four steps in computing the average deviation for grouped data.
1. Find the arithmetic mean $\bar{X}$ (or median if desired).
2. Find the absolute value of the deviation of the mid-point of each class from the arithmetic mean, or $|x| = |X - \bar{X}|$, and multiply the deviation in each class by the frequency in the class, or $f|x|$.
3. Find the sum of the absolute values of the deviations, or $\Sigma(f|x|)$.

4. Find the average deviation by dividing the sum by the number of values (n or $\Sigma\, f$) in the data.

$$A.D. = \frac{\Sigma\,(f\,|x|)}{n} \qquad\qquad (7\text{-}6)$$

Example 7

The miles traveled by 20 students in coming to Swan College from their homes are given in the first three columns of Table 7–3 (same as the values given in Example 5). Find the $A.D.$

Solution.   The $A.D.$ is computed as follows:

TABLE 7–3

COMPUTATION FOR EXAMPLE 7

Miles traveled (class interval)	Number of students	Average miles (mid-point)	Deviations from mean	Total deviations (disregard signs)
	f	X	$x = X - \bar{X}$	$f\,\lvert x\rvert$
0 and under 2	2	1	-4.1	8.2
2 and under 4	5	3	-2.1	10.5
4 and under 6	4	5	-0.1	0.4
6 and under 8	8	7	1.9	15.2
8 and under 10	1	9	3.9	3.9
Total	$n = 20$			$\Sigma(f\lvert x\rvert) = 38.2$

$$\bar{X} = 5.1 \text{ (page 122)}$$
$$A.D. = \frac{\Sigma(f\,\lvert x\rvert)}{n} = \frac{38.2}{20} = 1.91 \text{ miles}$$

Chief Characteristics of the Average Deviation

1. The average deviation is based on every value in the data. It therefore gives a better description of the dispersion than the range and the quartile deviation.

2. The average deviation is computed from an average, either the arithmetic mean or the median. It measures the dispersion around the average rather than the dispersion within certain values, as measured by the range and the quartile deviation.

3. The average deviation is the arithmetic mean of the absolute values of the deviations. It ignores the positive and negative signs of the deviations. This weakness creates the demand for a more reliable measure of dispersion: the standard deviation.

7.4 Standard Deviation (s) and Variance (s^2)

The standard deviation is a refined form of the average deviation. It is computed in the same manner as the average deviation, except that the positive and negative signs of the individual deviations are taken into consideration. The methods of computing the standard deviation for ungrouped data and for grouped data are given below.

Ungrouped Data

Basic Method

The standard deviation of a set of values is the square root of the arithmetic mean of the individual deviations squared. The individual deviations are based on the arithmetic mean of the values in the set.

The procedure of computing the standard deviation for ungrouped data is given below:

1. Find the arithmetic mean of the given data, $\bar{X}$.
2. Find the deviation of each value from the arithmetic mean, or $x = X - \bar{X}$.
3. Square each deviation to make it positive, or x^2.
4. Find the sum of the deviations squared, or $\sum x^2$.
5. Find the variance (s^2) by dividing the sum by the number of values (n) in the data.

$$s^2 = \frac{\sum x^2}{n}$$

6. Extract the square root of the variance to find the standard deviation (s).

$$s = \sqrt{\frac{\sum x^2}{n}} \qquad (7\text{–}7\text{a})$$

Example 8

Find the standard deviation of values $2, 3, 5, 7, and 10.

Solution. The standard deviation is computed by using formula (7–7a).

TABLE 7–4

COMPUTATION FOR EXAMPLE 8 (BASIC METHOD)

Values X	Deviations $x = X - \bar{X}$	Deviations Squared x^2
$ 2	−3.4	11.56
3	−2.4	5.76
5	−0.4	0.16
7	1.6	2.56
10	4.6	21.16
Total 27	0	41.20

$$\bar{X} = \frac{27}{5} = 5.4$$

$$s^2 = \frac{41.20}{5} = 8.24$$

$$s = \sqrt{8.24} = \$2.87$$

Short Method

The short method of computing the standard deviation for ungrouped data is based on formula (7–7b) as follows:

Let A = an assumed mean or an arbitrarily selected value (including zero)

v = the deviation of each value from the assumed mean, or

$= X - A$

Formula (7–7a) may be written

$$s^2 = \frac{\sum v^2}{n} - \left(\frac{\sum v}{n}\right)^2$$

$$s = \sqrt{\frac{\sum v^2}{n} - \left(\frac{\sum v}{n}\right)^2} \qquad (7\text{–}7b)*$$

Example 8 may be computed in a different way when formula (7–7b) is used: Let $A = 5$.

---

*Proof of Formula (7–7b).

Let
$$\frac{\sum v}{n} = \bar{v}$$

Then, from formula (5–1b),

$$\bar{X} = A + \frac{\sum v}{n} = A + \bar{v}$$

$$x = X - \bar{X} = X - (A + \bar{v}) = v - \bar{v}$$

or
$$v = x + \bar{v}$$

$$v^2 = (x + \bar{v})^2 = x^2 + 2x\bar{v} + \bar{v}^2$$

$$\sum v^2 = \sum x^2 + 2\bar{v}\sum x + \sum \bar{v}^2$$

Since $\bar{v}$ is a constant, $\sum \bar{v}^2 = n\bar{v}^2$. The sum of the deviations from the true mean is zero, or $\sum x = 0$ and $2\bar{v}\sum x = 0$.

$$\sum v^2 = \sum x^2 + n\bar{v}^2$$

$$\sum x^2 = \sum v^2 - n\bar{v}^2$$

Divide both sides by n.

$$\frac{\sum x^2}{n} = \frac{\sum v^2}{n} - \frac{n\bar{v}^2}{n}$$

Thus,
$$s^2 = \frac{\sum v^2}{n} - \left(\frac{\sum v}{n}\right)^2$$

TABLE 7-5

Computation for Example 8 (Short Method)

Values X	Deviations $v = X - A$	Deviations Squared v^2
$ 2	−3	9
3	−2	4
5	0	0
7	2	4
10	5	25
Total 27	2	42

$$s^2 = \frac{\sum v^2}{n} - \left(\frac{\sum v}{n}\right)^2$$

$$= \frac{42}{5} - \left(\frac{2}{5}\right)^2 = 8.4 - 0.16 = 8.24$$

$$s = \sqrt{8.24} = \$2.87$$

When zero is selected as the assumed mean, the deviation of each value from the assumed mean is the value unchanged, or $v = X - 0 = X$. Formula (7–7b) becomes

$$s^2 = \frac{\sum X^2}{n} - \left(\frac{\sum X}{n}\right)^2$$

or

$$s = \sqrt{\frac{\sum X^2}{n} - \left(\frac{\sum X}{n}\right)^2} = \sqrt{\frac{\sum X^2}{n} - (\bar{X})^2} \qquad (7\text{–}7c)$$

The value of X^2 may be obtained from the table of squares in the appendix. The above expression thus is convenient for computing the mean and the standard deviation at the same time. Example 8 may be computed as follows when the above expression is used.

Let $A = 0$.

Values X	X^2
$ 2	4
3	9
5	25
7	49
10	100
Total 27	187

$$\bar{X} = \frac{27}{5} = 5.4$$

$$s^2 = \frac{187}{5} - \left(\frac{27}{5}\right)^2 = 37.4 - 5.4^2$$

$$= 8.24$$

$$s = \sqrt{8.24} = \$2.87$$

Grouped Data

Basic Method

The procedure for computing the standard deviation for grouped data is given below:

  1. Find the arithmetic mean of the given data, $\bar{X}$.

  2. Find the deviation of the mid-point of each class from the arithmetic mean, or $x = X - \bar{X}$.

  3. Find the total of deviations squared for each class fx^2. This total can be obtained in two ways: (a) First, square the deviation (x^2). Then, multiply the squared deviation by the frequency of the class, $f(x^2)$. (b) First, multiply the deviation by the frequency of the class fx. Then, multiply the product by the deviation, $(fx)x = fx^2$. The second way is usually preferred.

  4. Find the sum of the deviations squared, or $\sum fx^2$.

  5. Find the variance (s^2) by dividing the sum by the number of values (n or $\sum f$) of the data.

$$s^2 = \frac{\sum fx^2}{n}$$

  6. Extract the square root of the variance to find the standard deviation (s).

$$s = \sqrt{\frac{\sum fx^2}{n}} \qquad (7\text{–}8a)$$

Example 9

The miles traveled by 20 students in coming to Swan College from their homes are given in the first three columns of Table 7–6 (same as the values given in Example 5, p. 177). Find the standard deviation.

Solution.   The standard deviation for the grouped data is computed by using formula (7–8a) as follows:

TABLE 7–6

Computation for Example 9 (Basic Method)

Miles Traveled	Average Miles (mid-point) X	Number of Students f	$x = X - \bar{X}$	fx	fx^2
0 and under 2	1	2	−4.1	− 8.2	33.62
2 and under 4	3	5	−2.1	−10.5	22.05
4 and under 6	5	4	−0.1	− 0.4	0.04
6 and under 8	7	8	1.9	15.2	28.88
8 and under 10	9	1	3.9	3.9	15.21
Total		20		0	99.80

$$\overline{X} = 5.1 \text{ miles (see page 122)}$$

$$s^2 = \frac{\Sigma fx^2}{n} = \frac{99.80}{20} = 4.99$$

$$s = \sqrt{4.99} = 2.23 \text{ miles}$$

Short Method—Deviations in Original Units (v)

The short method for computing the standard deviation for grouped data is basically the same as the short method for ungrouped data. When the deviations of the individual values from the assumed mean are expressed in the original units, formula (7–8b) below may be used to compute the standard deviation. The formula is analogous to formula (7–7b), except that v^2 and v values in each class are multiplied by the frequency in the class.

$$s^2 = \frac{\Sigma fv^2}{n} - \left(\frac{\Sigma fv}{n}\right)^2$$

or

$$s = \sqrt{\frac{\Sigma fv^2}{n} - \left(\frac{\Sigma fv}{n}\right)^2} \tag{7–8b}$$

When zero is selected as the assumed mean, the deviation of each value from the assumed mean is the value unchanged, or $v = X - 0 = X$. Formula (7–8b) becomes

$$s = \sqrt{\frac{\Sigma fX^2}{n} - \left(\frac{\Sigma fX}{n}\right)^2}$$

or

$$s = \sqrt{\frac{\Sigma fX^2}{n} - (\overline{X})^2} \tag{7–8c}$$

Example 9 may be computed in a different way when formulas (7–8b) and (7–8c) are used. The computations are presented in Table 7–7. Case I in the table shows the computation by using formula (7–8b), where the assumed mean is 5 miles, or $A = 5$. Case II shows the application of formula (7–8c), the assumed mean being zero mile, or $A = 0$.

Short Method—Deviations in Class-interval Units (d)

When the deviations of the individual values from the assumed mean are expressed in class-interval units, or $d = v/i$ and $v = di$, formula (7–8b) may be written

$$s^2 = i^2 \left[\frac{\Sigma fd^2}{n} - \left(\frac{\Sigma fd}{n}\right)^2 \right]$$

TABLE 7–7

COMPUTATION FOR EXAMPLE 9 (SHORT
METHOD—DEVIATIONS IN ORIGINAL UNITS)

Miles Traveled	Average Miles (mid-point) X	Number of Students f	Case I $A = 5$			Case II $A = 0$	
			$v = X - A$	fv	fv^2	fX	fX^2
0 and under 2	1	2	-4	-8	32	2	2
2 and under 4	3	5	-2	-10	20	15	45
4 and under 6	5	4	0	0	0	20	100
6 and under 8	7	8	2	16	32	56	392
8 and under 10	9	1	4	4	16	9	81
Total		20		2	100	102	620

Case I:
$$s^2 = \frac{100}{20} - \left(\frac{2}{20}\right)^2$$
$$= 5 - 0.01 = 4.99$$
$$s = \sqrt{4.99} = 2.23 \text{ miles}$$

Case II:
$$s^2 = \frac{620}{20} - \left(\frac{102}{20}\right)^2$$
$$= 31 - (5.1^2) = 4.99$$
$$s = \sqrt{4.99} = 2.23 \text{ miles}$$

or

$$s = i\sqrt{\frac{\sum fd^2}{n} - \left(\frac{\sum fd}{n}\right)^2} \qquad (7\text{–}8d)*$$

Example 9 may be computed by using formula (7–8d) as follows (see Table 7–8). (Note that the arrangement in the table is identical to the arrangement of computing the mean for Example 6 on page 125 except that the last column is added.)

$$s^2 = 2^2\left[\frac{25}{20} - \left(\frac{1}{20}\right)^2\right] = 4\,(1.25 - 0.0025) = 4\,(1.2475) = 4.99$$
$$s = \sqrt{4.99} = 2.23 \text{ miles}$$

The mid-point values (X's) are usually not shown in practical computation, since the d value for each class can easily be obtained (see page 126.)

Example 10 is an additional illustration of the use of formula (7–8d). The computation is in the most simplified form. This form is usually used in practice.

---

*Proof—formula (7–8d)

$$d = \frac{v}{i} \qquad v = di$$

Substitute $v = di$ in formula (7–8b).

$$s^2 = \frac{\sum f(di)^2}{n} - \left(\frac{\sum f(di)}{n}\right)^2$$

Since i is a constant, it can be factored out in each of the above summations. Thus,

$$s^2 = \frac{i^2 \cdot \sum f(d)^2}{n} - \left(\frac{i \cdot \sum fd}{n}\right)^2 = i^2\left[\frac{\sum f(d)^2}{n} - \left(\frac{\sum fd}{n}\right)^2\right]$$

TABLE 7-8

COMPUTATION FOR EXAMPLE 9 (SHORT
METHOD—DEVIATIONS IN CLASS-INTERVAL UNITS)

Miles Traveled	Average Miles (mid-point) X	Number of Students f	$A = 5$ miles $d = \dfrac{X - A}{i}$	fd	fd^2
0 and under 2	1	2	-2	-4	8
2 and under 4	3	5	-1	-5	5
4 and ander 6	5	4	0	0	0
6 and under 8	7	8	1	8	8
8 and under 10	9	1	2	2	4
Total		20		1	25

Example 10

The salaries earned by a group of 25 employees in the Barnes Company in a given period are given in the first two columns of Table 7–9. Compute the standard deviation.

Solution. Let $A = 8$, or "7–9" as the assumed mean class. The computation

TABLE 7-9

COMPUTATION FOR EXAMPLE 10

Salaries (class interval)	Number of Employees f	d	fd	fd^2
\$ 1- 3	1	-2	-2	4
4- 6	4	-1	-4	4
7- 9	9	0	0	0
10-12	6	1	6	6
13-15	2	2	4	8
16-18	3	3	9	27
Total	25		13	49

$$s^2 = 3^2\left[\frac{49}{25} - \left(\frac{13}{25}\right)^2\right] = 9[1.96 - 0.2704] = 9(1.6896) = 15.2064$$

$$s = \sqrt{15.2064} = \$3.9$$

of the standard deviation is given above. The arrangement in the table is the same as that of Example 7 on page 125 in computing the mean by the short method, except that the last column is added.

Chief Characteristics of the Standard Deviation

1. The standard deviation is based on every value in the data. Like the average deviation, it therefore gives a better description of the dispersion than the range and the quartile deviation.

2. The standard deviation is computed from the arithmetic mean of the values in the data. It measures the dispersion around the mean, not the dispersion within certain values as measured by the range and the quartile deviation.

3. The standard deviation is mathematically logical, since its computation does not neglect the positive and negative signs of the individual deviations. This fact increases the use of the standard deviation in further mathematical operations.

4. When each value of the given data is increased (or decreased) by a fixed number, the standard deviation is not affected. This is true because the mean, like each value, is also increased by the fixed amount. The deviation of each value from the mean thus is not affected (see Illustration I in Table 7–10). However, when each value of the data is multiplied (or divided) by a fixed number, the standard deviation is also multiplied (or divided) by the fixed number (see Illustration II in Table 7–10).

TABLE 7–10

THE FOURTH CHARACTERISTIC OF THE STANDARD DEVIATION

Original Data			I. Each value is increased by 2.			II. Each value is multiplied by 2.		
Value X	$x = X - 3$	x^2	$X + 2$	$x = (X + 2) - 5$	x^2	$X(2)$	$x = X(2) - 6$	x^2
1	-2	4	3	-2	4	2	-4	16
2	-1	1	4	-1	1	4	-2	4
6	3	9	8	3	9	12	6	36
Total 9	0	14	15	0	14	18	0	56

$$\bar{X} = \frac{9}{3} = 3 \qquad\qquad \bar{X} = \frac{15}{3} = 5 \qquad\qquad \bar{X} = \frac{18}{3} = 6$$

$$s = \sqrt{\frac{14}{3}} \qquad\qquad s = \sqrt{\frac{14}{3}} \qquad\qquad s = \sqrt{\frac{56}{3}} = 2\sqrt{\frac{14}{3}}$$

7.5 Relative Dispersion

The measures of dispersion expressed in absolute values as presented in the preceding sections are convenient in describing the dispersion for a single set of values. If two sets of values are being compared, the absolute values are convenient only when the averages of the two sets are about the same size and the units of measurement of the sets are alike. It is obvious that the comparison of two different units, such as the number of miles compared with the number of dollars, is meaningless.

When the averages are distinctively different, although the units may be the same, the task of comparing the degrees of dispersion based on the absolute values of the different sets is still difficult. For example, the weights of college students are generally greater than the weights of elementary school students. The average and the standard deviation of the weights of a group of college students, therefore, are expected to be greater than those of a group of elementary school students.

Examine the two sets of values in Table 7–11. The means of the two sets are distinctively different (200 pounds for one and 50 pounds for another). We cannot conclude that the higher standard deviation (32.7 pounds) gives the higher degree of dispersion in the weights of the college students. The standard deviation is significant only in relation to the mean from which it is computed.

A measure of dispersion expressed in a relative value is thus required for this type of comparison. In general, a relative dispersion is the quotient of a given measure of dispersion divided by the average from which the deviations were measured.

TABLE 7–11

COMPARISON OF STANDARD DEVIATIONS IN ABSOLUTE VALUES

Set 1—Weights of 3 College Students			Set 2—Weights of 3 Elementary School Students		
Weight in Pounds X	x	x^2	Weight in Pounds X	x	x^2
160	−40	1,600	40	−10	100
200	0	0	45	− 5	25
240	40	1,600	65	15	225
600		3,200	150		350

$$\bar{X} = \frac{600}{3} = 200$$

$$s = \sqrt{\frac{3,200}{3}} = 32.7 \text{ pounds}$$

$$\bar{X} = \frac{150}{3} = 50$$

$$s = \sqrt{\frac{350}{3}} = 10.8 \text{ pounds}$$

The most commonly used measure of dispersion expressed in a relative value is the *coefficient of variation*, represented by V. It is the quotient of the standard deviation divided by the arithmetic mean, or

$$V = \frac{s}{\bar{X}} \qquad (7\text{-}9)$$

The coefficient of variation of the weights of the three college students is

$$V = \frac{32.7}{200} = 0.1635 \text{ or } 16.35\%$$

The coefficient of variation of the weights of the three elementary school students is

$$V = \frac{10.8}{50} = 0.216 \text{ or } 21.6\%$$

The relative dispersion of the weights of the elementary school students is larger than that of the weights of the college students, although the absolute dispersion is much smaller for the elementary school students.

Similarly other measures of dispersion expressed in relative values may be obtained as follows:

coefficient of range: $V_r = \dfrac{\text{range}}{(\text{lowest value} + \text{largest value})/2}$

coefficient of average deviation: $V_{ad} = \dfrac{\text{average deviation}}{\text{mean (or median)}}$

coefficient of quartile deviation: $V_{qd} = \dfrac{\text{quartile deviation}}{(Q_1 + Q_3)/2}$

Obviously, if one of the measures of relative dispersion is used in describing one set of data, the same measure must be used in another set of data for comparison.

7.6 Summary

The variation of the values included in a series of data is called dispersion. A measure of dispersion can be expressed either in an absolute value or in a relative value. The most common types of dispersions expressed in absolute values are (1) range, (2) quartile deviation, (3) average deviation, and (4) standard deviation.

TABLE 7–12

SUMMARY OF THE CHIEF CHARACTERISTICS OF THE
ABSOLUTE DISPERSIONS

Characteristics	*Range* (R)	*Quartile Deviation* (Q.D.)	*Average Deviation* (A.D.)	*Standard Deviation* (s)
Computation—based on	Lowest and highest values	Q_1 and Q_3	Every value	Every value
Affected by extreme values	Greatest	Not by values smaller than Q_1 or larger than Q_3	Affected by every value	Affected by every value
Degree of precision as a measure of dispersion	Rough estimate	Better than the range only	Good, but only measures absolute deviations from the mean or median	Excellent, measures squared deviations from the mean
Mathematical advantages	Easy to compute	Can be used to measure asymmetrical distribution	Easier to compute than standard deviation	Hard to compute, but suitable for further mathematical calculation

The methods of computing the four types of dispersions are summarized in the formulas at the end of this section. Their chief characteristics are summarized in Table 7–12.

The measures of dispersion expressed in absolute values are convenient either in showing the degree of variation or in supplementing an average to describe the distribution for a single set of values. If two sets of values are being compared, the absolute values are convenient only when the units of measurement of the sets are alike and the averages of the sets are about the same size. When the units are not the same and when the averages are distinctively different although the units may be the same, a measure of dispersion expressed in a relative value is more significant. The most commonly used relative dispersion is the coefficient of variation.

SUMMARY OF FORMULAS

Application	Formula	Formula number	Reference page
Range	$R = X_n - X_1$	(7-1)	174
Quartile deviation	$Q.D. = \dfrac{Q_3 - Q_1}{2}$	(7-2)	174
	$Q_1 = L + \dfrac{n/4 - C}{f}(i)$	(7-3)	176
	$Q_3 = L + \dfrac{3n/4 - C}{f}(i)$	(7-4)	177
Average deviation			
Ungrouped data	$A.D. = \dfrac{\sum |x|}{n}$	(7-5)	179
Grouped data	$A.D. = \dfrac{\sum (f|x|)}{n}$	(7-6)	180
Standard deviation			
Ungrouped data Basic method	$s = \sqrt{\dfrac{\sum x^2}{n}}$	(7-7a)	181
Short method A = any assumed mean	$s = \sqrt{\dfrac{\sum v^2}{n} - \left(\dfrac{\sum v}{n}\right)^2}$	(7-7b)	182
$A = 0$	$s = \sqrt{\dfrac{\sum X^2}{n} - (\bar{X})^2}$	(7-7c)	183
Grouped data Basic method	$s = \sqrt{\dfrac{\sum fx^2}{n}}$	(7-8a)	184

Short method (original
    units)
  A = any assumed mean
	$s = \sqrt{\dfrac{\sum fv^2}{n} - \left(\dfrac{\sum fv}{n}\right)^2}$	(7-8b)	185

$A = 0$	$s = \sqrt{\dfrac{\sum fX^2}{n} - (\bar{X})^2}$	(7-8c)	185
Short method (class- interval units)	$s = i\sqrt{\dfrac{\sum fd^2}{n} - \left(\dfrac{\sum fd}{n}\right)^2}$	(7-8d)	186
Relative dispersion			
Coefficient of variation	$V = \dfrac{s}{\bar{X}}$	(7-9)	189

Exercises 7

1. Explain the term *dispersion*. Why is a measure of dispersion important in analyzing statistical data?

2. What are the differences and similarities, if any, between the range and the quartile deviation?

3. What are the differences and similarities, if any, between the average deviation and the standard deviation?

4. What are absolute dispersion and relative dispersion? Under what condition or conditions should we express a measure of dispersion in relative value?

5. The following values represent numbers of visits to a city by a group of eight farmers during one month (same values as given in Problem 3, Exercise 5).

$$1, 2, 2, 4, 5, 11, 11, 20$$

    (a) Compute:
        (1) Range.
        (2) Quartile deviation.
        (3) Average deviation from (I) the median and (II) the mean.
        (4) Standard deviation by using (I) the basic method and (II) the short method (let $A = 0$).
    (b) Interpret the findings in (a).

6. The weights of the 11 players in each of the two college football teams are given below (in pounds):

    Granby College: 160, 180, 190, 200, 210, 170, 250, 220, 180, 200, 240
    Norview College: 160, 190, 210, 230, 240, 220, 150, 190, 210, 160, 240

    (a) Compute the following to two decimal places for each team:
        (1) Range.
        (2) Quartile deviation.
        (3) Average deviation from the mean.
        (4) Standard deviation by the short method (let $A = 160$ pounds for the Granby team and let $A = 210$ pounds for the Norview team).
    (b) Interpret the answers in (a).
    (c) Does the standard deviation give a better description of the weights of each team than the the average deviation? If not, why should we compute the standard deviation?

7. Compute the variance (s^2) and the standard deviation (s) from Table 7–13 (compute s to two decimal places).

TABLE 7–13

Class Interval	f	d	fd	fd^2
4– 8	1			
9–13	9			
14–18	12	0		
19–23	15			
24–28	13			

8. Compute the variance (s^2 to two decimal places) and the standard deviation (s to one decimal place) from Table 7–14.

TABLE 7–14

Class Interval	f	d	fd	fd²
0.00– 9.99	4			
10.00–19.99	12			
20.00–29.99	18			
30.00–39.99	42			
40.00–49.99	58			
50.00–59.99	84	0		
60.00–69.99	45			
70.00–79.99	33			
80.00–89.99	16			
90.00–99.99	8			

9. The following frequency distribution table shows the heights of the 18 sales-men in Bayside Company (same as the values given in Problem 13, Exercise 5.)

Height (inches)	Number of salesmen
60 and under 62	2
62 and under 64	3
64 and under 66	7
66 and under 68	5
68 and under 70	1

Compute to two decimal place for the following:
  (a) The quartile deviation.
  (b) The average deviation from the mean ($\bar{X} = 65$).
  (c) The standard deviation.
    (1) Use the basic method.
    (2) Use the short method (let $A = 63$ inches).
      (a) Express the deviations in original units (v).
      (b) Express the deviations in class-interval units (d).

10. The grades of the 80 students in a statistics class are given below (same as the values given in Problem 14, Exercise 5).

Grades (Class interval)	Number of students
20–29	3
30–39	6
40–49	5
50–59	7
60–69	10
70–79	29
80–89	12
90–99	8
Total	80

Compute to two decimal places:

    (a) Quartile deviation ($Q.D.$).

    (b) Average deviation from the mean ($\bar{X} = 68.25$).

    (c) Standard deviation by using the short method. Express deviations in class-interval units (d). Let $A = 54.5$.

11. Refer to Problem 6.

    (a) Compute the coefficient of variation V for each set of values. (Compute to one decimal place of per cent unit.)

    (b) Compare the value of V for the Granby team with that for the Norview team. Which team has the higher variation among the weights of the players? Can you compare the variations between the two teams without finding the V values?

12. Refer to Problems 9 and 11.

    (a) Compute the coefficient of variation V for the heights of the 18 salesmen in Bayside Company given in Problem 9. (Compute to one decimal place of per cent unit.)

    (b) Compare the value V for the weights of the 11 players in the Granby team obtained in Problem 11(a) with the value V for the heights of the 18 salesmen in Bayside Company. Which group has the higher variation in the distribution? Can you compare the variations between the two groups without finding the V values?

8 Skewness and Kurtosis

In addition to the averages and the dispersions, there are two other measures used in describing the characteristics of a group of data. The two measures are the *skewness* and the *kurtosis*, which are presented in Sections 8.1 and 8.2, respectively. The two measures are especially useful in indicating the shapes of frequency distributions. A measure of skewness can indicate the direction of the distribution, either skewed to higher values or lower values. A measure of kurtosis can indicate the degree of concentration of the distribution, either peaked (values concentrated around the mode) or flat-topped (values decentralized from the mode). The two measures can be expressed by the use of moments. The meaning of a moment and its applications to skewness and kurtosis are explained in Section 8.3.

8.1 Measure of Skewness

As pointed out in Chapter 5, in a symmetrical frequency distribution, the values of mean, median, and mode will coincide under the frequency curve; that is, $\bar{X} = M_d = M_0$. When a frequency distribution is asymmetrical, the three values depart from each other. The more the mean departs from the mode, the greater the skewness. The frequency curve may skew either to the right side on the X scale (positively skewed) or to the left on the X scale (negatively skewed). (See Chart 5–8, p. 143.) In either case, the median is between the mode and the mean. When the difference between the mean and the mode is divided by the standard deviation, the quotient is called a *coefficient of skewness* (*Sk*) and is used by Karl Pearson for measuring the degree of skewness, or

$$Sk = \frac{\bar{X} - M_0}{s} \tag{8-1}$$

Example 1

The following information concerning the distribution of miles traveled by the 20 students in coming to Swan College is obtained from Table 5–15, page 122. Compute the coefficient of skewness by using formula (8–1).

Miles Traveled	Number of Students
0 and under 2	2
2 and under 4	5
4 and under 6	4
6 and under 8	8
8 and under 10	1
Total	20

Solution. $\bar{X} = 5.1$ (page 121); $M_0 = 6.7$ (page 139); $s = 2.23$ (page 186). Thus,

$$Sk = \frac{5.1 - 6.7}{2.23} = -0.72 \text{ or } -72\%$$

Since the coefficient is a negative value, the distribution is skewed to the left, or at smaller values on the X scale. The negative value also indicates that the mode is greater than the mean by an amount equal to 72% of the value of the standard deviation. Those facts are clearly shown in Chart 8–1. The chart shows

CHART 8–1

A Negatively Skewed Distribution with $\bar{X}$ and M_0

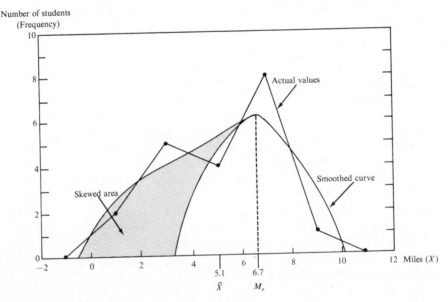

Source: Example 1.

the broken line representing the given values in Example 1 and a smoothed curve drawn through the broken line with the mode at 6.7. The smoothed curve gives a better indication of the direction of skewness than the broken line.

The above mode is obtained by using formula (5–5). For moderately skewed distributions, the empirical mode (formula 5–6) may be used, and formula 8–1 may be written

$$Sk = \frac{\bar{X} - EM_0}{s}$$

or

$$Sk = \frac{3(\bar{X} - M_d)}{s} \tag{8–2}$$

The coefficient of skewness for the distribution in Example 1 may thus be computed by formula (8–2) as follows:

$$Sk = \frac{5.1 - 6.3}{2.23} = -0.54 \text{ or } -54\%$$

or

$$Sk = \frac{3(5.1 - 5.5)}{2.23} = -0.54 \text{ or } -54\%$$

where $EM_0 = 6.3$ and $M_d = 5.5$ (page 142).

A different measure of skewness may be obtained by employing the quartiles. If the distribution is symmetrical, Q_3 and Q_1 will be equidistant from Q_2 (or the median); that is, $Q_3 - Q_2 = Q_2 - Q_1$. However, if the distribution is asymmetrical, the distance from Q_3 to Q_2 is not equal to the distance from Q_2 to Q_1. The difference between the two distances thus can be used as a basis for measuring the skewness of the distribution. For practical use, as suggested by Arthur Bowley, the difference is expressed in the units of interquartile range, or

$$Sk_q = \frac{(Q_3 - Q_2) - (Q_2 - Q_1)}{Q_3 - Q_1} \tag{8–3}$$

The value of Sk_q will vary from -1 to $+1$. Bowley pointed out that an absolute value (disregard $\pm$ signs) of 0.1 may be considered as a moderate degree of skewness and 0.3 as a marked skewness.

When formula (8–3) is applied, the coefficient of skewness for the distribution of miles traveled by the 20 college students in Example 1 is computed as follows:

$$Sk_q = \frac{(7 - 5.5) - (5.5 - 3.2)}{7 - 3.2} = \frac{1.5 - 2.3}{3.8} = -0.21 \text{ or } -21\%$$

where $Q_3 = 7$, $Q_1 = 3.2$ (page 177) and Q_2 (or M_d) = 5.5 (page 130).

The negative coefficient, -0.21, indicates that the distance between Q_3 and Q_2 is smaller than that between Q_2 and Q_1. Thus, the distribution is skewed to the left, or at smaller values on the X scale. The skewed distribution, represented by the smoothed curve, and the quartiles are also shown graphically in Chart 8–2. The curve is reproduced from Chart 8–1.

CHART 8–2

A Negatively Skewed Distribution with Quartiles and M_o

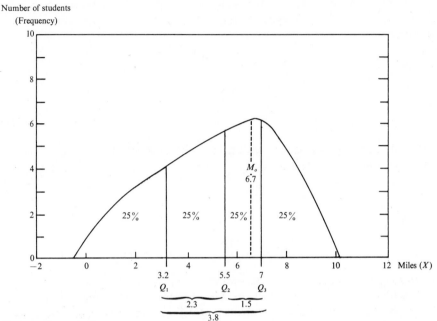

Source: Example 1.

The use of Bowley's formula thus is analogous to that of Pearson's formula. Obviously, if one formula is used in finding the skewness of one set of data, the same formula must be used in another set for comparison.

8.2 Measure of Kurtosis

In describing a frequency distribution a person can use an average to show the typical value or the central tendency in the distribution, a measure of dispersion to show the variation of values either within certain values (such as the range and the quartile deviation) or around an average of the distribution (such as the average deviation and the standard deviation), and a measure of skewness to show the direction of the distribution either skewed to the higher values (the right side on the X scale) or to the lower values (the left side on the X scale). Further, the measure of kurtosis, the fourth device in describing a frequency distribution, can be used to show the degree of concentration, with the values either concentrated in the area around the mode (a peaked curve) or decentralized from the mode to both tails of the frequency curve (a flat-topped curve),

A measure of kurtosis, or peakedness, can be obtained by using formulas presented in Section 8.3. However, a simple method to find the degree of peakedness or flatness of each distribution is by observing the frequency curve of the data. Chart 8–3 shows three types of curves: (A) the highest peaked curve, also called *leptokurtic*, (B) the intermediate peaked curve, called *mesokurtic*, and (C) the flat-topped curve, called *platykurtic*. It is assumed that the three distributions represented by the curves are symmetrical and have the same mean and the same dispersion as measured by the ranges.

CHART 8–3

DIFFERENT TYPES OF KURTOSIS

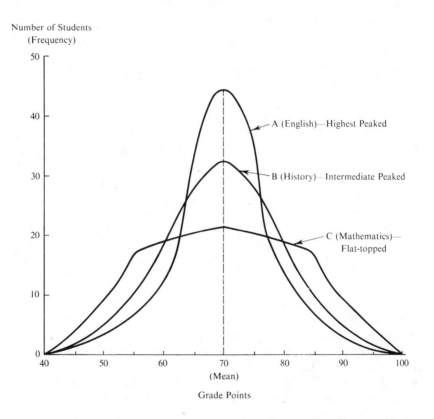

Curve A indicates that most of the students received about the same grades in the English test, between 65 to 75 points; curve B shows the *normal distribution* of the history grades, that about 2/3 of the students received 60 to 80 points; curve C indicates a wide variation of the mathematics grades, mostly between 55 to 85 points, among the group of students. The normal distribution, which is neither very peaked nor very flat-topped, is the intermediate peaked curve and is usually used as a standard for measuring the peakedness of a distribution. We shall discuss the normal distribution in greater detail in Chapter 11.

In addition to the various shapes of frequency distribution curves mentioned in this and the previous chapters, there are many other shapes of curves. Among them, the common types are J-shaped, reverse J-shaped, and U-shaped frequency curves. They can easily be recognized when they are plotted.

★8.3 Measuring Skewness and Kurtosis by Moments

Let $x = X - \bar{X}$ be the deviation of each X value from the arithmetic mean $\bar{X}$. Then, we define the rth moment about the mean $\bar{X}$ as

$$m_r = \frac{\sum x^r}{n} \qquad (8\text{–}4)$$

The first four moments about the mean and their applications are listed below:

Moment about the mean $\bar{X}$	For ungrouped data	For grouped data	Application
First moment, m_1	$\dfrac{\sum x}{n} = 0$	$\dfrac{\sum fx}{n} = 0$	Finding the mean by short methods
Second moment, m_2	$\dfrac{\sum x^2}{n} = s^2$	$\dfrac{\sum fx^2}{n} = s^2$	Finding the variance s^2 and the standard deviation s, since $m_2 = s^2$
Third moment, m_3	$\dfrac{\sum x^3}{n}$	$\dfrac{\sum fx^3}{n}$	Measuring absolute skewness
Fourth moment, m_4	$\dfrac{\sum x^4}{n}$	$\dfrac{\sum fx^4}{n}$	Measuring absolute kurtosis

The third power of each deviation (x^3) retains the original sign of the deviation. When the sum of the negative cubed-deviations is larger than the sum of the positive cubed-deviations, the value of $\sum x^3$ or $\sum fx^3$ must be negative and the third moment about the mean m_3 is also negative. A negative m_3 indicates the skewness of the distribution of X values toward the smaller values or the left side on the X scale. On the other hand, a positive m_3 indicates the skewness toward the larger values or the right side on the X scale. When the distribution is symmetrical, the value of m_3 is zero.

A measure of relative skewness (a_3) can be obtained by dividing the third moment about the mean by the third power of standard deviation, or

$$a_3 = \frac{m_3}{s^3} \qquad (8\text{–}5)$$

When the distribution is fairly symmetrical, the value of a_3 will be within the range $-\frac{1}{2}$ to $+\frac{1}{2}$. When a_3 is between $+\frac{1}{2}$ and $+1$ or between $-\frac{1}{2}$ and -1, the distribution is moderately skewed. A value of a_3 greater than $+1$ or smaller than -1 generally indicates a highly skewed distribution. These statements are diagrammed:

a_3 value	-1	$-\frac{1}{2}$	0	$\frac{1}{2}$	1
Distribution shape	highly skewed	moderately skewed	fairly symmetrical	moderately skewed	highly skewed

The fourth power of each deviation (x^4) is always positive. Thus, the value of the fourth moment about the mean m_4 is also positive. In general, a distribution with X values highly concentrated around the mean $\bar{X}$ will have smaller deviations since $x = X - \bar{X}$. The smaller deviations will give a smaller value of $\sum x^4$, which in turn will give a smaller value of m_4.

A measure of relative kurtosis (a_4) can be obtained by dividing the fourth moment by the fourth power of standard deviation, or

$$a_4 = \frac{m_4}{s^4} \tag{8-6a}$$

where $s^4 = (s^2)^2 = (m_2)^2$.

Or compute a_4 directly from deviations x, for ungrouped data:

$$a_4 = \frac{n \cdot \sum x^4}{(\sum x^2)^2} \tag{8-6b}$$

For grouped data:

$$a_4 = \frac{n \cdot \sum fx^4}{(\sum fx^2)^2} \tag{8-6c}$$

However, when the relative kurtosis is used, a distribution with X values highly concentrated around the mean $\bar{X}$ will have higher value of a_4. (See Example 4.) The smallest possible value of a_4 is 1.* The value of a_4 for a normal distribution is 3. Using the normal distribution as a standard reference, the measure of relative kurtosis may be expressed as follows:

* Proof:

$$a_4 = \frac{m_4}{s^4} = \frac{\sum x^4}{n} \cdot \frac{1}{s^4} = \frac{1}{n} \cdot \frac{\sum x^4}{s^4} = \frac{1}{n} \cdot \sum \left(\frac{x}{s}\right)^4$$

Let $z = \frac{x}{s}$

$$a_4 = \frac{1}{n} \cdot \sum z^4$$

The mean of z^2 values $= \dfrac{\sum \left(\frac{x}{s}\right)^2}{n} = \dfrac{\frac{\sum x^2}{s^2}}{n} = \dfrac{\sum x^2}{\sum x^2/n} = 1$

The variance of z^2 values $= \dfrac{\sum (z^2)^2}{n} - (\text{mean of } z^2 \text{ values})^2$

$$= \frac{\sum z^4}{n} - 1^2 = a_4 - 1 \qquad \text{(Formula 7-7c)}$$

Since a variance cannot be negative, the lowest value being zero, a_4 cannot be less than 1. The value of a_4 can be used as a measure of dispersion of z^2 values around their mean of 1.

$a_4 - 3 = 0$. This is the normal distribution.

$a_4 - 3 > 0$. The higher the difference $(a_4 - 3)$ above zero, the more peaked is the distribution.

$a_4 - 3 < 0$. The lower the difference below zero, the flatter is the distribution.

The applications of formulas (8–5) and (8–6) are presented in Examples 2 (for ungrouped data) and 3 (for grouped data). In addition, Example 4 compares the degrees of concentration of two distributions with the same number of values and the same range of values.

Example 2

Refer to the distribution of values $2, 3, 5, 7$, and 10 as given in Example 8, Chapter 7. Find (a) the relative skewness based on the third moment, and (b) the relative kurtosis based on the fourth moment.

Solution. The required values for the computation are presented in Table 8–1. Note that the first three columns of Table 8–1 are the same as Table 7–4, page 181.

TABLE 8–1

COMPUTATION FOR EXAMPLE 2

Values X	Deviations $x = X - \bar{X}$	x^2	x^3	x^4
$2	-3.4	11.56	-39.304	133.6336
3	-2.4	5.76	-13.824	33.1776
5	-0.4	0.16	-0.064	0.0256
7	1.6	2.56	4.096	6.5536
10	4.6	21.16	97.336	447.7456
Total 27	0	41.20	48.240	621.1360

$$\bar{X} = \frac{27}{5} = 5.4$$

$$m_2 = \frac{\Sigma x^2}{n} = \frac{41.20}{5} = 8.24$$

$$s^2 = m_2 = 8.24$$

$$s = \sqrt{8.24} = 2.87$$

(a) $$m_3 = \frac{\Sigma x^3}{n} = \frac{48.240}{5} = 9.648$$

$$a_3 = \frac{m_3}{s^3} = \frac{9.648}{2.87^3} = \frac{9.648}{23.648} = 0.41 \qquad \text{(Formula 8–5)}$$

Since the value of a_3 is positive (because m_3 is positive), it indicates the skewness toward the larger values or the right side on the X scale.

(b) $m_4 = \dfrac{\sum x^4}{n} = \dfrac{621.1360}{5} = 124.2272$

$a_4 = \dfrac{m_4}{(s^2)^2} = \dfrac{124.2272}{8.24^2} = \dfrac{124.2272}{67.8976} = 1.83$ (Formula 8–6a)

or $a_4 = \dfrac{n \cdot \sum x^4}{(\sum x^2)^2} = \dfrac{5(621.1360)}{41.20^2} = \dfrac{3,105.68}{1,697.44} = 1.83$ (Formula 8–6b)

Since the value of a_4 is smaller than 3, the distribution is relatively flat-topped.

The answers obtained above can also be checked graphically as shown in Chart 8–4. The mean ($\bar{X} = 5.4$) is on the right side of the median ($M_d = 5$) in the chart. Thus this distribution is skewed to the right or toward the higher values. This distribution has no mode since each value occurs only once. A smoothed curve is drawn through the given values. The curve also shows a right-sided skewness and a rather flat-topped distribution.

CHART 8–4

A POSITIVELY SKEWED AND FLAT-TOPPED DISTRIBUTION

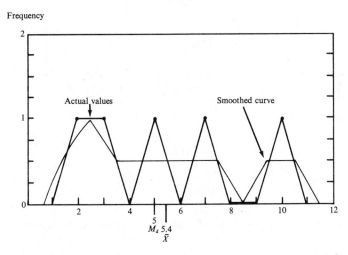

Source: Example 2.

Example 3

Refer to the distribution of miles traveled by the 20 students in coming to Swan College as given in Example 1. Find (a) the relative skewness based on the third moment, and (b) the relative kurtosis based on the fourth moment.

Solution.   The required values for the computation are presented in Table 8–2. Note that the first six columns of Table 8–2 are the same as Table 7–6 of Example 9 in Chapter 7, page 184.

TABLE 8–2

COMPUTATION FOR EXAMPLE 3

Miles Traveled	Average Miles (Mid-point) X	Number of Students f	$x = X - \bar{X}$	fx	fx^2	fx^3	fx^4
0 and under 2	1	2	−4.1	−8.2	33.62	−137.842	565.1522
2 and under 4	3	5	−2.1	−10.5	22.05	−46.305	97.2405
4 and under 6	5	4	−0.1	−0.4	0.04	−0.004	0.0004
6 and under 8	7	8	1.9	15.2	28.88	54.872	104.2568
8 and under 10	9	1	3.9	3.9	15.21	59.319	231.3441
Total		20		0	99.80	−69.960	997.9940

$$\bar{X} = 5.1 \text{ miles (See Table 7–6)}$$

$$m_2 = \frac{\Sigma fx^2}{n} = \frac{99.80}{20} = 4.99$$

$$s^2 = m_2 = 4.99$$

$$s = \sqrt{4.99} = 2.23$$

(a)
$$m_3 = \frac{\Sigma fx^3}{n} = \frac{-69.960}{20} = -3.498$$

$$a_3 = \frac{m_3}{s^3} = \frac{-3.498}{2.23^3} = \frac{-3.498}{11.090} = -0.32 \qquad \text{(Formula 8–5)}$$

Since the value of a_3 is negative (because m_3 is negative), it indicates the skewness toward the smaller values or the left side on the X scale of the distribution curve.

(b) $$m_4 = \frac{\Sigma fx^4}{n} = \frac{997.9940}{20} = 49.8997$$

$$a_4 = \frac{m_4}{(s^2)^2} = \frac{49.8997}{4.99^2} = \frac{49.8997}{24.9001} = 2.004 \qquad \text{(Formula 8–6a)}$$

or, $$a_4 = \frac{n \cdot \Sigma fx^4}{(\Sigma fx^2)^2} = \frac{20(997.9940)}{99.80^2} = \frac{19,959.88}{9,960.04} = 2.004 \quad \text{(Formula 8–6c)}$$

Since the value of a_4 is smaller than 3, the distribution is relatively flat-topped. Also see Chart 8–1 for the shape of the left-side skewed and flat-topped distribution.

The moments about the mean (m_2, m_3, and m_4) may also be obtained by the short method involving the deviations expressed in class-interval units (d) as follows:

$$m_2 = i^2\left[\frac{\Sigma fd^2}{n} - \left(\frac{\Sigma fd}{n}\right)^2\right] \qquad \text{[Same as formula (7–8d) for } s^2\text{]}$$

$$m_3 = i^3\left[\frac{\Sigma fd^3}{n} - 3\left(\frac{\Sigma fd}{n}\right)\left(\frac{\Sigma fd^2}{n}\right) + 2\left(\frac{\Sigma fd}{n}\right)^3\right] \qquad (8\text{–}7)$$

$$m_4 = i^4\left[\frac{\Sigma fd^4}{n} - 4\left(\frac{\Sigma fd}{n}\right)\left(\frac{\Sigma fd^3}{n}\right) + 6\left(\frac{\Sigma fd}{n}\right)^2\left(\frac{\Sigma fd^2}{n}\right) - 3\left(\frac{\Sigma fd}{n}\right)^4\right] \quad (8\text{–}8)$$

The required moments for the data given in Example 3 now may be computed by using the short method. First, add the two columns (fd^3) and (fd^4) to Table

7–8 on page 187 to form a new table, Table 8–3. Next, compute the moments from the values in Table 8–3 according to the short method formulas.

TABLE 8–3

COMPUTATION FOR EXAMPLE 3 BY SHORT METHOD

Miles Traveled	Average Miles (Mid-point) X	Number of Students f	($A = 5$ miles) d	fd	fd^2	fd^3	fd^4
0 and under 2	1	2	−2	−4	8	−16	32
2 and under 4	3	5	−1	−5	5	−5	5
4 and under 6	5	4	0	0	0	0	0
6 and under 8	7	8	1	8	8	8	8
8 and under 10	9	1	2	2	4	8	16
Total		20		1	25	−5	61

$$m_2 = s^2 = 2^2\left[\frac{25}{20} - \left(\frac{1}{20}\right)^2\right] = 4(1.25 - 0.0025) = 4(1.2475) = 4.99$$

(Formula 7–8d)

$$m_3 = 2^3\left[\frac{-5}{20} - 3\left(\frac{1}{20}\right)\left(\frac{25}{20}\right) + 2\left(\frac{1}{20}\right)^3\right] = 8\left[\frac{-1,749}{4,000}\right] = -3.498$$

(Formula 8–7)

$$m_4 = 2^4\left[\frac{61}{20} - 4\left(\frac{1}{20}\right)\left(\frac{-5}{20}\right) + 6\left(\frac{1}{20}\right)^2\left(\frac{25}{20}\right) - 3\left(\frac{1}{20}\right)^4\right] = 16\left[\frac{498,997}{160,000}\right]$$
$$= 49.8997$$

(Formula 8–8)

Example 4

The X values of Distribution I in Table 8–4 represent the mathematics grades of a group of 6 students and those of Distribution II represent the English grades of another group of 6 students. Compute a_4 for each group. Which one of the two distributions is more concentrated around the mean?

TABLE 8–4

DATA AND COMPUTATION FOR EXAMPLE 4

Distribution I—Mathematics Grades				Distribution II—English Grades			
Grade points X	$x = X - \bar{X}$	x^2	x^4	Grade points X	$x = X - \bar{X}$	x^2	x^4
50	−20	400	160,000	50	−20	400	160,000
60	−10	100	10,000	60	−10	100	10,000
60	−10	100	10,000	70	0	0	0
80	10	100	10,000	70	0	0	0
80	10	100	10,000	80	10	100	10,000
90	20	400	160,000	90	20	400	160,000
Total 420	0	1,200	360,000	420	0	1,000	340,000
$\bar{X} = 420/6 = 70$				$\bar{X} = 420/6 = 70$			

Solution.   Use the values computed in Table 8–4 and formula (8–6b) to compute a_4 as follows:

For Distribution I,

$$a_4 = \frac{n \cdot \sum x^4}{(\sum x^2)^2} = \frac{6(360,000)}{1,200^2} = 1.5$$

For Distribution II,

$$a_4 = \frac{6(340,000)}{1,000^2} = 2.04$$

Since Distribution II has a higher value of a_4, it is more concentrated around the mean than Distribution I. The answer is also shown clearly in Chart 8–5.

CHART 8–5

DIFFERENT DEGREES OF CONCENTRATIONS OF TWO DISTRIBUTIONS

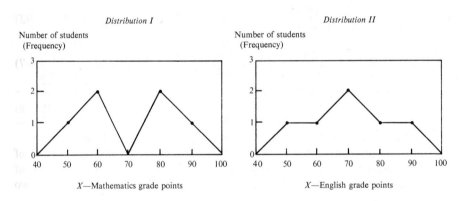

Source: Example 4.

Observe the value in Table 8–4. Distribution II has a lower value of $\sum x^4$ (= 340,000), but it has a higher value of a_4 (= 2.04). This fact is consistent with the statement made earlier concerning the value of the fourth moment m_4 (= $\sum x^4/n$). In general, a distribution with X values highly concentrated around the mean $\bar{X}$ will have a smaller value of m_4 or a higher value of a_4.

8.4 Summary

A measure of skewness is used to indicate the direction of a distribution, either skewed to the higher values (the right side on the X scale) or to the lower values (the left side on the X scale). There are two commonly used measures of skewness:

one developed by Pearson and the other developed by Bowley. The formulas for the two types of measures of skewness are listed at the end of this section.

The measure of kurtosis, in addition to the measures of central tendency (average), dispersion, and skewness, is the fourth device used in describing the characteristics of a frequency distribution. It can be used to show the degree of concentration of the distribution, either the values concentrated around the mode (peaked) or decentralized from the mode (flat). A simple method to find the degree of peakedness or flatness of a distribution is by observing the frequency curve of the data. The measures of skewness and kurtosis may also be obtained by the third and the fourth moments about the mean, respectively.

SUMMARY OF FORMULAS

Application	Formula	Formula number	Reference page
Measures of skewness			
Use mode (Formula 5-5)	$Sk = \dfrac{\bar{X} - M_0}{s}$	(8-1)	195
Use EM_0 (Formula 5-6)	$Sk = \dfrac{3(\bar{X} - M_d)}{s}$	(8-2)	197
Use quartiles	$Sk_q = \dfrac{(Q_3 - Q_2) - (Q_2 - Q_1)}{Q_3 - Q_1}$	(8-3)	197
Moment about mean			
rth moment	$m_r = \dfrac{\sum x^r}{n}$	(8-4)	200
As a measure of relative skewness	$a_3 = \dfrac{m_3}{s^3}$	(8-5)	200
As a measure of relative kurtosis General form	$a_4 = \dfrac{m_4}{s^4}$	(8-6a)	201
For ungrouped data	$a_4 = \dfrac{n \cdot \sum x^4}{(\sum x^2)^2}$	(8-6b)	201
For grouped data	$a_4 = \dfrac{n \cdot \sum fx^4}{(\sum fx^2)^2}$	(8-6c)	201
Short method for grouped data Second moment	$m_2 = i^2 \left[\dfrac{\sum fd^2}{n} - \left(\dfrac{\sum fd}{n} \right)^2 \right]$	(7-8d)	186
Third moment	$m_3 = i^3 \left[\dfrac{\sum fd^3}{n} - 3 \left(\dfrac{\sum fd}{n} \right)\left(\dfrac{\sum fd^2}{n} \right) + 2\left(\dfrac{\sum fd}{n} \right)^3 \right]$	(8-7)	204
Fourth moment	$m_4 = i^4 \left[\dfrac{\sum fd^4}{n} - 4 \left(\dfrac{\sum fd}{n} \right)\left(\dfrac{\sum fd^3}{n} \right) + 6\left(\dfrac{\sum fd}{n} \right)^2\left(\dfrac{\sum fd^2}{n} \right) - 3\left(\dfrac{\sum fd}{n} \right)^4 \right]$	(8-8)	204

Exercises 8

1. Explain briefly how the measures of skewness and kurtosis can be used in describing a frequency distribution.

2. Indicate the difference between the coefficient of skewness formulas as suggested by Pearson and Bowley. Why is the use of Bowley's formula analogous to that of Pearson's formula?

3. The following frequency distribution table shows the heights of the 18 salesmen in Bayside Company (same as the values given in Problem 9, Exercise 7.)

Height (inches)	Number of salesmen
60 and under 62	2
62 and under 64	3
64 and under 66	7
66 and under 68	5
68 and under 70	1

The values listed below are computed from the given distribution:

$$\bar{X} = 65 \quad M_o = 65.3333 \quad s = 2.11$$
$$Q_1 = 63.6667 \quad Q_2 = 65.1428 \quad Q_3 = 66.6$$

(a) Compute the coefficient of skewness for the distribution.
(1) Use Pearson's formula (to two decimal places).
(2) Use Bowley's formula (to three decimal places).
(b) Compare and interpret the findings in (a).

4. The grades of the 80 students in a statistics class are given below (same as the values given in Problem 10, Exercise 7).

Grades (Class interval)	Number of students
20–29	3
30–39	6
40–49	5
50–59	7
60–69	10
70–79	29
80–89	12
90–99	8
Total	80

The following values are computed from the given data:

$$\bar{X} = 68.25 \quad M_o = 74.78 \quad s = 18.26$$
$$Q_1 = 58.07 \quad Q_2 = 72.60 \quad Q_3 = 79.50$$

(a) Compute the coefficient of skewness for the given data.
(1) Use Pearson's formula (to two decimal places).
(2) Use Bowley's formula (to three decimal places).
(b) Compare and interpret the findings in (a).

★5. The following values represent numbers of visits to a city by a group of eight farmers during one month (same values as given in Problem 5, Exercise 7.)

$$1, 2, 2, 4, 5, 11, 11, 20$$

Compute and interpret:

(a) The relative skewness based on the third moment.

(b) The relative kurtosis based on the fourth moment.

(Use formula 8–6a to compute a_4.)

★6. Compute a_4 for each group of values: (Use formula 8–6b.)

(a) 1, 3, 3, 3, 3, 5

(b) 1, 1, 3, 3, 5, 5

Which one of the two groups is more concentrated around its mean?

★7. Refer to the data given in Problem 3. Find and interpret:

(a) The relative skewness based on the third moment.

(b) The relative kurtosis based on the fourth moment.

(Use formulas 8–6a and 8–6c to compute a_4.)

★8. Refer to the data given in Problem 4. Compute m_3 and m_4 by the short method. Let $A = 54.5$.

part 3

Probability and Probability Distributions

Part 3 provides the basic tools for studying statistical induction, which is presented in Part 4. Statistical induction, or statistical inference, concerns the generalization regarding the population or universe from a study of specific sample data based on probability. The theory of probability is presented in Chapter 9.

A set of values distributed according to the theory of probability is called a probability distribution. Probability distributions are of many types. This part will introduce five basic types. They are (1) the binomial distribution, (2) the hypergeometric distribution, (3) the multinomial distribution, (4) the Poisson distribution, and (5) the normal distribution. Among them, the most important type is the normal distribution. The first four types are presented in Chapter 10. The nature of the normal distribution is discussed in greater length in Chapter 11. The problems of sampling distributions can be solved by the use of the normal distribution. Sampling distributions are given in Chapter 12.

9 Probability

When a population is large or infinite, it is frequently difficult or sometimes impossible to find the true values of some characteristics of the population, such as the population mean and the population standard deviation. Fortunately, the characteristics of the population can be examined by the use of a sample drawn from the population. The relationship between a sample measure and a corresponding population measure is determined according to the theory of probability.

For example, if we wish to know the average annual family income in the United States, it would be impractical, although not impossible, to take a complete survey of the income of every family in this country each year. It is much simpler if we take only a sample from the population and then estimate the average of the population based on the average of the sample. Before a valid conclusion can be drawn from the sample about the population, however, the relationship between the average of the sample data and the average of the population must be clearly understood. In this example, the true average of the population is not known. The relationship is established by the use of the theory of probability.

This chapter will first discuss some mathematical concepts and operations which are basic to probability calculations. The subject of sets and subsets is presented in Section 9.1. The counting procedures for finding the number of possible arrangements of the objects in a set or sets are explained in Section 9.2. The probability calculations are given in Sections 9.3 and 9.4.

9.1 Sets and Subsets

Basic Concept

The word "set" has been used commonly in everyday life. We frequently speak of a set of chairs, a set of dishes, a set of pens, and so on, because we recognize that the chairs or dishes or pens are things of the same kind and are used together. Likewise, in mathematics, the word "set" is used to denote any collection of objects, things, or values, which are distinctly defined under our discussion. In other words, the individual objects of the collection must have common characteristics that cause them to belong to a given set. For example, in the preceding chapters the idea of a set has been frequently employed in illustrations, such as a set of values representing miles traveled by a group of 20 students in coming to Swan College from their homes during a period.

The concept of sets is basic in contemporary mathematics. In this section, only the terminology and idea of sets and the fundamental operations of sets that can be used in aiding the explanation of the use of probability in statistics are discussed.

A member of a set is called an *element* of the set. If the number of elements in a set is a positive number, the set is *finite*. If the number of elements in a set is unlimited, the set is *infinite*. A set which has no elements is called an *empty* set or a *null* set. The empty set plays a role in the theory of sets similar to that of zero in the decimal number system.

Example 1
Each of the following collections is regarded as a finite set:

1. The members of a family (father, mother, sons, and daughters).
2. The students in a statistics class in a college.
3. The hourly wages of a group of employees in a factory.
4. The counties in a state.
5. The numbers of dots on the faces of a die.
6. The odd numbers above 5 but less than 1501.

Example 2
Each of the following collections is regarded as an infinite set:

1. The people in the world (in the past, present, and future).
2. The units produced in the food industry (in the past, present, and future).
3. The integers above 3.
4. The decimal numbers below 0.95.

Example 3

Each of the following collections is regarded as an empty set:

1. The people over 30 feet tall.
2. The weights of dogs above 20 tons.
3. The living members of a family over 2,000 years old.
4. The even numbers between 6 and 8.

There are three common ways of expressing a set. The selection of a particular form of expression depends on the convenience under certain circumstances.

(a) Use a capital letter to represent the set and list the names of all elements of the set in braces.

Example 4

The ages of the five persons in the Smith family are expressed as set S:

$$S = \{6, 9, 12, 35, 40\}$$

(b) Use a capital letter to represent the set and state the conditions of the elements of the set in braces.

Example 5

Express the set of all odd numbers above 5 but less than 1501. Let S be the set:

$$S = \{x : x \text{ is an odd number and } 5 < x < 1501\}$$

This expression may be read "S is the set of all elements x such that x is an odd number and x is larger than 5 but smaller than 1501."

(c) Use a diagram, frequently called a *Venn diagram* or *Euler diagram*.

Example 6

U = the set of all students in the United States
A = the set of all students in College A
B = the set of all students in College B

Solution. The diagram is as follows:

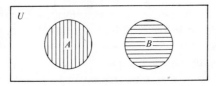

In the diagram, the total area is represented by the letter U and is called the *universal* or *population* set. Since every element of set A is also an element of set U, A is a *subset* of U. Likewise, B is a *subset* of the universal set U.

Fundamental Operations

The fundamental operations with sets are *union, intersection,* and *complement.* The three operations are illustrated below with A and B being two subsets of the universal set U.

Union

The union of sets A and B is denoted by the set $(A \cup B)$ (read "A union B," or "A cup B,") which contains all elements of U that belong *either* to A or to B or to both.

Observe in the three Venn diagrams shown below that every element of A belongs to $A \cup B$ and every element of B also belongs to $A \cup B$. (See shaded areas in the diagrams.)

Case a: $A \cup B$ Case b: $A \cup B = A$ Case c: $A \cup B$

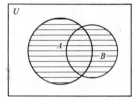 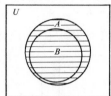 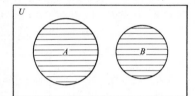

Example 7

Let
$$U = \{1, 2, 3, 4, 5\}$$
$$A = \{1, 2, 3\}$$
$$B = \{3, 5\}$$

Solution.

$$A \cup B = \{1, 2, 3, 5\}$$

(See Case a in the diagram above. Note that element 3 is not listed twice in set $A \cup B$.)

Example 8

Let
$$U = \{a, b, c, d, e\}$$
$$A = \{a, b, c\}$$
$$B = \{a, b\}$$

Solution.
$$A \cup B = \{a, b, c\} = A$$

(See Case b in the diagram above. Note that elements a and b are not listed twice in the set $A \cup B$).

Example 9

Let
$$U = \{x : x \text{ is an integer and } 3 < x < 15\}$$
$$A = \{4, 5, 6\}$$
$$B = \{8, 9\}$$

Solution.
$$A \cup B = \{4, 5, 6, 8, 9\}$$

(See Case c in the diagram above.)

Intersection

The intersection of sets A and B is denoted by the set $(A \cap B)$ (read "A intersection B," or "A cap B," which contains all elements of U that belong to *both* A and B.

Observe the shaded areas in the Venn diagrams below that every element which belongs to A *and* to B belongs to $A \cap B$.

Case a: $A \cap B$ Case b: $A \cap B = B$ Case c: $A \cap B = \emptyset$

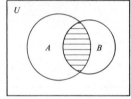 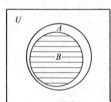 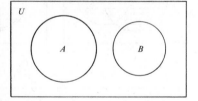

Example 10

Refer to Example 7. Find the set $A \cap B$.

Solution.
$$A \cap B = \{3\}$$

(See Case a in the diagram above. The number 3 is the only element of both sets A and B.

Example 11

Refer to Example 8. Find the set $A \cap B$.

Solution.
$$A \cap B = \{a, b\} = B$$

(See Case b in the diagram above.)

Example 12

Refer to Example 9. Find the set $A \cap B$.

Solution.

$$A \cap B = \varnothing \text{ (the empty set)}$$

(See Case c in the diagram above. Sets A and B in this case are said to be *disjoint* or *mutually exclusive*.)

Complement

The complement of set A is denoted by the set A' (read "A prime;" may also be written $\bar{A}$, $\sim A$, or $\tilde{A}$), which contains all elements in U that *do not* belong to A. Set A' is represented by the shaded area in the Venn diagram.

Complement of A ($= A'$)

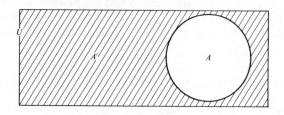

Example 13

Refer to Example 7. Find (a) set A', and (b) set B'.

Solution. (a) $A' = \{4, 5\}$

        (b) $B' = \{1, 2, 4\}$

9.2 Counting Possible Arrangements of Objects

The counting procedures for finding the numder of possible arrangements of the objects in a set or sets are essential in the study of probability. In counting the arrangements, it is helpful to list all the possible arrangements in the form of a tree, called a *tree diagram*.

Example 14

A lady has four coats (red, white, black, and yellow) and two hats (gray and pink). How many ways can she match one coat and one hat?

Solution.

A—If a coat is chosen first				*B—If a hat is chosen first*		

A—If a coat is chosen first

	Coat	Hat	Possible matches
	Red	Gray	R & G
		Pink	R & P
Lady (Origin)	White	Gray	W & G
		Pink	W & P
	Black	Gray	B & G
		Pink	B & P
	Yellow	Gray	Y & G
		Pink	Y & P

B—If a hat is chosen first

	Hat	Coat	Possible matches
	Gray	Red	G & R
		White	G & W
Lady (Origin)		Black	G & B
		Yellow	G & Y
	Pink	Red	P & R
		White	P & W
		Black	P & B
		Yellow	P & Y

There are eight ways of matching one coat and one hat at a time by actually counting the possible matches either in diagram A or in diagram B above.

In practice, it would be inconvenient to find the number of possible arrangements of the objects in every case by listing and counting, especially when the number of objects is large. The following three short-cut counting procedures are useful in various situations.

The Multiplication Principle

As shown in the diagrams of Example 14, any one of the four coats may be matched with any one of the two hats. Thus, the total number of possible ways of matching one coat and one hat may be computed as follows:

4 (ways to pick a coat) × 2 (ways to pick a hat) = 8 (total number of ways)

The multiplication principle is based on the same reasoning method as above and may be stated as follows:

> If one thing can be done in a ways, a second thing in b ways, a third thing in c ways, and so on for n things, then the n things can be done together in
>
> $$a \times b \times c \times \cdots (n \text{ factors}) \text{ ways}$$

Example 15

A salesman in City A wishes to drive his car to City D by passing through cities B and C. There are two highways from A to B, three from B to C, and two from C to D. How many different ways can the trip be made by the salesman?

Solution. According to the multiplication principle, the total number of ways is

$$a \times b \times c = 2 \times 3 \times 2 = 12 \text{ ways}$$

Permutations

The multiplication principle offers a general method for counting the number of possible arrangements of objects within a single set or among several sets. However, for a *single set* of objects, the formulas developed for permutations and combinations are more convenient in counting the number of possible arrangements. A *permutation* is an arrangement of all or part of the objects within a single set of objects in a *definite order*. The total number of permutations of a set of objects depends on the number of objects taken at a time for each permutation. The number of objects taken at a time for each permutation may be (a) all of the objects or (b) part of the objects.

Permutations of Different Objects Taken All at a Time

The total number of permutations of a set of objects taken all at a time may be obtained by reasoning similar to that used in Example 14 above. The procedure is illustrated in Example 16 below.

Example 16

Find the total number of permutations of the set of letters $\{a, b, c\}$ taken all at a time.

Solution. The possible orderly arrangements of three letters taken three at a time may be diagrammed as follows:

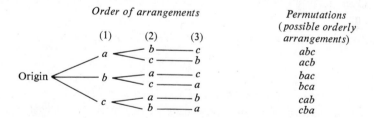

There are six permutations. Notice that the arrangement *abc* is different from *acb*, although each of the two arrangements consists of the same letters. The order of each letter is important in a permutation. The total number of permutations in Example 16 can be obtained by applying the multiplication principle as follows:

There are three ways to fill the first place in each permutation. Each of the three letters (a, b, c) can be used as the first place in a permutation. See column (1).

There are two ways to fill the second place in each permutation. For example, after the first place has been filled by the letter a, the second place can be filled by either b or c. See column (2).

There is only one way to fill the third place in each permutation. For example, after the first and second places have been filled by letters a and b respectively, the third place can be filled by only the letter c. See column (3).

Thus, the total number of permutations is

$$3 \times 2 \times 1 = 6$$

In general, let

$n =$ the number of objects in the given set

$n =$ the number of objects taken at a time for each permutation

${}_nP_n =$ the total number of permutation of n objects taken n at a time

Then, by the reasoning method similar to that used in Example 16

$${}_nP_n = n\,(n-1)\,(n-2)\,(n-3)\,\cdots\,(3)\,(2)\,(1) \qquad \text{(9–1a)}$$

For convenience, the right side of the above formula is written as $n!$ (Read n *factorial* or *factorial n.*) Thus the above formula may be written as

$${}_nP_n = n! \qquad \text{(9–1b)}$$

Note: If $n = 0$, then we define $0! = 1$.

By using formula (9–1b), Example 16 may be computed as follows: (here $n = 3$)

$${}_3P_3 = 3! = 3 \times 2 \times 1 = 6$$

Example 17

Find the total number of permutations of the set of digits $\{1, 3, 5, 7\}$ taken all at a time.

Solution.   Here $n = 4$ (number of objects in the given set)

$${}_4P_4 = 4! = 4 \times 3 \times 2 \times 1 = 24$$

Permutations of Different Objects Taken Part at a Time

The total number of permutations of a set of objects taken part at a time may also be obtained by either a tree diagram or a formula (9–2). The tree diagram is similar to that illustrated in Example 16 except that the number of columns in the present case is equal to the number of objects taken for each permutation. In general, let

$r =$ the number of objects taken at a time for each permutation

${}_nP_r =$ the total number of permutations of n objects taken r at a time.

Then,

$${}_nP_r = n\,(n-1)\,(n-2)\,(n-3)\,\cdots\,(n-r+1) \text{ for } r \text{ factors} \qquad \text{(9–2a)}$$

Note that the last factor $(n-r+1)$ is simplified from $[n-(r-1)]$. Also, when $r = n$, the last factor becomes $(n-n+1) = 1$. Thus, when $r = n$, formula (9–2a) is identical to formula (9–1). Formula (9–2a) may also be written:

$${}_nP_r = \frac{n!}{(n-r)!} \qquad \text{(9–2b)}$$

This formula is convenient for computation when tables of $n!$ and $(n - r)!$ are available.

Example 18

Find the total number of permutations of the set of letters $\{a, b, c, d\}$ taken (a) three at a time and (b) two at a time.

Solution. (a) Here

$n = 4$ (number of letters in the given set)
$r = 3$ (number of letters taken at a time for each permutation)

$${}_nP_r = {}_4P_3 = 4 \times 3 \times 2 = 24$$

or $\quad {}_nP_r = \dfrac{n!}{(n - r)!} = \dfrac{4 \times 3 \times 2 \times 1}{1} = 24$

(b) Here

$$n = 4$$
$$r = 2$$
$${}_nP_r = {}_4P_2 = 4 \times 3 = 12$$

or $\quad {}_nP_r = \dfrac{n!}{(n - r)!} = \dfrac{4 \times 3 \times 2 \times 1}{2 \times 1} = 12$

Example 19

Three officials—president, vice-president, and secretary—are to be elected from 20 members of a club. In how many ways can the three officials be elected?

Solution. This is a permutation problem. The order of arrangement is taken into consideration. For example, the set of president A, vice-president B, and secretary C is different from the set of president B, vice-president C, and secretary A.

Here

$$n = 20$$
$$r = 3$$
$${}_nP_r = {}_{20}P_3 = 20 \times 19 \times 18 = 6,840 \text{ ways}$$

Combinations

A combination is a subset or an arrangement of all or part of the objects of a single set *without regarding the order* of the objects. The total number of possible combinations of a set of objects taken all at a time is 1. For example, the possible arrangements from the set of letters $\{a, b\}$ are ab and ba. Since the order of arrangements is disregarded, arrangement ab is the same as ba. Thus, there is only one combination (a and b) possible for the set.

The total number of possible combinations of a set of different objects taken part at a time may be obtained by first finding the total number of permutations, and then counting the permutations with the same objects as one combination. This procedure is illustrated in Example 20 below.

Example 20

Find the total number of combinations of the set of letters $\{a, b, c\}$ taken two at a time.

Solution.   The total number of permutations of three letters taken two letters at a time is

$$_3P_2 = 3 \times 2 = 6$$

Each combination consists of two letters.

The total number of permutations of the same two letters taken all at a time is

$$_2P_2 = 2! = 2 \times 1 = 2$$

The two permutations consisting of the same letters are considered as only one combination. Thus, the total number of combinations is

$$\frac{_3P_2}{2!} = \frac{6}{2} = 3$$

The answer may be supported by the use of a tree diagram as follows:

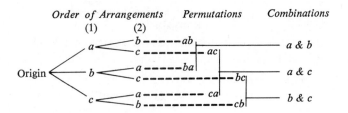

In general, let

    n = the number of objects in a given set
    r = the number of objects taken at a time for each combination
    $_nC_r$ = the total number of combinations of n objects taken r at a time

Then, by reasoning similar to that used in Example 20,

$$_nC_r = \frac{_nP_r}{r!} \tag{9–3a}$$

or

$$_nC_r = \frac{n!}{r!\,(n-r)!} \tag{9–3b}$$

Example 20 may be computed by using formula (9–3a) as follows:

$$_3C_2 = \frac{3 \times 2}{2 \times 1} = 3$$

or by using formula (9–3b) as follows:

$$_3C_2 = \frac{3!}{2!(3-2)!} = \frac{3 \times 2 \times 1}{(2 \times 1)(1)} = 3$$

Example 21

Three from 20 members are to be selected to form a committee. In how many ways can the committee be formed?

Solution. The committee formed by members A, B, C is the same as the committee formed by B, C, A, or other arrangements consisting of the three members. Thus, this is a type of combination problem, since we disregard the order of arrangement.

Here

$$n = 20$$
$$r = 3$$

$$_nC_r = {}_{20}C_3 = \frac{{}_{20}P_3}{3!} = \frac{20 \times 19 \times 18}{3 \times 2 \times 1} = 1,140$$

The following expression often simplifies computation of some combination problems:

$$_nC_r = {}_nC_{n-r}$$

Example 22

Compute (a) $_3C_2$ and (b) $_{100}C_{98}$

Solution. (a)

$$_3C_2 = {}_3C_{3-2} = {}_3C_1 = \frac{3}{1} = 3$$

(b)

$$_{100}C_{98} = {}_{100}C_{100-98} = {}_{100}C_2 = \frac{100 \times 99}{2 \times 1} = 4,950$$

9.3 Basic Concept of Probability

Probability and Trial

Probability deals with the problems of chance. The idea of probability first was applied in gambling games during the seventeenth century in France. Today the application of the theory of probability is one of the most important topics in mathematics and statistics.

An act which is performed to find out what is going to happen to an event involving chance is called a *trial*. The trial may be performed only once, called a single trial, or repeated a series of times under the same condition. The process of performing the trials is frequently referred to as a *random, stochastic,* or *chance process*. The probability of success or failure in a single trial is usually expressed in a ratio form:

$$P \text{ (successful outcome)} = \frac{\text{number of successful outcomes}}{\text{number of possible outcomes}}$$

$$P \begin{pmatrix} \text{failure or unsuc-} \\ \text{cessful outcome} \end{pmatrix} = \frac{\text{number of unsuccessful outcomes}}{\text{number of possible outcomes}}$$

where

$$\begin{pmatrix} \text{Number of} \\ \text{successful} \\ \text{outcomes} \end{pmatrix} + \begin{pmatrix} \text{Number of} \\ \text{unsuccessful} \\ \text{outcomes} \end{pmatrix} = \begin{pmatrix} \text{Number of} \\ \text{possible} \\ \text{outcomes} \end{pmatrix}$$

Sample Points and Sample Space

An outcome of a trial is also called a *sample point* or an *elementary event*. A *sample space* is formed by a set of sample points which represent all possible outcomes of the trial. In tossing a coin, for example, the possible outcomes are H (a head) and T (a tail). Thus there are two sample points. The two sample points form a sample space. Let S be the sample space, or the set of sample points; then

$$S = \{H, T\}$$

If two coins (a dime and a quarter) are tossed together, the possible outcomes can be obtained according to the following diagram:

		Quarter	
		H	T
Dime	H	HH	HT
	T	TH	TT

Possible outcomes (Sample points)

Thus, the sample space consists of four sample points, or

$$S = \{HH, HT, TH, TT\}$$

Theoretical Probability

In some cases, we can count exactly the different ways in which a given event may or may not happen and can assume that all the possible ways will occur on an *equally likely* basis. The probability obtained under such cases and the assumption is called *theoretical probability* or *mathematical probability*. Consider that an event (A) can happen in h ways out of a total of n possible "equally likely" ways. Then, the probability of the occurrence of the event A (success) is

$$P(A) = \frac{h}{n} \tag{9-4}$$

The probability of the nonoccurrence of the event A (failure) is

$$P(not\ A) = \frac{n - h}{n} = 1 - \frac{h}{n} = 1 - P(A)$$

or $$P(not\ A) = 1 - P(A) \tag{9-5}$$

The sum of the probability of success and the probability of failure is always equal to 1, or

$$P(A) + P(not\ A) = 1$$

Example 23

Find the probability that one throw of a die will (a) have "3" on top and (b) not have "3" on top.

Solution. A die (plural *dice*) is a cube-shaped gambling instrument with dots on each of the six sides: 1(·), 2(:), 3(·⋅), 4(: :), 5(:·:), 6(::). Assume that it is a well-balanced die; that is, each of the six sides will have an *equal chance* to turn on the top when the die stops in a throw. Then, the sample points are the six dots and the sample space is

$$S = \{1, 2, 3, 4, 5, 6\}$$

(a)

$$P(\text{having a "3"}) = \frac{1\,(\text{number of successful outcome})}{6\,(\text{number of possible outcomes})}$$

(b)

$$P(\text{not having a "3"}) = \frac{6 - 1}{6} = \frac{5\,(\text{number of unsuccessful outcomes})}{6\,(\text{number of possible outcomes})}$$

Empirical Probability

In some cases, although we may count exactly the different ways in which the given event may or may not happen, we can not assume, or we may doubt, that all the possible ways will happen or not happen on an "equally likely" basis. The computation of the probability of the event thus must be based on our experiences or experiments of what has happened on similar occasions in the past. The probability obtained from statistical data, recorded from our experiences or experiments, is called *empirical probability* or *statistical probability*. Since our experiences and the results of various experiments may be different, the value of empirical probability of an event may vary. This fact is illustrated in Example 24.

Example 24

Find the probability that a person who is now 20 years old will (a) live at age 21, (b) die before reaching age 21.

Solution. Although we can count exactly the two different ways in which the person will be at age 21, dead or alive, we cannot assume that the two ways will happen "equally likely." Thus, the statistical data concerning the similar occasions in the past must be used as a basis to predict the probability of the event that the person who is 20 will be living at 21.

1. According to the Commissioners 1941 Standard Ordinary Mortality Table, based on the experience of life insurance companies for the years from 1930 to 1940, 949,171 persons out of 951,483 men now aged 20 will be alive at age 21. Thus,

(a) $$P(\text{the person will live at age 21}) = \frac{949,171}{951,483} = 0.99757$$

(b) $$P(\text{the person will die before 21}) = \frac{951,483 - 949,171}{951,483} = 0.00243$$

Note that (a) + (b) = 0.99757 + 0.00243 = 1, the total probability.

2. According to the Commissioners 1958 Standard Ordinary Mortality Table, based on the experience of life insurance companies for the years from 1950 to 1954, 9,647,694 out of 9,664,994 men now age 20 will be alive at age 21. Thus,

(a) $$P(\text{the person will live at age 21}) = \frac{9,647,694}{9,664,994} = 0.99821$$

(b) $$P(\text{the person will die before 21}) = \frac{9,664,994 - 9,647,694}{9,664,994} = 0.00179$$

Here (a) + (b) = 0.99821 + 0.00179 = 1

Example 24 indicates that the two answers for each question are different because the experiences (based on the different years) are different, and that the empirical probability is usually based on a large number of occurrences in an experiment.

Relative Frequency

In an experiment of repeated trials, the ratio of the number of successful outcomes (h) to the number of trials (n) is also referred to as the *relative frequency*, such as the ratio in 1 (a) of Example 24,

$$\frac{h}{n} = \frac{949,171}{951,483}$$

The idea of relative frequency is important in understanding the concept of theoretical probability. For example, if we toss a well-balanced coin, we may get either one of the two sides: a head or a tail. If we wish to have a head, the chance of success is one out of two, giving the theoretical probability of the event $1/2 = 0.5$. If we toss a coin 20 times, the *expected* number of heads according to the theoretical probability is 10 (or $20 \times \frac{1}{2} = 10$). However, if we actually toss a coin 20 times, we may have a result somewhat different from the expected number, such as only 9 heads in 20 trials. Does the relative frequency $9/20$ (or $= 0.45$) disprove the result based on theoretical probability? Or should we give up the idea of theoretical probability in actual work? Most of the events in the real economic and business worlds occur differently from theoretical expectation. However, the relative frequency based on a large number of trials and the expectation based on theoretical probability are normally very close to each other.

If we continue to toss the coin many times, we should get relative frequencies closer and closer to the theoretical probability $1/2$ (or $= 0.5$). For example, if we toss the coin for 2,000 times, we may have 950 heads (or 1,050 tails). The relative frequency becomes $950/2,000 = 0.475$. If we further toss the coin for a total of 20,000 times, we may have 9,600 heads (or 10,400 tails), the relative frequency being 0.480 (or $9,600/20,000$), and so on. Thus, the theoretical probability may be defined as the limiting value of the relative frequency of the event when the number of trials becomes infinitely large, or

$$P(A) = \lim_{n \to \infty} \frac{h}{n}$$

Read "the probability of the occurrence of the event A is the limit of the relative frequency h/n as the number of trials n approaches infinity. The relationship between the theoretical probability and the relative frequency of an event enables us to use the theoretical probability to predict the relative frequency without actually carrying out the experiment, or to use the relative frequency of an experiment or of sample data to estimate the characteristics of the population or universe.

9.4 Computing Theoretical Probability

In the following computation, we assume that each of the possible outcomes of a trial will occur equally likely. In other words, we are computing the theoretical or mathematical probability. The idea of sets can be used conveniently in solving such probability problems. In the language of sets:

> All possible outcomes of the trial or experiment are elements of the universal set. For instance, the universal set (or the sample space) in Example 23 has six elements $\{1, 2, 3, 4, 5, 6\}$.
>
> All outcomes of a given event are the elements of the event set which is a subset of the universal set. The event set in Example 23(a) has only one element 3. The event set is a subset of the universal set since 3 is also an element of the universal set.

We denote a universal set by the letter U and subsets (events) by letters A, B, C, $\cdots$ and so on. Then,

$$P(A) = \frac{h}{n} = \frac{\text{number of elements in set } A}{\text{number of elements in set } U}$$

Probability of One Event

The computation of the probability of a single event has already been illustrated in Example 23. Example 25 illustrates the method of computing the probability of one event by using the idea of sets.

Example 25

One ball is drawn from a bag containing four black balls and three red balls. What is the probability (a) that it is a red ball, and (b) that it is a white ball?

Solution.   Here, the universe has seven elements:

$U = \{B, B, B, B, R, R, R\}$ (the possible outcomes in one drawing of one ball)

Thus

$$n = 7$$

(a) Let A = the event of drawing a red ball. Then the subset has three elements:

$$A = \{R, R, R\}$$

Thus

$$h = 3$$

The probability of event A is

$$P(A) = \frac{h}{n} = \frac{3}{7}$$

(b) Let A = the event of drawing a white ball. Since there is no white ball in the bag, the event set is an empty set, or contains no elements. Thus,

$$h = 0 \text{ and } P(A) = \frac{h}{n} = \frac{0}{7} = 0$$

Probability of Two or More Events

Two or more events may be (a) mutually exclusive, (b) partially overlapping, (c) independent, and (d) dependent. The probability of each of the different types of events is presented below.

Mutually Exclusive Events

Mutually exclusive events are also called *disjoint events*. Two or more events are considered mutually exclusive if the events cannot occur together; that is, the occurrence of any one of them excludes the occurrences of the others. Let A and B be mutually exclusive or disjoint events as shown in the Venn diagram below:

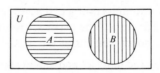

Then, the probability of event A or B is written

$$P(A \text{ or } B) = P(A \cup B) = P(A) + P(B) \tag{9-6}$$

Notice that events A and B are subsets of universal set U. The two subsets have no elements in common; that is, they are *disjoint subsets*. Thus, the probability that both events A and B will occur is zero since the intersection of the two event subsets is the empty set, or

$$P(A \text{ and } B) = P(A \cap B) = 0.$$

Example 26

Find the probability that one throw of a die will have 4 or 5.

Solution. Let $A =$ the event having 4

$B =$ the event having 5

Since one throw of a die can not have both sides on the top, events A and B are mutually exclusive.

or

$$U = \{1, 2, 3, 4, 5, 6\}$$
$$n = 6 \text{ (elements or possible outcomes)}$$
$$A = \{4\}, \text{ or } h = 1 \text{ (element)}$$
$$P(A) = \frac{h}{n} = \frac{1}{6}$$
$$B = \{5\}, \text{ or } h = 1 \text{ (element)}$$
$$P(B) = \frac{h}{n} = \frac{1}{6}$$

Thus

$$P(A \cup B) = P(A) + P(B) = \frac{1}{6} + \frac{1}{6} = \frac{2}{6} = \frac{1}{3}$$

Partially Overlapping Events

Partially overlapping events are called *intersected events*. Two or more events are considered partially overlapping if part of one event and part of another event occur together. Let A and B be partially overlapping or intersected events; that is, part of event A is also a part of event B as shown in the Venn diagram below:

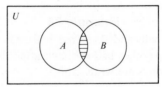

Then

$$P(A \text{ or } B) = P(A \cup B) = P(A) + P(B) - P(A \cap B) \qquad (9\text{–}7)$$

Notice that the elements in intersection of subsets A and B appeared twice; that is, the shaded area belongs to $P(A)$ and also $P(B)$. Thus, $P(A \cap B)$, the shaded area, must be subtracted from the sum of $P(A)$ and $P(B)$ to avoid the duplication.

Example 27

In drawing one card from a deck of 52 cards, find the probability that a single draw will be either a face card or a heart card.

Solution. There are 13 cards in each of the four suits: heart, spade, diamond, and club. Thus, the U set contains 52 elements (cards) or $n = 52$.

Let A = the event of drawing a face card. There are three face cards (jack, queen, king) in each of the four suits.

Then $A = \{J_h, Q_h, K_h, J_s, Q_s, K_s, J_d, Q_d, K_d, J_c, Q_c, K_c\}$
or $h = 12$ elements

Let B = the event of drawing a heart card. There are 13 cards with hearts

                      $B = \{J_h, Q_h, K_h, A_h, 2_h, 3_h, 4_h, 5_h, 6_h, 7_h, 8_h, 9_h, 10_h\}$
or $h = 13$

Notice that elements J_h, Q_h, K_h which are elements of set B are also elements of set A, or set

$$A \text{ and } B = A \cap B = \{J_h, Q_h, K_h\}$$

and $h = 3$

Thus,

$$P(A \cup B) = P(A) + P(B) - P(A \cap B) = \frac{12}{52} + \frac{13}{52} - \frac{3}{52} = \frac{22}{52} = \frac{11}{26}$$

Independent Events

Two or more events are considered to be independent if the events in no way affect each other. For example, A and B are independent events when the occurrence of event A has no effect on the occurrence of event B and vice versa. The probability that both independent events A and B will occur is

$$P(A \text{ and } B) = P(A \cap B) = P(A) \cdot P(B) \tag{9-8}$$

Example 28

A bag contains five white balls and three red balls. One ball is drawn from the bag and then is *replaced*. Another ball is drawn after the replacement. Find the probability that both drawings are white balls.

Solution.   The happening on the first draw certainly has nothing to do with the second draw, since the second ball is drawn after the first ball is replaced. Thus, the two drawings are independent events.

Let A = the event of the first draw that will have a white ball
      B = the event of the second draw that will have a white ball also

Since there are five white balls in the bag of eight balls in each draw,

$$P(A \cap B) = P(A) \cdot P(B) = \frac{5}{8} \times \frac{5}{8} = \frac{25}{64}$$

Dependent Events

If events A and B are so related that the occurrence of B depends on the occurrence of A, then A and B are called dependent events. The probability of event B depending on the occurrence of event A is called *conditional probability* and is written:

$P(B|A)$ or $P_A(B)$, which may be read, "the probability of B, given that event A has occurred," or simply "the probability of B, given A." The probability that both the dependent events A and B will occur is:

$$P(A \text{ and } B) = P(A \cap B) = P(A) \cdot P(B|A) \qquad (9\text{--}9)$$

Example 29

Refer to Example 28. Assume that the first ball is not returned to the bag when the second ball is drawn. Find the probability that both balls of the two drawings are white.

Solution.

Let $A =$ the event of the first draw that will have a white ball
          $B =$ the event of the second draw that will have a white ball also

Then,

$$P(A) = \frac{5}{8} \text{ (since there are five white balls in the bag of eight balls)}$$

Event B depends on the occurrence of event A. If the first ball is white, the probability of the second draw that will have a white ball is

$$P(B|A) = \frac{4}{7} \text{ (since there are only four white balls left in the bag}$$

of seven balls after the first white ball is drawn)

The probability that both balls of the two drawings are white is

$$P(A \cap B) = P(A) \cdot P(B|A) = \frac{5}{8} \times \frac{4}{7} = \frac{5}{14}$$

This answer agrees with the result obtained by the multiplication principle. There are eight ways to draw one ball in the first draw and only seven ways in the second draw. The draws can be made in $(8)(7) = 56$ possible ways. There are only $(5)(4) = 20$ ways that the two draws can have both white balls. Thus, the probability that both draws are white is $20/56 = 5/14$.

From Formula (9–9), the conditional probability of B, given that event A has occurred, is

$$P(B|A) = \frac{P(A \text{ and } B)}{P(A)}$$

In a like manner, the conditional probability of A, given that event B has occurred, is

$$P(A|B) = \frac{P(B \text{ and } A)}{P(B)}$$

The conditional probability formulas are required in developing *Bayes' theorem* in Chapter 20.

9.5 Summary

This chapter provides the basic tools for studying statistical induction, or statistical inference. In most cases, we do not know specific characteristics, such as the mean and the standard deviation, of the population from which a sample is drawn. Thus, we must estimate the characteristics of the population from a study of specific sample data based on the theory of probability.

The concept of sets and the counting procedures for finding the number of possible arrangements of the objects in a set or sets are basic to probability calculations. The word "set" is used to denote any collection of objects, things, or values that are of the same kind and are used together. A member of a set is called an element of the set. If the number of elements in a set is a positive number, the set is finite. If the number of elements in a set is an unlimited number, the set is infinite. A set which has no elements is called an empty set or a null set. There are also universal or population set and subsets of the universal set.

In counting the number of possible arrangements of the objects in a set or sets, it is helpful to list all the arrangements in the form of a tree, called tree diagram. In practice, however, it would be inconvenient to find the number of possible

SUMMARY OF FORMULAS

Application	Formula	Formula Number	Reference Page
Permutations			
n objects taken n at a time	$_nP_n = n(n-1)(n-2)(n-3)\cdots$ $(3)(2)(1)$	9–1a	220
	$_nP_n = n!$	9–1b	220
n objects taken r at a time	$_nP_r = n(n-1)(n-2)(n-3)\cdots$ $(n-r+1)$ for r factors	9–2a	220
	$_nP_r = \dfrac{n!}{(n-r)!}$	9–2b	220
Combinations			
n objects taken r at a time	$_nC_r = \dfrac{_nP_r}{r!}$	9–3a	222
	$_nC_r = \dfrac{n!}{r!(n-r)!}$	9–3b	222
Probability			
One event			
occurrence of event A	$P(A) = \dfrac{h}{n}$	9–4	225
nonoccurrence of event A	$P(not\ A) = \dfrac{n-h}{n} = 1 - \dfrac{h}{n}$ $= 1 - P(A)$	9–5	225
Two (or more) events A, B			
mutually exclusive events (disjoint)	$P(A \cup B) = P(A) + P(B)$	9–6	228
partially overlapping events (intersected)	$P(A \cup B) = P(A) + P(B)$ $- P(A \cap B)$	9–7	229
independent events	$P(A \cap B) = P(A) \cdot P(B)$	9–8	230
dependent events	$P(A \cap B) = P(A) \cdot P(B|A)$	9–9	231

arrangements by listing and counting, especially when the number of objects is large. The following three short-cut counting procedures are useful in various situations: (1) the multiplication principle, (2) permutations of different objects, including those taken (a) all at a time and (b) part at a time, and (3) combinations of different objects, including those taken (a) all at a time and (b) part at a time.

The probability of success or failure is usually expressed in a ratio form. The ratio is based on the number of successful or nonsuccessful outcomes and the number of possible outcomes. An outcome of a trial is also called a sample point or an elementary event. A sample space is formed by a set of sample points which represent all possible outcomes of the trial.

There are theoretical probability and empirical probability. The theoretical probability or mathematical probability is obtained under the assumption that all the possible ways will occur on an "equally likely" basis. The empirical probability or statistical probability is obtained from statistical data, which are recorded from our experiences or experiments. The ratio of the number of occurrences of an event to the number of possible occurrences in an experiment is also referred to as the relative frequency. When the theoretical probability can be applied in such an experiment, and when the number of trials becomes infinitely large, the theoretical probability may be defined as the limiting value of the relative frequency of the event.

Exercises 9–1

Reference: Sections 9.1 and 9.2

1. Indicate whether each of the following collections is (a) a finite set? (b) an infinite set? or (c) an empty set?
  1. The books in the college library.
  2. The integers above 5 but below 180.
  3. The numbers between 6 and 10.
  4. The people over 20 feet tall in a city.
  5. The families in the United States in 1975.

2. Let all (5) students in an English class be the universal set, or

$$U = \{\text{Abel, Campball, Griffin, Smith, Watson}\}$$

Also, let all (3) students who come from Virginia be set A, or

$$A = \{\text{Campball, Griffin, Smith}\}$$

and all (2) students who are freshmen be set B, or

$$B = \{\text{Abel, Campball}\}$$

Find (a) $A \cup B$, (b) $A \cap B$, and (c) A'.

3. Let $U = \{f, g, h, i, j, k, l\}$
        $A = \{f, h, i, k, l\}$
        $B = \{f, h, j\}$

Find sets: (a) $A \cup B$, (b) $A \cap B$, (c) A', and (d) B'.

4. Two letters followed by a five-digit number are used to make up license plates. How many license numbers can be made? (Numbers below 10,000, such as 00001, 00002,. . . 09999, are not considered five-digit numbers.)

5. A restaurant offers onion, catsup, mustard, and pickle as flavors added to a plain hamburger. How many different kinds of hamburgers can be made if the flavors can be taken 0 (no flavor), 1, 2, 3, or 4 at a time?

6. Find the total number of permutations of the set of letters $\{a, b, c, d, e, f\}$ taken (a) all at a time, and (b) three at a time.

7. Two officials, president and vice-president, are to be elected from four candidates: $A, B, C,$ and D. In how many ways can the two officials be elected?

8. Find the total number of combinations of the set of letters $\{a, b, c, d, e, f\}$ taken (a) all at a time, and (b) three at a time.

9. Five from nine basketball players are to be selected to form a team. In how many ways can the team be made?

10. There are 1,000 people in city X. If we wish to take the ages of three people in the city as a sample, in how many ways can the sample be taken?

11. How many four-letter words can possibly be formed from the letters of the word *universal*?

12. In how many ways can we select five cards from a deck of 52 cards?

Exercises 9–2

Reference: Sections 9.3 and 9.4

1. One student is chosen by lot from a class with 15 boys and 20 girls. What is the probability that a girl is chosen?

2. According to the Commissioners 1958 Standard Ordinary Mortality Table, 63,037 out of 97,165 men now aged 95 will be alive at age 96. Find the probability that a person who is now 95 will (a) be living at age 96 and (b) die before reaching age 96. (Compute the probability to two decimal places.)

3. One chip is drawn from a bag containing five black chips and four pink chips. What is the probability that (a) it is a black chip, (b) it is a pink chip, and (c) it is a quarter?

4. One card is drawn from a deck of 52 cards. What is the probability of drawing a spade? An ace? A spade or an ace?

5. A grocery store offers 500 first prizes and 500 second prizes but gives away 10,000 tickets. (a) What is the probability that a person who receives one ticket will win either a first prize or a second prize? (b) Not win a prize?

6. A bag contains two red and three white balls. One ball is drawn from the bag and is replaced for the next draw. Find the probability that two drawings are red balls.

7. Refer to Problem 6. If the first ball is not returned to the bag for the second draw, what is the probability that the two drawings are red balls?

8. Find the probability that one throw of two dice (one red and one white) will have (a) the sum 7, (b) the sum 4 or the sum 9, (c) 2 on either red or white die, (d) 2 or less on the red and 3 or more on the white die, and (e) 2 or less on the red and the sum 4 or less. Show the formula in each case.

Hint: Write down all sample points or all possible outcomes of the two dice in a single throw as suggested below before computing the probabilities. For example, the entry (2, 1) in the second row of the following arrangement represents the outcome: red die shows 2 and white die shows 1.

Red Die \ White Die	1	2	3	4	5	6
1	(1, 1)	(1, 2)	(1, 3)	...	...	...
2	(2, 1)	...	...			
3	(3, 1)					
4	(4, 1)					
5	(5, 1)					
6	(6, 1)					

10 *Discrete Probability Distributions

There are many types of probability distributions. In general, a probability distribution is a set of values distributed according to the theory of probability. The values in the set are the results of all possible outcomes of repeated trials in an experiment and are also termed a *random variable*. When the values are discrete data, the distribution is a *discrete probability distribution*. When the values are continuous data, it is a *continuous probability distribution*.

The idea of repeated trials is fundamental to the derivation of a probability distribution (Section 10.1). The method of computing the repeated trials is the extension of the method used for a single trial presented in the previous chapter. Discrete probability distributions introduced in this chapter are *binomial* (10.2), *hypergeometric* (10.3), *multinomial* (10.5), and *Poisson* (10.6). An application of the formulas derived from the binomial and the hypergeometric distributions is also presented (10.4). Tchebycheff's inequality theorem can be used to show the relationship among the mean, the standard deviation, and the amount of probability of a distribution of any type (10.7).

One of the most important types of continuous probability distributions is the *normal distribution*. Details concerning the normal distribution are presented in the next chapter. Other important types of continuous probability distributions—Student's (t), chi-square (χ^2), and variance ratio (F) distributions—are discussed in Chapters 15, 17, and 18, respectively.

10.1 Repeated Trials— Bernoulli Process

Instead of a single trial, an experiment may include repeated trials. Each trial in the experiment is performed independently; that is, the outcome of a trial has no influence on the next trial. When each trial has only *two possible outcomes*, such as success or failure, yes or no, accepted or rejected, and the *probability of either outcome remains the same* throughout the experiment, the repeated trials are called the *Bernoulli process*, in honor of the Swiss mathematician Jacob Bernoulli (1654–1705).

Let the probabilities of the two possible outcomes of each trial be:

$P = P(A)$, the probability that event A will happen in a single trial and
$Q = P$ (not A), the probability that event A will not happen or
$Q = 1 - P$

Then, the probability that event A will happen exactly X times in n repeated trials with the probability of success in a single trial being P, denoted by the *combined symbol $P(X; n, P)$*, is:

$$P(X; n, P) = {}_nC_X \cdot P^X \cdot Q^{n-X} \tag{10-1}$$

The capital letter P outside the parentheses in the symbol $P(X; n, P)$ represents *the probability of*, whereas the capital letter P inside the parentheses represents *the probability that event A will happen in a single trial*. Thus, in 3 repeated trials ($n = 3$), if we wish that event A will happen exactly 2 times ($X = 2$), with the probability of success in a single trial being 1/5 (or $P = 1/5$), the combined symbol becomes $P(2; 3, 1/5)$. When formula (10–1) is applied, the computation is

$$P(2; 3, 1/5) = {}_3C_2 \cdot \left(\frac{1}{5}\right)^2 \cdot \left(\frac{4}{5}\right)^{3-2} = \frac{3 \times 2}{2 \times 1} \cdot \left(\frac{1}{5}\right)^2 \cdot \left(\frac{4}{5}\right) = 0.096$$

Example 1 illustrates the derivation of formula (10–1) from a complete discrete probability distribution.

Example 1

A bag contains five balls: one white and four black. One ball is drawn and is replaced after each drawing. In three repeated drawings, determine the probability that the drawings will have (a) exactly three white balls, (b) exactly two white balls, (c) exactly one white ball, and (d) no white ball.

Solution.

Let $A =$ the event of drawing the white ball in each drawing
          $B =$ the event of not drawing the white ball in each drawing

Then,

$P(A) = P = 1/5$, and
$P(B) = P(\text{not } A) = Q = 1 - 1/5 = 4/5$

Since the result of each draw has nothing to do with each successive drawing, the three drawings are independent events.

When the formula (9–8) for independent events is used, the probability that the first draw is white, the second is white, and the third is white is

$$P(A \cap A \cap A) = P(A) \cdot P(A) \cdot P(A) = P^3 = (1/5)^3 = 0.008$$

The probability that the first draw is white, the second is black, and the third is white is

$$P(A \cap B \cap A) = P(A) \cdot P(B) \cdot P(A) = P \cdot Q \cdot P = P^2 Q$$
$$= (1/5)^2(4/5) = 0.032$$

and so on. The possible outcomes and their probabilities are shown in the following tree diagram.

TABLE 10–1

COMPUTATION FOR EXAMPLE 1

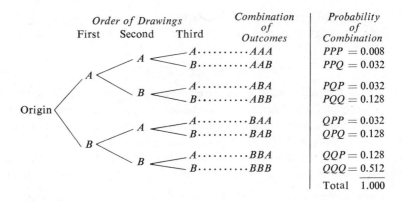

Order of Drawings			Combination of Outcomes	Probability of Combination
First	Second	Third		
			$A \cdots AAA$	$PPP = 0.008$
			$B \cdots AAB$	$PPQ = 0.032$
			$A \cdots ABA$	$PQP = 0.032$
			$B \cdots ABB$	$PQQ = 0.128$
			$A \cdots BAA$	$QPP = 0.032$
			$B \cdots BAB$	$QPQ = 0.128$
			$A \cdots BBA$	$QQP = 0.128$
			$B \cdots BBB$	$QQQ = 0.512$
				Total 1.000

The answers to the questions of Example 1 may be obtained by adding the probabilities of the combinations with the same number of occurrences of event A as follows:

(a) The probability that event A will occur exactly three times (or three white balls) in the three repeated drawings is

$$P(A \cap A \cap A) = P^3 = (1/5)^3 = 1/125 = 0.008$$

(b) The probability that event A will occur exactly two times (or two white balls) in the three repeated drawings is

$$P(A \cap A \cap B) + P(A \cap B \cap A) + P(B \cap A \cap A)$$
$$= P^2 Q + P^2 Q + P^2 Q = 3(P^2 Q) = 3(1/5)^2(4/5) = 3(0.032) = 0.096$$

The combinations (AAB, ABA, BAA) are mutually exclusive, since the combinations cannot occur together. For example, if combination AAB occurs in the three repeated drawings, the other combinations cannot occur. Thus, the probabilities of the combinations with the same number of occurrences of event A may be added (Formula 9–6).

(c) The probability that event A will occur exactly one time (or one white ball) in the three repeated drawings is

$$P(A \cap B \cap B) + P(B \cap A \cap B) + P(B \cap B \cap A)$$
$$= PQ^2 + PQ^2 + PQ^2 = 3(PQ^2) = 3(1/5)(4/5)^2 = 3(0.128) = 0.384$$

(d) The probability that event A will occur exactly zero time (or no white ball) in the three repeated drawings is

$$P(B \cap B \cap B) = Q^3 = (4/5)^3 = 0.512$$

The answers may be written in a probability distribution table as shown in Table 10–2. The X values, called X random variable, represent all possible successful outcomes (0, 1, 2, and 3 white balls) in three repeated trials. The sum of the probabilities of all possible X values is always equal to 1. This discrete probability distribution is also shown graphically in Chart 10–1.

TABLE 10–2

ANSWERS TO EXAMPLE 1

Number of White Balls (or successes) X	Probability* (or theoretical frequency)
0	0.512
1	0.384
2	0.096
3	0.008
Total	1.000

*The probabilities for these and other figures may also be obtained from Table 3 in the Appendix.

CHART 10–1

A DISCRETE PROBABILITY DISTRIBUTION

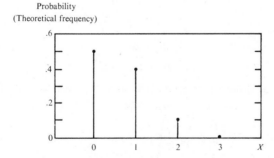

Source: Table 10–2.

Observe Example 1. Notice that the number of combinations with the same number of A's corresponds to the number computed by the combination formula

$$_nC_r = \frac{_nP_r}{r!}$$

where $n = 3$ (drawings or repeated trials) and
           $r = $ number of occurrences of event A in 3 drawings (or X)

Thus, for example, when $r = 2$, or exactly two white balls in the three repeated trials, the number of combinations is

$$_3C_2 = \frac{3 \times 2}{2 \times 1} = 3$$

Check: (AAB), (ABA), (BAA)

The probability for Example 1(b) that event A will occur exactly two times in the three repeated trials thus can be written as

$$P(2; 3, 1/5) = {}_3C_2 \cdot P^2 Q = 3(1/5)^2(4/5) = 3(0.032) = 0.096$$

Extending this idea, the probability that event A will happen exactly X times in n repeated trials with the probability of a single trial being P is

$$P(X; n, P) = {}_nC_X \cdot P^X \cdot Q^{n-X}$$

$P(X; n, P)$ is also called the *binomial probability density function* since it represents the general form of each term in a binomial expansion for computing a probability distribution. This fact is illustrated in the next section. Binomial probability distributions for selected values of n and P are listed in Table 3 of the Appendix.

10.2 Binomial Distribution (from Infinite Population)

The Binomial Theorem

By actual multiplication, the product of $(a + b)(a + b)$ may be written

$$
\begin{array}{r}
a + b \\
(\times)\, a + b \\
\hline
ab + b^2 \\
a^2 + ab \\
\hline
a^2 + 2ab + b^2
\end{array}
$$

Likewise if the binomial $(a + b)$, where a and b are real numbers, is raised to successive positive integer powers, the expanded expressions are as follows:

$$(a + b)^1 = a + b$$
$$(a + b)^2 = a^2 + 2ab + b^2$$
$$(a + b)^3 = a^3 + 3a^2b + 3ab^2 + b^3$$
$$(a + b)^4 = a^4 + 4a^3b + 6a^2b^2 + 4ab^3 + b^4$$
. . . .

In general, when the binomial is raised to nth power, n being a positive integer, the expansion known as the *binomial theorem* can be written:

$$(a + b)^n = {}_nC_0 a^n + {}_nC_1 a^{n-1}b + {}_nC_2 a^{n-2}b^2 + \cdots + {}_nC_{n-1} ab^{n-1} + {}_nC_n b^n$$

Note the generalized binomial expansion:

1. The number of terms in the expansion is $(n + 1)$.

2. The letter a in the first term of the expansion has the exponent n, in the second term has the exponent $(n - 1)$, in the third term has the exponent $(n - 2)$, and so on.

3. The letter b in the second term of the expansion has the exponent 1, in the third term has the exponent 2, in the fourth term has the exponent 3, and so on.

4. The sum of the exponents of a and b in any term is n.

5. The coefficients of the terms which are equidistant from the ends are the same. For example, the coefficient of the first term is $1(= {}_nC_0)$ and the coefficient of the last term is also $1(= {}_nC_n)$.

6. The value of ${}_nC_r$ is the coefficient of $(r+1)$th term of the expansion.

Since a and b are any real numbers, we let $b = P$ (the probability of an event in a single trial), and $a = 1 - P = Q$. Then, the binomial theorem becomes

$$(Q + P)^n = {}_nC_0 Q^n + {}_nC_1 PQ^{n-1} + {}_nC_2 P^2 Q^{n-2}$$
$$+ \cdots + {}_nC_X P^X Q^{n-X} \cdots + {}_nC_n P^n \qquad (10\text{--}2)$$

Each of the terms at the right side of the above equation is identical to that of the probability formula $P(X; n, P)$; that is, $X = 0$ in the first term, $X = 1$ in the second term, and so on. Here, X again represents the exact number of times that an event, having probability P, will happen in n trials. Also, since $1 - P = Q$, or $Q + P = 1$, the left side becomes $1^n = 1$. Thus, formula (10–2) becomes

$$1 = P(0; n, P) + P(1; n, P) + P(2; n, P)$$
$$+ \cdots + P(X; n, P) + \cdots P(n; n, P)$$

which indicates that the sum of the probabilities of all the possible combinations $(X = 0, 1, 2, 3, \cdots, n)$ must be equal to 1. The values of X's are integers and are of a discrete nature. The probabilities obtained by the binomial theorem thus form a *discrete probability distribution*.

When Example 1 is computed by the binomial formula (10–2), the probabilities of all possible numbers of successes (X's) in the three repeated drawings may be computed systematically as follows:

$$P = 1/5, Q = 4/5, X = 0, 1, 2, 3, \text{ and } n = 3$$
$$(Q + P)^3 = {}_3C_0 Q^3 + {}_3C_1 PQ^2 + {}_3C_2 P^2 Q + {}_3C_3 P^3$$
$$= 1(4/5)^3 + 3(1/5)(4/5)^2 + 3(1/5)^2(4/5) + 1(1/5)^3$$
$$= (64/125) + (48/125) + (12/125) + (1/125)$$
$$= 125/125 = 1$$

or

$$= 0.512 + 3(0.128) + 3(0.032) + 0.008$$
$$= 0.512 + 0.384 + 0.096 + 0.008 = 1$$

Thus, the probabilities that event A will occur exactly 0, 1, 2, and 3 times (white balls) in the 3 repeated drawings are 0.512, 0.384, 0.096, and 0.008, respectively. The answers obtained above are similar to those obtained in Example 1 in the preceding section.

Chief Characteristics of Binomial Distributions

(1) A binomial distribution is obtained from an infinite population, such as the population in Example 1. The five balls given in the example form an infinite population since we replace each ball after each drawing. The number of drawings (n) thus may also be infinitely large.

(2)   When P (the probability of an event that will happen in a single trial) is smaller than Q ($= 1 - P$), such as $P = 1/5 = .20$ and $Q = 4/5 = .80$, the binomial distribution is skewed to the right. (See Chart 10–2 of Example 2.) When P is larger than Q, the direction of the skewness is reversed; that is, the distribution is skewed to the left.

(3)   A binomial distribution is more moderately skewed if the value of n (number of trials) increases; that is, if $P = .20$, a distribution having $n = 20$ is more moderately skewed than one having $n = 8$. (See Chart 10–2.) When n is large, the distribution tends to approach the normal curve (Chart 10–4).

(4)   When P deviates away from 0.5 (either above or below 0.5, or $P \neq Q$), the distribution curve tends to be more and more skewed. When P is equal to 0.5, the curve is symmetrical. (See Chart 10–3 of Example 3.)

(5)   A binomial distribution is the limit of a frequency distribution obtained by observing the actual outcomes of an experiment with a large number (n) of trials. On the other hand, the normal curve is the limit of binomial distributions when n is infinitely large. (See Chart 10–4 of Example 4.)

Examples 2, 3, and 4 illustrate the chief characteristics of binomial distributions.

Example 2

Refer to Example 1 concerning the five balls (one white and four black.) Find the probabilities of having the various possible numbers of white balls in (a) eight repeated drawings, and (b) 20 repeated drawings. Again, assume that one ball is drawn and is replaced after each drawing.

Solution.

(a) $X = 0, 1, 2, 3, \cdots, 8$ (exact number of successes or white balls)
  $n = 8$ (repeated trials)
  $P = 1/5$ (one white out from 5 balls) $= 0.2$
  $Q = 1 - 1/5 = 4/5 = 0.8$

The probabilities are obtained by using formula $P(X; n, P)$ and are shown in Table 10–3 and Chart 10–2.

Example of the computation of the probabilities:
When $X = 5$, or exactly five white balls in eight repeated drawings, then

$$P(5 \, ; 8, 0.2) = {}_8C_5(0.2)^5(0.8)^3$$
$$= \frac{8 \times 7 \times 6 \times 5 \times 4}{5 \times 4 \times 3 \times 2 \times 1}(0.00032)(0.512)$$
$$= 0.00917504, \text{ rounded to } 0.0092$$

(b) $X = 0, 1, 2, 3, \cdots, 20$ (exact number of white balls)
  $n = 20$ (repeated trials)
  $P = 1/5 = 0.2$
  $Q = 0.8$

The probabilities are shown in Table 10–4 and also on Chart 10–2. Example of the computation of the probabilities:

When $X = 4$, or exactly four white balls in twenty repeated drawings, then

$$P(4; 20, 0.2) = {}_{20}C_4(0.2)^4(0.8)^{16} = 0.2181994, \text{ rounded to } 0.2182$$

TABLE 10–3

ANSWERS TO EXAMPLE 2a

Number of white balls (or successes) X	Probability (or theoretical frequency) $P(X; 8, 0.2)$
0	0.1678
1	0.3355
2	0.2936
3	0.1468
4	0.0459
5	0.0092
6	0.0011
7	0.0001
8	0.0000
Total	1.0000

TABLE 10–4

ANSWERS TO EXAMPLE 2b

Number of White Balls (or successes) X	Probability (or theoretical frequency) $P(X; 20, 0.2)$	Number of White Balls (or successes) X	Probability (or theoretical frequency) $P(X; 20, 0.2)$
0	0.0115	11	0.0005
1	0.0576	12	0.0001
2	0.1369	13	0.0000
3	0.2054	14	0.0000
4	0.2182	15	0.0000
5	0.1746	16	0.0000
6	0.1091	17	0.0000
7	0.0545	18	0.0000
8	0.0222	19	0.0000
9	0.0074	20	0.0000
10	0.0020		

Chart 10–2 illustrates two facts: (1) The three curves are skewed to the right because the value of $P (= .20)$ is smaller than the value of $Q (= .80)$. (2) When the value of n gets larger and larger (increased from 3 to 8 to 20), the curve representing the probability distribution tends to be less and less skewed. If the value of n becomes infinitely large, the curve will approach a symmetrical or bell-shaped smooth curve, as shown on Chart 10–4.

CHART 10–2

BINOMIAL PROBABILITY DISTRIBUTIONS

FOR n (NUMBER OF TRIALS OR SAMPLE SIZE) $= 3, 8,$ AND 20, WHERE $P = 1/5 = .20$

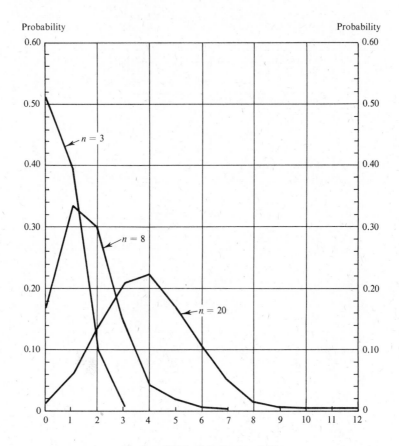

Number of White Balls (Successes), X

Source: Examples 1 and 2.

   In Example 2, the probability of success of the event, P, is constant, but the values of n increase. When n is large, the probability distribution tends to approach the normal curve. When n is constant, the changes of P also affect the shape of the binomial probability distributions. The four distribution curves shown in Chart 10–3 of Example 3 are used to illustrate the change.

Example 3

Let $n = 8$, and $P = 0.1, 0.2, 0.4,$ and 0.5. Construct the binomial probability distribution curves.

Solution.   The curves representing the given P values are shown in Chart 10–3. The method of computing the probabilities for each of the curves in the chart is similar to that of Examples 1 and 2. For example, when $n = 8$, $P = 0.5$, and $X = 4$, the probability is

$$P(4; 8, 0.5) = {}_8C_4(0.5)^4(0.5)^4 = \frac{70}{256} = 0.27343750$$

The result of the computation may be stated as follows:

> A bag contains ten balls: five white and five black. One ball is drawn and is replaced after each drawing. In eight repeated drawings, the probability of having exactly four white balls is 0.27343750.

Actually, it makes no difference if the bag contains 100 balls, 50 being white and 50 being black. As long as the probability of drawing a white ball in a single trial is not changed, such as $5/10 = 50/100 = 0.5$, the binomial probability distribution is not affected.

CHART 10–3

BINOMIAL PROBABILITY DISTRIBUTIONS

FOR n (NUMBER OF TRIALS OR SAMPLE SIZE) $= 8$, WHERE $P = 0.1, 0.2, 0.4,$ AND 0.5

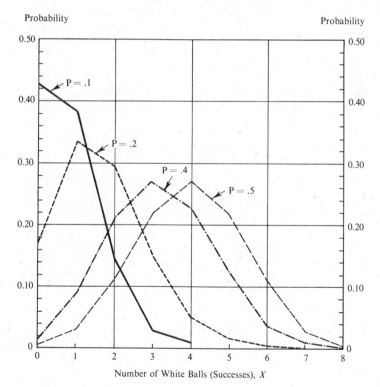

Number of White Balls (Successes), X

Source: Example 2 for $P = 0.2$, and Example 3 for $P = 0.5$.

Observe Chart 10–3. When *P* is equal to 0.5, the shape of the curve is symmetrical, or is close to the shape of the normal curve. When *P* deviates away from 0.5 (decreased from .4 to .2 to .1), the distribution curve tends to be more and more skewed.

Example 4 gives a different situation in applying the binomial formula when $n = 8$ and $P = 0.5$. The computed probability distribution is the same as that obtained above. Example 4 also shows the comparison of the probability distribution and the actual frequencies obtained by an experiment.

Example 4

Eight coins are actually tossed 256 times. The first toss shows four heads:

The second toss shows six heads:

(H) (H) (H) (T) (H) (H) (H) (T)

The results of the 256 tosses are given in the last column of Table 10–5. What are the corresponding theoretical occurrences of the various numbers of heads?

TABLE 10–5

COMPUTATION FOR EXAMPLE 4

Number of heads (X)	Probability of having X heads in a single toss of 8 coins P(X ; 8, 1/2)	Theoretical Frequency Based on Probability in 256 tosses (256)·P(X; 8, 1/2)	Observed Frequency (actual tosses of 256 times)
0	P(0 ; 8, 1/2) = 1/256	1	0
1	P(1 ; 8, 1/2) = 8/256	8	6
2	P(2 ; 8, 1/2) = 28/256	28	28
3	P(3 ; 8, 1/2) = 56/256	56	48
4	P(4 ; 8, 1/2) = 70/256	70	77
5	P(5 ; 8, 1/2) = 56/256	56	50
6	P(6 ; 8, 1/2) = 28/256	28	32
7	P(7 ; 8, 1/2) = 8/256	8	14
8	P(8 ; 8, 1/2) = 1/256	1	1
Total	256/256 = 1	256	256

Solution. The theoretical occurrences in 256 tosses are listed in the third column of Table 10–5. The theoretical frequency or probability of having X heads in a single toss of 8 coins is computed by formula (10–1). Here each trial consists of tossing 8 coins. Since the outcome of each coin is independent of others, each trial may be thought of as if it were a single coin tossed 8 times. For example, the probability that will have 3 heads in a single toss of 8 coins is the same as the probability that will have 3 heads in 8 tosses of a single coin. The probability that will have 3 heads in 8 tosses of a single coin (or a single toss of 8 coins) is computed below.

$n = 8$ (number of repeated trials)
$X = 3$ (number of occurrences of the event that will have a head
  in a toss)
$P = 1/2$ (the probability of the event in a single toss of a coin)
$Q = 1 - 1/2 = 1/2$

Substitute the values in formula (10–1) as follows:

$$P(3; 8, 1/2) = {}_8C_3 \cdot P^3 Q^{8-3} = \frac{(8)(7)(6)}{(3)(2)(1)}(1/2)^3(1/2)^5 = \frac{56}{256}$$

The theoretical frequency, or *expected frequency*, of having exactly 3 heads in 256 tosses is

$$(256) \cdot P(3; 8, 1/2) = 256(56/256) = 56$$

The expected frequency and the actual frequency for each possible occurrence (number of heads) are plotted on Chart 10–4.

Notice in Chart 10–4 that the theoretical (expected) frequency distribution curve and the observed (actual experiment or sample data) frequency distribution curve are fairly close to each other. Also, there are eight straight lines showing the theoretical frequencies in the form of a perfect symmetrical frequency polygon. The number of sides of the polygon corresponds to the number of coins tossed. If the number of coins increases, the number of sides of the polygon will correspondingly increase and the polygon will more and more approach a smooth curve. The smooth curve, which is the limit of the theoretical frequency distribution, is called the *normal distribution curve* or simply *normal curve*. The relationship between the observed frequency distribution and the normal distribution is fundamental to the study of inductive statistics.

In Example 4, the probability of success of the event is equal to the probability of failure of the event, or $P = 1/2 = Q$. If P and Q are not equal, the theoretical frequency distribution curve also approaches the normal curve when n is large. This fact has been illustrated in Example 2 and Chart 10–2.

The binomial probability distribution also has a mean and a standard deviation. The mean and the standard deviation of Example 1 may be computed by the methods presented in Chapters 5 and 7 as shown in Table 10–6 and the accompanying paragraph.

CHART 10–4

A COMPARISON OF NORMAL CURVE, THEORETICAL FREQUENCIES,
AND ACTUAL FREQUENCIES AS THE RESULTS OF 256 TOSSES OF 8 COINS

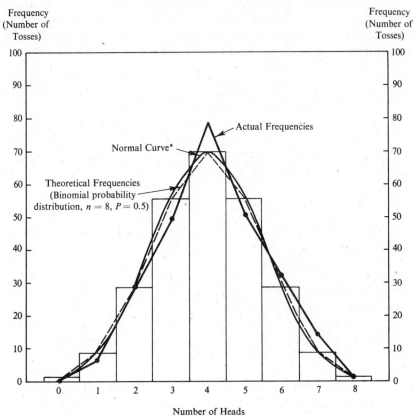

Source: Example 4.

*The method of constructing a normal curve is presented in Chapter 11.

However, a simple way to compute the mean and the standard deviation of a
binomial probability distribution for an infinite population is to use the formulas
below:

$$\mu \text{ (of numbers of successes)} = nP \qquad (10\text{–}3)$$

$$\sigma \binom{\text{of numbers of successes drawn}}{\text{from an infinite population}} = \sqrt{nPQ} \qquad (10\text{–}4)$$

The Greek symbols μ (mu) and σ (sigma) are customarily used to denote the
mean and the standard deviation of the population or a complete set of data,
whereas the symbols $\bar{X}$ and s are used to denote the two values of a sample. When
the two formulas are used, the same answers can be obtained:

TABLE 10–6

COMPUTATION OF MEAN AND STANDARD
DEVIATION FROM THE PROBABILITY DISTRIBUTION OF EXAMPLE 1

Number of Successes X	$P(X; n, P)$ or Theoretical Frequency f	fX	$X - \bar{X}$ ($\bar{X} = 0.6$)	$f(X - \bar{X})$	$f(X - \bar{X})^2$
0	0.512	0	-0.6	-0.3072	0.18432
1	0.384	0.384	0.4	0.1536	0.06144
2	0.096	0.192	1.4	0.1344	0.18816
3	0.008	0.024	2.4	0.0192	0.04608
Total	1.000	0.600		0.0000	0.48000

$$\mu \text{ (or } \bar{X}) = \frac{\Sigma fX}{\Sigma f} = \frac{0.600}{1} = 0.6$$

$$\sigma \text{ (or } s) = \sqrt{\frac{\Sigma f(X - \bar{X})^2}{\Sigma f}} = \sqrt{\frac{0.48}{1}} = \sqrt{0.48} = 0.69$$

$$\mu = nP = 3(1/5) = 0.6$$

$$\sigma = \sqrt{nPQ} = \sqrt{3(1/5)(4/5)} = \sqrt{12/25} = \sqrt{0.48} = 0.69$$

Formulas (10–3) and (10–4) are obtained under the assumption that the probabilities are computed from an infinite population. The five balls given in Example 1 form an infinite population since we replace each ball after each drawing. The number of drawings (n) can be any positive integer from zero to infinity.

Note that when n is large, the computation of a binomial probability distribution is tedious. Since the binomial distribution approaches the normal distribution under such conditions, we may conveniently use the latter to estimate the former for practical purposes. The method of computing the probability for the normal distribution curve is discussed in the next chapter.

10.3   Hypergeometric Distribution (from Finite Population)

The hypergeometric probability distribution is obtained from a finite population. The size of the population is reduced each time after a trial. Thus, the maximum number of repeated trials is limited to the size of the population. Also, the probability of a successful event changes if the population size changes, even though the number of trials remains unchanged. The computation of a hypergeometric distribution is illustrated in Example 5. (Example 5 is a revision of Example 1.) The revision is made so that the infinite population now is changed to the finite population.*

---

*Detailed values of the hypergeometric distribution are tabulated by Gerald J. Lieberman and Donald B. Owen. See their publication, *Tables of the Hypergeometric Probability Distribution*, (Stanford, Calif.: Stanford University Press, 1961).

Example 5

A bag contains five balls, one white and four black. One ball is drawn and is *not* replaced into the bag. (1) In three repeated drawings, what is the probability that the drawings will have exactly (a) three white balls, (b) two white balls, (c) one white ball, and (d) zero white ball? (2) Compute the mean and the standard deviation of the probability distribution.

Solution. (1) The total number of possible outcomes, or the number of ways of taking 3 out of 5 balls in the bag, is

$$_5C_3 = \frac{5 \times 4 \times 3}{3 \times 2 \times 1} = 10$$

(a) There is only one white ball. Thus, the total number of successful outcomes, or having 3 white balls, is zero.

$$P(\text{having 3 white balls}) = \frac{0}{10} = 0$$

(b) The total number of successful outcomes, or having 2 white balls, is also zero.

$$P(\text{having 2 white balls}) = \frac{0}{10} = 0$$

(c) The total number of successful outcomes, or having 1 white ball and 2 black balls, is 6.

$$_1C_1(\text{1 out of 1 white ball}) \times {_4C_2}(\text{2 out of 4 black balls}) = 1 \times \frac{4 \times 3}{2 \times 1} = 6$$

Check: (W, B_1, B_2); (W, B_1, B_3); (W, B_1, B_4); (W, B_2, B_3); (W, B_2, B_4); (W, B_3, B_4). Each of the 4 black balls is labeled for counting.

$$P(\text{having 1 white and 2 black balls}) = \frac{6}{10} = 0.6$$

(d) The total number of successful outcomes, or having 0 white ball and 3 black balls, is 4.

$$_1C_0(\text{0 out of 1 white ball}) \times {_4C_3}(\text{3 out of 4 black balls}) = 1 \times 4 = 4$$

Check: (B_1, B_2, B_3); (B_1, B_2, B_4); (B_1, B_3, B_4); (B_2, B_3, B_4)

$$P(\text{having 0 white and 3 black balls}) = \frac{4}{10} = 0.4$$

(2) The mean and the standard deviation may be computed from Table 10–7.

The computation of probability for each question in solution (1) of Example 5 may be written symbolically as follows:

TABLE 10–7

Computation for example 5 ($N = 5$)

Number of Successes (white balls) X	Probability (Theoretical frequencies) f	fX	$X - \bar{X}$ ($\bar{X} = 0.6$)	$f(X - \bar{X})$	$f(X - \bar{X})^2$
0	0.4	0.0	− 0.6	− 0.24	0.144
1	0.6	0.6	0.4	0.24	0.096
Total	1.0	0.6		0.00	0.240

$$\mu(\text{or } \bar{X}) = \frac{\Sigma fx}{\Sigma f} = \frac{0.6}{1} = 0.6$$

$$\sigma(\text{or } s) = \sqrt{\frac{\Sigma f(X - \bar{X})^2}{\Sigma f}} = \sqrt{\frac{0.24}{1}} = \sqrt{0.24} = 0.49$$

$$P(X) = \frac{{}_aC_X \cdot {}_bC_{n-X}}{{}_NC_n} \tag{10-5}$$

where a = number of successful items in population
               b = number of unsuccessful items in population
               $N = a + b$ = population size
               n = number of repeated trials and
               X = number of possible successes in n trials

The computation of the mean and standard deviation in solution (2) of Example 5 can be simplified when formulas are used. The formula for the mean of the binomial distribution (10–3) may also be used for a probability distribution obtained from a finite population. However, the standard deviation formula (10–4) should be corrected before it is used for a distribution from a finite population.

$$\sigma\begin{pmatrix}\text{of numbers of successes drawn} \\ \text{from a finite population}\end{pmatrix} = \sqrt{nPQ} \cdot \sqrt{\frac{N - n}{N - 1}} \tag{10-6}$$

The factor $\sqrt{(N - n)/(N - 1)}$ in the above formula is called *finite correction factor.*

When N is small, the finite correction factor is important to the computation of the standard deviation of the probability distribution. When formulas (10–3) and (10–6) are used, the mean and the standard deviation of Example 5 are computed as follows:

         $P = 1/5 = 0.2$ (the probability of having a white ball, or success-
                        ful event, in a single drawing; it is also the proportion
                        of the number of white balls to the population)
         $n = 3$ (number of drawings, or sample size)

Use formula (10–3) to compute the mean.

$$\mu = nP = 3(0.2) = 0.6 \quad \text{(same as the mean of Example 1)}$$

Use formula (10–6) to compute the standard deviation.

$$N = 5 \text{ (size of the population)}, Q = 1 - P = 1 - 0.2 = 0.8$$

$$\sigma = \sqrt{nPQ}\sqrt{\frac{N-n}{N-1}} = \sqrt{3(0.2)(0.8)}\sqrt{\frac{5-3}{5-1}} = \sqrt{0.48}\cdot\sqrt{\frac{2}{4}}$$

$$= \sqrt{0.48 \times 0.5} = \sqrt{0.24} = 0.49$$

The population size $N (= 5)$ in Example 5 is very small. The probability distribution (the X's and their f's) of the finite population differs greatly from that of the infinite population of Example 1. When the population size increases and P is constant, however, the probability distribution of the finite population will approach that of the infinite population.

Thus, if there were 100 balls, 20 being white (or $1/5$ of the population) and 80 being black in Example 5, the probability distribution for 3 ($= n$) repeated trials is computed by using formula (10–5) as shown in Table 10–8.

The probability distribution (the X's and their f's) of the finite population ($N = 100$) in the table is very close to that of the infinite population (page 239.)

TABLE 10–8

PROBABILITY DISTRIBUTION FROM A FINITE POPULATION ($N = 100$)

(1) Number of Successes (white balls) X	(2) Total Number of Successful Outcomes X out of 20 white balls $_{20}C_X$	(3) $(3 - X)$ out of 80 black balls $_{80}C_{3-X}$	(4) Total	(5) Probability, f (4) ÷ Total number of possible outcomes
0	$_{20}C_0$	× $_{80}C_3$	= 82,160	0.5081
1	$_{20}C_1$	× $_{80}C_2$	= 63,200	0.3908
2	$_{20}C_2$	× $_{80}C_1$	= 15,200	0.0940
3	$_{20}C_3$	× $_{80}C_0$	= 1,140	0.0071
Total number of possible outcomes $_{100}C_3$			= 161,700	
Total probability				1.0000

TABLE 10–9

PROBABILITY DISTRIBUTION FROM A FINITE POPULATION ($N = 500$)

(1) Number of Successes (white balls) X	(2) Total Number of Successful Outcomes X out of 100 white balls $_{100}C_X$	(3) $(3 - X)$ out of 400 black balls $_{400}C_{3-X}$	(4) Total (2) × (3)	(5) Probability, f (4) ÷ Total number of possible outcomes
0	$_{100}C_0$	× $_{400}C_3$	= 10,586,800	0.5113
1	$_{100}C_1$	× $_{400}C_2$	= 7,980,000	0.3853
2	$_{100}C_2$	× $_{400}C_1$	= 1,980,000	0.0956
3	$_{100}C_3$	× $_{400}C_0$	= 161,700	0.0078
Total number of possible outcomes $_{500}C_3$			= 20,708,500	
Total probability				1.0000

Further, if there were 500 balls, 100 being white (or 1/5 of the population) and 400 being black in Example 5, the probability distribution for 3 ($= n$) repeated trials is practically identical to that of the infinite population. The distribution is computed in Table 10–9.

The probability distributions obtained from various population sizes in Examples 1 and 5 are compared in Table 10–10. Note that the values of n ($= 3$) and P ($=1/5$) are constant in each of the distributions.

TABLE 10–10

COMPARISON OF PROBABILITY
DISTRIBUTIONS FROM VARIOUS FINITE POPULATIONS

Number of Successes (white balls) X	*Probability* (Theoretical frequency)			
	From finite population			From infinite population, $N = \infty$
	$N = 5$	$N = 100$	$N = 500$	
0	0.4	0.5081	0.5113	0.512
1	0.6	0.3908	0.3853	0.384
2	0	0.0940	0.0956	0.096
3	0	0.0071	0.0078	0.008

When the population size N is large and the number of trials (n, which is frequently referred to as sample size in a sampling study) is relatively small with respect to the population size, the finite correction factor will be close to unity or 1, such as

$$N = 1000,\ n = 30, \text{ the finite correction factor being}$$

$$\sqrt{\frac{N-n}{N-1}} = \sqrt{\frac{1{,}000 - 30}{1{,}000 - 1}} = \sqrt{\frac{970}{999}} = \sqrt{0.970971} = 0.985$$

When N is very large, or approaches infinity, the factor approaches 1 even if n is large, such as

$$N = 1{,}000{,}000,\ n = 1000, \text{ the factor being}$$

$$\sqrt{\frac{1{,}000{,}000 - 1{,}000}{1{,}000{,}000 - 1}} = \sqrt{\frac{999{,}000}{999{,}999}} = \sqrt{0.999001} = 0.9995$$

Since a number will not change its value when it is multiplied by 1, formula (10–6) becomes formula (10–4) when N is very large. This fact indicates an important relationship between the probability distribution obtained from a finite population and the binomial distribution obtained from an infinite population. This relationship will be emphasized in the sampling study in Chapter 12.

10.4   An Application— Probability Distribution of Proportions (p)

The formulas for the mean and the standard deviation derived from the binomial distribution (10–3 and 10–4) and from the hypergeometric distribution (10–6) may be adjusted for use in a probability distribution of proportions and its population. The basic concept of a probability distribution of proportions and the adjustments are discussed in this section. The discussions concerning an infinite population and a finite population are based on Example 1 and Example 5 respectively. More detailed discussion concerning the distribution of sample proportions is presented in Chapter 12.

In sampling study, a number of success (X) is frequently expressed in proportion form. Thus, instead of saying draw one white ball $(X = 1)$ in three repeated drawings $(n = 3)$, we may say that the proportion of the successful event (white balls) in three drawings is $1/3$ (or X/n).

The probability distribution of Example 1 for the proportions of successful event thus may be written as shown in Table 10–11.

TABLE 10–11

PROBABILITY DISTRIBUTION OF
PROPORTIONS FROM AN INFINITE POPULATION (EXAMPLE 1)

Number of Successes X	Proportion $p = X/n$	Probability (Theoretical frequency) f	fp
0	0	0.512	0
1	1/3	0.384	0.128
2	2/3	0.096	0.064
3	3/3=1	0.008	0.008
Total		1.000	0.200

The mean of the distribution of the proportions may be computed from Table 10–11.

$$\mu \text{ (of proportions)} = \frac{\Sigma fp}{\Sigma f} = \frac{0.200}{1} = 0.2$$

The mean may also be obtained by dividing formula (10–3) by n, since each p is obtained by dividing X by n, or

$$\mu \text{ (of proportions)} = \frac{\mu \text{ (of numbers of successes)}}{n, \text{ number of repeated trials}} = \frac{nP}{n} = P$$

or

$$\mu \text{ (of proportions)} = P \text{ (proportion of the population)} \qquad (10\text{–}7)$$

We have known that $P = 1/5 = 0.2$ in Example 1.

The standard deviation of the distribution of the proportions drawn from the infinite population may also be computed from Table 10–11. However, it can be obtained in a simpler manner by dividing formula (10–4) by n, or

$$\sigma \text{ (of proportions)} = \frac{\sigma \text{ (of numbers of successes)}}{n, \text{ number of repeated trials}} = \frac{\sqrt{nPQ}}{n} = \sqrt{\frac{PQ}{n}}$$

Or written:

$$\sigma \text{ (of proportions from infinite population)} = \sigma_p = \sqrt{\frac{PQ}{n}} \qquad (10\text{–}8)^*$$

Thus, the standard deviation of the distribution of proportions of Example 1 is

$$\sigma \text{ (of proportions)} = \sqrt{\frac{0.2(0.8)}{3}} = \sqrt{0.053333} = 0.23$$

Likewise, for a finite population,

$$\mu \text{ (of proportions)} = P \text{ (proportion of the population)}$$

which is formula (10–7);

$$\sigma \text{ (of proportions from finite population)} = \sigma_p = \sqrt{\frac{PQ}{n}}\sqrt{\frac{N-n}{N-1}} \quad (10\text{–}9)$$

which is obtained by dividing formula (10–6) by n.

The probability distribution of Example 5 for the proportions may be written as shown in Table 10–12.

TABLE 10–12

PROBABILITY DISTRIBUTION OF
PROPORTIONS FROM A FINITE POPULATION (EXAMPLE 5)

Number of successes X	Proportion $p = X/n$	Probability (Theoretical frequency) f	fp
0	0	0.4	0
1	1/3	0.6	0.2
Total		1.0	0.2

$$\mu \text{ (of proportions)} = \frac{\Sigma fp}{\Sigma f} = \frac{0.2}{1} = 0.2 = P$$

$$\sigma \text{ (of proportions)} = \sqrt{\frac{PQ}{n}}\sqrt{\frac{N-n}{N-1}} = \sqrt{\frac{0.2(0.8)}{3}}\sqrt{\frac{5-3}{5-1}}$$

$$= \sqrt{\frac{0.16}{3} \times \frac{2}{4}} = \sqrt{0.026667} = 0.16$$

Note that the mean and the standard deviation of a population (infinite or finite), from which the probability distribution (or the proportions) is obtained,

---

*Since each p is obtained by dividing X by n, the standard deviation σ of p values may be obtained by dividing σ of X values by n. See the 4th characteristic of the standard deviation on page 188.

can be computed respectively as follows:

$$\mu \text{ (of population)} = P \tag{10-10}$$
$$\sigma \text{ (of population)} = \sqrt{PQ} \tag{10-11}$$

The applications of the formulas can be illustrated by assigning 1 to each successful element and 0 to each unsuccessful element of the population set. This assignment will make the population proportion of the successful event equal to the probability of the event in a single trial, or $n = 1$. When $n = 1$, the possible outcomes are 0 and 1 success. Also, formula (10-4) becomes

$$\sigma \begin{pmatrix} \text{of 0 and 1 success drawn from} \\ \text{an infinite population} \end{pmatrix} = \sqrt{nPQ} = \sqrt{PQ}$$

and formula (10-6) becomes

$$\sigma \begin{pmatrix} \text{of 0 and 1 success drawn from} \\ \text{a finite population} \end{pmatrix} = \sqrt{nPQ} \cdot \sqrt{\frac{N-n}{N-1}} = \sqrt{PQ}$$

These results are the same as formula (10-11). Using the elements of the population set given in Example 1, we now have Table 10-13.

TABLE 10-13

POPULATION DISTRIBUTION (EXAMPLE 1)

Element in Population	X (Value Assigned)	f	fX	x (X − μ)	fx	fx²
Black ball	0 (unsuccessful)	4/5 = 0.8	0	−0.2	−0.16	0.032
White ball	1 (successful)	1/5 = 0.2	0.2	0.8	0.16	0.128
Total		5/5 = 1	0.2		0	0.160

The mean computed from Table 10-13 is

$$\mu = \frac{\Sigma fX}{\Sigma f} = \frac{0.2}{1} = 0.2$$

and from the formula (10-10) is

$$\mu \text{ (population)} = P = 1/5 = 0.2$$

Obviously, the mean of the population (unsuccessful element being 0 and successful element being 1), the population proportion (P), and the mean of sample proportions (p's) are equal to each other. In other words, formula (10-7) is the same as formula (10-10).

The standard deviation computed from Table 10-13 is

$$\sigma = \sqrt{\frac{\Sigma fx^2}{\Sigma f}} = \sqrt{\frac{0.16}{1}} = 0.4$$

and from formula (10-11) is

$$\sigma \text{ (population)} = \sqrt{PQ} = \sqrt{0.2(0.8)} = \sqrt{0.16} = 0.4$$

10.5 Multinomial Distribution

The multinomial distribution is an extension of the binomial distribution. Instead of handling only two possible outcomes, the method of computing the multinomial distribution deals with two or more possible outcomes in each trial.

Example 6

A bag contains five balls: one white, two black, and two red. One ball is drawn and is replaced after each drawing. Find the probabilities of having the various possible numbers of balls in two repeated drawings.

Solution.   Let the events and probabilities of the events in *each* drawing be as follows:

X_1 = the event of drawing a white ball
X_2 = the event of drawing a black ball
X_3 = the event of drawing a red ball
P_1 = the probability of drawing a white ball = $1/5$ = .2
P_2 = the probability of drawing a black ball = $2/5$ = .4
P_3 = the probability of drawing red ball = $2/5$ = .4

The required probabilities can be obtained by taking a square of the trinomial $(P_1 + P_2 + P_3)$, or

$$
\begin{aligned}
(P_1 + P_2 + P_3)^2 &= P_1^2 + P_2^2 + P_3^2 + 2P_1P_2 + 2P_1P_3 + 2P_2P_3 \\
&= .2^2 + .4^2 + .4^2 + 2(.2)(.4) + 2(.2)(.4) + 2(.4)(.4) \\
&= .04 + .16 + .16 + .16 + .16 + .32 \\
&= 1.00
\end{aligned}
$$

The exponent 2 (square) on the left side of the above equation is equal to the number of repeated drawings. The answers obtained on the right side may be written in a probability distribution table as shown in Table 10–14. (Let W = white, B = black, and R = red ball in the table.)

The mean and standard deviation for a multinomial distribution are computed on the basis of the individual event. For example, the mean and the standard deviation of X_1 values, or the possible occurrences of the event of drawing white balls in two repeated drawings, are computed in the following manner: First, find the probability distribution of X_1 values based on the results shown in Table 10–14. The X_1 distribution is shown in the first three columns of Table 10–15. Next, compute the mean and the standard deviation from the distribution in the usual manner as also shown in Table 10–15.

Likewise, the probability distribution of X_2 values and that of X_3 values can be obtained from Table 10–14. The mean and the standard deviation of each distribution can be computed in a similar manner to the computation in Table 10–15.

The work of computing a multinomial distribution involving a large number

TABLE 10–14

MULTINOMIAL DISTRIBUTION
(EXAMPLE 6)

Number of balls in two repeated drawings		Probability	
First	Second	Symbol	Value
W	W	P_1^2	.04
B	B	P_2^2	.16
R	R	P_3^2	.16
W	B	$2P_1P_2$	.16
B	W		
W	R	$2P_1P_3$	.16
R	W		
B	R	$2P_2P_3$	.32
R	B		
Total			1.00

TABLE 10–15

COMPUTATION OF MEAN AND STANDARD DEVIATION FOR THE PROBABILITY
DISTRIBUTION OF X_1 VALUES OBTAINED FROM TABLE 10–14

Number of white balls X_1	Result of two repeated drawings	Probability f	fX_1	$X_1 - \bar{X}_1$ ($\bar{X}_1 = .4$)	$f(X_1 - \bar{X}_1)$	$f(X_1 - \bar{X}_1)^2$
	BB	.16				
	RR	.16				
	BR	.32				
	RB					
0		.64	.00	−.4	−.256	.1024
	WB	.16				
	BW					
	WR	.16				
	RW					
1		.32	.32	.6	.192	.1152
2	WW	.04	.08	1.6	.064	.1024
Total		1.00	.40		.000	.3200

$$\mu \text{ (or } \bar{X}_1) = \frac{\Sigma fX_1}{\Sigma f} = \frac{.40}{1} = .40$$

$$\sigma = \sqrt{\frac{\Sigma f(X_1 - \bar{X}_1)^2}{\Sigma f}} = \sqrt{\frac{.32}{1}} = \sqrt{.32} = .57$$

of repeated trials is very tedious. Fortunately, the binomial formula $P(X; n, P)$ can be extended for computing any term of a multinomial expansion. The extended formula is

$$P(X_1, X_2, \ldots, X_k; n; P_1, P_2, \ldots, P_k) = \frac{n!}{X_1! X_2! \ldots X_k!}(P_1)^{X_1}(P_2)^{X_2} \ldots (P_k)^{X_k}$$

$$(10\text{-}12)$$

where $X_1, X_2, \ldots, X_k$ represent outcomes of the k events in n repeated trials with probabilities $P_1, P_2, \ldots, P_k$ respectively, and $X_1 + X_2 + \ldots + X_k = n$.

The formulas of the mean and the standard deviation of a particular set of X values are:

$$\mu_i \text{ (of } X_i \text{ values)} = nP_i \qquad\qquad (10\text{-}13)$$
$$\sigma_i \text{ (of } X_i \text{ values)} = \sqrt{nP_i(1 - P_i)} \qquad\qquad (10\text{-}14)$$

Any term of the multinomial expansion in Example 6 thus may be obtained directly by using formula (10–12). For example, the first term P_1^2 represents the probability of drawing two white balls in two repeated drawings. Thus,

$$X_1 = 2 \text{ (white)}, X_2 = 0 \text{ (black)}, X_3 = 0 \text{ (red)}, \text{ and}$$
$$P_1 = .2, P_2 = .4, P_3 = .4$$

Substitute the above values in formula (10–12),

$$P(2, 0, 0; 2; .2, .4, .4) = \frac{2!}{2! 0! 0!} (.2)^2(.4)^0(.4)^0$$

$$= \frac{2}{(2)(1)(1)}(.04)(1)(1) = .04$$

Also, the fourth term $2P_1P_2$ represents the probability of drawing one white ball and one black ball in two repeated drawings. Thus,

$$X_1 = 1 \text{ (white)}, X_2 = 1 \text{ (black)}, X_3 = 0 \text{ (red)}, \text{ and}$$
$$P_1 = .2, P_2 = .4, P_3 = .4$$

Substitute the values in formula (10–12),

$$P(1, 1, 0; 2; .2, .4, .4) = \frac{2!}{1! 1! 0!}(.2)^1(.4)^1(.4)^0 = .16$$

The mean and the standard deviation of X_1 values in Example 6 can be obtained by using formulas (10–13) and (10–14) respectively as follows:

$$\mu_1 \text{ (of } X_1 \text{ values)} = nP_1 = 2(.2) = .4$$
$$\sigma_1 \text{ (of } X_1 \text{ values)} = \sqrt{nP_1(1 - P_1)} = \sqrt{2(.2)(1 - .2)} = \sqrt{.32} = .57$$

The answers are the same as those obtained from Table 10–15.

The fraction on the right side of the equation in formula (10–12) is the coefficient of a multinomial term. The manner in which the multinomial coefficient can be derived is presented in Example 7 below.

Example 7

Refer to Example 6. Find the probability of having two white balls, one black ball, and three red balls in six repeated drawings.

Solution.   Here $X_1 = 2$ white balls, $X_2 = 1$ black ball, $X_3 = 3$ red balls, and

$$X_1 + X_2 + X_3 = 2 + 1 + 3 = 6 = n$$

There are

$$_6C_2 = \frac{6!}{2!(6-2)!} = \frac{6!}{2!4!} = 15 \qquad \text{(formula 9–3b)}$$

different ways of drawing 2 white balls in 6 drawings.

After the white balls are drawn, the black ball is to be drawn in the remaining $6 - 2 = 4$ drawings. There are

$$_4C_1 = \frac{4!}{1!(4-1)!} = \frac{4!}{1!3!} = 4$$

different ways of drawing 1 black ball in 4 drawings.

Next, the 3 red balls are to be drawn from the remaining $6 - 2 - 1 = 3$ drawings. There is only

$$_3C_3 = 1$$

way of drawing the three red balls in three drawings. The total number of ways of drawing six balls of three groups is

$$_6C_2 \cdot {}_4C_1 \cdot {}_3C_3 = \frac{6!}{2!4!} \cdot \frac{4!}{1!3!} \cdot 1 = \frac{6!}{2!1!3!} = 15(4)(1) = 60$$

The probability of drawing 2 white, 1 black, and 3 red balls in any way of 6 repeated drawings is

$$P_1 \cdot P_1 \cdot P_2 \cdot P_3 \cdot P_3 \cdot P_3 = P_1^2 \cdot P_2^1 \cdot P_3^3 = (.2)^2(.4)^1(.4)^3 = .001024$$

The total probability of the 60 ways is

$$60(.001024) = .06144$$

In summary, the computation of Example 7 may be arranged as follows:

$$P(2, 1, 3; 6; .2, .4, .4) = \frac{6!}{2!1!3!}(.2)^2(.4)^1(.4)^3 = .06144$$

Formula (10–12) can be derived by extending the computation method in Example 7 to include k different groups of X values and substituting the values by symbols.

Example 8 is used to illustrate an additional application of formula (10–12).

Example 8

A factory manufactures four different grades of products: 10% of the products manufactured by the company are grade A; 20% are grade B; 30% are grade C; and 40% are grade D. Find the probability of drawing a sample of eight products including five A products, one B product, two C products, and no D product.

Solution.   Here $X_1 = 5$, $X_2 = 1$, $X_3 = 2$, $X_4 = 0$, $P_1 = 10\% = .1$, $P_2 = .2$, $P_3 = .3$, $P_4 = .4$, and $X_1 + X_2 + X_3 + X_4 = 5 + 1 + 2 + 0 = 8 = n$.

Substituting the given values in formula (10–12), the required probability is

$$P(5, 1, 2, 0; 8; .1, .2, .3, .4) = \frac{8!}{5!\,1!\,2!\,0!}\,(.1)^5(.2)^1(.3)^2(.4)^0$$

$$= 168(.00000018) = .00003024$$

10.6   Poisson Distribution

This section will present the applications of the Poisson distribution in three ways. (1) The Poisson distribution may be used to approximate the binomial distribution when n (the number of repeated trials) is fairly large and P (the probability that a given event will happen in a single trial) is very small. (2) It may be used independently as a distribution when only the mean μ of the population is known. (3) It can be used to show the close relationship between the Poisson distribution and the normal curve.

The Poisson distribution formula is

$$P(X) = \frac{\mu^X \cdot e^{-\mu}}{X!} \tag{10–15}$$

where $P(X) =$ the probability that exactly X $(= 0, 1, 2, \ldots, n)$ times
a given event will happen in n trials

$\mu = nP$, the mean of occurrences of the event in n trials, and
$e = 2.71828$, the base of natural logarithms, or expressed as
$[1 + (1/n)]^n$, where n approaches infinity.

For convenience, values of $e^{-\mu}$ for selected values of μ computed to four (for $\mu = 0.9$ or less) and five (for $\mu = 1$ or more) decimal places are listed in Table 10–16.

TABLE 10–16

VALUES OF $e^{-\mu}$ FOR SELECTED VALUES OF μ $(= nP)$*

μ	$e^{-\mu}$	μ	$e^{-\mu}$	μ	$e^{-\mu}$
0.00	1.0000				
0.01	0.9900	0.1	0.9048	1	0.36788
0.02	0.9802	0.2	0.8187	2	0.13534
0.03	0.9704	0.3	0.7408	3	0.04979
0.04	0.9608	0.4	0.6703	4	0.01832
0.05	0.9512	0.5	0.6065	5	0.00674
0.06	0.9418	0.6	0.5488	6	0.00248
0.07	0.9324	0.7	0.4966	7	0.00091
0.08	0.9231	0.8	0.4493	8	0.00034
0.09	0.9139	0.9	0.4066	9	0.00012
				10	0.00005

*The values of $e^{-\mu}$ for other values of μ may be obtained by using the common logarithms or the laws of exponents as in the following example:

$$e^{-2.6} = (e^{-2})(e^{-0.6}) = (0.13534)(0.5488) = 0.0743$$

(1) An Approximation of the Binomial

As presented in Sections 10.2 and 10.3, the binomial formulas are appropriate for use in computing a probability distribution involving two possible outcomes in repeated trials from both infinite population and large finite population. The sections also pointed out that when n is large, the computation of a binomial probability distribution is tedious. Since the binomial distribution approaches the normal distribution when n is large, it is more convenient to use the latter to estimate the former for practical purposes.

The normal distribution to be discussed in the next chapter is a symmetrical and continuous probability distribution. It therefore should be used to estimate a binomial distribution which is or at least approaches a symmetrical distribution. Section 10.2 indicates that when P is not too far away from 0.5, such as 0.2 in Chart 10–2, and n is large, the binomial distribution approaches a symmetrical shape. Also, when P is close to 0.5, the binomial distribution is nearly symmetrical even if n is not too large (Chart 10–3).

However, when P is too far away from 0.5, either close to 0 or close to 1, and n is not large enough, the binomial distribution is skewed. The skewed distribution can not be estimated very well by the normal distribution. In such a case, the probability that exactly $X(= 0, 1, 2, \cdots, n)$ times a given event will happen in n trials should be computed by the binomial formula (10–1 or 10–2), or can be conveniently approximated by the following Poisson distribution formula.

The Poisson formula is usually used to approximate binomial distribution when $P = 0.1$ or less and nP or $\mu = 5$ or less. When μ is above 5, the binomial distribution is increasingly closer to normal distribution than it is to Poisson distribution, although the probability distribution obtained by the Poisson formula also approaches the normal distribution as shown in Table 10–18 and Chart 10–5.

The method of computing a probability distribution by using the Poisson formula to approximate the binomial result is illustrated in Example 9(b).

Example 9

Assume that a manufacturing process turns out 10 percent of the units produced defective. Find the probability that a sample of ten units chosen from the process will have exacty three defects by using (a) the binomial formula, and (b) the Poisson formula.

Solution. The population size N (total of units produced) is very large. The probability of a defective unit drawn from the population is $P = 10\% = 0.1$, $n = 10$, and $X = 3$.

(a) $P(X; n, P) = P(3; 10, 0.1) = {}_{10}C_3(0.1)^3(0.9)^7$

$$= \frac{10 \times 9 \times 8}{3 \times 2 \times 1}(0.001)(0.4782969)$$

$$= 0.057395628 \text{ or } 0.06$$

(b) $\mu = nP = 10(0.1) = 1$

$$P(X) = P(3) = \frac{1^3 e^{-1}}{3!} = \frac{1(0.36788)}{3 \times 2 \times 1}$$

$$= 0.0613133 \text{ or } 0.06 \qquad \text{(Table 10–16)}$$

Note that the computation in (b) is simpler than that in (a). The approximated value by the Poisson formula is very close to the answer given by the binomial formula. The value of $nP(=1)$ in this example is less than 5.

(2) Independent Applications

Formula (10–15) indicates that if only the mean μ is known, the distribution can be computed. Thus, the Poisson distribution formula is useful when neither n nor P is known but the mean μ is known (or can be approximated) and the population is large. Examples 10 and 11 are used to illustrate this type of computation.

Example 10

Assume that the number of defects per square yard of a certain type of cloth manufactured by a mill is measured as no defect, one defect, two defects, and so on. In the average, the number of defects is 0.5. Compute the probabilities that a square yard will have (a) any number (X) of defects, and (b) two or less defects.

Solution. Here, $\mu = 0.5$ defects and $n =$ sample size or the sum of possible defects and nondefects per square yard, which is unknown but large. The population, or the number of square yards of the type of cloth manufactured by the mill, is also large. The probabilities are computed as follows by using formula (10–15):

(a) The probability that a square yard of cloth will have 0 defect is:

$$P(0) = \frac{0.5^0 e^{-0.5}}{0!} = \frac{1(0.6065)}{1} = 0.6065 \qquad \text{(Table 10–16)}$$

The probability that it will have 1 defect is:

$$P(1) = \frac{0.5^1 e^{-0.5}}{1!} = \frac{0.5(0.6065)}{1} = 0.3033$$

The probability that it will have 2 defects is:

$$P(2) = \frac{0.5^2 e^{-0.5}}{2!} = \frac{0.25(0.6065)}{2 \times 1} = 0.0758$$

In a similar manner, the other probabilities are computed and are shown in column (2) of Table 10–17.

TABLE 10–17

COMPUTATION FOR EXAMPLE 10
POISSON PROBABILITY DISTRIBUTION ($\mu = 0.5$)

(1) Number of Defects X	(2) Probability P(X)	(3) Cumulative Probability for X or Less Defects P(X or less)
0	0.6065	0.6065
1	0.3033	0.9098
2	0.0758	0.9856
3	0.0126	0.9982
4	0.0016	0.9998
5	0.0002	1.0000
Total	1.0000	

It is not likely that there would be five or more defects in this case since the probability of having 5 or more defects is only 0.0002 (or 2 out of 10,000 cases).

Note that the total of the probabilities for all possible numbers of defects is always equal to 1, although in some cases the total may not be 1 because of rounding.

(b) The probability that a square yard of cloth will have 2 or less defects is the sum of the probabilities of having 0, 1, and 2 defects:

$$P(2 \text{ or less}) = P(0) + P(1) + P(2) = 0.6065 + 0.3033 + 0.0758$$
$$= 0.9856$$

which is, along with other cumulative probabilities, shown in column (3) of Table 10–17.

TABLE 10–18

POISSON PROBABILITIES FOR SELECTED VALUES OF μ ($= nP$)

Number of Defects X	$\mu = 1$		$\mu = 5$		$\mu = 10$	
	$P(X)$	$P(X$ or less)	$P(X)$	$P(X$ or less)	$P(X)$	$P(X$ or less)
0	0.3679	0.3679	0.0067	0.0067	0.0000	0.0000
1	0.3679	0.7358	0.0337	0.0404	0.0005	0.0005
2	0.1839	0.9197	0.0842	0.1246	0.0023	0.0028
3	0.0613	0.9810	0.1404	0.2650	0.0076	0.0104
4	0.0153	0.9963	0.1755	0.4405	0.0189	0.0293
5	0.0031	0.9994	0.1755	0.6160	0.0378	0.0671
6	0.0005	0.9999	0.1462	0.7622	0.0631	0.1302
7	0.0001	1.0000	0.1044	0.8666	0.0901	0.2203
8			0.0653	0.9319	0.1126	0.3329
9			0.0363	0.9682	0.1251	0.4580
10			0.0181	0.9863	0.1251	0.5831
11			0.0082	0.9945	0.1137	0.6968
12			0.0034	0.9979	0.0948	0.7916
13			0.0013	0.9992	0.0729	0.8645
14			0.0005	0.9997	0.0521	0.9166
15			0.0002	0.9999	0.0347	0.9513
16			0.0001	1.0000	0.0217	0.9730
17					0.0128	0.9858
18					0.0071	0.9929
19					0.0037	0.9966
20					0.0019	0.9985
21					0.0009	0.9994
22					0.0004	0.9998
23					0.0002	1.0000
24					0.0001	

In practical use, it is not necessary to compute the Poisson distribution as is done in Example 10. Extensive tables of the Poisson distribution for selected values of μ and their cumulative probabilities can be obtained from published sources, such as the well-known *Poisson's Exponential Binomial Limit*, computed by E. C.

Molina and published by D. Van Nostrand Company in 1942, which gives probabilities for values of μ from 0.001 to 100. Table 10–18 shows the Poisson probabilities of $\mu = 1$, 5, and 10 for illustrative purposes. Table 4 of the Appendix is a more detailed table.

The standard deviation of a Poisson probability distribution may be computed by a method similar to that used for the binomial probability distribution (see Table 10–6). However, a simple way to compute the standard deviation is by using the following formula:

$$\sigma = \sqrt{\mu} \qquad (10\text{--}16)$$

where σ represents the standard deviation of the Poisson distribution.

By formula (10–16), the standard deviation of Example 10 is:

$$\sigma = \sqrt{\mu} = \sqrt{0.5} = 0.707 \text{ defects}$$

Another important type of problem that can best be described by the Poisson distribution is the *waiting line* or *queuing* formation. There are many occasions of forming waiting lines in business activities, for example, at a supermarket checkout counter, a bank teller's window, or a theater ticket office. Customers come in randomly and wait for services. Example 11 is used to illustrate only the form of arrivals involved in a waiting line problem.

Example 11

Assume that the customers of a jewelry store arrive randomly at the average rate of 1.5 customers per hour. The average is computed from the store's record of recent 100 business hours, as shown in Table 10–19. Examine whether or not the arrivals follow the Poisson distribution.

TABLE 10–19

DATA FOR EXAMPLE 11

Number of Customers Arriving Per Hour X	Number of Hours Recorded f	fX
0	20	0
1	35	35
2	26	52
3	14	42
4	4	16
5	1	5
6 or more	0	0
Total	100	150

Mean, μ or $\bar{X} = \frac{150}{100} = 1.5$ customers per hour

Solution. The Poisson probability distribution based on $\mu = 1.5$ can be obtained directly from Table 4 in the Appendix and is shown in Table 10–20, column (2). Each expected number of hours shown in column (3) in the table is obtained by multiplying individual probability (or relative frequency) by 100 (the total hours).

TABLE 10–20

COMPUTATION FOR EXAMPLE 11
POISSON PROBABILITY DISTRIBUTION ($\mu = 1.5$)

(1) *Number of Customers* *Arriving Per Hour* X	(2) *Relative Frequency,* *Probability P(X)* *(Table 4, Appendix)*	(3) *Expected Number* *of Hours* *100 × (2)*
0	.2231	22.31
1	.3347	33.47
2	.2510	25.01
3	.1255	12.55
4	.0471	4.71
5	.0141	1.41
6 or more	.0035	0.35
Total	1.0000	100.00

The expected number of hours based on the Poisson probability distribution and the observed number of hours given in the problem are plotted in Chart 10–5 for comparison. The chart shows that the arrivals to the jewelry store follow the Poisson distribution closely.

CHART 10–5

A COMPARISON BETWEEN OBSERVED FREQUENCIES
AND POISSON DISTRIBUTION OF ARRIVALS, EXAMPLE 11

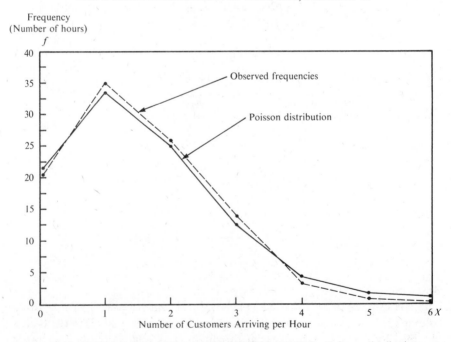

Source: Table 10–19 for observed frequencies and Table 10–20 for Poisson distribution.

(3) Relation to the Normal Curve

Generally, there is a close relationship between the Poisson distribution and the normal curve when P is small, n is very large, and thus μ is large or $\mu = nP = 5$ or more as shown in Chart 10–6.

The Poisson probability distribution for $\mu = 0.5$ in Example 10 is shown graphically on Chart 10–6. The chart also shows the Poisson distribution for $\mu = 1$, $\mu = 5$, and $\mu = 10$ obtained from Table 10–18. It can be seen that as μ increases, the probability curve becomes more symmetrical; that is, it approaches the shape of the normal curve.

CHART 10–6

POISSON PROBABILITY DISTRIBUTIONS FOR $\mu = 0.5$, $\mu = 1$, $\mu = 5$, AND $\mu = 10$

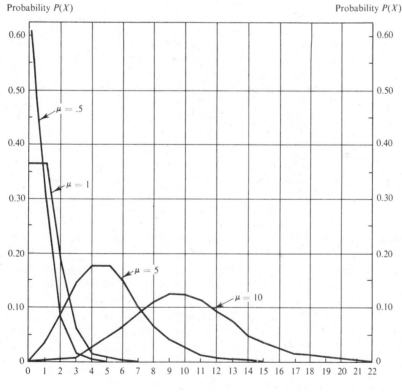

Number of Defects (X)

Source: Example 10 and Table 10–18.

10.7 Tchebycheff's Inequality

The methods of computing means and standard deviations for the different types of probability distributions have been introduced in the previous sections. An interesting relationship among the mean, the standard deviation, and the amount of probability of X variable can be described by the following expression, called Tchebycheff's (also commonly spelled Chebyshev's) inequality. This expression can be applied to a distribution of any type, including the probability distributions to be introduced in later chapters.

$$P(|x| \geq k\sigma) \leq \frac{1}{k^2} \qquad (10\text{--}17)$$

where $x = X - \mu$, which is positive when X is larger than μ, or negative when X is smaller than μ. However, the $+$ and $-$ signs are disregarded in $|x|$.

This expression states that the probability of X values which differ from the

CHART 10–7

TCHEBYCHEFF'S INEQUALITY

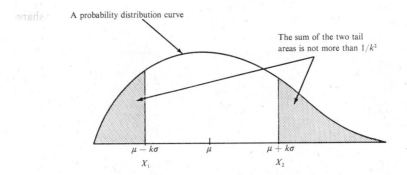

mean by k or more than k standard deviations will be equal to or less than $1/k^2$. Stated in a different way, based on the distribution curve shown in Chart 10–7, the sum of the two tail-areas representing probabilities of X values outside the range from $X_1 (= \mu - k\sigma)$ to $X_2 (= \mu + k\sigma)$ is not more than $1/k^2$.

Example 12

Refer to the binomial probability distribution of Example 1 as shown in Table 10–6, page 249. $\mu = 0.6$ and $\sigma = .69$. Let $k = 2$. Find the probability of X values which differ from the mean μ by $k\sigma$.

Solution. $k\sigma = 2(.69) = 1.38$

$$X_1 = \mu - k\sigma = .6 - 1.38 = -.78$$
$$X_2 = \mu + k\sigma = .6 + 1.38 = 1.98$$

$$P\binom{\text{outside the range,}}{-\,.78 \text{ to } 1.98} = P(2) + P(3)$$

$$= .096 + .008$$
$$= .104 \text{ (See Table 10–6).}$$

The answer .104 is smaller than $1/k^2 = 1/2^2 = .25$.

Example 13

Refer to the Poisson Probability distribution of Example 10 as shown in Table 10–17. $\mu = 0.5$ and $\sigma = .707$. Let $k = 3$. Find the probability of X values deviated from the mean μ by $k\sigma$.

Solution.
$$k\sigma = 3(.707) = 2.121$$
$$X_1 = .5 - 2.121 = -1.621$$
$$X_2 = .5 + 2.121 = 2.621$$

$$P\begin{pmatrix}\text{outside the range,}\\ -1.621 \text{ to } 2.621\end{pmatrix} = P(3) + P(4) + P(5)$$
$$= 0.0126 + 0.0016 + 0.0002$$
$$= 0.0144 \text{ (See Table 10–17)}$$

The answer 0.0144 is smaller than $1/k^2 = 1/3^2 = .1111$

Observe that the X_1 values in Examples 12 and 13 are negative. Since the lowest X value in each of the two probability distributions is 0, only the right-tail-area under the distribution curve will represent the probability in each answer.

10.8   Summary

A set of values distributed according to the theory of probability is called a probability distribution. This chapter introduces four basic discrete probability distributions: (1) binomial, (2) hypergeometric, (3) multinomial, and (4) Poisson.

The binomial distribution can be computed by the probability formula for repeated trials or by the binomial theorem for an infinite population. When n is very large and P is equal to or close to $0.5(= \frac{1}{2})$, the symmetrical binomial distribution will approach the smooth curve, called the normal distribution curve (Charts 10–3, –4). When n is small and P deviates from 0.5, the binomial distribution is skewed. However, when the value of n increases, although P is significantly different from 0.5, the binomial distribution gets closer and closer to the normal curve (Chart 10–2).

The hypergeometric probability distribution is obtained from a finite population. The size of the population is reduced each time after a trial. Also, the probability of a successful event changes if the population size changes, even though the number of trials remains unchanged. The formulas for the mean and the standard deviation derived from the binomial distribution and the hypergeometric distribution may be adjusted for use in a probability distribution of proportions and its population.

The multinomial distribution is an extension of the binomial distribution. It is extended to deal with two or more possible outcomes in each trial.

The Poisson distribution, like the binomial distribution, is a discrete probability distribution. It may be used in three ways: (1) It may be used to approximate the binomial distribution when n is fairly large and P is very small ($P = 0.1$ or less and $nP = 5$ or less) as shown in Example 9. (2) It may be used independently as a distribution to show the probabilities of various events in an experiment or action when neither n nor P is known but the mean μ ($= nP$) is known or can be approximated, such as shown in Example 10. (3) The Poisson distribution is closely related to the normal curve when μ increases indefinitely (5 or more) as shown in Chart 10–5.

Exercises 10

1.   What are discrete probability distributions and continuous probability distributions? Give an example for each type.

2.   What is the difference between the process of performing repeated trials for the binomial distribution and that for the multinomial distribution?

3.   Expand the binomial $(x + y)^5$.

4.   Expand the trinomial $(x + y + z)^3$.

5.   A bag contains two red balls and three white balls. One ball is drawn and is *replaced* for the next draw. (a) Find the probability distribution of having zero, one, two, and three red balls in three repeated drawings. Also find (b) the mean, and (c) the standard deviation of the distribution.

6.   Refer to Problem 5. Find the answers to (a), (b), and (c) for four repeated drawings. (d) Plot the probability distributions obtained from Problem 5 and this problem on a chart.

7.   Refer to Problem 5. Find the answers to (a), (b), and (c) for four repeated drawings if the numbers of successes (having red balls) are expressed in proportion form.

8.   A bag contains two red balls and three white balls. One ball is drawn and is *not replaced* for the next draw. (a) Find the probability distribution of having zero, one, two, and three red balls in three repeated drawings. Also, find (b) the mean, and (c) the standard deviation of the distribution.

9.   Refer to Problem 8. Find the answers to (a), (b), and (c) for four repeated drawings.

10. Refer to Problem 8. Find the answers to (a), (b), and (c) for four repeated drawings if the numbers of successes (having red balls) are expressed in proportion form.

11.   What are (a) the population mean, and (b) the population standard deviation in Problem 5? In Problem 8?

12. Refer to Problem 5. Find the probability that will have exactly four red balls in six repeated drawings.

13.   Twelve boys are to be devided into three groups of four each to play a game. How many different groups can possibly be made?

SUMMARY OF FORMULAS

Application	Formula	Formula Number	Reference Page
n repeated trials			
Successful event happens X times	$P(X; n, P) = {}_nC_X \cdot P^X \cdot Q^{n-X}$	10-1	237
Binomial distribution (*from infinite population*)			
Probabilities of X's (number of successes)	$(Q + P)^n = {}_nC_0 Q^n + {}_nC_1 P Q^{n-1} + \cdots$ $+ {}_nC_X P^X Q^{n-X} + \cdots + {}_nC_n P^n$	10-2	241
Mean of the distribution of X's	$\mu = nP$	10-3	248
Standard deviation of the distribution of X's	$\sigma = \sqrt{nPQ}$	10-4	248
Hypergeometric distribution (*from finite population*)			
Probability of X	$P(X) = \dfrac{{}_aC_X \cdot {}_bC_{n-X}}{{}_NC_n}$	10-5	251
Standard deviation of the distribution of X's	$\sigma = \sqrt{nPQ} \cdot \sqrt{\dfrac{N-n}{N-1}}$	10-6	251
Probability distribution of p's (*proportions*)			
Mean of the distribution of p's	$\mu = P$	10-7	254
Standard deviation of the distribution of p's from an infinite population	$\sigma_p = \sqrt{\dfrac{PQ}{n}}$	10-8	255
Standard deviation of the distribution of p's from a finite population	$\sigma_p = \sqrt{\dfrac{PQ}{n}} \sqrt{\dfrac{N-n}{N-1}}$	10-9	255
Mean of population	$\mu = P$	10-10	256
Standard deviation of population	$\sigma = \sqrt{PQ}$	10-11	256
Multinomial distribution			
Probability of $X_1, X_2 \cdots X_k$ times	$P(X_1, X_2, \ldots X_k; n; P_1, P_2, \ldots P_k) =$ $\dfrac{n!}{X_1! X_2! \cdots X_k!}(P_1)^{X_1}(P_2)^{X_2}\cdots(P_k)^{X_k}$	10-12	259
Mean of the distribution of X_i's	$\mu_i = nP_i$	10-13	259
Standard deviation of the distribution of X_i's	$\sigma_i = \sqrt{nP_i(1 - P_i)}$	10-14	259
Poisson Distribution			
Probability of X	$P(X) = \dfrac{\mu^X e^{-\mu}}{X!}$	10-15	261
Standard deviation of the distribution of X's	$\sigma = \sqrt{\mu}$	10-16	265
Tchebycheff's Inequality	$P(|x| \geq k\sigma) \leq \dfrac{1}{k^2}$	10-17	268

14. Refer to Example 6, page 257. Find the mean and the standard deviation of (a) X_2 values, aud (b) X_3 values.

15. Refer to Example 8, page 260. Find the probability of drawing a sample of (a) eight products, including four A products, one B product, two C products, and one D product, (b) ten products, including four A products, three B products, two C products, and one D product.

16. Assume that 20 percent of TV viewers prefer network L, 30 percent prefer network M, and 50 percent prefer network N. Find the probability of selecting a sample of six viewers of which two prefer L, three M, and one N.

17. A factory finds 2 percent of the units produced by a machine to be defective. Find the probability that a sample of 100 units produced by the machine will have exactly (a) two defects and (b) four defects by the Poisson formula.

18. Assume that the average number of defects per lot of a certain shipment is one defect. Find the probability of two defects or less for a given lot by the Poisson formula.

19. Let $k = 1$. Use the answers obtained in Problem 5 to prove the Tchebycheff's inequality.

20. Let $k = 2$. Use the answers obtained in Problem 6 to prove the Tchebycheff's inequality.

11 The Normal Distribution

Statistical data concerning business and economic problems are frequently displayed in the form of the normal distribution or close to the normal distribution. The *normal distribution*, also called the *normal probability distribution*, is thus regarded as the most important type among various probability distributions. When the normal distribution is shown graphically, the curve representing the distribution, called the *normal curve*, is symmetrical or bell-shaped. Since the normal distribution is symmetrical, the mid-point under the curve is the mean of the distribution. The shape of the normal curve indicates that the frequencies in a normal distribution are concentrated in the center portion of the distribution and the values above and below the mean are equally distributed. The construction of the normal curve is presented in Section 11.1.

The normal distribution is a continuous distribution. The number of cases in it may be infinitely large. The probability or relative frequency of the occurrence of a certain event is therefore measured according to the size of the area representing the event under the normal curve. There are unlimited shapes of normal curves. However, all shapes can be converted to the standard normal curve, shown in Section 11.2. The method of finding the size of an area under the standard normal curve is discussed in Section 11.3.

The normal curve of a population may be approximated by using a sample. The approximated normal curve thus permits us to make statistical inferences concerning the population based on the sample. The method of obtaining the normal curve fitted to the frequency distribution of a sample is illustrated in Section 11.4.

11.1 The Construction of Normal Curves

Basically, the normal curve is constructed according to the heights of the ordinates (Y variable) computed from the formula below:

$$Y_x = \frac{N}{\sigma} f(z)$$ (11-1)

where $x = X - \mu$, the distance between X and μ on the X axis
μ = the mean of the population or universe (X's)
Y_x = the height of ordinate Y at point x on the X axis
N = number of items or cases in the population
σ = the standard deviation of the population
$z = x/\sigma$, the distance x expressed in units of σ
$f(z)$ = function of z (not f times z). The value of $f(z)$ depends on the value of z and can be obtained from Table 11–1*

TABLE 11–1

VALUES OF $f(z)$ FOR $z = 0$ TO 4 AT AN INTERVAL OF .25†

$z\left(=\dfrac{x}{\sigma}\right)$	$f(z)$	$z\left(=\dfrac{x}{\sigma}\right)$	$f(z)$
0.00	0.39894	2.25	0.03174
0.25	0.38667	2.50	0.01753
0.50*	0.35207	2.75	0.00909
0.75	0.30114	3.00	0.00443
1.00	0.24197	3.25	0.00203
1.25	0.18265	3.50	0.00087
1.50	0.12952	3.75	0.00035
1.75	0.08628	4.00	0.00014
2.00	0.05399		

†For more detailed values of $f(z)$, see Table 5 in the Appendix.

When formula (11–1) is used, three values must be known. They are: N, μ, and σ. Examples 1 to 4 are used to illustrate the method of constructing a normal curve.

Example 1
Assume that the average monthly income of 10,000 workers in King City is $500 and the standard deviation is $100. Construct a normal distribution curve.

---

*The value of $f(z)$ can be expressed mathematically as follows:

$$f(z) = \frac{1}{\sqrt{2\pi}} e^{-x^2/2\sigma^2} = \frac{1}{\sqrt{2\pi}} e^{-z^2/2} \text{ where } \pi = 3.14159$$

Thus, $f(z) = [1/\sqrt{2(3.14159)}]e^{-z^2/2} = (0.398942)e^{-z^2/2}$. The values of $e^{-z^2/2}$ may be computed by logarithms. $e = 2.71828$. For example, when $z = 0.50$
$$e^{-z^2/2} = 2.71828^{-(0.5)^2/2} = 2.71828^{-0.125}$$

By logarithms,
$$(-0.125) \log 2.71828 = (-0.125)(0.434293) = -0.0542866 = 9.9457134 - 10$$
Find antilog, $e^{-z^2/2} = 0.8825$.
$$f(z) = f(0.50) = 0.398942(0.8825) = 0.35207$$

Solution.   Here, $N = 10,000$, $\mu = \$500$, $\sigma = \$100$, and $N/\sigma = 10,000/100 = 100$. The height of an ordinate at point x is Y_x. The ordinates at an interval of $\frac{1}{2}\sigma$ $(= \$100[\frac{1}{2}] = \$50)$ from the mean $(\mu = \$500)$ up to 3σ (or $z = 3$) are computed in Table 11–2 so that enough coordinates on the graph will enable us to draw the normal curve.

TABLE 11–2

COMPUTATION FOR EXAMPLE 1

X	$x = X - \mu$	$z = x/\sigma$	$f(z)$	$Y_x = \dfrac{N}{\sigma} f(z)$
$500 (= \mu)$	0	0	0.39894	39.894
550	50	0.50	0.35207	35.207
600	100	1.00	0.24197	24.197
650	150	1.50	0.12952	12.952
700	200	2.00	0.05399	5.399
750	250	2.50	0.01753	1.753
800	300	3.00	0.00443	0.443

Since the normal curve is symmetrical, the height of the ordinate on the left side of the mean must be the same as the one on the right side of the mean if they are at the same distance but opposite direction from the mean. Thus, Y_x at point $\$550$ $(x = 550 - 500 = 50 = 0.5\sigma)$ is the same as that at point $\$450$ $(x = 450 - 500 = -50 = -0.5\sigma)$ since their distance (x's) from the mean are the same; Y_x at point $\$600$ $(x = 600 - 500 = 100 = 1\sigma)$ is the same as that at point $\$400$ $(x = 400 - 500 = -100 = -1\sigma)$; and so on. The points are plotted in Chart 11–1. When a smooth curve, A, is drawn through all points, the curve is called a normal curve.

Example 2

Refer to Example 1. Assume that the average monthly income of the 10,000 workers in King City is $600 and the standard deviation is also $100. Construct the normal curve.

Solution.   Here, $N = 10,000$, $\mu = \$600$, $\sigma = \$100$, and $N/\sigma = 10,000/100 = 100$. The ordinates at an interval of $\frac{1}{2}\sigma$ $(= \$100[\frac{1}{2}] = \$50)$ from the mean (μ) up to 3σ are computed in Table 11–3.

Note that the computations, except those in the first column (X), are the same as those in Table 11–2. The results are also plotted in Chart 11–1, curve B. Notice that the shape of curve A and the size of the area under the curve are the same as those of curve B. The only difference between the two curves is the location of the ordinates at the X scale. For example, the maximum ordinate (Y_0) of curve A is at point $X = \$500$, (the mean in Example 1), whereas that of curve B is at point $X = \$600$ (the mean in Example 2).

TABLE 11–3

COMPUTATION FOR EXAMPLE 2

X	$x = X - \mu$	$z = x/\sigma$	$f(z)$	$Y_x = \dfrac{N}{\sigma} f(z)$
600 $(= \mu)$	0	0	0.39894	39.894
650	50	0.50	0.35207	35.207
700	100	1.00	0.24197	24.197
750	150	1.50	0.12952	12.952
800	200	2.00	0.05399	5.399
850	250	2.50	0.01753	1.753
900	300	3.00	0.00443	0.443

CHART 11–1

COMPARISON OF TWO NORMAL DISTRIBUTIONS

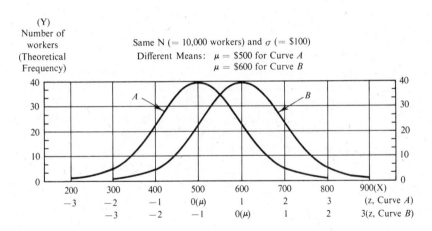

Source: Example 1, Curve *A*. Example 2, Curve *B*.

Example 3

Refer to Example 1. Assume that there are only 5,000 workers in King City and that the average monthly income of the workers and the standard deviation of the income distribution are the same. Construct a normal distribution curve.

Solution. Here, $N = 5,000$, $\mu = \$500$, $\sigma = \$100$, and $N/\sigma = 5,000/100 = 50$. The ordinates at an interval of $\frac{1}{2}\sigma$ from the mean up to 3σ (or $z = 3$) can be obtained in a way similar to that used in Example 1. The values in the four columns, X, x, z, and $f(z)$, in Table 11–4 are identical to those in Table 11–2 since the values of μ and σ are the same.

TABLE 11–4

COMPUTATION FOR EXAMPLE 3

X	$x = X - \mu$	$z = x/\sigma$	$f(z)$	$Y_x = \dfrac{N}{\sigma} f(z)$
500 $(= \mu)$	0	0	0.39894	19.9470
550	50	0.50	0.35207	17.6035
600	100	1.00	0.24197	12.0985
650	150	1.50	0.12952	6.4760
700	200	2.00	0.05399	2.6995
750	250	2.50	0.01753	0.8765
800	300	3.00	0.00443	0.2215

The results are plotted on Chart 11–2, curve C. The chart also shows curve A for comparison. Notice that when the N value decreases, the curve (C) is flat-topped.

CHART 11–2

COMPARISON OF TWO NORMAL DISTRIBUTIONS

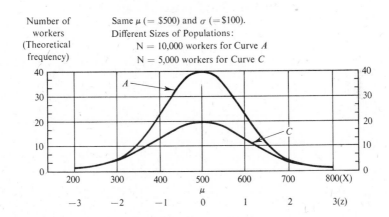

Number of workers (Theoretical frequency)

Same μ (= \$500) and σ (=\$100).
Different Sizes of Populations:
    N = 10,000 workers for Curve A
    N = 5,000 workers for Curve C

Source: Example 1, Curve A. Example 3, Curve C.

Example 4

Refer to Example 1. Assume that the average monthly income of the 10,000 workers in King City is also \$500 but the standard deviation is only \$50. Construct a normal distribution curve.

Solution. Here, $N = 10,000$, $\mu = \$500$, $\sigma = \$50$, and $N/\sigma = 10,000/50 = 200$. The ordinates at an interval of $\frac{1}{2}\sigma$ $(= \frac{1}{2} \cdot \$50 = \$25)$ from the mean up to 3σ (or $z = 3$) are computed in Table 11–5.

TABLE 11–5

COMPUTATION FOR EXAMPLE 4

X	$x = X - \mu$	$z = x/\sigma$	$f(z)$	$Y_x = \dfrac{N}{\sigma} f(z)$
\$500 $(= \mu)$	\$ 0	0	0.39894	79.788
525	25	0.50	0.35207	70.414
550	50	1.00	0.24197	48.394
575	75	1.50	0.12952	25.904
600	100	2.00	0.05399	10.798
625	125	2.50	0.01753	3.506
650	150	3.00	0.00443	0.886

The results are plotted on Chart 11–3, curve D. The chart also shows curve A for comparison. Notice that when σ decreases, the curve (D) is highly peaked.

CHART 11–3

COMPARISON OF TWO NORMAL DISTRIBUTIONS

Same N (= 10,000 workers) and μ (= \$500)

Different Standard Deviations: $\sigma =$ \$100 for Curve A

$\sigma =$ \$50 for Curve D

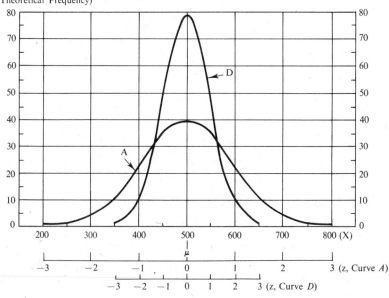

Source: Example 1, Curve A. Example 4, Curve D.

11.2   The Standard Normal Curve

Let $N = 1$ (the absolute frequencies of the distribution are considered as one unit),

$\sigma = 1$ (the standard deviation is expressed as one unit), and

$\mu = 0$ (the mean of the population distribution is expressed as zero, the origin on the X scale).

Then, $z = x/\sigma = x$. Formula 11–1 can be written:

$$Y_z = \frac{1}{1} f(z) \quad \text{or} \quad Y_z = f(z)$$

The values of $f(z)$ for $z = 0$ to 4 at an interval of 0.25 are listed in Table 11–1. When the values in the table are plotted (Chart 11–4), the smooth curve drawn through all the points is called the *standard normal curve*. The standard normal curve has unit population ($N = 1$), unit standard deviation ($\sigma = 1$), and zero mean ($\mu = 0$).

The standard normal curve has the following properties:

1. The curve is symmetrical with respect to the Y_0 ordinate.

2. The curve approaches the X-axis more and more closely when z is greater than ± 3. However, the curve extends without ending in both directions. Since $f(z)$ can never be zero for all values of z, the curve will not intercept the X-axis.

3. The curve has a maximum point when $z = 0$; that is, $f(z) = 0.39894$.

4. The curve has points of inflection when $z = \pm 1$; that is, the curve is concave downward between $z = -1$ and $z = +1$, and is concave upward to the right of $z = +1$ and to the left of $z = -1$.

5. Any point on the curve is not the value of probability; that is, the ordinate at the point is a relative height to the maximum ordinate. However, the total area under the normal curve and above the X-axis is the total probability of the distribution, or equal to 1.

6. The unit population may represent a finite population as well as an infinite population, or $N = 1 = 100\%$. Thus, if the mean and the standard deviation of an infinite population are known, the population distribution is determined. Stated in a different way, if two infinite populations have the same mean and the same standard deviation, the two populations are identical.

All shapes of normal curves can be converted to the shape of the standard normal curve. The conversion may be made by changing the designation of the unit width on both the X and Y scales of the original normal curve to coincide with that of the standard normal curve. The ordinates (Y_x) of a normal curve are obtained by multiplying $f(z)$ by the constant value N/σ (formula 11–1). When the key numbers labeled on the Y scale of the normal curve are divided by the constant value N/σ, the quotients will coincide with the values on the Y scale of the standard normal curve. For example, the number 20 on the Y scale for curve C (where $N = 5,000$ and $\sigma = 100$) coincides with the number 0.40 for the standard normal curve since

$$20 \div \frac{N}{\sigma} = 20 \div \frac{5,000}{100} = 0.40$$

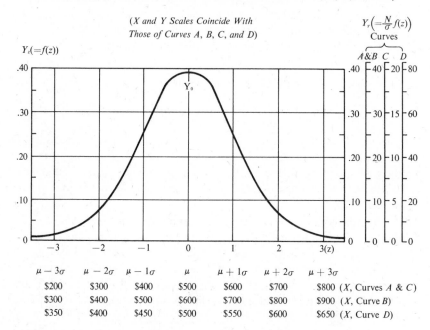

CHART 11–4

THE STANDARD NORMAL CURVE

(X AND Y SCALES COINCIDE WITH THOSE OF CURVES $A, B, C,$ AND D)

(X and Y Scales Coincide With Those of Curves A, B, C, and D)

$$Y_z\left(=\frac{N}{\sigma}f(z)\right)$$

Curves

$Y_z(=f(z))$

	$\mu - 3\sigma$	$\mu - 2\sigma$	$\mu - 1\sigma$	μ	$\mu + 1\sigma$	$\mu + 2\sigma$	$\mu + 3\sigma$	
	$200	$300	$400	$500	$600	$700	$800	(X, Curves A & C)
	$300	$400	$500	$600	$700	$800	$900	(X, Curve B)
	$350	$400	$450	$500	$550	$600	$650	(X, Curve D)

Source: Table 11-1 and Examples 1, 2, 3, and 4 of Section 11.1.

Similarly, the number 60 on the Y scale for curve D (where $N = 10,000$ and $\sigma = 50$) coincides with the number 0.30 for the standard normal curve since

$$60 \div \frac{N}{\sigma} = 60 \div \frac{10,000}{50} = 0.30$$

The values on the X scale can easily be designated according to the number (z) of standard deviations (σ); that is, $z\sigma$. When $\sigma = 1$, $z\sigma = z$. Thus, z is also called the *standard normal deviate*. Chart 11–4 also shows that curves A, B, C, D, and the standard normal curve are actually the same shape.

11.3　Areas Under the Normal Curve

The probabilities represented by areas of various sizes under the standard normal curve have been computed. A portion of the computed probabilities is listed in Table 11–6 and shown in Chart 11–5.

Each value in column $A(z)$ of the table represents the area between the maximum ordinate at the mean (Y_0) and the ordinate at the point on X axis, expressed in units of $z(= x/\sigma)$ away from the mean (or Y_z). The table gives z values at an

TABLE 11–6

AREAS UNDER THE NORMAL CURVE—VALUES OF $A(z)$*
BETWEEN ORDINATE AT MEAN (Y_0) AND ORDINATE AT z (OR Y_z)

$z\left(=\dfrac{x}{\sigma}\right)$	$A(z)$ (area)	$z\left(=\dfrac{x}{\sigma}\right)$	$A(z)$ (area)
0.00	0.00000	2.25	0.48778
0.25	0.09871	2.50	0.49379
0.50	0.19146	2.75	0.49702
0.75	0.27337	3.00	0.49865
1.00	0.34134	3.25	0.49942
1.25	0.39435	3.50	0.49977
1.50	0.43319	3.75	0.49991
1.75	0.45994	4.00	0.49997
2.00	0.47725		

*For more detailed values of $A(z)$, see Table 6 in the Appendix.

interval of 0.25, from $z = 0$ to $z = 4$. The total area under the normal curve and above the X-axis is the total probability which is equal to 1 or 100%. Chart 11–5 shows that there are 68.268% ($= 0.34134[2] = 0.68286$ or 68.268%) of the area within the range $\mu \pm 1\sigma$, 95.45% ($= 0.47725[2] = 0.95450$ or 95.45%) within the range $\mu \pm 2\sigma$, and 99.73% ($= 0.49865[2] = 0.99730$) within the range of $\mu \pm 3\sigma$. If the total area is used to represent the total absolute frequency N of a distribution, a portion of the area expressed in terms of probability (or relative

CHART 11–5

AREAS UNDER THE NORMAL CURVE
FOR SELECTED VALUES OF z FROM TABLE 11–6

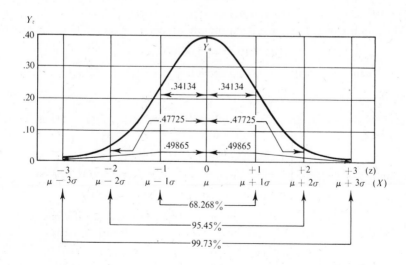

frequency) obtained from Table 11–6 must be multiplied by N to find the absolute frequency of that portion.

Because any shape of normal curve can be converted to the shape of the standard normal curve, the table showing the areas under the standard normal curve is the only one required in finding the probability of a certain area under a normal curve. The method of computing the probability based on Table 11–6 is illustrated in the following two different cases.

When the X Value is Given, the Area A(z) is to be Determined

Example 5

Assume that the average monthly income of 10,000 workers in King City is $500 and the standard deviation is $100. If the distribution is normal, find the number of workers having a monthly income (a) below $500, (b) above $500 but below $600, and (c) above $600.

Solution. The related information is shown in Chart 11–6. Before using the normal curve area table (Table 11–6), the value of X must be converted to z by using formula

$$z = \frac{x}{\sigma} = \frac{X - \mu}{\sigma}$$

This example gives that $\mu = \$500$ and $\sigma = \$100$.

(a) The required area is below $500, which is equivalent to the point $z = 0$, since

$$z = \frac{X - \mu}{\sigma} = \frac{500 - 500}{100} = 0$$

where $X = \$500$ and $\mu = \$500$, or $X = \mu$.

Since the maximum ordinate Y_0 is located at the point where $z = 0$, the area representing below $z = 0$ is the entire area to the left side of Y_0, or $0.5 (= 50\%)$. (Note: When $z = -\infty$, the value of $A(z) = 0.50000$.)

The number of workers having a monthly income below $500 is

$$10,000(0.5) = 5,000 \text{ (workers)}$$

(b) When $X = \$500$, $z = 0$ (see above).

When $X = \$600$, $z = \left(\dfrac{600 - 500}{100}\right) = 1$

In Table 11–6, the area (or probability) between $z = 0$ and $z = 1$ is 0.34134 or 34.134%.

$$A(1) = 0.34134 \text{ or } 34.134\%$$

The number of workers having a monthly income above $500 but below $600 is

$$10,000 \times 0.34134 = 3,413.4 \text{ or } 3,413 \text{ (workers)}$$

(c) When $X = \$600$, $z = 1$. The area above $z = 1$ is the difference between the area above $z = 0$ (or 0.5) and the area between $z = 0$ and $z = 1$ (or 0.34134).

$$0.5 - 0.34134 = 0.15866 \text{ or } 15.866\%$$

The number of workers having a monthly income above $600 is

$$10,000 \times 0.15866 = 1,586.6 \text{ or } 1,587 \text{ (workers)}$$

CHART 11–6

AREAS UNDER THE NORMAL CURVE FOR VALUES OF EXAMPLE 5

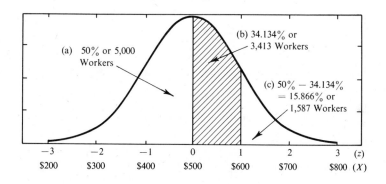

Note that the total probability is equal to 1 or 100%. Thus, the total number of workers obtained in above is 10,000, which is 100% of the distribution.

$$\text{Total workers} = 5,000 + 3,413 + 1,587 = 10,000$$

The following examples further illustrate the use of Table 11–6. For simplicity, however, only the areas of probability are computed, not the absolute frequencies.

Example 6

If $\mu = 400$, and $\sigma = 100$, what is the probability (area) of values (a) between 250 and 500 and (b) less than 250?

Solution. This example is analyzed in Chart 11–7 and is computed below:

(a) The area under the curve between 250 and 500 is the sum of the area between 250 and 400 and the area between 400 and 500. The area between 250 and 400, where $400 = \mu$, is computed as follows:

$$\text{When } X = 250, z = \frac{X - \mu}{\sigma} = \frac{250 - 400}{100} = -1.5$$

The area between Y_0 and the ordinate at $z = -1.5$ is the same as the area between Y_0 and the ordinate at $z = +1.5$ in the table, since the normal curve shows a perfect symmetrical distribution.

$$A(z) = A(-1.5) = 0.43319$$

The area between 400 and 500 is computed as follows:

$$\text{When } X = 500, z = \frac{500 - 400}{100} = 1$$

CHART 11–7

AREAS UNDER THE NORMAL CURVE FOR VALUES OF EXAMPLE 6

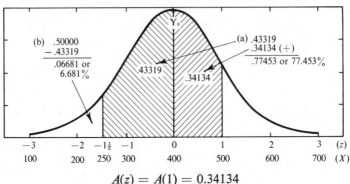

$$A(z) = A(1) = 0.34134$$

Thus, the area between 250 and 500 is

$$\begin{array}{r} 0.43319 \\ +0.34134 \\ \hline 0.77453 \end{array} \quad \text{or} \quad 77.453\%$$

(b) When $X = 250$, $z = -1.5$ (See (a) above.) The area below $z = -1.5$ is the difference between the area below $z = 0$ (or 0.50000) and the area between Y_0 and the ordinate at $z = -1.5$ (or 0.43319).

$$\begin{array}{r} 0.50000 \\ -0.43319 \\ \hline 0.06681 \end{array} \quad \text{or} \quad 6.681\%$$

Example 7

If $\mu = 300$ and $\sigma = 50$, find the probability (area) of values (a) between 200 and 225 and (b) between 350 and 400.

Solution.   This example is analyzed in Chart 11–8 and is computed below:

(a) When $X = 200$, $z = (200 - 300)/50 = -2$. $A(-2) = 0.47725$.

When $X = 225$, $z = (225 - 300)/50 = -1.5$. $A(-1.5) = 0.43319$.

The area between 200 and 225 = the area between 200 and 300 —

the area between 225 and 300

$$= A(-2) - A(-1.5)$$
$$= 0.47725 - 0.43319$$
$$= 0.04406 \text{ or } 4.406\%$$

(b) When $X = 350$, $z = (350 - 300)/50 = 1$. $A(1) = 0.34134$.

When $X = 400$, $z = (400 - 300)/50 = 2$. $A(2) = 0.47725$.

The area between 350 and 400 = the area between 300 and 400 —

the area between 300 and 350

$$= A(2) - A(1)$$
$$= 0.47725 - 0.34134$$
$$= 0.13591 \text{ or } 13.591\%$$

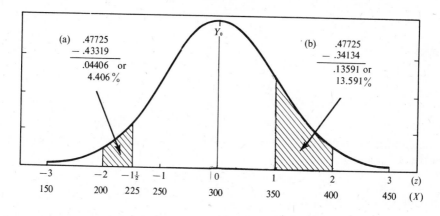

CHART 11–8

AREAS UNDER THE NORMAL CURVE FOR VALUES OF EXAMPLE 7

When the Area A(z) is Given, the X Value is to be Determined

Example 8

Assume that the average monthly income of 10,000 workers in King City is $500 and the standard deviation is $100. If the distribution is normal, what is the amount of the income above which are the earnings of 60% of the workers?

Solution. This problem is analyzed in Chart 11–9 and is computed as follows: Since the area to the right side of the maximum ordinate Y_0 is 50%, we must find the 10% (or 60% − 50%) area to the left of the maximum ordinate, or $A(-z) = 10\%$. Table 11–6 shows that

CHART 11–9

AREAS UNDER THE NORMAL CURVE FOR VALUES OF EXAMPLE 8

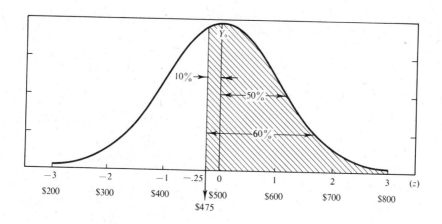

$$A(z) = A(0.25) = 0.09871 \text{ or } 9.871\%$$

which is the closest value to the required 10%. Since the area required is to the left side of Y_0, the value of z must be negative, or

$$z = -0.25$$

Since

$$z = \frac{x}{\sigma}$$

then

$$x = z(\sigma) = (-0.25)(100) = -\$25$$
$$X = \mu + x = 500 - 25 = \$475$$

Thus, above $475 on the X scale, there will be 60% of the area under the normal curve. In other words, there are

$$10{,}000 \times 60\% = 6{,}000 \text{ workers}$$

making a monthly income above $475.

Example 9

Let $\mu = 1{,}000$ miles and $\sigma = 200$ miles. Find (a) the point X_1 above which there will be 10% of the area under the curve and (b) the point X_2 below which there will be 10% of the area under the curve.

Solution. This problem is analyzed in Chart 11–10 and is computed as follows: First, find the 40% ($=50\% - 10\%$) area between the maximum ordinate and the ordinate at z, or $A(z) = 0.40$ or 40%. The area table, Table 6 in the Appendix (Table 11–6 is not detailed enough for this example) shows:

$$A(z) = A(1.28) = 0.39973 \text{ or } 39.973\%$$

which is the closest value to the required 40%.

CHART 11–10

AREAS UNDER THE NORMAL CURVE FOR VALUES OF EXAMPLE 9

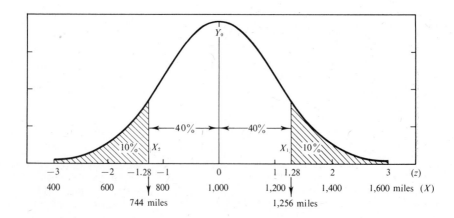

(a) Since the area required represents the top 10% of the values, the value of z must be on the right side of Y_0 and is positive, or

$$z = +1.28$$

Since

$$z = \frac{x}{\sigma}$$

then
$$x = z(\sigma) = 1.28(200) = 256 \text{ miles}$$
$$X_1 = \mu + x = 1{,}000 + 256 = 1{,}256 \text{ miles}$$

Thus, above 1,256 miles on the X scale, there will be 10% area under the normal curve.

(b) Since the area required represents the low 10% of the values, the value of z must be on the left side of Y_0 and is negative, or

$$z = -1.28$$
$$x = z(\sigma) = (-1.28)(200) = -256 \text{ miles}$$
$$X_2 = \mu + x = 1{,}000 + (-256) = 744 \text{ miles}$$

Thus, below 744 miles on the X scale, there will be 10% area under the normal curve. Note that there will be 80% values above 744 miles but below 1,256 miles (or 256 miles deviate from the mean, 1,000 miles, in both directions).

★11.4   Fitting a Normal Curve to Sample Data

There are at least two purposes for fitting a normal curve to sample data:

1. To provide a visual device for judging whether or not the normal curve is a good fit to the sample data.

A population usually follows the form of normal distribution if the individual values of the population are influenced by random forces, which must operate independently, on an *equal* basis. If the normal curve and the curve representing the sample data are close to each other, we have an additional reason to believe that the population is normally distributed.

2. To use the smoothed normal curve, instead of the irregular curve representing the sample data, to estimate the characteristics of the population.

If the population is normally distributed, the smoothed normal curve should serve as a basis for estimating the characteristics of the population. The curve representing sample data usually shows a lack of uniformity and, therefore, is not a good basis for estimating the population. The method of fitting the normal curve to sample data is illustrated in Example 10.

Example 10

The salaries earned by 25 employees in the Barnes Company during a given period are shown below. Assume that the salaries are selected as a sample. Use the sam-

TABLE 11–7

DATA FOR EXAMPLE 10

Salaries (Class interval)	Number of Employees (Actual frequency)
$ 1 — 3	1
4 — 6	4
7 — 9	9
10 — 12	6
13 — 15	2
16 — 18	3
Total	25

TABLE 11–8

COMPUTATION OF FITTING A NORMAL CURVE TO SAMPLE DATA

Salaries (Class interval)	Real Class Limits X	x $(X-\bar{X})$	z (x/s)	Area Between Y_0 and z	Theoretical Frequencies		Number of Employees (Actual)
					Between Y_0 and z	Between Class Limits	
(1)	(2)	(3)	(4)	(5)	(6)	(7)	(8)
	$-\infty$	$-\infty$	$-\infty$	0.50000	12.50		
Under 0.5						0.25	...
	0.5	−9.06	−2.32	0.48983	12.25		
1 — 3						1.26	1
	3.5	−6.06	−1.55	0.43943	10.99		
4 — 6						3.93	4
	6.5	−3.06	−0.78	0.28230	7.06		
7 — 9						6.86	9
	9.5	−0.06	−0.02	0.00798	0.20 ⎫		
10 — 12					⎬ 7.03		6
	12.5	2.94	0.75	0.27337	6.83 ⎭		
13 — 15						4.06	2
	15.5	5.94	1.52	0.43574	10.89		
16 — 18						1.33	3
	18.5	8.94	2.29	0.48899	12.22		
Above 18.5						0.28	...
	$+\infty$	∞	∞	0.50000	12.50		
Total						25.00	25

$\bar{X} = 9.56$ (see page 125); $s = 3.9$ (see page 187).

ple data to find the theoretical number of employees in each class interval. (Same as the values given in Example 7, Chapter 5, and in Example 10, Chapter 7.)

Solution. The theoretical frequency of each class is computed in Table 11–8, based on the table of Areas Under the Normal Curve (Table 6 in the Appendix). The procedure given in Table 11–8 is explained below.

Column (1) gives the original class intervals and two additional classes: (a) "Under 0.5," where 0.5 is the real lower class limit of the first class "1 − 3." (b) "Above 18.5," where 18.5 is the real upper class limit of the last class "16 − 18." The two classes are added for recording the theoretical frequencies computed in column (7).

Column (2) gives the real class limits for the class intervals in column (1). The real class limits (X's) are to be used for computing the theoretical frequencies from the table showing the areas under the normal curve. Since the lower and upper limits of the normal curve are infinite in both directions, the real limits in this column are extended from $-\infty$ to $+\infty$. We understand that the actual lower limit will not be below zero since no employee will make a monthly salary below nothing, or zero.

Column (3) shows the deviations from the mean, or $x = X - \bar{X}$. We should use the mean of population μ in constructing a normal curve. However, the population mean is not known in this case. The sample mean is used as an estimate of the population mean. The sample mean is computed on page 125 and is 9.56. Thus, for example, $-9.06 = 0.5 - 9.56$.

Column (4) expresses the value of each deviation (x) in units of the sample standard deviation (s), or $z = x/s$. The sample standard deviation is computed on page 187 and is 3.90. For example, $-2.32 = (-9.06)/3.9$. The normal curve calls for σ, the standard deviation of the population. However, the true σ is not known in the present case and is therefore estimated from the standard deviation of the sample.*

Column (5) shows the areas between the maximum ordinate Y_0 and the ordinates at z listed in column (4). The areas are obtained from the table of Areas Under the Normal Curve in the Appendix. (Table 11–6 is not detailed enough for the present case.) For example, the area between Y_0 and the ordinate at $z = -2.32$ is 0.48983. The area between Y_0 and the ordinate at $z = +\infty$ or $-\infty$ is equal to 50% of the total area under the curve, or 0.50000.

---

*When the sample standard deviation is used for estimating the population standard deviation, the value of the sample standard deviation is usually increased by applying formula $\hat{s} = \sqrt{s^2(n)/(n-1)}$. (See Chapter 13.) The value of s^2 for the above example is 15.2064. Substitute the value in the formula. We have $\hat{s} = \sqrt{(15.2064)(25)/(25-1)} = \sqrt{15.84} = 3.98$. When 3.98 is used instead of 3.9 for the estimated standard deviation of the population, the final answers to the theoretical frequencies in column (7) for classes from "Under 0.5" upward are 0.28, 1.33, 3.91, 6.78, 6.96, 4.04, 1.39, and 0.31 respectively. The changes of the individual frequencies in the column are rather small in the present case. Thus, the value of 3.9 (s) is used here for simplicity. When a sample has a small size, $\hat{s}$ is definitely superior to s. For example, when $n = 5$, the value of $n/(n-1) = 5/4 = 1.25$, which increases a variance s^2 by 25%.

Column (6) gives the theoretical (expected) frequencies based on the areas listed in column (5). Each area listed in column (5) is multiplied by 25, the total number of employees. Thus, $12.50 = 0.5 \times 25$; $12.25 = 0.48983 \times 25$; and so on. The computed theoretical frequencies can be interpreted in the same way as the areas under the normal curve. For example, the theoretical frequency 12.25 indicates that the monthly salaries of 12.25 employees are between $0.5 (which is $9.06 or 2.32 standard deviations [$z = -2.32$] smaller than the mean) and $9.56 (which is the mean or the location of the maximum ordinate Y_0). Likewise, the theoretical frequency 10.99 indicates that the monthly salaries of 10.99 employees are between $3.5 (which is $6.06 or 1.55 standard deviations small than the mean) and $9.56.

Column (7) shows the differences between the theoretical frequencies listed in column (6). For example, $12.25 - 10.99 = 1.26$, which represents the frequency between $0.5 and $3.5 (or class interval $1 - $3). One exception is the frequency 7.03, which is equal to $(0.20 + 6.83)$. The addition is necessary since the mean $9.56 is within the real class limits $9.5 and $12.5, or the class interval "$10 - 12$." This class covers both the area between Y_0 and $z = -0.02$ (or frequency $= 0.20$) and the area between Y_0 and $z = +0.75$ (or frequency $= 6.83$).

Column (8) gives the actual (observed) frequencies, or numbers of employees in the classes of the given example.

The results shown in columns (7) and (8) are plotted on Chart 11–11. Each value in the two columns is centered respectively within the real class limits labeled on the horizontal scale. The value of the maximum ordinate Y_0 for the normal curve is computed by formula (11–2).

$$Y_x = \frac{Ni}{\sigma} f(z) \tag{11--2}$$

The formula is obtained by multiplying the right side of formula (11–1) by i, the size of the class interval. The multiplication is necessary because each theoretical frequency in column (7) is the ordinate of an area with a width of class interval. In the present case,

$$Y_0 = \frac{25(3)}{3.9} f(0) = \frac{75}{3.9} (0.39894) = 7.67$$

where $i = \$3$.

Chart 11–11 shows the theoretical distribution (the normal curve) compared with the actual frequencies (sample data), class by class. The comparison reveals the fact that the normal curve does not fit at all points representing the actual frequencies. There are two possible reasons for the fact. (a) The population distribution *is* in the form of the normal curve. Thus, the departure of the sample data from the theoretical distribution is due to sample fluctuations. Ordinarily, there are many samples that may be drawn from the same population. A sample selected from the population usually does not fit the form of the population distribution perfectly, although a large size of sample may fit the population better. The problem of determining the proper size of a sample and other problems

CHART 11–11

FITTING A NORMAL CURVE TO SAMPLE DATA

(*Salaries Earned by 25 Employees in the Barnes Company*)

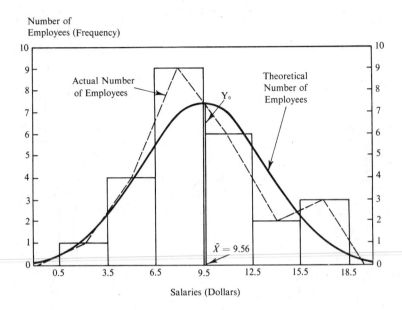

Source: Table 11-8.

between a sample and its population are discussed in detail in Chapters 12 through 18. (b) The population distribution *is* actually *not* in the form of the normal curve. In such cases, the normal curve should not be used for describing the characteristics of the population. A test of whether or not the departure of the sample data from the normal curve is too large to be considered sample fluctuation is given in Chapter 17 when the chi-square distribution is presented. The chi-square distribution may be used to test the fit of the normal curve to sample data. (See Example 2 of Chapter 17.)

At this stage, we may rely on our observation of the normal curve and the curve representing the sample data in the chart to draw a conclusion. Since the two curves are fairly close to each other, it is reasonable to conclude that the population distribution of the salaries earned by all employees in the Barnes Company during the given period is in the normal form.

11.5 Summary

The normal distribution, also called the normal probability distribution, can be shown graphically as a symmetrical or bell-shaped curve. The shape of the normal curve indicates that the frequencies in a normal distribution are concentrated in the center portion of the distribution and the values above and below the mean are equally distributed.

Basically, the normal curve is constructed according to the heights of the ordinates (Y variable) computed from formula (11–1), or $Y_x = (N/\sigma)f(z)$. In the application of this formula, three values, N, μ, and σ, must be known. When any one of the three values of a distribution is different from the one of another distribution, the shapes of the two normal curves are different from each other. This fact is illustrated in Examples 1 to 4 of Section 11.1. However, all shapes of normal curves can be converted to the shape of the standard normal curve (Chart 11–4). The standard normal curve has unit population ($N = 1$), unit standard deviation ($\sigma = 1$), and zero mean ($\mu = 0$).

The probabilities represented by areas of various sizes under the standard normal curve have been computed, and a portion of the computed probabilities is listed in Table 11–6. This table shows the areas under the normal curve between the ordinate at mean (maximum ordinate) and ordinates at various values of z. Because any shape of normal curve can be converted to the shape of the standard normal curve, the table is the only one required in finding the probability of a certain area under a normal curve. The method of computing the probability based on Table 11–6 is illustrated in Examples 5 to 9 of Section 11.2.

There are at least two purposes for fitting a normal curve to sample data:

1. To provide a visual device for judging whether or not the normal curve is a good fit to the sample data. Two possible reasons for the normal curve not completely fitting the sample data are: (a) sample fluctuation, (b) population distribution not actually in the form of the normal curve. In such cases, the normal curve should not be used for describing the characteristics of the population. A test of whether or not the departure of the sample data from the normal curve is too large to be considered sample fluctuation is given in Chapter 17 when the chi-square distribution is presented.

SUMMARY OF FORMULAS

Application	Formula	Formula number	Reference page
Constructing a normal curve			
General form	$Y_x = \dfrac{N}{\sigma} f(z)$	11–1	274
Standard form	where $z = x/\sigma$ $Y_z = f(z)$ where $N = 1$, $\sigma = 1$, and $\mu = 0$	11–1	279
Fitting a normal curve to sample data			
Grouped data	$Y_x = \dfrac{Ni}{\sigma} f(z)$ where i = class interval	11–2	290

2. To use the smooth normal curve, instead of the irregular curve representing the sample data, to estimate the characteristics of the population if the population is normally distributed.

Exercises 11

1. What is the normal probability distribution? Why is it regarded as the most important type among various probability distributions?

2. What values must be known in constructing a normal curve for (a) a finite population and (b) an infinite population?

3. What are the properties of the standard normal curve? Why is the table showing the areas under the standard normal curve the only one required in finding the probability of a certain area under a normal curve of any shape?

4. What is the standard normal deviate? How would it be used in locating an area under the normal curve?

5. Assume that the average typing speed of 800 students in College X is 40 words per minute and the standard deviation is 10 words. Construct a normal curve. Show the computation of the ordinates at an interval of $\frac{1}{2}\sigma$ from the mean up to 3σ. Also, explain the method for converting the normal curve constructed to the shape of the standard normal curve.

6. Refer to Problem 5. Assume that the average typing speed of the 800 students is 55 words per minute and the standard deviation is also 10 words. Construct a normal curve. The computation of the ordinates is not required to be shown.

7. Refer to Problem 5. Find the number of students having a typing speed:
    (a) From 20 to 60 words.
    (b) Above 40 but below 50 words.
    (c) From 25 to 50 words.
    (d) Above 20 but below 35 words.

8. Refer to Problem 5. Find the point on the X scale:
    (a) Above the point are the speeds of 65% of the students.
    (b) Below the point are the speeds of 20% of the students.
    (c) Above the point there will be 5% of the area under the normal curve.
    (d) Below the point there will be 5% of the area under the normal curve.

9. The mean and the standard deviation of all cash sale amounts in Abertson Company in the last year are $60 and $10, respectively. Assume that the amounts are distributed normally.
    (a) What proportion of the amounts is
        (1) between $55 and $72.50?
        (2) above $72.50?
        (3) below $55?
    (b) What is the amount
        (1) above which the proportion of the amounts is 20%?
        (2) below which the proportion of the amounts is 20%?

10. Refer to Problem 9(a). Assume that Abertson Company has 10,000 cash sale transactions during January of this year. Estimate the number of transactions in each case based on the found proportion.

★11. What are the purposes of fitting a normal curve to sample data? What are the possible reasons for the normal curve not fitting all points representing the sample data?

★12. The grades of the 80 students in a statistics class are given below: (Same as the values given in Problem 10. Exercises 7.)

Grades (Class interval)	Number of Students
20—29	3
30—39	6
40—49	5
50—59	7
60—69	10
70—79	29
80—89	12
90—99	8
Total	80
$\bar{X} = 68.25$	$s = 18.3$

Required:

(a) Compute the theoretical number of students of each class interval. (Compute to two decimal places.)

(b) Draw a normal curve to fit the actual data. Show the value of Y_0.

(c) If 100 students are expected to take this course in the coming year, how many students are estimated to make grades from 69.5 to 89.5 according to the normal curve?

(d) Examine the normal curve and the curve representing the sample data. Do you believe that the grades are normally distributed? Would you use the normal curve instead of the actual data to estimate the grades of the students? Why?

12 Sampling Distributions

A sampling distribution involves all possible samples of the same size drawn from a given population. This chapter will illustrate the basic relationship between a population and the samples drawn from the population. A knowledge of the relationship is important because it aids in understanding the process of statistical inference. Discussion of the process of statistical inference is presented in Part 4 (Chapters 13 through 16). In this chapter, the new terms concerning the population and the samples are first introduced (12.1). The methods of selecting a sample from a given population are then discussed (12.2). The mean and the proportion of a population are the most frequently used figures in statistical inference. Thus, the mean and the proportion of a population and those of all possible samples of same size drawn from the population are analyzed in detail (12.3 to 12.5).

12.1 Terminology

The frequently used terms in statistical inference are as follows:

Statistic

A statistic is a measure used to describe some characteristic of a *sample*, such as an arithmetic mean, a median, or a standard deviation of a sample.

Parameter

A parameter is a measure used to describe some characteristic of a *population*, such as an arithmetic mean, a median, or a standard deviation of a population.

When these two terms are used, the process of estimation in statistical inference may be described as the process of estimating a parameter from the corresponding statistic, such as using a sample mean (a statistic) to estimate the population mean (a parameter).

The symbols used to represent the *statistics* and *parameters* in this and the following chapters are summarized in Table 12–1.

TABLE 12-1

SYMBOLS FOR CORRESPONDING STATISTICS AND PARAMETERS

Measure	Symbol for Statistic (Sample)	Symbol for Parameter (Population)
Mean	$\bar{X}$	μ
Standard deviation	s	σ
Number of items	n	N
Proportion	p	P

Sampling Distribution

When the sample size (n) is smaller than the population size (N), two or more samples may be drawn from the same population. A certain statistic can be computed for each of the samples possibly drawn from the population. A distribution of the *statistics* obtained from the samples is called the *sampling distribution of the statistic*.

For example, if the sample size is two and the population size is three (items A, B, C), it is possible to draw three samples (AB, BC, and AC) from the population. We may compute the mean for each sample. Thus, we have three sample means for the three samples. The three sample means form a distribution. The distribution of the means is called the *distribution of sample means*, or the *sampling distribution of the mean*. Likewise, the distribution of the proportions (or percent rates) obtained from all possible samples of the same size drawn from a population is called the *sampling distribution of the proportion*.

Standard Error

The standard deviation of a sampling distribution of a statistic is frequently called the *standard error of the statistic*. For example, the standard deviation of the means of all possible samples of the same size drawn from a population is called the *standard error of the mean*. Likewise, the standard deviation of the proportions of all possible samples of the same size drawn from a population is called the *standard error of the proportion*. The difference between the terms "standard deviation" and "standard error" is that the former concerns original values, whereas the latter concerns computed values. A statistic is a computed value, computed from the items included in a sample.

Sampling Error

The difference between the result obtained from a sample (a statistic) and the result which would have been obtained from the population (the corresponding parameter) is called the *sampling error*. A sampling error usually occurs when the complete survey of the population is not carried out, but a sample is taken for estimating the characteristic of the population. The sampling error is measured by the standard error of the statistic in terms of probability under the normal curve. The result of the measurement indicates the *precision* of the estimate of the population based on the sample study. The smaller the sampling error, the greater is the precision of the estimate. For example, the difference between the mean of a sample and the mean of the population, if it were obtained, is a type of sampling error and is measured by the standard error of the mean. It should be noted that the errors made in a sample survey, such as answers being inconsistent, incomplete, or not determinable (as discussed in Section 2.1), are not considered sampling errors. The nonsampling errors may also occur in a complete survey of the population.

12.2   Methods of Selecting Samples

A sample must be a representative one if it is to be used for estimating the characteristics of the population. The procedure of selecting a representative sample depends on numerous factors, such as time, money, skill available for taking a sample, and the nature of the individual items of the population. A large volume would be required to include all types of the sampling methods. In this section, however, only the most common ones are discussed.

The methods of selecting samples may be classified according to (1) the number of samples taken from a given population for a study and (2) the manner used in selecting the sample items. The sampling methods based on the two types of classifications are discussed below.

Classification According to the Number of Samples Taken

Under this classification, there are three common types of sampling methods. They are single, double, and multiple sampling.

Single Sampling
This type of sampling uses only one sample from a given population for the purpose of statistical inference. Since only one sample is taken, the sample size must be large enough for drawing a conclusion. A large sample often costs more money and time than necessary.

Double Sampling
Under this type of sampling, when the result of the study of the first sample is not decisive, a second sample is drawn from the same population. The two samples are then combined in analyzing the results. This method permits the person to begin with a relatively small sample to save costs and time. If the first sample shows a definite result, the second sample may not be needed.

For example, in testing the quality of a lot of manufactured products, if the first sample shows a very high quality, the lot is accepted; if it shows a very poor quality; the lot is rejected. Only if the first sample shows an intermediate quality will the second sample be required. A typical double sampling plan may be obtained from the *Military Standard Sampling Procedures and Tables for Inspection by Attributes*, published by the Department of Defense and also used by many private industries. In testing the quality of a lot consisting of 3,000 manufactured units, when the number of defects found in the first sample of 80 units is 5 or less, the lot is considered good and is accepted; if the number of defects is 9 or more, the lot is considered poor and is rejected; if the number is between 5 and 9, a decision cannot be reached and the second sample of 80 units is drawn from the lot. If the number of defects in the two samples combined (including 80 + 80 = 160 units) is 12 or less. the lot is accepted; if the combined number is 13 or more, the lot is rejected. (See Table 16–8, page 410).

Multiple Sampling

The procedure under this method is similar to that discussed in double sampling except that the number of successive samples required to reach a decision is more than two samples. Table 16–8 also shows a multiple sampling plan.

Classification According to the Method Used in Selecting the Items

The items of a sample may be selected in two different manners: based on the judgment of a person and selection at random. The sampling methods classified according to the two different manners are presented below.

Judgment Sampling

A sample is called a judgment sample when the items of the sample are selected by personal judgment. The person who selects the items of the sample usually is an expert on the given subject. A judgment sample is also called a *nonprobability sample* since this method is based on a person's subjective views and the *theoretical probability* cannot be employed in measuring the sampling error. The chief advantages of a judgment sample are the ease in obtaining it and the usual low cost.

Random Sampling

A sample is said to be drawn *at random* when the manner of selection is such that each element of the population has an *equal chance* of being selected. A random sample is also called a *probability sample* since each element has a known chance. Probability samples are generally preferred by statisticians because the selection of the samples is objective and the sampling error may be measured in terms of probability under the normal curve. The common types of random sampling are simple random sampling, systematic sampling, stratified sampling, and cluster sampling.

Simple Random Sampling.   A simple random sample is selected in such a way that each possible *sample* of the same size has an *equal probability* of being selected from the population. For example, in a population of three items (A, B, C) it is possible to draw three samples of two items each (AB, BC, and AC). If each of the three samples has the same probability ($\frac{1}{3}$) of being selected, a selected sample is a simple random sample.

To obtain a simple random sample, each *element* in the population must have an *equal chance* of being selected. However, it is also possible that although each element in the population has the same chance of being selected, the sampling plan may not yield a simple random sample. For example, if we wish to estimate the heights of all college students by using a sample of five students, it is not a simple random sample if we ask only the students in a boy's dormitory. Although each boy living in the dormitory has an equal chance of being selected, not all the possible samples of the same size, such as a sample consisting of three boys and two girls, have the same probability of being selected.

A simple method to obtain a simple random sample is first to write the name or a code number of each item in the population on a card. The cards are dropped in a box. A sample is then drawn from the box after the cards are thoroughly mixed. For convenience, this method may be replaced by a table of random numbers, such as shown in Table 12–2.

The table is constructed by drawing each of the digits from 0 to 9 on an "equally likely" basis; that is, each of the ten digits has the same chance ($\frac{1}{10}$) of being selected. The ten digits are written on separate cards and are then mixed in a box. A card is drawn and the digit on the card is recorded. A second card is drawn after the first card is returned to the box and the ten digits in the box are again thoroughly mixed. When five digits are recorded, the next digit is recorded in a separate group until a large number of groups is obtained. Table 12–2 shows only 320 groups of five digits each for the purpose of illustration. Many extensive random tables are published and are available for practical uses.

Application of the random table in drawing a sample is simple. For example, if we wish to select four students from a group of 30 students at random, we may first assign each student a number from 01 to 30. Next, select a starting point in the table at random, such as by dropping a finger on the table after closing eyes. Then move from the starting point in any direction, such as moving down, up, right, left, or diagonally, to get the desired numbers on the table. Assume that the starting point is on the fourth column and thirty-first row in the table and the moving direction is from the starting point down. Only the first two digits in each group of five digits are needed for our purpose since our numbers are of two digits (from 01 to 30). The numbers are ignored if they are larger than the largest number in the population (30) or are repeated. Thus, the numbers representing the four students are 22 (80 and 64 are ignored since they are larger than 30), 23 (the second 23 is ignored), 28, and 01.

To obtain a simple random sample is not an easy or practical task under many circumstances. It may be time-consuming or costly and sometimes is theoretically impossible. For example, if we wish to take a simple random sample from a large finite population of one million families, although it is possible, it is not a simple task to assign a number to each of the families and then to draw a sample at random from the numbers. When a population is infinite, it is obvious that the task of numbering each element of the population is impossible. Therefore, certain modifications of simple random sampling are necessary. The most common types of the modified random samples are systematic, stratified, and cluster samples.

Systematic Sampling. A systematic sample is obtained when the items are selected in an orderly manner. The manner of the selection depends on the number of items included in the population and the size of the sample. The number of items in the population is first divided by the number desired in the sample. The quo-

TABLE 12-2

RANDOM NUMBERS

Line	(1)	(2)	(3)	(4)	(5)	(6)	(7)	(8)
1	24571	23165	39407	60614	99692	53643	15237	75497
2	13670	32919	85543	04891	95940	36404	76575	21672
3	30051	56205	28399	57818	50250	64143	21454	43778
4	17977	66365	89867	29215	16767	78664	61052	20792
5	92178	52766	05531	89370	29936	73564	89039	87520
6	55136	12504	50905	63482	77089	16116	69540	42617
7	56292	65313	87697	77362	25261	41434	53533	78662
8	28852	23758	99995	89994	80072	16037	09242	02476
9	36575	00384	56044	71864	37692	93583	67871	55693
10	73548	41988	41754	77623	74789	47006	71348	08856
11	72077	69908	26013	89159	29262	21100	13848	26884
12	62212	59442	56691	84042	17000	40994	90372	92380
13	15835	87145	62164	00392	48946	85269	43931	84040
14	11418	34412	57620	27362	40064	36081	14038	56486
15	75712	00893	75595	99815	16218	04983	25848	55323
16	82781	43481	65187	25236	97977	79008	07923	01439
17	11341	31929	46669	91080	53736	98034	03093	82350
18	67303	67361	40344	20562	65616	94776	74398	73140
19	15563	55883	75112	51585	85050	45388	82118	85799
20	24516	67385	30307	70874	29955	71904	07820	24392
21	29072	24881	51692	76856	70769	91318	33021	56804
22	98398	84599	80434	23325	57478	09349	70354	51570
23	76460	07180	17586	89669	51861	94388	38384	55181
24	74656	38853	70503	48664	67571	54981	12207	02247
25	44671	32847	91009	54934	52683	77609	50212	30923
26	27906	28645	45676	35369	65227	48617	74535	12413
27	40996	11863	78692	95991	14621	30490	84343	91097
28	21917	59181	36117	04700	17593	03176	08479	02393
29	71416	97913	42120	91634	87433	42675	21008	72725
30	06475	16974	50487	28922	50273	34026	12902	41020
31	46714	35033	60626	22039	28633	10540	89239	59295
32	44155	67697	73377	80482	52186	90008	20379	21490
33	64558	40620	56281	64703	21641	93937	48274	05923
34	48437	75697	68647	23526	89468	08261	92414	84681
35	48573	61866	08920	23339	55006	41144	53299	19156
36	15915	75596	87992	28897	85916	53472	02117	10983
37	65319	21980	74852	01770	82811	57641	40814	03221
38	16521	15881	35674	05940	79340	40810	89367	85602
39	83865	33163	21158	89532	62634	05451	83992	28510
40	47653	32290	48778	87661	56275	33849	92014	20928

tient will indicate whether every tenth, every eleventh, or every hundredth item in the population is to be selected.

The first item of the sample is selected at random. Thus, a systematic sample may yield the same precision of estimate about the population as a simple random sample when the items in the population are in random order. For example, assume that a college registration office has a file of 10,000 students, filed according to alphabetical order. If a sample of the ages of 200 students is desired, the ages may be selected from the file for every 50th student (10,000 ÷ 200 = 50), since the order of the file has nothing to do with the ages of the students. The first age may be selected from the first group of 50 students in the file at random. If the first item is 18th in the population, the second will be 68th (= 18 + 50), the third will be 118th, the fourth will be 168th, and so on.

The systematic sampling may also be carried out by using physical measurement. If the information is filed by means of cards of uniform thickness, such as IBM cards, the linear space of the file is measured and the length is divided by the number of items needed for the sample. Assume that the file occupies 67 inches and that 200 items are required in the sample. A card is selected at random in the first 1/3 inch since 67 inches ÷ 200 = .335 or approximately 1/3 inch. Thereafter, one card is selected for every 1/3 inch in the file, starting from the first card.

Stratified Sampling. The first step in obtaining a stratified random sample is to divide of the population into groups, called *strata*, that are more homogeneous than the population as a whole. The items of the sample are then selected at random or by a systematic method from each stratum. Estimates of the population based on the stratified sample usually have greater precision (or smaller sampling error) than if the whole population were sampled by simple random sampling.

The number of items selected from each stratum may be proportionate or disproportionate to the size of the stratum in relation to the population. Under the proportionate method, for example, if the size of stratum A is 40 percent of the population, the same rate is used to select the number of items of the sample from the stratum. Thus, if the sample size is 200 items, 40 percent of the sample size, or 80 items, are to be selected from stratum A. When the selection is disproportionate, however, it is relatively difficult to weigh the results from individual strata properly.

Cluster Sampling. A cluster sample is obtained by dividing the population into groups that are convenient for sampling. Next, a portion of the groups is selected at random or by a systematic method. Finally, all of the items or a part of the items are taken at random or by a systematic method from the selected groups to obtain a sample. Under this method, although not all groups are sampled, every group does have an equal chance of being selected. Thus the sample is a random one.

For example, when it is impractical to take a simple random sample from a population of one million families in a city, the city may be divided into small areas according to the city map. A number is assigned to each area. A part of the areas is selected at random or by a systematic method from the numbers representing the areas. The families within the selected areas are interviewed, either all of the families or families selected at random or by a systematic method. This kind of cluster sample is also called an *area sample*.

A cluster sample usually yields a greater sampling error (thus gives less precision of the estimate about the population) than a simple random sample of the same size. Individual items within each "cluster" usually tend to be alike. For example, rich people may live in the one neighborhood whereas poor people may live in another area. Not all of the areas are sampled in area sampling. The variation among the items obtained from the selected areas is therefore frequently greater than that if the whole population is sampled at one time by simple random sampling. This weakness may be reduced when the size of the area sample is increased. The increase of the sample size can easily be done in an area sample. The interviewers do not have to walk too far in a small area to interview more families. Thus a large area sample may be obtained within a short period of time and at a low cost.

On the other hand, a cluster sample may yield the same precision of the estimate as a simple random sample if the variation of the individual items within each cluster is as great as that of the population.

12.3 Sampling Distribution of the Mean ($\overline{X}$)

The sampling distribution of the mean, or the distribution of the means obtained from all possible samples of the same size drawn from the population, provides a cornerstone of statistical inference. The relationship between the mean of the population and the sampling distribution of the mean may be illustrated by using a small finite population. In actual work, a sample may be drawn from a large finite population or from a population which is infinitely large. However, the basic relationship is the same regardless of the size of a population. For convenience, only simple random samples are used in the illustration of this chapter.

Example 1

The hourly wages of the six workers in the Small Shop are given in Table 12–3 below. The table also shows the computation of the mean and the standard deviation of the wages. Consider the wages earned in the shop as a population. Find: (a) the means of the samples of two items (wages of two workers), (b) the mean of the sample means, (c) the standard error of the mean, and (d) the probabilities of the sample means.

Solution. (a) The means of the samples of the hourly wages of two workers are shown in the last column of Table 12–4. The table shows that for a simple random sample of two items from the population of six items, the number of the

TABLE 12-3

HOURLY WAGES IN THE SMALL SHOP AND THE
COMPUTATION OF THE MEAN AND THE STANDARD
DEVIATION OF THE POPULATION

Workers	Hourly Wages X	x $(= X - \mu)$	x^2
A	\$1	-2	4
B	2	-1	1
C	3	0	0
D	3	0	0
E	4	1	1
F	5	2	4
Total	\$18	0	10

$$\mu = \frac{18}{6} = \$3$$

$$\sigma^2 = \frac{10}{6} = \frac{5}{3}$$

$$\sigma = \sqrt{\frac{5}{3}} = \$1.29$$

possible samples is 15.* They are A and B, A and C, . . . and so on as listed in the fourth column. There are 15 sample means ($\bar{X}$), one for each sample. The sample mean of sample (A and B), for example, is computed by dividing the sum of the hourly wages of workers A and B by two, or $\bar{X} = (1 + 2)/2 = 1.5$.

(b) The total of the 15 sample means is \$45 as shown in the last column of Table 12–4. Thus, the mean of the sample means, denoted by $\bar{\bar{X}}$, is

$$\bar{\bar{X}} = \frac{45}{15} = \$3$$

which is equal to the mean of the population, or $\bar{\bar{X}} = \mu$.

(c) The standard error of the mean, or the standard deviation of all possible sample means, denoted by $\sigma_{\bar{x}}$, is computed in Table 12–5. $\sigma_{\bar{x}} = \$0.816$.

(d) The probabilities of the 15 sample means are shown in the last column of Table 12–5. When the 15 samples are selected at random, each sample will have the probability 1/15 of being selected. Since there are three samples with the mean \$3.5 (Table 12–4), for example, the probability of a sample being selected to have the mean \$3.5 is $1/15 + 1/15 + 1/15 = 3/15$.

The method of computing the standard error of the mean illustrated in Example 1 involves the tedious work of finding all possible sample means. When a population is very large, it is almost impossible to do the work. For example, the

---

*The number of possible samples of size two taken from six items may be computed as follows.

$$_6C_2 = \frac{6 \times 5}{2 \times 1} = 15$$

TABLE 12-4

COMPUTATION OF SAMPLE MEANS FOR EXAMPLE 1

Workers (1)	(2)	Possible Arrangements (Permutations) (3)	Samples (Combinations) (4)	Total Wage in Each Sample (5)	Sample Means (Total × $\frac{1}{2}$) (6)
A	B	AB	A,B	$1+2=3$	1.5
	C	AC	A,C	$1+3=4$	2.0
	D	AD	A,D	$1+3=4$	2.0
	E	AE	A,E	$1+4=5$	2.5
	F	AF	A,F	$1+5=6$	3.0
B	A	BA	*		
	C	BC	B,C	$2+3=5$	2.5
	D	BD	B,D	$2+3=5$	2.5
	E	BE	B,E	$2+4=6$	3.0
	F	BF	B,F	$2+5=7$	3.5
C	A	CA	*		
	B	CB	*		
	D	CD	C,D	$3+3=6$	3.0
	E	CE	C,E	$3+4=7$	3.5
	F	CF	C,F	$3+5=8$	4.0
D	A	DA	*		
	B	DB	*		
	C	DC	*		
	E	DE	D,E	$3+4=7$	3.5
	F	DF	D,F	$3+5=8$	4.0
E	A	EA	*		
	B	EB	*		
	C	EC	*		
	D	ED	*		
	F	EF	E,F	$4+5=9$	4.5
F	A	FA	*		
	B	FB	*		
	C	FC	*		
	D	FD	*		
	E	FE	*		
Total					45.0

*The arrangements with the identical letters are counted as one sample. For example, the group of worker A and worker B is same as the group consisting of worker B and worker A. Thus, AB and BA are considered as one sample, or "A, B."

number of possible samples of five items drawn from a population of only 1,000 items is more than 990 trillion.† In practice, the following formula is usually employed in computing the standard error of the mean:

$$\sigma_{\bar{x}} = \sqrt{\frac{\sigma^2}{n} \cdot \frac{N-n}{N-1}} \quad \text{or} \quad \sigma_{\bar{x}} = \frac{\sigma}{\sqrt{n}} \sqrt{\frac{N-n}{N-1}} \qquad (12\text{-}1)$$

---

† $_{1,000}C_5 = \dfrac{1,000 \times 999 \times 998 \times 997 \times 996}{5 \times 4 \times 3 \times 2 \times 1} = 990{,}034{,}950{,}024{,}000$ samples

TABLE 12-5

COMPUTATION OF THE STANDARD ERROR OF THE MEAN AND THE
PROBABILITY DISTRIBUTION OF SAMPLE MEANS (EXAMPLE 1)

Sample Means $\bar{X}$	Number of Sample Means f	d $(A = 3.0)$	fd	$f(d)^2$	Probability of Sample Means
$1.5	1	-3	-3	9	1/15
2.0	2	-2	-4	8	2/15
2.5	3	-1	-3	3	3/15
3.0	3	0	0	0	3/15
3.5	3	1	3	3	3/15
4.0	2	2	4	8	2/15
4.5	1	3	3	9	1/15
Total	15		0	40	15/15

$$\sigma_{\bar{x}}^2 = (1/2)^2 \left[\frac{40}{15} - \left(\frac{0}{15} \right)^2 \right] = \frac{2}{3}$$

$$\sigma_{\bar{x}} = \sqrt{2/3} = \$0.816 \qquad \text{(Formula 7-8d where } i = 0.5 = 1/2)$$

Note: The mean of the sample means may also be computed from Table 12-5 by using formula (5-3c).

$$\bar{\bar{X}} = 3 + \frac{0}{15}(0.5) = 3 \text{ (where A or the assumed mean} = \$3)$$

where $\sigma_{\bar{x}}$ = the standard error of the mean
              σ = the standard deviation of the population
              N = the size of the population
              n = the size of the sample

When this formula is used, the standard error of the mean for Example 1 is computed as follows:

$$\sigma_{\bar{x}} = \sqrt{\frac{5/3}{2} \cdot \frac{6-2}{6-1}} = \sqrt{\frac{2}{3}} = \$0.816$$

The size of the samples drawn from a population affects the values of the sampling distribution of the mean and the standard error of the mean. Example 2 is used to illustrate the effects of larger size samples drawn from the same population of Example 1.

Example 2

Refer to Example 1. Find (a) the means of the samples of four items (the wages of four workers) from the population, (b) the mean of the sample means, (c) the standard error of the mean, and (d) the probabilities of the sample means.

Solution. (a) The means of the samples of the hourly wages of four workers are shown in Table 12–6. The possible samples listed on the table are obtained by a method similar to the diagram used in Table 12–4. Table 12–6 shows that for a simple random sample of four items from the population of six items, the number of the possible samples is 15.* The sample mean of the first sample A, B, C, D, for example, is obtained by dividing the total of the hourly wages of the workers by four, or

$$\bar{X} = (1 + 2 + 3 + 3)/4 = \$2.25$$

(b) The total of the 15 sample means is \$45 as shown in the table. Thus, the mean of the sample means is

$$\bar{\bar{X}} = \frac{45}{15} = \$3$$

which is also equal to the mean of the population, or $\bar{\bar{X}} = \mu$.

TABLE 12-6

COMPUTATION OF SAMPLE MEANS FOR EXAMPLE 2

Workers in Each Sample (Combination)	Total Wage in Each Sample	Sample Means (Total $\times \frac{1}{4}$)
A,B,C,D,	$1+2+3+3=\ 9$	\$2.25
A,B,C,E,	$1+2+3+4=10$	2.50
A,B,C,F,	$1+2+3+5=11$	2.75
A,B,D,E,	$1+2+3+4=10$	2.50
A,B,D,F,	$1+2+3+5=11$	2.75
A,B,E,F,	$1+2+4+5=12$	3.00
A,C,D,E,	$1+3+3+4=11$	2.75
A,C,D,F,	$1+3+3+5=12$	3.00
A,C,E,F,	$1+3+4+5=13$	3.25
A,D,E,F,	$1+3+4+5=13$	3.25
B,C,D,E,	$2+3+3+4=12$	3.00
B,C,D,F,	$2+3+3+5=13$	3.25
B,C,E,F,	$2+3+4+5=14$	3.50
B,D,E,F,	$2+3+4+5=14$	3.50
C,D,E,F,	$3+3+4+5=15$	3.75
Total 15 samples		\$45.00

$$\bar{X} = \frac{45}{15} = \$3 = \mu$$

(c) The standard error of the mean is computed by formula (12–1) since the standard deviation of the population is known (from Example 1).

$$\sigma_{\bar{x}} = \sqrt{\frac{\sigma^2}{n} \cdot \frac{N-n}{N-1}} = \sqrt{\frac{5/3}{4} \cdot \frac{6-4}{6-1}} = \sqrt{\frac{1}{6}} = \$0.408$$

(d) The probabilities of the 15 sample means are shown in the last column of Table 12–7. When the 15 samples are selected at random, each sample will have the probability 1/15 of being selected. Since there are two samples with the mean \$3.5, for example, the probabiltiy of a sample being selected to have the mean \$3.5 is 1/15 + 1/15 = 2/15.

---

*The number of possible samples of size four taken from six items may be computed as follows:

$$_6C_4 = \frac{6 \times 5 \times 4 \times 3}{4 \times 3 \times 2 \times 1} = 15$$

TABLE 12-7

THE PROBABILITY DISTRIBUTION OF SAMPLE MEANS
FOR EXAMPLE 2

Sample Means $\bar{X}$	Number of Sample Means f	Probability (*From Table* 12-6)
$2.25	1	1/15
2.50	2	2/15
2.75	3	3/15
3.00	3	3/15
3.25	3	3/15
3.50	2	2/15
3.75	1	1/15
Total	15	15/15

The sampling distributions of the means obtained in Examples 1 and 2 are plotted on Charts 12–1 and 12–2, respectively. The charts indicate:

(1) The mean of the sample means is always equal to the population mean. That is,

$$\bar{\bar{X}} = \mu = \$3$$

The symbol $\bar{\bar{X}}$ is also frequently written as $E(\bar{X})$, which represents the *expected* value of $\bar{X}$.

(2) The class interval of the means (or the width of the bar in the histogram) is equal to $1/n$. When the sample size is two (Chart 12–1), $1/n = 1/2 = 0.5$; when the sample size is four (Chart 12–2), $1/n = 1/4 = 0.25$. Thus, when the sample size becomes larger and larger, the class interval gets smaller and smaller. If the middle points on the tops of the bars are connected by straight lines to form a frequency polygon, the polygon will be closer and closer to the smoothed normal curve.

(3) When the size of the samples drawn from a population is increased, the dispersion of the sample means from the population mean is decreased. In the two examples, when $n = 2$, the standard error of the mean is 0.816; however, when $n = 4$, the standard error of the mean is only 0.408. Also, the range of the sample means is $3 (= 4.5 - 1.5)$ for $n = 2$, and is only $1.5 (= 3.75 - 2.25)$ for $n = 4$. This fact is called the *law of large numbers* and is important in finding a proper sample size for a given population by the use of formula (12–1) as discussed in the next chapter.

A normal curve fitted to the sampling distribution of the mean is constructed in each of the two charts (12–1 and 12–2).* The curve in each chart shows a

---

*The normal curves are constructed by applying formula (11-2), where σ is replaced by $\sigma_{\bar{x}}$. For example, Y_o for Chart 12-1 is:

$$\frac{Ni}{\sigma_{\bar{x}}} f(z) = \frac{15(0.5)}{0.816}(0.39894) = 3.7$$

Y_o for Chart 12-2 is:

$$\frac{Ni}{\sigma_{\bar{x}}} f(z) = \frac{15(0.25)}{0.408}(0.39894) = 3.7$$

CHART 12-1

SAMPLING DISTRIBUTION OF THE MEAN, SAMPLE SIZE $n = 2$ (EXAMPLE 1)

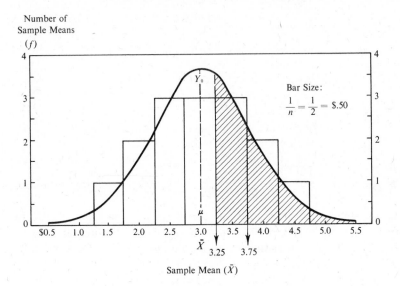

Source: Table 12-5.

CHART 12-2

SAMPLING DISTRIBUTION OF THE MEAN, SAMPLE SIZE $n = 4$ (EXAMPLE 2)

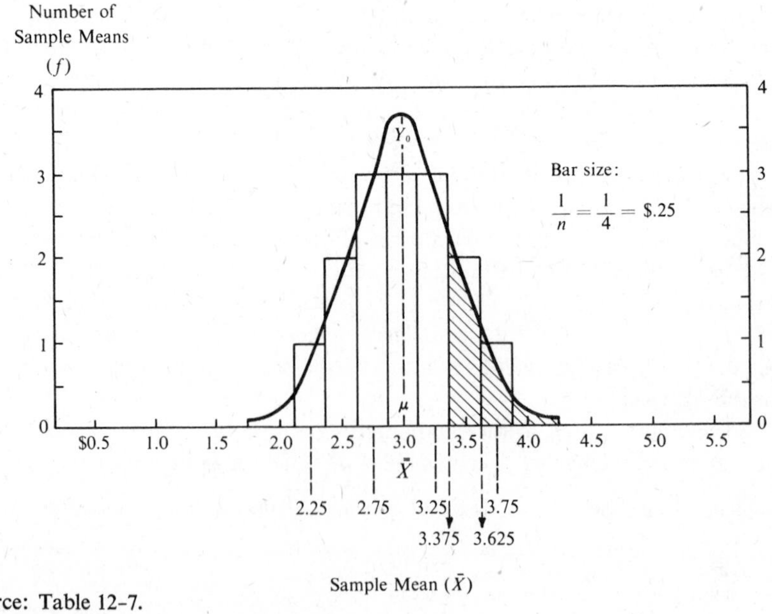

Source: Table 12-7.

close fitting to the sampling distribution. Since it is impractical to obtain all sample means in every study, the normal curve is usually used to approximate the probabilities of the sample means in a sampling distribution. In general the normality of a sampling distribution of the mean is called *central limit theorem* and may be stated as follows:

(1) When the population is fairly large and is normally distributed, the distribution of the sample means will be normal.

(2) When the population is not normally distributed, the distribution of the sample means will approach a normal distribution if the sample size is large enough, usually 30 or more.

(3) The normal distribution of sample means has the mean equal to the expected mean $E(\bar{X})$ and the standard error $\sigma_{\bar{x}}$. The values of $E(\bar{X})$ and $\sigma_{\bar{x}}$ can theoretically be computed from the population mean μ and the population standard deviation σ, respectively.

In finding the probability of a sample mean from a sampling distribution by the normal curve, we must know the following three values:

$$\mu = \text{the mean of the population}$$
$$\bar{X} = \text{the sample mean, and}$$
$$\sigma_{\bar{x}} = \text{the standard error of the mean}$$

If the true values of μ and $\sigma_{\bar{x}}$ are unknown, the values may be estimated from a sample. The process of the estimation is discussed in the next chapter. Examples 3 and 4 are used to illustrate the method of finding the probabilities of the sample means by using the normal curve.

Example 3

Refer to Example 1. Find the probability of selecting a sample of two items with a mean of $3.50 or more by using the normal curve.

Solution. The distribution of the sample means in Table 12–5 is a discrete variable. However, when it is plotted on Chart 12–1 in the form of a histogram, the distribution is assumed to be a continuous variable. The width of each bar is $1/n = 1/2 = 0.5$. Thus, the value of $\bar{X} = \$3.5$ on the $\bar{X}$ scale is represented by the area from $3.25 (\frac{1}{2}$ of the width of the bar, $\frac{1}{2}(0.5) = 0.25$, smaller than 3.5, or the midpoint of 3 and 3.5) to $3.75 (\frac{1}{2}$ of the width of the bar larger than 3.5, or the midpoint of 3.5 and 4). The area representing $\bar{X} = \$3.25$ or more based on the normal curve can be obtained from the Areas Under the Normal Curve Table, where

$$z = \frac{\bar{X} - \mu}{\sigma_{\bar{x}}} = \frac{3.25 - 3}{0.816} = 0.30637 \text{ or } 0.31$$
$$A(0.31) = 0.12172$$

The area greater than $A(0.31)$ is

$$0.50000 - 0.12172 = 0.37828 \text{ or } 37.828\%$$

From Table 12–5, the actual probability of selecting a sample of two items with a mean $3.5 or more (including $4.0 and $4.5) is

$$\frac{3}{15} + \frac{2}{15} + \frac{1}{15} = \frac{6}{15} = 0.4$$

The discrepancy between the approximated probability based on the normal curve and the actual probability is only

$$0.40000 - 0.37828 = 0.02172 \text{ or } 2.172\%$$

Example 4

Refer to Example 2. Find the probability of selecting a sample of four items with a mean $3.50 or more by using the normal curve.

Solution. The width of each bar is $1/n = \frac{1}{4} = 0.25$ as shown in Chart 12–2. The value of $\bar{X} = \$3.5$ on the $\bar{X}$ scale is represented by the area from $3.375 [= 3.5 - \frac{1}{2}(1/n) = 3.5 - \frac{1}{2}(0.25)]$ to $\$3.625 [= 3.5 + \frac{1}{2}(1/n) = 3.5 + \frac{1}{2}(0.25)]$. The area representing $\bar{X} = \$3.375$ or more based on the normal curve is obtained from the Areas Under the Normal Curve Table, where

$$z = \frac{\bar{X} - \mu}{\sigma_{\bar{x}}} = \frac{3.375 - 3}{0.408} = 0.91911 \text{ or } 0.92$$
$$A(0.92) = 0.32121$$

The area greater than $A(0.92)$ is

$$0.50000 - 0.32121 = 0.17879 \text{ or } 17.879\%$$

From Table 12–7, the actual probability of selecting a sample of four items with a mean $3.5 or more (including $3.75) is

$$\frac{2}{15} + \frac{1}{15} = \frac{3}{15} = 0.2$$

The discrepancy between the approximated probability based on the normal curve and the actual probability is

$$0.20000 - 0.17879 = 0.02121 \text{ or } 2.121\%$$

Compare the discrepancies obtained in Examples 3 and 4. The discrepancy between the approximated probability and the actual probability is smaller when the sample size is increased. The discrepancy is 2.172% when $n = 2$ and is only 2.121% when $n = 4$. In other words, the normal curve shows a better fit to the sampling distribution of the mean when the sample size is increased. Thus, for a sample of a large size, $n = 30$ or more, the normal curve is a proper device in finding the probability of the sample mean. If $n = 30$, for example, the width of each bar is only 1/30 of the unit of X variable, such as dollar unit, and the distribution of the sample means will be very close to the smoothed normal curve.

12.4 Sampling Distribution of the Proportion (p)

The proportion of a given thing in a population probably is needed as frequently as the arithmetic mean. For example, we may wish to know the proportion (or percentage) of smokers or nonsmokers in the United States, the proportion of voters or nonvoters in a certain election, the proportion of users or nonusers of a given product, and so on. The relationship between the proportion of the population and the sampling distribution of the proportion is again illustrated by using a small finite population. The sampling distribution of the proportion is a set of proportions of all possible samples of the same size drawn from a given population.

Example 5

There are six salesmen in the May Company. Salesmen A, B, C are smokers and the others, X, Y, Z, are nonsmokers. Consider the six salesmen as a population and let 1 represent the value of a smoker and 0 a nonsmoker. The proportion of the number of smokers (P) and the standard deviation of the values (σ) are computed in Table 12-8. Find (a) the proportions of smokers of the possible samples of four salesmen, (b) the mean of sample proportions, (c) the standard error of the proportion, and (d) the probabilities of the sample proportions.

TABLE 12-8

Smokers (1) and Nonsmokers (0) in May Company, and Computation of P and $\sigma(=\sqrt{PQ})$

Salesmen	X	$x(=X-P)$	x^2
A	1 (smoker)	0.5	0.25
B	1	0.5	0.25
C	1	0.5	0.25
X	0 (nonsmoker)	−0.5	0.25
Y	0	−0.5	0.25
Z	0	−0.5	0.25
Total	3	0	1.50

$$P = \frac{3}{6} = 0.5 \text{ or } 50\% \text{ (parameter)}$$

Or, use formula (10–10),

$$\mu = P = \frac{3}{6} \quad \text{(3 out of 6 are smokers)}$$

$$\sigma^2 = \frac{1.50}{6} = 0.25; \ \sigma = \sqrt{0.25} = 0.5$$

Or, use formula (10–11),

$$\sigma = \sqrt{PQ} = \sqrt{P(1-P)} = \sqrt{0.5(0.5)} = 0.5$$

Solution. (a) The proportions of smokers of the possible samples of four persons are shown in Table 12–9. The proportion of a sample is denoted by p and is a statistic. The table shows that for a simple random sample of four persons chosen from the population of six salesmen, the number of possible samples is 15.*

(b) The mean of sample proportions, denoted by $\bar{p}$, is obtained from Table 12–9.

TABLE 12-9

COMPUTATION OF SAMPLE PROPORTIONS FOR EXAMPLE 5

4 Salesmen in Each Sample		Sample Proportion (p)
Smokers	*Nonsmokers*	(= *number of smokers/number of salesmen*)
A	X,Y,Z	0.25 (= 1/4)
B	X,Y,Z	0.25
C	X,Y,Z	0.25
A,B	X,Y	0.50 (= 2/4)
A,B	X,Z	0.50
A,B	Y,Z	0.50
A,C	X,Y	0.50 (= 2/4)
A,C	X,Z	0.50
A,C	Y,Z	0.50
B,C	X,Y	0.50 (= 2/4)
B,C	X,Z	0.50
B,C	Y,Z	0.50
A,B,C	X	0.75 (= 3/4)
A,B,C	Y	0.75
A,B,C	Z	0.75
Total, 15 samples		7.50

$$\bar{p} = \frac{7.50}{15} = 0.50$$

which is equal to the proportion of the population or $\bar{p} = P$. The symbol $\bar{p}$ may also be written as $E(p)$, which represents the *expected* value of p.

(c) The standard error of the proportion, or the standard deviation of all possible sample proportions, denoted by σ_p, is computed in Table 12–10.

$$\sigma_p = 0.158$$

---

*The number of possible samples may be computed as follows:
1 smoker and 3 nonsmokers
$\quad$ $_3C_1$ (1 out of 3 smokers) $\times$ $_3C_3$ (3 out of 3 nonsmokers) = 3 $\times$ 1 = 3
2 smokers and 2 nonsmokers
$\quad$ $_3C_2$ (2 out of 3 smokers) $\times$ $_3C_2$ (2 out of 3 nonsmokers) = 3 $\times$ 3 = 9
3 smokers and 1 nonsmoker
$\quad$ $_3C_3$ (3 out of 3 smokers) $\times$ $_3C_1$ (1 out of 3 nonsmokers) = 1 $\times$ 3 = 3

Total, or $_6C_4$ (4 out of 6 salesmen). 15

(d) The probabilities of the 15 sample proportions are shown in the last column of Table 12–10. When the 15 samples are selected at random, each sample will have the probability $1/15$ of being selected. Since there are three samples with the proportion 0.75, for example, the probability of a sample being selected to have the proportion 0.75 is $1/15 + 1/15 + 1/15 = 3/15$.

This method of computing the standard error of the proportion involves the tedious work of finding all possible sample proportions. For a finite population, formula (10–9) will simplify the computation and yield the same result. The formula is renumbered (12–2) here for easy reference.

$$\sigma_p = \sqrt{\frac{PQ}{n} \cdot \frac{N-n}{N-1}} \qquad (12\text{--}2)$$

where σ_p = the standard error of the proportion
   P = the proportion of a given event (or success) of the population
   $Q = 1 - P$, the proportion of not a given event (or failure) of the population
   N = the size of the population
   n = the size of the sample

TABLE 12-10

THE STANDARD ERROR OF THE PROPORTION AND THE PROBABILITY
DISTRIBUTION OF THE SAMPLE PROPORTIONS (EXAMPLE 5)

Possible samples (See Table 12-9)	Sample proportion p	Number of samples f	$p - \bar{p}$ ($\bar{p} = .50$)	$f(p - \bar{p})$	$f(p - \bar{p})^2$	Probability of sample proportion
1(s), 3(nons)	0.25	3	-0.25	-0.75	0.1875	3/15
2(s), 2(nons)	0.50	9	0	0	0	9/15
3(s), 1(nons)	0.75	3	0.25	0.75	0.1875	3/15
Total		15		0.00	0.3750	15/15

$$\sigma_p{}^2 = \frac{0.3750}{15} = 0.025$$

$$\sigma_p = \sqrt{0.025} = 0.158$$

When formula (12–2) is used, the standard error of the proportion for Example 5 is computed as follows:

$$\sigma_p = \sqrt{\frac{(0.50)(1 - 0.50)}{4} \cdot \frac{6-4}{6-1}} = \sqrt{0.025} = 0.158$$

The size of a sample affects the standard error of the proportion in a similar manner as it affects the standard error of the mean. When the size is increased, the standard error of the proportion is decreased. For example, by using the population given in Example 5, when $n = 5$, the standard error of the proportion is only 0.1 as computed below:

$$\sigma_p = \sqrt{\frac{PQ}{n} \cdot \frac{N-n}{N-1}} = \sqrt{\frac{(0.50)(0.5)}{5} \cdot \frac{6-5}{6-1}} = \sqrt{\frac{1}{100}} = 0.1$$

The distribution of the sample proportions listed in Table 12–10 is shown graphically in Chart 12–3. The normal curve fitted to the sampling distribution is also constructed in the chart.* The curve in the chart shows a fairly close fit to the sampling distribution.

In finding the proportion from a sampling distribution by the normal curve, we must know the following three values:

$$P = \text{the proportion of the population}$$
$$p = \text{the sample proportion}$$
$$\sigma_p = \text{the standard error of the proportion}$$

If the true values of P and σ_p are unknown, the values may be estimated from a sample. The processs of the estimation is discussed in the next chapter. Example 6 is used to illustrate the method of finding the probabilities of the sample proportions by using the normal curve.

Example 6

Refer to Example 5. Find the probability of selecting a sample of four items (salesmen) with a proportion of smokers 0.75 (or 75%) or more by the normal curve.

Solution. The width of each bar in Chart 12–3 is $1/n = 1/4 = 0.25$. The value of $p = 0.75$ on the p scale is represented by the area from 0.625 (the midpoint 0.5 and 0.75, or $\frac{1}{2}$ of the width of the bar smaller than 0.75, or $0.75 - \frac{1}{2}(0.25) = 0.625$) to 0.875 ($\frac{1}{2}$ of the width of the bar larger than 0.75, or $0.75 + \frac{1}{2}(0.25) = 0.875$). The area representing $p = 0.625$ or more based on the normal curve can be obtained from Table 6 in the Appendix. where

$$z = \frac{p - P}{\sigma_p} = \frac{0.625 - 0.5}{0.158} = 0.79$$
$$A(0.79) = 0.28524$$

The area greater than $A(0.79)$ is

$$0.50000 - 0.28524 = 0.21476 \text{ or } 21.476\%$$

From Table 12–10, the actual probability of selecting a sample of four items with a proportion of smokers 0.75 or more is $3/15 = 0.2$ or 20%. (There is no proportion larger than 0.75.) The discrepancy between the approximated probability based on the normal curve and the actual probability is

$$0.21476 - 0.20000 = 0.01476 \text{ or } 1.476\%$$

---

*The normal curve is constructed by applying formula (11–2), where σ is replaced by σ_p. For example, Y_0 is computed as follows:

$$\frac{Ni}{\sigma_p} \cdot f(z) = \frac{15(0.25)}{0.158}(0.39894) = 9.47$$

As in a sampling distribution of the mean, the discrepancy will tend to be smaller when n becomes larger. In other words, the normal curve shows a better fit to the sampling distribution of the proportion when sample size is increased. The statement is also true when the value of P is not equal to 0.5 or 50% but the sample size is large.

The shape of a sampling distribution curve depends upon the value of the population proportion. When the population proportion is 0.5, such as $P = 0.5$ in Example 5, the sampling distribution of the proportion is symmetrical. When the population proportion deviates away from 0.5, the shape of the distribution curve tends more and more asymmetrical. When P is larger than 0.5, the distribution is skewed to the left; whereas, when P is less than 0.5, the distribution is skewed to the right. However, when the sample size is increased, the distribution curve becomes less skewed. This fact has been shown in Chart 10–2, which shows that when the sample size is increased from 3 to 20, the distribution curve more and more approaches the normal curve. It is obvious that when the value of P departs from 0.5 to a great extent and at the same time the sample size is small, the approximation of the probability for a sampling distribution of the proportion by the normal curve will be less meaningful.

CHART 12-3

SAMPLING DISTRIBUTION OF THE PROPORTION, SAMPLE SIZE $n = 4$ (EXAMPLE 5)

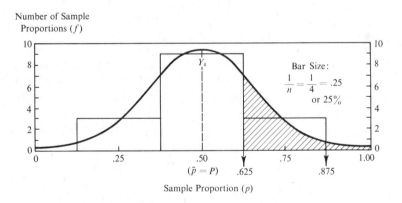

Sample Proportion (p)

12.5   Simplified Standard Error Formulas

In practice, the following two formulas are used to compute the standard error of sampling distributions obtained from either infinite populations or large finite populations:

$$\sigma_{\bar{x}} = \frac{\sigma}{\sqrt{n}} \qquad (12\text{-}3)$$

$$\sigma_p = \sqrt{\frac{PQ}{n}} \qquad\qquad (12\text{--}4)$$

The discussions presented in Examples 1 to 6 are based on the samples drawn from small finite populations. When the population size N is large and the sample size n is relatively small with respect to the population size, the finite correction factor $\sqrt{(N-n)/(N-1)}$ in formulas (12–1) and (12–2) will approach unity or 1 (as illustrated on page 253). Since a number will not change its value when it is multiplied by 1, the two formulas may be simplified:

Formula (12–1) $\sigma_{\bar{x}} = \dfrac{\sigma}{\sqrt{n}} \cdot \sqrt{\dfrac{N-n}{N-1}}$ is close to $\dfrac{\sigma}{\sqrt{n}}$

Formula (12–2) $\sigma_p = \sqrt{\dfrac{PQ}{n}} \cdot \sqrt{\dfrac{N-n}{N-1}}$ is close to $= \sqrt{\dfrac{PQ}{n}}$

In general, when the ratio of the sample size to the population size does not exceed 5% (or $n/N = 5\%$ or less), the finite correction factor is ignored in computing the standard error of the mean or of the proportion for a finite population.

Also, when discrete data are shown continuously by a histogram, the data are represented by bars, as shown in Charts 12–1 and 12–3. Each bar is centered at the value represented on the horizontal scale. However, when the normal curve is used to approximate the probability of a sampling distribution with a large sample size, the continuity correction factor, $\frac{1}{2}$ of the width of the bar or $\frac{1}{2}(1/n)$, is usually ignored. When n is large, the value of $\frac{1}{2}(1/n)$ gives only small effect to the final answers.

A sample may also be drawn from an infinite population. Note that formula (12–4) is identical to formula (10–8), which is used to compute the standard deviation (or standard error) of the distribution of proportions (or sample proportions) obtained from an infinite population. Similarly, formula (12–3) is originally derived for an infinite population.

The following two examples are used to illustrate the use of the simplified formulas.

Example 7

Assume that the average monthly income of 10,000 workers in King City is $500 and the standard deviation is $100. What is the probability of selecting a simple random sample of 400 workers that will have a mean of $510 or more per month?

Solution. Here, $N = 10,000$, $\mu = \$500$, $\sigma = \$100$, $n = 400$, and $\bar{X} = \$510$. By applying the simplified formula (12–3),

$$\sigma_{\bar{x}} = \frac{\sigma}{\sqrt{n}} = \frac{100}{\sqrt{400}} = \frac{100}{20} = 5$$

$$z = \frac{\bar{X} - \mu}{\sigma_{\bar{x}}} = \frac{510 - 500}{5} = 2$$

$A(z) = A(2) = 0.47725$ (From Table 6).

The area which is *larger* than $A(2)$ is

$$0.50000 - 0.47725 = 0.02275 = 2.275\%, \text{ rounded to } 2\%$$

Thus, there are about 2 chances in 100 of selecting samples of size 400 (workers) that will have a mean of $510 or more per month. The reasoning process may become clearer when the solution is diagramed in Chart 12–4.

CHART 12-4

DIAGRAM FOR EXAMPLE 7

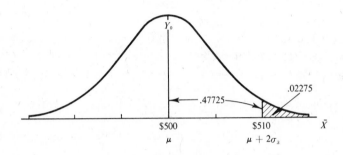

Example 8

Assume that 40% of the 5,000 students in a stadium are girls. What is the probability of selecting a simple random sample of 150 students that the number of girls will be 35% or less?

Solution. Here, $N = 5,000$, $P = 40\% = 0.4$, $Q = 1 - P = 1 - 0.4 = 0.6$, $n = 150$, and $p = 35\% = 0.35$. Substitute values P, Q, and n in the simplified formula (12–4) as follows:

$$\sigma_p = \sqrt{\frac{PQ}{n}} = \sqrt{\frac{(0.4)(0.6)}{150}} = \sqrt{0.0016} = 0.04$$

CHART 12-5

DIAGRAM FOR EXAMPLE 8

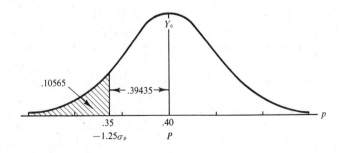

$$z = \frac{p - P}{\sigma_p} = \frac{0.35 - 0.4}{0.04} = -1.25$$

$A(z) = A(-1.25) = 0.39435$ (From Table 6).

The area *smaller* than $A(-1.25)$ is

$$0.50000 - 0.39435 = 0.10565 = 10.565\%, \text{ rounded to } 11\%$$

Thus, there are about 11 chances in 100 of selecting samples of size 150 (students) that the number of girls will be 35% or less. The reasoning process may become clearer when the solution is diagramed in Chart 12–5.

Notes to Examples 7 and 8.

1. The finite correction factor $\sqrt{(N-n)/(N-1)}$ is ignored in the computation of each example. The ratio in each of the two examples is less than 5%; that is, $400/10,000 = 0.04$ or 4% (Example 7) and $150/5,000 = 0.03$ or 3% (Example 8).

2. The continuity correction factor $\frac{1}{2}(1/n)$ is also ignored in the computation of each example. The factor for Example 7 is $\frac{1}{2}(1/400) = 0.00125$ and for Example 8 is $\frac{1}{2}(1/150) = 0.003$. The factor would not affect the final answers greatly if they were included in the computations.

12.6   Summary

This chapter indicates the basic relationship between a population and samples drawn from the population. The relationship is illustrated by using a parameter and its corresponding statistic. A measure, such as a mean, a median, or a standard deviation, used to describe some characteristic of a population is called a *parameter*, whereas a measure of a sample is called a *statistic*. A distribution of the *statistics* obtained from all possible samples of the same size drawn from the population is called the *sampling distribution* of the statistic.

The standard deviation of a sampling distribution of a statistic is frequently called the *standard error of the statistic*. Thus, the standard deviation of the means obtained from all possible samples is called the *standard error of the mean*, and the standard deviation of the proportions obtained from all possible samples is called the *standard error of the proportion*.

The difference between a statistic and a corresponding parameter is called a *sampling error*. The sampling error is measured by the standard error of the statistic in terms of probability under the normal curve. The result of the measurement indicates the precision of the estimate of the population based on the sample study.

The methods of selecting a representative sample are numerous. The most common methods may be classified in two types: (1) according to the number of samples taken from a population for a study, they are single sampling, double sampling, and multiple sampling; and (2) according to the manner used in selecting the items of a sample, they are judgment sampling and random sampling. A random sample is also called a probability sample since each element of the population has an equal chance of being selected. The common types of random sampling are simple random, systematic, stratified, and cluster sampling. Among them, the simple random sampling is a basic type and is used in illustrating the basic relationship between a parameter and its corresponding statistic in a sampling distribution. The basic relationships may be summarized as follows:

(1) The mean of the sample means (or sample proportions) is always equal to the population mean (or population proportion).

(2) When the size of the samples drawn from a population is increased, the dispersion of the sample means (or sample proportions) from the population mean (or population proportion) is decreased.

(3) When the population is fairly large and the sample size is large enough (larger than 30 items), the distribution of the sample means (or sample proportions) will be a normal one. The probability of a given statistic may be obtained from Table 6 in the Appendix when the distribution of the statistic is normal.

SUMMARY OF FORMULAS

Application	Formula	Formula Number	Reference Page
Standard error of the mean From finite population	$\sigma_{\bar{x}} = \sqrt{\dfrac{\sigma^2}{n} \cdot \dfrac{N-n}{N-1}}$	12-1	304
or	$\sigma_{\bar{x}} = \dfrac{\sigma}{\sqrt{n}} \sqrt{\dfrac{N-n}{N-1}}$		
Standard error of the proportion From finite population	$\sigma_p = \sqrt{\dfrac{PQ}{n} \cdot \dfrac{N-n}{N-1}}$	12-2	313
Standard error of the mean From infinite population or from large finite population when n/N is 5% or less	$\sigma_{\bar{x}} = \dfrac{\sigma}{\sqrt{n}}$	12-3	315
Standard error of the proportion From infinite population or from large finite population when n/N is 5% or less	$\sigma_p = \sqrt{\dfrac{PQ}{n}}$	12-4	316

Exercises 12

1. Explain the difference between (a) a statistic and a parameter and (b) a standard error and a sampling error.

2. Explain briefly (a) single sampling, (b) double sampling, and (c) multiple sampling.

3. Indicate the difference between (a) judgment sampling and random sampling, (b) simple random sampling and systematic sampling, and (c) stratified sampling and cluster sampling.

4. Assume that a manager wishes to select ten items from a group of 200 items for a sampling inventory. How would he use the Table of Random Numbers to select the ten items? (Let your starting point in the table be column (2) on line 6; then move from the point down.)

5. Find the total number of possible samples drawn from six items for all possible sample sizes ($n = 0, 1, 2, 3, 4, 5,$ and 6). (Hint: You may use the tree diagram, the combination formula, or the multiplication principle as presented in this chapter or in Chapter 9 to find the answer.)

6. The units produced by the five employees in Edwards Company during a given period are listed below:

Name of employee	Units produced
Brook	1
Dennis	2
Horton	3
Jones	4
Kent	5

Consider the production by the employees as a population. Find the total number of possible samples drawn from the population for all possible sample sizes ($n = 0, 1, 2, 3, 4, 5$).

7. Refer to the data given in Problem 6. If the size of the possible samples is two (units produced by any two employees), find:
    (a) The means of the possible samples
    (b) The mean of the sample means
    (c) The standard error of the mean by using
        (1) the sample means
        (2) Formula 12–1
    (d) The probability of selecting a sample of two employees that will have a mean of four units or more by using
        (1) the normal curve method
        (2) the actual counting method
    (e) The probability of selecting a sample of two employees that will have a mean of two units or less by using methods (1) and (2) listed in (d) above.

8. Refer to Problem 6. Find the standard error of the mean by using the formula method if the sample size is four. What is the probability of selecting a sample of four employees that will have a mean of four units or more by (a) the normal curve method and (b) the actual counting method? Compare the probabilities of having a mean of four units or more by the normal curve method obtained in this problem and in Problem 7(d). Which probability is larger? Why?

9. Assume that there are 20 senior students in a high school. Among them, eight students indicated that they are going to college. If the possible samples of five students each are drawn from the 20 students, what are (a) the mean of the sample proportions that the students are going to college? (b) the standard error of the proportion? and (c) the probability that a sample will have a proportion of 60% or more? It is not necessary to list all possible samples.

10. Refer to Problem 9. Find the answers if the sample size is ten students. Compare the answers to question (c) in the two problems. How can you interpret the difference?

11. Of 50,000 families in a city, 70 percent subscribed to the evening newspaper. What is the probability of selecting a simple random sample of 500 families with a proportion of 72 percent or more? Use the simplified method.

12. The average deposit of the 15,000 depositors of a bank on a given date is $325 and the standard deviation of the deposits is $20. What is the probability of selecting a simple random sample of 200 depositors with a mean between $320 and $330? Use the simplified method.

part 4

Statistical Induction— Process and Applications

The process of statistical induction may take one of the two forms: estimation and test of hypothesis. The process of estimation is discussed in Chapter 13. Basic aspects of tests of hypotheses are presented in Chapter 14. Tests of hypotheses by z and t distributions are illustrated in Chapter 15. An application of statistical induction—statistical quality control—is presented in Chapter 16.

13 Estimation of Parameters

This chapter introduces the basic concept as well as operations of estimation. Estimation is a process of using a sample statistic to estimate the corresponding unknown population parameter. The new terms frequently used in statistical estimation will be introduced first, Section 13.1. The methods of estimating a population mean from a sample mean are discussed in Section 13.2, and those of estimating a population proportion from a sample proportion are discussed in Section 13.3. The determination of a proper sample size for the process of estimation is presented in Section 13.4.

13.1 Terminology

In addition to the terms introduced in Chapter 12, the understanding of the following terms will be helpful in statistical estimation.

Point Estimate and Interval Estimate

An estimate of a parameter may be expressed in two ways: a point estimate and an interval estimate. A point estimate is a *single number* that is used to represent the estimate of the parameter. An interval estimate is a *stated range* within which we may expect the parameter to lie. For example, if the estimate of the average hourly wage in a city is expressed as $4, it is a point estimate; if it is expressed as between $3.5 to $4.5, it is an interval estimate.

As we shall see later, when an estimate is expressed in an interval specified by two numbers, we may use the term *probability* to indicate the *confidence* of our estimation.

Confidence Intervals, Limits, and Coefficients

When a sampling distribution of the mean (or the proportion) is normal, the probability that the sample means (or proportions) will lie between the maximum ordinate (Y_0) and the ordinate at z may be obtained from Table 6. For example, the probability that the sample means lie within a range

$$\text{from } \mu + 1\sigma_{\bar{x}} \text{ to } \mu - 1\sigma_{\bar{x}}$$

is represented by the area under the normal curve between Y_0 and $z = \pm 1$ and is

$$0.34134 + 0.34134 = 0.68268 \text{ or } 68.268\%$$

The 68.268% probability indicates that about 68 out of every 100 sample means will lie within the range of $\mu \pm 1\sigma_{\bar{x}}$.

Let $\mu = \$6$, $\sigma_{\bar{x}} = \$0.50$, and the number of possible samples drawn from a given population $= 100$ million. Each sample, of course, has a mean. Then, 68.268 million sample means, out of the 100 million sample means, will have their values between \$5.50 and \$6.50, or *within* the range

$$\text{from } \mu - 1\sigma_{\bar{x}} = 6 - 0.50 = \$5.50 \text{ to}$$
$$\mu + 1\sigma_{\bar{x}} = 6 + 0.50 = \$6.50$$

This range is shown graphically in Illustration I of Chart 13–1.

Conversely, if we select a sample mean within the range, such as $\bar{X}_1$ or $\bar{X}_2$ as shown in Illustration II of Chart 13–1, the *interval* represented by the sample mean $\pm$ one standard error, such as

$$\bar{X}_1 \pm 1\sigma_{\bar{x}} = 5.62 \pm 0.50 = \$5.12 \text{ to } \$6.12$$

or

$$\bar{X}_2 \pm 1\sigma_{\bar{x}} = 6.38 \pm 0.50 = \$5.88 \text{ to } \$6.88$$

will include the population mean $\mu\ (=\$6)$. We have 68.268 million such intervals. Stated in a different way, we have 31.732 million $(=100$ million $- 68.268$ million) sample means, such as $\bar{X}_3$ and $\bar{X}_4$ shown in the chart, having their values *outside* the range from \$5.50 to \$6.50. The intervals based on $\bar{X}_3$ and $\bar{X}_4$ values do not include the population mean μ. Thus, we have the *confidence* to make a statement that if we select a sample at random to estimate the population mean, we will have 68 out of 100 chances to have the true population mean within the interval of the sample mean $\pm$ one standard error of the mean.

Again, let $\mu = \$6$, and $\sigma_{\bar{x}} = \$0.50$, the range $\mu \pm 1\sigma_{\bar{x}}$ may be written in different forms as follows:

$$\mu \pm 1\sigma_{\bar{x}}, \text{ or } \$6 \pm \$0.50, \text{ or } \$5.50 \text{ to } \$6.50$$

The range as written in these three forms is called the *confidence interval;* the two values (\$5.50 and \$6.50) which specify the ranges are called the *confidence limits;* and the probability, 68.268%, is called the *confidence coefficient* or *confidence level*. Likewise, a 95% confidence interval may be expressed:

$$\mu \pm 1.96\sigma_{\bar{x}}, \text{ where } z = \pm 1.96$$

The area under the normal curve between Y_0 and $z = \pm 1.96$ is

$$0.47500 + 0.47500 = 0.95 \text{ or } 95\%$$

CHART 13–1

A 68% Confidence Interval for a Sampling Distribution of the Mean $(\bar{X})$

Illustration I: From the population mean $(\mu = \bar{\bar{X}})$ to locate sample means $(\bar{X})$
Illustration II: From sample means to locate the population mean

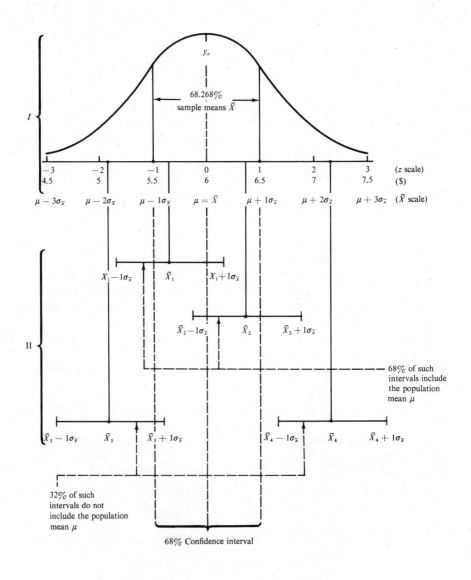

The most frequently used confidence coefficients and their z values are tabulated in Table 13–1.

TABLE 13–1

COMMON CONFIDENCE COEFFICIENTS AND THEIR z VALUES

Confidence Coefficient	50%	68.27%	90%	95%	95.45%	99%	99.73%
z	0.6745	1.00	1.645	1.96	2.00	2.58	3.00

Where $z = \dfrac{\bar{X} - \mu}{\sigma_{\bar{x}}}$ for a sampling distribution of the mean, $\bar{X}$; or $z = \dfrac{p - P}{\sigma_p}$ for a sampling distribution of the proportion, p.

The concept of confidence intervals is important when an interval estimate is used in the estimation of a parameter.

Unbiased Estimator

A statistic that is used for estimating a parameter is called an *estimator*, such as a sample mean being used for estimating the population mean. An estimator is *unbiased* when the *expected value* of the statistic used as the estimator is equal to the value of the parameter. The expected value of the statistic (or expressed symbolically, E [statistic]) is the arithmetic mean of the sampling distribution of the statistic. Since the mean of the sampling distribution of the mean (or the mean of the means of all possible samples of the same size drawn from the same population, or the expected value of sample mean) is equal to the population mean (see page 306), a sample mean is an unbiased estimator; or stated in a different way,

$\bar{X}$ is an *unbiased estimate* of μ, since $E(\bar{X}) = \mu$

Likewise, since the mean of the sampling distribution of the proportion is equal to the population proportion (see page 312, Table 12–9), a sample proportion is an unbiased estimator, or

p is an *unbiased estimate* of P, since $E(p) = P$

However, the mean of the sampling distribution of the variance (s^2) is not equal to the population variance (σ^2), or

s^2 is a *biased estimate* of σ^2 since $E(s^2) \neq \sigma^2$

(See exception below.)

The value of the variance s^2 or σ^2 as discussed in Chapter 7, page 181, is computed by dividing the sum of the squared deviations from the mean $\sum x^2$ by n (sample size) or N (population size); that is,

$$s^2 = \frac{\sum x^2}{n} = \frac{\sum (X - \bar{X})^2}{n} \quad \text{and} \quad \sigma^2 = \frac{\sum x^2}{N} = \frac{\sum (X - \mu)^2}{N} \qquad (7\text{–}7a)$$

On the other hand, if the sum of the squared deviations from the mean $\sum x^2$ is divided by $(n - 1)$ for a sample, denoted by $\hat{s}^2$, or by $(N - 1)$ for a population, denoted by $\hat{\sigma}^2$, the mean of the sampling distribution of the modified variance $\hat{s}^2$ is equal to the modified population variance $\hat{\sigma}^2$, or

$\hat{s}^2$ is an *unbiased estimate* of $\hat{\sigma}^2$, since $E(\hat{s}^2) = \hat{\sigma}^2$

$$\hat{s}^2 = \frac{\Sigma x^2}{n-1} = \frac{\Sigma (X - \bar{X})^2}{n-1}$$

and

$$\hat{\sigma}^2 = \frac{\Sigma x^2}{N-1} = \frac{\Sigma (X - \mu)^2}{N-1} \qquad (13\text{--}1a)$$

(Note: Read the "estimate sign ^" as "hat," such as "s hat" for $\hat{s}$ and "σ hat" for $\hat{\sigma}$.)

Tables 13–2 and 13–3, based on the information in Example 2, page 305, are used to illustrate the biased and the unbiased sample variances. When the variances are computed by using formula (7–7a), the population variance σ^2 is $10/6 =$ 1.67, whereas the mean of the sample variances s^2 is \$1.5 (Table 13–3, column 3). However, when the modified variances are computed by using formula (13–1a), the population variance $\hat{\sigma}^2$ is \$2, which is the same as the mean of the sample variances $\hat{s}^2$ (Table 13–3, column 4.)

TABLE 13–2

COMPUTATION OF THE POPULATION VARIANCE BY TWO
DIFFERENT METHODS (σ^2 AND $\hat{\sigma}^2$)

Worker	Hourly Wage (X)	x	x²
A	$1	−2	4
B	2	−1	1
C	3	0	0
D	3	0	0
E	4	1	1
F	5	2	4
Total	18	0	10

$$\mu = 18/6 = \$3$$
$$\sigma^2 = \frac{10}{6} = \frac{5}{3} = \$1.67 \qquad \text{[Formula (7–7a)]}$$
$$\hat{\sigma}^2 = \frac{10}{6-1} = \$2 \qquad \text{[Formula (13–1a)]}$$

The values of $\hat{s}^2$ and $\hat{\sigma}^2$ may also be obtained from s^2 and σ^2 respectively from formulas below:

$$\hat{s}^2 = s^2 \left(\frac{n}{n-1}\right) \quad \text{and} \quad \hat{\sigma}^2 = \sigma^2 \left(\frac{N}{N-1}\right) \qquad (13\text{--}1b)^*$$

$* \quad s^2 \left(\frac{n}{n-1}\right) = \frac{\Sigma x^2}{n} \left(\frac{n}{n-1}\right) = \frac{\Sigma x^2}{n-1} = \hat{s}^2.$

Similarly, this method may also be used to prove the formula for $\hat{\sigma}^2$.

TABLE 13–3

BIASED AND UNBIASED VARIANCES OF THE SAMPLING DISTRIBUTION OF THE MEAN

Sample (Wages of 4 Workers)	Sample Mean	Biased Variance $s^2 = \sum x^2/n$	Unbiased Variance $\hat{s}^2 = \sum x^2/(n-1)$
$1, 2, 3, 3	$2.25	2.75/4	2.75/3
1, 2, 3, 4	2.50	5.00/4	5.00/3
1, 2, 3, 5	2.75	8.75/4	8.75/3
1, 2, 3, 4	2.50	5.00/4	5.00/3
1, 2, 3, 5	2.75	8.75/4	8.75/3
1, 2, 4, 5	3.00	10.00/4	10.00/3
1, 3, 3, 4	2.75	4.75/4	4.75/3
1, 3, 3, 5	3.00	8.00/4	8.00/3
1, 3, 4, 5	3.25	8.75/4	8.75/3
1, 3, 4, 5	3.25	8.75/4	8.75/3
2, 3, 3, 4	3.00	2.00/4	2.00/3
2, 3, 3, 5	3.25	4.75/4	4.75/3
2, 3, 4, 5	3.50	5.00/4	5.00/3
2, 3, 4, 5	3.50	5.00/4	5.00/3
3, 3, 4, 5	3.75	2.75/4	2.75/3
Total, 15 samples	45.00	90.00/4 = 22.50	90.00/3 = 30.00
Mean of Statistics (Expected Value)	$\dfrac{45.00}{15} = \$3$	$\dfrac{22.50}{15} = \$1.5$	$\dfrac{30.00}{15} = \$2$
Population Parameter (Table 13–2)	$\mu = \dfrac{18}{6} = \$3$	$\sigma^2 = \dfrac{10}{6} = \$1.67$	$\hat{\sigma}^2 = \dfrac{10}{5} = \$2$

Source: Table 12–6.

For example, the biased variance of the first sample of 4 in Table 13–3 is computed in Table 13–4, $s^2 = 0.6875$. By using formula (13–1b), the value of $\hat{s}^2$ is obtained as follows:

$$\hat{s}^2 = s^2 \left(\frac{n}{n-1} \right) = 0.6875 \left(\frac{4}{4-1} \right) = \frac{\$2.75}{3}$$

TABLE 13–4

COMPUTING VARIANCE OF THE FIRST SAMPLE IN TABLE 13–3

Worker	Hourly Wage (X)	x	x^2
A	$1	−1.25	1.5625
B	2	−0.25	0.0625
C	3	0.75	0.5625
D	3	0.75	0.5625
Total	9	0.00	2.7500

$$\bar{X} = 9/4 = \$2.25$$
$$s^2 = 2.75/4 = 0.6875$$

Likewise, the value of $\hat{\sigma}^2$ in Table 13–2 may be obtained as follows:

$$\hat{\sigma}^2 = \sigma^2\left(\frac{N}{N-1}\right) = \frac{10}{6}\left(\frac{6}{6-1}\right) = \$2$$

Formula (13–1b) is convenient for those who are familiar with the short method of computing a standard deviation (as discussed in Chapter 7).

Thus, the statement "$\hat{s}^2$ is an unbiased estimate of $\hat{\sigma}^2$" may be written as:

$$s^2\left(\frac{n}{n-1}\right) \text{ or } \hat{s}^2 \text{ is an } \textit{unbiased estimate} \text{ of } \sigma^2\left(\frac{N}{N-1}\right)$$

When the population size (N) is large or infinite, the factor $N/(N-1)$ approaches 1, and $\sigma^2[N/(N-1)]$ approaches σ^2. Thus,

$\hat{s}^2$ is an unbiased estimate of σ^2 when the population is large.

In addition, when the population is large and the sample size is also large, both factors $n/(n-1)$ and $N/(N-1)$ approach 1. By similar reasoning, we may state:

s^2 is an *unbiased estimate* of σ^2 when both the population and sample size are large.

Similarly, when a population represents the proportion of an event and the individual value of the population is 1 for "the event" (such as 1 representing the value of a smoker) but is 0 for "not the event" (such as 0 representing the value of a nonsmoker), we may state the biasness or unbiasness of a sample variance for a sampling distribution of the proportion as follows:

pq ($= s^2$) is a *biased estimate* of PQ ($=\sigma^2$) in general, but

$$pq\left(\frac{n}{n-1}\right), (= \hat{s}^2), \text{ is an } \textit{unbiased estimate} \text{ of } PQ\left(\frac{N}{N-1}\right), (=\hat{\sigma}^2)$$

Also,

$pq[n/(n-1)]$ is an *unbiased estimate* of PQ when the population is large, since $N/(N-1)$ approaches 1;

and

pq is an *unbiased estimate* of PQ when both the population size and the sample size are large, since both $N/(N-1)$ and $n/(n-1)$ approach 1.

Degrees of Freedom

The number of variables that can vary freely in a set of variables under certain conditions is frequently referred to as the number of *degrees of freedom*. Assume that we have a set of three variables A, B, C, and the sum of the variables is ten, or

$$A + B + C = 10$$

If we select a value for A and a value for B, such as $A = 1$ and $B = 3$, then the value of the third variable is automatically determined since

$$1 + 3 + C = 10, \quad C = 10 - 4 = 6$$

Thus, the number of variables that can vary freely in the set is two. We then say that there are two (or 3 minus 1) degrees of freedom when the sum of the three variables is known. Similarly, if we have n variables and the sum of the variables is a fixed value (or a constant), we say that the number of degrees of freedom is $n - 1$.

The term *degrees of freedom* is frequently used in statistical work concerning inference. For example, in finding an unbiased sample variance as discussed above, one must divide the sum of the squared deviation $\sum x^2$ by the sample size minus one, or $(n - 1)$. The quantity $(n - 1)$ is also referred to as the degrees of freedom. The reason for calling $(n - 1)$ the number of degrees of freedom in such cases is explained below.

Suppose we let X_1, X_2, X_3, and X_4 be the elements of a sample and the sample mean be five. If we select freely a value for each of any three of the four elements, the value of the variance is automatically determined since the fourth value is fixed. The algebraic sum of the deviations of the individual elements from the mean is always equal to zero, or $\sum (X - \bar{X}) = 0$. If we select $X_1 = 1$, $X_2 = 6$, and $X_3 = 4$, then

$$(1 - 5) + (6 - 5) + (4 - 5) + (X_4 - 5) = 0$$
$$(-4) + (1) + (-1) + (X_4 - 5) = 0$$
$$X_4 = 9$$

Thus, when we measure the variance, we are concerned with the deviations of only $n - 1$ elements. In other words, when the sample size is four, three of the four values may be selected freely; that is, the number of degrees of freedom is $4 - 1 = 3$. The sum of the squared deviations from the mean is therefore divided by the number of degrees of freedom $n - 1$ to obtain the unbiased variance, or

$$s^2 = \frac{(-4)^2 + (1)^2 + (-1)^2 + (9 - 5)^2}{3} = \frac{16 + 1 + 1 + 16}{3} = 11\frac{1}{3}$$

13.2 Estimating a Population Mean (μ) from a Sample Mean ($\bar{X}$)

In estimating a population mean from a sample mean, it is usually best to find both the point estimate and the interval estimate. When an interval estimate is found, the precision of the estimate of the parameter, the population mean, may be expressed in terms of probability. The following procedure should be followed in finding an interval estimate:

(1) *Find the sample mean, $\bar{X}$.* The sample mean is a point estimate. If an interval estimate is not used in an estimation, the sample mean is the best estimate of the population mean.

(2) *Compute the standard error of the mean, $\sigma_{\bar{x}}$.* The standard error of the mean may be obtained by using formula (12–1) in the complete form

$$\sigma_{\bar{x}} = \frac{\sigma}{\sqrt{n}} \sqrt{\frac{N - n}{N - 1}}$$

or the simplified form

$$\sigma_{\bar{x}} = \frac{\sigma}{\sqrt{n}}$$

In this section, the simplified form is used; that is, the population size N is assumed to be large finite or infinite.

(3) *Compute the confidence limits.* The confidence limits are computed as follows:

$$\text{Upper confidence limit} = \bar{X} + z\sigma_{\bar{x}}$$
$$\text{Lower confidence limit} = \bar{X} - z\sigma_{\bar{x}}$$

The value of z indicates the confidence coefficient or the probability as tabulated on page 327. The upper and lower confidence limits specify the confidence interval. When the confidence interval is expressed by two numbers, it is the value of the *interval estimate*.

This procedure is illustrated in two different cases below.

The Standard Deviation of the Population σ Is Known

When the standard deviation of the population is known, the size of a sample will affect the value of the standard error of the mean. In the formula $\sigma_{\bar{x}} = \sigma/\sqrt{n}$, the standard deviation of the population σ is constant. The increase of the sample size n will reduce the value of the standard error of the mean $\sigma_{\bar{x}}$. Thus, if an investigator wishes to have a small range of the confidence limits, a large sample size is probably required. Details concerning the determination of a proper sample size will be discussed in Section 13.4. Examples 1 and 2 below are used to illustrate the procedure of estimating a population mean from a sample mean when σ is known.

Example 1

Assume that we have selected the heights of 100 students in a large university as a simple random sample. The mean of the sample is 68 inches. The standard deviation of the heights of all students in the university is 2 inches. Find the mean of the population (the mean height of all students in the university) at (a) 68.27%, (b) 95.45%, and (c) 99.73% confidence intervals.

Solution.   (1) Find the sample mean. $\bar{X} = 68$ inches. (The point estimate)

(2) Compute the standard error of the mean. $\sigma = 2$ inches, $n = 100$ heights.

$$\sigma_{\bar{x}} = \frac{\sigma}{\sqrt{n}} = \frac{2}{\sqrt{100}} = 0.2 \text{ inches.}$$

(3) Compute the confidence limits. (The interval estimate)

(a) The 68.27% confidence interval—$z = \pm 1$—is

$$\bar{X} \pm 1\sigma_{\bar{x}} = 68 \pm 1(0.2) = 67.8 \text{ to } 68.2 \text{ inches}$$

(b) The 95.45% confidence interval—$z = \pm 2$—is

$$\bar{X} \pm 2\sigma_{\bar{x}} = 68 \pm 2(0.2) = 67.6 \text{ to } 68.4 \text{ inches}$$

(c) The 99.73% confidence interval—$z = \pm 3$—is

$$\bar{X} \pm 3\sigma_{\bar{x}} = 68 \pm 3(0.2) = 67.4 \text{ to } 68.6 \text{ inches}$$

The answers obtained above may be stated as follows: (a) there are 68 out of 100 chances that the population mean would fall within the range 67.8 to 68.2 inches; (b) there are about 95 out of 100 chances that it would fall within 67.6 to 68.4 inches; and (c) it is practically certain, or 99.73 out of 100 chances, that the population mean would fall within the range 67.4 to 68.6 inches.

Example 2

Refer to Example 1. Assume that we have selected the heights of only 16 students as a simple random sample. Find the population mean at the different levels of confidence intervals, (a) to (c).

Solution. (1) $\bar{X} = 68$ inches.

(2) $\sigma = 2$ inches, and $n = 16$ heights.

$$\sigma_{\bar{x}} = \frac{\sigma}{\sqrt{n}} = \frac{2}{\sqrt{16}} = \frac{2}{4} = 0.5 \text{ inches.}$$

(3) (a) The 68.27% confidence interval is

$$68 \pm 1(0.5) = 67.5 \text{ to } 68.5 \text{ inches}$$

(b) The 95.45% confidence interval is

$$68 \pm 2(0.5) = 67 \text{ to } 69 \text{ inches}$$

(c) The 99.73% confidence interval is

$$68 \pm 3(0.5) = 66.5 \text{ to } 69.5 \text{ inches}$$

The Standard Deviation of the Population σ Is Unknown

When the standard deviation of the population σ is unknown, the value of the standard error of the mean $\sigma_{\bar{x}}$ can not be obtained directly from σ. However, the value of $\sigma_{\bar{x}}$ may be estimated by using a sample standard deviation $\hat{s}$. Whenever it is possible, an unbiased estimator instead of a biased estimator should be used in estimating a required parameter. Section 13.1 indicated that, in general, the sample variance s^2 is a biased estimate of the population variance σ^2, but the modified sample variance $\hat{s}^2$ is an unbiased estimate of σ^2 when the population is large. Thus the modified sample standard deviation $\hat{s}$, instead of s, is usually preferred as an estimatior for estimating the value of σ, although neither s nor $\hat{s}$ is an unbiased estimate of σ.

When $\hat{s}$ is used as an estimated population standard deviation, the standard error formula $\sigma_{\bar{x}} = \sigma/\sqrt{n}$ may be written

$$s_{\bar{x}} = \frac{\hat{s}}{\sqrt{n}} \qquad (13\text{-}2a)$$

where $\hat{s}$ is a point estimate of σ, and $s_{\bar{x}}$ is a point estimate of $\sigma_{\bar{x}}$. Or, the estimated standard error of the mean may be computed from s as follows:

$$S_{\bar{x}} = \frac{s}{\sqrt{n-1}} \qquad (13\text{-}2b)^*$$

Formula (13–2b) is convenient for those who are familiar with the short method in computing the standard deviation s as discussed in Chapter 7.

It should be noted that when $s_{\bar{x}}$ instead of $\sigma_{\bar{x}}$ is used in obtaining an interval estimate of a population mean, the sample size n must be sufficiently large. When n is small, $s_{\bar{x}}$ may become a poor estimate of $\sigma_{\bar{x}}$ since $\acute{s}$ may differ greatly from σ. When n is large, $s_{\bar{x}}$ usually is a good estimate of $\sigma_{\bar{x}}$ since $\acute{s}$ approaches σ. Thus, the normal curve area table is appropriate only for a large sample in estimating the population mean. In general, the sample size should not be less than 30, preferably 100 or more. If it is less than 30, the method as discussed in the t-distribution (Chapter 15, page 367) should be used. On the other hand, if it is 100 or more, formula (13–2b) may be simplified as follows:

$$S_{\bar{x}} = \frac{s}{\sqrt{n}} \qquad (13\text{-}3)$$

When the sample size is large, 100 or more, the value of $\sqrt{n-1}$ will approach the value of $\sqrt{n}$. Note that $\sqrt{100-1} = \sqrt{99} = 9.94987$, and $\sqrt{100} = 10$. Thus, when 1 is not subtracted from n, the effect to the final answer for $s_{\bar{x}}$ is small. In the following illustrations, the simplified formula (13–3) will be used when a sample size (n) is 100 or more.

Example 3

The entrance examination grades of 200 students in a large university during a given period are selected as a simple random sample and are shown in the first two columns of the table below. Estimate the mean of the examination grades of all students in the university at a 95% confidence interval.

Solution. (1) *Find the sample mean.* $\bar{X} = 51.5$. The value of $\bar{X}$ is computed by using the short method, formula (5–3c), page 124. The assumed mean in the computation is 50 points, the midpoint of class "40 and under 60."

(2) *Compute the estimated standard error of the mean, $s_{\bar{x}}$.* First, compute the sample variance s^2 (or the sample standard deviation s) by using formula (7–8d).

$$s^2 = (20)^2\left[\frac{235}{200} - \left(\frac{15}{200}\right)^2\right] = 400(1.175 - 0.075^2) = 467.75$$

Then, compute the value of $s_{\bar{x}}$ by using formula (13–3) since the sample size is larger than 100, or $n = 200$.

$$s_{\bar{x}} = \frac{s}{\sqrt{n}} = \sqrt{\frac{s^2}{n}} = \sqrt{\frac{467.75}{200}} = \sqrt{2.33875} = 1.529$$

or rounded to 1.5 points.

Note that if formula (13–2b) is used in computing the value of $s_{\bar{x}}$, the answer is also 1.5 points, or

---

* $s_{\bar{x}} = \dfrac{\acute{s}}{\sqrt{n}} = \dfrac{\sqrt{s^2[n/(n-1)]}}{\sqrt{n}} = \sqrt{\dfrac{s^2}{n-1}} = \dfrac{s}{\sqrt{n-1}}$ (see formula 13–1b)

TABLE 13–5

DATA AND COMPUTATION FOR EXAMPLE 3

Grade Points (Class interval)	Number of Students f	d	fd	$f(d)^2$
0 and under 20	15	−2	−30	60
20 and under 40	45	−1	−45	45
40 and under 60	70	0	0	0
60 and under 80	50	1	50	50
80 and under 100	20	2	40	80
Total	200		15	235

$$\bar{X} = 50 + \frac{15}{200}(20) = 51.5$$

$$s_{\bar{x}} = \frac{s}{\sqrt{n-1}} = \sqrt{\frac{s^2}{n-1}} = \sqrt{\frac{467.75}{200-1}} = \sqrt{2.3505} = 1.533$$

or rounded to 1.5 points.

(3) *Compute the confidence limits.* At a 95% confidence interval, $z = \pm 1.96$.

$$\bar{X} \pm 1.96\, s_{\bar{x}} = 51.5 \pm 1.96(1.5) = 51.5 \pm 2.94 = 48.56 \text{ to } 54.44 \text{ points.}$$

13.3 Estimating a Population Proportion (P) from a Sample Proportion (p)

The procedure of estimating a population proportion is analogous to that of estimating a population mean.

(1) *Find the sample proportion, p.* The sample proportion is a point estimate. If an interval estimate is not used in an estimation, the sample proportion p is the best estimate of the population proportion P.

(2) *Compute the standard error of the proportion, σ_p.* The standard error of the proportion may be obtained by using formula (12–2) in the complete form

$$\sigma_p = \sqrt{\frac{PQ}{n} \cdot \frac{N-n}{N-1}}$$

or formula (12–4) in the simplified form

$$\sigma_p = \sqrt{\frac{PQ}{n}}$$

In this section, the simplified form is used; that is, the population size N is assumed to be large finite or infinite. However, in computing σ_p by using formula (12–4) above, the product of PQ must be known. Since we do not know the value of P, we will have no knowledge of the product of PQ. Thus, the product must be estimated from a sample. Section 13.1, page 330, indicated that in a population representing the proportion of an event (1 for the event and 0 for not having the event), the sample variance pq (or s^2) is a biased estimate of the population variance PQ (or σ^2). But, $pq[n/(n-1)]$ (or $\hat{s}^2$) is an unbiased estimate of PQ (or σ^2) when the population is large. When $pq[n/(n-1)]$ is used as an estimated population variance, the simplified standard error formula (12–4) may be written:

$$s_p = \sqrt{\frac{pq[n/(n-1)]}{n}}$$

or simply

$$s_p = \sqrt{\frac{pq}{n-1}} \qquad (13\text{–}4)$$

Also, it should be noted that when s_p, a point estimate of σ_p, is used in obtaining an interval estimate of a population proportion P, the sample size should be sufficiently large, 30 or more, in order to use the normal curve table. Further, if the sample size is 100 or more, formula (13–4) may be simplified as follows:

$$s_p = \sqrt{\frac{pq}{n}} \qquad (13\text{–}5)$$

The largest possible value of the product pq is $(0.5)(0.5) = 0.25$. When $n = 100$, the values of s_p based on the largest possible product and the two formulas are computed as follows:

$$s_p = \sqrt{\frac{(0.5)(0.5)}{100-1}} = 0.05025 \qquad \text{(formula 13–4)}$$

$$s_p = \sqrt{\frac{(0.5)(0.5)}{100}} = 0.05 \qquad \text{(formula 13–5)}$$

The difference between the two values is only 0.00025 or 0.025%. In the following illustrations, the simplified formula (13–5) will be used when a sample size (n) is 100 or more.

(3) *Compute the confidence limits.* The confidence limits are computed as follows:

$$\text{Upper confidence limit} = p + zs_p$$
$$\text{Lower confidence limit} = p - zs_p$$

The two limits specify the interval estimate of the population proportion P.

Example 4

A political group in a large state wishes to know the percentage of voters who will vote for candidate Smith in a certain election. A sample of 400 voters has been taken at random. The sample shows that 140 voters will vote for Smith. Estimate the percentage of the population of voters in the state that will vote for the candidate at a 95% confidence interval.

Solution. (1) Find the sample proportion. $p = \dfrac{140}{400} = 0.35$ or 35%. (The point estimate.)

(2) Compute the standard error of the proportion. $n = 400$. Thus, formula (13–5) is used.

$$s_p = \sqrt{\frac{pq}{n}} = \sqrt{\frac{0.35(1 - 0.35)}{400}} = \sqrt{\frac{0.2275}{400}} = \sqrt{0.00056875} = 0.024$$

(3) Compute the confidence limits. At 95% confidence interval, $z = \pm 1.96$.

$$p \pm 1.96\, s_p = 0.35 \pm 1.96(0.024) = 0.35 \pm 0.04704$$
$$= 0.30296 \text{ to } 0.39704, \text{ or } 30.296\% \text{ to } 39.704\%$$

Thus, there are 95 out of 100 chances (95% confidence interval) that about 30% to 40% of the voters will vote for candidate Smith in the election.

13.4   Determining a Proper Sample Size (*n*)

The determination of a proper sample size is an important and practical problem in a sampling study. If the sample size is too large, more money and time are to be spent; but the result obtained from the large sample may not be more accurate than that from a smaller sample. On the other hand, if the sample size is too small, a valid conclusion may not be reached from the study. The method of determining a proper sample size is presented in the two cases below.

Sample Size for Estimating a Population Mean*

The relationship between the population mean μ and the sampling distribution of the mean $\bar{X}$ may be reviewed by sketching the normal curve as in Chart 13–2.

The normal curve shows the confidence interval

$$\mu \pm z\sigma_{\bar{x}} = \mu \pm E, \text{ and } E = z\sigma_{\bar{x}}$$

where $E =$ the sampling error, or the difference between a sample mean $\bar{X}$ and the population mean μ; that is, $E = \bar{X} - \mu$. It also shows that the range of the confidence interval is $2(E)$.

When $\sigma_{\bar{x}}$ is substituted by $\sigma/\sqrt{n}$, formula (12–3), the above equation becomes

$$E = z\frac{\sigma}{\sqrt{n}} \text{ and } \sqrt{n} = \frac{z\sigma}{E}$$

Solve for n by squaring both sides,

$$n = \left(\frac{z\sigma}{E}\right)^2 \text{ or } n = \frac{z^2\sigma^2}{E^2} \tag{13–6}$$

---

*Tchebycheff's inequality theorem can also be applied to the sampling distribution of the mean of any type, normally distributed or not normally distributed. According to the theorem (see page 268), the chance that a sample mean will fall farther than $k\sigma_{\bar{x}}$ from the population mean μ is less than $1/k^2$.

Let $a = k\sigma_{\bar{x}} = k(\sigma/\sqrt{n})$, then $k = a\sqrt{n}/\sigma$ and $1/k^2 = (\sigma/a\sqrt{n})^2 = \sigma^2/(a^2 n)$.

If the sample size n is very large, the maximum chance $1/k^2$ approaches zero. This result is called the *law of large numbers*.

CHART 13–2

THE NORMAL CURVE, SHOWING THE RELATIONSHIP BETWEEN μ AND $\bar{X}$

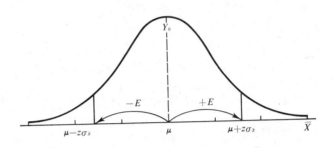

The values in the above formula may be obtained as follows:

(1) E: This is the maximum sampling error. An investigator must specify the maximum sampling error that he is willing to accept for his sampling study. For example, he may specify that if the mean obtained from the sample is $6 either above or below the true (population) mean, he will consider that the sampling study is satisfactory. Thus, $E = \$6$, and the confidence interval is $\mu \pm \$6$.

(2) z: This is set by the level of the confidence interval as shown in Table 13–1. For example, if the investigator wishes that the result of the estimation should be practically certain, or 99.73% (that is, the estimated population mean from the sample will be within the range of the true population mean $\mu \pm \$6$), the value of z is found to be 3 in the Areas Under the Normal Curve Table, Table 6 in the Appendix.

(3) σ: The value of the standard deviation of the population σ may be actual, or estimated from the past experience, or estimated by using $\hat{s}$ which is the standard deviation of either a previous sample or a pilot study.

After a sample of the computed size is taken, the result of the sample should be evaluated. This can be done by finding the standard error of the mean $s_{\bar{x}}$ according to the standard deviation of the sample $\hat{s}$. If the product of $z(s_{\bar{x}})$ is less than the specified sampling error, the estimation of the sample is considered satisfactory. If the product is larger, the sample size should be revised and increased.

Example 5

A service station manager wants to sample the sales tickets in order to find out the average (mean) amount per sale during a given period. He indicates that (1) the maximum sampling error should not be more than 20¢ above or below the true mean, (2) the level of the confidence interval should be 99.73%, and (3) the standard deviation of the population based on his experience is estimated at 80¢. Find a proper sample size according to his indications.

Solution.   (1) The confidence interval is $\mu \pm \$0.20$. Thus, $E = \$0.20$.
(2) At the 99.73% confidence interval, $z = 3$.

(3) The standard deviation of the population is $\sigma = 80\cent = \$0.80$. Substitute the above three values in formula (13–6).

$$n = \left(\frac{z\sigma}{E}\right)^2 = \left(\frac{3(0.80)}{0.20}\right)^2 = 12^2 = 144 \text{ (sample size)}$$

Note that the units of σ and E are the same, namely, dollars.

Evaluation of the Sample Size 144—Case I.   Assume that after a sample of 144 sales tickets is taken at random, the sample statistics are found as follows: the sample mean $\bar{X} = \$2.70$, and the sample standard deviation $\hat{s} = \$0.72$. Then,

$$s_{\bar{x}} = \frac{\hat{s}}{\sqrt{n}} = \frac{0.72}{\sqrt{144}} = \frac{0.72}{12} = \$0.06$$

At the 99.73% confidence interval, the population mean is within the range

$$\bar{X} \pm 3s_{\bar{x}} = 2.70 \pm 3(0.06) = \$2.70 \pm \$0.18 = \$2.52 \text{ to } \$2.88$$

Since the computed sampling error $0.18 or 18\cent does not exceed the specification $20\cent$, the sample size is a proper one.

Evaluation of the Sample Size 144—Case II.   Assume that the standard deviation of the above sample of 144 tickets is $84\cent$, or $\hat{s} = \$0.84$, then

$$s_{\bar{x}} = \frac{0.84}{\sqrt{144}} = \frac{0.84}{12} = 0.07$$

At the 99.73% confidence interval, the population mean is then within the range

$$\bar{X} \pm 3s_{\bar{x}} = 2.70 \pm 3(0.07) = \$2.70 \pm \$0.21$$

Thus, the computed sampling error $21\cent$ exceeds the maximum specification $20\cent$. Based on the sampling study, the sample size must be revised as follows:

$$n = \left(\frac{z\sigma}{E}\right)^2 = \left(\frac{3(0.84)}{0.20}\right)^2 = 12.6^2 = 158.76 \text{ or rounded to } 159$$

Thus the proper sample size is increased to 159.

Example 6

Refer to Example 5. If the level of the confidence interval is 95.45%, what is the proper sample size?

Solution.   Here $z = 2$. Thus,

$$n = \left(\frac{z\sigma}{E}\right)^2 = \left(\frac{2(0.80)}{0.20}\right)^2 = 8^2 = 64 \text{ (sample size)}$$

The computed sample size should also be evaluated before a valid conclusion is drawn from the sampling study.

Sample Size for Estimating a Population Proportion

The relationship between the population proportion P and the sampling distribution of the proportion p may be reviewed by sketching the normal curve in Chart 13–3.

CHART 13–3

THE NORMAL CURVE, SHOWING THE RELATIONSHIP BETWEEN P AND p

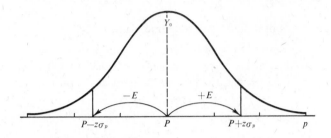

The normal curve shows the confidence interval

$$P \pm z\sigma_p = P \pm E \text{ and } E = z\sigma_p$$

where E = the sampling error, or the difference between a sample proportion p and the population proportion P; that is, $E = p - P$. When σ_p is substituted by $\sqrt{PQ/n}$ (Formula 12–4), the above equation becomes

$$E = z\sigma_p = z\sqrt{\frac{PQ}{n}} \text{ or } \sqrt{\frac{PQ}{n}} = \frac{E}{z}$$

Square both sides,

$$\frac{PQ}{n} = \left(\frac{E}{z}\right)^2 = \frac{E^2}{z^2}$$

Solve for n,

$$n = \frac{z^2 PQ}{E^2} \tag{13–7}$$

The values in the formula are obtained in a similar manner as those in formula (13–6) except a few changes noted below:

(1) E: This is specified in a proportional form, such as a percentage. For example, an investigator may specify that if the proportion obtained from the sample is 3% either above or below the population proportion, he will consider that the sampling study is satisfactory. Thus, $E = 3\%$, and the confidence interval is $P \pm 3\%$.

(2) PQ: The value of P, or the population proportion, may be estimated from the past experience, a previous sampling study, or a pilot study. $Q = 1 - P$. Since the maximum value of the product of PQ is $(0.5)(0.5) = 0.25$, for safety, 50% (or 0.5) is usually assigned to both factors P and Q in finding a maximum sample size.

Example 7

The business manager of a large company wants to check the inventory records against the physical inventories by a sampling study. He indicates that (1) the maximum sampling error should not be more than 5% above or below the true proportion of the inaccurate records, (2) the level of the confidence interval should be 99.73%, and (3) the proportion of the inaccurate records is estimated at 35% according to the past experience. Find the sample size.

Solution. (1) The confidence interval is $P \pm 5\%$. Thus, $E = 5\% = 0.05$.

(2) At the 99.73% confidence interval, $z = 3$.

(3) The estimated population proportion $P = 35\% = 0.35$, and $Q = 1 - P = 1 - 0.35 = 0.65$.

Substitute the values in formula (13–7).

$$n = \frac{z^2 PQ}{E^2} = \frac{3^2(0.35)(1 - 0.35)}{0.05^2} = \frac{9(0.2275)}{0.0025} = 819 \text{ (sample size)}$$

Example 8

Refer to Example 7. What is the maximum sample size if the population proportion is not given?

Solution. Here, $P = 50\% = 0.50$, $Q = 1 - P = 1 - 0.50 = 0.50$, $z = 3$ (same), and $E = 5\% = 0.05$ (same).

Substitute the above values in formula (13–7).

$$n = \frac{3^2(0.50)(0.50)}{0.05^2} = \frac{9(0.2500)}{0.0025} = 900 \text{ (sample size)}$$

Example 9

Refer to Example 7. What is the maximum sample size if the level of confidence interval is set on 95.45%?

Solution. Here, $z = 2$, $P = 0.50$, $Q = 0.50$, and $E = 0.05$.

$$n = \frac{2^2(0.50)(0.50)}{0.05^2} = \frac{4(0.2500)}{0.0025} = 400 \text{ (sample size)}$$

It is not necessary to evaluate or to revise the sample size if the maximum sample size is used.

13.5 Summary

The process of statistical inference may take one of the two forms: (1) estimation—use a sample statistic to estimate the corresponding unknown population parameter, and (2) tests of hypotheses—test the significance of the difference between a sample statistic and the corresponding hypothetical population parameter in order to make a decision.

An estimate of a parameter may be expressed in two ways: a point estimate which is a single number, and an interval estimate which is a stated range. The area (such as under a normal curve) which represents the probability of having the true population parameter (such as population mean) within interval estimates (such as based on sample means) is called a confidence interval. The two values which specify the confidence interval are called the confidence limits, and the probability is called the confidence coefficient or confidence level.

A statistic that is used for estimating a parameter is called an estimator. An estimator is unbiased when the expected value of the statistic used as the estimator is equal to the value of the parameter. Since the expected value of a sample mean $\bar{X}$ (or the mean of the sampling distribution of $\bar{X}$) is equal to the population mean μ, $\bar{X}$ is an unbiased estimate of μ. Likewise, a sample proportion p is an unbiased estimate of the population proportion P.

Generally, a sample variance $s^2 (= \sum x^2/n)$ is not an unbiased estimate of the population variance $\sigma^2 (= \sum x^2/N)$. However, when the sample size n is large (30 or above) and the population size N is also large, s^2 is considered as an unbiased estimate of σ^2. A modified sample variance $\hat{s}^2 [= \sum x^2/(n-1)]$ is an unbiased estimated of the modified population variance $\hat{\sigma}^2 [= \sum x^2/(N-1)]$. When N is large, $\hat{\sigma}^2$ approaches σ^2; $\hat{s}^2$ thus becomes an unbiased estimate of σ^2, regardless of the sample size n being large or small.

Similarly, a sample variance pq is not an unbiased estimate of the population variance PQ. However, when the sample size n and the population size N are large, pq is considered as an unbiased estimate of PQ. A modified sample variance $pq[n/(n-1)]$ is an unbiased estimate of the modified population variance $PQ[N/(N-1)]$. When N is large, $PQ[N/(N-1)]$ approaches PQ; and $pq[n/(n-1)]$ becomes an unbiased estimate of PQ, regardless of the sample size n being large or small.

SUMMARY OF FORMULAS

Application	Formula	Formula Number	Reference Page
Estimation			
Unbiased estimate of σ^2. n is small (If n is large, use s^2 as the estimator.)	$\hat{s}^2 = \dfrac{\sum x^2}{n-1} = s^2\left(\dfrac{n}{n-1}\right)$	13–1 a & b	328
Estimated standard error of the mean			
n is small	$s_{\bar{x}} = \dfrac{\hat{s}}{\sqrt{n}} = \dfrac{s}{\sqrt{n-1}}$	13–2 a & b	333, 334
n is large	$s_{\bar{x}} = \dfrac{s}{\sqrt{n}}$	13–3	334
Estimated standard error of the proportion			
n is small	$s_p = \sqrt{\dfrac{pq}{n-1}}$	13–4	336
n is large	$s_p = \sqrt{\dfrac{pq}{n}}$	13–5	336
Sample size			
Estimating population mean	$n = \left(\dfrac{z\sigma}{E}\right)^2$	13–6	337
Estimating population proportion	$n = \dfrac{z^2 PQ}{E^2}$	13–7	340

The quantity $(n - 1)$ given above is also called the number of degrees of freedom. The number of degrees of freedom (D) is the number of variables that can vary freely in a set of variables under certain conditions.

The steps used in finding an interval estimate of population mean μ from a sample mean $\bar{X}$ are: (1) find $\bar{X}$, (2) compute $\sigma_{\bar{x}}$ if σ is known (or compute $s_{\bar{x}}$ if σ is to be estimated from s), and (3) compute $\bar{X} \pm z\sigma_{\bar{x}}$ (or $\bar{X} \pm zs_{\bar{X}}$).

Likewise, the steps used in finding an interval estimate of population proportion P from a sample proportion p are: (1) find p, (2) find σ_p if the product PQ is known (or s_p if the product is to be estimated from pq), and (3) compute $p \pm z\sigma_p$ (or $p \pm zs_p$).

The formulas used in estimating various population parameters from sample statistics and used in finding proper sample sizes for the estimations are summarized at the end of this section.

Exercises 13

1. Assume that $\mu = \$100$, $z = \pm 2.58$, and $\sigma_{\bar{x}} = \$10$. Find:
   (a) The confidence interval.
   (b) The confidence limits.
   (c) The confidence coefficient.

2. Refer to Problem 1. Assume that μ is unknown and $\bar{X} = \$105$. Find:
   (a) The point estimate.
   (b) The interval estimate.
   (c) Interpret the findings of the two types of estimates. If μ is known to be $\$100$, is the true population mean within the interval estimate?

3. Assume that $P = 45\%$, $z = \pm 1.645$, and $\sigma_p = 6\%$. Find:
   (a) The confidence interval.
   (b) The confidence limits.
   (c) The confidence coefficient.

4. Refer to Problem 3. Assume that P is unknown and $p = 48\%$. Find:
   (a) The point estimate.
   (b) The interval estimate.
   (c) Interpret the findings of the two types of estimates. If P is known to be 45%, is the true population proportion within the interval estimate?

5. Take all possible samples of two students from the following population which represents the hours spent by three students in their part-time jobs during a week. Prove that $\hat{s}^2$ is an unbiased estimate of $\hat{\sigma}^2$.

Student	Hours Worked
L	2
M	3
N	7

6. Give the reasons and the conditions, if any, to support the following statements:

(a) $\bar{X}$ is an unbiased estimate of μ.
(b) s^2 is a biased estimate of σ^2.
(c) $\hat{s}^2$ is an unbiased estimate of σ^2.
(d) s^2 is an unbiased estimate of σ^2.

7. Take all possible samples of three men from the following population of four employees in an office. Find the proportion of married men for each sample and prove that

$$pq\left(\frac{n}{n-1}\right) \text{ is an unbiased estimate of } PQ\left(\frac{N}{N-1}\right)$$

Employee	Marriage Status
Adams	Married
Bonner	Married
Casey	Married
Denton	Unmarried

8. Give the reasons and the conditions, if any, to support the following statements:

(a) p is an unbiased estimate of P.
(b) pq is a biased estimate of PQ.
(c) $pq[n/(n-1)]$ is an unbiased estimate of PQ.
(d) pq is an unbiased estimate of PQ.

9. What is the number of degrees of freedom if we have a set of 20 values and the sum of the values is 500?

10. Assume that you have selected at random the weights of 400 employees from the personnel file of a large company. The mean of the sample is 165 pounds. The standard deviation of the weights of all employees in the company is 8 pounds. Find the population mean at (a) 68.27%, (b) 95.45%, and (c) 99.73% confidence intervals.

11. Assume that you have asked 350 workers regarding their weekly wages as a random sample in a factory of 40,000 workers. The results of your inquiries are tabulated below. Estimate the population mean of the wages earned by all workers at a 95% confidence interval. (Compute to two decimal places.)

Weekly wages	Number of workers
$ 50 and under $ 70	50
70 and under 90	70
90 and under 110	125
110 and under 130	75
130 and under 150	30
Total	350

12. Assume that you are one of the 80,000 football fans in a large stadium. You have taken a random sample of 500 fans and found that 200 of them are girls.

Estimate the population proportion of the number of girls at a 99.73% confidence interval. (Compute to two decimal places of a percent rate.)

13. A research division of a state highway department wishes to know the average number of miles travelled during a week by trucks. The division head states that (a) the maximum sampling error should not be more than 15 miles above or below the true mean, (b) the level of confidence interval should be 95.45%, and (c) the standard deviation of the population based on a previous sampling study is 120 miles. What is the proper sample size in order to meet his requirements?

14. Assume that you have taken a random sample of the size obtained from Problem 13. Would you consider the sample size as satisfactory if the standard deviation of the sample (s) is (a) 150 miles, and (b) 96 miles? If yes, why? If not, what should the sample size be then?

15. The research project given in Problems 13 and 14 was made in an area having 500,000 trucks traveling. Assume that the same conditions and requirements are to be applied in an area of 2,000,000 trucks. What is the proper sample size?

16. A radio station manager wants to know the proportion of the people who like a particular program. He specifies that (a) the maximum sampling error should not be more than 2% from the true proportion, (b) the level of the confidence interval should be 95%, and (c) the proportion of the people who like the program is about 60%.

  (1) What is the proper sample size according to his specification?
  (2) What is the maximum sample size if he specifies only (a) and (b)?

14 Basic Aspects of Tests of Hypotheses

This chapter explains the basic idea of tests of hypotheses. The actual work involving the tests will be presented in the next chapter. The new terms used in tests of hypotheses are first explained in Section 14.1. The basic concept of testing hypotheses by the use of the normal curve is discussed in Section 14.2. The operating characteristic and power curves are probability curves that are constructed under hypotheses. The constructions are presented in Section 14.3. The summary of this chapter is given in Section 14.4.

14.1 Terminology

Statistical Hypotheses

A *statistical hypothesis*, or simply *hypothesis*, is an assumption or a guess concerning the population. Before accepting or rejecting a hypothesis, an investigator should test the validity of the hypothesis since it may or may not be true. Clearly, a sure way of testing the hypothesis would be an examination of the entire population. However, the examination may become impractical or impossible. A practical way is to test the hypothesis by using a sample according to the theory of probability. The result of the test will guide a statistician either to accept the hypothesis or to reject it. The acceptance or rejection will lead an investigator to make a decision.

For example, a large army plans to order uniforms for its soldiers. Before a decision concerning the quantity of different sizes of uniforms is reached, it is desirable to know the average height of the soldiers in the army. A hypothesis or an assumption that the average height of the soldiers in the army is 68 inches is first made. The hypothesis is made according to the past experience which may or may not be true today. In order to find out the validity of the hypothesis, we may take a sample of 400 soldiers and then test the hypothesis based on the sample mean. The test is conducted according to the probability of the sampling distribution of the mean. If the test shows a high probability that the population mean is 68 inches, we then state that the fact found in the sample is *not inconsistent* with the hypothesis and, therefore, we *accept* the hypothesis. Unless additional information shows otherwise, the army may reach a decision that the quantity of different sizes of uniforms is to be distributed according to the mean of 68 inches. On the other hand, if the test shows a low probability, we may state that the fact found in the sample is *inconsistent* with the hypothesis and thus we *reject* or *nullify* the hypothesis. In such a case, a different decision concerning the quantity of different sizes of uniforms may be reached.

Another common way of stating a hypothesis is that there is *no difference* between two values, such as a population mean being compared with a sample mean. The term "no difference" does not mean that the actual value of the difference is zero. Instead, the term means that the difference is merely due to sampling fluctuation; therefore, the difference is considered as "no difference" or "zero." If the mean height of the sample of 400 soldiers is found to be $67\frac{1}{2}$ inches, we may state the hypothesis: There is no difference between the population mean and the sample mean. When the test shows that the difference, $68 - 67\frac{1}{2} = \frac{1}{2}$ inches, is *not significant* (that is, only due to sampling fluctuation), we accept the hypothesis and conclude that the population mean is 68 inches. On the other hand, if the statistical test shows that the difference is *significant*, the hypothesis is rejected or nullified.

Statistical hypotheses which are stated for the purpose of possible rejection or nullification are called *null hypotheses*. A null hypothesis is usually denoted by the symbol H_0. In the above illustrations, the first hypothesis that the average height is 68 inches may be expressed symbolically:

$$H_0: \quad \mu = 68 \text{ inches}$$

and the second hypothesis that there is no difference (or the difference is not significent) between the population mean and the sample mean may be written:

$$H_0: \quad \bar{X} - \mu = 0$$

Any hypothesis which differs from the null hypothesis is called an *alternative hypothesis* and is denoted by H_1. In a given test, there is usually only one null hypothesis but there may be many alternative hypotheses. For example, concerning the average height of the soldiers, we may state the null and alternative hypotheses as follows:

$$H_0: \quad \mu = 68 \text{ inches}$$
$$H_1: \quad \mu \neq 68 \text{ inches (or } \mu \text{ is "not equal to" 68 inches)}$$

The statistical methods which are used to decide whether to accept or reject hypotheses are called *tests of hypotheses, tests of significance*, or *methods of making decision rules*.

Type I Error (α) and Type II Error (β)

There are always two possible *states of nature*, or *states of world*, of the happening in a given problem. In the example concerning the average height of the soldiers in the army, the first state of nature of the happening will be:

The average height of the soldiers *is* 68 inches

The second state of nature will be:

The average height of the soldiers *is not* 68 inches

Again, let the null hypothesis be:

The average height of the soldiers is 68 inches

When one makes a decision concerning the hypothesis, his decision may be correct or incorrect, depending on the true state of nature of the happening. In general, the results of making such decisions can be classified in the following manner:

1. Two types of correct decisions:
   I. He *rejects* a hypothesis when it is *not true.*
   II. He *accepts* a hypothesis when it is *true.*
2. Two types of incorrect decisions or errors:
   Type I error. He *rejects* a hypothesis but actually it is *true.*
   Type II error. He *accepts* a hypothesis but actually it is *not true.*

Thus, in the above example, if the hypothesis of 68 inches is rejected according to the result of the test but actually the average is 68 inches, a Type I error has been made. On the other hand, if the hypothesis of 68 inches is accepted according to the test result but actually the average is 70 inches (or not 68 inches), a Type II error has been made by the investigator.

The probability of making a Type I error is usually denoted by α and the probability of making a Type II error is denoted by β. The possible decisions and the results from the decisions concerning the above example may be summarized as follows:

Possible decision	State of nature	
	H_0 is true (*68* inches)	H_0 is not true (*70* inches)
Reject H_0	Incorrect decision — Type I error (α)	Correct decision
Accept H_0	Correct decision	Incorrect decision — Type II error (β)

Level of Significance Specifying Type I Error (α)

The maximum probability of making a Type I error specified in a test of hypothesis is called the *level of significance.* The level of significance is usually specified before a test is made. Otherwise the result obtained from the test may influence the decision concerning the hypothesis. In practice, the value of 5% ($\alpha = 0.05$) or 1% ($\alpha = 0.01$) is frequently used to set the level of significance, although other values may also be used. For example, in the illustration concerning the average height, if we select a 5% level of significance, we will expect that the probability of making an error of rejecting the hypothesis (assumed to be 68 inches) when it is true (actually it is 68 inches) is 5%. In other words, we are about 95% confident that we will make a correct decision, although we could be wrong with a probability of 5%, or about 5 out of 100 chances of being wrong.

14.2   Two-Tailed Tests and One-Tailed Tests

The level of significance may be represented by a portion of the area under the normal curve in two ways: (a) two "tails" or sides under the curve (Chart 14–1), and (b) one "tail" or side under the curve—either the right tail (Chart 14–2A) or the left tail (Chart 14–2B). The tests of hypotheses which are based on the level of significance represented by both tails under the normal curve are called *two-tailed tests* or *two-sided tests*. If the level of significance is represented by only one tail, the tests are called *one-tailed tests* or *one-sided tests*.

The fundamental concepts of the two kinds of tests are illustrated by using the information as follows:

Null hypothesis: The mean of a certain population is $500, or H_0: $\mu = \$500$.

The standard error of the sampling distribution of the mean drawn from the population is $5, or $\sigma_{\bar{x}} = \$5$.

The level of significance is 5%, or $\alpha = 5\%$.

Illustration of a Two-Tailed Test

This kind of test will answer the question: What should be the amount of a sample mean, which may be *below* or *above* the hypothetical mean ($500), that will enable us to decide to reject the hypothesis?

We know that the mean of a sampling distribution of the mean (or the mean of the means of all possible samples of the same size drawn from the population) is equal to the population mean. Thus, if the hypothesis is true, the mean of a sampling distribution of the mean of a given sample size should be $500 also. Further, if the hypothesis is true, 95% (or $1 - 5\%$) of the sample means of the distribution will fall within the range or the confidence interval

$$\mu \pm z\sigma_{\bar{x}} = 500 \pm 1.96(5) = 500 \pm 9.80 = \$490.20 \text{ to } \$509.80$$

Now if we actually take a simple random sample and find that the sample mean is *less* than $490.20 or *more* than $509.80, we would have reason to conclude that the difference between the hypothetical population mean and the sample mean is significant. If the hypothesis is true, such a sample mean (an extreme value) could happen with a probability of only 5%. Since the sample gives better information (observed facts) than the hypothesis, which may not be more than a rough estimate, we thus reject the hypothesis based on the sampling study. However, if the sample mean has a value within the range of $490.20 to $509.80, we will accept the hypothesis since a sample mean with such a value has a higher probability of being selected if the hypothesis is true.

The sampling distribution of the mean based on the hypothetical population mean is shown in Chart 14–1. The 5% level of significance ($\alpha = 5\%$) is repre-

CHART 14–1

TWO-TAILED TEST, LEVEL OF SIGNIFICANCE: $\alpha = 5\%$

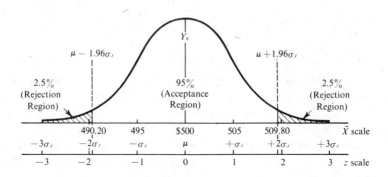

CHART 14–2

ONE-TAILED TEST, LEVEL OF SIGNIFICANCE: $\alpha = 5\%$

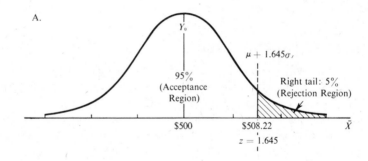

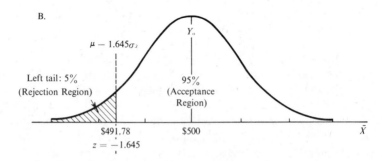

sented by the shaded areas located on both tails of the normal curve in the chart. It is split into two equal areas: 2.5% on each tail of the normal curve. The z values which indicate the two equal areas are called *critical values* and are ± 1.96 according to Table 6. The critical values may also be expressed in the actual units: \$490.20 and \$509.80. The shaded areas which are outside the range of z

$= \pm 1.96$ are called the *regions of rejection of the hypothesis* or simply *rejection region*. The area inside the range is then called the *region of acceptance of the hypothesis* or simply *acceptance region*.

Illustration of One-Tailed Tests

The hypothesis, the 5% level of significance, and the value of the standard error of the mean stated above are again used in the illustration for the following two cases.

A Test Involving Extreme High Values (A Right-Tailed Test)
This kind of test will answer the question: What should be the amount of a sample mean, which is *higher than* the hypothetical population mean ($500), that will enable us to decide to reject the hypothesis? Chart 14–2A shows that the 5% extreme high values far above the population mean of $500 are represented by the single shaded area located on the right tail of the normal curve. The critical value of z is $+ 1.645$ which is obtained from the normal curve area table. It separates the normal curve into two parts: the 5% rejection region and the 95% acceptance region. The critical value may also be expressed in dollar value, or

$$500 + 1.645(5) = \$508.225 \text{ or round to } \$508.22$$

If we actually take a simple random sample and find that the sample mean is *more* than $508.22, we would conclude that the difference is significant and reject the hypothesis that the population mean is $500.

A Test Involving Extreme Low Values (A Left-Tailed Test)
This kind of test will answer the question: What should be the amount of a sample mean, which is *smaller than* the hypothetical population mean ($500), that will enable us to decide to reject the hypothesis? Chart 14–2B shows that the 5% extreme low values far below the hypothetical population mean of $500 are represented by the single shaded area located on the left tail of the normal curve. The critical value of z is $- 1.645$ in the units of the standard error of the mean, or expressed in dollar value:

$$500 - 1.645(5) = \$491.775 \text{ or round to } \$491.78$$

If we actually take a simple random sample and find that the sample mean is *less* than $491.78, we would conclude that the difference is significant and reject the hypothesis that the population mean is $500.

TABLE 14–1

SELECTED CRITICAL VALUES OF z

α, *Level of Significance*	1%	5%	10%
z, Critical value for one-tailed tests	+2.33 or −2.33	+1.645 or −1.645	+1.28 or −1.28
z, Critical values for two-tailed tests	+2.58 and −2.58	+1.96 and −1.96	+1.645 and −1.645

Some frequently used critical values of z for both two-tailed and one-tailed tests at various levels of significance are given above. The critical values of z for other levels of significance may be obtained from the Areas Under the Normal Curve Table.

★14.3 Operating Characteristic Curve and Power Curve—Showing Type II Errors

An operating characteristic curve (or OC curve) shows the probabilities of making Type II errors (β). Under various alternative hypotheses, the probabilities of making Type II errors may be computed according to a fixed Type I error (α) assigned to the null hypothesis. A power curve can be derived from an OC curve. It is simply a reversed OC curve. It shows the values of $(1 - \beta)$, which is the probability of making the correct decision of rejecting the null hypothesis when in fact the hypothesis is not true. The OC curve or the power curve thus may be conveniently used to illustrate the relationship between a specified or fixed Type I error and a series of Type II errors.

Construction of Curves from Sampling Distribution of the Mean

Examples 1 and 2 are used to illustrate the method of constructing operating characteristic curves and power curves from a sampling distribution of the mean. Examples 3 and 4 are used to illustrate the method of constructing the curves from a sampling distribution of the proportion.

Example 1

The average monthly family income in a large city is assumed to be $500, or H_0: $\mu_0 = \$500$. The standard deviation of the population according to the past experience is $100. In order to test the validity of the hypothesis, we plan to take a sample of 400 families. The level of significance is set at 5%; that is, the probability of making a Type I error should not exceed 5% (or $\alpha = 5\%$). Let the alternative hypotheses be that the means are not equal to $500, or H_1: $\mu \neq \$500$. What are the probabilities of making Type II errors under various alternative hypotheses?

Solution.   The definitions of the two types of errors are first reviewed for this example.

Null Hypothesis: The average is $500.

Commit a Type I error: Reject the hypothesis, but in fact the hypothesis is true or the average is $500.

Commit a Type II error: Accept the hypothesis, but in fact the hypothesis is not true; or, the average is not $500.

Next, we specify the acceptance region and the rejection region based on the null hypothesis, $\mu_0 = \$500$. If the true population mean is $500, the mean of the sampling distribution of the mean of sample size 400 families should be $500 also. The standard error of the mean of the sampling distribution is

$$\sigma_{\bar{x}} = \frac{\sigma}{\sqrt{n}} = \frac{100}{\sqrt{400}} = \$5$$

When we set the level of significance at 5%, we will expect to have 95% of all possible means of the distribution within the range of critical values $490.20 to $509.80, or

$$\mu \pm z\sigma_{\bar{x}} = 500 \pm 1.96(5) = 500 \pm 9.80 = \$490.20 \text{ to } \$509.80$$

The area within the range is the region of acceptance of the null hypothesis and the areas outside the range are the regions of rejection of the null hypothesis as shown in Chart 14–3(a). Note that the chart is based on a *two-tailed test* since the values of the sample means which are used as a basis to reject the null hypothesis in this type of problem may be *above* or *below* the hypothetical mean $500.

Let the alternative hypotheses (H_1) be $485 and $505 for illustration purpose.

(1) H_1: $\mu = \$485$. If the mean $\mu = \$485$ is true, the probability of drawing a sample of 400 with a mean within the range of $490.20 to $509.80 from the population is 0.14917, as shown in Chart 14–3(b), and is computed as follows:

The value of σ is assumed to be unchanged. The assumption is reasonable since the standard deviation is based on the deviations of individual values from the mean. When individual values are increased (or decreased), the mean is also increased (or decreased). Therefore, the deviations will not be changed greatly when the mean of a given population is changed.

Since σ is not changed, the value of $\sigma_{\bar{x}}$ should still be $5.

The probability of having a sample mean within values from $485 to $490.20 is 0.35083:

When $\bar{X} = \$490.20$, the lower critical value,

$$z = \frac{\bar{X} - \mu}{\sigma_{\bar{x}}} = \frac{490.20 - 485}{5} = 1.04 \qquad A(1.04) = 0.35083$$

When $\bar{X} = \$509.80$, the upper critical value,

$$z = \frac{509.80 - 485}{5} = 4.96 \qquad A(4.96) = 0.50000$$

The required probability $= 0.50000 - 0.35083 = 0.14917$

Thus, when $\mu = \$485$, we will expect 14.917% of the sample means of the sampling distribution to fall within the range of $490.20 to $509.80. Since the sample means are within the acceptance region, we may state that if H_1 (or $\mu = \$485$) is true, the probability of accepting H_0 (or $\mu_0 = \$500$) is 0.14917. The acceptance of the hypothesis when in fact it is not true is the Type II error, or $\beta = 0.14917$, which is listed in Table 14–2.

(2) H_1: $\mu = \$505$. If the mean $\mu = \$505$ is true, the probability of having a sample mean within the values from $490.20 to $509.80 in the sampling distribution is 0.82993, as shown in Chart 14–3(c), and is computed as follows:

The probability of having a sample mean within values from $505 to $509.80 is 0.33147:

$$z = \frac{509.80 - 505}{5} = 0.96 \qquad A(0.96) = 0.33147$$

CHART 14-3

PROBABILITIES OF MAKING TYPE II ERRORS (β VALUES)
UNDER VARIOUS HYPOTHESES FOR TWO-TAILED TESTS
H_0: $\mu_0 = \$500$, H_1: $\mu \neq \$500$, $\alpha = 5\%$

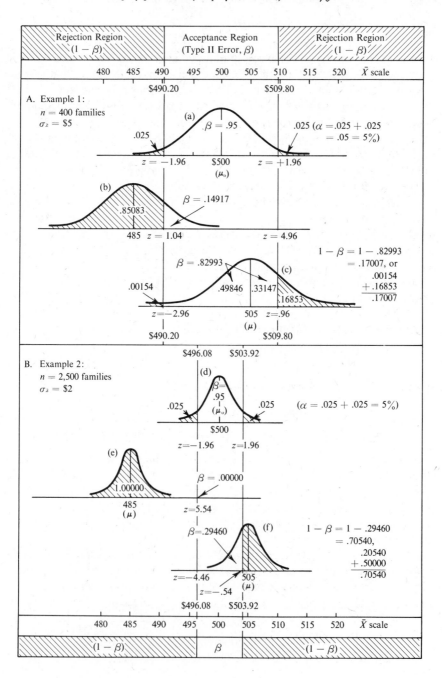

The probability of having a sample mean within the values from \$490.20 to \$505 is 0.49846:

$$z = \frac{490.20 - 505}{5} = -2.96 \qquad A(-2.96) = 0.49846$$

The total probability $\beta = 0.33147 + 0.49846 = 0.82993$, which is also listed in Table 14–2.

Other values of β as listed in Table 14–2 are computed in the same manner.

Examine curve (a) in part A of Chart 14–3. It can be seen that if the acceptance region (β value or Type II error) is narrowed, the rejection region (α value

TABLE 14–2

Operating Characteristic and Power Functions
$H_0 : \mu_0 = \$500, H_1 : \mu \neq \$500,$ and $\alpha = 5\%$
Two-Tailed Tests

| Hypothetical Values of μ | A. $n = 400$ | | B. $n = 2{,}500$ | |
	OC Functions (Type II Errors) β	Power Functions $1 - \beta$	OC Functions (Type II Errors) β	Powers Function $1 - \beta$
480	0.02068	0.97932	0.00000	1.00000
485	0.14917	0.85083	0.00000	1.00000
490	0.48401	0.51599	0.00118	0.99882
495	0.82993	0.17007	0.29460	0.70540
500	0.95000	0.05000	0.95000	0.05000
505	0.82993	0.17007	0.29460	0.70540
510	0.48401	0.51599	0.00118	0.99882
515	0.14917	0.85083	0.00000	1.00000
520	0.02068	0.97932	0.00000	1.00000

or Type I error) is increased. For example, if the acceptance region is reduced from the range of *\$490.20 to \$509.80* to the range of *\$495 to \$505* (or $\mu_0 \pm 1\sigma_x$ = 500 ± 5), the rejection region in curve (a) is increased from 5% to 32% (or 1 − 68%, since the area under the normal curve is 68% when $z = \pm 1$). That is to say: When we wish to reduce Type II error, we are forced to increase Type I error at the same time.

However, when Type I error is specified or fixed, such as 5%, there is one way to reduce Type II error under a given alternative hypothesis. The only way is to increase the sample size, which may or may not be possible because of time, money, and the type of population. The following example is used to illustrate the effect of increasing the sample size to Type II errors under the various alternative hypotheses.

Example 2

Refer to Example 1. Suppose that we plan to take a sample of 2,500 families. What are the probabilities of making Type II errors under various alternative hyptheses?

Solution.   The answers to this question are listed in part B of Table 14–2 and shown in Chart 14–3B. They are computed as follows:

$$\sigma_{\bar{x}} = \frac{\sigma}{\sqrt{n}} = \frac{100}{\sqrt{2,500}} = \frac{100}{50} = \$2$$

The value of α is still 5%. Thus,

$$\mu \pm 1.96\,\sigma_{\bar{x}} = 500 \pm 1.96(2) = 500 \pm 3.92 = \$496.08 \text{ to } \$503.92$$

The range of \$496.08 to \$503.92 forms the region of acceptance of the hypothesis that the average family income is \$500 per month (curve d). Also let the alternative hypotheses (H_1) be \$485 and \$505 for illustration purpose.

(1) H_1: $\mu = \$485$. If the mean $\mu = \$485$ is true, the probability of having a sample mean within values from \$496.08 to \$503.92 in the sampling distribution of the mean (curve e) is 0 and is computed as follows:

The probability of having a sample mean within values from \$485 to \$496.08 is 0.5:

$$z = \frac{496.08 - 485}{2} = 5.54 \qquad A(5.54) = 0.50000$$

Thus, $\beta = 0.50000 - 0.50000 = 0$.

(2) H_1: $\mu = \$505$. If the mean $\mu = \$505$ is true, the probability of having a sample mean within values from \$496.08 to \$503.92 in the sampling distribution of the mean (curve f) is 0.29460 and is computed as follows:

The probability of having a sample mean within values from \$496.08 to \$505 is 0.5:

$$z = \frac{496.08 - 505}{2} = -4.46 \qquad A(-4.46) = 0.50000$$

The probability of having a sample mean within values from \$503.92 to \$505 in the distribution is 0.20540:

$$z = \frac{503.92 - 505}{2} = -0.54 \qquad A(-0.54) = 0.20540$$

Thus, $\beta = 0.50000 - 0.20540 = 0.29460$.

Other values listed in part B of Table 14–2 are also computed in the same manner.

Compare the β values in part A with those in part B in Chart 14–3. It can be seen that the values of β are reduced under the same alternative hypotheses when the sample size is increased from 400 to 2,500 families. The β values are also called the *operating characteristic functions*. When the functions are shown by a curve, it is called an *operating characteristic curve* as shown in Chart 14–4. The values of $1 - \beta$ are called *power functions*. When the power functions are shown by a curve, the curve is called a *power curve* as shown on Chart 14–5.

CHART 14–4

OPERATING CHARACTERISTIC CURVE

$H_0: \mu_0 = \$500, H_1: \mu \neq \$500,$ AND $\alpha = 5\%$

TWO-TAILED TESTS

Probability of Accepting
H_0 under Given μ
(β)

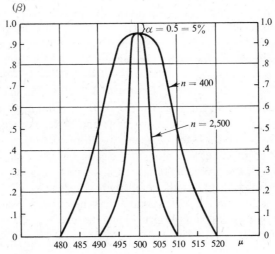

Source: Table 14–2.

CHART 14–5

POWER CURVE

$H_0: \mu_0 = \$500, H_1: \mu \neq \$500,$ AND $\alpha = 5\%$

TWO-TAILED TESTS

Probability of Rejecting
H_0 under Given μ
$(1 - \beta)$

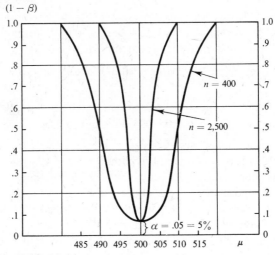

Source: Table 14–2.

Construction of Curves from a Sampling Distribution of the Proportion

The OC and power curves may also be drawn for sampling distributions of the proportions under various alternative hypotheses in a similar manner to that for sampling distributions of the means. However, the standard error of the proportion σ_p should be computed for each sampling distribution. The value of the standard error of a sampling distribution of the proportion depends upon the value of P of a population from which the samples of size n are drawn. Since the values of P are specified differently in the hypotheses, the standard errors can not be assumed to be the same for each distribution.

Example 3

A large company has an agreement with its supplier that it will accept a shipment of products with 4% or less defective units and reject the shipment with more than 4% defective units. If the company would inspect every unit in each shipment, a decision of either accepting or rejecting a given shipment is a simple matter. However, the procedure of inspecting all units in a large shipment may be too costly or impossible. Thus, the company has decided to take a sample in a large shipment for inspection purpose. Assume that the company has decided that the sample size should be 300 and the level of significance 5%. Construct an OC curve and a power curve.

Solution.   Since the agreement states that the company will accept a shipment with 4% or less defective units, we need only to find the highest defective rate which is due to the sampling fluctuation. Thus, one-tailed tests (the right tail) are used in this example. The example is diagrammed in Chart 14–6. The null and alternative hypotheses are:

$$H_0: \qquad P_0 = 4\% \text{ or less}$$
$$H_1: \qquad P = \text{more than } 4\%$$

Here $z = 1.645$. If the shipment has exactly 4% defectives, the 5% (α) sample proportions of the sampling distribution of size 300 will lie above the critical value 0.06 (or 6%). The critical value is computed as follows:

$$\sigma_p = \sqrt{\frac{PQ}{n}} = \sqrt{\frac{(0.04)(0.96)}{300}} = 0.011$$

$$\text{Critical value} = P_0 + 1.645\,\sigma_p = 0.04 + 1.645(0.011)$$
$$= 0.058095 \text{ or round to } 0.06$$

Curve (a) in Chart 14–6 shows the acceptance region ($\beta = 95\%$) and the rejection region ($\alpha = 5\%$) based on the null hypothesis: $P_0 = 4\%$ for the distribution of sample proportions. Thus, if $P_0 = 4\%$ is true, there are only 5 out of 100 chances that a sample proportion of 6% or above will occur in the distribution.

The following decision rules now may be established for taking a simple random sample of 300 units for a large shipment.

(1) If the sample has above 6% defective units, we will reject the shipment; that is, we will reject the hypothesis that the population (shipment) has 4% or less defective units.

CHART 14–6

PROBABILITIES OF MAKING TYPE II ERRORS (β VALUES) UNDER VARIOUS HYPOTHESES FOR ONE-TAILED TESTS; $H_0: P_0 = 4\%$ OR LESS, $H_1: P =$ ABOVE 4%, $\alpha = 5\%$

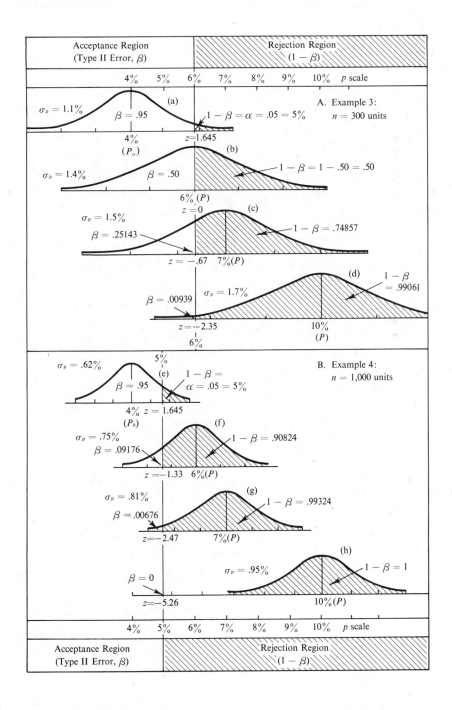

(2) If the sample has 6% or less defective units, we will accept the shipment; that is, we will accept the hypothesis that the population has 4% or less defective units. Note that there will be 95% of the sample proportions fluctuating below 6% on the p scale of the normal curve when the population proportion is 4%.

Let the alternative hypotheses be $P = 6\%$, 7%, and 10% defective units for illustration purpose.

(1) $H_1: P = 6\% = 0.06$. If the shipment has a proportion of 6% defective units (or $P = 6\%$ is true), the probability of having a sample proportion of 6% or less (within the acceptance region) is 0.5:

$$z = \frac{\text{Critical value} - P}{\sigma_p} = \frac{0.06 - 0.06}{\sigma_p} = \frac{0}{\sigma_p} = 0$$

where $\qquad \sigma_p = \sqrt{\dfrac{PQ}{n}} = \sqrt{\dfrac{(0.06)(0.94)}{300}} = 0.014$ or 1.4%

$$A(z) = A(0) = 0. \quad \beta = 0.50000 - 0 = 0.50000 \ [\text{curve(b)}]$$

(2) $H_1: P = 7\% = 0.07$. If the shipment has a proportion of 7% defective units, the sampling distribution of the proportion will have a standard error 0.015 as computed below:

$$\sigma_p = \sqrt{\frac{PQ}{n}} = \sqrt{\frac{(0.07)(0.93)}{300}} = 0.015 \text{ or } 1.5\%$$

The probability of having a sample proportion 6% or less from the distribution, if the hypothesis ($P = 7\%$) is true, is 0.25143:

$$z = \frac{0.06 - 0.07}{0.015} = -0.67 \qquad A(-0.67) = 0.24857$$

$$\beta = 0.50000 - 0.24857 = 0.25143 \ [\text{curve (c)}]$$

Thus, if the population proportion of 7% defective is true, the probability of accepting the shipment (or $H_0: P_0 = 4\%$, the false hypothesis) is about 25%.

(3) $H_1: P = 10\% = 0.10$. If the shipment has a proportion of 10% defective units, the sampling distribution of the proportion will have a standard error 0.017 as computed below:

$$\sigma_p = \sqrt{\frac{PQ}{n}} = \sqrt{\frac{(0.10)(0.90)}{300}} = 0.017 \text{ or } 1.7\%$$

The probability of having a sample proportion 6% or less, when $P = 10\%$ is 0.00939:

$$z = \frac{0.06 - 0.10}{0.017} = -2.35 \qquad A(-2.35) = 0.49061$$

$$\beta = 0.50000 - 0.49061 = 0.00939 \ [\text{curve (d)}]$$

Thus, if the shipment has a true proportion of 10% defective units, the probability of accepting the null hypothesis, $H_0: P_0 = 4\%$, is about 0.9%.

The above values of β and values $(1 - \beta)$ are listed in Table 14–3A and are shown in Charts 14–7A and 14–8A respectively. The β values under other alternative hypotheses as listed in the table are obtained in the same manner.

The sample size also affects the β values for the sampling distributions of the proportions. This is illustrated in the following example.

Example 4

Refer to Example 3. Construct the OC curve and the power curve if the sample size for inspection is 1,000 units in a large shipment.

Solution.   The OC curve is shown in Chart 14–7B and the power curve is shown in Chart 14–8B. The computation of β values is based on the diagram shown in Chart 14–6B. The β values are listed in part B of Table 14–3 and are computed below.

The critical value is computed as follows:

$$\sigma_p = \sqrt{\frac{PQ}{n}} = \sqrt{\frac{(0.04)(0.96)}{1,000}} = 0.0062 \text{ or } 0.62\%$$

Critical value $= P_0 + 1.645\,\sigma_p = 0.04 + 1.645(0.0062) = 0.050199$, or round to 5% [see Chart 14–6B, curve (e)]

Let the alternative hypotheses also be $P = 6\%$, 7%, and 10% defective units for illustration purpose.

(1) H_1: $P = 6\% = 0.06$ [curve (f).]

$$\sigma_p = \sqrt{\frac{PQ}{n}} = \sqrt{\frac{(0.06)(0.94)}{1,000}} = 0.0075 \text{ or } 0.75\%$$

$$z = \frac{0.05 - 0.06}{0.0075} = -1.33 \quad A(-1.33) = 0.40824$$

$$\beta = 0.50000 - 0.40824 = 0.09176$$

(2) H_1: $P = 7\% = 0.07$ [curve (g).]

$$\sigma_p = \sqrt{\frac{PQ}{n}} = \sqrt{\frac{(0.07)(0.93)}{1.000}} = 0.0081 \text{ or } 0.81\%$$

$$z = \frac{0.05 - 0.07}{0.0081} = -2.47 \quad A(-2.47) = 0.49324$$

$$\beta = 0.50000 - 0.49324 = 0.00676$$

(3) H_1: $P = 10\% = 0.10$ [curve (h).]

$$\sigma_p = \sqrt{\frac{PQ}{n}} = \sqrt{\frac{(0.10)(0.90)}{1,000}} = 0.0095 \text{ or } 0.95\%$$

$$z = \frac{0.05 - 0.10}{0.0095} = -5.26 \quad A(-5.26) = 0.50000$$

$$\beta = 0.50000 - 0.50000 = 0$$

Charts 14–7 and 14–8 also show the OC and power curves when $n = 100$. The values of β and $(1 - \beta)$ for the two curves can be obtained in the same manner as above.

Remarks on Charts Constructed

In conclusion, the following remarks are made about Charts 14–3 through 14–8:

Charts 14–3 and 14–6 [*Normal Curves Showing β and* (1 − β) *Values*]

1. In order to obtain a good result from a test of hypothesis, we should minimize the two types of errors. However, it is impossible to reduce both errors at one time, since a decrease of one type of error will be accompanied by an increase of another type of error. For example, when we specify Type I error (α) to be 5%,

TABLE 14–3

OPERATING CHARACTERISTIC AND POWER FUNCTIONS
$H_0: P_0 = 4\%$ OR LESS, $H_1: P =$ ABOVE 4%, $\alpha = 5\%$
ONE TAILED TEST

	A. $n = 300$		*B.* $n = 1000$	
Hypothetical values of P	*OC functions (Type II errors)* β	*Power functions* $1 - \beta$	*OC functions (Type II errors)* β	*Power functions* $1 - \beta$
2%	1.00000	0.00000	1.00000	0.00000
3%	0.99865	0.00135	0.99997	0.00003
4%	0.95000	0.05000	0.95000	0.05000
5%	0.77935	0.22065	0.50000	0.50000
6%	0.50000	0.50000	0.09176	0.90824
7%	0.25143	0.74857	0.00676	0.99324
8%	0.10565	0.89435	0.00024	0.99976
9%	0.03438	0.96562	0.00000	1.00000
10%	0.00939	0.99601	0.00000	1.00000

Source: Examples 3 and 4.

CHART 14–7

OPERATING CHARACTERISTIC CURVE
$H_0: P_0 = 4\%$ OR LESS, $H_1: P =$ ABOVE 4%, $\alpha = 5\%$
ONE-TAILED TESTS

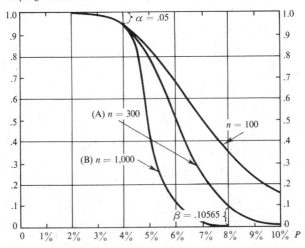

Source: Table 14–3.

CHART 14–8

POWER CURVE

$H_0\colon P_0 = 4\%$ OR LESS, $H_1\colon P =$ ABOVE 4%, $\alpha = 5\%$

ONE-TAILED TESTS

$1 - \beta$, Probability of
Rejecting H_0 under Given P

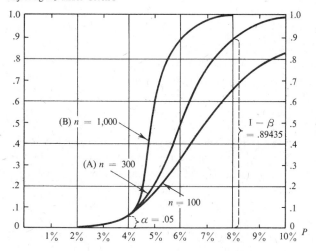

Source: Table 14–3.

the maximum Type II error (β) is 95% as shown in Chart 14–3 (a). If Type I error is decreased to 1%, the maximum Type II error will be increased to 99%. In practice, one should examine the seriousness of the outcomes of the two types of errors for a given problem. If Type I error is more serious than the other, a smaller value of the probability should be assigned to Type I error.

2. We cannot assign 0 or 1 to any type of error in a test of hypothesis. For example, if we assigned 0 to Type I error, the value of Type II error must be 1; that is, we have only the region of acceptance of the null hypothesis. Then, we would not be allowed to reject any null hypothesis but would be forced to accept the hypothesis whether it is true or not true.

3. Once the probability of making Type I error is specified or fixed, the probability of making Type II error under a given alternative hypothesis may be reduced by increasing the sample size. For example, under the alternative hypothesis: $\mu = \$505$, the value of β is 0.82993 when $n = 400$ [Chart 14–3 (c)], but it is only 0.29460 when $n = 2500$ [Chart 14–3(f)].

4. When the difference between the null hypothesis and an alternative hypothesis is decreased, the probability of making Type II error is increased. In other words, when the assumed population mean (or proportion) is very close to the true population mean, the chance of making an error by accepting the assumed mean as the true mean is increased. The limiting value of Type II error is $1 - \alpha$.

Charts 14–4 and 14–7 (Operating Characteristic Curves)

1. The OC curve shows the probability of accepting the null hypothesis H_0 when it is not true, but a given alternative hypothesis H_1 is true. For example, in

Chart 14–7, when $n = 300$, the OC curve shows that the probability of accepting the null hypothesis ($P_0 = 4\%$ or less) when the population proportion is actually 8% is 0.10565 (or about 10%). In other words, there are about 10 out of 100 chances that a shipment with actually 8% defectives will be accepted by a receiver as a shipment with only 4% defectives under the sampling method of inspection. The same curve also shows that the chance of accepting a shipment with 10% defectives is practically zero.

2. Compare the OC curves in Chart 14–4 or in Chart 14–7. The steeper the OC curve, or the larger the sample size, the better it is for decision-making. For example, if a shipment with 8% defectives is submitted for inspection, the inspector will not hesitate to reject the shipment when he is using curve B in Chart 14–7. The steepest curve, curve B, shows that when $P = 8\%$ the probability of accepting the null hypothesis ($P_0 = 4\%$ or less) is close to zero, or $\beta = 0.00024$.

3. The values of β vary with the values of H_1 for each curve. The maximum value of β in a chart (Chart 14–4) showing two-tailed tests is $1 - \alpha$, whereas in a chart (Chart 14–7) showing one-tailed tests is close to 1. (Theoretically speaking, it will never be 1 according to the Areas Under the Normal Curve Table.) In general, the minimum value of β is close to zero.

4. Although the values of β on the curves above a given H_1 are different, there is a common point of all curves in each chart. The common point is at $1 - \alpha$ just above H_0; that is, the point in Chart 14–4 is at 0.95 just above $\mu_0 = \$500$, and the point in Chart 14–7 is at 0.95 just above $P_0 = 4\%$. The Type I error, rejecting the H_0 when it is true, can be made only when H_0 is true. Thus, the value of α is indicated only above H_0 in each chart.

Charts 14–5 and 14–8 (Power Curves)

1. The power curve is used in a similar manner to that of the OC curve. The power curve shows the probability of rejecting the null hypothesis H_0 when it is not true, but a given alternative hypothesis H_1 is true. For example, in Chart 14–8, when $n = 300$, the power curve shows that the probability of rejecting the null hypothesis ($P_0 = 4\%$ or less) when the population proportion is actually 8% (or $P = 8\%$) is 0.89435 (or about 90%). In other words, there are about 90 out of 100 chances that we will reject a shipment with 8% defectives. Thus a power curve provides an investigator the *power* or *guide* to reject the null hypothesis when the curve shows a high probability of rejection at a given alternative hypothesis.

2. Compare the power curves in Chart 14–5 or in Chart 14–8. Again, the steeper the power curve, the better it is for decision-making. The curve on a higher position (or the steeper one) is called a more *powerful curve*.

3. The maximum value of $1 - \beta$ is close to 1 although, theoretically speaking, it will never be 1 according to the Areas Under the Normal Curve Table. The minimum value of $1 - \beta$ is α for two tailed tests and is close to zero for one-tailed tests.

4. The common point of all power curves in each chart is at α or 0.05 just above H_0.

14.4   Summary

A statistical hypothesis, or simply hypothesis, is an assumption or a guess concerning the population. The hypothesis can be tested statistically by using a sample according to the theory of probability. The result of the test will guide a statistician either to accept the hypothesis or to reject (nullify) it. The acceptance or rejection will lead an investigator to make a decision. Statistical hypotheses which are stated for the purpose of possible rejection or nullification are called null hypotheses (H_0). Any hypothesis which differs from the null hypothesis is called an alternative hypothesis (H_1).

When one rejects a hypothesis but actually it is true, one commits a Type I error. When one accepts a hypothesis but actually it is not true, one commits a Type II error. The maximum probability of making a Type I error specified in a test of hypothesis is called the level of significance, α. The test of hypotheses which are based on the level of significance represented by both tails under the normal curve are called two-tailed tests or two-sided tests. If the level of significance is represented by only one tail, the tests are called one-tailed tests or one-sided tests.

An operating characteristic curve (or OC curve) shows the probabilities of making Type II errors (β); that is, the probabilities of accepting H_0 when it is not true but various alternative hypotheses are true. The OC curve is constructed according to a fixed Type I error (α). A power curve is simply a reversed OC curve, which shows the probabilities of rejecting H_0 (or $1 - \beta$) when various alternative hypotheses are true. The basic aspects concerning the two kind of curves are summarized on pages 363-64.

Exercises 14

1. State the difference between: (a) a null hypothesis and an alternative hypothesis, (b) a Type I error and a Type II error.

2. Explain the following: (a) level of significance, (b) critical value, (c) region of rejection, and (d) region of acceptance.

3. Indicate the difference between: (a) a two-tailed test and a one-tailed test, and (b) a left-tailed test and a right-tailed test.

4. If the level of significance is 5%, what are z values under the different types of tests mentioned in Problem 3?

★**5.** When Type I error is specified, such as at 1%, can we reduce Type II error under a given alternative hypothesis? If we can, what is the minimum value of Type II error?

★**6.** What is the relationship between an operating characteristic curve and a power curve?

★**7.** Refer to Table 14-2. Prove the β and $(1 - \beta)$ values for H_1: $\mu = \$495$ when (a) $n = 400$, and (b) $n = 2,500$.

★**8.** Refer to Table 14-3. Prove the values of β and $(1 - \beta)$ for H_1: $P = 8\%$ when (a) $n = 300$, and (b) $n = 1,000$.

15 Methods of Testing Hypotheses

This chapter presents four selected methods of testing hypotheses. The first two sections discuss the fundamental procedure of testing hypotheses and the t-distributions (15.1 and 15.2). The selected methods are: (1) the tests concerning the difference between a sample mean and a population mean, Section 15.3; (2) the difference between two sample means, Section 15.4; (3) the difference between a sample proportion and a population proportion, Section 15.5; and (4) the difference between two sample proportions, Section 15.6. A summary of this chapter is given in Section 15.7.

The methods commonly used in testing hypotheses are based on z (standard normal deviate), t (Student's), χ^2 (Chi-square), and F (variance ratio) distributions and various nonparametric tests. This chapter will present the methods involving z and t distributions. Other special types of tests are discussed in later chapters.

15.1 Fundamental Procedure of Testing Hypotheses

The fundamental procedure for the various methods of testing hypotheses involving either z or t distribution is outlined below:

(1) State the null hypothesis as follows: There is *no difference* between the two given values, or the *difference is zero*, such as $\bar{X} - \mu = 0$. In other words, an assumption is made that the difference between the two given values is only due to sampling fluctuation or chance; therefore, the *difference* is considered as "no difference" or "not significant."

(2) Express the difference in units of the standard error of the statistic as follows:

> (a) When the true standard deviation of the population σ is known, the standard error of the statistic is determined. The difference is expressed in the value of the standard normal deviate z, such as

$$z = \frac{\bar{X} - \mu}{\sigma_{\bar{x}}} = \frac{\bar{X} - \mu}{\dfrac{\sigma}{\sqrt{n}}}$$

When the value of z is obtained, the Areas Under the Normal Curve, Table 6, can be used in the tests of hypotheses. The distribution of the values of z is normal and is called the standard normal distribution with mean 0 and standard deviation 1.

(b) When the true standard deviation of the population σ is unknown, the standard error of the statistic is estimated from a sample standard deviation. The difference is then expressed in the value of t, such as

$$t = \frac{\bar{X} - \mu}{s_{\bar{x}}} = \frac{\bar{X} - \mu}{\dfrac{s}{\sqrt{n-1}}}$$

When the value of t is obtained, the t-distribution table, Table 7 in the Appendix or Table 15–1, should be used in the tests of hypotheses.

(3)   Make a decision. The decision rule is based on the level of significance, either for a two-tailed test or for a one-tailed test, as follows:

(a) If the computed z or t value falls in the acceptance region, we accept the hypothesis.

(b) If the computed z or t value falls in the rejection region, we reject the hypothesis.

15.2   The t Distribution

The distribution of the values of t is not normal, but its use and the shape are somewhat analogous to those of the standard normal distribution of z. The t-distribution is also symmetrical about 0 on the t-scale. However, there is a series of t-distributions. The shape of each t-distribution is affected by the number of degrees of freedom, D, which is computed from the sample size n. For example, the value $n - 1$ in the t formula represents the number of degrees of freedom. If $n = 3$, $D = n - 1 = 3 - 1 = 2$.

Just as we may construct a table showing the areas under the standard normal curve, we may construct a table showing the areas under a t-distribution curve between the maximum ordinate Y_0 and the ordinate at t for every number of degrees of freedom that may possibly occur; that is, $D = 1, 2, 3, \ldots$ and so on. However, in practice only the most frequently used values of t are tabulated in a compact form. The t-distribution table gives the selected values of t for two-tailed tests. The probabilities of the t-distribution at various t values for the two-tails are listed in the captions of the table and the numbers of degrees of freedom are listed in the stubs. The t-distribution table can easily be adapted for one-tailed tests. For example, if $\alpha = 0.05$ is specified for a one-tailed test, the t value with $D = 4$ is 2.132 which is listed under $\alpha = 0.10$ (or 2×0.05) column in the t-distribution table, Table 15–1.

Compare the right sides of the equations representing z and t values above. When the sample size n is large, the value of $\sqrt{n-1}$ approaches the value of $\sqrt{n}$, and the value of s approaches the value of σ; therefore t approaches z.

TABLE 15–1

COMPARISON OF SELECTED VALUES OF t AND z REPRESENTING
THE SAME PROBABILITIES FOR TWO-TAILED TESTS

	Probabilities, α			
	(or Areas Under t and z Distribution Curves)			
	0.50	0.10	0.05	0.01
Number of Degrees of freedom (D)		*Values of t* *		
1	1.000	6.314	12.706	63.657
2	0.816	2.920	4.303	9.925
3	0.765	2.353	3.182	5.841
4	0.741	2.132	2.776	4.604
5	0.727	2.015	2.571	4.032
10	0.700	1.812	2.228	3.169
20	0.687	1.725	2.086	2.845
30	0.683	1.697	2.042	2.750
40	0.681	1.684	2.021	2.704
120	0.677	1.658	1.980	2.617
∞	0.674	1.645	1.960	2.576
	Value of z			
	0.674	1.645	1.960	2.576

*Notes: 1. For more detailed values of t, see the t-Distribution, Table 7 in the Appendix.
   2. For one-tailed tests, the probability listed in each column shoud be divided by two. For example,

   at 0.10 for a two-tailed test, $t = 2.132$ where $D = 4$
   at 0.05 for a one-tailed test, $t = 2.132$ where $D = 4$

CHART 15–1

COMPARISON OF VARIOUS t-DISTRIBUTIONS AND
THE STANDARD NORMAL DISTRIBUTION

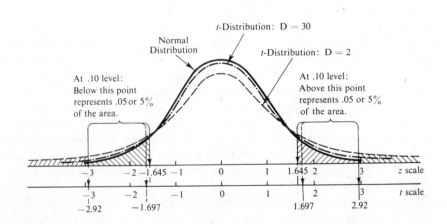

Chart 15–1 and Table 15–1 give the comparison of the t-distribution and the z-distribution (normal curve) for the various numbers of degrees of freedom based on two-tailed tests. Note that when n is small, the t-distribution curve is more spread out than the normal curve; whereas when the number of degrees of freedom is 30, the t-distribution curve is fairly close to the normal distribution curve.

The t-distribution is also called *Student's distribution*. The name Student is the pseudonym used by William S. Gosset for the publication of his work on t-distribution in 1908. Gosset's employer did not permit him to use his real name for the publication then.

15.3 The Difference Between a Sample Mean and a Population Mean

The value of z for the difference between a sample mean and a population mean, as mentioned before, is expressed in units of the standard error of the mean as follows:

$$z = \frac{\bar{X} - \mu}{\sigma_{\bar{x}}} \text{ where } \sigma_{\bar{x}} = \frac{\sigma}{\sqrt{n}} \cdot \sqrt{\frac{N - n}{N - 1}}$$

or

$$\sigma_{\bar{x}} = \frac{\sigma}{\sqrt{n}} \text{ (when } N \text{ is large)}$$

Thus,

$$z = \frac{\bar{X} - \mu}{\dfrac{\sigma}{\sqrt{n}}} \tag{15–1}$$

In a test of hypothesis, $\bar{X}$ represents a sample mean, μ represents a hypothetical population mean, and $\sigma_{\bar{x}}$ represents the standard error of the distribution of the sample means, or the values of $\bar{X}$. The sampling distribution of $\bar{X}$ is usually normal (see page 309). If the hypothetical population mean μ is true, the mean of the sampling distribution is equal to μ. Thus, the distribution of the values of z is normal if the hypothetical mean is true.

Example 1

Assume that Alan Company produces a certain type of thread. The production records of the last year showed that the mean breaking strength of the thread was 12.46 ounces and the standard deviation was 1.80 ounces. The production manager recently took a sample of 100 pieces of thread and found that the mean beaking strength of the sample was increased to 12.82 ounces. Can he conclude that the quality of the thread has become higher than before? Let the level of significance be (a) 0.05 and (b) 0.01.

Solution. (1) The null hypothesis, or H_0: There is no difference between the population mean and the sample mean. The mean breaking strength of the thread of the last year's production is the hypothetical population mean (μ) for the continous production. Thus,

$$\bar{X} = 12.82 \text{ ounces, } n = 100 \text{ and } \mu = 12.46 \text{ ounces}$$

(2) Find z according to formula (15–1). The true standard deviation σ is known: $\sigma = 1.80$ ounces (if the population has not been changed, or if H_0 is true).

Substitute the above values in the formula:

$$\sigma_{\bar{x}} = \frac{\sigma}{\sqrt{n}} = \frac{1.80}{\sqrt{100}} = 0.18$$

$$z = \frac{\bar{X} - \mu}{\sigma_{\bar{x}}} = \frac{12.82 - 12.46}{0.18} = 2$$

(3) Make a decision based on the level of significance. Here we are interested in testing whether or not there has been an increase of the mean breaking strength of the thread. Thus, a one-tailed (the right tail) test is used in locating the critical value of z (which is higher than the mean 0 and is positive.)

(a) At 0.05 level of significance (Chart 15–2A): The critical value of $z = +1.645$. The obtained value of $z = 2$ falls in the rejection region. The probability of having a sample mean as great as 12.82 ounces or greater, when the population mean is 12.46 ounces, is only 0.02275, or less than the 0.05 level. ($0.50000 - A(2) = 0.50000 - 0.47725 = 0.02275$.) Thus, we reject the hypothesis that there is *no difference* between the population mean and the sample mean. In other words, the difference is significant. The quality of the present production is higher than that of the last year.

(b) At 0.01 level of significance (Chart 15–2B): The critical value of $z = +2.33$. The obtained value of $z = 2$ falls in the acceptance region. Thus, we accept the hypothesis. The difference between the population mean and the sample mean is not significant; that is, the quality of the present production is the same as that of the last year. The difference is due to chance or sampling fluctuation.

CHART 15–2

ONE-TAILED TESTS OF HYPOTHESES
(EXAMPLE 1)

A. Level of Significance at .05   B. Level of Significance at .01

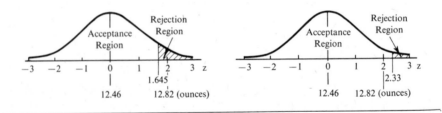

Sometimes, the result of a test of the difference showing significance at the 0.01 level is referred to as *highly significant* and significance at the 0.05 level but not at the 0.01 level as *probably significant*. The results obtained from Example 1 may then be called probably significant. Thus, a further investigation of the production may be suggested before a firm conclusion concerning the difference is made.

In practice, when the true population mean μ is unknown, the true population standard deviation σ is unknown. The value of σ thus is estimated by using the modified sample standard deviation $\hat{s}$, which is the square root of the unbiased estimate of the population variance $\hat{s}^2$. When σ is estimated by $\hat{s}$, formula (15–1) becomes formula (15–2), as mentioned above:

$$t = \frac{\bar{X} - \mu}{s_{\bar{x}}} \text{ where } s_{\bar{x}} = \frac{\hat{s}}{\sqrt{n}} = \frac{s}{\sqrt{n-1}}$$

Thus,

$$t = \frac{\bar{X} - \mu}{\dfrac{s}{\sqrt{n-1}}} \tag{15–2}$$

Formula (15–2) is preferred when the sample size (n) is small, usually less than 30.

When the sample size n is large, $n-1$ in formula (15–2) approaches n. The formula may then be written:

$$z = \frac{\bar{X} - \mu}{s_{\bar{x}}} \text{ where } s_{\bar{x}} = \frac{s}{\sqrt{n}}$$

Thus,

$$z = \frac{\bar{X} - \mu}{\dfrac{s}{\sqrt{n}}} \tag{15–3}$$

The value of z, instead of t, is used in the above formula because when n is large, s approaches σ. Therefore, the normal curve may be used to approximate the t-distribution when the sample size n becomes large, 30 or above.

Example 2

Assume that Dale Company buys a large shipment of the type of thread produced by Alan Company. The buyer took a sample of 31 pieces of thread and found that the mean breaking strength of the sample is 12.02 ounces and the standard deviation of the sample is 1.74 ounces. The desired mean breaking strength is 12.50 ounces. Should the buyer accept the shipment based on the sampling inspection? Assume that the company set the level of significance at 5% for a two-tailed test.

Solution. (1) H_0: There is no difference between the sample mean ($\bar{X} = 12.02$ ounces) and the assumed or hypothetical mean ($\mu = 12.50$ ounces).

(2) Find t from formula (15–2) since the true standard deviation of the population is unknown but is estimated from the sample.

$$s = 1.74 \text{ ounces, and } n = 31$$

$$s_{\bar{x}} = \frac{s}{\sqrt{n-1}} = \frac{1.74}{\sqrt{31-1}} = 0.32$$

$$t = \frac{\bar{X} - \mu}{s_{\bar{x}}} = \frac{12.02 - 12.50}{0.32} = \frac{-0.48}{0.32} = -1.50$$

(3) Make a decision (see Chart 15–3A): The value of t for 30 (or $n - 1 = 31 - 1 = 30$) degrees of freedom at the 0.05 level of significance is ± 2.042 as found from the table. The computed value of t (-1.50) is larger than -2.042, or falls in the acceptance region. Thus, we accept the hypothesis that there is no difference between the sample mean and the assumed mean. In other words, we accept the shipment since the difference between 12.02 and 12.50, or -0.48 ounces, results from sampling fluctuation. Note that the interpretation of the value of t is the same as that of z in Example 1 above.

CHART 15–3

TWO-TAILED TESTS OF HYPOTHESES
(EXAMPLE 2)

A. t-Distribution—D $= 30$　　　　　　　　　　B.　z-Distribution (Normal Curve)

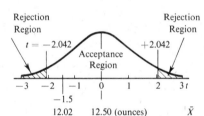　　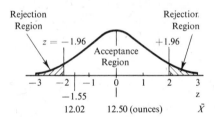

Example 3

Refer to Example 2. Use the approximation method to find the answer.

Solution. (1) H_0: There is no difference between the sample mean and the population mean. (Same as the null hypothesis in Example 2.)

(2) Find z from formula (15–3).

$$s = 1.74 \text{ ounces, and } n = 31$$

$$s_{\bar{x}} = \frac{s}{\sqrt{n}} = \frac{1.74}{\sqrt{31}} = 0.31$$

$$z = \frac{\bar{X} - \mu}{s_{\bar{x}}} = \frac{12.02 - 12.50}{0.31} = -1.55$$

(3) Make a decision (see Chart 15–3B): At 0.05 level of significance for a two-tailed test, the critical value of $z = \pm 1.96$. The computed value of z falls in the acceptance region. We thus accept the hypothesis. Note that the critical value of $z = \pm 1.96$ based on the normal curve area table is the same as the value of t on

the last line in the t-distribution table when $D = \infty$ (infinitely large). This fact indicates that when the sample size increases, the t-distribution approaches the the normal distribution.

15.4 The Difference Between Two Sample Means

In the following discussion the first population and its samples are identified by subscript 1, and the second population and its samples are identified by subscript 2. Thus, $\bar{X}_1$ represents the mean of the sample drawn from population 1, and $\bar{X}_2$ represents the mean of the sample drawn from population 2.

When the difference between two means of samples drawn from separate populations is involved in a test of hypothesis, the hypothesis may be stated in one of the following two forms:

I. The two populations have the same means but different variances.

II. The two populations have the same means and the same variances; that is, the two populations are essentially the same.

The methods of testing the hypotheses in the two forms are discussed below.

Hypothesis I: Two sample means ($\bar{X}_1$ and $\bar{X}_2$) are obtained from separate populations with the same means ($\mu_1 = \mu_2$) but different variances ($\sigma_1^2 \neq \sigma_2^2$).

Under this hypothesis, the value of z is obtained by expressing the difference between two sample means in units of the standard error of the difference as follows:

$$z = \frac{\bar{X}_1 - \bar{X}_2}{\sigma_{(\bar{x}_1 - \bar{x}_2)}} \quad \text{where} \quad \sigma_{(\bar{x}_1 - \bar{x}_2)} = \sqrt{\frac{\sigma_1^2}{n_1} + \frac{\sigma_2^2}{n_2}} \quad (15\text{-}4)$$

Formula (15-4) is illustrated by using the facts in Table 15-2. The table shows that the mean of sampling distribution of ($\bar{X}_1 - \bar{X}_2$) is equal to the difference between the means of the sampling distributions of $\bar{X}_1$ and $\bar{X}_2$. It is also equal to the difference between the two population means ($\mu_1 - \mu_2$), or symbolically,

$$\overline{\bar{X}_1 - \bar{X}_2} = \bar{\bar{X}}_1 - \bar{\bar{X}}_2 = \mu_1 - \mu_2 = 5.5$$

(Both short and long bars above symbol X represent "arithmetic mean of.")

When the sampling distribution of $\bar{X}_1$ and that of $\bar{X}_2$ are normal, the sampling distribution of the difference between two sample means ($\bar{X}_1 - \bar{X}_2$) is also normal. Since the mean of the sampling distribution of ($\bar{X}_1 - \bar{X}_2$) is equal to the difference between the two population means, the distribution of

$$z = \frac{(\bar{X}_1 - \bar{X}_2) - (\mu_1 - \mu_2)}{\sigma_{(\bar{x}_1 - \bar{x}_2)}}$$

is also normal. If the hypothesis is true (that is, $\mu_1 = \mu_2$, or $\mu_1 - \mu_2 = 0$), the numerator of the z equation becomes

$$(\bar{X}_1 - \bar{X}_2) - 0 = \bar{X}_1 - \bar{X}_2$$

which is the numerator of the z equation in formula (15-4) above.

TABLE 15–2

COMPUTATION OF VARIOUS MEANS AND VARIANCES
OF TWO POPULATIONS AND THEIR SAMPLES

Population		Possible Samples of 3 from Population 1	Possible Samples of 2 from Population 2	Differences Between All Possible Sample Means $\bar{X}_1 - \bar{X}_2$
1	2			
X_1	X_2	1st sample:	1st sample:	$6 - 1.5 = 4.5$
		3	1	$6 - 2.0 = 4.0$
3	1	6	2	$6 - 2.5 = 3.5$
6	2	9	3	
9	3	18	$\bar{X}_2 = 3/2 = 1.5$	$7 - 1.5 = 5.5$
12	6	$\bar{X}_1 = 18/3 = 6$		$7 - 2.0 = 5.0$
30		2nd sample:	2nd sample:	$7 - 2.5 = 4.5$
$\mu_1 = 30/4$	$\mu_2 = 6/3$	3	1	$8 - 1.5 = 6.5$
$= 7.5$	$= 2$	6	3	$8 - 2.0 = 6.0$
		12	4	$8 - 2.5 = 5.5$
		21	$\bar{X}_2 = 4/2 = 2.0$	$9 - 1.5 = 7.5$
		$\bar{X}_1 = 21/3 = 7$		$9 - 2.0 = 7.0$
		3rd sample:	3rd sample:	$9 - 2.5 = 6.5$
		3	2	66.0
		9	3	
		12	5	
		24	$\bar{X}_2 = 5/2 = 2.5$	
		$\bar{X}_1 = 24/3 = 8$		
		4th sample:		
		6		
		9		
		12		
		27		
		$\bar{X}_1 = 27/3 = 9$		
		$\Sigma \bar{X}_1 = 30$	$\Sigma \bar{X}_2 = 6.0$	
		$\bar{\bar{X}}_1 = 30/4 = 7.5$	$\bar{\bar{X}}_2 = 6.0/3 = 2.0$	
$\mu_1 - \mu_2 = 7.5 - 2$ $= 5.5$		$\bar{\bar{X}}_1 - \bar{\bar{X}}_2 = 7.5 - 2.0 = 5.5$		$\bar{X}_1 - \bar{X}_2 = 66.0/12$ $= 5.5$
		$\sigma^2_{\bar{x}_1} = 5/4$	$\sigma^2_{\bar{x}_2} = 1/6$	$\sigma^2_{(\bar{x}_1 - \bar{x}_2)}$ $= \dfrac{\Sigma[(\bar{X}_1 - \bar{X}_2) - 5.5]^2}{12}$
		$\sigma^2_{\bar{x}_1} + \sigma^2_{\bar{x}_2} = \dfrac{5}{4} + \dfrac{1}{6} = 1\dfrac{5}{12}$		$= \dfrac{17}{12} = 1\dfrac{5}{12}$

Table 15–2 also shows that the variance of the difference between two sample means is the sum of variances of the two sample means, or symbolically

$$\sigma^2_{(\bar{x}_1 - \bar{x}_2)} = \sigma^2_{\bar{x}_1} + \sigma^2_{\bar{x}_2} = \frac{17}{12} = 1\frac{5}{12}$$

When N is large

$$\sigma_{\bar{x}} = \frac{\sigma}{\sqrt{n}} \text{ or } \sigma^2_{\bar{x}} = \frac{\sigma^2}{n}$$

Thus,

$$\sigma^2_{(\bar{x}_1 - \bar{x}_2)} = \frac{\sigma^2_1}{n_1} + \frac{\sigma^2_2}{n_2}$$

Take the square roots of both sides of the above equation. We may write $\sigma_{(\bar{x}_1-\bar{x}_2)}$ in the form of formula (15-4).

If necessary, when N and n are large, we can use the sample variances s_1^2 and s_2^2 as estimates of σ_1^2 and σ_2^2. Then, formula (15-4) may be written:

$$z = \frac{\bar{X}_1 - \bar{X}_2}{s_{(\bar{x}_1-\bar{x}_2)}} \quad \text{where} \quad s_{(\bar{x}_1-\bar{x}_2)} = \sqrt{\frac{s_1^2}{n_1} + \frac{s_2^2}{n_2}} \qquad (15\text{-}5)$$

However, we should not use s_1^2 and s_2^2 as estimates of σ_1^2 and σ_2^2 in this case since the computation of the number of degrees of freedom is more involved when σ_1^2 is not equal to σ_2^2. Thus, when sample sizes (n) are small, we should avoid this method. Instead, we may use a nonparametric test to find an answer (see Chapter 16, Example 9.)

Example 4

An English instructor gave the same test to two classes. The first class has 48 students and the mean grade was 72 points with a standard deviation of 12 points. The second class has 36 students and the mean grade was 75 points with a standard deviation of 6 points. Is there a significant difference between the mean performances of the two classes based on the above information? Let the level of significance be 0.05.

Solution.   (1) Let μ_1 be the mean of the first population from which the first class comes, and μ_2 be the mean of the second population from which the second class comes.

H_0: The two population means are the same; that is, $\mu_1 = \mu_2$, or $\mu_1 - \mu_2 = 0$.

Under this hypothesis, the difference between $(\bar{X}_1 - \bar{X}_2)$ and $(\mu_1 - \mu_2)$, or simply the difference between $\bar{X}_1$ and $\bar{X}_2$, is merely due to chance.

(2) Find z by using formula (15-5). Since we do not know the true popula-

CHART 15-4

A Two-Tailed Test of Hypothesis
(Difference Between Two Sample Means)
Level of Significance: .05 (Example 4)

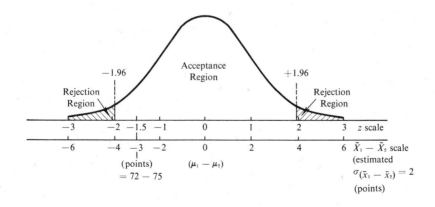

tion standard deviations, we must use the sample standard deviations as estimates of σ_1 and σ_2.

$$s_{(\bar{x}_1-\bar{x}_2)} = \sqrt{\frac{s_1^2}{n_1} + \frac{s_2^2}{n_2}} = \sqrt{\frac{12^2}{48} + \frac{6^2}{36}} = \sqrt{3+1} = 2$$

$$z = \frac{\bar{X}_1 - \bar{X}_2}{s_{(\bar{x}_1-\bar{x}_2)}} = \frac{72 - 75}{2} = -1.5$$

(3) Make a decision (see Chart 15–4). At a 0.05 level, the difference is significant for a two-tailed test if z falls outside the range ± 1.96. The computed z value is -1.5, which is inside of the range. Thus, we conclude that the two population means are the same. In other words, the two classes have the same mean performance.

Hypothesis II: Two sample means ($\bar{X}_1$ and $\bar{X}_2$) are obtained from separate populations with the same means ($\mu_1 = \mu_2$) and the same variances ($\sigma_1^2 = \sigma_2^2$).

When two populations have the same means and the same variances, they are identical populations. (See property 6 of the standard normal curve, page 279.) Thus, if the hypothesis is true, ($\mu_1 = \mu_2$) and ($\sigma_1^2 = \sigma_2^2$), the two samples may be thought of as drawn from the same population.

When the two equal population variances are known, there is no new problem. Formula (15–4) may still be used for finding the value of z in testing the hypothesis. Also, when the two equal population variances are unknown, they must be estimated by the sample variances. When sample sizes are large, use formula (15–5). When sample sizes are small, use the t-distribution. The value of t is obtained by replacing σ in formula (15–4) by $\hat{s}$, the estimated population standard deviation, as follows:

$$t = \frac{\bar{X}_1 - \bar{X}_2}{s_{(\bar{x}_1-\bar{x}_2)}}$$

where $\qquad s_{(\bar{x}_1-\bar{x}_2)} = \hat{s}\sqrt{\dfrac{1}{n_1} + \dfrac{1}{n_2}}$ and $\hat{s} = \sqrt{\dfrac{n_1 s_1^2 + n_2 s_2^2}{n_1 + n_2 - 2}}$ $\qquad$ (15–6)*

---

*The derivation of formula (15–6) is illustrated below.
When $\sigma_1^2 = \sigma_2^2$, formula (15–4) may be written:

$$\sigma_{(\bar{x}_1-\bar{x}_2)} = \sqrt{\sigma^2\left(\frac{1}{n_1} + \frac{1}{n_2}\right)} \quad \text{where } \sigma^2 \text{ may be } \sigma_1^2 \text{ or } \sigma_2^2$$

When σ_1^2 and σ_2^2 are unknown, the value of σ^2 must be estimated. The best estimate of the equal variance of each population is the weighted mean of the two unbiased estimates of the population variances

$$\hat{s}_1^2 = \frac{\sum x_1^2}{n_1 - 1} \text{ and } \hat{s}_2^2 = \frac{\sum x_2^2}{n_2 - 1}$$

The weights are based on the numbers of degrees of freedom $n_1 - 1$ and $n_2 - 1$ (the denominators of the two fractions), or

$$\hat{s}^2 = \frac{(n_1 - 1)\,\hat{s}_1^2 + (n_2 - 1)\,\hat{s}_2^2}{(n_1 - 1) + (n_2 - 1)} = \frac{n_1 s_1^2 + n_2 s_2^2}{n_1 + n_2 - 2}$$

since $\qquad (n-1)\,\hat{s}^2 = (n-1)\dfrac{\sum x^2}{(n-1)} = \sum x^2 = \dfrac{\sum x^2}{n} n = s^2 n$

When σ^2 is replaced by $\hat{s}^2$, the value of $\sigma_{(\bar{x}_1-\bar{x}_2)}$ may be written in forms of $s_{(\bar{x}_1-\bar{x}_2)}$ and $\hat{s}$ in formula (15–6).

The quantity $(n_1 + n_2 - 2)$ in formula (15–6) is the number of degrees of freedom of the t distribution.

Formula (15–6) provides a satisfactory method in testing a hypothesis concerning the difference between two sample means when the sizes of the two samples are small and the two population variances are believed to be equal.

Example 5

A young man wants to know whether the mean of the wages of electricians differs significantly from the mean of the wages of carpenters in a certain city. He took two samples of a given period and computed the mean and the sum of the squared deviations for each sample as shown in Table 15–3. Assume that the wages of each occupation are normally distributed. Based on the above information and the level of significance of 0.05, find the answer.

TABLE 15–3

COMPUTATION FOR EXAMPLE 5

Electricians (Name)	Monthly Income X_1	x_1	x_1^2	Carpenters (Name)	Monthly Income X_2	x_2	x_2^2
Adams	$74	4	16	Logan	$75	0	0
Baker	65	−5	25	Morton	78	3	9
Carter	72	2	4	North	74	−1	1
Davis	69	−1	1	Ornel	76	1	1
				Palmer	72	−3	9
$n_1 = 4$				$n_2 = 5$			
Total	280	0	46	Total	375	0	20
$\bar{X}_1 = 280/4 = \$70$				$\bar{X}_2 = 375/5 = \$75$			
$n_1 s_1^2 = \sum x_1^2 = 46$				$n_2 s_2^2 = \sum x_2^2 = 20$			

Solution. (1) Let μ_1 and μ_2 be the means of the periodical incomes of all electricians and all carpenters in the city, respectively. H_0: The two population means are the same; that is, $\mu_1 = \mu_2$, or $\mu_1 - \mu_2 = 0$.

(2) Find t by using formula (15–6) since the two samples are small. When the formula is used, an assumption is made that the two samples are drawn from two populations with the same variance. Substitute the above values in formula (15–6).

$$\hat{s} = \sqrt{\frac{46 + 20}{4 + 5 - 2}} = \sqrt{\frac{66}{7}}$$

$$s_{(\bar{x}_1 - \bar{x}_2)} = \sqrt{\frac{66}{7}}\sqrt{\frac{1}{4} + \frac{1}{5}} = \sqrt{\frac{66}{7} \times \frac{5 + 4}{20}} = 2.06$$

$$t = \frac{\bar{X}_1 - \bar{X}_2}{s_{(\bar{x}_1 - \bar{x}_2)}} = \frac{70 - 75}{2.06} = -2.427$$

(3) Make a decision. At a 0.05 level, the difference is significant for a two-tailed test if t, with 7 degrees of freedom $(n_1 + n_2 - 2 = 4 + 5 - 2 = 7)$, falls

outside the range $+2.365$. The computed t value is -2.427, which falls outside the range. We thus reject the hypothesis that the two population means are the same. In other words, the difference between the two population means is significant. The mean of carpenters' wages is higher than that of electricians in the city.

15.5   The Difference Between a Sample Proportion and a Population Proportion

The value of z for the difference between a sample proportion and a population proportion is obtained by expressing the difference in units of the standard error of the proportion as follows:

$$z = \frac{p - P}{\sigma_p} \text{ where } \sigma_p = \sqrt{\frac{PQ}{n}}\sqrt{\frac{N - n}{N - 1}}$$

or when N is large,
$$\sigma_p = \sqrt{\frac{PQ}{n}}$$

Thus,

$$z = \frac{p - P}{\sqrt{\dfrac{PQ}{n}}} \tag{15-7}$$

In a test of hypothesis, p represents a sample proportion, P represents a hypothetical population proportion, and σ_p represents the standard error of the sampling distribution of p with mean P. When P is equal to or nearly equal to 0.50, the sampling distribution of p is normal even for small samples. When P departs from 0.50, the distribution approaches normal if the sample size is large, large enough to make the product of nP equal to or larger than 5. When a sampling distribution of the proportion is not normal, the binomial or Poisson distribution as discussed in Chapter 10 may be used in finding the probability in a test of hypothesis. For simplification, only samples of large sizes are discussed in a test of hypothesis concerning a sampling distribution of the proportion in this text.

Example 6

Assume that a television station claimed that 70% of the TV sets in a certain city were turned on for its special program on a Monday evening. A competitor wants to challenge the claim. He thus took a random sample of 200 families and found that the sample proportion was only 65%. Can the competitor conclude that the claim was not valid if he let the level of significance be 0.05?

Solution.   (1) H_0: $P = 70\%$; that is, the population proportion of TV sets turned on for the program is 70%.

(2) Find z by using formula (15–7).

$$\sigma_p = \sqrt{\frac{PQ}{n}} = \sqrt{\frac{(0.70)(0.30)}{200}} = \sqrt{0.00105} = 0.0324$$

$$z = \frac{p - P}{\sigma_p} = \frac{0.65 - 0.70}{0.0324} = -1.54$$

(3) Make a decision (see Chart 15–5.). We are interested in finding the critical value which is below or smaller than $P = 70\%$ on the p scale. Thus, the one-tailed (left tail) test is used. At 0.05 level, the difference is significant for a one-tailed test if z is smaller than -1.645. The computed value of z is -1.54, which is larger than the critical value z and is inside the acceptance region. We thus accept the hypothesis that there are 70% of TV sets turned on for the special program.

CHART 15–5

ONE-TAILED TEST OF HYPOTHESIS

(DIFFERENCE BETWEEN A SAMPLE PROPORTION AND A POPULATION PROPORTION)

LEVEL OF SIGNIFICANCE: .05

(EXAMPLE 6)

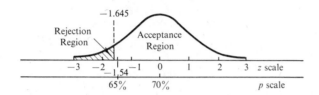

15.6 The Difference Between Two Sample Proportions

Let p_1 and p_2 be the proportions of two samples drawn from respective populations with proportions P_1 and P_2. The null hypothesis is that there is no difference between the two population proportions; that is, $P_1 = P_2$. If the null hypothesis is true, $P_1 = P_2$, the two populations are really the same population.

The basic concept concerning the difference between two sample proportions is analogous to that concerning the difference between two sample means. By a method of reasoning similar to that used in the illustration for Table 15–2, corresponding results for the sampling distribution of the difference between two sampling proportions may be obtained as follows:

(1) The mean of the sampling distribution of $(p_1 - p_2)$ is equal to the difference between the two population proportions, P_1 and P_2, or

$$\overline{p_1 - p_2} = P_1 - P_2$$

(2) The variance of the difference between two sample proportions is the sum of variances of the two sample proportions,

$$\sigma^2_{(p_1 - p_2)} = \sigma^2_{p_1} + \sigma^2_{p_2} = \frac{P_1 Q_1}{n_1} + \frac{P_2 Q_2}{n_2}$$

When the sampling distributions of p_1 and p_2 are normal, the distribution of the differences between p_1 and p_2 is also normal. Since the mean of the sampling distribution of $(p_1 - p_2)$ is equal to the difference between the two population proportions, the distribution that follows is normal.

$$z = \frac{(p_1 - p_2) - (P_1 - P_2)}{\sigma_{(p_1 - p_2)}}$$

When $P_1 = P_2$, $P_1 - P_2 = 0$ and $P_1Q_1 = P_2Q_2 = PQ$, where $Q = 1 - P$. Thus,

$$z = \frac{p_1 - p_2}{\sigma_{(p_1 - p_2)}}$$

where
$$\sigma_{(p_1 - p_2)} = \sqrt{\frac{P_1Q_1}{n_1} + \frac{P_2Q_2}{n_2}} = \sqrt{PQ\left(\frac{1}{n_1} + \frac{1}{n_2}\right)} \qquad (15\text{--}8)$$

When the value of P is to be estimated by using the two sample proportions and when the sample sizes of the two samples are large, z becomes an approximated value and is written:

$$z = \frac{p_1 - p_2}{s_{(p_1 - p_2)}}$$

where
$$s_{(p_1 - p_2)} = \sqrt{pq\left(\frac{1}{n_1} + \frac{1}{n_2}\right)} \quad \text{and} \quad p = \frac{n_1p_1 + n_2p_2}{n_1 + n_2} \qquad (15\text{--}9)$$

p is the best estimate of the population proportion P and $q = 1 - p$. p is the weighted mean of the two sample proportions, p_1 and p_2. The weights are based on sample sized n_1 and n_2.

Example 7

A sample of 500 voters taken from City A showed that 59.6% of the voters were in favor of a given candidate for a state office; whereas a sample of 300 voters taken from City B showed that 50% of the voters were in favor of the candidate. Assume that the candidate made a speech in City A but not in City B. Is there a real difference between the opinions of the two cities concerning the candidate? Let the level of significance be 0.05.

Solution. (1) Let P_1 and P_2 denote the population (city) proportions of the voters in favor of the candidate.

H_0: There is no difference between the two population proportions, or $P_1 = P_2$ and $P_1 - P_2 = 0$.

(2) Find z from formula (15–9).

$$n_1 = 500, \ p_1 = 59.6\% = 0.596, \ n_2 = 300, \ p_2 = 50\% = 0.5$$

$$p = \frac{500(0.596) + 300(0.5)}{500 + 300} = 0.56, \text{ and } q = 1 - 0.56 = 0.44$$

$$s_{(p_1 - p_2)} = \sqrt{(0.56)(0.44)\left(\frac{1}{500} + \frac{1}{300}\right)} = \sqrt{0.2464\left(\frac{3 + 5}{1,500}\right)} = 0.036$$

$$z = \frac{0.596 - 0.5}{0.036} = \frac{0.096}{0.036} = 2.67$$

(3) Make a decision (see Chart 15–6). We are interested in finding the critical values which may be above or below the mean 0 (or $P_1 - P_2 = 0$) on the $p_1 - p_2$ scale. Thus, a two-tailed test is used. At 0.05 level, the difference is significant

for a two-tailed test if z is outside the range ± 1.96. The computed value of z falls outside the range or in the rejection region. Thus, we reject the hypothesis. There is significant difference of the opinions between the two cities about the candidate.

CHART 15–6

A Two-Tailed Test of Hypothesis
(Difference Between Two Sample Proportions)
Level of Significance: .05 (Example 7)

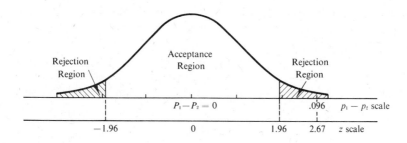

The fundamental procedure outlined in Section 15.1 for testing various types of hypotheses by using z (or normal) distribution may also be used in the tests concerning the sampling distributions other than those of means and proportions. However, care should be emphasized in computing the value of z. In general, the z value may be expressed as follows:

$$z = \frac{\left(\begin{array}{c}\text{Sample}\\\text{statistic}\end{array}\right) - \left(\begin{array}{c}\text{Corresponding hypothetical}\\\text{population parameter}\end{array}\right)}{\text{Standard error of the statistic}}$$

TABLE 15–4

Values Required in Computing z for Selected
Sampling Distributions

Sample Statistic	Corresponding * Population Parameter	Standard Error of the Statistic	Conditions of Applications
Median (m_d), or Second quartile (q_2)	M_d or $\mu\,(=\overline{m_d})$	$\sigma_{m_d} = \dfrac{1.2533\sigma}{\sqrt{n}}$	$n \geq 30$
First quartile (q_1) Third quartile (q_3)	$Q_1\,(=\bar{q}_1 \text{ very nearly})$ $Q_3\,(=\bar{q}_3 \text{ very nearly})$	$\sigma_{q_1} = \sigma_{q_3} = \dfrac{1.3626\sigma}{\sqrt{n}}$	$n \geq 30$
Standard deviation (s)	$\sigma\,(=\bar{s} \text{ very nearly})$	$\sigma_s = \dfrac{\sigma}{\sqrt{2n}}$	$n \geq 100$
Variance (s^2)	$\sigma^2\,(=\overline{s^2} \text{ very nearly})$	$\sigma_{s^2} = \sigma^2 \sqrt{\dfrac{2}{n}}$	$n \geq 100$

*Bars above symbols represent "arithmetic mean of."
$>$ represents "is larger than."

The corresponding hypothetical population parameter is assumed to be equal or very nearly equal to the mean of the sampling distribution of the statistic. The related values for computing z for the most commonly used statistics other than those mentioned before are listed in Table 15–4. The values are given under the condition that the populations from which the samples are drawn are normal or approximately normal. Other conditions in applying the values are listed for each statistic.

15.7 Summary

This chapter presented the methods of testing hypotheses involving z and t distributions. The fundamental steps employed in the procedure for the various methods of testing hypotheses are:

(1) State the null hypothesis in the form: There is no difference between the two given values—a sample statistic and the hypothetical population parameter. The parameter is assumed to be the mean of the sampling distribution of the statistic.

TABLE 15–5

SUMMARY OF SAMPLE STATISTICS AND CORRESPONDING PARAMETERS

Sample Statistic	Hypothetical Population Parameter (Mean of the Sampling Distribution of the Statistic)	Hypothesis (Difference = 0 or Due to Sampling Fluctuation)
$\bar{X}$	$\bar{X} = \mu$	$\bar{X} - \mu = 0$
$\bar{X}_1 - \bar{X}_2$	$\bar{X}_1 - \bar{X}_2 = \mu_1 - \mu_2 = 0$	$(\bar{X}_1 - \bar{X}_2) - (\mu_1 - \mu_2) = 0$ or simply $(\bar{X}_1 - \bar{X}_2) = 0$
p	$\bar{p} = P$	$p - P = 0$
$p_1 - p_2$	$p_1 - p_2 = P_1 - P_2 = 0$	$(p_1 - p_2) - (P_1 - P_2) = 0$ or simply $(p_1 - p_2) = 0$

(2) Express the difference in units of the standard error of the statistic, or

$$= \frac{\left(\begin{array}{c}\text{Sample} \\ \text{statistic}\end{array}\right) - \left(\begin{array}{c}\text{Corresponding hypothetical} \\ \text{population parameter}\end{array}\right)}{\text{Standard error of the statistic}}$$

The above expression may be applied under three different circumstances.

(a) When the true population standard deviation σ (or standard deviations of populations) is known, the difference can be expressed in the value of z, such as

$$z = \frac{\bar{X} - \mu}{\sigma / \sqrt{n}}$$

(b) When the the true population standard deviations σ (or standard deviations of populations) is unknown and it is estimated by using a modified sample standard deviation $\hat{s}$, the difference is expressed in the value of t, such as

$$ t = \frac{\bar{X} - \mu}{\hat{s}/\sqrt{n}} = \frac{\bar{X} - \mu}{s/\sqrt{n-1}} $$

The t-distribution table is usually used for small samples, or when the number of degrees of freedom, $n-1$, is 30 or less. When the sample size

SUMMARY OF FORMULAS

Application	Formula	Formula Number	Reference Page
Difference between sample mean and population mean σ is known	$z = \dfrac{\bar{X} - \mu}{\sigma/\sqrt{n}}$	15-1	369
σ is estimated n is small	$t = \dfrac{\bar{X} - \mu}{s/\sqrt{n-1}}$	15-2	371
n is large	$z = \dfrac{\bar{X} - \mu}{s/\sqrt{n}}$	15-3	371
Difference between two sample means σ_1^2 and σ_2^2 are known	$z = \dfrac{\bar{X}_1 - \bar{X}_2}{\sigma_{(\bar{x}_1 - \bar{x}_2)}}$, where $\sigma_{(\bar{x}_1 - \bar{x}_2)} = \sqrt{\dfrac{\sigma_1^2}{n_1} + \dfrac{\sigma_2^2}{n_2}}$	15-4	373
σ_1^2 and σ_2^2 are estimated. n_1 and n_2 are large	$z = \dfrac{\bar{X}_1 - \bar{X}_2}{s_{(\bar{x}_1 - \bar{x}_2)}}$, where $s_{(\bar{x}_1 - \bar{x}_2)} = \sqrt{\dfrac{s_1^2}{n_1} + \dfrac{s_2^2}{n_2}}$	15-5	375
$\sigma_1^2 = \sigma_2^2$, and both are estimated. n_1 and n_2 are small. [If n_1 and n_2 are large, use formula (15-5).]	$t = \dfrac{\bar{X}_1 - \bar{X}_2}{s_{(\bar{x}_1 - \bar{x}_2)}}$, where $s_{(\bar{x}_1 - \bar{x}_2)} = \hat{s}\sqrt{\dfrac{1}{n_1} + \dfrac{1}{n_2}}$, and $\hat{s} = \sqrt{\dfrac{n_1 s_1^2 + n_2 s_2^2}{n_1 + n_2 - 2}}$	15-6	376
Difference between sample proportion and population proportion	$z = \dfrac{p - P}{\sqrt{PQ/n}}$	15-7	378
Difference between two sample proportions $P_1 Q_1 = P_2 Q_2 = PQ$ are known	$z = \dfrac{p_1 - p_2}{\sigma_{(p_1 - p_2)}}$, where $\sigma_{(p_1 - p_2)} = \sqrt{PQ\left(\dfrac{1}{n_1} + \dfrac{1}{n_2}\right)}$	15-8	380
$P_1 Q_1 = P_2 Q_2 = PQ$ are estimated. n_1 and n_2 are large.	$z = \dfrac{p_1 - p_2}{s_{(p_1 - p_2)}}$, where $s_{(p_1 - p_2)} = \sqrt{pq\left(\dfrac{1}{n_1} + \dfrac{1}{n_2}\right)}$, and $p = \dfrac{n_1 p_1 + n_2 p_2}{n_1 + n_2}$	15-9	380

n is large, the t value can be approximated by the z value (see c. below). (c) Under the circumstances described in (b), the difference may be expressed in the value of z, provided the sample size is large, or the number of degrees of freedom is above 30, such as

$$z = \frac{\bar{X} - \mu}{s/\sqrt{n}}$$

since $n - 1$ approaches n, and s approaches σ when n is large. The detailed formulas are summarized on page 383.

(3) Make a decision. The decision rule is based on the level of significance, α, either for a two-tailed test or for a one-tailed test. If the computed z or t value falls in the acceptance region, we consider the difference between two given values is due to sampling fluctuation and then accept the hypothesis. If the computed z or t value falls in the rejection region, we consider that the difference is not likely due to sampling fluctuation and then reject the hypothesis.

Exercises 15

1. How would you express the difference between two given values in units of the standard error of the statistic (a) when the population σ is known and (b) when the population σ is unknown?

2. Discuss the similarities and dissimilarities, if any, between the normal curve and t-distribution curves.

3. The mean and the standard deviation of the breaking strengths of ropes produced by Company A were 600 pounds and 40 pounds respectively. Recently a new technique is applied in the manufacturing process. It is believed that the strengths of ropes can be increased by the new process. A sample of 64 ropes is taken by the production manager for testing the effectiveness of the new technique. The sample shows that the mean of breaking strengths is now 609 pounds. Can he conclude that there is an increase of the mean breaking strength at a 0.05 level of significance?

4. Refer to Problem 3. Suppose the production manager believes that the new technique may affect breaking strengths of ropes in either direction; that is, it may increase or may decrease the strengths. He now wishes to know whether there is a change of the mean breaking strength after the new technique is applied. Would the answer to the problem be different?

5. A company claimed that the mean lifetime of all car batteries produced by the company is 40 months. However, you have found that the mean lifetime of a sample of 100 batteries produced by the company is only 38.5 months with a standard deviation of 5 months. Determine whether or not the company's claim is overstated at a level of significance of (a) 0.01 and (b) 0.05.

6. Refer to Problem 5. Suppose you have found that the mean lifetime of a sample of 26 batteries produced by the company is also 38.5 months with a standard deviation of 5 months. Determine whether or not the company's claim is over-stated at a 0.05 level of significance.

7. A sample of 60 male students of the Physical Education Department in a college has a mean height of 69.5 inches with a standard deviation of 3 inches. Another sample of 40 male students of the Economics Department in the college has a mean height of 68.4 inches with a standard deviation of 2 inches. Is there a significant difference between the mean heights at (a) 0.01 and (b) 0.05 level of significance?

8. A man has two farms. The mean weight of 10 turkeys on one farm was 14.35 pounds with a standard deviation of 2.5 pounds, whereas the mean weight of 20 turkeys on another farm was 12.19 pounds with a standard deviation of 2 pounds. The variances of the weights of all turkeys on the two farms are believed to be the same. He wants to know whether the mean weight of the turkeys on one farm differs significantly from that of another farm. Based on the level of significance of (a) 0.01 and (b) 0.05, find the answer.

9. A medicine manufacturer claimed that his product, Rin, was 95% effective in relieving hay fever misery within a period of 5 hours. A sample of 150 persons who used the product showed that it provided such relief for 138 persons. Do you believe that the claim made by the manufacturer is valid at 0.10 level of significance?

10. Assume that a machine, after proper setting and adjustment, is expected to turn out an average of 4% defective units. A sample of 300 units showed that 15 units were defective. Find whether the higher defective rate of the sample differs significantly from the expected rate, based on 0.05 level of significance.

11. Mr. Jones has two classes: X and Y. A special teaching aid is given to class X but not to class Y; otherwise the two classes are treated equally. At the end of the semester, 108 out of 120 students in class X have passed the course, whereas 68 out of 80 students in class Y have passed the course. Test the hypothesis that the teaching aid did not give special help to the students to pass the course at a level of significance 0.05.

12. Assume that a sample of 200 units sold by Jack showed 4 units returned for refund, whereas a sample of 100 units sold by Bill showed 5 units returned for refund. Is there a real difference between the rates of returned units of the two samples? Let the level of significance be 0.01.

16 Statistical Quality Control

The techniques used in statistical quality control are basically sampling methods, as presented in the previous chapters. Statistical quality control was initiated in the United States in the 1920's. However, it has developed rapidly only since the outbreak of World War II. The war forced the United States government to quickly enlarge its military personnel and supply. More effective techniques were urgently needed to save manpower and time for inspecting the quality of the huge war supply. The urgent need thus made the armed forces play a leading role in the development and application of statistical quality control techniques during the war. In 1946 a national organization, known as the American Society for Quality Control, was organized by various interested groups and individuals to further develop the use of statistical quality control techniques in the United States. Today more and more governmental agencies and industries in the United States and in most other industrialized nations are making some attempt to use statistical quality control techniques.

The statistical quality control techniques are used initially for improving the quality of product from a manufacturing process. In recent years, however, the application of the techniques has been extended to many types of *nonmanufacturing* business activities, such as checking the efficiency of clerical performance and testing the accuracy of inventory records.

The quality of manufactured products may be expressed in two different ways: by *variables* and by *attributes*. When the quality is expressed by an actual measurement, the quality is said to be expressed by a variable, such as dimension in inches and weight in pounds. When the quality is expressed by either conforming or nonconforming to the specified requirements, good or bad, yes or no, accepted or rejected, defective or nondefective, the quality is said to be expressed by an attribute, such as a piece of cracked glass being listed as a bad or defective item and one not cracked being listed as a good or nondefective item. Frequently a quality characteristic may be expressed in either one of the two ways. For example, a dimension of a product may be inspected by using a measuring device and recorded in actual units of measurement (variable) or inspected by using a go and not-go gage and recorded as yes or no (attribute).

Statistical quality control techniques may be classified into two groups: (1) control charts and (2) specified acceptance sampling plans. Both groups can further be classified according to the two ways of quality expressions: by variables and by attributes. In this chapter, the major parts and the use of a control chart are described in Section 16.1. Details concerning the mechanics of constructing control charts for variables and for attributes are discussed in Sections 16.2 and 16.3 respectively. Selected acceptance sampling plans are discussed in Section 16.4.

16.1 General Description of a Control Chart

The major parts and the use of a control chart for variables are basically the same as those for attributes. Thus, it is convenient to discuss them together.

Major Parts of a Control Chart

A control chart generally includes the four major parts shown in Chart 16–1.

Quality Scale
This is a vertical scale. The scale is marked according to the quality characteristic (either in variables or attributes) of each sample.

Plotted Samples
The qualities of individual items of a sample are not shown in a control chart. Only the quality of the entire sample represented by a single value (a statistic) is plotted. The single value plotted on the chart is in the form of a dot (or sometimes a small circle or a cross). For example, if the qualities of the 4 items of a sample are expressed by variables 1, 2, 2, and 7 pounds, only the mean ($\bar{X}$) of the 4 items, 3 pounds (or $(1 + 2 + 2 + 7)/4 = 3$), is plotted on the chart by a single dot. Thus, the quality scale is marked in numbers of pounds according to the values of $\bar{X}$'s. The chart is then called an $\bar{X}$ *chart*.

Sample Numbers
The samples plotted on a control chart are numbered individually and consecutively on a horizontal line. The line is usually placed at the bottom of the chart. The samples are also referred to as *subgroups* in statistical quality control. Whenever it is practical, at least 25 subgroups should be used in constructing a control chart.

The sample size for variables is usually equal and small, often 4 or 5 items in each sample. Samples of small sizes can be obtained more frequently and thus they are more informative than infrequent samples of larger sizes. Samples of an equal sizes are convenient in computing required values for constructing a chart.

The sample size for attributes should be much larger, preferably 100 or more items, in order to meet sampling theory requirements.

Three Horizontal Lines

The central solid line represents the average quality of the samples plotted on the chart. For example, the line may represent the mean of the sample means ($\bar{\bar{X}}$) if it is an $\bar{X}$ chart. The line above the central line shows the upper control limit (UCL), which is commonly obtained by adding 3 sigmas (σ with proper subscripts) to the average, such as $\bar{\bar{X}} + 3\sigma_{\bar{x}}$. The line below the central line is the lower control limit (LCL), which is obtained by subtracting 3 sigmas from the average, such as $\bar{\bar{X}} - 3\sigma_{\bar{x}}$. The upper and lower control limits are usually drawn as dotted lines.

CHART 16–1

MAJOR PARTS OF A CONTROL CHART

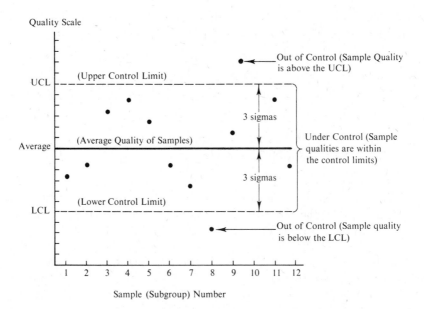

Sample (Subgroup) Number

It should be noted that in some type of process there may be upper control limit only, such as a dimension not exceeding a given number of inches; or there may be lower control limit only, such as the strength of a rope being above a minimum number of pounds.

Use of a Control Chart

A control chart provides the important information of the following three kinds, any one of which may be used as a basis for management to take action.

The Quality Variation of the Samples

The quality variation of the samples drawn from the process is clearly shown by the scattered dots plotted on the chart. In practice, the past data are first analyzed. The result of the analysis is then used as a guide in checking the quality variation of current products.

It is seldom, if ever, that a manufacturing process can produce all products exactly alike. The variation of the product quality is due to numerous causes that affect the process, such as the structure of the machine, the quality of raw material, the skill of the operator, and so on. The variation is unavoidable and must be recognized by the management in setting up specifications. Specification is the required quality value of products produced by a process. It may be expressed in many different ways. For example, a specification may be expressed in terms of the maximum and minimum quality values, within the two extreme values the product quality is allowed to vary. A dimensional tolerance for a certain type of electrical device thus may be specified as 0.5000 ± 0.0025 inches, or a maximum value of 0.5025 inches and a minimum value of 0.4975 inches.

Although a control chart does not show the quality variation of individual products, it does show the quality variation of the samples included in the chart. If the variation of the samples shown in the chart were so great that it would be impossible to produce all the products by the process within the specified tolerance range limits, the manufacturing process should be improved in order to meet the specification. The improvement may be made in various ways, such as replacing certain parts of the machine, using a better grade of material, or hiring a more skilled worker. Otherwise, the management may face the more costly alternative that a 100% inspection (by checking all products instead of sampling inspection) must be carried out so that the good products can be sorted from the bad ones to meet the specification.

Under Control or Out of Control of a Process

This kind of information is provided by the two lines representing the upper and the lower control limits. When samples (dots) plotted on the chart are within the two control limits, the manufacturing process is considered to be *under control*. On the other hand, if any sample is outside the two control limits, the process is considered to be *out of control*.

The two control limits cover the range of "average $\pm$ 3 sigmas." The probability obtained from the Areas Under the Normal Curve, Table 6 in the Appendix, for this range indicates that 99.73% of the possible samples drawn from the process (a large or an infinite population) will fall within the two limits.

Thus, under the normal operation, the samples plotted on the chart should be shown in a stable pattern (or the process is under control—dots are within the control limits). The quality variation is therefore considered as the result of the interactions among numerous minor causes. The minor causes are inherent in the process and are called *chance* or *random causes*. As long as the samples are within the control limits the management should leave the process alone in order to avoid unnecessary adjustment of the process and to save money and time. When any sample is outside the control limits the stable pattern of the variation is interrupted. We thus may suspect that some major and unusual cause or causes has entered into the process (or the process is out of control). This type of unusual and major cause is called *assignable*, such as the breakdown of the machine, poor quality of material in a certain lot, and performance by an unskilled worker. The assignable cause should be found and corrected. The control limits thus provide the information which will tell when the manager should leave a process alone (i.e., if the process is under control) and when he should take a corrective action or look for the causes of trouble (i.e., if the process is out of control).

It should be noted that since the control limits are established on the theory of probability, the use of the control chart may cause us to commit two types of errors (mentioned in Chapter 14). When a sample plotted on the chart is outside the control limits, we will reject the hypothesis that the process is under control. When we reject the hypothesis but in fact it is true or the process is under control, we are committing a Type I error. The result of making such an error is to stop the process to look for *nonexisting* assignable causes. The unnecessary stoppage will cost money and time. On the other hand, when a sample as shown in the chart is within the control limits, we will accept the hypothesis that the process is under control. When we accept the hypothesis but in fact it is not true or the process is out of control, we are committing a Type II error. Thus, we fail to look for *existing* assignable causes. The result of committing a Type II error is that we may produce some products which are not meeting the required quality. Of course, if a 100% inspection method is used, the two types of errors can be avoided.

The Average Quality Level

The central line in the control chart provides the information concerning the average quality of the samples plotted on the chart. This type of information may help the management to maintain or to improve the quality level in order to save money and time. Frequently an average quality level may be too high or too low although the process is under control. In filling orange juice in containers, for example, if the average of the fillings is much higher than the required level, the management may reduce the average, thus saving material and time.

16.2 Control Charts for Variables

The most common types of control charts for variables are the $\bar{X}$ chart and R chart. The symbol $\bar{X}$ represents the mean of the values included in a sample, whereas R represents the range (the difference between the largest value and the smallest value) of a sample. The $\bar{X}$ chart gives more detailed information than the R chart. However, the R chart is easy in computation.

$\bar{X}$ Chart ($\bar{X}$ = Mean of a Sample)

The following values must first be computed before an $\bar{X}$ chart is constructed:

(1) *The Mean of Each Sample, $\bar{X}$*

This is obtained by dividing the sum of the values included in a sample ($\sum X$) by the number of items in the sample (n or sample size).

$$\bar{X} = \frac{\sum X}{n}$$

where X = the value of each item in a sample

(2) *The Mean of the Sample Means, $\bar{\bar{X}}$*

This is obtained by dividing the sum of the sample means ($\sum \bar{X}$) by the number of samples to be included in the chart.

$$\bar{\bar{X}} = \frac{\sum \bar{X}}{\text{Number of samples}}$$

(3) *Upper Control Limit ($UCL_{\bar{X}}$) and Lower Control Limit ($LCL_{\bar{X}}$) for $\bar{X}$ Chart*

The upper control limit may be obtained by adding $3\sigma_{\bar{X}}$ (the standard error of $\bar{X}$, based on all possible sample means of size n: $\sigma_{\bar{X}} = \sigma/\sqrt{n}$) to μ (the mean of means of all possible samples of the same size n = the population mean). The lower control limit may be obtained by subtracting $3\sigma_{\bar{X}}$ from μ. However, in practice the value of μ is usually estimated from samples. The best estimate of μ is $\bar{\bar{X}}$ since the mean of all sample means to be included in the chart ($\bar{\bar{X}}$) can give a more accurate estimate than a single sample mean ($\bar{X}$).

The control limits based on the mean of all sample means can be expressed symbolically:

$$UCL_{\bar{X}} = \bar{\bar{X}} + 3\sigma_{\bar{X}}$$

and
$$LCL_{\bar{X}} = \bar{\bar{X}} - 3\sigma_{\bar{X}}$$

where $3\sigma_{\bar{X}} = 3\dfrac{\sigma}{\sqrt{n}}$ and σ = the process (population) standard deviation

The process standard deviation is usually estimated by using $\hat{s}$ since $\hat{s}^2$ is the unbiased estimate of σ^2. The control limits based on $\bar{\bar{X}}$ and the value of $\hat{s}$ may be obtained by various methods. We shall introduce two basic methods below.

Use the Average of Sample Variance, $\overline{s^2}$

When σ is replaced by $\hat{s}$, $\sigma_{\bar{X}}$ becomes $s_{\bar{X}} = \dfrac{\hat{s}}{\sqrt{n}} = \dfrac{s}{\sqrt{n-1}}$. The control limits now may be written:

$$UCL_{\bar{X}} = \bar{\bar{X}} + 3\frac{s}{\sqrt{n-1}}$$

and
$$LCL_{\bar{X}} = \bar{\bar{X}} - 3\frac{s}{\sqrt{n-1}} \qquad \text{(16-1a)}$$

where $s = \sqrt{\overline{s^2}}$; that is, instead of using a *single sample*, we use the average of variances of *all samples* to find the value of s.

The computation of a large number of sample variances is tedious work. Example 1 includes only 5 samples for the sake of simplicity in illustration.

Example 1

The individual measurements made on 5 samples of 4 items taken from a manufacturing process at random are shown in columns (1) and (2) of Table 16–1. (a) Compute the upper and lower control limits of the $\bar{X}$ chart by using formula (16–1a). (b) Construct the $\bar{X}$ chart.

Solution. (a) The required values for computing the control limits are shown in columns (3), (4), and (5) of Table 16–1. The computations are given below.

TABLE 16–1

COMPUTATION OF THE CONTROL LIMITS (EXAMPLE 1)
(UNIT OF MEASUREMENT: 0.001 INCHES IN EXCESS OF
0.600 INCHES. THUS, 12 REPRESENTS 0.612 INCHES.)

(1) Sample number	(2) Measurement of Each Item in a sample, X				(3) $\sum X$	(4) $\bar{X}$	(5) s^2
	1st	2nd	3rd	4th			
1	12	14	16	6	48	12	14.0
2	5	9	8	10	32	8	3.5
3	3	13	5	7	28	7	14.0
4	20	18	18	16	72	18	2.0
5	4	5	1	10	20	5	10.5
Total					200	50	44.0

(1) The mean of each sample, $\bar{X}$. The computation of each $\bar{X}$ is shown in columns (3) and (4). For example, the $\bar{X}$ for the first sample is 12, or

$$\bar{X} = \frac{\sum X}{n} = \frac{48}{4} = 12$$

(2) The mean of the sample means, $\bar{\bar{X}}$. The sum of 5 sample means ($\sum \bar{X}$) is also obtained from column (4).

$$\bar{\bar{X}} = \frac{\sum \bar{X}}{n} = \frac{50}{5} = 10$$

Note: It can also be computed from the total of all items, or

$$\bar{\bar{X}} = \frac{200[\text{total of all items, column (3)}]}{20(\text{number of items of 5 samples})} = 10$$

(3) Upper and lower control limits. First, compute s^2 for each sample. The method of computing each sample variance s^2 was discussed in Chapter 7, page 181. The value of s^2 for the first sample, for example, is computed as shown in Table 16–2.

Next, compute s from the values of s^2 shown in column (5). The average of the 5 sample variances is

$$\bar{s^2} = \frac{44}{5} = 8.8, \text{ and } s = \sqrt{8.8}$$

TABLE 16–2

COMPUTATION OF s^2 FOR TABLE 16–1

X	$x = X - \bar{X}$ $(\bar{X} = 12)$	x^2
12	0	0
14	2	4
16	4	16
6	-6	36
48	0	56

$$s^2 = \frac{56}{4} = 14.0$$

The control limits based on formula (16–1a) are:

$$UCL_{\bar{x}} = \bar{\bar{X}} + 3\frac{s}{\sqrt{n-1}} = 10 + 3\frac{\sqrt{8.8}}{\sqrt{4-1}} = 10 + 3\sqrt{\frac{8.8}{3}}$$
$$= 10 + 3(1.71) = 10 + 5.13 = 15.13$$

$$LCL_{\bar{x}} = \bar{\bar{X}} - 3\frac{s}{\sqrt{n-1}} = 10 - 5.13 = 4.87$$

CHART 16–2

$\bar{X}$ CHART—MEASUREMENTS OF MANUFACTURED ITEMS
(UNIT OF MEASUREMENT: .001 INCHES IN EXCESS OF .600 INCHES)

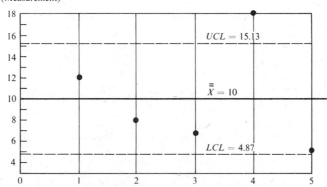

Source: Example 1 and Table 16–1.

Note that the values in this computation are expressed in units of 0.001 inches in excess of 0.600 inches. The actual value for the $UCL_{\bar{x}}$ thus is 0.61513 inches and that for the $LCL_{\bar{x}}$ is 0.60487 inches.

(b) The control chart for this example is shown on Chart 16–2. The chart shows that the mean of sample number 4 (or $\bar{X} = 18$) is above the control limit 15.13. Thus, the process is out of control. The assignable cause of this process should be discovered and corrected immediately. A new control chart is needed for controlling the process in the future.

Use the Average of Sample Ranges, $\bar{R}$

This method simplifies the computation of the control limits. It is especially useful when the number of samples and the size of each sample are large. The symbol $\bar{R}$ represents the mean of the ranges of the samples to be included in the chart. When this method is used, the control limits may be written:

$$UCL_{\bar{x}} = \bar{\bar{X}} + A_2\bar{R}$$

and

$$LCL_{\bar{x}} = \bar{\bar{X}} - A_2\bar{R} \qquad (16\text{--}1b)^*$$

A_2 is a control limit factor. The values of A_2 for selected sample sizes ($n = 2$ to 20) are listed in Table 16–3. Example 2 is used to illustrate this method.

Example 2

A food company puts orange juice into cans advertised as containing 10 ounces of the juice. The weights of the juice drained from cans immediately after filling for 25 samples are taken by a random method (at an interval of every 20 minutes). Each of the samples includes 4 cans. The samples are tabulated in columns (1) and (2) of Table 16–4. The weights in the table are given in units of 0.01 ounces in excess of 10 ounces. For example, the weight of juice drained from the first can of the first sample is 10.12 ounces, which is in excess of 10 ounces for 0.12 ounces (or $10.12 - 10 = 0.12$). Since the unit in the table is 0.01 ounces, the excess is recorded as 12 units in the table. Construct an $\bar{X}$ chart to control the weights of orange juice for the filling.

Solution. The required values for constructing the $\bar{X}$ chart are shown in columns (3) and (4) and are computed as follows:

(1) The mean of each sample, $\bar{X}$. The computation of each $\bar{X}$ is shown in column (3). For example, the $\bar{X}$ for the first sample is 19.75, or

$$\bar{X} = \frac{\sum X}{n} = \frac{79}{4} = 19.75$$

---

*Proof for formula (16–1b).

Let $\bar{R}' =$ the mean of ranges of all possible samples of the same size n drawn from a population

$\sigma =$ the standard deviation of the population from which the samples are drawn

$d_2 = \bar{R}'/\sigma$. The values of d_2 for various sample size (n) from a normal population can be computed. (See the *Manual on Quality Control of Materials,* published by the American Society for Testing and Materials.)

From the d_2 ratio equation, $\sigma = \bar{R}'/d_2$. The best estimate of $\bar{R}'$ is $\bar{R}$. Thus

$$\hat{s} \text{ (or the estimate of } \sigma) = \frac{\bar{R}}{d_2}$$

Substitute the value of σ by $\hat{s}$ in the 3 sigma formula $3\sigma_{\bar{x}} = 3\dfrac{\sigma}{\sqrt{n}}$. Then

$$3\frac{\hat{s}}{\sqrt{n}} = 3\frac{\bar{R}/d_2}{\sqrt{n}} = \left(\frac{3}{d_2\sqrt{n}}\right)\bar{R}$$

Let $A_2 = \dfrac{3}{d_2\sqrt{n}}$. The estimate of $3\sigma_{\bar{x}}$ can be written: $A_2\bar{R}$.

TABLE 16–3

FACTORS FOR COMPUTING 3-SIGMA CONTROL LIMITS FOR $\bar{X}$ AND R CHARTS

Sample Size n	For $\bar{X}$ Chart Factor for Lower or Upper Control Limit A_2	For R Chart Factor for Lower Control Limit D_3	For R Chart Factor for Upper Control Limit D_4
2	1.880	0	3.267
3	1.023	0	2.575
4	0.729	0	2.282
5	0.577	0	2.115
6	0.483	0	2.004
7	0.419	0.076	1.924
8	0.373	0.136	1.864
9	0.337	0.184	1.816
10	0.308	0.223	1.777
11	0.285	0.256	1.744
12	0.266	0.284	1.716
13	0.249	0.308	1.692
14	0.235	0.329	1.671
15	0.223	0.348	1.652
16	0.212	0.364	1.636
17	0.203	0.379	1.621
18	0.194	0.392	1.608
19	0.187	0.404	1.596
20	0.180	0.414	1.586

Source : American Society for Testing and Materials, *A. S. T. M. Manual on Quality Control of Materials, Table B2* (Philadelphia, 1951) p. 115. (Reprinted with permission.)

(2) The mean of the sample mean, $\bar{X}$. The sum of 25 sample means ($\sum \bar{X}$) is obtained from column (3).

$$\bar{X} = \frac{\sum \bar{X}}{25} = \frac{425}{25} = 17$$

(3) Upper and lower control limits. The value of $\bar{R}$ is computed from the values of R shown in column (4). For example, the value of R for the first sample is computed as follows:

$$R = 26 - 12 = 14$$

The $\bar{R}$ represents the mean of the values of R, or

$$\bar{R} = \frac{\sum R}{25} = \frac{305}{25} = 12.20$$

Substitute values of $\bar{X}$, $\bar{R}$, and A_2 ($n = 4$, Table 16–3) in formula (16–1b):

$$UCL_{\bar{X}} = \bar{X} + A_2\bar{R} = 17 + 0.729(12.20)$$
$$= 17 + 8.8938 = 25.8938, \text{ or round to } 25.9$$

TABLE 16–4

WEIGHT OF ORANGE JUICE OF 25 SAMPLES—EXAMPLE 2
WITH CALCULATION FOR $\bar{X}$ CHART (CHART 16–3) AND R CHART (CHART 16–4)
(WEIGHT OF EACH CAN IS EXPRESSED IN UNITS OF 0.01 OUNCES
IN EXCESS OF 10 OUNCES)

(1) Sample Number	(2) Weight of Each Can (4 cans in each sample, $n = 4$) X				(3) Total Weight of 4 Cans $\sum X$	Sample Mean $\bar{X}$	(4) Sample Range R
1	12	19	22	26	79	19.75	14
2	24	6	18	16	64	16.00	18
3	7	10	17	23	57	14.25	16
4	16	23	28	13	80	20.00	15
5	20	14	19	8	61	15.25	12
6	18	24	16	15	73	18.25	9
7	24	3	12	21	60	15.00	21
8	25	14	12	18	69	17.25	13
9	12	23	17	16	68	17.00	11
10	14	21	23	14	72	18.00	9
11	22	17	19	17	75	18.75	5
12	9	8	10	13	40	10.00	5
13	8	14	21	25	68	17.00	17
14	13	23	18	20	74	18.50	10
15	1	12	14	7	34	8.50	13
16	4	15	19	19	57	14.25	15
17	19	22	24	18	83	20.75	6
18	17	24	20	21	82	20.50	7
19	26	23	27	22	98	24.50	5
20	28	19	17	8	72	18.00	20
21	24	13	18	24	79	19.75	11
22	27	21	17	4	69	17.25	23
23	14	5	16	17	52	13.00	12
24	17	19	24	18	78	19.50	7
25	15	8	14	19	56	14.00	11
Total						425.00	305

$$LCL_{\bar{X}} = \bar{\bar{X}} - A_2\bar{R} = 17 - 8.8938$$
$$= 8.1062, \text{ or round to } 8.1$$

Note that the values in the above computation are expressed in units of 0.01 ounces in excess of 10 ounces. The actual value for the $UCL_{\bar{X}}$ thus is 10.259 ounces and that for the $LCL_{\bar{X}}$ is 10.081 ounces. The control chart for this example is shown in Chart 16–3.

Observe Chart 16–3. All of the sample means (dots) plotted on the chart are within the two control limits. Thus, the process is considered as under control and the contol limits are extended on the chart for checking the quality produced

CHART 16–3

$\bar{X}$ Chart—Drained Weights of Orange Juice in Cans
(Unit of Weight: .01 Ounces in Excess of 10 Ounces)

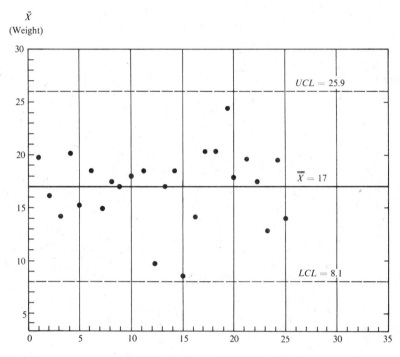

Sample Number (Each sample includes 4 cans)

Source: Example 2 and Table 16–4.

by the process in the future. It should be noted here that if any of the sample means is outside the control limits, either above or below the two control limits, the values of $\bar{\bar{X}}$, $\bar{R}$, and the two control limits should be recomputed. The sample mean that is out of control is excluded in the recomputation. If necessary, the procedure of recomputation may be repeated again and again until all of the sample means are within the new control limits. The new limits are then extended for checking the quality of future products.

R Chart (R = Range of a Sample)

The R chart is used to show the variability or dispersion of the quality produced by a given process. In general, the procedure of constructing the R chart is similar to that for the $\bar{X}$ chart. The required values for constructing the R chart are:

(1) *The range of each sample, R.*
(2) *The mean of the sample ranges, $\bar{R}$.*
(3) *Upper control limit (UCL_R) and lower control limit (LCL_R) for R chart. The*

mean of the sample ranges $\bar{R}$ is used as the estimate of the mean of the ranges of all possible samples of the same size n drawn from the process (population). Thus,

$$UCL_R = \bar{R} + 3\sigma_R$$

and

$$LCL_R = \bar{R} - 3\sigma_R$$

where $\sigma_R =$ the standard error of the range (or the standard deviation of the ranges of all possible samples of the same size n drawn from a given population)

The value of σ_R may be estimated by finding the standard deviation of the ranges of the samples included in a chart. In practice, however, it is rather convenient to compute the upper and lower control limits by using the values D_4 and D_3 as provided in Table 16–3 according to various sample size ($n = 2$ to 20). When the tabulated values are used, the two limits may be written as follows:

$$UCL_R = D_4\bar{R}$$

and

$$LCL_R = D_3\bar{R} \qquad\qquad (16\text{–}2)^*$$

Note that the distribution of the ranges of all possible samples of a small size drawn from a normal population is not normal. However, the fact that a sample range falls outside the 3-sigma control limits is unusual and should be regarded as the presence of assignable causes. The lower control limit for sample size of 6 or less may become negative. Since the value of range cannot be negative, the lower control limit is set at zero under such a situation. The value of D_3 is zero when $n = 6$ or less in Table 16–3.

Example 3

Refer to Example 2 and Table 16–4. Construct the R chart.

Solution.   The required values for the chart are presented below.
  (1) The range of each sample, R. See column (4) of Table 16–4.
  (2) The mean of the sample ranges, $\bar{R}$.

$$\bar{R} = \frac{305}{25} = 12.20$$

  (3) The upper and lower control limits. Use D_4 and D_3 values for $n = 4$ in Table 16–3 and formula (16–2).

$$UCL_R = D_4\bar{R} = 2.282(12.20) = 27.8404, \text{ or round to } 27.8$$
$$LCL_R = D_3\bar{R} = 0(12.20) = 0$$

The control chart for R is shown in Chart 16–4. The chart shows that the process

---

*Let $D_4 = 1 + \dfrac{3\sigma_R}{\bar{R}}$ $UCL_R = \bar{R} + 3\sigma_R = \left(1 + \dfrac{3\sigma_R}{\bar{R}}\right)\bar{R} = D_4\bar{R}$

Let $D_3 = 1 - \dfrac{3\sigma_R}{\bar{R}}$ $LCL_R = \bar{R} - 3\sigma_R = \left(1 - \dfrac{3\sigma_R}{\bar{R}}\right)\bar{R} = D_3\bar{R}$

is under control since all of the R values plotted on the chart are within the two control limits. The three lines representing the average value $\bar{R}$ and the two control limits are thus extended for checking the quality variation of subsequent production by the process.

The choice between the $\bar{X}$ chart and the R chart is a managerial problem. The $\bar{X}$ chart is used to show the quality averages of the samples drawn from a given process, whereas the R chart is used to show quality dispersion (variabilities) of the samples. If the presence of an assignable cause will show on both types of charts that the process is out of control, only one type of chart probably is sufficient for statistical quality control purpose. On the other hand, it is possible that the increase of quality dispersion of a sample may cause the process to be out of control in the R chart, but keeps the process under control in the $\bar{X}$ chart. Conversely, the increase of a sample mean may cause the process to be out of control in the $\bar{X}$ chart, but keeps the process under control in the R chart. Under such circumstances, the management may wish to keep both types of control charts.

In practice, the R charts should be constructed first. If the R chart indicates that the dispersion of the quality by the process is out of control generally, it is better not to construct an $\bar{X}$ chart until the quality dispersion is brought under control. Note that the computation of the control limits of an $\bar{X}$ chart depends on the value of $\bar{R}$.

CHART 16–4

R Chart—Drained Weights of Orange Juice in Cans
(Unit of Weight: .01 Ounces in Excess of 10 Ounces)

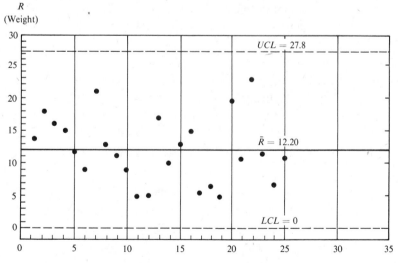

Sample Number (Each sample includes 4 cans)

Source: Example 3 and Table 16–4.

16.3 Control Charts
for Attributes

The control charts for variables show the quality characteristics that are measured and expressed in units by numbers. The control charts for attributes, on the other hand, deal with the quality characteristics that are observed only by conforming or nonconforming to the specified requirements and expressed by two opposite words, such as yes and no, good and bad, and nondefective and defective. The common types of control charts for attributes are: (1) p (fraction defective) chart, (2) np (number of defectives) chart, and (3) c (number of defects) chart. The procedure of constructing a control chart for attributes is basically the same as that for variables. Details for the three different types of charts are presented individually below.

p Chart (p = Fraction Defective of a Sample)

The following values are required for constructing a p chart.

(1) *The Fraction Defective of Each Sample, p*
This can be expressed:

$$p = \frac{\text{Number of defectives of a sample}}{\text{Number of items inspected (sample size)}} = \frac{np}{n}$$

The number of items inspected for each sample, or sample size n, for a p chart should be relatively larger than that for a control chart for variables. The reason is obvious. For example, if only 0.1% of the process is defective, the sample size must be 1,000 items if we expect to find an average of one defective per sample. However, if we measure less than 1,000 items, say 200 items, as a sample drawn from the process and record the measurements by numbers of units (variables), we certainly will have a good indication from the sampling study for the quality by the process.

(2) *The Average Fraction Defective of Samples, $\bar{p}$*
This can be written:

$$\bar{p} = \frac{\text{Total number of defectives of all samples}}{\text{Total number of items inspected}} = \frac{\Sigma\, np}{\Sigma\, n}$$

(3) *Upper Control Limit (UCL_p) and Lower Control Limit (LCL_p) for p Chart*
The control limits are again based on 3 sigmas and may be expressed as follows:

$$UCL_p = P + 3\sigma_p$$
and
$$LCL_p = P - 3\sigma_p$$

where P = the true process (population) fraction defective and
$\sigma_p = \sqrt{PQ/n}$, the standard error of p

The value of P is frequently unknown. When the process is under control, $\bar{p}$ is usually used as the estimate of P. The estimate of $\sigma_p = \sqrt{PQ/n}$ thus is $\sqrt{\bar{p}(1-\bar{p})/n}$ since $Q = 1 - P$. When the estimated values are used, the two limits may be written:

$$UCL_p = \bar{p} + 3\sqrt{\frac{\bar{p}(1-\bar{p})}{n}}$$

and
$$LCL_p = \bar{p} - 3\sqrt{\frac{\bar{p}(1-\bar{p})}{n}} \tag{16-3}$$

The sample size (n) for a p chart should be constant; that is, the sample size should be the same for all samples included in the chart. However, in many cases, the sample size may vary. Observe formula (16–3). The control limits will change when n varies. The procedure of constructing a p chart for samples of a constant size thus is different from that for samples of various sizes.

Constructing a p Chart for Samples of Constant Size

Example 4

Certain television parts produced by a process are inspected by a random method for a single quality characteristic. The results of inspection are given in columns (1), (2), and (3) of Table 16–5. Construct a p chart.

Solution.   The required values for constructing the p chart are first computed:
(1) The fraction defective of each sample (p) is shown in column (4). For example, the value of p of the first sample is 0.08, which is computed:

$$p = \frac{np}{n} = \frac{16}{200} = 0.08$$

(2) The average fraction defective:

$$\bar{p} = \frac{\Sigma\, np}{\Sigma\, n} = \frac{460}{5,000} = 0.092$$

(3) The upper and lower control limits. Use formula (16–3).

$$3\sqrt{\frac{\bar{p}(1-\bar{p})}{n}} = 3\sqrt{\frac{0.092(1-0.092)}{200}} = 0.061$$
$$UCL_p = 0.092 + 0.061 = 0.153$$
$$LCL_p = 0.092 - 0.061 = 0.031$$

The above computed values are plotted on Chart 16–5 as the *trial values*. The trial values may be revised when the process is out of control.

The trial values should be revised in the present case. Chart 16–5 shows that four samples are outside the two control limits, sample number 21 being below the *LCL* and samples 6, 11, and 24 being above the *UCL*. Since the goal of keeping a p chart is to maintain a better quality level or lower values of p, investigation on sample 21 may not be needed. However, the assignable causes for the other three samples should be discovered and corrected. Assume that the investigation reveals that the three samples were taken from the lots produced by a group of new workers. A training program may be introduced for the workers. The values of $\bar{p}$ and the two control limits should be revised for checking the quality of future

TABLE 16–5

INSPECTION RESULTS FOR TELEVISION PARTS
(WITH CALCULATION FOR p CHART—CHART 16–5)

(1) Sample Number	(2) Number of Units Inspected n	(3) Number of Defectives np	(4) Fraction Defective p
1	200	16	0.08
2	200	14	0.07
3	200	8	0.04
4	200	20	0.10
5	200	10	0.05
6	200	34	0.17
7	200	20	0.10
8	200	16	0.08
9	200	18	0.09
10	200	12	0.06
11	200	36	0.18
12	200	20	0.10
13	200	22	0.11
14	200	18	0.09
15	200	26	0.13
16	200	8	0.04
17	200	16	0.08
18	200	20	0.10
19	200	22	0.11
20	200	14	0.07
21	200	6	0.03
22	200	12	0.06
23	200	22	0.11
24	200	38	0.19
25	200	12	0.06
Total	5,000	460	—

Source: Example 4.

production. The revised values are computed below after the three samples of 600 items ($= 200 + 200 + 200$) with 108 defectives ($= 34 + 36 + 38$) are eliminated.

$$\bar{p} = \frac{460 - 108}{5,000 - 600} = \frac{352}{4,400} = 0.080$$

$$3\sqrt{\frac{\bar{p}(1 - \bar{p})}{n}} = 3\sqrt{\frac{0.08(1 - 0.08)}{200}} = 0.058$$

$$UCL_p = 0.080 + 0.058 = 0.138$$

$$LCL_p = 0.080 - 0.058 = 0.022$$

The revised values are also plotted on Chart 16–5. Excluding the three samples, there is no sample outside the revised control limits. The lines representing the new values ($\bar{p}$ and two control limits) are thus extended for checking the future quality by the process.

CHART 16–5

p CHART—INSPECTION RESULTS FOR TELEVISION PARTS

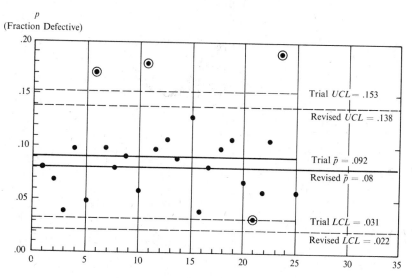

Sample Number (Each sample includes 200 units.)

Source: Example 4 and Table 16–5.

Constructing a p Chart for Samples of Varying Sizes

Example 5

Columns (1) to (3) of Table 16–6 reported the daily inspection results for radio tubes in May of this year. There are 22 working days during the month. The sample size for each day varies. Construct a p chart.

Solution. The required values for constructing the p chart are computed below.

(1) The fraction defective of each sample p is first computed. The computed values of p are shown in column (4) of Table 16–6.

(2) The average fraction defective $\bar{p}$ is computed from the totals shown in columns (2) and (3):

$$\bar{p} = \frac{\Sigma\, np}{\Sigma\, n} = \frac{445}{4{,}356} = 0.102$$

(3) The upper and lower control limits. The control limits are obtained by using formula (16–3), which may be written in different forms as follows:

$$UCL_p = \bar{p} + \frac{3\sqrt{\bar{p}(1-\bar{p})}}{\sqrt{n}}$$

$$LCL_p = \bar{p} - \frac{3\sqrt{\bar{p}(1-\bar{p})}}{\sqrt{n}}$$

TABLE 16-6

DAILY INSPECTION RESULTS FOR RADIO TUBES IN MAY
(WITH CALCULATION FOR p CHART—CHART 16-6)

(1) Date (Sample Order)	(2) Number Inspected n	(3) Number of Defectives np	(4) Fraction Defective p	(5) $\sqrt{n}$	(6) 3 sigmas $=\dfrac{0.90795}{\sqrt{n}}$	(7) UCL$=$ $\bar{p}+3sigmas=$ $0.102+col.(6)$	(8) LCL$=$ $\bar{p}-3sigmas=$ $0.102-col.(6)$
May 2	225	27	0.12	15.0	0.061	0.163	0.041
3	210	21	0.10	14.5	0.063	0.165	0.039
6	196	14	0.07	14.0	0.065	0.167	0.037
7	180	14	0.08	13.4	0.068	0.170	0.034
8	176	16	0.09	13.3	0.068	0.170	0.034
9	190	23	0.12	13.8	0.066	0.168	0.036
10	215	30	0.14	14.7	0.062	0.164	0.040
13	220	24	0.11	14.8	0.061	0.163	0.041
14	182	16	0.09	13.5	0.067	0.169	0.035
15	185	21	0.11	13.6	0.067	0.169	0.035
16	204	16	0.08	14.3	0.063	0.165	0.039
17	205	31	0.15	14.3	0.063	0.165	0.039
20	196	24	0.12	14.0	0.065	0.167	0.037
21	215	17	0.08	14.7	0.062	0.164	0.040
22	230	28	0.12	15.2	0.060	0.162	0.042
23	174	17	0.10	13.2	0.069	0.171	0.033
24	188	23	0.12	13.7	0.066	0.168	0.036
27	194	17	0.09	13.9	0.065	0.167	0.037
28	170	14	0.08	13.0	0.070	0.172	0.032
29	185	16	0.09	13.6	0.067	0.169	0.035
30	210	21	0.10	14.5	0.063	0.165	0.039
31	206	15	0.07	14.4	0.063	0.165	0.039
Total	4,356	445					

Source: Example 5.

The value of 3 sigmas is different for each sample since the sample size n is different for each sample in this example. The numerator of the 3 sigmas, however, is the same for all samples, or

$$3\sqrt{\bar{p}(1-\bar{p})} = 3\sqrt{0.102(1-0.102)} = 0.90795$$

The 3 sigmas for each sample thus may be written as

$$\frac{3\sqrt{\bar{p}(1-\bar{p})}}{\sqrt{n}} = \frac{0.90795}{\sqrt{n}}$$

The values of $\sqrt{n}$ and the 3 sigmas for each sample are shown in columns (5) and (6) respectively. The value of the 3 sigmas of the first sample, for example, is computed:

$$3 \text{ sigmas} = \frac{0.90795}{\sqrt{225}} = \frac{0.90795}{15} = 0.061$$

CHART 16–6

p CHART—DAILY INSPECTION RESULTS FOR RADIO TUBES

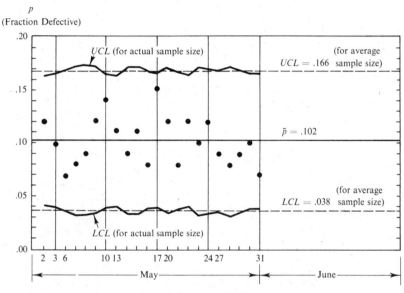

Sample Order (Sample size varies.)

Source: Example 5 and Table 16–6.

The upper and lower control limits for each sample are shown in columns (7) and (8) respectively. The values of the limits for the first sample are computed as follows:

$$UCL_p = \bar{p} + 3 \text{ sigmas} = 0.102 + 0.061 = 0.163$$
$$LCL_p = \bar{p} - 3 \text{ sigmas} = 0.102 - 0.061 = 0.041$$

The above computed values are plotted on Chart 16–6. Note that the control limits are connected by broken lines.

The computation of control limits for every sample included in a p chart is time consuming. In practice, whenever the sample size is expected to vary moderately, only a single set of limits is used for checking the future production. The average size of the samples to be taken in the future is first estimated. The single set of control limits is then computed for the average sample size. For example, the estimated average size of the samples to be taken in June based on the total items inspected in May of Example 5 is

$$n = \frac{4,356 \text{ items inspected}}{22 \text{ days}} = 198 \text{ items per day}$$

The control limits for all samples included in the chart based on the average size are:

$$3 \text{ sigmas} = \frac{3\sqrt{\bar{p}(1-\bar{p})}}{\sqrt{n}} = \frac{3\sqrt{0.102(1-0.102)}}{\sqrt{198}} = \frac{0.90795}{14.1} = 0.064$$

$$UCL_p = 0.102 + 0.064 = 0.166$$
$$LCL_p = 0.102 - 0.064 = 0.038$$

The single set of control limits is drawn on Chart 16–6 as two straight horizontal lines. Observe the p values on the chart. All of the p values are within the two control limits. Thus, the process is considered under control and the two control lines as well as the average fraction defective line are extended for inspecting the products to be completed in June. However, a user of the two straight control limit lines should observe the following facts:

(1) When the actual sample size is larger than the average sample size, the true control limits for the sample are *inside* the two straight lines, such as samples taken on May 2 and 3.

(2) When the actual sample size is smaller than the average sample size, the true control limits for the sample are *outside* the two straight lines, such as samples taken on May 8 and 9.

The understanding of the above facts sometimes can help the user to detect whether the process is under or out of control based on a sample which is substantially larger or smaller than the average sample size, without computing a set of new control limits. Of course, if it cannot help the user to make a judgment, a set of new control limits should be computed for the sample of unusual size.

np Chart (np = Number of Defectives of a Sample)

An np chart shows the actual number of defectives found in each sample. The chart is applied only when the samples to be included are of constant size. When the sample size is varying, the control chart for fraction defectives (p chart) should be used to show the product quality of a process. The following values are required for constructing an np chart:

(1) *The number of defectives of each sample, np.*

(2) *The average number of defectives per sample of a constant size, $n\bar{p}$.* This is obtained by dividing the total number of defectives of all samples ($\sum np$) by the number of samples. If $\bar{p}$ is known, it can be obtained by multiplying $\bar{p}$ by n. Thus, $\overline{np} = n\bar{p}$.

(3) *Upper control limit (UCL_{np}) and lower control limit (LCL_{np}) for np chart.* The control limits based on 3 sigmas may be written:

$$UCL_{np} = nP + 3\sigma_{np}$$

and

$$LCL_{np} = nP - 3\sigma_{np}$$

where P = the true process (population) fraction defective
         nP = the average number of defectives per sample based on all possible samples of size n from the process
         $\sigma_{np} = \sqrt{nP(1-P)}$, the standard error of np derived from formula (10–4), page 248.

When the process is under control, $\bar{p}$ is usually used as the estimate of P. When P is replaced by $\bar{p}$, the control limits may be written:

$$UCL_{np} = n\bar{p} + 3\sqrt{n\bar{p}(1 - \bar{p})}$$

and

$$LCL_{np} = n\bar{p} - 3\sqrt{n\bar{p}(1 - \bar{p})} \qquad \textbf{(16-4)}$$

Example 6

Refer to Example 4 and Table 16–5, which gives 25 samples of the constant size 200 items. Compute the required values for constructing an np chart.

Solution. The required values for constructing the np chart are given below.

(1) The number of defectives of each sample, np, is shown in column (3) of Table 16–5.

(2) The average number of defectives per sample:

$$n\bar{p} = \overline{np} = \frac{\Sigma\, np}{\text{number of samples}} = \frac{460}{25} = 18.4$$

Or, $n\bar{p} = 200(0.092) = 18.4$. $\bar{p} = 0.092$ (see Example 4)

(3) The upper and lower control limits. Use formula (16–4).

$$3\sqrt{n\bar{p}(1 - \bar{p})} = 3\sqrt{200(0.092)(1 - 0.092)} = 12.3$$
$$UCL_{np} = 18.4 + 12.3 = 30.7$$
$$LCL_{np} = 18.4 - 12.3 = 6.1$$

If these computed values are plotted on an np chart with a vertical scale 200 ($= n$) times the scale of Chart 16–5 (the p chart), the appearance of the np chart should be the same as that of the p chart. For example, the position of the central line on the np chart ($n\bar{p} = 18.4$) is the same as that on the p chart ($\bar{p} = 0.092$ and $0.092 \times 200 = 18.4$). The p and the np charts thus provide the same type of information regarding the process; either it is under or out of control. When the process is out of control, the samples, which are outside the control limits on a p chart, will likewise be outside the control limits of an np chart. The procedure of revising the central line and the two control limits on an np chart for an out of control process is similar to that discussed in Example 4 above.

When sample size is constant, the np chart has certain advantages over the p chart. The np chart does not require the computation of the values of p for all samples. The actual number of defectives plotted on the np chart is more readily understandable than those relative fractions on a p chart. However, when sample size changes, it is rather difficult to construct and to read an np chart; because the average number of defectives per sample, $n\bar{p}$, as well as the control limits, would need to be computed with every change in sample size, n. The lines representing $n\bar{p}$ values and limits would be broken lines due to the changes. Since both p and np charts provide the same type of information for quality control purpose, one should select only one of the two charts for defectives in order to avoid confusion.

c Chart (c = Number of Defects Per Unit [Sample])

The samples included in a c chart are individual products of constant size. The number of defects in each product, represented by letter c, is counted and recorded as the value of a sample. A defect is different from a defective. A defective is a product that fails to conform to one or more specified requirements; whereas, a defect is a single nonconformance which causes the product to be a defective. A defective may have one or more defects.

In applying a c chart, the probability distribution of the numbers of defects of a product (values of c) from a process should follow the pattern of Poisson distribution. (See Section 10.6, page 261.) Examples of such products are 10-ounce glass milk bottles (c = number of air bubbles inside the glass of each bottle), painted metal sheets (c = number of surface defects on each sheet), electric wire (c = number of weak spots in a given length of the wire), and bolts of cotton cloth (c = number of imperfections observed in a square yard of the cloth). The examples indicated that it is not necessary that the samples for a c chart must be individual manufactured products. Each sample may be any unit of constant size that has equal opportunity for the occurrence of defects, such as a length and an area. Furthermore, the application of a c chart may be extended to many types of operations, such as in an inventory operation, c = number of errors made by a worker during a given period of time.

The required values for constructing a c chart are as follows:

(1) *The number of defects in each sample, c.* The defects in a sample are counted individually. Therefore, the number of defects is expressed in a whole number. The samples are plotted on a c chart according to the c scale (vertical) and the order of sample numbers on the horizontal scale.

(2) *The average number of defects of samples $\bar{c}$.* This can be written:

$$\bar{c} = \frac{\text{Total number of defects of all samples}}{\text{Number of samples inspected}}$$

The value of $\bar{c}$ is represented by the central line in a c chart.

(3) *Upper control limit (UCL_c) and lower control limit (LCL_c) for c chart.* The control limits based on 3 sigmas may be written:

$$UCL_c = \hat{c} + 3\sqrt{\hat{c}}$$

and

$$LCL_c = \hat{c} - 3\sqrt{\hat{c}}$$

where $\hat{c}$ = the true average number of defects per product (or sample unit) of the process, and

$\sqrt{\hat{c}}$ = the standard error of c (or the standard deviation of the Poission distribution of values of c). See formula (10–16), page 265.

The value of $\hat{c}$ is usually unknown. When the process is under control, $\bar{c}$ is used as the estimate of $\hat{c}$. When the estimated value is used, the two limits may be written:

$$UCL_c = \bar{c} + 3\sqrt{\bar{c}}$$

and

$$LCL_c = \bar{c} - 3\sqrt{\bar{c}} \tag{16-5}$$

Example 7

Assume that 25 10-ounce glass milk bottles are selected at random from a process. The numbers of air bubbles (defects) observed from the bottles are given in Table 16–7. Compute the required values for constructing a c chart.

TABLE 16–7

OBSERVED DEFECTS IN 25 GLASS MILK BOTTLES
(c = NUMBER OF AIR BUBBLES [DEFECTS] IN EACH BOTTLE)

Bottle Number (Sample order)	Defects c	Bottle Number (Sample order)	Defects c	Bottle Number (Sample order)	Defects c
1	8	10	4	19	9
2	7	11	5	20	8
3	6	12	6	21	7
4	5	13	7	22	6
5	3	14	3	23	5
6	4	15	2	24	4
7	8	16	4	25	3
8	5	17	10		
9	4	18	11	Total	144

Source: Example 7.

Solution. (1) The number of defects (c) for each sample (bottle). The values of c are already shown in Table 16–7.

(2) The average number of defects, $\bar{c}$. The total number of defects of the 25 samples is 144. Thus,

$$\bar{c} = \frac{144}{25} = 5.76$$

(3) The upper and lower control limits. Use formula (16–5).

$$3\sqrt{\bar{c}} = 3\sqrt{5.76} = 3(2.4) = 7.2$$
$$UCL_c = \bar{c} + 3\sqrt{\bar{c}} = 5.76 + 7.2 = 12.96$$
$$LCL_c = \bar{c} - 3\sqrt{\bar{c}} = 5.76 - 7.2 = -1.44 \text{ or recorded as zero,}$$
since the number of defects cannot be negative.

Observe each of the values of c in Table 16–7. All of the c values are below 12.96 defects, the UCL_c. Thus, the process is under control. We therefore learned the nature of the process without having a c chart. Note that a c chart may be constructed for Example 7 by following the steps used in constructing previous control charts.

16.4 Acceptance Sampling

Acceptance sampling is the process of sampling inspection used by a purchaser to decide whether he should accept or reject a shipment of products on the basis of predetermined standards. If the shipment is rejected, it may be returned to the supplier or it may be inspected completely (100 percent inspection). All items below standard found by 100 percent inspection must then be replaced or corrected for acceptance.

Many types of acceptance sampling plans have been developed by government agencies, private industries, and individual statisticians. In general, the plans are designed for either variables or attributes and are based on sampling theory as presented in previous chapters. The application of acceptance sampling by attributes is much simpler and less costly than that by variables. It is easier to record and compute the conformances or nonconformances to specifications than the actual measured values. Moreover, many quality characteristics are not measurable but are observable as attributes. Acceptance sampling by attributes thus is more popularly used than that by variables.

The acceptance sampling plans for attributes presented in Table 16–8 are selected from a government publication for illustration purposes. In those types of plans the purchaser first designates the acceptable quality level (AQL) which is the required quality of the products submitted by the supplier. There are various AQL's available for different types of products and suppliers. The AQL is expressed in terms of maximum percent defective and is 4 percent for the selected sampling plans. Thus, if an inspection lot has 2,000 items, the purchaser will accept the lot if there are 80 (= 2,000 × 4%) or less defectives. However, the actual number of defectives in a lot is not known to the purchaser since the products are inspected by sampling, not by 100 percent inspection. The selected sampling plans provide the lot which consists of not more than the specified maximum number of defectives (AQL) a high probability of acceptance.

TABLE 16–8

SELECTED SAMPLING PLANS

INSPECTION LOT SIZE: 1,201 TO 3,200 ITEMS

ACCETABLE QUALITY LEVEL (AQL): 4%

Type of Sampling Plan	Sample Size	Cumulative Sample Size	Number of Defectives Cumulated in Samples	
			Acceptance Number	Rejection Number
Single Sampling (One sample only)	125	125	10	11
Double Sampling				
1st sample	80	80	5	9
2nd sample	80	160	12	13
Multiple Sampling				
1st sample	32	32	0	5
2nd sample	32	64	3	8
3rd sample	32	96	6	10
4th sample	32	128	8	13
5th sample	32	160	11	15
6th sample	32	192	14	17
7th sample	32	224	18	19

Source: U.S. Department of Defense, *Sampling Procedures and Tables for Inspection by Attributes (MIL-STD-105D)*, April 29, 1963, p. 49.

The sample size for inspection is determined by the lot size to be inspected and the type of sampling plan used. When a shipment of products is received, it may be divided into a number of lots for inspection. Each inspection lot is thus a population from which a sample or samples are drawn. Table 16–8 gives three sampling plans for the lot size ranging from 1,201 to 3,200 items: (1) the single sampling plan, (2) the double sampling plan, and (3) the multiple sampling plan.

Under the single sampling plan, the sample size is 125 items. If 10 or less defectives are found in the single sample, the lot is accepted; if 11 or more defectives are found, the lot is rejected. Under the double sampling plan, the sample size for the first sample is 80 items. If 5 or less defectives are found in the first sample, the lot is accepted; if 9 more defectives are found, the lot is rejected; and there is no need to take a second sample. If more than 5 but less than 9 de-

CHART 16-7

OPERATING CHARACTERISTIC CURVE FOR THE SINGLE
SAMPLING PLAN GIVEN IN TABLE 16–8
(CURVES FOR DOUBLE AND MULTIPLE SAMPLING ARE VERY CLOSE
TO THE CURVE FOR SINGLE SAMPLING)

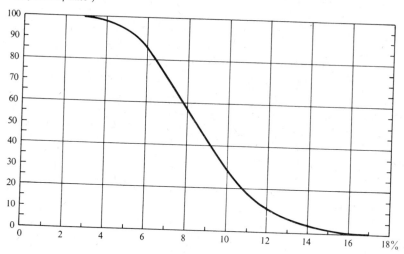

Percent of lots expected
to be accepted
(probability of acceptance*)

Quality of Submitted Lots (p = percent defective)

*The values of probability are based on Poisson distribution as an approximation to the Binomial. The Poisson probability can be obtained from published table such as table 10–18 and Table 4 in the Appendix. For example, when $p = 4\% = .04$, $np = 125(.04) = 5 = \mu$. The probability is .9863 or 98.63% for 10 defects or less. When $p = 8\% = .08$, $np = 125(.08) = 10 = \mu$. The probability is .5831 or 58.31% for 10 defects or less.

Source: U.S. Department of Defence, *Sampling Procedures and Tables for Inspection by Attributes* (MIL-STD-105D), Aril 29, 1963, p. 48.

fectives are found, a second sample of 80 items is taken. The defectives found in the first and the second samples are then accumulated. If 12 or less defectives are found in the two samples, that lot is accepted; if 13 or more defectives are found, the lot is rejected. The double sampling plan may save money and time for the inspection since at some times, especially when the lot quality is either very good or very poor, the second sample is not needed. Note that the size of the first sample under the double sampling plan (80) is smaller than the size of the sample under the single sampling plan (125). The inspection procedure of the multiple sampling plan is similar to that of the double sampling plan, except that after the first sample, the number of successive samples required to reach a decision may be more than one.

The probability of accepting a submitted lot of any quality level can be obtained from the operating characteristic curve shown in Chart 16–7. The horizontal scale of the chart indicates the true percent defective of the lot submitted for inspection. The vertical scale gives the probability that the lot will be accepted. For example, if the submitted lot has 4 percent defective, the required quality level, the probability that the lot will be accepted under the single sampling plan is 98.63 percent. The acceptance and rejection numbers of defectives for the sampling plan thus make a purchaser accept the great majority of the lots that the supplier submits, provided that the true percent defective in these lots be no greater than the designated AQL of 4 percent. If the lot has 8 percent defectives, the probability of its being accepted is only 58.31 percent. Further, if the submitted lot has more than 16 percent defectives, the lot is expected to be not accepted at all since the probability of acceptance is practically zero as read from the OC curve. Thus, a supplier will maintain his products on the required AQL if he wishes to have low rejection rates under the sampling plans.

Again we note the two types of errors by the use of sampling. The null hypothesis is that the inspection lot has 4 percent or less defectives. When we reject the hypothesis based on the findings of a sampling plan but in fact it is true, we are committing a Type I error. The result of making such an error is to reject the lot of acceptable quality. When we accept the hypothesis but in fact it is not true, we are committing a Type II error. The result of making such an error is to accept the lot of below specified quality level.

16.5   Summary

The techniques used in statistical quality control are basically sampling techniques as presented in the previous chapters. They are initially used for improving the quality of products from a manufacturing process. In recent years, however, the application of the techniques has been extended to many types of nonmanufacturing business activities.

The qualities of manufactured products may be expressed in two different ways. When the quality is expressed by an actual measurement, it is said to be expressed by a variable. When the quality is expressed by either conforming or nonconforming to the specified requirements, it is said to be expressed by an attribute. Frequently, a quality characteristic may be expressed in either of the two ways.

Statistical quality control techniques may be classified into two groups: (1) control charts and (2) specified acceptance sampling plans. Both groups can further be classified according to the quality expressions: by variables and by attributes. The most common types of control charts for variables are the $\bar{X}$ (mean) chart and the R (range) chart. Control charts for attributes are the p (fraction defective) chart, the np (number of defectives) chart, and the c (number of defects) chart. The formulas of the control limits for the various types of control charts are summarized at the end of this section.

Acceptance sampling is the process of sampling inspection used by a purchaser to decide whether he should accept or reject a shipment of products on the basis of predetermined standards. Many types of acceptance sampling plans have been developed. Acceptance sampling by attributes is more popularly used than that by variables. The acceptance sampling plans for attributes in this chapter are selected from a governmental publication. In those types of plans, various acceptable quality levels (AQL) are available. Only the $AQL = 4$ percent is chosen for the selected sampling plans in the illustration. The sample size for inspection is determined by the inspection lot size and the type of sampling plan used. The illustration gives three sampling plans, including simple, double, and multiple sampling, for the lot size stated in a range from 1,201 to 3,200 items. The specified acceptance and rejection numbers of defectives for the sampling plans will make the purchaser accept the great majority of the lots if the supplier submits the products with a required quality level (AQL). The probability of accepting a submitted lot of any quality level can be obtained from the operating characteristic curve. Since the inspection is based on samples, the two types of errors may occur in the plans.

SUMMARY OF FORMULAS

Application	Formula	Formula Number	Reference Page
Control limits for $\bar{X}$ chart	$UCL_{\bar{X}} = \bar{\bar{X}} + 3\dfrac{s}{\sqrt{n-1}}$ $LCL_{\bar{X}} = \bar{\bar{X}} - 3\dfrac{s}{\sqrt{n-1}}$	16–1a	391
	$UCL_{\bar{X}} = \bar{\bar{X}} + A_2\bar{R}$ $LCL_{\bar{X}} = \bar{\bar{X}} - A_2\bar{R}$	16–1b	394
Control limits for R chart	$UCL_R = D_4\bar{R}$ $LCL_R = D_3\bar{R}$	16–2	398
Control limits for p chart	$UCL_p = \bar{p} + 3\sqrt{\dfrac{\bar{p}(1-\bar{p})}{n}}$ $LCL_p = \bar{p} - 3\sqrt{\dfrac{\bar{p}(1-\bar{p})}{n}}$	16–3	401
Control limits for np chart	$UCL_{np} = n\bar{p} + 3\sqrt{n\bar{p}(1-\bar{p})}$ $LCL_{np} = n\bar{p} - 3\sqrt{n\bar{p}(1-\bar{p})}$	16–4	407
Control limits for c chart	$UCL_c = \bar{c} + 3\sqrt{\bar{c}}$ $LCL_c = \bar{c} - 3\sqrt{\bar{c}}$	16–5	408

Exercises 16

1. Basically, what are the techniques used in statistical quality control? Can the techniques be used for improving the qualities of manufactured products as well as nonmanufacturing business activities?

2. What are variables? attributes? Give examples in your answers.

3. What are the major parts of a control chart? What type of sample size is preferred for the control chart for variables? for attributes?

4. State briefly the use of a control chart. Explain the difference between random causes and assignable causes.

5. What are $\bar{X}$ and R charts? By comparison, what are the advantages and disadvantages of the two charts?

6. Under what circumstances may the management wish to keep both $\bar{X}$ and R charts?

7. State the similarities and differences between the p chart, np chart, and c chart.

8. When the fraction defective (p) of a sample falls below the lower control limit, is the process considered as out of control? Should the control limit be revised by excluding the sample?

9. What is meant by acceptance sampling? Explain the term "acceptable quality level."

10. Explain the reasons for using the double and multiple sampling plans.

11. The average life of the batteries manufactured from a process and the standard deviation of the lives are estimated at 52 hours and 8 hours respectively. Compute the 3-sigma upper and lower control limits of an $\bar{X}$ chart if the samples are of 4 items each.

12. Assume that 28 samples of 6 products each were taken from a manufacturing process. The mean of the 28 sample means ($\bar{\bar{X}}$) was 2.54 pounds and the mean of the ranges of the samples ($\bar{R}$) was 0.86 pounds. Find the upper and lower control limits for (a) the $\bar{X}$ chart, and (b) the R chart.

13. The average fraction defective $\bar{p}$ of a group of 35 samples of various sizes is 0.084. (a) Compute the 3-sigma upper and lower control limits for the sample sizes from the group as follows:
    (1) The average size of 225 items.
    (2) The largest sample size of 400 items.
    (3) The smallest sample size of 100 items.
    (b) Assume that the control limits for the above three sample sizes are extended on the p chart for checking future production quality. Explain the significance of the control limits if
    (1) The size of the 36th sample is 350 items, and
    (2) The size of the 37th sample is 120 items.

14. The following information in units of inches is obtained from a group of 32 samples of 10 items each for an $\bar{X}$ chart:

$$UCL_{\bar{x}} = 7.04 \qquad \bar{\bar{X}} = 6.27$$
$$LCL_{\bar{x}} = 5.50 \qquad \bar{R} = 2.50$$

Two of the 32 sample means are out of control limits. The two sample means are: 9.25 inches (with $R = 5.60$ inches for the sample) and 3.59 inches (with $R = 2.40$ inches). Compute the revised upper and lower control limits.

15. The following table shows the measurements of length of valves manufactured by a process for 20 samples of 5 valves each. The measurements are expressed in units of 0.001 inches in excess of 0.500 inches. For example, the length of the first valve of the first sample is actually 0.507 inches. Compute and plot the trial upper and lower control limits for (a) the $\bar{X}$ chart (use the $\bar{R}$ method), and (b) the R chart. Explain if the control limits should be extended for checking the quality of future production.

MEASUREMENTS OF LENGTH OF VALVES FOR 20 SAMPLES
(VALUES ARE EXPRESSED IN UNITS OF 0.001 INCHES
IN EXCESS OF 0.500 INCHES)

Sample Number	Measurement on Each Value, X (5 valves in each sample)					Total Measurement of 5 values, $\sum X$
1	7	20	32	25	41	125
2	35	60	18	46	41	200
3	56	57	22	53	62	250
4	10	27	42	38	33	150
5	6	18	29	43	4	100
6	16	48	56	47	58	225
7	12	27	45	49	17	150
8	57	42	43	39	19	200
9	38	23	16	34	14	125
10	5	14	28	26	27	100
11	17	65	67	61	65	275
12	56	52	14	42	36	200
13	48	32	22	7	16	125
14	58	60	46	47	14	225
15	45	11	35	18	16	125
16	54	53	63	23	32	225
17	63	53	59	54	21	250
18	10	12	18	14	46	100
19	6	10	40	11	8	75
20	66	22	63	59	65	275
Total						3,500

16. Refer to the $\bar{X}$ and R charts obtained in Problem 15. Compute the $\bar{X}$ and R values of the three subsequent samples of 5 items each and plot them on the charts respectively.

Sample 21: $X = 23, 59, 45, 40, 43$
Sample 22: $X = 4, 78, 86, 53, 29$
Sample 23: $X = 64, 65, 58, 60, 63$

Is the process under control in each case? If not, what should the plant manager do?

17. The following table shows the number of defectives for 16 samples of 500 items each. Compute the required values for constructing (a) the p chart, and (b) the np chart. Explain the difference of the two charts. Is the process under control according to the two charts?

NUMBER OF DEFECTIVES OF 16 SAMPLES OF 500 ITEMS

Sample Number	Number of Items Inspected	Number of Defectives
1	500	15
2	500	30
3	500	35
4	500	10
5	500	15
6	500	20
7	500	35
8	500	5
9	500	10
10	500	40
11	500	15
12	500	15
13	500	35
14	500	45
15	500	40
16	500	35
Total	8,000	400

18. Refer to Example 7, page 409. Construct a c chart.

19. A shipment of 3000 items is received by a purchaser. If he uses the double sampling plan of Table 16–8 to inspect the shipment, what is the size of the first sample? If he finds 7 defectives in the sample, what should he do then?

20. If the seller of the items stated in Problem 19 knows that the shipment has 10 percent defectives, what is the chance that the shipment is expected to be accepted?

part 5

Other Tests of

Hypotheses and

Decision Theory

This part has two chapters presenting other tests of hypotheses. Chi-square (χ^2) tests and various nonparametric tests are presented in Chapter 17. Tests of hypotheses by the use of the F distribution are introduced in Chapter 18. The remaining two chapters (19 and 20) explain the decision theory, which is also referred to as Bayesian statistics.

17 Chi-Square Distribution and Nonparametric Tests

During the process of testing hypotheses as discussed in Chapter 15, we first made some assumptions about a population from which a sample was drawn. For example, the population was usually assumed as a normal distribution; and the population parameters, such as the population mean and the population variance, were used either directly or indirectly in the tests.

However, there are situations in which it is difficult or impossible for us to make some or all of the required assumptions about the shape of distribution of the population. *Nonparametric*, or *distribution-free*, methods thus have been developed. The methods of testing hypotheses employed in the area of nonparametric statistics are not concerned with population parameters. This provides a distinct advantage since we do not have to know the shape of the population (either distributed normally or not normally), or make assumptions about the population for the purpose of estimating population parameters.

Other advantages of nonparametric techniques are the relative ease of computation and understanding. Also, they are useful with ranked information (see rank correlations in Chapter 28) and are applicable to tests involving small samples. However, their applications are limited to certain types of information, and they are less precise and efficient than parametric methods.

This chapter first presents the chi-square (χ^2) distribution (17.1) and the methods of testing hypotheses based on the χ^2 distribution (17.2). Next, it introduces other selected nonparametric tests which have become more popular in recent years. The other tests are the sign test (17.3), the rank-sum tests (17.4), and the test of randomness (17.5).

17.1 Chi-Square Distribution

The chi-square, denoted by the Greek letter χ^2, is frequently used in testing a hypothesis concerning the difference between a set of observed frequencies of a sample and a corresponding set of expected or theoretical frequencies. A chi-square is a sample statistic. It is computed as follows:

$$\chi^2 = \Sigma \left[\frac{(O - E)^2}{E} \right] \tag{17-1}*$$

where O = observed frequency
E = expected or theoretical frequency

If we throw 8 coins, we may have 6 heads and 2 tails in the throw. However, the expected number of heads in the single throw is 4 (or $8 \times \frac{1}{2} = 4$ heads, $\frac{1}{2}$ being the probability of having a head in a single throw of a coin). The expected number of tails in the single throw is also 4 (or $8 \times \frac{1}{2} = 4$ tails, $\frac{1}{2}$ being the probability of having a tail in a single throw of a coin). The value of χ^2 in the single experiment is computed according to formula (17–1) as shown in Table 17–1 or written:

$$\chi^2 = \frac{(6 - 4)^2}{4} + \frac{(2 - 4)^2}{4} = 1 + 1 = 2$$

Likewise, if we have 1 head and 7 tails in the throw, the value of χ^2 is 4.5:

$$\chi^2 = \frac{(1 - 4)^2}{4} + \frac{(7 - 4)^2}{4} = 4.5$$

There are 9 possible outcomes in throwing 8 coins at a time, as shown in the first two columns of Table 17–2. The values of χ^2 of other possible outcomes in the table are computed in a similar manner.

---

*The value of chi-square may also be defined as

$$\chi^2 = \Sigma \left(\frac{X - \mu}{\sigma} \right)^2$$

where X variable is normally distributed with mean μ and standard deviation σ; or

$$\chi^2 = \Sigma \left(\frac{x}{\sigma} \right)^2 = \frac{\Sigma x^2}{\sigma^2}$$

where $x = X - \mu$ and $x/\sigma = z$, which is normally distributed with mean $= 0$ and standard deviation $= 1$.

The number of x^2 values in the above expression is the number of degrees of freedom, D. When $D = 1$, $\chi^2 = x^2/\sigma^2$. Let the subscripts s and f represent *success* and *failure* respectively. By using formula (10–3), then, formula (17–1) may be written:

$$\chi^2 = \Sigma \left[\frac{(O - E)^2}{E} \right] = \frac{(O_s - E_s)^2}{E_s} + \frac{(O_f - E_f)^2}{E_f}$$

$$= \frac{(O_s - nP)^2}{nP} + \frac{[(n - O_s) - n(1 - P)]^2}{n(1 - P)} = \frac{(O_s - nP)^2}{nP(1 - P)}$$

$$= \frac{(O_s - E_s)^2}{nPQ} = \frac{x^2}{\sigma^2}$$

Thus, if we have 6 heads (successes) and 2 tails (failures) in throwing 8 ($= n$) coins

$$\chi^2 = \frac{(O_s - E_s)^2}{nPQ} = \frac{(6 - 4)^2}{8 \times 1/2 \times 1/2} = \frac{4}{2} = 2$$

TABLE 17–1

COMPUTATION OF A CHI-SQUARE (χ^2) VALUE

Observed Frequency (O)	Expected Frequency (E)	$O - E$	$(O - E)^2$	$\dfrac{(O - E)^2}{E}$
6 (heads)	4 (heads)	2	4	$4/4 = 1$
2 (tails)	4 (tails)	−2	4	$4/4 = 1$
8 (sides)	8 (sides)	0		$\chi^2 = 2$

We may consider each outcome as a sample drawn from the infinite population. The relative frequencies of the possible samples can be obtained from Example 4, Chapter 10. Table 17–2 thus shows the sampling distribution of χ^2, or the distribution of values of χ^2 of the possible samples of the experiment of throwing 8 coins.

TABLE 17–2

SAMPLING DISTRIBUTION OF χ^2 FOR THROWING EIGHT COINS
(NUMBER OF DEGREES OF FREEDOM = 1)

Possible Outcomes		χ^2	Relative Frequency
(Number of Heads)	(Number of Tails)		
0	8	8.0	1/256
1	7	4.5	8/256
2	6	2.0	28/256
3	5	0.5	56/256
4	4	0.0	70/256
5	3	0.5	56/256
6	2	2.0	28/256
7	1	4.5	8/256
8	0	8.0	1/256
Total			$256/256 = 1$

When the like values of χ^2 are combined, the frequency distribution of χ^2 is arranged as shown in Table 17–3.

TABLE 17–3

SAMPLING DISTRIBUTION OF χ^2, TABLE 17–2 REARRANGED

χ^2	Relative Frequency	Cumulative Frequency
8.0	0.0078	0.0078
4.5	0.0625	0.0703
2.0	0.2188	0.2891
0.5	0.4375	0.7266
0.0	0.2734	1.0000
Total	1.0000	

For example, when $\chi^2 = 8$, the relative frequency $= 1/256 + 1/256 = 0.0078$. The relative frequency on each line is the probability of having the χ^2 value on the same line. Thus, the probability of having $\chi^2 = 8$ is only 0.0078 out of 1, or 78 out of 10,000 cases. Now, if we actually throw 8 coins and find that the χ^2 value of the single throw is 8, we may suspect whether all coins are fair. The sampling distribution of χ^2 is a discrete series. In practice, it may be conveniently approximated by a tabulated continuous series of χ^2 distribution. The discrete series is arranged in a continuous scale shown by class intervals in Table 17–4.

TABLE 17–4

COMPARISON OF SAMPLING DISTRIBUTION OF
χ^2 AND THE χ^2 DISTRIBUTION
(NUMBER OF DEGREES OF FREEDOM $= 1$)

	Relative frequency (or probability)	
Values of χ^2 (class intervals)	Sampling distribution of χ^2 (discrete series) (Table 17–3)	χ^2 distribution (continuous series) (Table 17–5)
0 and above	1.0000	1.00
1.074 and above	0.2891	0.30*
2.706 and above	0.0703	0.10
3.841 and above	0.0703	0.05
5.412 and above	0.0078	0.02*
6.635 and above	0.0078	0.01

*See Table 8 in the Appendix.

The table also shows the continuous χ^2 distribution obtained from Table 17–5 and Table 8 in the Appendix in the same class intervals. The approximation of the sampling distribution of χ^2 by the continuous χ^2 distribution is fairly good in most

TABLE 17–5

SELECTED VALUES OF χ^2 DISTRIBUTION
(FOR ONE-TAILED [RIGHT TAIL] TESTS)

Number of Degrees of Freedom (D)	Probability (or Area Under χ^2 Distribution Curve)			
	0.50	0.10	0.05	0.01
	Value of χ^2			
1	0.455	2.706	3.841	6.635
2	1.386	4.605	5.991	9.210
3	2.366	6.251	7.815	11.345
4	3.357	7.779	9.488	13.277
5	4.351	9.236	11.070	15.086
10	9.342	15.987	18.307	23.209
20	19.337	28.412	31.410	37.566
30	29.336	40.256	43.773	50.892

Note: For more detailed values of χ^2, see the χ^2 distribution table, Table 8 in the Appendix.

class intervals. If the sample size (or the number of coins) is increased, the approximation will be improved.

In general, the χ^2 distribution (Table 17–5) is used to approximate the sampling distribution of χ^2 when the expected frequency in each case is 5 or above. Note that the expected number of heads or tails in Table 17–2 is only 4. If 10 coins were thrown each time, the expected frequency of heads or tails would be 5 (or $10 \times \frac{1}{2} = 5$).

In each throw of eight coins, the total number of occurrences of heads and tails is known. If the number of heads is known, the number of tails is automatically determined. Thus, the number of degrees of freedom is 1. Likewise, if we use a 3-sided die in our experiment, we may obtain a sampling distribution of χ^2 with $D = 2$, and so on. Thus, there is a χ^2 distribution for every number of degrees of freedom. Chart 17–1 is constructed to show various χ^2 distribution curves corresponding to the numbers of degrees of freedom 2, 5, and 10. The Y scale represents the relative height of any point on each curve. The height, like an

CHART 17–1

CHI-SQUARE DISTRIBUTIONS FOR NUMBERS
OF DEGREES OF FREEDOM $= 2, 5,$ AND 10

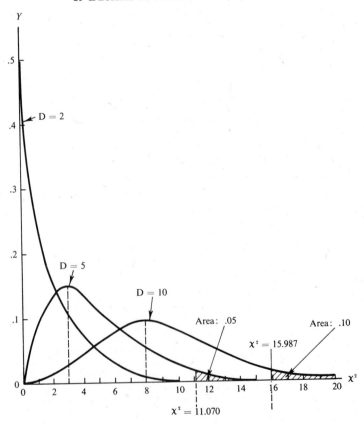

ordinate of a normal curve, may be computed by a formula.* The horizontal scale represents the values of χ^2. Some important properties of the χ^2 distribution curves are as follows:

(1) The mode of each distribution is equal to $D - 2$ on the χ^2 scale when D is equal to or larger than 2. For example, the maximum value of Y for the curve with $D = 5$ is at the point where $\chi^2 = 3$ (or $5 - 2 = 3$).

(2) The total area under each curve is 1 or 100%. The median of a χ^2 distribution divides the area into two equal parts, each part being 0.50 or 50%. The mean of a χ^2 distribution is equal to the number of degrees of freedom. For example, the median for curve with $D = 5$ is 4.351 as shown in Table 17–5 and the mean for the curve is 5.

(3) The curves show a fairly fast approach to symmetry as the number of degrees of freedom increases.

(4) The chi-square is obtained from squared numbers. Thus, it can never be negative. The smallest possible value for chi-square is 0, and the largest number is infinite.

17.2 Chi-Square Tests

In testing a hypothesis by using χ^2 distribution, we may determine whether the differences between the two sets of frequencies are significant, or whether the differences are too great to be attributable to sampling fluctuation. Observe the methods of computing the χ^2 values in Table 17–1. If $\chi^2 = 0$, the observed frequencies will agree exactly with the expected or theoretical frequencies. The larger the value of χ^2, the greater is the difference between observed and expected or theoretical frequencies.

The χ^2 test can be applied to many different types of problems. If a set of observed frequencies (or sample data) is obtained and a set of corresponding expected or theoretical frequencies can be derived, the χ^2 test may be applied. The most frequently used types of tests of hypotheses by χ^2 distributions are: (1) tests for goodness of fit and (2) tests for contingency tables. Both types of tests involve the comparison between observed frequencies and expected or theoretical frequencies.

Note that the selected values of χ^2 in Table 17–5 and in the more detailed χ^2 distribution table, Table 8 in the Appendix, are limited to the distributions with D from 1 to 30. The use of the χ^2 distribution table is analogous to that of the t-distribution table as discussed in the previous chapter, except that χ^2 Distribution Table is primarily designed for one-tailed tests.

$$* Y = \frac{1}{\left(\frac{D}{2} - 1\right)!} 2^{(-D/2)} \cdot (\chi^2)^{[(D/2)-1]} \cdot e^{-\chi^2/2}$$

where $D =$ the number of degrees of freedom, and
$e = 2.71828$, the value of the base of natural logarithms.

Tests for Goodness of Fit

The chi-square test can be used to determine whether a set of theoretical or expected frequencies, such as frequencies obtained from normal or binomial distribution or from other rational, uniform or ideal methods, fits a corresponding set of observed frequencies of a sample. The number of degrees of freedom for this type of test can be obtained as follows:

$$D = g - m$$

where g = the number of groups, or classes, or components of the observed or expected frequencies in a sample, and

m = the number of known constant values, which are used as constraints for finding the expected frequencies of the sample.

Example 1

Last year, the sales record of Edwards Automobile Company showed that the cars sold in Districts A, B, C, and D were 20%, 10%, 30%, and 40% respectively of the total cars sold by the Company. This year, the cars sold in the districts are: A—85; B—60; C—175; and D—180 cars. Does the sales distribution in this year differ significantly from that of last year at 0.05 level?

Solution. Testing the hypothesis:

(1) H_0: There is no difference between the distributions of the sales percents of the two years; that is, the sales percents of this year in the districts are the same as those of last year.

(2) Find the value of χ^2 by using formula (17–1). If the hypothesis is true, the sales percents of the districts in this year should be the same as the percents in the last year. Thus, the expected sales of this year are:

A: 500 (total sales) × 20% = 100 cars
B: 500 × 10% = 50 cars
C: 500 × 30% = 150 cars
D: 500 × 40% = 200 cars

The computation is arranged in Table 17–6 for the value of χ^2.

TABLE 17–6

COMPUTATION FOR EXAMPLE 1

| Districts | Cars Sold in This Year | | $O - E$ | $(O - E)^2$ | $\dfrac{(O - E)^2}{E}$ |
	Actual or Observed O	Expected E			
A	85	100	−15	225	2.250
B	60	50	10	100	2.000
C	175	150	25	625	4.167
D	180	200	−20	400	2.000
Total	500	500	0		$\chi^2 = 10.417$

$g = 4$, number of groups consisting of expected frequencies

$m = 1$, number of constant value (the total or $n = 500$ cars) used in estimating the expected frequencies

$D = 4 - 1 = 3$. [If any 3 expected frequencies from the 4 groups are freely selected, the frequency of the remaining group is automatically determined from the known total 500 cars.]

(3) Make a decision (see Chart 17–2). At a 0.05 level, the difference is significant if χ^2 with 3 degrees of freedom is above 7.815. The computed value of $\chi^2 = 10.417$ is larger than the critical value 7.815, or falls in the rejection region. Thus, we reject the hypothesis. In conclusion, the study shows that there has been a change in the distribution of sales percents in the districts in this year.

CHART 17–2

TEST OF HYPOTHESIS BY χ^2 DISTRIBUTION
$D = 3$, LEVEL OF SIGNIFICANCE $= .05$
(EXAMPLE 1)

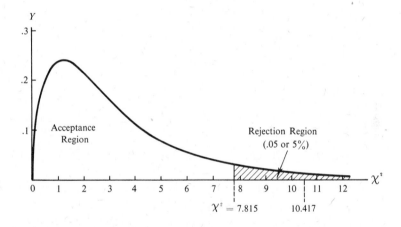

Example 2

A sample of the salaries earned by 25 employees in the Barnes Company during a given period is shown in columns (1) and (2) in Table 17–7. The theoretical number of employees [column (3)] for each class interval is obtained by using the normal curve as presented in Table 11–8, page 288. Determine whether or not the normal curve fits the sample data at a 0.05 level of significance.

Solution.   Testing the hypothesis by using χ^2 distribution:

(1) H_0: There is no difference between the actual distribution of the salaries among the employees of the sample and the theoretical distribution in the company.

(2) Find the value of χ^2 by using formula (17–1) as shown in Table 17–8.

Note: As a rule of safety for applying the chi-square distribution table, the expected frequency in each class should at least be 5. When there are small ex-

TABLE 17-7

DATA FOR EXAMPLE 2

Salaries (Class Interval)	Number of Employees (Actual)	Number of Employees (Theoretical)
Under $0.5	0	$\left.\begin{array}{l}0.25 \\ 1.26 \\ 3.93\end{array}\right\} = 5.44$
$1 − 3	1	
4 − 6	4	
7 − 9	9	6.86
10 − 12	6	7.03
13 − 15	2	$\left.\begin{array}{l}4.06 \\ 1.33 \\ 0.28\end{array}\right\} = 5.67$
16 − 18	3	
Above 18.5	0	
Total	25	25.00

TABLE 17-8

COMPUTATION FOR EXAMPLE 2

Salaries (Class Interval)	Number of Employees		$O - E$	$(O - E)^2$	$\dfrac{(O - E)^2}{E}$
	Actual or Observed O	Expected E			
Under $6.5	5	5.44	−0.44	0.1936	0.0356
$7 − 9	9	6.86	2.14	4.5796	0.6676
10 − 12	6	7.03	−1.03	1.0609	0.1509
Above 12.5	5	5.67	−0.67	0.4489	0.0792
Total	25	25.00	0.00		$\chi^2 = 0.9333$

pected frequencies in several classes, the classes should be combined to meet the requirement. Thus, the combined expected frequency for the three low-valued classes is $0.25 + 1.26 + 3.93 = 5.44$. The combined expected frequency for the three high-valued classes is $4.06 + 1.33 + 0.28 = 5.67$.

$g = 4$, number of classes consisting of expected frequencies

$m = 3$, number of constant values which are used to estimate the population parameters in order to derive the expected frequencies. They are sample statistics: $n = 25$, $\bar{X} = 9.56$, and $s = 3.9$ (see Table 11-8)

$D = 4 - 3 = 1$

(3) Make a decision. At a 0.05 level, the difference is significant if χ^2 with 1 degree of freedom is above 3.841, or falls in the rejection region. The computed value of $\chi^2 = 0.9333$ is less than 3.841, or falls in the acceptance region. Thus, we accept the hypothesis. There is no difference between the actual distribution of salaries among the employees and the theoretical distribution in the company.

Example 3

Assume that a television station claimed that 70% of the TV sets in a certain city were turned on for its special program on a Monday evening. A competitor wants to challenge the claim. He thus took a random sample of 200 families who were watching TV during that evening, and found that 130 sets were turned on for the special program. Can the competitor conclude that the claim was not valid if he let the level of significance be 0.05?

Solution.   Testing the hypothesis:

(1) H_0: The population proportion of TV sets turned on for the program is 70%.

(2) Find the value of χ^2 by using formula (17–1). If the hypothesis is true, the expected number of families whose TV sets were turned on for the special program is

$$200 \times 70\% = 140$$

and the expected number of families who were watching TV but who were not watching the special program is

$$200 \times (100\% - 70\%) = 60$$

The computation of χ^2 is arranged in Table 17–9.

TABLE 17–9

COMPUTATION FOR EXAMPLE 3

Groups	Families		$O - E$	$(O - E)^2$	$\dfrac{(O - E)^2}{E}$
	Observed or Actual O	Expected E			
For the program	130	140	−10	100	0.714
Not for the program	70	60	10	100	1.667
Total	200	200	0		$\chi^2 = 2.381$

$g = 2$, number of groups consisting of expected frequencies

$m = 1$, number of constant value (the total or 200 families) required in estimating the expected frequencies

$D = 2 - 1 = 1$

(3) Make a decision. At a 0.05 level, the difference is significant if χ^2 with 1 degree of freedom is above 3.841, or falls in the rejection region. The computed value of $\chi^2 = 2.381$ is less than the critical value 3.841, or falls in the acceptance region. We thus accept the hypothesis that there are 70% of TV sets turned on for the special program.

Note: This problem is restated here from Example 6 of Chapter 15. $\chi^2 = 2.381$ obtained above is equal to the square of the value of $z = -1.54$ obtained on that page. In general, in a χ^2 test *involving only two categories*, such as yes and no, for the program and not for the program, a rejection of a hypothesis at the 0.05 level of significance by using χ^2 distribution is equivalent to a rejection by using z distribution for a two-tailed test at the 0.05 level or for a one-tailed test at the 0.025 level. It can be proved that when the value of $z = +1.96$ or -1.96 is obtained from the data of this type of problem, the value of $\chi^2 = 3.841$ can also be obtained from the same data. Again, observe the relationship between the values of χ^2 and z:

$$\chi^2 = 3.841 = (\pm 1.96)^2 = z^2$$

Contingency Tables—Tests of Independence

A contingency table is a cross-classified table showing observed frequencies of a sample. When there are r rows and c columns in the table, it is called an $r \times c$ contingency table. For example, a 2×2 contingency table has two rows and two columns. There are four cells in a 2×2 table. The frequencies in the cells are called *cell frequencies*. The total of the frequencies in each row or each column is called the *marginal frequency*.

In testing a hypothesis involving a contingency table, corresponding expected or theoretical cell frequencies are first computed according to the hypothesis based on the rules of probability. The sum of all expected frequencies must be equal to the sum of all observed frequencies.

Next, Formula (17-1) is applied to compute the value of χ^2. The χ^2 distribution table is used in the usual manner, provided each expected cell frequency is not too small—preferably five or above.

The number of degrees of freedom of an $r \times c$ contingency table is $(r - 1)(c - 1)$. There are r values in each column of the table. When any $(r - 1)$ of the r values are assigned, the remaining 1 value is automatically determined. Similarly, there are c values in each row of the table. When any $(c - 1)$ of the c values are assigned, the remaining 1 value is automatically determined. Thus, for a 2×2 contingency table, there is $(2 - 1)(2 - 1) = 1$ degree of freedom. When the expected frequency in any one of the four cells is known, the frequencies in other three cells are automatically determined by the known row and column totals (the marginal frequencies). Likewise, for a 3×4 contingency table, there are $(3 - 1)(4 - 1) = 2 \times 3 = 6$ degrees of freedom; and so on.

Contingency tables are frequently used in tests of independence. This type of test will tell us whether or not the two bases of classification used respectively in rows and columns of a contingency table are independent (or not related). Examples 4 and 5 are used to illustrate the tests.

Example 4

Assume that we are interested in the relationship between adults who wear glasses and their educational levels in a city. A sample of 100 adults is taken. The results are shown in Table 17–10.

Find the answer if the level of significance is at 0.05.

TABLE 17–10

DATA FOR EXAMPLE 4

Education level	Adults wearing glasses	Adults not wearing glasses	Total (marginal frequency)
Elementary	7	3	10
High school	14	16	30
College	39	21	60
Totals (marginal frequency)	60	40	100

Solution. Testing the hypothesis:

(1) H_0: There is no relationship between wearing glasses and educational levels of the adults in the city; that is, wearing glasses and educational levels are *independent*.

(2) Find the value of χ^2 by using formula (17–1). First, compute the expected

TABLE 17–11

COMPUTATION OF THE EXPECTED CELL FREQUENCIES

(EXAMPLE 4)

Educational Level	Expected Numbers of Adults		Totals
	Wearing Glasses	Not Wearing Glasses	
Elementary	$60 \times (10/100) = 6$	$40 \times (10/100) = 4$	10
High School	$60 \times (30/100) = 18$	$40 \times (30/100) = 12$	30
College	$60 \times (60/100) = 36$	$40 \times (60/100) = 24$	60
Totals	60	40	100

TABLE 17–12

COMPUTATION OF THE VALUE OF χ^2

(EXAMPLE 4)

Types of Adults*	Numbers of Adults		$O - E$	$(O - E)^2$	$\dfrac{(O - E)^2}{E}$
	Observed O	Expected E			
Elementary, G	7	6	1	1	0.1667
Elementary, NG	3	4	−1	1	0.2500
High School, G	14	18	−4	16	0.8889
High School, NG	16	12	4	16	1.3333
College, G	39	36	3	9	0.2500
College, NG	21	24	−3	9	0.3750
Totals	100	100	0		$\chi^2 = 3.2639$

*G = wearing glasses
NG = not wearing glasses

cell frequencies. The expected frequencies are computed on the basis of the hypothesis. If the hypothesis is true that educational levels have nothing to do with wearing glasses, the same ratio of educational levels should be applied to both types of adults, wearing or not wearing glasses. Thus, 10 out of every 100 adults in the city should have had elementary education, 30 out of every 100 adults should have had high school education, and 60 out of every 100 adults should have had college education. The computation of the expected cell frequencies is shown in Table 17–11.

Next, arrange the corresponding observed and theoretical frequencies in an orderly form, then compute the χ^2 value as shown in Table 17–12.

$$D = (3 - 1)(2 - 1) = 2$$

Check: (Use the 3 × 2 table below.)

			Totals
			10
			30
			60
Totals	60	40	100

If the expected frequencies in 2 [= 2 (or 3 − 1 values in a column) × 1 (or 2 − 1 values in a row)] of the 6 (= 3 × 2) cells are filled, the frequencies of the remaining 4 cells are automatically determined from the known totals (marginal frequencies).

(3) Make a decision. (See Chart 17–3). At a 0.05 level, the difference is significant if χ^2 with 2 degrees of freedom is above 5.991. The computed value of $\chi^2 = 3.2639$ is smaller than the critical value, or falls in the acceptance region.

CHART 17–3

TEST OF HYPOTHESIS BY χ^2 DISTRIBUTION
$D = 2$, LEVEL OF SIGNIFICANCE = .05
(EXAMPLE 4)

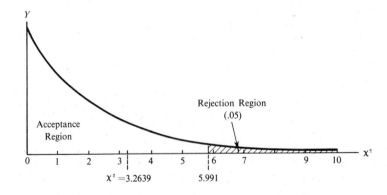

Thus, we accept the hypothesis. The conclusion is that there is no relationship between wearing glasses and educational levels of the adults in the city. Note that the basis of classifications used in columns is wearing glasses and that used in rows is educational level. The test shows that the two bases are independent (or not related). In other words, it is not necessary that a person who wears glasses has higher (or lower) education.

Example 5

A sample of 500 voters taken from City A showed that 59.6% of the voters were in favor of a given candidate for a state office; whereas a sample of 300 voters taken from City B showed that 50% of the voters were in favor of the candidate. Assume that the candidate made a speech in City A but not in City B. Is there a real difference between the opinions of the two cities concerning the candidate? Let the level of significance be 0.05.

Solution.   Testing the hypothesis:

(1) H_0: There is no relationship between speech making and voting in the two cities. In other words, speech making and actual voting are independent (or not related).

(2) Find the value of χ^2 by using formula (17–1). The actual cell frequencies are obtained from the given problem.

TABLE 17–13

COMPUTATION OF THE ACTUAL CELL FREQUENCIES
(EXAMPLE 5)

Speech making	Voters are in favor of the candidate	Voters are not in favor of the candidate	Totals
Made a speech (in city A)	(Cell 1) 298 (59.6% of 500)	(Cell 2) 202 (40.4% of 500)	500
Made no speech (in city B)	(Cell 3) 150 (50% of 300)	(Cell 4) 150 (50% of 300)	300
Totals	448 (56% of 800)	352 (44% of 800)	800

The expected cell frequencies are obtained according to the hypothesis that speech making has nothing to do with voting in the two cities. Thus, the percentages (56% in favor and 44% not in favor) applied to the total of the two samples (800 voters) should also be applied to each total of the two samples.

$$\text{Cell 1: } 500 \times 56\% = 280 \quad \text{Cell 3: } 300 \times 56\% = 168$$
$$\text{Cell 2: } 500 \times 44\% = 220 \quad \text{Cell 4: } 300 \times 44\% = 132$$
$$\text{Total} = 500 \qquad\qquad\qquad \text{Total} = 300$$

Now, arrange corresponding observed (actual) and expected (theoretical) frequencies in the orderly form and compute the χ^2 value.

TABLE 17–14

COMPUTATION OF THE VALUE OF χ^2
(EXAMPLE 5)

Speech Making	Voters		$O - E$	$(O - E)^2$	$\dfrac{(O - E)^2}{E}$
	Observed O	Expected E			
Made a speech (cell 1)	298	280	18	324	1.157
Made a speech (cell 2)	202	220	−18	324	1.473
Made no speech (cell 3)	150	168	−18	324	1.929
Made no speech (cell 4)	150	132	18	324	2.455
Total	800	800	0		$\chi^2 = 7.014$

$$D = (r - 1)(c - 1) = (2 - 1)(2 - 1) = 1$$

Check: (Use the 2 × 2 table below.)

			Totals
			500
			300
Totals	448	352	800

If the expected frequency of any one of the 4 ($=2 \times 2$) cells is filled, the frequencies of the remaining 3 cells are automatically determined.

(3) Make a decision. At a 0.05 level, the difference is significant if χ^2 with 1 degree of freedom is above 3.841. The computed value of $\chi^2 = 7.014$ is larger than the critical value, or falls in the rejection region. Thus, we reject the hypothesis. The conclusion is that there is a relationship between speech making and voting in the two cities. In other words, voting depends on speech making in the election.

Note: This problem is restated here from Example 7 of Chapter 15. $\chi^2 = 7.014$ obtained above is equal to the square of the value of $z = 2.67$ obtained in Example 7. Also see the note for Example 3. p. 428.

Yates' Correction for Continuity

When the continuous χ^2 distributions are applied to discrete distributions of χ^2, a correction for continuity called *Yates' correction*, which is analogous to that used in normal distribution for discrete data as shown in Chart 12–1, is available. The correction is made simply by reducing each *absolute* difference (disregarding + and − signs) between O and E by 0.5, or

$$\chi^2 = \Sigma \left[\frac{(|O - E| - 0.5)^2}{E} \right] \tag{17-2}$$

Example 6
Refer to Example 5. Use Yates' correction to find the answer.

Solution.

$$\chi^2 = \frac{(|298 - 280| - 0.5)^2}{280} + \frac{(|202 - 220| - 0.5)^2}{220}$$

$$+ \frac{(|150 - 168| - 0.5)^2}{168} + \frac{(|150 - 132| - 0.5)^2}{132}$$

$$= \frac{(17.5)^2}{280} + \frac{(17.5)^2}{220} + \frac{(17.5)^2}{168} + \frac{(17.5)^2}{132} = 6.629$$

The corrected χ^2 value leads to the same conclusion: rejection of the hypothesis (H_0 as stated in Example 5).

Yates' correction always reduces the value of χ^2. However, when each expected cell frequency is large, this correction will give a negligible amount of reduction to the value of χ^2. In general, Yates' correction is applied only when the number of degrees of freedom is equal to 1. If the uncorrected and the corrected χ^2 values give the same conclusion regarding a hypothesis, such as rejection of the hypothesis at the 0.05 level in the above two examples, there is no problem in decision-making. If the two values give different conclusions, the investigator should either increase the sample sizes or use the exact methods of computing probability.

17.3 The Sign Test

Frequently two sets of values are related and the individual values can be matched to pairs at random or according to a common object. The sign test is used to determine whether the differences between all paired values are significantly large. However, instead of using the actual magnitudes, only the simple plus (+) and minus (−) signs are used to show the differences in the test. If the difference between two values in a pair is zero, the pair is excluded and the number of the remaining pairs of values will be used as the sample size n in the computation.

The null hypothesis assumes that the two populations, from which the two samples (or two sets of values) to be analyzed are drawn respectively, are the same. If the hypothesis is true, the positive and negative differences of the paired values should be distributed evenly. In other words, the probability of having a + sign and the probability of having a − sign for each difference should be equal, or

$$P = P(\text{having a + sign}) = P(\text{having a − sign}) = \tfrac{1}{2}$$

Also, the total number of + signs and the total number of − signs in the population should be equal, or the expected value of the + signs (or − signs) is

$$E(+ \text{ or } - \text{ signs}) = nP, \text{ which is formula (10–3)}$$

The probability of having a desired X number of + or − signs from a sample size n with $P = \tfrac{1}{2} = 0.5$ can be computed by the binomial formula

$$P(X; n, P) = {}_nC_X \cdot P^X \cdot Q^{n-X} \tag{10–1}$$

or can be approximated by the normal curve with

$$\sigma = \sqrt{nPQ} \tag{10-4}$$

and

$$z = \frac{X - 0.5 - nP}{\sigma} \tag{17-3}$$

where 0.5 is the correction for continuity under the normal curve, and X is the number of *less frequent* signs.

Example 7

A cab company used 20 cars driven for two weeks, using brand A gasoline for the first week and brand B gasoline for the second week. A record of average miles per gallon by each brand was kept for each car as shown in Table 17–15. The manager of the company wishes to know whether or not there is a significant difference between the mileage yields of the two kinds of gasoline. Let the level of significance be 0.05.

TABLE 17–15

DATA AND COMPUTATION FOR EXAMPLE 7

Car number	Miles per gallon by using A brand (1)	Miles per gallon by using B brand (2)	d (1) − (2)	Sign of d
1	15	18	− 3	−
2	13	12	+ 1	+
3	14	16	− 2	−
4	18	22	− 4	−
5	19	24	− 5	−
6	12	18	− 6	−
7	20	13	+ 7	+
8	16	13	+ 3	+
9	15	23	− 8	−
10	21	21	0	excluded
11	18	27	− 9	−
12	25	15	+10	+
13	23	11	+12	+
14	11	24	−13	−
15	12	27	−15	−
16	12	26	−14	−
17	20	20	0	excluded
18	16	11	+ 5	+
19	28	12	+16	+
20	13	24	−11	−

Total: 7 + signs
$\underline{11 - \text{signs}}$
18 is the sample size n

Solution. (1) H_0: There is no difference between the two kinds of gasoline in mileage yields.

(2) Find z by using formula (17–3).

Here X = number of the *less frequent* sign = 7(+ signs)

n = total number of + and − signs = 18

$P = .5$

$$\sigma = \sqrt{nPQ} = \sqrt{18(0.5)(1 - 0.5)} = \sqrt{4.5} = 2.12$$

$$z = \frac{7 - 0.5 - 18(0.5)}{2.12} = \frac{-2.5}{2.12} = -1.18$$

(3) Make a decision based on the level of significance. Here we are interested in testing whether or not the number of less frequent signs (X or 7 + signs) is significantly different from (or smaller than) the expected number of the signs (or $nP = 18(.5) = 9$). Thus, a one-tailed (the left tail) test is used in locating the critical value of z. At 0.05 level of significance, the critical value of $z = -1.645$. The obtained value of $z = -1.18$ falls in the acceptance region. Thus, we accept the hypothesis. Brand A gasoline is equally good as brand B gasoline in mileage yields.

Notes:

(1) The normal curve approximation to the binomial probability distribution is adequate for n as small as 10. This is due to the symmetrical binomial distribution when $P = 0.5$. (See the curve on Chart 10–3, page 245, where $n = 8$ and $P = 0.5$.) However, if a more accurate answer is desired, the binomial formula should be used. The computation is as follows:

$$X = 7, \text{ or } X = 7, 6, 5, 4, 3, 2, 1, 0; n = 18; P = 0.5$$

$$P(X \le 7; 18, 0.5) = .1214 + .0708 + .0327 + .0117 + .0031$$
$$+ .0006 + .0001 + .0000$$
$$= .2404 \text{ (see the binomial table, Table 3 in}$$
$$\text{the Appendix)}$$

which is more than the desired level of .05. Thus, we accept the hypothesis.

(2) Section 15.3 indicated that in testing the difference between two sample means, the t test should not be used when the sample sizes are small unless we can assume that the populations from which the samples are drawn are normally distributed and their population variances are the same. If the two assumptions cannot be made, the sign test can be used for testing the hypothesis that the two population means are equal.

(3) The sample size n should not be too small in using the sign test. At 5% level for a one-tailed test, for example, the sample size should be at least 5. If it is 4, we can never reject the null hypothesis at the 5% level. Because, even in the extreme case, where $X = 0$, or all differences are either + or −,

$$P(X = 0 \text{ or less}) = P(0; 4, .5) = 0.5^4 = 0.0625$$

which is more than the desired level of .05. Thus, we have to accept the hypothesis that the total number of + signs and the total number of − signs are equal.

(4) The following formula can also be used for finding the critical values of X for the sign test:

Let X' be the critical value,

$$X' = \frac{n-1}{2} - k\sqrt{n+1}$$

where k is 1.2879, 0.9800, and 0.8224 for the two-tailed test at 1%, 5%, and 10% level of significance respectively. Reject H_0 if $X \leq X'$, but accept H_0 if $X > X'$ for the sign test. In Example 7, where the level of significance is 5% for the one-tailed test or 10% for the two-tailed test,

$$X' = \frac{18-1}{2} - 0.8224\sqrt{18+1} = 8.5 - 3.58 = 4.92$$

Accept the H_0 since $X = 7$, which is larger than X'.

17.4 Rank-Sum Tests

There are various ways of making rank-sum tests. In this section, only the basic ones are discussed: (1) the Wilcoxon test for matched pairs, (2) the Mann-Whitney (U) test, and (3) the Kruskal-Wallis (H) test.

The Wilcoxon Test for Matched Pairs

The *Wilcoxon test* is also called the *signed-rank* test. It is more detailed in procedure and more powerful than the sign test. In addition to the $+$ and $-$ signs of the paired values, the Wilcoxon test considers the magnitudes of the differences.

Let n be the number of paired values and d be the difference of each pair of values. First, the n differences are ranked according to their absolute values ($|d|$, or disregarding $+$ and $-$ signs of d) from the smallest (ranked number 1) to the largest (ranked number n). If two or more differences are tied in ranks, each difference is given the mean of the ranks. Thus, if two differences are tied in the 3rd and the 4th ranks, each difference is ranked as $3.5 = (3 + 4)/2$. Next, each rank is prefixed the original sign of the difference represented. The sum of all positive ranks and the sum of all negative ranks are then computed.

The null hypothesis again assumes that the two populations, from which the two samples (or two sets of values) to be analyzed are drawn respectively, are the same. Thus, the expected rank sums for the positive and negative differences should be equal, or

$$\left(\begin{array}{c}\text{Sum of positive}\\ \text{ranks}\end{array}\right) = \left(\begin{array}{c}\text{Sum of negative}\\ \text{ranks}\end{array}\right) = \left(\frac{\text{Sum of all ranks}}{2}\right)$$

The sum of all ranks, 1, 2, 3, ..., n (an arithmetic progression) can be obtained by the formula

$$\frac{n}{2}(1+n)$$

If $n = 18$, the sum of all ranks 1 through 18 is

$$\frac{18}{2}(1 + 18) = 171$$

and the expected sum of either positive ranks or negative ranks is

$$\frac{171}{2} = 85.5$$

The purpose of the Wilcoxon test is to decide whether the difference between a computed signed-rank sum and the expected rank sum of the same sign is significantly large.

Let $T =$ the computed total of either the positive ranks or the negative ranks, whichever is smaller. The expected value of T is $\frac{1}{2}$ of the sum of all ranks, or

$$E(T) = \frac{n(1 + n)}{4} \tag{17-4}$$

When the sample size is large, preferably 10 or more, the sampling distribution of T is approximately normal. The standard error of the statistic T is

$$\sigma_T = \sqrt{\frac{n(n + 1)(2n + 1)}{24}} \tag{17-5}$$

and the standard normal deviate z is

$$z = \frac{T - E(T)}{\sigma_T} \tag{17-6}$$

Example 8

Refer to Example 7. Find the answer by using the Wilcoxon test.

Solution. (1) H_0: There is no difference between the two kinds of gasoline in mileage yields.

(2) Find z.

$$T = +64 \qquad \text{(obtained from Table 17-16)}$$

$$E(T) = \frac{n(1 + n)}{4} = \frac{18(1 + 18)}{4} = \frac{171}{2} = 85.5 \qquad \text{(Formula 17-4)}$$

$$\sigma_T = \sqrt{\frac{n(n + 1)(2n + 1)}{24}} = \sqrt{\frac{18(18 + 1)(2 \times 18 + 1)}{24}}$$

$$= \sqrt{527.25} = 22.96 \qquad \text{(Formula 17-5)}$$

$$z = \frac{64 - 85.5}{22.96} = -.94 \qquad \text{(Formula 17-6)}$$

(3) Make a decision based on the level of significance. At 0.05 level for a one-tailed (the left tail) test, the critical value of $z = -1.645$. The computed value of $z = -.94$ falls in the acceptance region. Thus, we accept the hypothesis, the same conclusion as Example 7.

TABLE 17–16

DATA AND COMPUTATION FOR EXAMPLE 8
(BASED ON TABLE 17–15)

Car number	Miles per gallon by A (1)	Miles per gallon by B (2)	d (1) – (2)	Rank of |d|	Signed rank	
					Positive rank	Negative rank
1	15	18	– 3	3.5		– 3.5
2	13	12	+ 1	1	+ 1	
3	14	16	– 2	2		– 2
4	18	22	– 4	5		– 5
5	19	24	– 5	6.5		– 6.5
6	12	18	– 6	8		– 8
7	20	13	+ 7	9	+ 9	
8	16	13	+ 3	3.5	+ 3.5	
9	15	23	– 8	10		–10
10	21	21	0	excluded		
11	18	27	– 9	11		–11
12	25	15	+10	12	+12	
13	23	11	+12	14	+14	
14	11	24	–13	15		–15
15	12	27	–15	17		–17
16	12	26	–14	16		–16
17	20	20	0	excluded		
18	16	11	+ 5	6.5	+ 6.5	
19	28	12	+16	18	+18	
20	13	24	–11	13		–13
Total: 7 + signs 11 – signs 18 signs = sample size n				171	+64 = T, smaller total	–107

The Mann-Whitney (U) Test

The *Mann-Whitney test*, also called the U test, is another type of rank-sum test. This test can be used to determine whether two independent samples are drawn from identical populations or from two populations with the same mean.

   Let n_1 = the number of items in the first sample, A; and
      n_2 = the number of items in the second sample, B.

   Next, arrange all $(n_1 + n_2)$ items into one group according to their magnitudes and rank them. If two or more items of different samples are tied in ranks, each item is given the mean of the ranks.

   Then, the sum of cumulative items of the first sample, denoted by the letter U, can be obtained by two methods:

   (1) Count the number of A items that precedes each B item. The sum of the numbers counted is the value of U.

(2) Use formula

$$U = n_1 n_2 + \frac{n_1(n_1 + 1)}{2} - R_1 \qquad (17\text{--}7)$$

where R_1 = total of ranks of A items.

Thus, if sample A has 4 items (60, 64, 72, and 78) and sample B has 5 items (62, 67, 69, 72, and 74), or $n_1 = 4$ and $n_2 = 5$, the value of U for the combined sample of 9 items is 10.5 by Method (1) as shown in Table 17–17.

TABLE 17–17

COMPUTING U VALUE

Items arranged	Rank	Method (1) Number of A items precedes each B*	Method (2) Rank of A items
60A	1		1
62B	2	1	
64A	3		3
67B	4	2*	
69B	5	2	
72A	6.5		6.5
72B	6.5	2.5*	
74B	8	3	
78A	9		9
Total		10.5 = U	19.5 = R_1

*For example, the 2 A items before 67B are 60A and 64A. The 2.5 items before 72B are 60A, 64A, and 72A (72A is counted as $\frac{1}{2}$ item since it ties with 72B.)

The value of U is computed by using Method (2) as follows: Substituting $n_1 = 4$, $n_2 = 5$, and $R_1 = 19.5$ into formula (17–7),

$$U = (4)(5) + \frac{4(4 + 1)}{2} - 19.5 = 10.5$$

Method (1) explains the meaning of U value while Method (2) gives a simpler way to compute U. In the following discussion, Method (2) is used.

When the sample sizes are large, preferably both n_1 and n_2 larger than 10, the sampling distribution of U is approximately normal. The expected value of U is

$$E(U) = \frac{n_1 n_2}{2} \qquad (17\text{--}8)$$

the standard error of the statistic U is

$$\sigma_U = \sqrt{\frac{n_1 n_2 (n_1 + n_2 + 1)}{12}} \qquad (17\text{--}9)$$

and the standard normal deviate z is

$$z = \frac{U - E(U)}{\sigma_U} \qquad (17\text{--}10)$$

Example 9

An English instructor gave the same test to two classes. The first class, or Class A, has 12 students and the second class, or Class B, has 10 students. The grade points of the 22 students are ranked and listed in Table 17–18. The instructor wishes to know whether there is a significant difference between the mean grades of the two classes. Let the level of significance be 0.05.

TABLE 17–18

DATA AND COMPUTATION FOR EXAMPLE 9

Student number	Class A		Class B	
	Grade	Rank	Grade	Rank
1	60	1	62	2
2	64	3	67	4
3	72	6.5	69	5
4	78	9	72	6.5
5	80	10	74	8
6	88	14	82	11
7	88	15	84	12
8	91	17	86	13
9	93	18	90	16
10	96	20	95	19
11	97	21		
12	100	22		
Total	1,007	$156.5 = R_1$	781	
Mean	$\frac{1,007}{12} = 84$		$\frac{781}{10} = 78$	

Solution. (1) H_0: The two population means are the same. There is no difference between the two class means.

(2) Find z.

$$U = (12)(10) + \frac{12(12 + 1)}{2} - 156.5 = 41.5$$

(Formula 17–7 and Table 17–18)

$$E(U) = \frac{n_1 n_2}{2} = \frac{(12)(10)}{2} = 60$$

(Formula 17–8)

$$\sigma_U = \sqrt{\frac{(12)(10)(12 + 10 + 1)}{12}} = \sqrt{230} = 15.16$$

(Formula 17–9)

$$z = \frac{41.5 - 60}{15.16} = -1.18$$

(Formula 17–10)

(3) Make a decision. Here we are interested in testing whether or not the sum U obtained from the ranks of the items of the first sample is significantly *larger* or *smaller* than the expected sum $E(U)$. Thus, a two-tailed test is used in locating

the critical value of z. At a 0.05 level, the difference is significant if z falls outside of the range ± 1.96. The computed z value is -1.18, which is inside of the range. Thus, accept the hypothesis.

Compare the nonparametric test method used here for Example 9 with the parametric test methods used for Examples 4 and 5 in Section 15.4. The nonparametric test method requires no assumptions about the populations from which the samples are drawn and offers much simpler computation.

The Kruskal-Wallis (H) Test

The *Kruskal-Wallis test* is also a type of rank-sum test. This test can be used to determine whether k independent samples are drawn from identical populations or from k populations with the same mean. The nonparametric test may be used to substitute for the method of one-way analysis of variance to be presented in the next chapter.

> Let n_i = the number of items in the ith sample,
> $i = 1, 2, 3, \ldots, k$,
> n = the number of all items in the combined large sample
> $= n_1 + n_2 + n_3 + \cdots + n_k$, and
> R_i = the total of ranks of n_i items in the ith sample after the k samples are combined into one large sample and all items are ranked according to their magnitudes. The method of computing R_i is similar to that of computing R_1 in the Mann-Whitney test.

Then, the statistic H can be used in the test,

$$H = \frac{12}{n(n+1)} \cdot \Sigma \left[\frac{R_i^2}{n_i} \right] - 3(n+1) \qquad (17\text{--}11)$$

The Kruskal-Wallis test thus is also called the H test. If the null hypothesis that k samples are drawn from identical populations is true and each sample size is five or more, the sampling distribution of the statistic H can be approximated by the χ^2 distribution with D (degrees of freedom) $= k - 1$. Thus the χ^2 distribution table in the appendix can be used in the H test process.

Example 10

Assume that a young man took a sample of the wages of 13 workers in a large city during a given period. He classified the wages into three classes according to occupations: electricians, carpenters, and painters as shown in Table 17–19. Use the H test to determine whether or not the means of the wages classified by the three different occupations are significantly different. Let the level of significance be 0.05.

Solution. (1) H_0: The population means of the wages of the three different occupations are the same.

TABLE 17–19

DATA AND COMPUTATION FOR EXAMPLE 10

Sample item (worker)	Sample 1 (electricians)		Sample 2 (carpenters)		Sample 3 (painters)	
	Wages	Rank	Wages	Rank	Wages	Rank
1	$65	5	$72	7.5	$52	1
2	69	6	74	9.5	53	2
3	72	7.5	75	11	55	3
4	74	9.5	76	12	56	4
5			78	13		
Total	$280	$28.0 = R_1$	$375	$53.0 = R_2$	$216	$10 = R_3$
Mean	$70		$75		$54	

(2) Find H. $n_1 = 4$, $n_2 = 5$, $n_3 = 4$, $n = 4 + 5 + 4 = 13$, $R_1 = 28$, $R_2 = 53$, and $R_3 = 10$. (See Table 17–19.)

Substitute the values into formula (17–11),

$$H = \frac{12}{13(13 + 1)} \left(\frac{28^2}{4} + \frac{53^2}{5} + \frac{10^2}{4}\right) - 3(13 + 1) = 9.61$$

Note: The sum of all ranks 1 through 13 is

$$R_1 + R_2 + R_3 = 28 + 53 + 10 = 91$$

which can be checked by using formula

$$\frac{n}{2}(1 + n) = \frac{13}{2}(1 + 13) = 91$$

(3) Make a decision. At a 0.05 level, the difference is significant if χ^2 with $D = k - 1 = 3 - 1 = 2$ degrees of freedom is above 5.991. The computed $H = 9.61$ is larger than the critical value of 5.991. Thus, we reject the hypothesis. We may state that the means of the wages of the three different occupations in the populations are not equal. The three populations are different ones. This answer is the same as the answer to Example 6 on page 461. Of course, if each sample size is five or more, the approximation to H value by χ^2 distribution would be closer. Here, the sample sizes are four, five, and four respectively in the test.

17.5 Test of Randomness

In the previous discussion, we frequently made the assumption that a sample was drawn *randomly* from a population. However, we could not be certain if the assumption of randomness was a reasonable one. This section will introduce the methods of testing the randomness based on the theory of runs.

A *run* is a sequence of identical expressions uninterrupted by other kind of expression in a sample. Thus, if a sample consists of the series of expressions "$+ + + - - + + + +$," it has three runs. The first run has three $+$'s, the second run has two $-$'s, and the third run has four $+$'s.

Let $n_1 =$ the number of expressions of one kind, such as 7 $+$'s in the above illustration,

$n_2 =$ the number of expressions of the other kind, such as 2 $-$'s in the above illustration, and

$R =$ the number of runs in a given sample, such as 3 runs above.

The expected value of R, or $E(R)$, and the standard error of the sampling distribution of the statistic R, or σ_R, may be computed under two cases.

Case I. $n_1 \neq n_2$ (General Case)
This is the general case in a sample: the number of expressions of one kind is usually not equal to that of another kind.

When $n_1 \neq n_2$, the expected value of runs is

$$E(R) = \frac{2n_1 n_2}{n_1 + n_2} + 1 \tag{17–12a}$$

the standard error of the statistic R is

$$\sigma_R = \sqrt{\frac{2n_1 n_2(2n_1 n_2 - n_1 - n_2)}{(n_1 + n_2)^2(n_1 + n_2 - 1)}} \tag{17–13a}$$

and the standard normal deviate is

$$z = \frac{R - E(R)}{\sigma_R} \tag{17–14}$$

When the sample sizes are large, preferably both n_1 and n_2 are larger than 10, the sampling distribution of the statistic R is approximately normal. Thus, the normal curve can be used in the tests of randomness.

Example 11
Suppose that a sample of 25 customers was taken in a large department store. The question asked each of the customers was "Did you buy anything in this store today?" Let $Y =$ yes and $N =$ no. The 25 answers were: (arranged in order of time asked)

YY   NNN   YYYY   NN   YYYYY   NNNNN   YYYY

Use the test of randomness to determine whether this sample was taken at random. Let the level of significance be 0.05.

Solution.   (1) H_0: The yes-customers and no-customers are coming to the store at random. There is no difference between the number of runs obtained in the sample and the expected number of runs.

(2) Find z. Here $n_1 = 15$ (yes), $n_2 = 10$ (no), and $R = 7$ (runs).

$$E(R) = \frac{2(15)(10)}{15 + 10} + 1 = 13 \qquad \text{(Formula 17–12a)}$$

$$\sigma_R = \sqrt{\frac{2(15)(10)[2(15)(10) - 15 - 10]}{(15 + 10)^2(15 + 10 - 1)}} = 2.35 \qquad \text{(Formula 17–13a)}$$

$$z = \frac{7 - 13}{2.35} = -2.55 \qquad \text{(Formula 17–14)}$$

(3) Make a decision. Here we are interested in testing whether or not the number of runs of the sample is significantly larger or smaller than the expected number of runs $E(R)$. Thus, a two-tailed test is used in locating the critical value of z. At a 0.05 level, the difference is significant if z falls outside the range ± 1.96. The computed z value is -2.55, which is outside of the range. Thus, reject the hypothesis. We suspect that definite grouping or clustering yes-customers and no-customers were coming to the store during the sampling period.

Case II. $n_1 = n_2$ (Runs Above and Below the Median)
This is a special case in a sample: the number of expressions of one kind is equal to that of another kind.

When a sample is a set of numerical values, a median of the values can be obtained. Let a and b denote the values falling *above* and *below* the median respectively. The values which are equal to the median of the sample are disregarded. Also, let $n_1 =$ the number of a's and $n_2 =$ the number of b's. Then, $n_1 = n_2$ since the median divided the values in the sample into two equal groups. The total number of runs above and below the median now can be used to test the randomness of the sample values based on the simplified formulas shown below.

Let $n = n_1 + n_2$. Then, $n_1 = n_2 = n/2$. Substitute $n/2$ for n_1 and n_2 into formulas (17–12a) and (17–13a). The two formulas can be simplified:
The expected value of runs is

$$E(R) = \frac{n}{2} + 1 \qquad \text{(17–12b)}$$

and the standard error of the statistic R is

$$\sigma_R = \sqrt{\frac{n(n - 2)}{4(n - 1)}} \qquad \text{(17–13b)}$$

Example 12
A grocery store manager took a sample of 30 sales tickets. The amounts of the tickets are: (arranged in order of time sold)

$16, 14, 17, 18, 13, 28, 39, 12, 11, 19, 8, 12, 10, 30, 10, 40, 2, 9, 25, 45, 7, 50, 63, 6, 5, 49, 20, 4, 1, 23.

The median is $15 (= (14 + 16)/2)$. Let a and b represent the values above and below $15 respectively. The sales tickets, again arrranged in order of time sold, are indicated by runs below:

$$a, b, aa, b, aa, bb, a, bbb, a, b, a, bb, aa, b, aa, bb, aa, bb, a$$

There is a total of 19 runs. Use the test of randomness to determine whether this sample was taken at random. Let the level of significance be 0.05.

Solution.   (1) H_0: The sales tickets are taken at random. There is no difference between the number of runs obtained in the sample and the expected number of runs.

(2) Find z. Here $n_1 = n_2 = 15$, $n = n_1 + n_2 = 15 + 15 = 30$, and $R = 19$.

$$E(R) = \frac{30}{2} + 1 = 16 \qquad\qquad\qquad \text{(Formula 17–12b)}$$

$$\sigma_R = \sqrt{\frac{30(30-2)}{4(30-1)}} = \sqrt{\frac{210}{29}} = \sqrt{7.24} = 2.69 \qquad \text{(Formula 17–13b)}$$

$$z = \frac{19 - 16}{2.69} = 1.12 \qquad\qquad\qquad \text{(Formula 17–14)}$$

(3) Make a decision. At a 0.05 level for a two-tailed test, the difference is significant if z falls outside the range ± 1.96. The computed z value is 1.12, which is inside the range. Thus, accept the hypothesis. We may state that the sample was taken at random.

17.6 Summary

The methods of testing hypotheses employed in the area of nonparametric statistics do not concern population parameters. Thus we do not have to know the shape of the population (either distributed normally or not normally) or make assumptions about the population for the purpose of estimating population parameters. There are many types of nonparametric tests. Only selected types are introduced here.

The chi-square, denoted by the Greek letter χ^2, is frequently used in testing a hypothesis concerning the difference between a set of observed values of a sample and a corresponding set of expected or theoretical values. The χ^2 is a sample statistic. The χ^2 distribution is usually used to approximate the sampling distribution of χ^2 when the expected frequency in each class is 5 or more. Like t-distributions, there is a χ^2 distribution for every number of degrees of freedom. Unlike t-distributions, the χ^2 distributions are not symmetrical about the mean when the numbers of degrees of freedom are small.

There are two most frequently used types of tests of hypotheses by χ^2 distributions. Both types of tests involve the comparison between observed frequencies and expected or theoretical frequencies.

(1) Tests for goodness of fit. This type of test can determine whether a set of theoretical or expected values, such as values obtained from normal or binomial distribution or from other rational, uniform, or ideal methods, fits a corresponding set of observed values of a sample. The number of degrees of freedom for this type of test is the difference between the number of groups of the expected frequencies minus the number of known constants which are used for finding the expected frequencies.

(2) Tests for contingency tables. A contingency table is a cross-classified table showing observed frequencies of a sample. When there are r rows and c columns

SUMMARY OF FORMULAS

Application	Formula	Formula Number	Reference Page
Chi Square Tests			
Generally	$\chi^2 = \Sigma \left[\dfrac{(O - E)^2}{E} \right]$	17–1	419
Yates' correction	$\chi^2 = \Sigma \left[\dfrac{(\lvert O - E \rvert - 0.5)^2}{E} \right]$	17–2	432
The sign test	$P(X; n, P) = {}_nC_X \cdot P^X \cdot Q^{n-X}$	10–1	433
	$\sigma = \sqrt{nPQ}$	10–4	434
	$z = \dfrac{X - 0.5 - nP}{\sigma}$	17–3	434
The Wilcoxon test	$E(T) = \dfrac{n(1 + n)}{4}$	17–4	436
	$\sigma_T = \sqrt{\dfrac{n(n + 1)(2n + 1)}{24}}$	17–4	437
	$z = \dfrac{T - E(T)}{\sigma_T}$	17–6	437
The Mann-Whitney test	$U = n_1 n_2 + \dfrac{n_1(n_1 + 1)}{2} - R_1$	17–7	439
	$E(U) = \dfrac{n_1 n_2}{2}$	17–8	439
	$\sigma_U = \sqrt{\dfrac{n_1 n_2(n_1 + n_2 + 1)}{12}}$	17–9	439
	$z = \dfrac{U - E(U)}{\sigma_U}$	17–10	439
The Kruskal-Wallis test	$H = \dfrac{12}{n(n + 1)} \cdot \Sigma \left[\dfrac{R_i^2}{n_i} \right] - 3(n + 1)$	17–11	441
Test of Randomness			
$n_1 \neq n_2$	$E(R) = \dfrac{2n_1 n_2}{n_1 + n_2} + 1$	17–12a	443
$n_1 = n_2$	$E(R) = \dfrac{n}{2} + 1$	17–12b	444
$n_1 \neq n_2$	$\sigma_R = \sqrt{\dfrac{2n_1 n_2(2n_1 n_2 - n_1 - n_2)}{(n_1 + n_2)^2(n_1 + n_2 - 1)}}$	17–13a	443
$n_1 = n_2$	$\sigma_R = \sqrt{\dfrac{n(n - 2)}{4(n - 1)}}$	17–13b	444
either $n_1 \neq n_2$ or $n_1 = n_2$	$z = \dfrac{R - E(R)}{\sigma_R}$	17–14	443

in the table, it is called an $r \times c$ contingency table. The frequencies in the cells are called cell frequencies, and the total of the frequencies in each row or each column is called the marginal frequency. In testing a hypothesis involving a contingency table, corresponding expected or theoretical cell frequencies are first computed. Next, formula (17–1) is applied to compute the value of χ^2. The number of degrees of freedom of an $r \times c$ contingency table is $(r - 1)(c - 1)$. Contingency tables are frequently used in tests of independence. This type of test will tell us whether or not the two bases of classification used respectively in rows and columns of a contingency table are independent (or not related).

When the continuous χ^2 distributions are applied to discrete distribution of χ^2, a correction for continuity, called Yates' correction, can be used. The correction is made simply by reducing each absolute difference between O and E by 0.5.

Frequently two sets of values are related and the individual values can be matched to pairs according to a common object. The sign test is used to test whether or not the differences between all paired values are significantly large. However, instead of using the actual magnitudes, only the simple plus $(+)$ and minus $(-)$ signs are used to show the differences in the test.

Three types of rank-sum tests are introduced in this chapter. The Wilcoxon test, also called the signed-rank test, is more detailed in procedure and more powerful than the sign test. In addition to the $+$ and $-$ signs of the differences of the paired values as in the sign test, the Wilcoxon test considers the ranks of the magnitudes of the differences. The Mann-Whitney (U) test is another type of rank-sum test. This test can be used to determine whether two independent samples are drawn from identical populations or from two populations with the same mean. The third rank-sum test is the Kruskal-Wallis (H) test. This test can be used to determine whether k independent samples are drawn from identical populations or from k populations with the same mean.

The methods of testing the randomness introduced in this chapter are based on the theory of runs. A run is a sequence of identical expressions uninterrupted by an other kind of expression in a sample. The formulas used in the test of randomness and the formulas for the other tests presented in this chapter are summarized on the preceding page.

Exercises 17-1

Reference: Sections 17.1 and 17.2

1. Coin X was tossed 400 times; 214 heads and 186 tails were recorded. Coin Y was also tossed 400 times; 220 heads and 180 tails were recorded. Test the hypothesis that each coin is fair. Use a level of significance of 5%.

2. Eight coins were tossed 256 times. The results of the tosses and the theoretical occurrences of having heads are given below:

Number of heads	Theoretical occurrences	Observed occurrences
0	1	0
1	8	6
2	28	28
3	56	48
4	70	77
5	56	50
6	28	32
7	8	14
8	1	1
Total	256	256

The theoretical occurrences are obtained from the binomial distribution by assuming that the coins are fair (or $P = \frac{1}{2}$; see page 246). Use the chi-square test to determine whether the difference between the observed and theoretical (or expected) occurrences is too great to be attributed to chance. Use a level of significance of 5%. (Hint: the number of known constant value which is used as the constraint for finding the expected occurrences is 1, the *total* or 256 occurrences.)

3. The following information concerning grade distribution of the students in a college is obtained from the office of registration:

LAST YEAR		THIS YEAR	
Grades	Percent distribution	Grades	Number of students
A	15%	A	120
B	18%	B	190
C	30%	C	330
D	25%	D	270
F	12%	F	90
Total	100%	Total	1,000

Use the chi-square test to determine whether or not the grade distribution of this year differs significantly from that of last year at 0.01 level.

4. A sample of grades of 80 students in a statistics class are given in columns (1) and (2) of the following table. The theoretical number of students for each class shown in column (3) is obtained by using the normal curve (Problem 12, Exercises 11, page 294. Determine whether or not the normal curve fits the sample data at a 0.05 level of significance.

(1)	(2)	(3)
Grades (Class interval)	Number of students (actual)	Number of students (theoretical)
20–29	3	1
30–39	6	3
40–49	5	8
50–59	7	13
60–69	10	17
70–79	29	16
80–89	12	12
90–99	8	6
Above 99.5	0	4
Total	80	80

5. Work Problem 9 of Exercises 15, page 385. Use the chi-square test to determine whether or not the claim made by the manufacturer is valid at 0.10 level of significance.

6. The sales records of Palo Company indicated that Mr. Myers sold 652 times out of 1,000 calls and Mr. Logan sold 1,048 times out of 1,500 calls. Does the sales record made by Myers differ significantly from that by Logan? Use the chi-square test at (a) 0.05 and (b) 0.01 level of significance.

7. The following table shows the quality of a sample of 500 items produced by two different operators. Test the hypothesis that the quality distribution of the products is the same for both operators. Use the level of significance of 0.05.

Quality of product	Operator 1	Operator 2	Total
High	75	100	175
Medium	98	102	200
Low	47	78	125
Total items	220	280	500

8. Work Problem 11 of Exercises 15. (a) Use the chi-square to test the hypothesis that the teaching aid did not give special help to the students to pass the course at a level of significance of 0.05. (b) Use Yates' correction to find the answer.

Exercises 17-2

Reference: Sections 17.3 to 17.5

1. Two types of corn, A and B, were planted in separate sections of 18 plots. The yields from the plots are shown below. Use the sign test to determine whether there is a significant difference between the average yields of the two types of corn. Let the level of significance be (a) 0.01 and (b) 0.05. Use the normal curve method only.

Plot number	Corn A Yield (bushels)	Corn B Yield (bushels)
1	36	20
2	45	34
3	28	32
4	37	27
5	42	33
6	29	35
7	31	24
8	30	30
9	38	43
10	25	39
11	40	23
12	44	41
13	42	40
14	38	38
15	26	27
16	35	47
17	46	33
18	50	35

2. The weights of the 15 players in each of the two teams, A and B, are arranged randomly as presented below. Match the weights of corresponding players into 15 pairs based on the order presented.

Team A: 183, 194, 152, 178, 180, 170, 196, 204, 174, 186, 193, 173, 191, 164, 184 (pounds)

Team B: 210, 171, 165, 179, 155, 175, 203, 195, 182, 205, 193, 176, 197, 168, 189 (pounds)

Perform the sign test of the null hypothesis that there is no difference in the average weights between the two teams. Let the level of significance be 0.05. Use (a) the normal curve method, and (b) the binomial formula.

3. Refer to Problem 1. Find the answer by using the Wilcoxon test.

4. Refer to Problem 2. Find the answer by using the Wilcoxon test.

5. Use the yields of corn A from the first 11 plots and corn B from the first 10 plots given in Problem 1. Determine whether there is a significant difference between the average yields of the two types of corn by the Mann-Whitney test. Let the level of significance be 0.05.

6. Perform the Mann-Whitney test of the null hypothesis given in Problem 2.

7. The numbers of sales made by the 18 salesmen of three different realty companies during last year are given on the following page.

Salesman	Number of Sales		
	Company A	Company B	Company C
1	124	140	133
2	150	152	146
3	168	160	155
4	179	184	165
5	180	190	176
6	195		187
7			193
Total	996	826	1,155
Mean	166	165.2	165

Use the H test to determine whether the means of the numbers of sales of the three companies are significantly different. Let the level of significance be 0.05.

8. Assume that the dean of a college took a sample of freshman English grades of 15 students. He classified the grades on the basis of teachers who taught the students as shown in the following table. Use the H test to test the hypothesis that the means of the grades of the three classes are the same. Let the level of significance be 5%.

Student	English Grades		
	Teacher A	Teacher B	Teacher C
1	2	3	4
2	3	7	5
3	6	3	9
4	10	10	8
5	4	7	9

9. Assume that a sample of 28 units produced by a certain machine was taken in a factory. The units are classified into defective (D) and nondefective (N) and are arranged in order of time taken:

DD NNN DDD NNNN D NN DDD

NN DDDD N DDD

Use the test of randomness to determine whether this sample was taken at random. Let the level of significance be 0.05.

10. Refer to the yields of corn A given in Problem 1. Assuume that there are 20 plots. The yields of corn A from the 19th and 20th plots are 24 and 32 bushels respectively. Use the runs above and below the median to test the randomness of the yields of the 20 plots. Let the level of significance be 0.05.

18 The F Distribution and Analysis of Variance

This chapter will introduce the concept and the uses of the F distribution. The concept of the F distribution is explained in Section 18.1. The F distribution can be used for testing hypotheses concerning: (1) the equality of two estimated population variances and (2) the equality of three or more estimated population means. The technique used in (1) is presented in Section 18.2. The technique used in (2) is generally referred to as the method of *analysis of variance*. The fundamental concept and procedure of the analysis of variance are introduced in Section 18.3. Finally, this chapter will extend the procedure to present the two basic models: (1) *one-way* analysis of variance and (2) *two-way* analysis of variance in Sections 18.4 and 18.5, respectively.

18.1   The *F* Distribution

The letter F, chosen in honor of R. A. Fisher, represents the *variance ratio* showing the relationship between two independently estimated population variances as follows:

$$F = \frac{\hat{s}_1^2}{\hat{s}_2^2} \qquad (18\text{--}1)$$

where the subscripts 1 (in the numerator) and 2 (in the denominator) indicate the sample numbers and each $\hat{s}^2$ represents the estimate of the population variance based on the sample. The value of $\hat{s}^2$ can be obtained by the use of formula (13–1), or

$$\hat{s}^2 = \frac{\sum x^2}{n-1} = s^2 \left(\frac{n}{n-1} \right)$$

The F ratio may also be defined as the ratio of two independent chi-square variables, each divided by the corresponding degrees of freedom, or

$$F = \frac{\chi_1^2/D_1}{\chi_2^2/D_2} \qquad (18\text{--}2)^*$$

Observing formula (18–1), when the two estimated population variances are equal, the F ratio should be unity or 1. When F is not equal to 1, the difference from 1 may be attributed to chance (that is, the difference is not significant), or may not be attributed to chance (that is, the difference is significant—either too large or too small). To make such decisions, we must rely on the distribution of the statistic F.

Some important properties of the F distribution are listed below:

(1) The range of the values of F is from 0 to infinity. The value of F can not be negative since both terms of the F ratio are squared values.

(2) The shape of F distribution curve depends upon the number of degrees of freedom for the first term (D_1 for $\hat{s}_1^2$) and that for the second term (D_2 for $\hat{s}_2^2$) of the F ratio. In general, the F curve is skewed to the right as shown schematically in Chart 18–1. When the degrees of freedom increase, the F distribution tends to symmetrical shape.**

(3) For the same probability value, such as 5 percent (or 5 percent of the area under the F distribution curve), the critical value of F for the lower area (left tail) is the reciprocal of F for the upper area (right tail) with D_1 and D_2 interchanged. Tables 9A and 9B in the Appendix show detailed F values for upper 5 percent and 1 percent probabilities respectively. Table 18–1 shows selected values of F for upper 5 percent probabilty. When $D_1 = 5$ and $D_2 = 14$, for example, the right tail above $F = 2.96$ represents 0.05 or 5 percent of the area under the curve. When the numbers of degrees of freedom are interchanged, that is, $D_1 = 14$ and

---

*We know from the footnote on page 419

$$\chi^2 = \frac{\Sigma(X - \mu)^2}{\sigma^2}$$

with N (population size or number of X values) degrees of freedom.
Now, if we take n values of X to be a sample with the sample mean $\bar{X}$, then

$$\chi^2 = \frac{\Sigma(X - \bar{X})^2}{\sigma^2} = \frac{\Sigma x^2}{\sigma^2}, \text{ or } \Sigma x^2 = \chi^2 \sigma^2$$

with $(n - 1)$ degrees of freedom.
Likewise, if we take two samples of sizes n_1 and n_2 independently, we should have

$$\hat{s}_1^2 = \frac{\Sigma x_1^2}{n_1 - 1} = \frac{\chi_1^2 \sigma^2}{D_1} \text{ and } \hat{s}_2^2 = \frac{\Sigma x_2^2}{n_2 - 1} = \frac{\chi_2^2 \sigma^2}{D_2}$$

The population variance σ^2 is the same for both samples. Substitute the two values of $\hat{s}^2$ in formula (18–1),

$$F = \frac{\chi_1^2 \sigma^2}{D_1} \div \frac{\chi_2^2 \sigma^2}{D_2} = \frac{\chi_1^2/D_1}{\chi_2^2/D_2}$$

**The density function of the F distribution is

$$g(F) = \frac{\left(\dfrac{D_1 + D_2 - 2}{2}\right)!}{\left(\dfrac{D_1 - 2}{2}\right)! \left(\dfrac{D_2 - 2}{2}\right)!} \left(\frac{D_1}{D_2}\right)^{D_1/2} \frac{F^{(D_1 - 2)/2}}{\left(1 + \dfrac{D_1 F}{D_2}\right)^{(D_1 + D_2)/2}} \geq 0$$

where $D_1 = n_1 - 1$ and $D_2 = n_2 - 1$. The value of $g(F)$ will vary when the degrees of freedom D_1 and D_2 vary.

$D_2 = 5$, the value of F is 4.64. The reciprocal of 4.64, or $1/4.64 = 0.22$, is the critical value of F for the left tail of 5 percent probability.

CHART 18–1

EXAMPLE OF F DISTRIBUTION

$D_1 = 5$, $D_2 = 14$ AT 10% PROBABILITY LEVEL:
RIGHT TAIL 5%—$F = 2.96$, LEFT TAIL 5%—$F = .22$

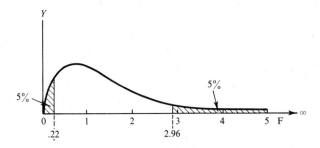

TABLE 18–1

SELECTED VALUES OF F FOR UPPER 5% PROBABILITY
(OR 5% AREA UNDER F DISTRIBUTION CURVE AT RIGHT TAIL)

D_1 D_2	1	2	3	4	5	6	10	12	14	∞
1	161	200	216	225	230	234	242	244	245	254
2	18.51	19.00	19.16	19.25	19.30	19.33	19.39	19.41	19.42	19.50
3	10.13	9.55	9.28	9.12	9.01	8.94	8.78	8.74	8.71	8.53
4	7.71	6.94	6.59	6.39	6.26	6.16	5.96	5·91	5.87	5.63
5	6.61	5.79	5.41	5.19	5.05	4.95	4.74	4.68	4.64	4.36
6	5.99	5.14	4.76	4.53	4.39	4.28	4.06	4.00	3.96	3.67
7	5.59	4.74	4.35	4.12	3.97	3.87	3.63	3.57	3.52	3.23
8	5.32	4.46	4.07	3.84	3.69	3.58	3.34	3.28	3.23	2.93
10	4.96	4.10	3.71	3.48	3.33	3.22	2.97	2.91	2.86	2.54
12	4.75	3.88	3.49	3.26	3.11	3.00	2.76	2.69	2.64	2.30
14	4.60	3.74	3.34	3.11	2.96	2.85	2.60	2.53	2.48	2.13
16	4.49	3.63	3.24	3.01	2.85	2.74	2.49	2.42	2.37	2.01
20	4.35	3.49	3.10	2.87	2.71	2.60	2.35	2.28	2.23	1.84
24	4.26	3.40	3.01	2.78	2.62	2.51	2.26	2.18	2.13	1.73
30	4.17	3.32	2.92	2.69	2.53	2.42	2.16	2.09	2.04	1.62
40	4.08	3.23	2.84	2.61	2.45	2.34	2.07	2.00	1.95	1.51
50	4.03	3.18	2.79	2.56	2.40	2.29	2.02	1.95	1.90	1.44
100	3.94	3.09	2.70	2.46	2.30	2.19	1.92	1.85	1.79	1.28
200	3.89	3.04	2.65	2.41	2.26	2.14	1.87	1.80	1.74	1.19
∞	3.84	2.99	2.60	2.37	2.21	2.09	1.83	1.75	1.69	1.00

Note: For more detailed values of F, see the F distribution tables in the Appendix—Table 9A for 5% and Table 9B for 1% probability (pages 791 and 792).

(4) When D_1 appears in the F tables but D_2 does not, the interpolation in the tables is based on direct proportions among the reciprocals of D_2 values and the two nearest values of F in the D_1 column as shown in Example 1.

Example 1

Given $D_2 = 6$ and $D_2 = 18$. Find F by using Table 18–1.

Solution.   The two nearest F values in $D_1 = 6$ column are 2.74 and 2.60 on the $D_2 = 16$ and 20 rows respectively. The interpolation is as follows:

Reciprocals of D_2	F	
$\dfrac{1}{16} = 0.0625$	2.74	(1)
$\dfrac{1}{18} = 0.0556$	x	(2)
$\dfrac{1}{20} = 0.0500$	2.60	(3)

$$\frac{(2) - (3)}{(1) - (3)} \qquad \frac{0.0056}{0.0125} = \frac{x - 2.60}{0.14}$$

$$x = 2.60 + 0.14\left(\frac{56}{125}\right) = 2.66$$

Thus, when $D_1 = 6$ and $D_2 = 18$, for 5% probability, $F = 2.66$.

Similarly, this method may be used when D_2 appears in the F tables but D_1 does not.

---

(5) F, χ^2, and t are related at the same probability point α as follows:

I. χ^2 (at α with D) $= D \cdot F$ (at α with $D_1 = D$ and $D_2 = \infty$)

Example 2

Given: $\alpha = 0.05$, $D = 6$. Find χ^2 value based on (a) F Table and (b) χ^2 Table.

Solution.   (a) From F Table, when $D_1 = 6$ and $D_2 = \infty$, at .05, $F = 2.09$. Thus,

$$\chi^2 = D \cdot F = 6(2.09) = 12.54$$

(b) From χ^2 Table, when $D = 6$, at .05, $\chi^2 = 12.592$. The discrepancy between the two answers is due to rounding in the tables.

II. t (at α with D) $= \sqrt{F}$ (at α with $D_1 = 1$ and $D_2 = D$)

Example 3

Given: $\alpha = 0.05$, $D = 6$. Find the t value based on (a) F Table, and (b) t Table.

Solution.   (a) From F Table, $t = \sqrt{F} = \sqrt{5.99} = 2.447$.

(b) From t Table, $t = 2.447$(for two tails)

18.2 The Equality of Two Population Variances

Chapter 11 indicated that a normally distributed population can be defined if the population mean μ and the population standard deviation σ or variance σ^2 are known (page 279). If we took two samples from separate populations, we would have two sample means and two sample variances. Now suppose that by the methods of testing hypotheses on the samples, we found that the two samples have the same mean and the same variance; that is, there is no significant difference either between the sample means or between the sample variances. We then have reason to believe that the two populations are essentially the same population. The methods of testing the hypotheses concerning the difference between two sample means are based on z tests for large samples and t tests for small samples (Section 15.2). The methods of testing the hypotheses concerning the equality of two population variances estimated from samples are based on F ratio tests which are illustrated in Examples 4 and 5.

Example 4

Refer to Example 5, page 377, concerning the mean wage of electricians and that of carpenters used for a t test. It was assumed in solution (2) of the example that the two samples are drawn from two populations with the same variance. Find out if the assumption is valid. Let the level of significance be 5%.

Solution. (1) H_0: The variance of the distribution of the wages of all electricians (σ_1^2) and that of all carpenters (σ_2^2) in the city are the same. Or, the two population variances are the same, $\sigma_1^2 = \sigma_2^2$.

(2) Find F by using formula (18–1).

$$\hat{s}_1^2 = \frac{\sum x_1^2}{n_1 - 1} = \frac{46}{4 - 1} = \frac{46}{3} = 15.33$$

$$\hat{s}_2^2 = \frac{\sum x_2^2}{n_2 - 1} = \frac{20}{5 - 1} = \frac{20}{4} = 5$$

$$F = \frac{15.33}{5} = 3.07$$

(3) Make a decision. Since the F ratio may be above 1 or below 1 when the two estimated population variances are not equal, the two-tailed test is used in this case. At 5% level for a two-tailed test, or 2.5% for each tail, the F value is significantly large if F, with $D_1 = 4 - 1 = 3$ and $D_2 = 5 - 1 = 4$, is above 9.98 (see Chart 18–2). The computed F value, 3.07, is smaller than 9.98 at 2.5% probability level. We thus conclude that $F = 3.07$ is not significantly large, or the difference of the F value from 1 is due to chance. We accept the hypothesis that the two population *variances are the same.*

Note that the 2.5% probability value, $F = 9.98$, is not tabulated in the Appendix in this text. Actually, we do not have to know the F value at 2.5% probability level in this case since it is between 6.59 (at 5% level) and 16.69 (at 1% level) as shown in the chart. Since the computed F value, 3.07, is even smaller than 6.59, we may draw the same conclusion either at 5% or at 2.5% level that $F = 3.07$ is not significantly large.

Note also that the above computation of F ratio is made by letting the larger $\hat{s}^2$ be the numerator of the ratio so that the ratio may be larger than 1. This practice is convenient since the values of F in the F tables are greater than 1 (except when $D_1 = \infty$ and $D_2 = \infty$, $F = 1$) and are for the probabilities of right tails. It is not necessary to find the F value at 2.5% level for the left tail in this case since the conclusion will be the same. However, when the left tail is applied in the test, the larger $\hat{s}^2$ must be used as the denominator. Thus, the F value now is smaller than 1.

CHART 18–2

A TWO-TAILED TEST OF HYPOTHESIS BY F RATIO (EXAMPLE 4)
AT 5% LEVEL OF SIGNIFICANCE OR 2.5% FOR EACH TAIL $D_1 = 3$, $D_2 = 4$

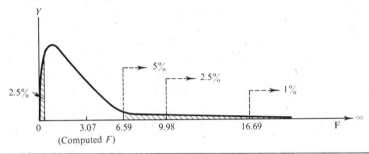

Example 5

Assume that two different methods A and B are used in teaching statistics for two groups of students. A sample of final grade points is taken for each of the two groups. The samples are shown:

	Group A	Group B
Sample size	$n_1 = 41$ students	$n_2 = 25$ students
Standard deviation	$s_1 = 12$ points	$s_2 = 10$ points

Find out if the variance of the grade points for Group A is significantly larger than that for Group B at 5% level.

Solution.   (1) H_0: The variance of the grade points for Group A and that for Group B are the same, or $\sigma_1^2 = \sigma_2^2$.

(2) Find F by using formula (18–1).

$$\hat{s}_1^2 = \left(\frac{n_1}{n_1 - 1}\right)s_1^2 = \left(\frac{41}{41 - 1}\right)12^2 = 147.60$$

$$\hat{s}_2^2 = \left(\frac{n_2}{n_2 - 1}\right)s_2^2 = \left(\frac{25}{25 - 1}\right)10^2 = 104.17$$

$$F = \frac{147.60}{104.17} = 1.42$$

(3) Make a decision. Here we are interested in whether or not the variance for Group A is greater than that for Group B. Thus, a one-tailed test is used. At 5% level the F value is significantly large if F, with $D_1 = 41 - 1 = 40$ and D_2

$= 25 - 1 = 24$, is above 1.89. The computed F value, 1.42, is smaller than 1.89. Thus, we conclude that $F = 1.42$ is not significantly large, or the difference of the F value from 1 is attributed to chance. We accept the hypothesis that the two population variances are the same.

18.3   Fundamental Concept and Procedure of the Analysis of Variance

The technique of *analysis of variance* was originated in agricultural research. In recent years, however, it has been developed as a powerful tool in analyzing different types of scientific problems as well as problems in business and economics. This section introduces the fundamental concept and procedure of analysis of variance. Only simple data are used in the illustration (Example 6). More complicated data will be used in the next two sections.

The hypothesis for the analysis of variance is that the *means* of normally distributed populations, such as three populations a, b, and c, are equal, or $\mu_a = \mu_b = \mu_c$. If we further assume that the variances of the populations are equal, we will have three equal populations. When the three populations are combined into a single large population, it is reasonable to expect that the mean and the variance of the large population (μ and σ^2) will be equal to those of the original populations, or

$$\mu = \mu_a = \mu_b = \mu_c$$

and
$$\sigma^2 = \sigma_a^2 = \sigma_b^2 = \sigma_c^2$$

Now if we take a random sample from each of the three original populations, we may consider the three samples as *classes* of a single large sample drawn from the single large population. The sample data may be arranged in columns (Table 18–2) or rows (Table 18–3).

TABLE 18–2

SAMPLE DATA PRESENTATION IN COLUMNS

Observations in each sample	Sample		
	a	b	c
1	X_a	X_b	X_c
2	X_a	X_b	X_c
3	X_a	X_b	X_c
4	X_a	X_b	X_c
5	—	X_b	—
Total	$\sum X_a$	$\sum X_b$	$\sum X_c$
Sample mean	$\bar{X}_a$	$\bar{X}_b$	$\bar{X}_c$

TABLE 18–3

SAMPLE DATA PRESENTATION IN ROWS

Sample	Observations in each sample					Total	Sample mean
	1	2	3	4	5		
a	X_a	X_a	X_a	X_a	—	$\sum X_a$	$\bar{X}_a$
b	X_b	X_b	X_b	X_b	X_b	$\sum X_b$	$\bar{X}_b$
c	X_c	X_c	X_c	X_c	—	$\sum X_c$	$\bar{X}_c$

$$\bar{X} = \text{Grand mean} = \frac{\sum X_a + \sum X_b + \sum X_c}{4 + 5 + 4} = \frac{\sum X}{13}$$

The unbiased estimate of the large population variance (σ^2) based on the above samples may be obtained by any one of the three methods below.

(1) Use the variance between classes (or between samples).

$$\hat{s}_1^2 = \frac{\text{Sum of } \textit{variations} \text{ of class means from grand mean}}{\text{Degrees of freedom between classes}}$$

$$= \frac{n_a(\bar{X}_a - \bar{X})^2 + n_b(\bar{X}_b - \bar{X})^2 + n_c(\bar{X}_c - \bar{X})^2}{3 \text{ (number of classes)} - 1}$$

or simply written in a general form: $$\hat{s}_1^2 = \frac{\sum n_i(\bar{X}_i - \bar{X})^2}{r - 1}$$ (18–3)*

where i = individual classes or samples a, b, c, . . .

   n_i = size of class i, or size of sample drawn from population i, such as $n_a = 4$, $n_b = 5$, $n_c = 4$ in the above
   illustration

   $\bar{X}_i$ = mean of the items in class or sample i

   $\bar{X}$ = the grand mean, or the mean of all items in the single
   large sample

   $\bar{X}_i - \bar{X}$ = deviation

   $(\bar{X}_i - \bar{X})^2$ = *variation*, or squared deviation (The term "variation"
   has been used loosely in previous discussions. Here,
   the term is limited to represent the squared deviation.)

   r = number of classes or samples, such as 3 classes in
   the above illustration

(2) Use the variance within classes (or within individual samples).

$$\hat{s}_2^2 = \frac{\text{Sum of } \textit{variations} \text{ of class items from class means}}{\text{Degrees of freedom within classes}}$$

$$= \frac{\sum (X_a - \bar{X}_a)^2 + \sum (X_b - \bar{X}_b)^2 + \sum (X_c - \bar{X}_c)^2}{n_a + n_b + n_c - 3}$$

or simply written in a general form: $$\hat{s}_2^2 = \frac{\sum [\sum (X_i - \bar{X}_i)^2]}{n - r}$$ (18–4)**

---

*Formula (18–3) is derived from the relation

$$\sigma_{\bar{x}}^2 = \frac{\sigma^2}{n} \text{ and } \sigma^2 = n\sigma_{\bar{x}}^2$$

According to the expression used in the sampling distribution of the mean $\bar{X}_i$, page 305, $\sigma_{\bar{x}}^2 = \frac{\sum f(\bar{X}_i - \mu)^2}{\sum f}$. When there is only one sample mean in the distribution, $\sum f = 1$, $\sigma_{\bar{x}}^2 = (\bar{X}_i - \mu)^2$, and $\sigma^2 = n_i(\bar{X}_i - \mu)^2$. Here we use $\bar{X}$ as the estimate of μ. The estimated popula-tion variance thus is $\hat{s}_i^2 = n_i(\bar{X}_i - \bar{X})^2$. The value of $\hat{s}_1^2$ is the average of the three individually estimated population variances $\hat{s}_a^2$, $\hat{s}_b^2$, and $\hat{s}_c^2$. Since the estimation is based on samples, instead of dividing by 3, use the degrees of freedom for the unbiased estimate $3 - 1 (= 2)$ as the divisor.

**Since samples a, b, and c are assumed to be drawn from the populations with the same va-riance σ^2, we could use any one of the three sample variances as an estimate of σ^2 in the form

$$\hat{s}_i^2 = \frac{\sum (X_i - \bar{X}_i)^2}{n_i - 1}$$

However, a better estimate is the weighted average of the estimated population variances $\hat{s}_a^2$, $\hat{s}_b^2$, and $\hat{s}_c^2$. The value of $\hat{s}_2^2$ is the weighted average, the weights being the numbers of degrees of freedom, $n_a - 1$, $n_b - 1$, and $n_c - 1$. (Also see footnote page 376.)

where X_i = individual items in class i

$n = n_a + n_b + n_c$ = number of items in the single large sample

(3) Use the variance of the single large sample.

$$\hat{s}_3^2 = \frac{\text{Sum of } \textit{variations} \text{ of all items from grand mean}}{\text{Degrees of freedom of all items}}$$

$$= \frac{\Sigma (X - \bar{X})^2}{n - 1} \qquad \qquad \text{which is formula (13–1)}$$

There are only two terms in the F ratio. For the purposes of testing hypotheses, we shall use only the estimated population variances by the first two methods; that is,

$$F = \frac{\text{Variance } \textit{between} \text{ classes (samples)}}{\text{Variance } \textit{within} \text{ classes (samples)}} = \frac{\hat{s}_1^2}{\hat{s}_2^2}$$

The reasons for selecting the two terms are given below.

(1) If the hypothesis is true (or $\mu = \mu_a = \mu_b = \mu_c$) and the assumption regarding the same variance is valid (or $\sigma^2 = \sigma_a^2 = \sigma_b^2 = \sigma_c^2$), the two independent estimates of the large population variance should not be greatly different; that is, $\hat{s}_1^2$ is close to $\hat{s}_2^2$. The value of F ratio, therefore, should be close to unity or 1.

(2) If the hypothesis is true, the means of the samples drawn from the populations a, b, and c respectively should not vary significantly from each other and also from the grand mean (except variation attributed to chance), or $\bar{X}_a$, $\bar{X}_b$, $\bar{X}_c$, and $\bar{X}$ are close to each other. Thus, the variance between classes should be small. In fact, the variance is zero if the sample means and the grand mean are the same. Therefore, the value of F ratio should be small or close to zero if the hypothesis is true.

(3) If the hypothesis is not true, the sample means will differ from each other and also from the grand mean by more than the extent attributed to chance. Thus, the variance between classes becomes large. On the other hand, the variance within classes is not affected by the differences of the sample means since it is obtained from the deviations within individual classes. Therefore, the value of F ratio is large if the hypothesis is not true.

Then, how *large* should the F ratio be in order to reject the hypothesis? This decision is usually based on the *right-tail* test according to the F distribution. For this reason, the values given in the F tables cover only the upper probability points in the analysis of variance.

However, the estimation by the third method is also useful in the process of the analysis of variance since

$$\left(\begin{array}{c} \text{variations} \\ \text{between classes} \end{array} \right) + \left(\begin{array}{c} \text{variations} \\ \text{within classes} \end{array} \right) = \left(\begin{array}{c} \text{variations of} \\ \text{all items} \end{array} \right)$$

or expressed in a different way,

$$\text{numerator for } \hat{s}_1^2 + \text{numerator for } \hat{s}_2^2 = \text{numerator for } \hat{s}_3^2$$

Also,

$$\left(\begin{array}{c}\text{degrees of freedom}\\\text{between classes}\end{array}\right) + \left(\begin{array}{c}\text{degrees of freedom}\\\text{within classes}\end{array}\right) = \left(\begin{array}{c}\text{degrees of freedom}\\\text{of all items}\end{array}\right)$$

or expressed in a different way,

$$\text{denominator for } \hat{s}_1^2 + \text{denominator for } \hat{s}_2^2 = \text{denominator for } \hat{s}_3^2$$

The relationships may serve as a checking point in the process as shown in an analysis of variance table, such as Tables 18–5 and 18–6.

Example 6

Assume that a young man took a sample of the wages of 13 workers in a large city during a given period. He classified the wages into three classes according to occupations: electricians, carpenters, and painters as shown on the rows in Table 18–4 [also shown in column (2) of Table 18–5]. He wishes to know whether or not the means of the wages classified by the three different occupations are significantly different. Let the level of significance be (a) 5% and (b) 1%. Also see Example 5, page 377.

TABLE 18–4

DATA FOR EXAMPLE 6

| Occupation | Amount of wages during the given period | | | | | Total |
| | Sample item (worker) | | | | | |
	1	2	3	4	5	
Electricians	$74	65	72	69	—	$280
Carpenters	75	78	74	76	72	375
Painters	56	55	53	52	—	216
Grand Total						$871

Solution. (1) H_0: The means of the wages of the three different occupations in the city are equal, or $\mu_a = \mu_b = \mu_c$, where class a represents the wages of electricians, class b the carpenters, and class c the painters. We also assume that the three populations (or wages of the occupations) are normally distributed.

(2) Find the F ratio. The detailed computation is shown in Table 18–5.

By formula (18–3), the variance between classes is

$$\hat{s}_1^2 = \frac{4(70 - 67)^2 + 5(75 - 67)^2 + 4(54 - 67)^2}{3 - 1}$$

$$= \frac{36 + 320 + 676}{2} = \frac{1,032}{2} = 516$$

By formula (18–4), the variance within classes is

TABLE 18–5

COMPUTATION FOR THE ANALYSIS OF VARIANCE
(EXAMPLE 6)

Item in Each Class	Amount of Wages X_i or X	Between Classes		Within Classes		Total	
		$\bar{X}_i - \bar{X}$	$n_i(\bar{X}_i - \bar{X})^2$	$X_i - \bar{X}_i$	$(X_i - \bar{X}_i)^2$	$X - \bar{X}$	$(X - \bar{X})^2$
		$i = a$, or Class a (electricians); $n_a = 4$					
1	$74			4	16	7	49
2	65			−5	25	−2	4
3	72			2	4	5	25
4	69			−1	1	2	4
Total	280			0	46	12	82
	$\bar{X}_a = \frac{280}{4} = 70$	$\begin{array}{l}70-67\\=3\end{array}$	$4(3)^2 = 36$				
		$i = b$, or Class b (carpenters); $n_b = 5$					
1	75			0	0	8	64
2	78			3	9	11	121
3	74			−1	1	7	49
4	76			1	1	9	81
5	72			−3	9	5	25
Total	375			0	20	40	340
	$\bar{X}_b = \frac{375}{5} = 75$	$\begin{array}{l}75-67\\=8\end{array}$	$5(8)^2 = 320$				
		$i = c$, or Class c (painters); $n_c = 4$					
1	56			2	4	−11	121
2	55			1	1	−12	144
3	53			−1	1	−14	196
4	52			−2	4	−15	225
Total	216			0	10	−52	686
	$\bar{X}_c = \frac{216}{4} = 54$	$\begin{array}{l}54-67\\=-13\end{array}$	$4(-13)^2 = 676$				
Grand Total	871		1,032	0	76	0	1,108
	$n = 13$	$\begin{array}{l}D_1 = 3-1\\=2\end{array}$		$\begin{array}{l}D_2 = 13-3\\=10\end{array}$		$\begin{array}{l}D = 13-1\\=12\end{array}$	
	$\bar{X} = \frac{871}{13} = 67$	$\begin{array}{l}\hat{s}_1^2 = \frac{1,032}{2}\\=516\end{array}$		$\hat{s}_2^2 = \frac{76}{10} = 7.6$		$\begin{array}{l}\hat{s}^2 = \frac{1,108}{12}\\=92.33\end{array}$	
		$F = \frac{\hat{s}_1^2}{\hat{s}_2^2} = \frac{516}{7.6} = 67.89$					

*See formula 18–8 for short-cut method.

$$\hat{s}_2^2 = \frac{46 + 20 + 10}{(4 + 5 + 4) - 3} = \frac{76}{13 - 3} = \frac{76}{10} = 7.6$$

By formula (18–1), the F ratio is

$$F = \frac{516}{7.6} = 67.89$$

Check the variations:

1032 (between classes) + 76 (within classes) = 1,108 (total)

Check the degrees of freedom:

$(3 - 1)$ for between classes + $(13 - 3)$ for within classes = 12
$(13 - 1)$ for all items or total = 12

The total variation and the total of degrees of freedom are shown in the last column of Table 18–5.

(3) Make a decision. With $D_1 = 3 - 1 = 2$ and $D_2 = 13 - 3 = 10$, the F value is significantly large if, based on the right-tailed test,

F is above 4.10 at the 5% level
F is above 7.56 at the 1% level

Since the computed F value, 67.89, is far above 7.56, we may draw the same conclusion either at 5% or at 1% level that $F = 67.89$ is significantly large. We thus reject the hypothesis. We may state that the means of the wages of the three different occupations in the city are not equal. The three populations, wages of electricians, carpenters, and painters, are different ones.

The entire procedure of testing the hypothesis for Example 6 is summarized in the analysis of variance table, Table 18–6.

TABLE 18–6

ANALYSIS OF VARIANCE TABLE

Source of variation	Variation (Sum of squares)	Degrees of freedom	Variance (Mean square)	F Ratio Computed	F Ratio at 5%	F Ratio at 1%
Between classes	$\sum n_i(\bar{X}_i - \bar{X})^2$ = 1,032	$3 - 1 = 2$	$\frac{1,032}{2} = 516$			
				$\frac{516}{7.6} = 67.89$	4.10	7.56
Within classes	$\sum[\sum(X_i - \bar{X}_i)^2]$ = 76	$n - 3 = 10$	$\frac{76}{10} = 7.6$			
Total	$\sum(X - \bar{X})^2$ = 1,108	$n - 1 = 12$		Test result: Reject H_0		

Note that if the wages in Example 6 are classified on another basis, such as by different education levels, the method of the analysis of variance used above may also be used to determine whether or not the means of the new classes are significantly different.

**18.4 One-Way Analysis
    of Variance Model**

In the *one-way* analysis of variance, the given data are grouped according to one-way classification. There is only one variable in the analysis. The illustration in Example 6 of the previous section was based on one way of classifying the given data, the classification being three different occupations or samples. The variable in the example was the set of wages earned by the workers of the three occupations.

The one-way analysis of variance to be discussed in this section, however, will extend the procedure used previously to include any number of samples. Also, the sizes of the samples, or classes, were not equal in Example 6. When the sample sizes, n_i, are equal, the method of computing the required variances may be simplified. The simplified method will be introduced in the illustrations below. Further, for consistency, the one-way classification of a given set of values hereafter is written as the stubs in the rows of a table. (See Tables 18–7 and 18–8).

TABLE 18–7

SAMPLE DATA CLASSIFIED IN ROWS FOR
ONE-WAY ANALYSIS OF VARIANCE

Samples, *ith sample* ($i = 1, 2, 3, \ldots r$)	Observations in each sample, *jth observation* ($j = 1, 2, 3, \ldots c$)				Sample *(row)* total $\sum X_i$	Sample *(row)* mean $\bar{X}_i$
	1	2	...	c		
1	X_1	X_1	...	X_1	$\sum X_1$	$\bar{X}_1$
2	X_2	X_2	...	X_2	$\sum X_2$	$\bar{X}_2$
⋮	⋮	⋮	⋮	⋮	⋮	⋮
r	X_r	X_r	...	X_r	$\sum X_r$	$\bar{X}_r$

The symbols used in Table 18–7 are defined as follows:

i = individual row classes, or samples, $1, 2, 3, \ldots, r$
j = individual columns, $1, 2, 3, \ldots, c$
r = number of rows, or number of samples ($r = 3$ in Table 18–8)
c = number of columns, or the size of each sample ($c = 4$ in Table 18–8.)

Note: The sample sizes c now are assumed to be equal. The unequal sample sizes were represented in previous illustrations, such as in Example 6, by n_i.

$n = cr$ = number of X values in the large combined sample
    ($cr = (4)(3) = 12$ in Table 18–8)
X_i = X value in the ith row, $X_1, X_2, X_3, \ldots, X_r$

The *model of the one-way analysis of variance* is based on the X_i values and is expressed as follows:

$$X_i = \mu + (\mu_i - \mu) + (X_i - \mu_i)$$

Or, graphically:

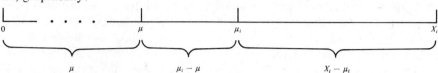

where μ = the large population mean
    μ_i = the ith population mean
    $\mu_i - \mu$ = the deviation *between* ith population mean and the large
        population mean
    $X_i - \mu_i$ = the deviation *within* the ith population

Thus, if the hypothesis that $\mu = \mu_i$ $(i = 1, 2, 3, \ldots, r)$ is true, the deviation between μ_i and μ should be zero.

When we use the sample mean $\bar{X}_i$ and the large sample mean $\bar{X}$ (or the grand mean) as estimators of μ_i and μ respectively, the model becomes

$$X_i = \bar{X} + (\bar{X}_i - \bar{X}) + (X_i - \bar{X}_i)$$

Based on the model, the total deviation of an X_i value from the grand mean $\bar{X}$ may be separated into a set of deviations:

$$(X_i - \bar{X}) = (\bar{X}_i - \bar{X}) + (X_i - \bar{X}_i)$$

or $\begin{pmatrix} \text{Total} \\ \text{deviation} \end{pmatrix} = \begin{pmatrix} \text{Deviation } \textit{between } i\text{th row} \\ \text{mean and grand mean} \end{pmatrix} + \begin{pmatrix} \text{Deviation } \textit{within} \\ i\text{th row} \end{pmatrix}$

Thus, if $X_i = 10$, $\bar{X} = 7$, and $\bar{X}_i = 8$

(see Table 18–8, A_1 row), the total deviation is

$$X_i - \bar{X} = 10 - 7 = 3$$

and its equivalent deviations are

$$\bar{X}_i - \bar{X} = 8 - 7 = 1$$

and $$X_i - \bar{X}_i = 10 - 8 = 2$$

The total variation and its equivalent deviations are diagrammed in Chart 18–3.

CHART 18–3

DEVIATIONS FOR THE ONE-WAY ANALYSIS OF VARIANCE

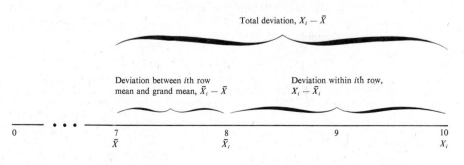

The deviation $(\bar{X}_i - \bar{X})$ is also called the *effect* and the nature of the sample i is called the *treatment*. An effect is due to a particular type of treatment, just as applying different types of fertilizers (treatments) to different sections of a piece of land will produce different grades of corns (effects) in an agricultural experiment. Likewise, the deviation $(X_i - \bar{X}_i)$ is the effect due to random chance or *error* obtained within i row class.

From the above equation of deviations, it can be proved that the corresponding sums of variations, or sums of squared deviations, are also equal, or*

$$\sum [\sum (X_i - \bar{X})^2] = \sum [\sum (\bar{X}_i - \bar{X})^2] + \sum [\sum (X_i - \bar{X}_i)^2]$$

Each term inside a bracket [ ] is the sum of c squared deviations since there are c X_i values in each row. Each final term is the sum of $cr = n$ squared deviations since there are r rows in the model. Further, the X_i values are identical with the X values when all i's $(i = 1, 2, 3, \ldots, r)$ are included in the equation of sums. The sum on the left side of the equation may be simplified to $\sum (X - \bar{X})^2$. Also, since the number of X_i's in each row, or the sample size c, is equal, the first term on the right side of the equation may be simplified to $\sum c(\bar{X}_i - \bar{X})^2$. Thus, the equation may be rewritten:

$$\sum (X - \bar{X})^2 = \sum c\,(\bar{X}_i - \bar{X})^2 + \sum [\sum (X_i - \bar{X}_i)^2]$$

or

$$\begin{pmatrix} Total \\ variation \end{pmatrix} = \begin{pmatrix} \text{Sum of variations} \\ \text{between } row \text{ classes} \end{pmatrix} + \begin{pmatrix} \text{Sum of variations} \\ within \text{ row classes} \end{pmatrix}$$

or expressed by simplified symbols, respectively, $S_t = S_r + S_w$

Formula (18–3) now becomes, $$\hat{s}_1^2 = \frac{S_r}{r - 1} \qquad \text{(18–5)}$$

Formula (18–4) becomes, $$\hat{s}_2^2 = \frac{S_w}{n - r} \qquad \text{(18–6)}$$

The denominators $(r - 1)$ and $(n - r)$ are the degrees of freedom for computing the unbiased estimates of the population variance $\hat{s}_1^2$ and $\hat{s}_2^2$ respectively.

The relationship among the sums of variations also gives:

$$S_w = S_t - S_r \qquad \text{(18–7)}$$

Formula (18–7) provides a simplified method to compute the variations within classes (S_w) since the fomulas of computing the total variation (S_t) and the variations between classes (S_r) may be simplified as follows:

---

*Proof. Square both sides of the equation of deviations,
$$(X_i - \bar{X})^2 = [(\bar{X}_i - \bar{X}) + (X_i - \bar{X}_i)]^2$$
$$(X_i - \bar{X})^2 = (\bar{X}_i - \bar{X})^2 + 2\,(\bar{X}_i - \bar{X})(X_i - \bar{X}_i) + (X_i - \bar{X}_i)^2$$
There are c X_i values in each row, r rows and a total of $cr = n$ individual X_i values in the large sample. Thus, there are n such equations. First sum each group of c equations containing the X_i values in the ith row. Next, sum the r sums obtained in the first step. Thus,
$$\sum [\sum (X_i - \bar{X})^2] = \sum [\sum (\bar{X}_i - \bar{X})^2] + 2 \sum [(\bar{X}_i - \bar{X}) \cdot \sum (X_i - \bar{X}_i)] + \sum [\sum (X_i - \bar{X}_i)^2]$$
The second term in the right side of the above equation is zero, since $\sum (X_i - \bar{X}_i)$ is the sum of the deviations within the ith row and is equal to zero. Thus,
$$\sum [\sum (X_i - \bar{X})^2] = \sum [\sum (\bar{X}_i - \bar{X})^2] + \sum [\sum (X_i - \bar{X}_i)^2]$$

Formula (7-7c) indicated that when A or the assumed mean $= 0$,

$$\frac{\Sigma (X - \bar{X})^2}{n} = \frac{\Sigma X^2}{n} - \left(\frac{\Sigma X}{n}\right)^2$$

Multiply each side by n,

$$\Sigma (X - \bar{X})^2 = \Sigma X^2 - \frac{(\Sigma X)^2}{n}$$

The left side of the above equation is the total variation, S_t. Thus

$$S_t = \Sigma X^2 - \frac{(\Sigma X)^2}{n} \tag{18-8}$$

Similarly, by applying formula (7-7c), we have the sum of variations between classes, S_r.

$$S_r = \frac{\Sigma (\Sigma X_i)^2}{c} - \frac{(\Sigma X)^2}{n} \tag{18-9}*$$

The application of formulas (18-7), (18-8), and (18-9) in the one-way analysis of variance is illustrated in Example 7.

Example 7

A large company offered a management training course to its employees. The employees were divided into three groups and taught by three different instructors,

TABLE 18-8

DATA FOR EXAMPLE 7

Instructors	Employees 1	2	3	4	Total ΣX_i	Mean $\bar{X}_i$
	Grade points					
A_1	6	7	9	10	32	8
A_2	2	2	5	7	16	4
A_3	7	9	10	10	36	9
Total ΣX_j	15	18	24	27	$\Sigma X = 84$	
Mean, $\bar{X}_j$	5	6	8	9		$\bar{X} = 7$

---

*Since c is constant,

$$S_r = \Sigma c(\bar{X}_i - \bar{X})^2 = c \Sigma (\bar{X}_i - \bar{X})^2$$

By applying formula (7-7c),

$$S_r = c \left[\Sigma (\bar{X}_i)^2 - \frac{(\Sigma \bar{X}_i)^2}{r} \right] = c \left[\Sigma \left(\frac{\Sigma X_i}{c}\right)^2 - \frac{\left[\Sigma \left(\frac{\Sigma X_i}{c}\right)\right]^2}{r} \right]$$

$$= \frac{\Sigma (\Sigma X_i)^2}{c} - \frac{c\left[\frac{1}{c}\Sigma (\Sigma X_i)\right]^2}{r} = \frac{\Sigma (\Sigma X_i)^2}{c} - \frac{(\Sigma X)^2}{cr}$$

$$S_r = \frac{\Sigma (\Sigma X_i)^2}{c} - \frac{(\Sigma X)^2}{n}$$

A_1, A_2, and A_3. However, the material covered in the course was the same for each group. At the end of the training, a uniform test was given to the employees. A sample of four employees was taken randomly from each of the three groups. The test grades, 10 points being the maximum, of the 12 employees in the three samples are listed in Table 18–8.

The education director of the company wishes to know whether or not the means of the grades ($\bar{X}_i$'s) classified by the three different instructors are significantly different. Let the level of significance be (a) 5% and (b) 1%.

Solution. (1) H_0: The means of the grade points classified by the three different instructors are equal.

(2) Find the F ratio. First, compute the required sums in Table 18–9. Next, substitute the sums obtained from Table 18–9 in formula (18–8),

$$S_t = \sum X^2 - \frac{(\sum X)^2}{n} = 678 - \frac{84^2}{12} = 678 - 588 = 90$$

and in formula (18–9),

$$S_r = \frac{\sum (\sum X_i)^2}{c} - \frac{(\sum X)^2}{n} = \frac{2,576}{4} - \frac{84^2}{12} = 644 - 588 = 56$$

where c = sample size in each row = 4,
r = number of rows or samples = 3, and
$n = cr = (4)(3) = 12$ (X values)

TABLE 18–9

COMPUTATION FOR EXAMPLE 7

A_i	X or (X_i)	$\sum X_i$	$(\sum X_i)^2$	X^2
A_1	6			36
	7			49
	9			81
	10			100
		32	1,024	
A_2	2			4
	2			4
	5			25
	7			49
		16	256	
A_3	7			49
	9			81
	10			100
	10			100
		36	1,296	
Sum	$\sum X = \sum (\sum X_i) = 84$		$\sum (\sum X_i)^2 = 2,576$	$\sum X^2 = 678$

By formula (18–7),

$$S_w = S_t - S_r = 90 - 56 = 34$$

Third, compute the terms in the F ratio.

By formula (18–5),

$$\hat{s}_1^2 = \frac{S_r}{r-1} = \frac{56}{3-1} = 28$$

By formula (18–6),

$$\hat{s}_2^2 = \frac{S_w}{n-r} = \frac{34}{12-3} = \frac{34}{9} = 3\frac{7}{9}$$

Thus,

$$F = \frac{28}{34/9} = \frac{126}{17} = 7.41$$

with $D_1 = 3 - 1 = 2$ and $D_2 = 12 - 3 = 9$.

(3) Make a decision. $D_1 = 2$ appears in the F table, but not $D_2 = 9$. In the 5% F Table the two nearest values of F in $D_1 = 2$ column are 4.10 on $D_2 = 10$ line and 4.46 on $D_2 = 8$ line. Since the computed F value is *larger* than both 4.10 and 4.46, the interpolation method as shown in Example 1 is not required and the hypothesis should be *rejected* at the 5% level. In the 1% F Table the two nearest F values are 7.56 and 8.65. Since the computed F is *lower* than both tabulated values, the hypothesis should be *accepted*. The test of the differences among the means shows significance at the 5% level but not at the 1% level. Thus, the result obtained from Example 7 may be considered *probably significant*. We may state that the means of the grade points classified by the instructors are probably not equal. The performances of the three instructors, A_1, A_2, and A_3, are probably different in effectiveness.

The entire procedure of testing the hypothesis for Example 7 is summarized in the analysis of variance table, Table 18–10.

TABLE 18–10

ANALYSIS OF VARIANCE TABLE

Source of variation	Variation (Sum of squares)	Degrees of freedom	Variance (Mean square)	F Ratio		
				Computed	at 5%	at 1%
Between row classes	$S_r = 56$	$D_1 = r - 1 =$ $3 - 1 = 2$	$\hat{s}_1^2 = \frac{56}{2} = 28$	$F = \frac{28}{34/9}$ $= \frac{126}{17}$ $= 7.41$	4.10 to 4.46	7.56 to 8.56
Within row classes	$S_w = 34$	$D_2 = n - r =$ $12 - 3 = 9$	$\hat{s}_2^2 = \frac{34}{9} = 3\frac{7}{9}$			
Total	$S_t = 90$	$n - 1 =$ $12 - 1 = 11$		Test result: Reject H_0 at 5% level. Accept H_0 at 1% level.		

18.5 Two-Way Analysis of Variance Model

A group of values often may be cross-classified in a table as shown in Table 18–11 in two ways: one classification for rows and another for columns. The two-way classifications give two independent variables A_i and B_j in the table. When we wish to analyze the two variables simultaneously, the technique used in one-way analysis of variance must be extended to investigate the variations *between row classes* as well as the variations *between column classes*. However, the total variation, S_t, remains unchanged, regardless of the number of classifications applied to the group of values.

TABLE 18–11

SAMPLE DATA CLASSIFIED IN ROWS AND COLUMNS FOR
TWO-WAY ANALYSIS OF VARIANCE

Row classes (Samples): A_i, ith row $(i = 1, 2, 3, \ldots, r,)$	Column classes: B_j, jth column $(j = 1, 2, 3, \ldots, c)$				Row total $\sum X_{i\cdot}$	Row mean $\bar{X}_{i\cdot}$
	B_1	B_2	$\cdots$	B_c		
A_1	X_{11}	X_{12}	$\cdots$	X_{1c}	$\sum X_{1\cdot}$	$\bar{X}_{1\cdot}$
A_2	X_{21}	X_{22}	$\cdots$	X_{2c}	$\sum X_{2\cdot}$	$\bar{X}_{2\cdot}$
.	.	.	$\cdots$	.	.	.
.	.	.	$\cdots$	.	.	.
.	.	.	$\cdots$	.	.	.
A_r	X_{r1}	X_{r2}	$\cdots$	X_{rc}	$\sum X_{r\cdot}$	$\bar{X}_{r\cdot}$
Column Total $\sum X_{\cdot j}$	$\sum X_{\cdot 1}$	$\sum X_{\cdot 2}$	$\cdots$	$\sum X_{\cdot c}$	$\sum X$	
Column Mean $\bar{X}_{\cdot j}$	$\bar{X}_{\cdot 1}$	$\bar{X}_{\cdot 2}$	$\cdots$	$\bar{X}_{\cdot c}$		$\bar{X}$

The new symbols used in Table 18–11 are defined as follows:

A_i = the ith row class

B_j = the jth column class

X_{ij} = X value located in the ith row and jth column, such as X_{12} representing the X value in the 1st row and 2nd column—the first subscript always indicates the location of the row and the second indicates the column

$i\cdot$ = the subscripts i and $\cdot$ representing only the value in the ith row, such as $\sum X_{1\cdot}$ being the sum of the X values in the first row and $\bar{X}_{2\cdot}$ being the mean of the X values in the second row

$\cdot j$ = the subscripts $\cdot$ and j representing only the value in the jth column, such as $\sum X_{\cdot 1}$ being the sum of the X values in the first column and $\bar{X}_{\cdot 2}$ being the mean of the X values in the second column

$$\Sigma X = \Sigma X_{ij} = \text{the sum of all } X \text{ values} = \Sigma\, [\Sigma\, X_{i\cdot}] = \Sigma\, [\Sigma X_{\cdot j}]$$

$$\bar{X} = \text{the grand mean} = \bar{\bar{X}}_{i\cdot} = \bar{\bar{X}}_{\cdot j}$$

The *model of the two-way analysis of variance* is expressed according to the X_{ij} values in a manner similar to the illustration used in obtaining the model of the one-way analysis of variance and can be written:

$$X_{ij} = \bar{X} + (\bar{X}_{i\cdot} - \bar{X}) + (\bar{X}_{\cdot j} - \bar{X}) + [(X_{ij} - \bar{X}_{i\cdot}) - (\bar{X}_{\cdot j} - \bar{X})]$$

Based on the model, the total deviation of an X_{ij} value from the grand mean $\bar{X}$ may be separated into a set of deviations:

$$(X_{ij} - \bar{X}) = (\bar{X}_{i\cdot} - \bar{X}) + (\bar{X}_{\cdot j} - \bar{X}) + [(X_{ij} - \bar{X}_{i\cdot}) - (\bar{X}_{\cdot j} - \bar{X})]$$

or $\left(\begin{array}{c}\text{Total}\\\text{deviation}\end{array}\right) = \left(\begin{array}{c}\text{Deviation }between\\\text{ith }row\text{ mean}\\\text{and grand mean}\end{array}\right) + \left(\begin{array}{c}\text{Deviation }between\\\text{jth }column\text{ mean}\\\text{and grand mean}\end{array}\right) + \left(\begin{array}{c}\text{Residual}\\\text{or error}\\\text{deviation}\end{array}\right)$

or identified as

$$(\underline{1}) = (\underline{2}) + (\underline{3}) + [(\underline{4}) - (\underline{3})]$$

in the diagrams on Chart 18–4. The numerical illustrations in the diagram on the chart are based on the deviations of X_{14} (= 10, the X value in the 1st row

CHART 18–4

DEVIATIONS FOR THE TWO-WAY ANALYSIS OF VARIANCE

Based on the value $X_{14} = 10$ (in A_1 row and B_4 column in Table 18-12); with $\bar{X}_{1\cdot} = 8$ (A_1 row mean), $\bar{X}_{\cdot 4} = 9$ (B_4 column mean), and $\bar{X} = 7$ (grand mean)

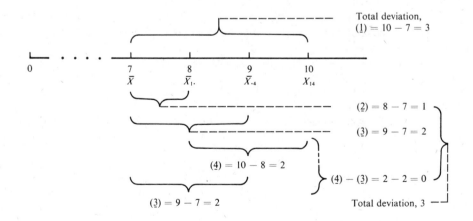

and 4th column) from the grand mean $\bar{X}$ (= 7) in Table 18–12.

From this equation of deviations, it can be proved that the corresponding sums of squared deviations are also equal.* The sums can be obtained as follows: first, add each group of c equations containing the X_{ij} values in the ith row. Next, add the r sums obtained in the first step, or

$$\Sigma \: [\Sigma \: (X_{ij} - \bar{X})^2] = \Sigma \: [\Sigma \: (\bar{X}_{i\cdot} - \bar{X})^2] + \Sigma \: [\Sigma \: (\bar{X}_{\cdot j} - \bar{X})^2]$$
$$+ \: \Sigma \: [\Sigma \: \{(X_{ij} - \bar{X}_{i\cdot}) - (\bar{X}_{\cdot j} - \bar{X})\}^2]$$

which may be written,

$$\left(\begin{matrix} Total \\ variation \end{matrix} \right) = \left(\begin{matrix} \text{Sum of variations} \\ \text{between} \\ row \text{ classes} \end{matrix} \right) + \left(\begin{matrix} \text{Sum of variations} \\ \text{between} \\ column \text{ classes} \end{matrix} \right) + \left(\begin{matrix} \text{Sum of residual} \\ \text{or } error \\ \text{variations} \end{matrix} \right)$$

or expressed by simplified symbols,

$$S_t = S_r + S_c + S_e$$

The individual sums of variations may be expressed in the following formulas for practical computations.

$$S_t = \Sigma \: (X - \bar{X})^2 = \Sigma X^2 - \frac{(\Sigma \: X)^2}{n} \qquad \text{(Formula 18–8)}$$

$$S_r = \Sigma \: c \: (\bar{X}_{i\cdot} - \bar{X})^2 = \frac{\Sigma \: (\Sigma \: X_{i\cdot})^2}{c} - \frac{(\Sigma \: X)^2}{n} \qquad \text{(Formula 18–9)}$$

$$S_c = \Sigma \: r \: (\bar{X}_{\cdot j} - \bar{X})^2 = \frac{\Sigma \: (\Sigma \: X_{\cdot j})^2}{r} - \frac{(\Sigma \: X)^2}{n} \qquad \text{(18–10)}$$

and $\qquad\qquad S_e = \Sigma \: [\Sigma \: \{(X_{ij} - \bar{X}_{i\cdot}) - (\bar{X}_{\cdot j} - \bar{X})\}^2]$

or simply $\qquad\qquad S_e = S_t - S_r - S_c \qquad\qquad\qquad\qquad\quad$ (18–11)

Formula (18–10) corresponding to formula (18–9) after exchanging c (number of columns) and $\bar{X}_{i\cdot}$ (row mean, or $\bar{X}_i$ in the one-way analysis of variance) with r (number of rows) and $\bar{X}_{\cdot j}$ (column mean, or $\bar{X}_j$ in the one-way analysis) respectively in the formulas.

When each of the four sums of variations is divided by its degrees of freedom, the quotient becomes an unbiased estimator of the population variance. The four estimated population variances are:

Estimated from the variations *between row* means,

$$\hat{s}_r^2 = \frac{S_r}{r - 1} \qquad\qquad \text{(Formula 18–5)}$$

Estimated from the variations *between column* means,

$$\hat{s}_c^2 = \frac{S_c}{c - 1} \qquad\qquad \text{(18–12)}$$

Estimated from the variations due to random chance or *error*,

---

*The method of proving the equation of variations is analogous to the method used in the one-way analysis of variance in the footnote of page 466.

$$\hat{s}_e^2 = \frac{S_e}{(c-1)(r-1)} \qquad (18\text{-}13)$$

Estimated from the *total* variation,

$$\hat{s}_t^2 = \frac{S_t}{n-1} \qquad \text{(Formula 13-1)}$$

There are two null hypotheses in the two-way analysis of variance:

I. H_0: The *row means* are equal, or $\mu = \mu_i$, $i = 1, 2, 3, \ldots, r$. Stated in a different way, there is no difference between the row means and the grand mean, or

$$\bar{X}_{i\cdot} - \bar{X} = 0 \text{ for all } i \text{ values}$$

The F ratio used for testing whether or not the *row* means are significantly different is

$$F_r = \frac{\hat{s}_r^2}{\hat{s}_e^2} \qquad (18\text{-}14)$$

If the hypothesis is true, the two independent estimates of the large population variance should not be greatly different; that is, $\hat{s}_r^2$ is close to $\hat{s}_e^2$. The value of F_r ratio thus should be close to unity or 1.

If the hypothesis is not true, the row means ($\bar{X}_{i\cdot}$) will differ from each other and also from the the grand mean $\bar{X}$ by more than the extent attributed to chance. Thus, the variance between row classes, $\hat{s}_r^2$, becomes large. On the other hand, the denominator $\hat{s}_e^2$ in the F_r formula is based on the sum of the error variations, or formula (18-11). The error deviations $(X_{ij} - \bar{X}_{i\cdot})$ and $(\bar{X}_{\cdot j} - \bar{X})$ are not affected by the differences of the row means $(\bar{X}_{i\cdot} - \bar{X})$, since $(X_{ij} - \bar{X}_{i\cdot})$ is obtained *within* the ith row class and $(\bar{X}_{\cdot j} - \bar{X})$ is obtained between the jth *column* mean and the grand mean. Therefore F_r is large if the hypothesis is not true.

II. H_0: The *column means* are equal, or $\mu = \mu_j$, $j = 1, 2, 3, \ldots, c$. Stated in a different way, there is no difference between the column means and the grand mean, or

$$\bar{X}_{\cdot j} - \bar{X} = 0 \text{ for all } j \text{ values}$$

The F ratio used for testing whether or not the column means are significantly different is

$$F_c = \frac{\hat{s}_c^2}{\hat{s}_e^2} \qquad (18\text{-}15)$$

If the hypothesis is true, the two independent estimates of the large population variance, $\hat{s}_c^2$ and $\hat{s}_e^2$ should not differ greatly. Thus the value of F_c ratio should be close to 1.

If the hypothesis is not true, the column means ($\bar{X}_{\cdot j}$) will differ significantly from each other and also from the grand mean $\bar{X}$. Thus, the variation between column means ($\bar{X}_{\cdot j} - \bar{X}$) becomes large and the numerator $\hat{s}_c^2$ also becomes large. Again, the denominator $\hat{s}_e^2$ in the F_c formula is based on the sum of the error variations. The first error deviation $(X_{ij} - \bar{X}_{i\cdot})$ is not affected by the differences of the column means. The second error deviation $(\bar{X}_{\cdot j} - \bar{X})$, which is also used in the numerator, is large. The denominator s_e^2 thus becomes relatively smaller than tne numerator s_c^2. The value of F_c therefore is large if the hypothesis is not true.

Example 8

Refer to Example 7. The employees were also divided into four groups based on numbers of years of college education: B_1, B_2, B_3, and B_4, representing one, two, three, and four years respectively. The test grades of the 12 employees in the sample are cross-classified in Table 18–12. The education director of the company wishes to know whether or not the means of the grades classified by (1) the three different instructors ($\bar{X}_i$'s) and (2) the four different years of college education of the employees ($\bar{X}_j$'s) are significantly different. Also, let the level of significance be (a) 5%, and (b) 1%.

TABLE 18–12

DATA FOR EXAMPLE 8*

Instructors	Employees				Total $\sum X_i.$	Mean $\bar{X}_i.$
	B_1	B_2	B_3	B_4		
	Grade points					
A_1	6	7	9	10	32	8
A_2	2	2	5	7	16	4
A_3	7	9	10	10	36	9
Total, $\sum X._j$	15	18	24	27	$\sum X = 84$	
Mean, $\bar{X}._j$	5	6	8	9		$\bar{X} = 7$

*Also see Table 18–8.

Solution.   The procedure of testing the hypothesis for Example 8 is presented in two parts below. In making the decisions in the two parts, we should be aware of the fact that each grade X_{ij} is the result of the *combined treatment* A_iB_j in the two-way classification, such as $X_{14} = 10$ points being the result of the treatment A_1B_4, or the employee taught by the first instructor (A_1) and having four years of college education (B_4). All variations thus are the *effects* of the combined treatments.

Part (1)—Testing the *row means* of variable A_i.

(1) H_0: The means of the grade points classified by the three different instructors are equal.

(2) Find the F_r ratio. The required sums S_t and S_r have already been computed on page 468 (where X_i has the same value as $X_i.$):

$$S_t = 90 \text{ and } S_r = 56$$

The other required sums S_c and S_e must be computed here.

$$S_c = \frac{\sum (\sum X._j)^2}{r} - \frac{(\sum X)^2}{n} = \frac{15^2 + 18^2 + 24^2 + 27^2}{3} - \frac{84^2}{12}$$

$$= 618 - 588 = 30 \qquad\qquad \text{(Formula 18–10)}$$

$$S_e = S_t - S_r - S_c = 90 - 56 - 30 = 4 \qquad \text{(Formula 18–11)}$$

The two terms of F_r ratio are:

$$\hat{s}_r^2 = \frac{S_r}{r-1} = \frac{56}{3-1} = \frac{56}{2} = 28 \qquad \text{(Formula 18–5)}$$

$$\hat{s}_e^2 = \frac{S_e}{(c-1)(r-1)} = \frac{4}{(4-1)(3-1)} = \frac{4}{6} \qquad \text{(Formula 18–13)}$$

Thus,

$$F_r = \frac{\hat{s}_r^2}{\hat{s}_e^2} = \frac{28}{4/6} = 42 \qquad \text{(Formula 18–14)}$$

(3) Make a decision. With $D_1 = 2$ and $D_2 = 6$, the F value is significantly large if it is above 5.14 at the 5% level or above 10.92 at the 1% level. The computed $F_r = 42$ is far above both 5.14 and 10.92. Thus, we reject the hypothesis at either 5% or 1% level. We may state the conclusion that the means of the grade points classified by the three different instructors are *not* equal. Different instructors (treatments) do produce different grades (effects). (Note: the computed $F = 7.41$ in Example 7 was concluded as probably significant. We were interested in only the effect from the single treatment A_i on the grade then.)

Part (2)—Testing the *column* means of variable B_j.

(1) H_0: The means of the grade points classified by the four different educational backgrounds (years of college education) are equal.

(2) Find the F_c ratio.

$$\hat{s}_c^2 = \frac{S_c}{c-1} = \frac{30}{4-1} = \frac{30}{3} = 10 \qquad \text{(Formula 18–12)}$$

$$F_c = \frac{\hat{s}_c^2}{\hat{s}_e^2} = \frac{10}{4/6} = 15$$

(3) Make a decision. With $D_1 = 3$ and $D_2 = 6$, the F value is significantly large if it is above 4.76 at the 5% level or above 9.78 at the 1% level. The computed $F_c = 15$ is far above both 4.76 and 9.78. Thus, we reject the hypothesis at either 5% or 1% level. We may state the conclusion that the means of the grade points classified by the four different educational backgrounds are *not* equal. Different educational backgrounds (treatments) do produce different grades (effects).

The entire procedure of testing the hypothesis for Example 8 is summarized in the analysis of variance table, Table 18–13.

The models introduced above are classified according to the number of ways of grouping data. There are other methods of classifying the models which can be used in the analysis of variance. An understanding of various types of the models is important to a researcher, since it enables him to design, conduct, and analyze an experiment more efficiently and inexpensively.

One of the other methods is to classify the models into *fixed effects, random effects,* and *mixed effects* according to our interest in particular information. A fixed model has *both* row and column classes selected for particular groups of information. In Example 8, we selected the three instructors A_1, A_2, and A_3 because we are particularly interested in the instructors in the company, and four

TABLE 18–13

ANALYSIS OF VARIANCE TABLE

(EXAMPLE 8)

Source of variation	Variation (Sum of squares)	Degrees of freedom	Variance (Mean square)	F Ratio Computed	F Ratio at 5%	F Ratio at 1%
Between row classes	$S_r = 56$	$D_1 =$ $r - 1 =$ $3 - 1 = 2$	$\hat{s}_r^2 = \dfrac{56}{2} = 28$	$F_r = \dfrac{\hat{s}_r^2}{\hat{s}_e^2}$ $= \dfrac{28}{4/6} = 42$	5.14	10.92
Between column classes	$S_c = 30$	$D_1 =$ $c - 1 =$ $4 - 1 = 3$	$\hat{s}_c^2 = \dfrac{30}{3} = 10$	$F_c = \dfrac{\hat{s}_c^2}{\hat{s}_e^2}$ $= \dfrac{10}{4/6} = 15$	4.76	9.78
Error	$S_e = 4$	$D_2 =$ $(c-1)(r-1) =$ $(4-1)(3-1) =$ 6	$\hat{s}_e^2 = \dfrac{4}{6}$			
Total	$S_t = 90$	$n - 1 = 11$		Test result—F_r: Reject H_0 at 5% and 1% levels. Test result—F_c: Reject H_0 at 5% and 1% levels.		

types of employees B_1, B_2, B_3, and B_4 because we are particularly interested in these types. The example illustrated a fixed model. Thus, the results of the analysis of variance can be applied only to the selected instructors and selected types of employees.

A random model has *both* row and column classes selected at random from available information. Refer to the data in Example 8. If we selected three instructors at random (assuming the company had instructors A_1, A_2, A_3, A_4, ..., and many others) and four types of employees at random (assuming the company had employees B_1, B_2, B_3, B_4, B_5, ..., and many others), we would have a random model. The results of the analysis of variance would be applicable to all instructors and all employees in the company.

A mixed model has *either* row *or* column classes (not both) selected at random and the other selected for particular information.

In one-way classification, the fixed effects and random effects models can further be classified into (1) *completely randomized design* and (2) *randomized block design*. Example 7 is a *fixed effects, completely randomized design, and one-way classification model*, since the three instructors are selected in a fixed manner and the four employees in each class are selected completely at random. The results of analysis of variance are applicable to only the instructors selected but to all employees in the company. If the four employees in each class represented particular types, called *blocks*, such as B_1, B_2, B_3, and B_4 as defined in Example 8, and the employees within each block had been assigned to instructors (or treatments) A_1, A_2, and A_3 by a random procedure, Example 7 would be a *fixed effects, randomized block design, and one-way classification model.* This one-way classification model, although identical to Example 8 in format, analyzes the effects only due to the single treatments A_1, A_2, and A_3.

Also, we may classify the models into types of having *one* observation per cell and n observations per cell in a table. In Example 8, there is only one grade in each cell, such as $X_{14} = 10$ points being the result of the treatment A_1B_4. Thus, it is a *one observation per cell model.* We could select n grades that are made by the employees who are taught by the first instructor (A_1) and who have four years of college education (B_4). If we continued to select n grades for each of other cells, we then would have an *n observations per cell model.* The different types of models will not affect the basic procedure of analyzing the variances. However, they do affect the methods of separating the total variations into equivalent groups and the interpretation of the results of the analysis.

18.6   Summary

This chapter first explained the nature of the F distribution. The letter F represents the variance ratio showing the relationship between two independently estimated population variances. The F ratio can be used in testing two types of hypotheses. (1) The hypothesis says that two estimated population variances based on samples are the same. When the two population variances are equal, the F ratio should be unity or 1. When F is not equal to 1, the difference from 1 may or may not be attributable to chance. How large should be the F ratio in order to determine that the difference is not due to chance (or reject the hypothesis)? We must rely on the F distribution. (2) The hypothesis says that three or more estimated population means based on samples are the same. The technique used in this test is called the analysis of variance. In this test, we also assume that the population variances are the same. When several populations have the same mean and the same variance, the populations are essentially the same population. Further, when we combine the populations into a single large poulation, we will expect the mean and the variance of the large population to be equal to those of the original populations. Then, if we take a random sample from each of three original populations, we may consider the three samples to be classes of a single large sample drawn from the single large population. We may now estimate the variance of the large population based on the samples by any one of the three methods:

1. Use the variance between classes (or between samples).
2. Use the variance within classes (or within individual samples).
3. Use the variance of the single large sample.

In computing the F ratio, we use the first two methods, or F = variance between classes ÷ variance within classes. The estimator by the third method may serve as a checking point in the process of the analysis of variance. The F ratio may be large (the variance between classes is greater than the variance within classes) or

SUMMARY OF FORMULAS

Application	Formula	Formula number	Reference page
Variance ratio (F) test			
General	$F = \dfrac{\hat{s}_1^2}{\hat{s}_2^2}$	18–1	452
or	$F = \dfrac{\chi_1^2/D_1}{\chi_2^2/D_2}$	18–2	453
Analysis of variance— *fundamental procedure*			
Variance between classes	$\hat{s}_1^2 = \dfrac{\sum n_i(\bar{X}_i - \bar{X})^2}{r - 1}$	18–3	459
Variance within classes	$\hat{s}_2^2 = \dfrac{\sum [\sum (X_i - \bar{X}_i)^2]}{n - r}$	18–4	459
One-way analysis of variance	$\hat{s}_1^2 = \dfrac{S_r}{r - 1}$	18–5	466
	$\hat{s}_2^2 = \dfrac{S_w}{n - r}$	18–6	466
	$S_w = S_t - S_r$	18–7	466
	$S_t = \sum X^2 - \dfrac{(\sum X)^2}{n}$	18–8	467
	$S_r = \dfrac{\sum (\sum X_i)^2}{c} - \dfrac{(\sum X)^2}{n}$	18–9	467
Two-way analysis of variance	$S_c = \dfrac{\sum (\sum X_{.j})^2}{r} - \dfrac{(\sum X)^2}{n}$	18–10	472
	$S_e = S_t - S_r - S_c$ (For S_r, use formula 18–9, where $X_t = X_{i.}$)	18–11	472
	$\hat{s}_r^2 = \hat{s}_1^2 = \dfrac{S_r}{r - 1}$	18–5	466
	$\hat{s}_c^2 = \dfrac{S_c}{c - 1}$	18–12	472
	$\hat{s}_e^2 = \dfrac{S_e}{(c - 1)(r - 1)}$	18–13	473
	$F_r = \dfrac{\hat{s}_r^2}{\hat{s}_e^2}$	18–14	473
	$F_c = \dfrac{\hat{s}_c^2}{\hat{s}_e^2}$	18–15	473

small (the variance between classes is close to the variance within classes). A large F ratio may indicate that the estimated population means based on samples are significantly different. How large should the F ratio be in order to reject the hypothesis? This decision is usually based on the right-tail test according to the F distribution.

The fundamental procedure summarized above is also used in the one-way analysis of variance. In the one-way analysis of variance, the given data is grouped according to one-way classification. There is only one variable in the analysis. This chapter also introduced the model of the two-way analysis of variance. In the two-way table, the given data are cross-classified: one classification for rows and another for columns. The two-way classifications give two independent variables A_i and B_j in the table. When we analyze the two variables simultaneously, we must investigate the variations between row classes as well as the variations between column classes.

Exercises 18

1. What is the F ratio? What are the hypotheses which can be tested by the use of the F distribution?

2. What are the assumptions that have been made for the analysis of variance?

3. What is the difference between one-way analysis of variance and two-way analysis of variance? Why should we select the variance between classes and the variance within classes as the two terms of the F ratio in the one-way analysis of variance?

4. What are the fixed, random, and mixed types of models which can be used in the analysis of variance?

5. Given: $\Sigma x_1^2 = 64$, $n_1 = 5$; $\Sigma x_2^2 = 12$, $n_2 = 7$. Compute the F ratio. Test the hypothesis that the two variances are the same at (a) 5%, and (b) 1% level of significance.

6. Given: $s_1 = 21$, $n_1 = 8$; $s_2 = 15$, $n_2 = 31$. Compute the F ratio. Test the hypothesis that the two variances are the same at (a) 5%, and (b) 1% level of significance.

7. Refer to Problem 8, Exercises 15, page 385. Test the hypothesis that the variances of the weights of the turkeys in the two farms are the same. Let the level of significance be (a) 0.01 and (b) 0.05.

8. Two different methods, X and Y, are used in producing a certain type of product by the workers in a factory. A sample of units produced per hour is taken for each method. The samples are shown in the following table. Test the hypothesis that the variances of the two populations from which the two samples are drawn are the same. Let the level of significance be 0.05.

SAMPLE No. 1—METHOD X		SAMPLE No. 2—METHOD Y	
Worker number	Units produced per hour	Worker number	Units produced per hour
1	21	1	10
2	12	2	13
3	14	3	15
4	15	4	17
5	18	5	20
		6	21
		7	23

9. A personnel manager took a sample of letters typed per hour by 15 typists. He classified the results into four groups according to years of experience of the typists. Test the hypothesis that the means of the letters typed of the four groups are the same. Let the level of significance be 5%.

Typist number	Years of experience			
	1	2	3	4
	Number of letters typed per hour			
1	2	4	10	7
2	3	5	7	8
3	3	3	10	9
4	4	6	9	

10. Assume that the dean of a college took a sample of freshman English grades of 15 students. He classified the grades on the basis of teachers who taught the students as shown in the following table. Test the hypothesis that the means of the grades of the three classes are the same. Let the level of significance be 5%.

Teacher	Student number				
	1	2	3	4	5
	Grade				
A	2	3	6	10	4
B	3	7	3	10	7
C	4	5	9	8	9

11. Refer to Problem 10. Suppose that the dean also classified the grades on the basis of high schools from which the students were graduated. He reconstructed the table as follows:

Teacher	High school				
	U	V	W	Y	Z
	Student's grade				
A	2	3	6	10	4
B	3	7	3	10	7
C	4	5	9	8	9

The dean wishes to know whether or not the means of the grades classified by (1) the three different teachers and (2) the five different high schools are significantly different. Let the level of significance also be 5%.

12. Four types of corn were planted in separate plots and fertilized by three kinds of fertilizer individually. A sample of the yields in pounds per plot for 12 plots is taken and shown in the table below. Test (a) the null hypothesis that there is no difference between the means of the yields of the four types of corn, and (b) the null hypothesis that there is no difference between the means of the yields by applying the three kinds of fertilizer. Let the level of significance be 5%.

Types of corn	Kinds of fertilizer		
	B_1	B_2	B_3
	Yields in pounds		
A_1	2	3	4
A_2	4	3	5
A_3	8	7	9
A_4	2	7	6

19 Decision-Making under Conditions of Uncertainty

The process of decision-making for problems under conditions of uncertainty is discussed in this chapter. This type of problem involves various possible *events*, the exact occurrence of any one of which is uncertain. The uncertainty provides many *alternative actions* for the decision-maker. The question is how to establish the criteria upon which the *best* action can be selected from the various alternatives. In general, the selection is accomplished by finding the *highest expected value* from all available information based on the application of probability concepts.

19.1 The Payoff Table

In making a decision for a problem under conditions of uncertainty, the decision-maker should list the following related items for analysis:

1. List *all* possible *states of nature*, called events, in the given problem. The list of events thus is *exhaustive*. The events are also *mutually exclusive* since the occurrence of any one of them excludes the occurrences of the other; that is, two or more of the events cannot occur together. The events are *uncertain* since we do not know which one of them will occur.
2. List *all* possible *acts* that the decision-maker may take.
3. List all *consequences*. Each consequence is the outcome or result of an act under a given event.

These items may be presented systematically in a table, such as the one shown in Table 19–1 below. A table which shows the relationships among all possible events, all possible acts, and all related consequences (or sometimes the values associated with the consequences) is called a *payoff table*, *decision matrix*, or *payoff matrix*. A "payoff" is simply a consequence, regardless of whether it is favorable or

482

unfavorable. For example, the payoffs may be profits, either positive numbers (representing profits) or negative numbers (representing losses).

Example 1

Assume that a faculty club in a university purchases steak dinners for resale. According to past experience, the numbers of dinners sold vary from 11 to 14. The cost of each dinner is $4 and the selling price of each dinner is $6. If the dinner is not sold during the day, it cannot be returned and thus has no value. Construct a payoff table to show all possible profits.

Solution.   The payoff table for Example 1 is shown in Table 19–1.

TABLE 19–1

A Payoff Table (Example 1)

Possible daily sales (Events) in number of dinners	Possible number of dinners stocked (Acts)			
	11	12	13	14
	Possible profit by each act (Consequences)			
11	(11×6) $-(11 \times 4) = \$22$	(11×6) $-(12 \times 4) = \$18$	(11×6) $-(13 \times 4) = \$14$	(11×6) $-(14 \times 4) = \$10$
12	(11×6) $-(11 \times 4) = 22$	(12×6) $-(12 \times 4) = 24$	(12×6) $-(13 \times 4) = 20$	(12×6) $-(14 \times 4) = 16$
13	(11×6) $-(11 \times 4) = 22$	(12×6) $-(12 \times 4) = 24$	(13×6) $-(13 \times 4) = 26$	(13×6) $-(14 \times 4) = 22$
14	(11×6) $-(11 \times 4) = 22$	(12×6) $-(12 \times 4) = 24$	(13×6) $-(13 \times 4) = 26$	(14×6) $-(14 \times 4) = 28$

The profit by each act is computed by the following expression:

$$\left(\begin{array}{c}\text{Number of dinners possibly sold}\\ \text{(limited by number stocked)}\\ \times\\ \text{Selling price per dinner}\end{array}\right) - \left(\begin{array}{c}\text{Number of dinners stocked}\\ \times\\ \text{Cost per dinner}\end{array}\right)$$

For example, if the demand event (in the possible daily sales column) is 13 dinners and the number of dinners stocked is 12, the profit is computed as follows:

$$[12 \text{ (dinners sold)} \times \$6] - [12 \text{ (dinners stocked)} \times \$4] = \$72 - \$48 = \$24$$

When 12 are stocked, only 12 can be sold, even if the demand event is 13 dinners or more. On the other hand, if the demand event (such as 11) is less than the number stocked (such as 12), the profit is reduced by the cost of overstocked dinners at $4 each.

19.2   Conditional and Expected Value

The profits shown in Table 19–1 are frequently referred to as *conditional values* since each profit is computed under certain conditions; that is, a specific number of dinners is sold (certain event occurs) and a specific number of dinners is stocked (certain act is taken). However, there is still uncertainty because the exact event which is going to occur is unknown. A decision-maker thus cannot choose the *best* or *optimum* act based only on the conditional values.

Although the exact occurrence of each event is not known, the probability of occurrence of each event can be assigned. A general way to assign the probability is based on historical data as shown in Example 2. The sum of the probabilities of all events is 1. The conditional value of each event under a given act is weighted by the probability of the event occurring. The sum of the weighted conditional values is called the *expected value* for the act. The optimum act is the one with the *highest* expected value (or profit).

The term *expected value* is generally used to represent an average of a set of values. The average used in this chapter is a weighted arithmetic mean, the weights being the probabilities of the individual values. Thus, the term will be used not only for conditional values but also for other types of values, such as the expected sales in Example 2 and the expected losses in Example 4.

Example 2

Refer to Example 1. The following additional information concerning daily sales during the recent 200 days is obtained from the faculty club:

Daily Sales (Number of Dinners)	Number of Days
11	20
12	60
13	80
14	40
Total	200

Compute: (a) The probability of each event of daily sales. (b) The expected sales for each day. (c) The expected profit for each act.

Solution. (a) The probability of each event is obtained by dividing the number of days for the event by the total number of days, 200. Thus, the probability of

---

TABLE 19–2

SOLUTION TO EXAMPLE 2(a) AND (b)

(1) Daily Sales (Event)	(2) Number of Days	(3) Probability of Each Event Col. (2) ÷ 200	(4) Expected Daily Sales Col. (1) × Col. (3)
11 *dinners*	20	0.10	1.10
12	60	0.30	3.60
13	80	0.40	5.20
14	40	0.20	2.80
Total	200	1.00	12.70

the event of 11 dinners is $20 \div 200 = 0.10$. The probabilities of all events are computed in the same manner and are shown in the third column of Table 19-2.

(b) The expected sales for each day based on the given information are computed in column (4) in Table 19-2. The daily sales for each event are multiplied by the probability of the event. The sum of the products, 12.70 dinners, is the expected sales per day.

(c) The expected profit for each act is computed in Table 19-3. The probability of each event is obtained from Table 19-2 and the conditional profits by each act are obtained from Table 19-1. The expected profit of the act of stocking 12 dinners is the highest value, $23.40. Thus, the optimum act is to stock 12 dinners each day.

TABLE 19-3

COMPUTING EXPECTED PROFITS (EXAMPLE 2(c))

Event (Dinners)	Probability of Event	Act (number of dinners stocked)							
		11		12		13		14	
		Calculations of Expected Profit (Conditional Profit × Probability of Event)							
		Profit		Profit		Profit		Profit	
		Conditional	Expected	Conditional	Expected	Conditional	Expected	Conditional	Expected
11	0.10	$22	$2.20	$18	$1.80	$14	$1.40	$10	$1.00
12	0.30	22	6.60	24	7.20	20	6.00	16	4.80
13	0.40	22	8.80	24	9.60	26	10.40	22	8.80
14	0.20	22	4.40	24	4.80	26	5.20	28	5.60
Total	1.00		$22.00		$23.40		$23.00		$20.20

Highest
(Optimum Act)

19.3   Expected Value of Perfect Information

Information which can be used by a business manager to change occurrences of events from uncertainty to certainty is called the *perfect information* or *perfect predictor*. The manager is usually interested in the cost of obtaining the perfect information. If the cost is higher than the additional profit derived from the information, he should not seek the information. The decision of whether or not to seek more information is commonly based on the *expected (or average) value of perfect information*.

Referring to Table 19-3, the expected profits were computed under the conditions of uncertainty—the exact occurrence of each event is unknown to the manager of the faculty club. The expected profit of the optimum act (stocking 12 dinners) under uncertainty is $23.40. Now, suppose that with perfect information in advance the manager can predict with certainty the demand of dinners in each day. The *best* act for each event is to stock the number of dinners in each day exactly equal to the demand. By doing so, the profit of each act is maximized. For example, if the manager knew tomorrow's demand would be 12 dinners, he would stock 12 dinners for the highest possible profit of $24. If he took any other action, such as only stocking 11 or 13 dinners, he would receive a smaller profit. The highest profit for each act can be obtained from Table 19–3. For convenience, the profits (or conditional values) are reproduced in Table 19–4.

When each conditional profit in Table 19–4 is multiplied, or weighted, by the probability of the corresponding event and the products are added, the *expected profit under certainty* is obtained. The expected profit under certainty is the highest *average* profit that the manager can receive from the varying demand (11, 12, 13, and 14 dinners) if he can get the perfect information daily in advance. The difference between the expected profit under certainty and the expected profit of the optimum act under uncertainty thus sets the limit for the manager to pay for the perfect information. The limit is called *expected value of perfect information.*

Example 3

Refer to Example 2. Compute: (a) the expected profit under certainty and (b) the expected value of perfect information.

Solution.   (a) The expected profit under certainty is $25.40. The computation is shown in the last column of Table 19–4.

(b) Expected profit under certainty $25.40
    Less: Expected profit of the optimum act (stocking 12
    dinners) under uncertainty   23.40
    Expected value of perfect information (See Tables 19–3
    and 19–4.) $ 2.00

TABLE 19–4

CONDITIONAL AND EXPECTED PROFIT UNDER CERTAINTY (EXAMPLE 3)

(1) Event (Dinners)	(2) Probability of Event	Act (number of dinners stocked)				(4) Expected Profit Under Certainty (2) × (3)
		11	12	13	14	
		(3) Conditional Profit Under Certainty (See Table 19–3)				
11	0.10	$22				$ 2.20
12	0.30		$24			7.20
13	0.40			$26		10.40
14	0.20				$28	5.60
Total	1.00					$25.40

19.4   The Conditional Loss Table

Instead of using conditional profits to find the optimum act and the expected value of perfect information, we may use *conditional losses* to find the same answers. However, in finding the optimum act by using conditional profits, we *maximize* the expected profit (see Table 19–3). Conversely, by using conditional losses, we must *minimize* the expected loss. The optimum act is the act with the highest expected profit, and also the smallest expected loss. The function of perfect information is to reduce the loss for any act under uncertainty to zero. Thus, the expected loss of the optimum act equals the expected value of perfect information.

The conditional losses which are incurred due to the uncertainty of event occurrences when a given action is taken can be classified into two groups (see Table 19-5). (1) Cost losses—caused by stocking more units than demand. A cost loss = unit *cost* × number of units overstocked. (2) Opportunity losses—caused by stocking fewer units than demand. An opportunity loss = unit *profit* × number of units understocked.

The conditional loss is zero if the action is the best for that event, or number of units stocked = number of units demanded.

Example 4

Refer to Example 1 (page 483) and Example 2 (page 484). (a) Construct a conditional loss table, (b) compute the expected loss for each act, and (c) indicate the optimum act based on the lowest expected loss.

Solution.   (a) The conditional loss table is Table 19-5. In constructing the loss table, first place zeros in the *diagonal*. Each zero indicates the best action for a given event, such as stocking 11 if the demand is 11 and stocking 12 if the demand is 12 dinners (the loss in either case is zero.)

The losses above the diagonal of zeros are cost losses. Each cost loss = \$4 (unit cost) × units overstocked. For example, if we stocked 13 units and the demand for the day is 11 units, the cost loss = \$4 × (13 − 11) = \$4 × 2 = \$8.

The losses below the diagonal of zeros are opportunity losses. Each opportunity loss = \$2 (unit profit: selling price \$6 − cost \$4) × units understocked. For example, if we stocked 11 units and the demand for the day is 14 units, the opportunity loss = \$2 × (14 − 11) = \$2 × 3 = \$6. We have lost the *opportunity* to sell 3 units, or make \$6 profit, since we have stocked 3 units less than the demand.

(b)   The expected loss for each act is computed in Table 19–6. The probability of each event is again obtained from Table 19-2 and the conditional losses by each act are obtained from Table 19-5. The conditional loss by a given act for each event is multiplied by the probability of the event for obtaining the expected loss.

(c)   The expected loss of the act of stocking 12 dinners is the lowest value, \$2.00. Thus, the optimum act is to stock 12 dinners each day in the faculty club. (The same answer as that of Example 2.)

TABLE 19–5

THE CONDITIONAL LOSS TABLE [EXAMPLE 4(a)]

Event (Dinners demand)	Act (number of dinners stocked)				
	11	12	13	14	
	Conditional Loss				
11	0	4	8	12	Cost losses:
12	2	0	4	8	$4 (unit cost) × units
13	4	2	0	4	overstocked.
14	6	4	2	0	

Opportunity losses: $2 (unit profit: selling price $6 − cost $4) × units understocked.

TABLE 19–6

COMPUTING EXPECTED LOSSES (EXAMPLE 4(b))

Event (Dinners)	Proba-bility of Event	Act (number of dinners stocked)							
		11		12		13		14	
		Calculations of Expected Loss (Conditional Loss × Probability of Event)							
		Loss		Loss		Loss		Loss	
		Condi-tional	Ex-pected	Condi-tional	Ex-pected	Condi-tional	Ex-pected	Condi-tional	Ex-pected
11	0.10	$0	$0.00	$4	$0.40	$8	$0.80	$12	$1.20
12	0.30	2	0.60	0	0.00	4	1.20	8	2.40
13	0.40	4	1.60	2	0.80	0	0.00	4	1.60
14	0.20	6	1.20	4	0.80	2	0.40	0	0.00
Total	1.00		$3.40		$2.00		$2.40		$5.20

↑
Lowest
(Optimum Act)

Note. (1) For a given event, the sum of each conditional profit in Table 19-3 and a corresponding conditional loss in Table 19–5 under uncertainty is equal to the conditional profit of the best act under certainty in Table 19-4. For example, for the event of selling 11 dinners a day, the following values may be checked:

	Act (number of dinners stocked)			
	11	12	13	14
Conditional profit under uncertainty (Table 19-3)	$22	$18	$14	$10
Add: Conditional loss under uncertainty (Table 19-5)	0	4	8	12
Conditional profit under certainty (Table 19-4)	$22	$22	$22	$22

(2)   The sum of expected profit and expected loss of any act under uncertainty is equal to the expected profit under certainty:

	Act (number of dinners stocked)			
	11	12	13	14
Expected profit under uncertainty (Table 19–3)	$22.00	$23.40	$23.00	$20.20
Add: Expected loss under uncertainty (Table 19–6)	3.40	2.00	2.40	5.20
Expected profit under certainty (Table 19–4)	$25.40	$25.40	$25.40	$25.40

(3)   The expected loss of the optimum act, stocking 12 dinners, is $2 which is equal to the expected value of perfect information obtained in Example 3(b).

(4)   In the previous illustrations, it was assumed that the overstocked dinners at the end of each day were completely worthless. However, if the overstocked dinners have some salvage value, such as $1 for each dinner, this value must be included in computing the conditional profits and losses for each action. For example, when the demand event is 11 dinners and the act is stocking 13 dinners, the conditional profit is:

$$11 \text{ (dinners sold)} \times \$6 \text{ (price per dinner)} = \$66$$
$$\text{Less: } 13 \text{ (dinners stocked)} \times \$4 \text{ (cost per dinner)} = \underline{52}$$
$$\text{Conditional profit without salvage value   } \$14 \text{ (Table 19–1)}$$
$$\text{Add: } 2 \text{ (dinners overstocked)} \times \$1 \text{ (salvage}$$
$$\text{value per dinner)} = \underline{2}$$
$$\text{Conditional profit with salvage value   } \$16$$

An alternative way to compute the conditional profit is:

$$\text{Profit per dinner} = \$6 - \$4 = \$2$$
$$\text{Net loss per dinner overstocked} = \$4 - \$1 = \$3$$
$$11 \text{ (dinners sold)} \times \$2 \qquad = \$22$$
$$2 \text{ (dinners overstocked)} \times \$3 = \underline{6\,(-)}$$
$$\text{Conditional profit} \qquad \$16$$

The conditional loss is:

$$\$3 \text{ (net loss per dinner)} \times 2 \text{ (overstocked)} = \$6$$

Check:

	Event 11, Act 13
Conditional profit under uncertainty	$16
Add: Conditional loss under uncertainty	6
Conditional profit under certainty (of the best act: stocking 11 since the demand event is 11—Table 19–4)	$22

The procedures of computing various expected values involving salvage values are the same as those illustrated above involving no salvage values.

19.5 Expected Utility

In the previous sections every consequence of an act under a given event was expressed in monetary value. The rule of finding the optimum act from all possible acts is to choose the act that has either the highest expected monetary gain (profit) or the lowest expected monetary loss. The decision-maker considers each dollar equally important under this rule. However, there are cases under conditions of uncertainty in which a decision-maker may consider each dollar not equally important. For example, a $1,000 potential loss may not be a serious problem to a person when he is rich, but it may become a very important matter to him when he is poor. The amount of risk that a person is willing to take, therefore, largely depends upon his personal opinions about money. In those cases, an expected monetary value is not an appropriate guide for making decisions. The *expected utility value*, derived from personal opinion, can be used as a guide instead.

Decisions Not Based on Expected Monetary Values

Example 5

A college student received $2000 from his father in May of this year. He may keep this money until September for the payment of tuition fee for the next semester, or he may invest his money in a restaurant at a beach during the summer. If the weather is good, he will be able to make a net profit of $5000. If the weather is bad, he will lose $2000. The chance of having a good weather during the summer is 30 percent and that of a bad weather is 70 percent. The expected monetary value of this investment is (also see Table 19–7)

$$(\$5000 \times 0.30) + (-\$2000 \times 0.70) = \$1500 - \$1400 = \$100 \text{ (profit)}$$

The expected monetary value of not making the investment is $0. However, the student *decided not to make* the investment since he did not wish to risk his education for the uncertain profit.

TABLE 19–7

COMPUTING EXPECTED MONETARY VALUES (EXAMPLE 5)

Event	Probability of Event	Act			
		Invest		Not Invest	
		Conditional Profit	Expected Profit	Conditional Profit	Expected Profit
Good weather	0.30	$5000	$1500	$0	$0
Bad weather	0.70	−$2000	−$1400	$0	$0
Total	1.00		$ 100		$0
			↑ Higher		

Example 6

A man owns a house valued at $20,000. Statistical records showed that the probability of having fire in his type of house is one out of 1000 (or $1/1000 = 0.001$). However, the insurance company said that the fire insurance premium on his house would be $100 per year. Should the man buy the fire insurance policy?

The expected monetary value of buying the insurance policy is $-\$100$ (or $100 loss) and that of not buying is $-\$20$ (or $20 loss). However, the owner *decided to buy* the fire insurance (having the larger loss) since he could not afford to risk having his house destroyed by fire and recovering nothing. (See Table 19–8.)

TABLE 19–8

COMPUTING EXPECTED MONETARY VALUES (EXAMPLE 6)

Event	Probability of Event	Act			
		Buy Insurance		Not Buy Insurance	
		Conditional Value	Expected Value	Conditional Value	Expected Value
No Fire	0.999	−$100	−$99.90	$0	$ 0
Fire	0.001	−$100	−$ 0.10	−$20,000	−$20
Total	1.000		−$100.00		−$20

↑
Higher Value
(Smaller loss)

These two examples indicate that the decision-makers did not use the expected monetary values as the bases for decision-making. The college student did not choose the act of investment although that act has the highest expected gain ($100 profit.) The home-owner did not choose the act of not buying fire insurance although that act has the lowest expected loss ($20 loss.) The examples illustrate cases under which an expected monetary value is not an appropriate guide for making decisions. However, the consequence of an act under a given event may be expressed in *utility* value instead of monetary value. When this is done, the rule of finding the optimum act from all possible acts is to choose the act that has the highest expected utility value. This rule will give answers consistent with the decisions made in the above examples.

Construction of Utility Values

Utility is a measurement of money based on the decision-maker's opinion of monetary value under conditions of uncertainty. A decision-maker may or may not consider every dollar to be equally important. Thus, utility values may or may not be proportional to monetary values represented, although a higher utility value usually represents a higher monetary value. When a set of utility values is proportional to a set of monetary values, the relationship is said to be *linear* (such as the representation listed below and shown in Chart 19–1).

Monetary value (Unit: dollar)	Utility value (Unit: utile)
$ 0	0
100	10
300	30
500	50

CHART 19–1

LINEAR RELATIONSHIP BETWEEN MONETARY VALUES
AND THEIR UTILITY VALUES

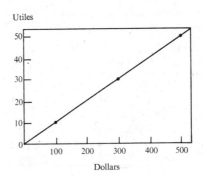

However, in many cases, the relationship between monetary values and their corresponding utility values is not linear. The procedure of constructing a set of utility values for a set of monetary values, when the relationship is not linear, is illustrated below.

(1)   Select two monetary values and assign their corresponding utility values. The two monetary values should have a fairly wide range, such as $0 and $5000. Their utility values are *assigned arbitrarily*, such as letting zero be the utility for $0 and 100 be the utility for $5000. For convenience, let the symbol

U (a monetary value) = the utility value of the given monetary value.

Then the two selected monetary values and their assigned utility values may be written symbolically as follows:

$$U (\$0) = 0$$
$$U (\$5,000) = 100$$

(2)   Compute the third utility for the decision-maker. The utility must reflect the maker's opinion of *indifference* between an amount of money with certainty and a lottery which will offer to him the two selected monetary values in (1) above with uncertainty but known probability distribution. The amount with certainty should be larger than $0 and smaller than $5000, preferably close to the lower value $0.

Assume that we are constructing a set of utility values for the decision-maker, Mr. Smith. He is indifferent between (A) $1000 for certain and (B) a lottery which will give $0 with 0.5 probability and $5000 with 0.5 probability also. The sum of the two probabilities must be equal to 1, or $0.5 + 0.5 = 1$. Since the amount in alternative (A) is not different than the amounts in alternative (B), their corresponding utilities must be equal also, or

$U(\$1,000) = U(\$0$ with 0.5 probability and $5,000$ with 0.5 probability)

The utility of a lottery is the expected utility of its component amounts. Thus,

$$U(\$1,000) = [0.5 \times U(\$0)] + [0.5 \times U(\$5,000)] \qquad \text{[See (1)]}$$
$$= (0.5 \times 0) + (0.5 \times 100) = 0 + 50 = 50$$

(3)   Compute the fourth utility for the decision-maker. The basic steps of computing the fourth utility are similar to those illustrated above. Assume that Mr. Smith is indifferent between (A) $2000 for certain and (B) a lottery which will give $1000 with 0.5 probability and $5000 with 0.5 probability. What is the utility value of $2000?

$$U(\$2000) = [0.5 \times U(\$1000)] + [0.5 \times U(\$5000)]$$
$$= (0.5 \times 50) + (0.5 \times 100)$$
$$= 25 + 50 = 75$$

Note: 1. The amounts included in the lottery are selected from the amounts whose utilities are already available. $U(\$1000) = 50$ is obtained from (2) and $U(\$5000) = 100$ is obtained from (1) above.

2. The utility of a higher monetary value, such as $U(\$2000)$ should be numerically larger than the utility of a lower monetary value, such as $U(\$1000)$. This concept is important in designing a lottery. The lottery in (3) is based on the lottery in (2) with the same probability distribution (0.5 and 0.5). Notice that the amount $0 in the old lottery is increased to $1000 in the new lottery.

(4) Compute the fifth utility for the decision-maker. Assume that Mr. Smith is indifferent between (A) $3000 for certain and (B) a lottery having $1000 with 0.2 probability and $5000 with 0.8 probability. What is the utility of $3000?

$$U(\$3000) = [0.2 \times U(\$1000)] + [0.8 \times U(\$5000)]$$
$$= (0.2 \times 50) + (0.8 \times 100) = 10 + 80 = 90$$

Note: The lottery in (4) is based on the lottery in (3) with the same amounts, $1000 and $5000. Notice that the probability of the larger amount, $5000, has been increased from 0.5 to 0.8. The increase is consistent with the expectation that the utility of $3000 is larger than the utility of $2000. When the probability of the larger amount is increased, the utility is also increased, from 75 in (3) to 90 in (4).

(5) Compute the sixth utility. The utilities from (2) to (4) were computed for the amounts within the range of the two selected monetary values, $0 and $5000. However, it is also possible to compute the utilities of the amounts either below

$0 or above $5000. The utility of an amount above $5000 is computed in this step.

Assume that Mr. Smith is indifferent between (A) $5000 for certain and (B) a lottery having $1000 with 0.2 probability and $10,000 with 0.8 probability. What is the utility of $10,000?

$$U(\$5000) = [0.2 \times U(\$1000)] + [0.8 \times U(\$10,000)]$$
$$100 = (0.2 \times 50) + [0.8 \times U(\$10,000)]$$
$$100 = 10 + [0.8 \times U(\$10,000)]$$
$$U(\$10,000) = (100 - 10)/0.8 = 112.5$$

Note: The lottery in (5) is based on the lottery in (4) with the same probability distribution (0.2 and 0.8). Notice that the amount $5000 in the old lottery is increased to $10,000 in the new lottery. The increase is consistent with the fact that the utility of $5000 is larger than the utility of $3000 [$U(\$5000) = 100$ and $U(\$3000) = 90$]. When the utility is increased, the amount (or amounts) in the new lottery should also be increased to a reasonable extent.

(6) Compute the seventh utility. The utility of an amount below $0 (a negative amount representing a loss) is computed in a similar manner.

Assume that Mr. Smith is indifferent between (A) $0 for certain and (B) a lottery which will give a loss of $2000 (or $-$2000) with 0.6 probability and a gain of $3000 with 0.4 probability. Find the utility of $2000 loss.

$$U(\$0) = [0.6 \times U(-\$2000)] + [0.4 \times U(\$3000)]$$
$$0 = [0.6 \times U(-\$2000)] + [0.4 \times 90]$$
$$U(-\$2000) = (0 - 36)/0.6 = -60.$$

Note: The lottery in (6) is different from any lottery illustrated in the previous steps, (2) to (5). It has a different probability distribution (0.6 and 0.4) and different amounts ($-$2000 and $3000). However, the answer must be consistent with the basic concept that a higher (or lower) utility value represents a higher (or lower) monetary value. The example in (6) illustrates another way to design a lottery for the decision-maker's approval.

(7) Construct a utility curve (Chart 19–2). The monetary values and their corresponding utility values, either assigned or computed in (1) to (6) above, are listed below and plotted on Chart 19–2. The points plotted on the chart give us enough indication to draw a smoothed curve which represents the utility function for the decision-maker, Mr. Smith. From this utility curve, we may read the utility representing a desired monetary value. For example, the utility of $4000 is 95, as shown by the broken line intersecting the curve.

The probability distribution of two monetary values in a lottery, which is equivalent to an amount for certain in the decision-maker's opinion, may also be derived from the utility curve. For example, a lottery which is equivalent to $4000 for certain has the two amounts $1000 (which is smaller than $4000) and $5000 (which is large than $4000). Find the probability distribution of the two amounts.

Let P be the probability of $1000. Then, the probability of $5000 must be $(1 - P)$.

$$U(\$4000) = [P \times U(\$1000)] + [(1 - P) \times U(\$5000)]$$
$$95 = [P \times 50] + [(1 - P) \times 100]$$
$$95 = 50P + 100 - 100P$$
$$50P = 5$$
$$P = 5/50 = 0.10 \text{ or } 10\%$$
$$1 - P = 1 - 10\% = 90\%$$

Monetary value (Unit: Dollar)	Utility value (Unit: Utile)
− $2,000	−60
$ 0	0
1,000	50
2,000	75
3,000	90
5,000	100
10,000	112.5

CHART 19–2

MR. SMITH'S UTILITY CURVE

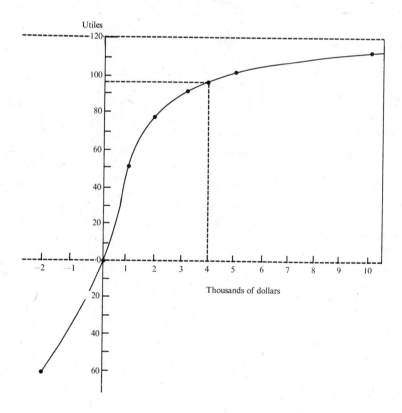

Thus Mr. Smith is indifferent between (A) $4000 for certain and (B) a lottery giving $1000 with 10 percent probability and $5000 with 90 percent probability.

As mentioned earlier, when the expected monetary value is not an appropriate guide for making decisions, we may use the expected utility value for the guidance. Examples 7 and 8 are used to illustrate the use of expected utility values.

Applications of Expected Utility Values

Example 7

Refer to Example 5, page 490. Suppose that Mr. Smith is the student mentioned in the example. What would be his decision if he uses expected utility values as a basis for decision-making?

Solution. The expected utility value of making the investment is -12 and the expected utility value of not making the investment is 0. (See Table 19–9.) Thus, Mr. Smith should not invest his money in the restaurant business. (Zero is larger than -12.)

TABLE 19–9

COMPUTING EXPECTED UTILITY VALUES (EXAMPLE 7)

Event	Proba-bility of Event	Act					
		Invest			Not Invest		
		Conditional Profit		Expected Utility	Conditional Profit		Expected Utility
		Dollars	Utiles		Dollars	Utiles	
Good weather	0.30	$5,000	100	30	$0	0	0
Bad weather	0.70	$-2,000	-60	-42	$0	0	0
Total	1.00			-12			0

↑
Higher

Example 8

Refer to Example 6, page 491. Assume that the home owner's utility values assigned to representing monetary values are as follows:

$$U(-\$20,000) = -2000$$
$$U(-\$100) = -1$$
$$U(\$0) = 0$$

Based on the given utility values, state the home owner's decision.

Solution. The expected utility value of buying the fire insurance is -1 utile and that of not buying the insurance is -2 utiles. Since -1 is larger than -2, the home owner should buy the fire insurance. See the computation in Table 19–10.

TABLE 19–10

COMPUTING EXPECTED UTILITY VALUES (EXAMPLE 8)

Event	Proba- bility of Event	Act					
		Buy Insurance			Not Buy Insurance		
		Conditional Value		Expected Utility	Conditional Value		Expected Utility
		Dollars	Utiles		Dollars	Utiles	
No Fire	0.999	−$100	−1	−0.999	$0	0	0
Fire	0.001	−$100	−1	−0.001	−$20,000	−2,000	−2
Total	1.000			−1.000			−2

$\uparrow$
Higher

The results obtained in Examples 7 and 8 are consistent with the decisions stated respectively in Examples 5 and 6, where the expected monetary values were not appropriate for decision-making. However, care must be exercised in using utility as a basis for decision-making. Utilities are derived from personal opinions on monetary values under conditions of uncertainty. A person's opinion may change from time to time and from one situation to another. Thus, the utility constructed for a person at one time for a particular type of problem involving risky acts may not be suitable to the same type of problem at another time. Different types of problems usually require the construction of different sets of utility values for the same person in making decisions. For example, a person's opinion on money used for college education, although the outcome is uncertain, is usually different from that used for gambling purposes. Furthermore, the scale of utility values of a person is arbitrarily selected. To compare the utility values of one person with those of another person is thus meaningless.

19.6 Decision Rules

There are many types of rules for making decisions. The common decision rules are:

(1) Choose the act with the *highest expected value*. The expected values are computed from subjective probability distributions. The values may be expressed either in terms of dollars (if appropriate) or in terms of utility (generally preferred). This rule has been applied in the above illustrations and is frequently called the *Bayesian decision rule*. The Bayesian decision rule is generally regarded as the superior one among the rules introduced in this section.

(2) Choose the act with the *maximum possible profit*. Thus the decision-maker ignores the probability distribution and the fact of possible losses. Refer to the information given in Table 19–7, page 490. The decision-maker will choose the act of investing since the act will yield the highest possible profit of $5000.

(3) Choose the act with the *minimum possible loss*. Under this rule, the decision-maker also ignores the probability distribution of event occurrences. This rule

tends to lead a decision-maker to do nothing since the minimum possible loss in a given problem is frequently zero dollars. The fact of making possible profit, either very high or small, is also ignored. Using the information given in Table 19–7, the decision is not to invest since the act has a minimum possible loss of $0.

(4) Choose the act with the *highest average profit*. Under this rule, the decision-maker assumes the occurrences of events are equally likely. Thus, if there are 2 possible events, the probability of the occurrence of each event is 1/2; for 3 possible events, the probability is 1/3; and so on. The method of computing the average profit for each event is the same as that used for computing the expected values. The computed average profit may be positive (gain) or negative (loss). If the highest average profit is positive, the rule maximizes the average profit; if it is negative, the rule minimizes the average loss.

Refer to Table 19-7. The average profit of the act of investing is

$$(1/2)(\$5000) + (1/2)(-\$2000) = 2500 - 1000 = \$1500$$

The average profit of the act of not investing is

$$(1/2)(\$0) + (1/2)(\$0) = \$0$$

Thus, the act of investing should be chosen under this rule.

(5) Choose the act with the *highest profit* for the event occurring *most likely*. Under this rule, the consequences of the less likely events are ignored. The profit may also be positive (gain) or negative (less). Refer to Table 19-7. The event which is most likely to occur is "bad weather," with 0.70 probability. The profit of not investing is $0, which is higher than the profit of investing, −$2000 or loss of $2000. Thus, the act of not investing is chosen under this decision rule.

19.7 Summary

This chapter discusses the process of decision-making for problems under conditions of uncertainty. The uncertainty provides many alternative actions available to the decision-maker. The question is how to establish the criteria upon which the best action can be selected from the various alternatives.

A table which shows the relationships among all possible events, all possible acts, and all related consequences, or sometimes the values associated with the consequences, is called a payoff table, decision matrix, or payoff matrix. A payoff is simply a consequence, such as a conditional profit as shown in Table 19–1.

The information which can be used by a business manager to change occurrences of events from uncertainty to certainty is called the perfect information or perfect predictor. The decision of whether or not to seek more information is commonly based on the expected value of perfect information, which is the difference between the expected profit under certainty and the expected profit of the best or optimum act under uncertainty. The optimum act is the one with the highest expected profit, or with the smallest expected loss if the conditional loss table is used. The expected loss of the optimum act also equals the expected value of perfect information. The conditional losses can be classified into two groups: cost losses and opportunity losses.

When expected monetary value is not an appropriate guide for making decisions, the expected utility value can be used as a guide instead. Utility is a measurement of money based on the decision-maker's opinion of monetary value under conditions of uncertainty. A decision-maker may or may not consider every dollar equally important. Thus utility values may or may not be proportional to monetary values represented, although a higher utility value usually represents a higher monetary value.

There are many types of rules for making decisions. The common ones are: (1) Choose the act with the highest expected value, either expressed in dollars or in utility. This rule is also called the Bayesian decision rule. (2) Choose the act with the maximum possible profit. (3) Choose the act with the minimum possible loss. (4) Choose the act with the highest average profit. (5) Choose the act with the highest profit for the most likely event.

Exercises 19–1

Reference: Sections 19.1 to 19.4

1. Explain briefly:
    (a) payoff table
    (b) conditional value
    (c) expected value
    (d) expected value of perfect information

2. What are:
    (a) conditional loss tables?
    (b) opportunity losses?

3. The following information is obtained from the record of recent weekly sales of Good magazine in the Bee Drug Store:

Weekly Sales (Number of Copies)	Number of Weeks
21	5
22	10
23	15
24	15
25	5
Total	50

The cost of each copy of the magazine is $0.25 and the selling price of each copy is $0.40. If the magazine is not sold during the week, it has no value. Construct a payoff table to show all possible profits.

4. Use the information given in Problem 3.
   (a) Find the probability of each event of weekly sales.
   (b) Compute the expected sales (in number of copies) for a week.

5. Use the information given in Problem 3. Compute the expected monetary profit for each possible act and indicate the optimum act.

6. Use the information given in Problem 3.
   (a) Compute the expected profit under certainty.
   (b) Compute the expected value of perfect information.

7. Use the information given in Problem 3 above.
   (a) Prepare a conditional loss table.
   (b) Compute the expected loss for each possible act.
   (c) Indicate the optimum act and the expected value of perfect information based on the expected loss.

8. Refer to Problem 3. Assume that a leftover magazine has a value of $0.03 per copy. Construct a payoff table to show all possible profits.

9. Use the assumption given in Problem 8.
   (a) Compute the expected monetary profit for each possible act and indicate the optimum act.
   (b) Compute the expected value of perfect information.

10. Use the assumption given in Problem 8. Find the answers to the questions listed in Problem 7.

Exercises 19–2

Reference: Sections 19.5 and 19.6

1. Mr. Jones has a utility value of 20 for a gain of $10 and 60 for a gain of $2000. He is indifferent between (A) having $500 for certain and (B) a lottery which will gain $10 with a 0.5 chance and $2000 with a 0.5 chance. Find his utility value for $500.

2. Ms. Brown has a utility value of 15 for a loss of $100 and 90 for a gain of $6000. She is indifferent between (A) receiving $1000 for certain and (B) a loss of $100 at a 0.3 chance and a gain of $6000 at a 0.7 chance. Find her utility value for $1000.

3. Refer to Problem 1. Assume that Mr. Jones is indifferent between (A) having $1200 for certain and (B) a lottery which will gain $10 with a 0.4 chance and $2000 with a 0.6 chance. Find his utility value for $1200.

4. Refer to Problem 2. Assume that Ms. Brown is indifferent between (A) receiving $2500 for certain and (B) a loss of $100 at a 0.2 chance and a gain of $6000 at an 0.8 chance. What is her utility value for $2500?

5. Refer to Problem 1. If Mr. Jones is indifferent between (A) having $2000 for certain and (B) gaining $10 at a 0.4 chance and $15,000 at 0.6 chance, what is his utility value for $15,000?

6. Refer to Problem 2. If Ms. Brown is indifferent between (A) receiving nothing for sure and (B) a loss of $100 at a 0.9 chance and a gain of $1000 at a 0.1 chance, what is her utility value for $0?

7. Refer to Problem 1. If Mr. Jones is indifferent between (A) having $10 for certain and (B) a loss of $3000 at a 0.2 chance and a gain of $2000 at an 0.8 chance, what is his utility value for a loss of $3000?

8. Refer to Problem 2. If Ms. Brown is indifferent between (A) receiving a loss of $100 for certain and (B) a loss of $4000 at a 0.6 chance and a gain of $6000 at a 0.4 chance, what is her utility value for a loss of $4000?

9. A lottery, which is equivalent to $2000 for certain, will give either $1000 or $3000 by chance. Based on the information given in Chart 19–2, find the probability distribution of the chance.

10. Suppose that the utility values of the student mentioned in Example 5, page 490, are: $U(\$5000) = 60$ and $U(-2000) = -10$. What would be his decision if he uses expected utility values as a basis for decision-making?

11. Use the dollar values given in Table 19–10, page 497. Indicate the decisions (buy or not buy the insurance) if the decision-maker chooses:
    (a)   the act with the maximum possible profit,
    (b)   the act with the minimum possible loss,
    (c)   the act with the highest average profit,
    (d)   the act with the highest profit of the event occurring most likely.

12. Use the utility values given in Table 19–10. What are the answers to the questions listed in Problem 11?

20 Bayesian Decision Theory

This chapter will first introduce Bayes' theorem, Section 20.1. The theorem can be used to revise a set of old probabilities, called *prior probabilities*, to a set of new probabilities, called *posterior probabilities*. The revision is usually made when additional information is available. The additional information can be obtained from past records of a business firm or from samples and is used to compute the conditional probabilities. The method of exact computation of the conditional probabilities is discussed in Section 20.2. The expected value of probability distribution is presented in Section 20.3. The method of obtaining the conditional probabilities by using the normal distribution tables is explained in Section 20.4. The computation of the posterior expected value and variance by formulas is shown in Section 20.5. A summary of this chapter is given in Section 20.6.

20.1 Bayes' Theorem

Bayes' theorem was created by the English clergyman and mathematician Thomas Bayes and published in 1763 after his death. The basic concept of this theorem is explained below.

Example 1

A deck of 10 cards includes:

The cards can be classified into two kinds:
   3 kings—1 printed in pink (unshaded) and 2 in gray (shaded)
   7 queens—3 in pink and 4 in gray.
The cards can also be classified into two colors:
   4 pink—1 king and 3 queens
   6 gray—2 kings and 4 queens
The cross classifications of the 10 cards are shown in Table 20–1.

TABLE 20–1

CROSS-CLASSIFIED CARDS

Color Kind	Pink	Gray	Row Total
King	1 (king and pink)	2 (king and gray)	3 (king)
Queen	3 (queen and pink)	4 (queen and gray)	7 (queen)
Column Total	4 (pink)	6 (gray)	10 cards

Case I. One card is drawn at random from the deck. What are the probabilities of all possible outcomes of the drawing?

Solution. The probabilities of all possible outcomes are shown in Table 20–2. The probabilities in the center four cells of the table [P (king and pink), P (king and gray), P (queen and pink), and P (queen and gray)] are known as *joint probabilities*. Each joint probability is obtained by dividing the number in the cell of Table 20–1 by the total number 10. For example, the probability of drawing a king-and-pink card, P (king and pink), is $1/10$. The sum of all joint probabilities is equal to 1. The probabilities in the right column (row totals) and in the bottom row (column totals) are called *marginal probabilities*. For example, the probability of drawing a king, ignoring the color, or P (king), is a row marginal probability and is $3/10$. The sum of the column marginal probabilities and that of the row marginal probabilities are also each equal to 1, as shown in Table 20–2.

TABLE 20–2

PROBABILITIES OF POSSIBLE OUTCOMES OF ONE DRAWING
EXAMPLE 1, CASE I.

Color Kind	Pink	Gray	Row Total
King	P(king and pink) $= 1/10$	P(king and gray) $= 2/10$	P(king) $= 3/10$
Queen	P(queen and pink) $= 3/10$	P(queen and gray) $= 4/10$	P(queen) $= 7/10$
Column Total	P(pink) $= 4/10$	P(gray) $= 6/10$	$10/10 = 1$

Case II. Suppose that the card which is to be drawn will be pink. What is then the probability that the card is a king? a queen?

Solution. Table 20–2 shows that the probability of drawing a king is 3/10. In the absence of any additional information, this would be the final answer. However, in the present case the additional information is available: the card to be drawn must be pink. Thus, the above question may be rephrased in another way. Suppose that the deck contained only four pink cards: 1 king and 3 queens. What is the probability of drawing a king? a queen?

Observe the "pink" column, Table 20–1.

$$\text{The probability of drawing a king} = \frac{1}{4}$$

$$\text{The probability of drawing a queen} = \frac{3}{4}$$

These questions may also be answered with conditional probability.

Observe the "pink" column, Table 20–2. The probability of drawing a king, given the card is pink, or

$$P(\text{king|pink}) = \frac{P(\text{king and pink})}{P(\text{pink})} = \frac{1/10}{4/10} = \frac{1}{4}$$

Similarly,

$$P(\text{queen|pink}) = \frac{P(\text{queen and pink})}{P(\text{pink})} = \frac{3/10}{4/10} = \frac{3}{4}$$

Observe: $\qquad P(\text{pink}) = P(\text{king and pink}) + P(\text{queen and pink})$

and $\qquad P(\text{king|pink}) + P(\text{queen|pink}) = \frac{1}{4} + \frac{3}{4} = 1$

Case III. Suppose that the card which is to be drawn will be gray. What is the probability that the card is a king? a queen?

Solution. Again this question may be rephrased in another manner. Suppose that the deck contained only six gray cards: 2 kings and 4 queens. What is the probability of drawing a king? a queen?

$$\text{The probability of drawing a king} = \frac{2}{6} = \frac{1}{3}$$

$$\text{The probability of drawing a queen} = \frac{4}{6} = \frac{2}{3}$$

This question may also be answered with conditional probability.

The probability of drawing a king, given the card is gray, or

$$P(\text{king | gray}) = \frac{P(\text{king and gray})}{P(\text{gray})} = \frac{2/10}{6/10} = \frac{2}{6} = \frac{1}{3}$$

The probability of drawing a queen, given the card is gray, or

$$P(\text{queen | gray}) = \frac{P(\text{queen and gray})}{P(\text{gray})} = \frac{4/10}{6/10} = \frac{4}{6} = \frac{2}{3}$$

Observe:

$$P(\text{gray}) = P(\text{king and gray}) + P(\text{queen and gray})$$
$$P(\text{king | gray}) + P(\text{queen | gray}) = 1/3 + 2/3 = 1$$

In general, let

A_1 and A_2 = the set of events which are mutually exclusive (the two events cannot occur together) and exhaustive (the conbination of the two events is the universal set), and

B = a single event which intersects each of the A events as shown in the diagram below:

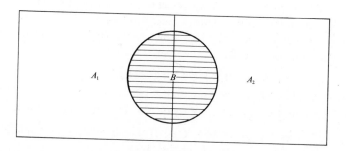

(Observe the diagram: The part of B which is within A_1 represents the area "A_1 and B," and the part of B within A_2 represents the area "A_2 and B.") Then the probability of event A_1, given event B, is

$$P(A_1|B) = \frac{P(A_1 \text{ and } B)}{P(B)}$$

and similarly, the probability of event A_2, given B, is

$$P(A_2|B) = \frac{P(A_2 \text{ and } B)}{P(B)}$$

where

$$P(B) = P(A_1 \text{ and } B) + P(A_2 \text{ and } B)$$

and, by applying the conditional probability formula (9–9),

$$P(A_1 \text{ and } B) = P(A_1) \cdot P(B|A_1)$$

and

$$P(A_2 \text{ and } B) = P(A_2) \cdot P(B|A_2)$$

Further, let $A_1, A_2, A_3, \ldots A_n$ be the set of n mutually exclusive and exhaustive events as shown in the following diagram ($n = 3$ in the diagram):

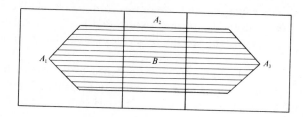

These expressions may be summarized and are called *Bayes' theorem*:

The probability of event A_i ($i = 1, 2, 3, \ldots n$), given event B, is

$$P(A_i|B) = \frac{P(A_i \text{ and } B)}{P(B)} \tag{20-1}$$

where $\quad P(B) = P(A_1 \text{ and } B) + P(A_2 \text{ and } B) + \cdots + P(A_n \text{ and } B),$

and $\qquad\qquad P(A_i \text{ and } B) = P(A_i) \cdot P(B|A_i)$

When Bayes' theorem is used to revise a set of old probabilities, called *prior probabilities*, to a set of new probabilities, celled *posterior probabilities*, the computation is usually done systematically in a table. Example 1 is now used to illustrate the computation employing Bayes' theorem. Let

$A_1 =$ the event of drawing a king card,

$A_2 =$ the event of drawing a queen card, and

$B =$ the event of drawing a *pink* card, either king or queen

The required values for Example 1, Case II, are computed in Table 20–3. The final answer is shown in column (5) of the table.

TABLE 20–3

COMPUTATION OF POSTERIOR PROBABILITIES
EXAMPLE 1, CASE II

(1) Event	(2) Prior Probability $P(A_i)$	(3) Conditional Probability of Event B (Pink) Given Event A $P(B\|A_i)$	(4) Joint Probability $P(A_i \text{ and } B)$ (2) × (3)	(5) Posterior (Revised) Probability $P(A_i\|B)$ (4) ÷ $P(B)$
A_1 (king)	3/10	1/3	1/10	$\frac{1/10}{4/10} = \frac{1}{4} = 0.25$
A_2 (queen)	7/10	3/7	3/10	$\frac{3/10}{4/10} = \frac{3}{4} = 0.75$
Total	10/10 = 1		$P(B) = 4/10$	1.00

Comments on Table 20–3:

Column (1)   The events for which the probabilities are required to be revised are listed in this column.

Column (2) $P(A_i)$ represents the prior (old or original) probability. Each card in the deck has equal chance to be drawn. Since 3 out of 10 cards are kings, the probability of the event of drawing a king, $P(A_1)$ is 3/10. Likewise, since 7 out of 10 cards are queens, the probability of the event of drawing a queen, $P(A_2)$, is 7/10.

Prior probabilties are the probabilities assigned to the events in column (1) before having the additional information that may be used to improve the quality of the probabilities. Prior probabilities may be assigned mathematically, such as the computation of $3/10$ and $7/10$ above, or subjectively, such as those based on past records of business activities (see Example 3 below.)

Column (3) $P(B|A_i)$ represents the conditional probability of B (event described by the additional information, drawing a pink card), given (or depending on) event A_i. For example, the additional information states that there are 1 pink king and 2 gray kings. Then the probability of the event of drawing a pink card, given the card is a king, or $P(B|A_1)$, is $1/3$. Also, there are 3 pink queens and 4 gray queens. Thus, the probability of the event of drawing a pink card, given the card is a queen, or $P(B|A_2)$, is $3/7$.

Column (4) $P(A_i$ and $B)$ represents the joint probability of event $(A_i$ and $B)$. The probability of the event of drawing a king and pink card is $P(A_1$ and $B) = P(A_1) \cdot P(B|A_1) = (3/10)(1/3) = 1/10$. The probability of the event of drawing a queen and pink card is $P(A_2$ and $B) = P(A_2) \cdot P(B|A_2) = (7/10)(3/7) = 3/10$.

Column (5) $P(A_i|B)$ represents the posterior (or revised prior) probability of A_i, given B. It is computed directly from the corresponding joint probability and the sum of the joint probabilities. For example, $P(A_1|B) = (1/10)/(4/10) = 0.25$, which represents the probability that the king card is pink. The sum of the posterior probabilities is equal to 1.

In a similar manner, let

A_1 = the event of drawing a king card,
A_2 = the event of drawing a queen card, and
B = the event of drawing a *gray* card, either king or queen.

Then, the required values for Example 1, Case III, by the use of Bayes' theorem can be computed in Table 20–4. The final answer is also shown in column (5) of the table.

TABLE 20–4

COMPUTATION OF POSTERIOR PROBABILITIES
EXAMPLE 1, CASE III

(1) Event	(2) Prior Probability $P(A_i)$	(3) Conditional Probability of Event B (Gray) Given Event A $P(B\|A_i)$	(4) Joint Probability $P(A_i$ and $B)$ $(2) \times (3)$	(5) Posterior (Revised) Probability $P(A_i\|B)$ $(4) \div P(B)$
A_1 (king)	$3/10$	$2/3$	$2/10$	$\frac{2/10}{6/10} = \frac{2}{6} = 0.33$
A_2 (queen)	$7/10$	$4/7$	$4/10$	$\frac{4/10}{6/10} = \frac{4}{6} = 0.67$
Total	$10/10 = 1$		$P(B) = 6/10$	1.00

The entire computation of Example 1 is summarized in the tree diagram shown in Chart 20–1.

CHART 20–1

COMPUTATION OF POSTERIOR PROBABILITIES
EXAMPLE 1

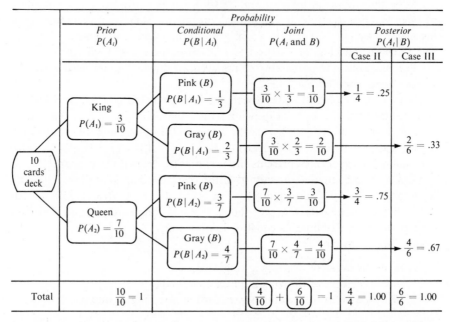

			Probability				
	Prior $P(A_i)$	Conditional $P(B \mid A_i)$		Joint $P(A_i \text{ and } B)$		Posterior $P(A_i \mid B)$	
						Case II	Case III
10 cards deck	King $P(A_1) = \frac{3}{10}$	Pink (B) $P(B \mid A_1) = \frac{1}{3}$	$\frac{3}{10} \times \frac{1}{3} = \frac{1}{10}$		$\frac{1}{4} = .25$	
		Gray (B) $P(B \mid A_1) = \frac{2}{3}$	$\frac{3}{10} \times \frac{2}{3} = \frac{2}{10}$			$\frac{2}{6} = .33$
	Queen $P(A_2) = \frac{7}{10}$	Pink (B) $P(B \mid A_2) = \frac{3}{7}$	$\frac{7}{10} \times \frac{3}{7} = \frac{3}{10}$		$\frac{3}{4} = .75$	
		Gray (B) $P(B \mid A_2) = \frac{4}{7}$	$\frac{7}{10} \times \frac{4}{7} = \frac{4}{10}$			$\frac{4}{6} = .67$
Total	$\frac{10}{10} = 1$		$\left(\frac{4}{10}\right) + \left(\frac{6}{10}\right) = 1$		$\frac{4}{4} = 1.00$	$\frac{6}{6} = 1.00$

Example 2 further illustrates the application of Bayes' theorem in a general case.

Example 2

Assume that a bag contains white and red balls marked X, Y, or Z as follows:

> Marked X — 8 balls, including 5 white and 3 red
> Marked Y — 8 balls, including 4 white and 4 red
> Marked Z — 8 balls, including 1 white and 7 red
> _____
> Total   —24 balls, including 10 white and 14 red

A white ball is drawn at random from the bag. Determine the probabilities that the white ball is marked X, Y, and Z.

Solution. Let $A_1 =$ the event of drawing an X ball
$\quad A_2 =$ the event of drawing a Y ball
$\quad A_3 =$ the event of drawing a Z ball and
$\quad B =$ the event of drawing a white ball, marked either X, Y, or Z. (additional information)

Events A_1, A_2, and A_3 are mutually exclusive since one drawing of one ball cannot produce two kinds of balls. The events are also exhaustive since there is no ball other than X, Y, and Z balls. The required values are computed in Table

20–5. The final answer is shown in column (5) of the table.

TABLE 20–5

COMPUTATION OF POSTERIOR PROBABILITIES
EXAMPLE 2

(1) Event	(2) Prior Probability $P(A_i)$	(3) Conditional Probability of Event B (White ball) Given Event A $P(B\|A_i)$	(4) Joint Probability $P(A_i \text{ and } B)$ (2) × (3)	(5) Posterior (Revised) Probability $P(A_i\|B)$ (4) ÷ P(B)
A_1 (X ball)	1/3	5/8	5/24	$\frac{5/24}{10/24} = \frac{5}{10} = 0.5$
A_2 (Y ball)	1/3	4/8	4/24	$\frac{4/24}{10/24} = \frac{4}{10} = 0.4$
A_3 (Z ball)	1/3	1/8	1/24	$\frac{1/24}{10/24} = \frac{1}{10} = 0.1$
Total	3/3 = 1		P(B) = 10/24	1.0

Comments on Table 20–5:

1. $P(A_i)$, the prior probability. Each ball in the bag has an equal chance to be drawn. Since 1/3 of all the balls in the bag (or 8 out of 24 balls) are marked X, the probability of the event of drawing an X ball, $P(A_1)$, is 1/3. The same probability value, 1/3, is assigned to $P(A_2)$ and $P(A_3)$.

2. $P(B\|A_i)$, the conditional probability of event B (described by the additional information), given event A_i. For example, the additional information states that there are 5 white and 3 red balls marked X. Then, the probability of the event of drawing a white ball, given balls marked X, or $P(B\|A_1)$, is 5/8.

3. $P(A_i \text{ and } B)$, the joint probability of event A_i and event B. For example, $P(A_1 \text{ and } B) = P(A_1) \cdot P(B\|A_1) = (1/3)(5/8) = 5/24$. Thus, the probability of the event of drawing an X and white ball from the bag is 5/24 (or 5 out of 24 balls in the bag).

4. $P(A_i\|B)$, the posterior (or revised prior) probability of A_i, given B. For example, $P(A_1\|B) = (5/24) \div (10/24) = 0.5$, which represents the probability that the white ball is marked X. The sum of the posterior probabilities is equal to 1.

20.2 Revising Probabilities Based on Past Records or Samples

This section illustrates the applications of Bayes' theorem in business problems. The original probability of a business event is revised according to historial records or samples. The applications are again illustrated in examples.

Example 3

Assume that a factory has two machines. Machine 1 produces 40 per cent of the items of total production and machine 2 produces 60 per cent of the items. Past records showd that 5 per cent of the items produced by machine 1 were defective and only 2 per cent produced by machine 2 were defective. If a defective item is drawn at random, what is the probability that the defective item was produced by machine 1? Machine 2?

Solution.   Let $A_1 =$ the event of drawing an item produced by machine 1

$A_2 =$ the event of drawing an item produced by machine 2

$B =$ the event of drawing a defective item produced either by machine 1 or machine 2

Then, from the first information, or prior probabilities

$$P(A_1) = 40\% = 0.40$$
$$P(A_2) = 60\% = 0.60$$

From the second information, or conditional probabilities.

$$P(B|A_1) = 5\% = 0.05$$
$$P(B|A_2) = 2\% = 0.02$$

The given information is diagrammed below and the required values are computed in Table 20–6. The final answer is shown in column (5) of the table.

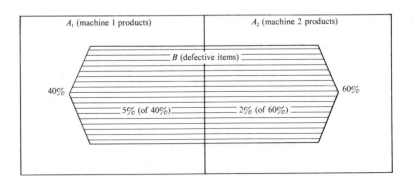

Observe the probabilities shown in Table 20–6. Without the additional information, we may be inclined to say that the defective item is drawn from machine 2 output since $P(A_2) = 60\%$ is larger than $P(A_1) = 40\%$. With the additional information, the accuracy of probability distribution is improved. The probability that the defective item was produced by machine 1 is 0.625 or 62.5% and that by machine 2 is only 0.375 or 37.5%. Thus, the defective item is most likely drawn

TABLE 20–6

COMPUTATION OF POSTERIOR PROBABILITIES
EXAMPLE 3

(1) Event	(2) Prior Probability $P(A_i)$	(3) Conditional Probability of Event B (defective item) Given Event A $P(B \mid A_i)$	(4) Joint Probability $P(A_i$ and $B)$ (2) × (3)	(5) Posterior (Revised) Probability $P(A_i \mid B)$ (4) ÷ $P(B)$
A_1 (Machine 1)	0.40	0.05	0.020	$\dfrac{0.020}{0.032} = 0.625$
A_2 (Machine 2)	0.60	0.02	0.012	$\dfrac{0.012}{0.032} = 0.375$
Total	1.00		$P(B) = 0.032$	1.000

from the output produced by machine 1.

The answer for Example 3 may be checked by using an actual number of items. Suppose that 10,000 items were produced by the two machines in a given period. The number of items produced by machine 1 is

$$10,000 \times 40\% = 4,000$$

and the number of items produced by machine 2 is

$$10,000 \times 60\% = 6,000$$

The number of defective items produced by machine 1 is

$$4,000 \times 5\% = 200$$

and by machine 2 is

$$6,000 \times 2\% = 120$$

The probability that a defective item was produced by machine 1 is

$$\frac{200}{200 + 120} = \frac{200}{320} = 0.625$$

and that by machine 2 is

$$\frac{120}{200 + 120} = \frac{120}{320} = 0.375$$

Example 4

Refer to Example 3. Suppose that the products of machine 1 and machine 2 are packed separately into many different lots. Two defective items instead of only one are drawn consecutively at random from an unidentified lot. What is the probability that the two defective items were produced by machine 1? By machine 2?

Solution. Let A_1 and A_2 represent the same events defined in Example 3, and

$B = $ the event of drawing two defective items produced either by machine 1 or machine 2.

The conditional probabilities based on the additional information are computed as follows:

$$P(B|A_1) = (0.05)^2 = 0.0025$$
$$P(B|A_2) = (0.02)^2 = 0.0004$$

The required values are computed in Table 20–7. The final answer is shown in column (5) of the table. The answer indicates that the probability of drawing the

TABLE 20–7

COMPUTATION OF POSTERIOR PROBABILITIES
EXAMPLE 4

(1) Event	(2) Prior Probability $P(A_i)$	(3) Conditional Probability of Event B (2 defective items), Given Event A $P(B|A_i)$	(4) Joint Probability $P(A_i \text{ and } B)$ (2) × (3)	(5) Posterior (Revised) Probability $P(A_i|B)$ (4) ÷ $P(B)$
A_1 (Machine 1)	0.40	0.0025	0.00100	$\dfrac{0.00100}{0.00124} = 0.806$
A_2 (Machine 2)	0.60	0.0004	0.00024	$\dfrac{0.00024}{0.00124} = 0.194$
Total	1.00		$P(B) = 0.00124$	1.000

defective items produced by machine 1 increases to 0.806 from 0.625 (Table 20–6). The increase is expected since machine 1 produces more defective items (5% of 40% of the total production) than machine 2 (2% of 60% of the total).

The sample size for obtaining the additional information in Example 3 is 1 (drawing one defective item) and that in Example 4 is 2 (drawing two defective items). Example 5 is expanded to use a larger sample size—20 items in one drawing. The basic procedure of applying Bayes' theorem for various sample sizes is the same. However, the binomial formula $P(X; n, P)$ is used in computing the conditional probabilities in Example 5.

Example 5

Assume that a factory has five machines. The production distribution and the defective prodution of the five machines are as follows:

Machine number	Production distribution	Defective items produced by each machine
M1	35%	2%
M2	30%	4%
M3	20%	5%
M4	10%	10%
M5	5%	20%
	100% (entire factory)	

A sample of 20 items is taken from a lot produced by an unidentified machine. Four of the 20 items are defective. Find the probability that the defective items were produced by each machine.

Solution. Let $A_1, A_2, \ldots A_5$ represent the event of drawing an item produced by machine 1, 2, ... 5 respectively, and

> $B =$ the event of drawing four defective items from 20 items produced by one of the five machines

The conditional probabilities based on the sample are computed according to the binomial formula, or

$$P(B|A_i) = P(X; n, P) = {}_nC_X \cdot P^X \cdot Q^{n-X}$$

where $X = 4$ (defective items), $n = 20$ (sample size), and $P =$ percent of defective items produced by each machine. ($Q = 1 - P$) The probability values can be obtained from Table 3 in the Appendix.

The conditional probabilities of a sample result for the given events are also called *likelihoods*. Thus, the likelihoods are:

$$P(4; 20, 0.02) = 0.0006 \quad \text{(Machine 1 with } P = 2\% = 0.02\text{)}$$
$$P(4; 20, 0.04) = 0.0065 \quad \text{(Machine 2 with } P = 4\% = 0.04\text{)}$$
$$P(4; 20, 0.05) = 0.0133 \quad \text{(Machine 3 with } P = 5\% = 0.05\text{)}$$
$$P(4; 20, 0.10) = 0.0898 \quad \text{(Machine 4 with } P = 10\% = 0.10\text{)}$$
$$P(4; 20, 0.20) = 0.2182 \quad \text{(Machine 5 with } P = 20\% = 0.20\text{)}$$

The required values for the answer to this example are computed in Table 20-8. The final answer is shown in column (5) of the table. The answer indicates that the highest revised probability of drawing the defective items is 0.442 by machine 5 and the lowest revised probabality is 0.008 by machine 1. Thus we may say that the given sample is most likely drawn from the lot produced by machine 5, but very unlikely by machine 1. According to the prior probabilities, however, the probability of drawing an item produced by machine 1 is the highest, 0.35.

The process of revising a set of prior probabilities may be repeated if more information can be obtained. Each new process must begin with a set of prior probabilities, which may be based on the set of posterior probabilities of the previous process. Thus, Bayes' theorem provides a powerful method in improving the quality of probability for aiding the management in decision-making under uncertainty.

TABLE 20–8

COMPUTATION OF POSTERIOR PROBABILITIES
EXAMPLE 5

(1) Event	(2) Prior Probability $P(A_i)$	(3) Conditional Probability, Likelihood $P(X = 4; n = 20, P)$	(4) Joint Probability $P(A_i \text{ and } B)$ (2) × (3)	(5) Posterior (Revised) Probability $P(A_i \mid B)$ (4) ÷ $P(B)$
$A_1(P = 0.02)$	0.35	0.0006	0.00021	0.008
$A_2(P = 0.04)$	0.30	0.0065	0.00195	0.079
$A_3(P = 0.05)$	0.20	0.0133	0.00266	0.108
$A_4(P = 0.10)$	0.10	0.0898	0.00898	0.363
$A_5(P = 0.20)$	0.05	0.2182	0.01091	0.442
Total	1.00		$P(B) = 0.02471$	1.000

20.3 The Expected Value of Probability Distribution

The method of computing the expected value of a prior probability distribution and that of a posterior probability distribution are illustrated in Table 20–9. The P values, the prior probabilities, and the posterior probabilities as shown in the table are obtained from columns (1), (2), and (5) of Table 20–8 respectively.

The expected value of P based on the posterior probability distribution (0.13342) is greater than that based on the prior probability distribution (0.0490). This reflects the greater weight that the posterior probability distribution gives to the larger values of P.

The posterior expected value of 0.13342 is obtained from the result of the sample of 20 items with the sample defective proportion of 0.20 (or 4 defective

TABLE 20–9

COMPUTATION OF EXPECTED VALUES OF PRIOR AND POSTERIOR
PROBABILITY DISTRIBUTIONS
EXAMPLE 5

(1) Values P	(2) Prior Probability	(3) Expected Value of P based on Prior Probability (1) × (2)	(4) Posterior Probability	(5) Expected Value of P based on Posterior Probability (1) × (4)
0.02	0.35	0.0070	0.008	0.00016
0.04	0.30	0.0120	0.079	0.00316
0.05	0.20	0.0100	0.108	0.00540
0.10	0.10	0.0100	0.363	0.03630
0.20	0.05	0.0100	0.442	0.08840
Total	1.00	0.0490	1.000	0.13342

items out from 20 items in the sample.) The posterior expected value is expected to be closer to 0.20 if the sample size is increased and the sample proportion remains at 0.20. This statement is supported by the answers to Example 6.

Example 6

Refer to Example 5. Suppose that the sample size is increased to 40 items. Eight of the 40 items are defective. Find the probability that the defective items were produced by each machine and the expected value of P.

Solution. Let B = the event of drawing 8 defective items from 40 items produced by one of the five machines.

The posterior probability distribution and its expected value are computed in Table 20–10.

TABLE 20–10

COMPUTATION OF POSTERIOR PROBABILITY AND EXPECTED VALUE
EXAMPLE 6

(1) Event A_i	(2) Prior Probability $P(A_i)$	(3) Conditional Probability, Likelihood $P(X = 8; n = 40, P)$	(4) Joint Probability $P(A_i \text{ and } B)$ $(2) \times (3)$	(5) Posterior (Revised) Probability $P(A_i \mid B)$ $(4) \div P(B)$	(6) Posterior Expected Value of P $(1) \times (5)$
$P = 0.02$	0.35	0.0000	0.000000	0.000	0.00000
$P = 0.04$	0.30	0.0001	0.000030	0.003	0.00012
$P = 0.05$	0.20	0.0006	0.000120	0.011	0.00055
$P = 0.10$	0.10	0.0264	0.002640	0.249	0.02490
$P = 0.20$	0.05	0.1560	0.007800	0.737	0.14740
Total	1.00		$P(B) = 0.010590$	1.000	0.17297

Observe column (6) of Table 20–10. The posterior expected value is increased to 0.17297, which is closer to the sample proportion 0.20 ($= 8/40$) than the expected value based on the smaller sample of 20 items. Of course, if the sample size is increased to be equal to the population size, the sample proportion becomes the population proportion and also the expected value of P.

Based on an expected value, which is an arithmetic mean weighted by probabilities, the standard deviation may be computed for a given probability distribution. The formula for the standard deviation involving weights is

$$s = \sqrt{\frac{\Sigma f x^2}{n}}, \text{ or simply } s = \sqrt{\Sigma f x^2}$$

where f = weight or probability
x = event value P − expected value $E(P) = P - E(P)$
n = the total of probabilities = 1

TABLE 20–11

COMPUTATION OF THE STANDARD DEVIATION OF POSTERIOR
PROBABILITY DISTRIBUTION—EXAMPLE 5, SAMPLE SIZE = 20 ITEMS

Event Value P	Posterior Probability f	$x = P - E(P)$ $= P - 0.13342$	fx	fx^2
0.02	0.008	− 0.11342	− 0.00090736	0.00010291
0.04	0.079	− 0.09342	− 0.00738018	0.00068946
0.05	0.108	− 0.08342	− 0.00900936	0.00075156
0.10	0.363	− 0.03342	− 0.01213146	0.00040543
0.20	0.442	0.06658	0.02942836	0.00195934
Total	1.000		0.00000000	0.00390870

$$s = \sqrt{0.00390870} = 0.063$$

TABLE 20–12

COMPUTATION OF THE STANDARD DEVIATION OF POSTERIOR
PROBABILITY DISTRIBUTION—EXAMPLE 6, SAMPLE SIZE = 40 ITEMS

Event Value P	Posterior Probability f	$x = P - E(P)$ $= P - 0.17297$	fx	fx^2
0.02	0.000	− 0.15297	0.00000000	0.00000000
0.04	0.003	− 0.13297	− 0.00039891	0.00005304
0.05	0.011	− 0.12297	− 0.00135267	0.00016634
0.10	0.249	− 0.07297	− 0.01816953	0.00132583
0.20	0.737	0.02703	0.01992111	0.00053847
Total	1.000		0.00000000	0.00208368

$$s = \sqrt{0.00208368} = 0.046$$

The standard deviations of the posterior probability distributions based on sample sizes 20 and 40 are computed above. (See Tables 20–11 and 20–12.) Observe that when sample size is increased (from 20 to 40), the standard deviation is decreased (from 0.063 to 0.046).

The value of each standard deviation obtained above can be used to measure the dispersion of each set of P values away from the expected value of P.

★20.4   Revising Normal Probabilities

The likelihoods (conditional probabilities) of a sample result for the given events

A_i were obtained by exact computation in Section 20.2, such as by using the binomial formula in Example 5, page 513. The likelihoods may also be obtained by using the normal probability distribution tables. When the likelihoods are obtained, the posterior probabilities, the posterior expected value, and the standard deviation of posterior probability distribution may be computed by the usual method or by using some developed formulas for short cuts. The computation involving the use of normal distribution tables is illustrated by the next two examples.

Example 7

A company manufactures G-D washing machines. The president of the company knows that during the past year, there were fifty million families in the U.S. and that the number of G-D washing machines sold was four million. The sales proportion, or the number of G-D washing machines sold per family, denoted by p, is thus

$$p = \frac{4,000,000 \text{ G-D washing machines sold}}{50,000,000 \text{ families in the U. S.}} = 0.08$$

This year the number of G-D washing machines to be sold will vary. Thus the value of p also will vary; p is a random variable. The president believes that the distribution of p is normal and that the range of sales in this year with 50 percent probability will be 500,000 above and below the sales of last year, or 4,000,000 $\pm$ 500,000 = 3,500,000 to 4,500,000. The president plans to take a sample for finding the sales proportion. (See Example 8.) He then will incorporate the sample result and the sales record of the last year into his decision-making analysis. Find the prior probability distribution of p based on the above information.

Solution. The given information suggests that the expected value of p, or $E(p)$, is 0.08. The probability of sales from 3,500,000 ($p = 3,500,000/50,000,000 = 0.07$) to 4,000,000 ($p = 0.08$) is 25 percent and that of sales from 4,000,000 to 4,500,000 ($p = 4,500,000/50,000,000 = 0.09$) is also 25 percent, since the total is 50 percent and p is normally distributed. The standardized normal deviate z, or

$$z = \frac{p - E(p)}{\sigma}$$

may be computed in a conventional manner by using the above suggested values. The value of σ may be found according to the information:

when $p = 0.09$ and $E(p) = 0.08$, $A(z) = 25\%$

This information is diagrammed in Chart 20–2.

CHART 20–2

PRIOR PROBABILITY DISTRIBUTION, FINDING σ

EXAMPLE 7

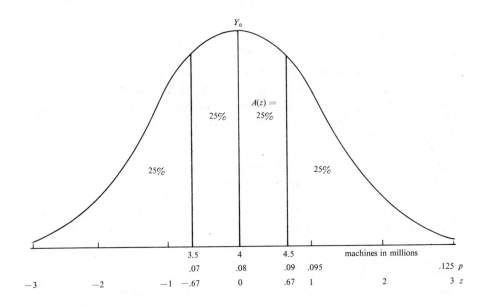

The Area Under the Normal Curve Table shows that

$$A(z) = 0.24857 \quad z = 0.67$$
$$A(z) = 0.25175 \quad z = 0.68$$

Select the nearest value, $z = 0.67$, since $A(z) = 25\% = 0.25$ is closer to 0.24857 than to 0.25175. Substitute the obtained values into the z formula and solve for σ.

$$0.67 = \frac{0.09 - 0.08}{\sigma}$$

$$\sigma = \frac{0.09 - 0.08}{0.67} = 0.015$$

When the mean and the standard deviation of a normally distributed random variable are known, the normal probability distribution of the variable is completely determined. Here, the random variable is p; the mean, or the expected value of $p = E(p) = 0.08$; and the standard deviation of the p values $= \sigma = 0.015$.

Thus, the area representing the probability between any two points under the curve can be obtained by using the normal curve table. However, most decision problems are concerned with discrete variables. In the present example, the president is interested in knowing the sales proprotion p expressed only in a single number, such as 0.08, 0.09, or 0.10 washing machine per family, not in a range specified by two numbers. The normal distribution is a continuous distribution. The probability of a single value of p, or at any point on the continuous p scale, is always equal to zero. Thus, when the normal curve is used to compute the probabililies of a discrete variable, each individual value of the variable must be considered as the midpoint of a selected interval on the continuous scale. The upper limit of an interval will be the lower limit of next interval on the scale. Although all of the intervals need not be of equal size, equal size is generally preferred for convenience. In Table 20–13, which is designed to compute the prior probability distribution of variable p, each interval is arbitrarily assigned to be 0.010 washing machine sold per family. Also, in selecting the intervals, care should be taken so that the mean and the standard deviation of the discrete prior probability distribution are the same as those for the original continuous normal distribution; that is, mean ($E(p)$) is 0.08 and standard deviation (σ) is 0.015. The final answer, prior probability, is shown in column (6) of Table 20–13.

Comments on Table 20–13:

1. Columns (1) and (2), p variable. First, determine in column (1) the upper and lower limits of the interval located in the center of the prior probability distribution. Since the distribution is normal, the midpoint of the central interval is the mean of the distribution and is equal to 0.08. The size of each interval is arbitrarily assigned to be 0.010. The upper and lower limits of each interval thus must be 1/2 of 0.010, or 0.005, above and below the midpoint of each interval. For the central interval,

$$\text{the upper limit} = 0.08 + 0.005 = 0.085$$
$$\text{the lower limit} = 0.08 - 0.005 = 0.075$$

Next, determine the limits of other intervals in column (1). The other limits can be obtained by either subtracting 0.010 consecutively from 0.075 for the intervals with values smaller than that of the central interval or adding 0.010 consecutively to 0.085 for the intervals with values larger than that of the central interval. Observe that the interval between any two midpoints in column (2) is also equal to 0.010.

2. Column (3)   The p values are the interval limits shown in column (1). The expected value of p, or $E(p)$, is the mean of the prior distribution, 0.08.

3. Column (4)   Each z value is the quotient of the difference obtained from column (3) divided by 0.015. The value of 0.015 is the standard error of propor-

TABLE 20–13

COMPUTATION OF PRIOR PROBABILITIES, USING AREAS UNDER THE
NORMAL CURVE TABLE—EXAMPLE 7

p Interval limits (1)	Midpoint (2)	$p - E(p) =$ Col. (1) $- 0.080$ (3)	$z = \dfrac{p-E(p)}{\sigma} = \dfrac{\text{Col. (3)}}{0.015}$ (4)	Area Between Y_o and z (5)	Prior Probability $P(p)$ (6)
Under 0.025		$-\infty$	$-\infty$	0.50000	
	$-\infty$				0.00012
0.025		-0.055	-3.67	0.49988	
	0.03				0.00123
0.035		-0.045	-3.00	0.49865	
	0.04				0.00855
0.045		-0.035	-2.33	0.49010	
	0.05				0.03756
0.055		-0.025	1.67	0.45254	
	0.06				0.11120
0.065		-0.015	1.00	0.34134	
	0.07				0.21204
0.075		-0.005	-0.33	0.12930	
	0.08			$+$	0.25860
0.085		0.005	0.33	0.12930	
	0.09				0.21204
0.095		0.015	1.00	0.34134	
	0.10				0.11120
0.105		0.025	1.67	0.45254	
	0.11				0.03756
0.115		0.035	2.33	0.49010	
	0.12				0.00855
0.125		0.045	3.00	0.49865	
	0.13				0.00123
0.135		0.055	3.67	0.49988	
	∞				0.00012
Above 0.135		∞	∞	0.50000	
Total					1.00000

TABLE 20–14

COMPUTATION OF PRIOR PROBABILITIES, USING ORDINATES
OF THE NORMAL CURVE TABLE
EXAMPLE 7

p		$p - E(p) =$ Col. (2) $-$ 0.08	$z = \dfrac{p - E(p)}{\sigma}$ $= \dfrac{\text{Col. (3)}}{0.015}$	Ordinate* at z	Prior Probability $P(p) = $ Col. (5) $\times$ 0.67
Interval	Mid-point				
(1)	(2)	(3)	(4)	(5)	(6)
0.025–0.035	0.03	$-$ 0.05	3.33	0.00146	0.00098
0.035–0.045	0.04	$-$ 0.04	2.67	0.01191	0.00798
0.045–0.055	0.05	$-$ 0.03	2.00	0.05399	0.03617
0.055–0.065	0.06	$-$ 0.02	1.33	0.16038	0.10745
0.065–0.075	0.07	$-$ 0.01	0.67	0.32297	0.21639
0.075–0.085	0.08	0.00	0.00	0.39894	0.26729
0.085–0.095	0.09	0.01	0.67	0.32297	0.21639
0.095–0.105	0.10	0.02	1.33	0.16038	0.10745
0.105–0.115	0.11	0.03	2.00	0.05399	0.03617
0.115–0.125	0.12	0.04	2.67	0.01191	0.00798
0.125–0.135	0.13	0.05	3.33	0.00146	0·00098
Total					1.00523

*See Table 5 in the Appendix.

tion of the prior distribution.

4. Column (5) The area, which represents the probability of the sales proportion for the range between a given limit (at z) to the mean (at Y_o), is obtained from the Areas Under the Normal Curve Table.

5. Column (6) Each prior probability is the difference between two adjoining values shown in column (5). For example, the prior probability of the interval 0.025 $-$ 0.035 with the midpoint 0.03 is

$$0.49988 - 0.49865 = 0.00123$$

One exception is the prior probability of the central interval, which is the sum of the two adjoining values in column (5), or

$$0.12930 + 0.12930 = 0.25860$$

The total of the prior probabilities is 1.

There is a second method to compute the prior normal probability distribution. This method uses the Ordinates of the Normal Curve Table. Generally, it yields approximately the same result if the size of each interval for p is relatively small. The computation in Table 20–14 illustrates the application of this method. Columns (1) and (2) of the table are the same as the first two columns of Table 20–13. Each z value in Table 20–14 is computed from the midpoint of an interval instead of from the interval limits as in Table 20–13. Each ordinate in column (5) is obtained from the Ordinates of the Normal Curve Table according to the nearest z value. Each prior probability in column (6) is the product of the ordinate at z multiplied by 0.67. The value of 0.67 is the size of each interval expressed in terms of the standardized normal deviate z, or the interval on the z scale $= \dfrac{0.010 \text{ washing machine per family}}{0.015 \text{ (the value of } \sigma \text{)}}$

$$= 0.67$$

The prior probability distribution obtained from Table 20–14 is shown graphically in Chart 20–3. Observe that the width of each bar on the z scale is 0.67. The area of each bar (the product of the ordinate at the midpoint and the width, 0.67) represents the prior probability of the interval. The chart also shows the prior probability distribution obtained from Table 20–13. Compare the areas (or probabilities) obtained from the two tables as shown in the chart. The difference between the area of the bar and the area under the normal curve for each interval is fairly small. It is obvious that the approximation by the bars would be closer to the areas under the normal curve if the width of each bar (or interval size) were reduced or the number of bars were increased.

CHART 20–3

PRIOR PROBABILITY DISTRIBUTION, SHOWING AREAS (PROBABILITIES) BY INTERVALS
EXAMPLE 7

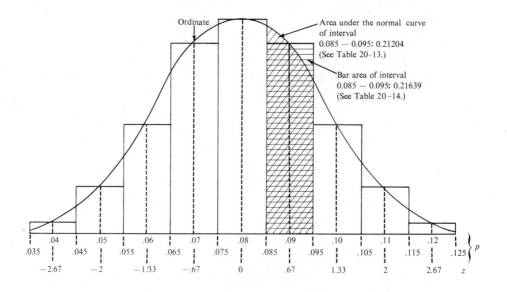

Example 8

Refer to Example 7. Assume that a sample of 100 families was taken in the early part of this year. Each family was asked how many G-D washing machines it will buy during this year. The number of G-D washing machines to be purchased by each family is represented by X. If a family will not buy, $X = 0$; if it will buy one, $X = 1$; if two, $X = 2$; and so on. The mean $\bar{X}$ and the standard deviation s of the 100 X's of the sample are computed as follows (details of 100 X's are omitted for simplicity):

$$\bar{X} = \frac{\Sigma X}{n} = \frac{9}{100} = 0.09 \text{ washing machine per family}$$

$$s = \sqrt{\frac{\Sigma (X - \bar{X})^2}{n}} = \sqrt{\frac{0.25}{100}} = \sqrt{0.0025} = 0.05$$

Revise the prior probability distribution obtained from Example 7 based on the above additional information. Also, compute the posterior expected value and the standard deviation.

Solution. The posterior or revised probabilities and the posterior expected value are shown in columns (5) and (6) respectively of Table 20–16. The standard deviation is computed in Table 20–17. The computations in the tables are explained below.

First, derive the likelihoods for the midpoints of the intervals listed in the prior probability distribution of Table 20–13. The derivation is done in Table 20–15. The midpoints and their respective intervals in columns (1) and (2) of the table are repeated from Table 20–13. Each likelihood shown in column (6) is the conditional probability of obtaining a sample mean of $\bar{X} = 0.09$ from a population which has a given mean p. For example, the probability of obtaining a sample with the mean 0.09 from a population with the given mean $p = 0.07$ is 0.00135 as shown in Table 20–15. The population mean $p = 0.07$ is specified by the interval limits 0.065 and 0.075 in the table. The difference between the sample mean and each limit, or $\bar{X} - p$ in column (3) in the table, is measured by the unit of standard error of the mean $s_{\bar{x}}$. When the sample size is large, 100 or more, the formula of $s_{\bar{x}}$ is

$$s_{\bar{x}} = \frac{s}{\sqrt{n}} \text{ where } s = \text{the standard deviation of the sample and}$$

$$n = \text{the sample size}$$

Substitute the s and n values in the above formula,

$$s_{\bar{x}} = \frac{0.05}{\sqrt{100}} = 0.005$$

When the difference $\bar{X} - p$ is expressed in the units of $s_{\bar{x}}$, the result is the value of z, or

$$z = \frac{\bar{X} - p}{s_{\bar{x}}} = \frac{0.09 - p}{0.005}$$

TABLE 20–15

COMPUTATION OF LIKELIHOODS, USING AREAS UNDER THE NORMAL
CURVE TABLE—EXAMPLE 8

Interval limits (1)	Midpoint (2)	$\bar{X}-p = 0.09 - Col.(1)$ (3)	$z = \dfrac{\bar{X}-p}{s_{\bar{x}}} = \dfrac{Col.(3)}{0.005}$ (4)	Area Between Y_0 and z (5)	Likelihood (6)
Under 0.025		$-\infty$	$-\infty$	0.50000	
	$-\infty$				0.00000
0.025		0.065	13.00	0.50000	
	0.03				0.00000
0.035		0.055	11.00	0.50000	
	0.04				0.00000
0.045		0.045	9.00	0.50000	
	0.05				0.00000
0.055		0.035	7.00	0.50000	
	0.06				0.00000
0.065		0.025	5.00	0.50000	
	0.07				0.00135
0.075		0.015	3.00	0.49865	
	0.08				0.15731
0.085		0.005	1.00	0.34134	
	0.09			$+$	0.68268
0.095		-0.005	-1.00	0.34134	
	0.10				0.15731
0.105		-0.015	-3.00	0.49865	
	0.11				0.00135
0.115		-0.025	-5.00	0.50000	
	0.12				0.00000
0.125		-0.035	-7.00	0.50000	
	0.13				0.00000
0.135		-0.045	-9.00	0.50000	
	∞				0.00000
Above 0.135		∞	∞	0.50000	

TABLE 20–16

COMPUTATION OF POSTERIOR PROBABILITY AND EXPECTED VALUE
EXAMPLE 8

(1) Event p		(2) Prior Probability (Table 20–13)	(3) Likelihood (Table 20–15)	(4) Joint Probability (2) × (3)	(5) Posterior Probability (4) ÷ 0.20327	(6) Posterior Expected Value of p (1) × (5)
Interval	Midpoint					
...	...	...	0	0	0	0
0.065–0.075	0.07	0.21204	0.00135	0.00029	0.00143	0.0001001
0.075–0.085	0.08	0.25860	0.15731	0.04068	0.20013	0.0160104
0.085–0.095	0.09	0.21204	0.68268	0.14476	0.71215	0.0640935
0.095–0.105	0.10	0.11120	0.15731	0.01794	0.08604	0.0086040
0.105–0.115	0.11	0.03756	0.00135	0.00005	0.00025	0.0000275
...	...	...	0	0	0	0
Total		1.00000		0.20327	1.00000	0.0888355

TABLE 20–17

COMPUTATION OF THE STANDARD DEVIATION OF POSTERIOR
PROBABILITY DISTRIBUTION
EXAMPLE 8

Event p (midpoint)	Posterior Probability f	$x = p - E(p)$ $= p - 0.089$	fx	fx^2
0.07	0.00143	− 0.019	− 0.00002717	0.00000052
0.08	0.20013	− 0.009	− 0.00180117	0.00001621
0.09	0.71215	0.001	0.00071215	0.00000071
0.10	0.08604	0.011	0.00094644	0.00001041
0.11	0.00025	0.021	0.00000525	0.00000011
Total	1.00000		− 0.00016450 (near to zero)	0.00002796

$$\sigma^2_1 = 0.00002796 \qquad \sigma_1 = \sqrt{0.00002796} = 0.0053$$

which is shown in column (4) in the table. The area between Y_o and z in column (5) is obtained from the Areas Under the Normal Curve Table. Each likelihood is the difference between two adjoining values shown in column (5). One exception is the likelihood of the central interval, which is the sum of the two adjoining values in column (5).

Next, compute the posterior probabilities and the expected value of the posterior probability distribution in the usual manner. The computation is shown in Table 20–16. The classes with zero likelihoods are omitted in the table since the omission does not change the result of the computation. The prior probabilities obtained from Table 20–13 are used in the computation. The highest prior probability in the table is 0.25860 for $p = 0.08$ washing machine per family. However, when the prior probablities are revised according to the sample information, the highest posterior probability is 0.71215 for $p = 0.09$. The expected value of the posterior probability distribution is 0.0888355 which is very close to the sample mean 0.09.

Finally, compute the standard deviation of the posterior distribution in a manner similar to that of Table 20–11. This is done in Table 20–17. The variance is 0.00002796 and the standard deviation is $\sqrt{0.00002796} = 0.0053(= \sigma_1)$.

★20.5   Computing the Posterior Expected Value and Variance by Formulas

When the interval size is reduced, or the number of midpoints is increased, the work of computing the expected value (or the mean) and the variance (or the standard deviation squared) of the posterior normal distribution as presented in Tables 20–16 and 20–17 is rather tedious. An easier way to compute the two values is to use the formulas listed below.

For computing the posterior expected value,

$$E(p)_1 = \frac{E(p)_o \cdot \left(\frac{1}{\sigma_o^2}\right) + \bar{X} \cdot \left(\frac{1}{\sigma_{\bar{x}}^2}\right)}{\frac{1}{\sigma_o^2} + \frac{1}{\sigma_{\bar{x}}^2}} \quad \text{or}$$

$$E(p)_1 = \frac{E(p)_o \cdot \sigma_{\bar{x}}^2 + \bar{X} \cdot \sigma_{po}^2}{\sigma_o^2 + \sigma_{\bar{x}}^2} \tag{20–2}$$

For computing the posterior variance,

$$\sigma_1^2 = \frac{\sigma_{po}^2 \cdot \sigma_{\bar{x}}^2}{\sigma_o^2 + \sigma_{\bar{x}}^2} \tag{20–3}$$

The subscript zero (o) on the far right side of a symbol indicates a prior distribution and the subscript one (1) indicates a posterior distribution. Thus, formula (20–2) indicates that the posterior expected value is a *weighted average* of the prior expected value $(E(p)_o)$ and the sample mean $(\bar{X})$, the weights being the reciprocals of the variances of the prior distribution $(1/\sigma_{po}^2)$ and the sampling distribution of the mean $(1/\sigma_{\bar{x}}^2)$ respectively.

Formula (20–3) is derived from the following expression (which can be proved by using calculus):

$$\frac{1}{\sigma_1^2} = \frac{1}{\sigma_o^2} + \frac{1}{\sigma_{\bar{x}}^2} \tag{20–4}$$

The expression indicates that the reciprocal of the posterior variance is the sum of the reciprocals of the variances of the prior distribution of p and the sampling distribution of the mean $\bar{X}$.

Substitute the values obtained previously.

$$E(p)_o = 0.08 \qquad \text{(Example 7, page 517)}$$
$$\sigma_o = 0.015 \qquad \text{(Example 7, page 518)}$$
$$\bar{X} = 0.09 \qquad \text{(Example 8, page 523) and}$$
$$\sigma_{\bar{x}} = 0.005 \quad \text{(Example 8, estimated by using } s_{\bar{x}}, \text{ page 523)}$$

into formulas (20–2) and (20–3) as follows:

$$E(p)_1 = \frac{0.08(0.005^2) + 0.09(0.015^2)}{0.015^2 + 0.005^2} = \frac{0.08(0.000025) + 0.09(0.000225)}{0.00025}$$

$$= \frac{0.00002225}{0.00025} = 0.089$$

$$\sigma_1^2 = \frac{0.015^2 \cdot 0.005^2}{0.015^2 + 0.005^2} = \frac{0.000000005625}{0.00025} = 0.0000225$$

$$\sigma_1 = \sqrt{0.0000225} = 0.0047$$

The two answers may be compared with those obtained by Tables 20–16 and 20–17 respectively.

Observe the weight attached to the sample mean $\bar{X}$ in the posterior expected value $E(p)_1$ formula. The weight is

$$\frac{1}{\sigma_{\bar{x}}^2} = \frac{n}{\sigma^2} \cdot \frac{N-1}{N-n}$$

The values of the population size N and the population variance σ^2 are fixed. Thus, when the sample size n becomes increasingly large, the relative weight attached to the sample mean also becomes increasingly large. Correspondingly, the relative weight $(1/\sigma_o^2)$ attached to the prior distribution becomes smaller since the value of σ_o^2 is not affected by the sample size.

From the above computed values, the estimate of the sales proportion at a 99.73% confidence interval is

$$E(p)_1 \pm 3(\sigma_1) = 0.089 \pm 3(0.0047) = 0.089 \pm 0.0141$$
$$= 0.0749 \text{ to } 0.1031 \text{ G-D washing machine per family}$$

If the Bayesian approach is not employed, that is, the prior information is completely disregarded, the estimate based on the sample information at a 99.73% confidence interval is computed as follows:

$$\bar{X} \pm 3s_{\bar{x}} = 0.09 \pm 3(0.005) = 0.09 \pm 0.015$$
$$= 0.075 \text{ to } 0.105 \text{ G-D washing machine per family}$$

Observe each term of Formula (20–4). A variance (or a standard deviation) measures the dispersion from the mean of a given distribution. When a variance is large, or the reciprocal of the variance is small, the quantity of information summarized by the mean of the distribution is small. Conversely, when a variance is small, or the reciprocal of the variance is large, the quantity of information summarized by the mean of the distribution is large. Now, let

I = the quantity of information summarized by a given mean and
    a subscript to I represent a probability distribution or

$$I_o = \frac{1}{\sigma_o^2} = \text{the quantity of information summarized by the prior}$$

mean $E(p)_o$

$$I_s = \frac{1}{\sigma_{\bar{x}}^2} = \text{the quantity of information summarized by the sample}$$

mean $\bar{X}$ and

$$I_1 = \frac{1}{\sigma_1^2} = \text{the quantity of information summarized by the poste-}$$

rior mean $E(p)_1$

Then Formula (20–4) may be written:

$$I_1 = I_o + I_s$$

which indicates that the posterior information is the sum of the prior information and the sample information. Also, I_o and I_s are the weights attached to the prior mean $E(p)_o$ and the sample mean $\bar{X}$ in Formula (20–2) respectively. Formula (20–2) may be written

$$E\,(p)_1 = \frac{I_o \cdot E(p)_o + I_s \cdot \bar{X}}{I_o + I_s} \tag{20–5}$$

In examples 7 and 8, the posterior expected value $(E(p)_1 = 0.089)$ is much closer to the sample mean $(\bar{X} = 0.09)$ than it is to the prior expected value $(E(p)_o = 0.08)$. It is evident that the sample information I_s gives greater weight $(I_s = 1/\sigma_{\bar{x}}^2 = 1/0.005^2 = 40,000)$ than the prior information I_o $(I_o = 1/\sigma_o^2 = 1/0.015^2 = 4,444)$ in computing the posterior expected value.

Finally, observe the terms in formula (20–3). If the two variances σ_o^2 and $\sigma_{\bar{x}}^2$ become increasingly large, the variance σ_1^2 also becomes increasingly large, because a product of two positive numbers increases at a greater rate than the sum of the two numbers when the two numbers are increasing. The quotient (σ_1^2) of a greater product $(\sigma_o^2 \cdot \sigma_{\bar{x}}^2)$ divided by a relatively smaller sum $(\sigma_o^2 + \sigma_{\bar{x}}^2)$ thus becomes larger. The effect is that the value of I_1, or the reciprocal of the larger σ_1^2, now becomes smaller. Thus, the quantity of information summarized by the posterior mean $E(p)_1$ is decreased. In such a case, the sample size n should be increased in order to reduce the value of $\sigma_{\bar{x}}^2$ $(\sigma_{\bar{x}}^2 = \sigma^2/n)$.

20.6 Summary

The assigned probabilities of all possible events in a given problem may be revised when additional information is available. The additional information can be obtained from past records of a business firm or samples. The revision process is usually based on Bayes' theorem, Formula (20–1). When Bayes' theorem is used to revise a set of old probabilities, called prior probabilities, to a set of new probabilities, called posterior probabilities, the computation is usually done systematically in a table, such as Tables 20–3 and 20–4.

The original probability of a business event can be revised according to historial records or samples. The basic procedure of applying Bayes' theorem for various sample sizes is the same. The conditional probabilities, or likelihoods, are computed by using the binomial formula. Instead of the exact computation method, the likelihoods may also be obtained by using the normal probability distribution tables.

The process of revising a set of prior probabilities may be repeated if more information can be obtained. Each new process must begin with a set of prior probabilities, which may be based on the set of posterior probabilities of the previous process.

The expected value of a probability distribution, either prior or posterior, is an arithmetic mean weighted by probabilities. Based on an expected value, the standard deviation may be computed for a given probability distribution.

There are two methods of computing the prior normal probability distribution: (1) use the Areas Under the Normal Curve Table, and (2) use the Ordinates of the Normal Curve Table. Generally the two methods yield approximately the same result if the size of each interval for p is relatively small. However, when the interval size is reduced, or the number of midpoints is increased, the work of computing the expected value and the standard deviation of the posterior normal distribution is rather tedious. An easier way to compute the two values is to use the formulas (20–2) and (20–3).

It can be proved that the reciprocal of the posterior variance is the sum of the reciprocals of the variances of the prior distribution of p and the sampling distribution of the mean $\bar{X}$ [formula (20–4)]. This relationship indicates the fact that the posterior information is the sum of the prior information and the sample information.

SUMMARY OF FORMULAS

Application	Formula	Formula number	Reference page
Bayes' theorem	$P(A_i \mid B) = \dfrac{P(A_i \text{ and } B)}{P(B)}$ where $P(B) = P(A_1 \text{ and } B) + P(A_2 \text{ and } B)$ $+ \cdots + P(A_n \text{ and } B)$ $P(A_i \text{ and } B) = P(A_i) \cdot P(B \mid A_i)$	20-1	506
Posterior expected value	$E(p)_1 = \dfrac{E(p)_o \cdot \left(\frac{1}{\sigma_o^2}\right) + \bar{X} \cdot \left(\frac{1}{\sigma_{\bar{x}}^2}\right)}{\frac{1}{\sigma_o^2} + \frac{1}{\sigma_{\bar{x}}^2}}$ $= \dfrac{E(p)_o \cdot \sigma_{\bar{x}}^2 + \bar{X} \cdot \sigma_{po}^2}{\sigma_o^2 + \sigma_{\bar{x}}^2}$	20-2	526
Posterior variance	$\sigma_{p1}^2 = \dfrac{\sigma_o^2 \cdot \sigma_{\bar{x}}^2}{\sigma_o^2 + \sigma_{\bar{x}}^2}$	20-3	526
Posterior variance	$\dfrac{1}{\sigma_1^2} = \dfrac{1}{\sigma_o^2} + \dfrac{1}{\sigma_{\bar{x}}^2}$	20-4	526
Posterior expected value	$E(p)_1 = \dfrac{I_o \cdot E(p)_o + I_s \cdot \bar{X}}{I_o + I_s}$	20-5	528

Exercises 20

1. An urn contains 30 marbles: 18 red and 12 blue. The marbles are also marked with either yellow or green stripes as follows:

18 red marbles—12 with yellow and 6 with green stripes

12 blue marbles—3 with yellow and 9 with green stripes

One marble is drawn randomly from the urn. (a) What are the probabilities of all possible outcomes of the drawing? (b) Suppose that the marble drawn from the urn has yellow stripes. What is the probability that the marble is red? blue?

2. Refer to Problem 1. Suppose that the marble drawn from the urn has green stripes. What is the probability that the marble is red? blue?

3. Assume that 40 students of a freshman class are divided into 4 groups as follows:

Group 1—6 girls and 4 boys

Group 2—8 girls and 2 boys

Group 3—5 girls and 5 boys

Group 4—3 girls and 7 boys

A student selected randomly from the class is a girl. What is the probability that the selected girl is from group 1? 2? 3? 4? From which group was she most likely selected?

4. Assume that a student randomly selected from the class stated in Problem 3 above is a boy. What is the probability that the selected boy is from group 1? 2? 3? 4?

5. In a manufacturing company, department X produces 20 percent of the output and department Y produces the remaining 80 percent. Based on the average, 1 in 20 items produced by department X is defective and 1 in 40 items produced by department Y is defective. The two departments produced 50,000 items in a single day. An item drawn at random from the day's output is found to be defective. What is the probability that the defective item was produced by department X? by Y?

6. Assume that an item drawn at random from the day's output given in Problem 5 above is found to be nondefective. What is the probability that the nondefective item was produced by department X? by Y?

7. Refer to Problem 5. Suppose that the 50,000 items produced in the company are packed into many bags according to departments. Two defective items are drawn randomly from an unidentified bag. What is the probability that the two defective items were produced by department X? by department Y?

8. Assume that a company has four departments. The production distribution and the defective production of the four departments are as follows:

Department number	Production distribution	Defective items produced by each department
1	10%	15%
2	40%	5%
3	20%	10%
4	30%	4%

A sample of 25 items is taken from a lot produced by an unidentified department. Five of the 25 items are defective. Find the probability that the defective items were produced by each department.

9. Refer to Problem 8. Compute the expected values of the prior and the posterior probability distributions and the standard deviation of the posterior distribution.

★10. Refer to Tables 20–13 and 20–14 of Example 7. Compute the prior probabilities of the midpoints 0.06 to 0.10 by letting the size of each interval of p be 0.005.

★11. Refer to Tables 20–15 and 20–16 of Example 8. Compute the posterior probabilities of the midpoints 0.07 to 0.09 by letting the size of each interval of p be 0.005. (Use the prior probabilities obtained from Problem 10.)

★12. Refer to Example 8, page 522. Instead of a sample of 100 families, a sample of 196 families was taken in the early part of this year. The mean $\bar{X}$ and the standard deviation s of the sample are: $\bar{X} = 0.07$ washing machine per family and $s = 0.084$. Compute the posterior expected value and the standard deviation by using formulas (20–2) and (20–3). Which information, I_o or I_s, gives a greater weight in computing the posterior expected value?

part 6

Time

Series

Analysis

This part presents business and economic changes based on a *time series*, which is a type of quantitative information classified on the basis of time. The study will include two major areas: (1) the construction of index numbers of time series (Chapter 21) and (2) the analysis of the basic patterns of time series movements (Chapters 22 to 24).

21 Index Numbers

An index number is a relative value with a base equal to 100 percent or a multiple of 100 percent, such as 10 or 100. It is used as an indicator for change of anything or any group of things. Important index numbers concerning business and economic activities may be classified into three types: (1) price indexes, (2) quantity indexes, and (3) value indexes. Practical examples of the three types of indexes (or *indices*) are shown in Table 21-1. The indexes on the table, published monthly in the *Survey of Current Business* by the U. S. Department of Commerce, provide important information to businessmen, economists, and government agencies for making decisions and analyzing economic changes.

Index numbers may be constructed for a single commodity, called *simple index numbers*, or for a group of commodities, called *composite index numbers*. The methods of computing simple index numbers are presented in Section 21.1 and those for composite index numbers, in Section 21.2. There are various formulas for constructing these index numbers. The criteria used in testing the quality of a formula are given in Section 21.3. There may be a need to shift the base of a set of index numbers previously constructed. The methods of shifting the base from one period to another are discussed in Section 21.4. The method of adjusting dollar values of a time series by price indexes is called *deflation*, which is presented in Section 21.5.

21.1 Simple Index Numbers (for a Single Commodity)

A simple index number referred to here is constructed from a time series concerning a single commodity. It is also called a *simple relative* since it is expressed in a form of ratio. A ratio has two terms: the first term (or the number mentioned first in the statement) and the second term (or the base used for comparison). The two terms may be written as a fraction as follows:

$$\text{The ratio of first term to second term} = \frac{\text{First Term}}{\text{Second Term}}$$

TABLE 21–1

EXAMPLES OF IMPORTANT INDEX NUMBERS CONCERNING BUSINESS
AND ECONOMIC ACTIVITIES, BY TYPES OF INDEXES

Type and Name	Primary Source	Current Base Period	Items Included in the Index
(A) *Price index*			
1. Prices received by farmers	Department of Commerce	1910–14	Crops, livestock and products
2. Prices paid by farmers	Department of Commerce	1910–14	Family living items and production items
3. Consumer prices	Department of Labor	1967	Food, housing, apparel and upkeep, health and recreation, etc.
4. Wholesale prices	Department of Labor	1967	Farm products, foods, and other commodities
(B) *Quantity index*			
1. Industrial production	Federal Reserve Board	1967	Manufacturing, mining, and utilities
2. Machinery and equipment	Department of Commerce	1967–69	Industrial supplies
3. Electrical equipment	Department of Commerce	1967	Motors and generators
4. Exports of U.S. merchandise	Department of Commerce	1967	Quantity
(C) *Value index*			
1. Construction contracts	F. W. Dodge Division	1967	Valuation
2. Advertising	McCann-Erickson	1957–59	Magazines, newspapers, television (network), and spot TV
3. Weekly payrolls	Department of Labor	1967	Construction, manufacturing, mining workers
4. Exports of U. S. merchandise	Department of Commerce	1967	Value

Source: U.S. Department of Commerce, *Survey of Current Business*, 1975 monthly issues.

Thus, if the price of a certain type of car was $3,000 in 1975 and $3,750 in 1976, the ratio of the 1976 price to the 1975 price is

$$\frac{\text{1976 Price}}{\text{1975 Price}} = \frac{\$3,750}{\$3,000} = 1.25 \text{ or } 125\%$$

and the ratio of the 1975 price to the 1976 price is

$$\frac{\text{1975 Price}}{\text{1976 Price}} = \frac{\$3,000}{\$3,750} = 0.80 \text{ or } 80\%$$

The percentage, obtained by moving the decimal point in the quotient two places to the right, now represents in each case the relative value of the first term and 1 or 100 percent represents the relative value of the second term, or the base. A base does not have to be 100 percent. When each relative value in a series is multiplied, or divided, by the same number (other than zero), the relationships among the values are not changed. Frequently index numbers are expressed in $100 \times$ each percent in a series. This can be done simply by dropping the percent sign ($\%$) in each percent. In this type of expression the base is 100, which is 100 times 100 percent. This is illustrated in Table 21–2.

TABLE 21–2

ILLUSTRATION OF RATIOS AS SIMPLE INDEX NUMBERS

Year	Unit Price of the Car	Index Numbers (or Simple Relatives)			
		1975 *price is base*		1976 *price is base*	
		1975 = 100%	1975 = 100	1976 = 100%	1976 = 100
1975	$3,000	100%	100	80%	80
1976	$3,750	125%	125	100%	100

In general, the simple price, quantity, and value relatives for a single commodity may be computed from the following formulas:

$$\text{Price Relative} = \frac{p_n}{p_0} \qquad (21\text{-}1)$$

$$\text{Quantity Relative} = \frac{q_n}{q_0} \qquad (21\text{-}2)$$

$$\text{Value Relative} = \frac{p_n q_n}{p_0 q_0} \qquad (21\text{-}3)$$

where p_n = price of a single commodity in the given year (or other unit of time),

p_0 = price of a single commodity in the base year,

q_n = quantity of a single commodity (produced, consumed, or sold) in the given year, and

q_0 = quantity of a single commodity in the base year.

The period of time in computing the simple relatives or index numbers is usually a year, although it may be a quarter, a month, or other unit of time. For many commodities, the unit prices in a year may not be the same all the time. In such cases, an appropriate average of the prices for the year may be used in the computation.

The applications of these three formulas are illustrated in Example 1.

Example 1

Assume that the unit prices and quantities produced of green coffee in the Moontie County in the years 1971 and 1976 are given below. Compute the indexes by using 1971 as the base year for (a) price, (b) quantity, and (c) value of the year 1976.

TABLE 21–3

DATA FOR EXAMPLE 1

Year	Price per pound p	Quantity produced q	Value pq
0—1971 (Base)	$0.25	200 lbs	$ 50
n—1976	0.40	250	100

Solution. The simple relatives are used as the index numbers as follows:

$$1976 \text{ Price relative} = \frac{p_n}{p_0} = \frac{0.40}{0.25} = 1.60 \text{ or } 160\% \qquad \text{(Formula 21–1)}$$

$$1976 \text{ Quantity relative} = \frac{q_n}{q_0} = \frac{250}{200} = 1.25 \text{ or } 125\% \qquad \text{(Formula 21–2)}$$

$$1976 \text{ Value relative} = \frac{p_n q_n}{p_0 q_0} = \frac{\$0.40 \times 250}{\$0.25 \times 200} = \frac{\$100}{\$ 50} = 2.00 \text{ or } 200\%$$

$$\text{(Formula 21–3)}$$

Example 1 illustrates the basic method of computing the simple relatives. When a time series includes information of more than two years, there are three ways to compute the relatives. The various ways of computation give different names for the simple relatives: (1) fixed-base relatives, (2) link relatives, and (3) chain relatives. Since the procedures of computing the three differently named simple relatives for price, quantity, and value of a single commodity are basically the same, only the time series of unit prices are used for illustration purposes.

Fixed-Base Relatives

Fixed-base relatives for unit prices are used to show the relative price changes during the years included in a time series. The series has a single number which is equal to 100 or a multiple of 100 percent (such as $100 \times 100\% = 100$) selected as the base. The base number may be the price of a single year or an average of the prices of several years. The criteria for selecting a base period depend upon the type and the use of the index numbers. For example, if the index numbers are frequently used to compare with published data, it is preferable if the base period agrees with that of the published data. In general, the base period should be normal with respect to the prices of other periods.

Fixed-base price relatives are easy to compute. The annual price in a series is divided by the same base number for each year. The computation is illustrated in Example 2.

Example 2

Assume that the unit prices of green coffee in the Moontie County in the years 1971 to 1976 are given in column (2) of Table 21–4.

TABLE 21–4

DATA AND ANSWERS FOR EXAMPLE 2

Year (1)	p_n Price per pound (2)	Index Numbers (Simple Relatives)	
		(a) 1971 = 100(%) (3)	(b) 1971–73 = 100(%) (4)
1971	$0.25	100.0(%)	71.4(%)
1972	0.30	120.0	85.7
1973	0.50	200.0	142.9
1974	0.20	80.0	57.1
1975	0.22	88.0	62.9
1976	0.40	160.0	114.3

Compute the price relative for each year by using (a) the year 1971 as the base, and (b) the average of the prices of the years 1971 to 1973 as the base.

Solution. (a) The price relative for each year by using 1971 as the base is computed according to formula (21–1), or

$$\text{Price relative of a given year} = \frac{\text{Price of a given year } (p_n)}{\text{Price of 1971, \$0.25 } (p_0)}$$

$$\text{Thus 1971 price relative} = \frac{0.25}{0.25} = 1 \text{ or } 100\%$$

$$\text{1972 price relative} = \frac{0.30}{0.25} = 1.20 \text{ or } 120\%$$

and so on. The answers are shown in column (3) of the above table.

(b) The average of the prices of the years 1971 to 1973 is obtained by using the arithmetic mean method, or

$$\text{the average price} = \frac{0.25 + 0.30 + 0.50}{3} = \frac{1.05}{3} = \$0.35$$

The price relative for each year by using the average price as the base is

$$\text{Price relative of a given year} = \frac{\text{Price of a given year } (p_n)}{\text{Average price, \$0.35 } (p_0)}$$

$$\text{Thus, 1971 price relative} = \frac{0.25}{0.35} = 0.714 \text{ or } 71.4\%$$

$$\text{1972 price relative} = \frac{0.30}{0.35} = 0.857 \text{ or } 85.7\%$$

and so on. The answers are shown in column (4) of Table 21–4.

Link Relatives

Link price relatives are used to show the relative price changes between two successive years in a time series. To obtain the link relative of a given year, divide the price for the given year by the price for the immediately preceding year (the base).

Example 3

Refer to the time series given in column (2) of the table in Example 2. Compute the link relative for each year.

Solution. There is no link relative of the first year, 1971. The link relatives of other years are computed as follows:

$$1972 \text{ link relative} = \frac{1972 \text{ price}}{1971 \text{ price}} = \frac{\$0.30}{\$0.25} = 1.20 \text{ or } 120\%$$

$$1973 \text{ link relative} = \frac{1973 \text{ price}}{1972 \text{ price}} = \frac{\$0.50}{\$0.30} = 1.667 \text{ or } 166.7\%$$

and so on. The answers are shown in column (3) of Table 21-5.

TABLE 21-5

DATA AND ANSWERS FOR EXAMPLE 3

Year (1)	Price per pound (2)	Link relatives (%) (3)
1971	$0.25	none
1972	0.30	120.0(%)
1973	0.50	166.7
1974	0.20	40.0
1975	0.22	110.0
1976	0.40	181.8

Chain Relatives

Chain price relatives, like the fixed-base price relatives, are used to show the relative price changes during the years included in a time series with a single base. However, chain relatives differ from fixed-base relatives in computation. Chain relatives are computed from link relatives, whereas fixed-base relatives are computed directly from the original data. The results obtained by the two different methods should be the same, but they may differ from each other slightly due to rounding decimal places.

In general, the *chain relative* of a given year is the product of the *link relatives* of the given year and the preceding years up to (not including) the base year. This statement is illustrated in Example 4 below.

Example 4

Refer to the link relatives given in Example 3. Compute the 1974 chain relative by using 1971 as the base year.

Solution.   1974 chain relative = 1974 link relative × 1973 link relative ×
                                1972 link relative
                        = 40% × 166.7% × 120%
                        = 0.4 × 1.667 × 1.2
                        = 0.80016 or round to 80.0%

Observe that before the rounding, the answer to the 1974 chain relative is 0.80016.
However, the 1974 fixed-base relative is exact 0.80, the base also being the 1971
price (see Example 2). The comparison shows the effect of computing the relatives
by the two different methods.

Note that the above computation may be written in the following form:

$$1974 \text{ chain relative} = \frac{1974 \text{ price}}{1973 \text{ price}} \times \frac{1973 \text{ price}}{1972 \text{ price}} \times \frac{1972 \text{ price}}{1971 \text{ price}}$$

$$= \frac{1974 \text{ price}}{1971 \text{ price}} = 1974 \text{ fixed-base relative, the base}$$
$$\text{being the 1971 price}$$

The chain relative of a given year may also be obtained by the following expression:

$$\begin{pmatrix} \text{Chain relative of} \\ \text{a given year} \end{pmatrix} = \begin{pmatrix} \text{Link relative of} \\ \text{the given year} \end{pmatrix} \times \begin{pmatrix} \text{Chain relative of} \\ \text{the preceding year} \end{pmatrix}$$

This expression is illustrated in Example 5.

Example 5

Refer to the link relatives given in Example 3. Compute the chain relatives for
the years 1971 to 1976 by using 1971 as the base year.

Solution.   1971 chain relative = 1 or 100%, since 1971 is the base year
                1972 chain relative = 1972 link relative × 1971 chain relative
                        = 1.2 × 1 = 1.2 or 120%

TABLE 21-6

DATA AND ANSWERS FOR EXAMPLE 5

Year	Link relative Previous year = base = 1.000	Chain relative	
		1971 = 1.000	1971 = 100%
(1)	(2)	(3)	(4)
1971	none	1.000	100.0(%)
1972	1.200	1.200	120.0
1973	1.667	2.000	200.0
1974	0.400	0.800	80.0
1975	1.100	0.880	88.0
1976	1.818	1.600	160.0

1973 chain relative = 1973 link relative × 1972 chain relative
= 1.667 × 1.2 = 2.0004, or
round to 200.0%

and so on. The answers are shown in column (4) of Table 21–6. Note that the link relatives in column (2) of the table are expressed in decimal numbers to facilitate the computation.

The chain relatives obtained from the link relatives above are the same as the fixed-base (1971 = 100%) relatives obtained from the original prices in Example 2. Chain relatives are useful when the original data are not available but the link relatives are.

21.2   Composite Index Numbers (for a Group of Commodities)

A composite index number is constructed from a group of time series concerning various commodities. Composite index numbers are used to show collectively the relative changes in prices, quantities, or values of the commodities included in the construction. Most of the index numbers in practical use are composites. For example, if we wish to know the relative changes (increase or decrease) of living expenses, we should not examine the prices of only a single item. We should include the prices of a group of living items, such as food, transportation, clothing, and housing in computing the living index numbers.

Composite index numbers may be computed from either original data or simple relatives. There are numerous formulas developed for the computation. The formulas presented below are the basic ones.

Computation of Original Data—Methods of Aggregates

The composite price (or quantity) index number of a given year may be computed by dividing the aggregate of the weighted prices (or quantities) of the given year by that of the base year. The weights assigned to a particular commodity for the given year and the base year should be the same. A weight represents the relative importance of the commodity with respect to other commodities included in the computation. Let w = weight, then

$$\text{Price index (by weighted aggregates)} = \frac{\sum p_n w}{\sum p_0 w} = \frac{\sum p_n q_0}{\sum p_0 q_0} \qquad (21\text{-}4)$$

The weights of a composite price index should be the *quantities* of a selected year. Frequently, the selected year is the base year, provided that the quantities sold or produced are normal in the base year. Hereafter, unless otherwise specified, we assume $w = q_0$ in formula (21–4).

$$\text{Quantity index (by weighted aggregates)} = \frac{\sum q_n w}{\sum q_0 w} = \frac{\sum q_n p_0}{\sum q_0 p_0} \qquad (21\text{-}5)$$

The weights of a composite quantity index should be the *prices* of a selected year. Hereafter we assume $w = p_0$ in formula (21–5). Note that when the weight is the number of the base year (that is, $w = q_0$ or $w = p_0$), the index is called *Laspeyres index*. On the other hand, when the weight is the number of the given year (that is, $w = q_n$ or $w = p_n$), the index is called *Paasche index*. Laspeyres and Paasche are the statisticians who first suggested to use the number of the base year and the number of the given year as the weight respectively.

The formula for the composite value index number by using the aggregative method is:

$$\text{Value index (by aggregates)} = \frac{\Sigma \, p_n q_n}{\Sigma \, p_0 q_0} \qquad (21–6)$$

A weight is not assigned to each commodity for the value index. Actually, both the price and the quantity of each commodity have been weighted in computing the value index since the value is the product of the price and the quantity.

Example 6

Assume that the Frost City Council wishes to construct the composite index numbers of (a) price, (b) quantity, and (c) value for 1976 food items by using the items in 1971 as the base. The average unit prices, quantities sold, and the values of sales of the selected food items during the years 1971 and 1976 in the city are shown in Table 21–7.

TABLE 21–7

DATA FOR EXAMPLE 6

Item	1971			1976		
	Unit Price	Quantity sold (1,000 units)	Value of Sales (in $1000)	Unit Price	Quantity sold (1,000 units)	Value of sales (in $1,000)
	p_0	q_0	$p_0 q_0$	p_n	q_n	$p_n q_n$
Eggs	$0.50 doz.	100 dozens	$50	$0.60 doz.	90 dozens	$54
Milk	$0.20 qt.	120 quarts	$24	$0.25 qt.	140 quarts	$35
Meat	$0.40 lb.	10 pounds	$ 4	$0.30 lb.	15 pounds	$ 4.5
Total	$1.10 unit	Cannot be added	$78	$1.15 unit	Cannot be added	$93.5

Solution.   (a) Use formula (21–4).

$$1976 \text{ price index} = \frac{\Sigma \, p_n q_0}{\Sigma \, p_0 q_0}$$

$$= \frac{(\$0.60 \times 100) + (\$0.25 \times 120) + (\$0.30 \times 10)}{\$78}$$

$$= \frac{\$60 + \$30 + \$3}{\$78} = \frac{\$93}{\$78}$$

$$= 1.192 \text{ or } 119.2\%$$

The common multiplier of 1,000 units given in the quantity column (q_0) is omitted in both numerator and denominator of the above computation. The omission does not affect the answer.

Note that if each price is unweighted, the aggregative price index would be

$$\frac{\Sigma\, p_n}{\Sigma\, p_0} = \frac{\$0.60 + \$0.25 + \$0.30}{\$0.50 + \$0.20 + \$0.40} = \frac{\$1.15}{\$1.10} = 1.045 \text{ or } 104.5\%$$

The unweighted aggregative price index is not commonly used. It is constructed under the assumption that each of the commodities is equally important. That is, if one dozen of eggs is sold, one quart of milk and one pound of meat are also sold during the same period. Furthermore, if a different unit is selected, such as milk being measured by gallons instead of quarts, the index would differ greatly.

(b) Use formula (21–5).

$$\begin{aligned}
1976 \text{ quantity index} &= \frac{\Sigma\, q_n p_0}{\Sigma\, q_0 p_0} \\
&= \frac{(90 \times \$0.50) + (140 \times \$0.20) + (15 \times \$0.40)}{\$78} \\
&= \frac{\$45 + \$28 + \$6}{\$78} = \frac{\$79}{\$78} \\
&= 1.013 \text{ or } 101.3\%
\end{aligned}$$

Note that an unweighted aggregative quantity index can not be obtained in this case since the aggregates of the quantities can not be computed. We can not add dozens, quarts, and pounds together. If the units of the selected food items were the same, such as all items being expressed in pounds, the unweighted aggregative quantity index could be computed by formula $\Sigma\, q_n / \Sigma\, q_0$. However, this type of quantity index is not commonly used. It is constructed under the assumption that the unit price of each commodity is the same. For example, if the quantity sold in a food store for cabbage was 500 pounds in 1971 and 1,900 pounds in 1976 and the quantity for steak was 1,000 pounds in 1971 and 1,100 pounds in 1976, the unweighted aggregative index of 1976 based on 1971 is

$$\frac{\Sigma\, q_n}{\Sigma\, q_0} = \frac{1,900 + 1,100}{500 + 1,000} = 2.00 \text{ or } 200\%$$

The index indicates that the quantity sold in 1976 is 200% of that in 1971. The fact of the high-price steak and the low-price cabbage is ignored. Thus, this type of index does not give enough information in analyzing quantity changes of a group of commodities during a period.

(c) Use formula (21–6).

$$1976 \text{ value index} = \frac{\Sigma\, p_n q_n}{\Sigma\, p_0 q_0} = \frac{\$93.5}{\$78.0} = 1.199 \text{ or } 119.9\%$$

A value is the product of unit price and quantity sold or produced. The value index thus shows the combined change of price and quantity. It is not used as frequently as the separated index numbers for price and quantity in decision-making.

Computation of Relatives—Methods of Averages

Sometimes the original data of prices and quantities are not available but the simple relatives and the actual values are. In such cases a composite index number of the price and the quantity may be obtained by averaging the relatives only or the relatives weighted by the values. When the relatives are unweighted, the divisor used in the averaging process is the number of the relatives, denoted by N, or

$$\text{Price index (average of unweighted relatives)} = \frac{\Sigma \left(\frac{p_n}{p_0}\right)}{N} \qquad \textbf{(21-7)}$$

$$\text{Quantity index (average of unweighted relatives)} = \frac{\Sigma \left(\frac{q_n}{q_0}\right)}{N} \qquad \textbf{(21-8)}$$

When the relatives are weighted, the divisor used in the averaging process is the sum of the weights. Let the actual values of the base year be the weights, $p_0 q_0$, then,

$$\text{Price index (average of weighted relatives)} = \frac{\Sigma \left(\frac{p_n}{p_0} \cdot p_0 q_0\right)}{\Sigma \left(p_0 q_0\right)} \qquad \textbf{(21-9)}$$

$$\text{Quantity index (average of weighted relatives)} = \frac{\Sigma \left(\frac{q_n}{q_0} \cdot p_0 q_0\right)}{\Sigma \left(p_0 q_0\right)} \qquad \textbf{(21-10)}$$

The application of the above formulas is illustrated in Example 7 below.

Example 7

The 1976 price relatives (p_n/p_0), 1976 quantity relatives (q_n/q_0), and the actual values of the year 1971, the base, $(p_0 q_0)$ given below are obtained from Example 6. Compute the unweighted and weighted price and quantity index numbers by the methods of averages.

TABLE 21–8

DATA FOR EXAMPLE 7

Item	Price relative, 1976 (1971 = 1.00) p_n/p_0	Quantity relative, 1976 (1971 = 1.00) q_n/q_0	Actual values, 1971 (base year) $p_0 q_0$
Eggs	1.20 (= 0.60/0.50)	0.900 (= 90/100)	$50
Milk	1.25 (= 0.25/0.20)	1.167 (= 140/120)	$24
Meat	0.75 (= 0.30/0.40)	1.500 (= 15/10)	$ 4
Total Σ	3.20	3.567	$78

Solution.   (a) Price index—average of unweighted relatives. Use formula (21–7).

$$\frac{\Sigma \left(\frac{p_n}{p_0}\right)}{N} = \frac{3.20}{3} = 1.067 \text{ or } 106.7\%$$

This type of price index is useful if we can assume that the quantity of each commodity is equal. Compare the unweighted price index by using original data with that by using relatives. The latter is better than the former, because if different units of an item are selected (such as milk being measured by gallons instead of quarts) the index will not be affected by using relatives.

(b) Quantity index—average of unweighted relatives. Use formula (21–8).

$$\frac{\Sigma \left(\frac{q_n}{q_0}\right)}{N} = \frac{3.567}{3} = 1.189 \text{ or } 118.9\%$$

This type of quantity index is useful if we can assume that the unit price of each commodity is equal. Compare the unweighted quantity index by using original data with that by using relatives. The latter is better than the former, because if various units of the quantities are included in a series, (such as dozens, quarts, and pounds which can not be added together) relatives can be used in constructing the index numbers.

(c) Price index—average of weighted relatives. Use formula (21–9).

$$\frac{\Sigma \left(\frac{p_n}{p_0} \cdot p_0 q_0\right)}{\Sigma (p_0 q_0)} = \frac{(1.20 \times 50) + (1.25 \times 24) + (0.75 \times 4)}{78}$$

$$= \frac{60 + 30 + 3}{78} = 1.192 \text{ or } 119.2\%$$

The result is the same as the answer to Example 6 (a) by using formula (21–4). Note that the numerator of formula (21–9) may be simplified to agree with that of formula (21–4) as follows:

$$\frac{p_n}{\cancel{p_0}} \cdot \cancel{p_0} q_0 = p_n q_0$$

(d) Quantity index—average of weighted relatives. Use formula (21–10).

$$\frac{\Sigma \left(\frac{q_n}{q_0} \cdot p_0 q_0\right)}{\Sigma (p_0 q_0)} = \frac{(0.9 \times 50) + (1.167 \times 24) + (1.5 \times 4)}{78}$$

$$= \frac{45 + 28.008 + 6}{78} = 1.013 \text{ or } 101.3\%$$

The result is the same as the answer to Example 6 (b) by using formula (21–5). Note that the numerator of formula (21–10) may be simplified to agree with that of formula (21–5) as follows:

$$\frac{q_n}{\cancel{q_0}} \cdot p_0 \cancel{q_0} = q_n p_0$$

21.3 Testing
Index Numbers

In addition to the formulas presented above, there are numerous formulas developed by statisticians for constructing index numbers. The adequacy of a formula may be tested in many ways. The most common ways of theoretical testing are (1) the time reversal test and (2) the factor reversal test.

It should be noted that a formula meeting one or both of the tests may not be suitable for practical use. It may have other weak points. Nevertheless, theoretical tests do provide logical criteria in selecting the index numbers for a particular purpose.

The Time Reversal Test

When the index numbers of any two years are constructed by the same method but with the bases reversed, the two index numbers should be the reciprocals of each other; the product of the two index numbers thus should be unity, or equal to 1. Thus, if a price index for 1977 is 200 with 1974 = 100, or

$$1977 \text{ index with 1974 as the base} = \frac{200}{100} = 200\%$$

then, the same index for 1974 should be 50 with 1977 = 100, or

$$1974 \text{ index with 1977 as the base} = \frac{100}{200} = \frac{50}{100} = 50\%$$

The product of the two index numbers is

$$200\% \times 50\% = 2.0 \times 0.5 = 1$$

The time reversal test is based on this theory. The formulas (21–1, 21–2, and 21–3) for the simple index numbers meet the test. However, many formulas for the composite index numbers can not satisfy this test very well.

Example 8

Refer to the information given in the table of Example 6 and formula (21–4). (a) Use the formula to compute the weighted composite price index for 1971 with 1976 as the base year. (b) Find the adequacy of the formula for the data by the time reversal test.

Solution. (a) Use formula (21–4). The weighted price index for 1971 is:

$$\frac{1971 \text{ weighted price aggregate}}{1976 \text{ weighted price aggregate}} = \frac{\sum p_n q_0}{\sum p_0 q_0}$$

$$\frac{(0.50 \times 90) + (0.20 \times 140) + (0.40 \times 15)}{93.5} = \frac{79}{93.5} = 0.845$$

(b) The time reversal test—The weighted composite price index for 1976 with 1971 as the base is 1.192. (See the answer to Example 6(a).) Thus, the product of the two index numbers with reversed bases is

$$1.192 \times 0.845 = 1.00724$$

which is not exactly equal to 1. Since the product is more than 1, formula (21–4) has an *upward bias* for the index numbers. However, the upward bias is not serious since the error is only 0.00724 or 0.724 percent. The test shows that the index numbers are reasonably correct in showing the relative changes of the prices during the two years. The bias is due to the application of the weights.

The Factor Reversal Test

In a given year, the value of a single commodity is the product of the quantity sold and the unit price, or

$$\text{price} \times \text{quantity} = \text{value}$$

Therefore, we should expect that in a given year

$$\text{price index} \times \text{quantity index} = \text{value index}$$

A test based on this expression is called factor reversal test. If the price and quantity indexes can not satisfy the test, there must be an error in one or both of the indexes. The error is also due to the application of the weights. The formulas (21–1, 21–2, and 21–3) for the simple index numbers can obviously meet the test. This is illustrated by using Example 1 as follows:

$$1976 \text{ price index} \times 1976 \text{ quantity index} = 1.60 \times 1.25 = 2.00$$
$$= 1976 \text{ value index, or}$$

$$\frac{p_n}{p_0} \times \frac{q_n}{q_0} = \frac{p_n q_n}{p_0 q_0}$$

However, many formulas for the composite index numbers do not meet the test, such as formulas (21–4) and (21–5) as applied in Example 6, or formulas (21–9) and (21–10) as applied in Example 7. In Example 6, 1976 price index $\times$ 1976 quantity index $= 1.192 \times 1.013 = 1.207$, which is not equal to the 1976 value index ($= 1.199$) although the error is small ($1.207 - 1.199 = 0.008$ or 0.8%).

There are many formulas that do meet both tests. For example, Professor Irving Fisher selected one from them as the "idea" index number since it is comparatively simple in computation.

$$\text{Ideal index of price} = \sqrt{\frac{\Sigma \, p_n q_0}{\Sigma \, p_0 q_0} \times \frac{\Sigma \, p_n q_n}{\Sigma \, p_0 q_n}} \qquad \textbf{(21–11)}$$

$$\text{Ideal index of quantity} = \sqrt{\frac{\Sigma \, q_n p_0}{\Sigma \, q_0 p_0} \times \frac{\Sigma \, q_n p_n}{\Sigma \, q_0 p_n}} \qquad \textbf{(21–12)}$$

Thus, the ideal index is the geometric mean of the two weighted aggregative indexes. That is, $w = q_0$ and $w = q_n$ in formula (21–4) for the ideal index of price, and $w = p_0$ and $w = p_n$ in formula (21–5) for the ideal index of quantity.

The Making of Index Numbers (Boston: Houghton Mifflin Co., 1927), pp. 220–21.

Example 9

Refer to Example 6. Compute (a) 1976 ideal price index and (b) 1976 ideal quantity index with 1971 as the base.

Solution. (a) Use formula (21–11). For the first factor,

$$\frac{\Sigma\, p_n q_0}{\Sigma\, p_0 q_0} = \frac{93}{78} \qquad \text{[See Example 6(a)]}$$

For the second factor,

$$\frac{\Sigma\, p_n q_n}{\Sigma\, p_0 q_n} = \frac{93.5}{(0.50 \times 90) + (0.20 \times 140) + (0.40 \times 15)} = \frac{93.5}{79}$$

Thus, 1976 ideal price index $= \sqrt{\dfrac{93}{78} \times \dfrac{93.5}{79}} = 1.188$

(b) Use formula (21–12). For the first factor,

$$\frac{\Sigma\, q_n p_0}{\Sigma\, q_0 p_0} = \frac{79}{78} \qquad \text{[See Example 6(b)]}$$

For the second factor,

$$\frac{\Sigma\, q_n p_n}{\Sigma\, q_0 p_n} = \frac{93.5}{(100 \times 0.60) + (120 \times 0.25) + (10 \times 0.30)} = \frac{93.5}{93}$$

Thus, 1976 ideal quantity index $= \sqrt{\dfrac{79}{78} \times \dfrac{93.5}{93}} = 1.009$ or 100.9%

The ideal index numbers do meet the time reversal test. From above,

$$\left(\begin{array}{l}\text{1976 ideal price index}\\\text{with 1971 as the base}\end{array}\right) \times \left(\begin{array}{l}\text{1971 ideal price index}\\\text{with 1976 as the base}\end{array}\right)$$

$$= \sqrt{\frac{93}{78} \times \frac{93.5}{79}} \times \sqrt{\frac{79}{93.5} \times \frac{78}{93}} = 1$$

Note that the 1971 ideal price index is obtained by interchanging the subscripts 0 and n in p and q in solution (a) above.

The ideal index numbers also meet the factor reversal test. That is,

$$\left(\begin{array}{l}\text{1976 ideal price index}\\\text{with 1971 as the base}\end{array}\right) \times \left(\begin{array}{l}\text{1976 ideal quantity index}\\\text{with 1971 as the base}\end{array}\right) = \left(\begin{array}{l}\text{1976 value}\\\text{index}\end{array}\right)$$

$$\sqrt{\frac{93}{78} \times \frac{93.5}{79}} \times \sqrt{\frac{79}{78} \times \frac{93.5}{93}} = \frac{93.5}{78} = 1.199 \text{ or } 119.9\%$$

[See Example 6(c)]

However, ideal indexes are too complex in calculation for practical use.

21.4 Shifting the Base of Index Numbers

There are occasions of shifting the base of index numbers from one period to another period. For example, it may be desirable to make the base more recent.

The U. S. Department of Labor changed the base period of the consumer price index (listed in Table 21–1) from 1957–59 = 100 to 1967 = 100 for the recent indexes. The recent base facilitates the analysis of recent price changes. The method of shifting the base is illustrated in Example 10 below.

Example 10

Assume that the price indexes of green coffee in the Moontie County in the years 1971 to 1976 are given in column (2) of Table 21–9. Compute new price indexes by shifting the base from 1971 = 100% to 1971–73 = 100%.

Solution. The new price index for each year is shown in column (4) of the same table. The average of the price indexes of the years 1971 to 1973 is obtained by using the arithmetic mean method, or

$$\text{the average price index} = \frac{100 + 120 + 200}{3} = 140 \text{ (in \%)}$$

The new price index for each year by using the average price index as the base is

$$\text{new price index of a given year} = \frac{\text{Old price index of a given year}}{\text{Average price index, 140}}$$

$$\text{thus, the new 1971 price index} = \frac{100}{140} = 0.714 \text{ or } 71.4\%$$

$$\text{the new 1972 price index} = \frac{120}{140} = 0.857 \text{ or } 85.7\%$$

and so on. Note that the answers shown in column (4), Table 21–9 are the same as those obtained by using the original prices in Example 2. The percent sign (%) is omitted in each index number for simplifying the computation, since, for example,

$$\frac{100\%}{140\%} = \frac{100}{140} = 0.714$$

TABLE 21–9

DATA AND ANSWERS FOR EXAMPLE 10

Year (1)	Price Index 1971 = 100% (2)	New Price Index	
		1971–73 = 1.000 (3)	1971–73 = 100% (4)
1971	100%	0.714	71.4%
1972	120	0.857	85.7
1973	200	1.429	142.9
1974	80	0.571	57.1
1975	88	0.629	62.9
1976	160	1.143	114.3

21.5 Deflating Time Series by Price Indexes

A time series expressed in dollar values may represent the combined changes in prices and quantities of a single commodity or a group of commodities. The process of removing the effects of price changes from the dollar values is called *deflation*. The deflated dollar values thus represent the quantity changes. The deflation can be done by the following expression:

$$\text{Deflated dollar value} = \frac{\text{Original dollar value}}{\text{A corresponding or an appropriate price index}}$$

In some cases, a dollar value can be divided by its corresponding price index in deflation. For example, if the price indexes of certain construction equipment are available, the dollar values of the construction equipment in a series can be divided by the corresponding index numbers to obtain the deflated dollar values. However, in some other cases, a corresponding price index is not available. Thus, we must select an appropriate price index for the deflation. For example, in deflating wage earnings, the consumer price index computed by the Department of Labor is appropriate since most of wages earned by workers are spent for the items (Table 21–1) included in computing the index.

The deflated dollar values are also called by various names, such as constant dollars, real wages or income, purchasing power of the dollar, and others.

Example 11

Annual weekly earnings in the mining industry in the United States for the years 1970 to 1974 are given in column (2) of Table 21–10. Compute the real wages based on the consumer price indexes listed in column (3) of the same table.

Solution.   The real wages in 1967 dollars are shown in column (4) of the table below. Each answer in column (4) is obtained by dividing the weekly earnings by the consumer price index of the year. Thus, the 1970 real wages in 1967 dollars are obtained:

$$\$162.64 \div 116.3\% = \$139.85$$

TABLE 21–10

DATA AND ANSWERS FOR EXAMPLE 11

Year (1)	Weekly wages* (2)	Consumer price* index 1967 = 100% (3)	Real wages 1967 dollars (4)
1970	$162.64	116.3%	$139.85
1971	171.74	121.3	141.58
1972	187.43	125.3	149.58
1973	200.60	133.1	150.71
1974	242.66	155.4	156.15

*Source: U. S. Department of Commerce, *Survey of Current Business.*

Note that the purchasing power of the dollar as measured by the consumer price index with $1967 = 100\%$ can also be computed for each year given in Example 11. For instance, the purchasing power of $1 in 1970 as measured by the consumer price index 116.3% is

$$\$1 \div 116.3\% = \$0.85985$$

Thus, the purchasing power of \$162.64 in 1970 is equivalent to

$$\$0.85985 \times 162.64 = \$139.85 \text{ in } 1967$$

21.6   Summary

An index number is a relative value with a base equal to 100 percent or a multiple of 100 percent, such as 10 and 100. Important index numbers concerning business and economic activities may be classified into three types: (1) price indexes, (2) quantity indexes, and (3) value indexes.

Index numbers may be constructed for a single commodity, called simple index numbers or simple relatives (ratios), or for a group of commodities, called composite index numbers. The length of period in computing the index numbers is usually a year, although it may be a quarter, a month, or other unit of time.

When a time series includes information of more than two years, there are three ways to compute the simple relatives: (1) fixed-base relatives, (2) link relatives, and (3) chain relatives.

Composite index numbers may be computed from either original data or simple relatives. The formulas presented in this chapter are basic ones and are summarized at the end of this section.

The most common ways of testing the quality of a formula for constructing index numbers are: (1) The time reversal test. This test is based on the theory that when the index numbers of any two years are constructed by the same method but with the bases reversed, the product of the two index numbers should be unity, or equal to 1. (2) The factor reversal test. This test is based on the theory that the product of the price index and the quantity index should equal the value index.

There are occasions of shifting the base of index numbers from one period to another period. The simplest way to shift the base is to divide each index number in the given series by the index number (or the average of the index numbers) of the new base. It is not necessary to use the original data in computing the index numbers with a new base.

The process of removing the effects of price changes from the dollar values is called deflation. A deflated dollar value is obtained by dividing the original dollar value by a corresponding or an appropriate price index.

SUMMARY OF FORMULAS

Application	Formula	Formula Number	Reference Page
Simple index numbers			
Price relative	$\dfrac{p_n}{p_0}$	21-1	536
Quantity relative	$\dfrac{q_n}{q_0}$	21-2	536
Value relative	$\dfrac{p_n q_n}{p_0 q_0}$	21-3	536
Composite index numbers			
Methods of aggregates (for original data)			
Price index	$\dfrac{\sum p_n q_0}{\sum p_0 q_0}$	21-4	541
Quantity index	$\dfrac{\sum q_n p_0}{\sum q_0 p_0}$	21-5	541
Value index	$\dfrac{\sum p_n q_n}{\sum p_0 q_0}$	21-6	542
Methods of averages (for relatives)			
Price index (unweighted)	$\dfrac{\sum \dfrac{p_n}{p_0}}{N}$	21-7	544
Quantity index (unweighted)	$\dfrac{\sum \dfrac{q_n}{q_0}}{N}$	21-8	544
Price index (weighted)	$\dfrac{\sum \left(\dfrac{p_n}{p_0} \cdot p_0 q_0\right)}{\sum (p_0 q_0)}$	21-9	544
Quantity index (weighted)	$\dfrac{\sum \left(\dfrac{q_n}{q_0} \cdot p_0 q_0\right)}{\sum (p_0 q_0)}$	21-10	544
Ideal index numbers			
Ideal price index	$\sqrt{\dfrac{\sum p_n q_0}{\sum p_0 q_0} \times \dfrac{\sum p_n q_n}{\sum p_0 q_n}}$	21-11	547
Ideal quantity index	$\sqrt{\dfrac{\sum q_n p_0}{\sum q_0 p_0} \times \dfrac{\sum q_n p_n}{\sum q_0 p_n}}$	21-12	547

Exercises 21

1. The unit prices and quantities sold of commodity X for the years 1975 and 1977 in a city are given below. Compute the indexes of (a) price, (b) quantity, and (c) value for 1977 with 1975 as the base.

Year	Price per unit	Units sold
1975	$2.80	160
1977	3.36	128

2. From the following data compute: (a) fixed-base relatives with 1973 as the base, (b) link relatives, and (c) chain relatives from the link relatives.

Year	Quantity produced (unit: 1,000 lbs)
1973	50
1974	75
1975	100
1976	120
1977	140

3. Use the information given in Problem 2 to compute fixed-base relatives with 1975–76 as the base.

4. From the following link relatives, compute the chain relatives for the years 1973 to 1977 with 1973 as the base year.

Year	Link relative
1973	none
1974	2.00
1975	1.20
1976	0.90
1977	0.80

5. Assume that the prices and quantities sold of the commodities during the years 1976 and 1977 in Pacific City are as follows:

Commodity	Price (per unit)		Quantity (in 1,000 units)	
	1976	1977	1976	1977
A	$ 0.60 per pound	$ 0.65 per pound	45	38
B	0.45 per pound	0.48 per pound	180	120
C	80.00 per ton	85.00 per ton	14	10
D	1.50 per bushel	1.42 per bushel	20	15

Use the methods of aggregates to construct the composite index numbers of (a) price, (b) quantity, and (c) value for 1977 items with 1976 as the base.

6. Apply the data given in Problem 5. Use the methods of averaging relatives to construct composite index numbers of (a) unweighted price, (b) unweighted quantity, (c) weighted price, and (d) weighted quantity. Also let the year 1976 be the base.

7. Refer to Problem 5. (a) Compute the weighted price index number for 1976 with 1977 as the base. (b) Find the adequacy of the method used in computing the price index by the time reversal test.

8. Refer to Problem 5. Determine if the computed composite index numbers meet the factor reversal test.

9. Refer to Problem 5. Compute (a) 1977 ideal price index and (b) 1977 ideal quantity index with 1976 as the base. Comment on the use of ideal indexes.

10. Refer to the fixed-base relatives obtained in Problem 2. Compute the new quantity indexes by shifting the base from 1973 = 100% to 1973–75 = 100%.

11. The wholesale price indexes with 1967 = 100% are as follows:

Year	Wholesale price index
1972	119.1
1973	134.7
1974	171.5

Compute the purchasing power of the dollar as measured by the wholesale price index for each year. Comment on your findings.

12. Refer to Example 11, page 550. Compute the real wages for each of the years 1972 to 1974 based on the wholesale price index given in Problem 11.

22 Secular Trend

This and the next two chapters will analyze business and economic changes of the past based on time series movements. The analysis of the past is important because it will enable the management to make a more accurate forecast of future activity. It also will increase the effectiveness of comparing different groups of data or different periods of the same data. The result of time series analysis thus improves efficiency in decision-making.

22.1 Basic Patterns of Time Series Movements

A time series representing a particular activity of an organization, such as sales activity in a firm or in an industry and an economic activity in a nation, is the result of interactions of many types of changing forces. The forces can be business, economic, political, and social influences as well as the forces of nature. The forces are usually investigated for decision-making after the movements of a time series are separated into the following four basic patterns or components. The graphs and sketches shown in Chart 22–1 are used to portray the patterns.

Secular Trend

Secular trend points out the direction of a time series movement over a long period of time. It may be an upward or a downward movement. When it is shown graphically, it is usually represented by a straight line or by a smooth curve. Many types of business and economic activities in the United States have upward trends. For example, Gross National Product from 1940 to 1975 as shown in Chart 22–1A has an upward trend. The chief forces that caused the upward long-term move-

CHART 22–1

FOUR BASIC PATTERNS OF TIMES SERIES MOVEMENTS

A. *Secular Trend*
    Gross National Product and Farm
    Employment in the United States

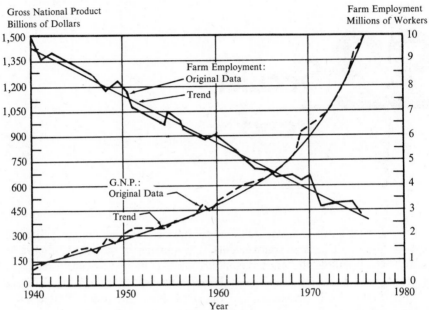

Source: Federal Reserve System, 1964 Historical Chart Book, pp. 73-105. Data for 1964–75 are obtained from *Survey of Current Business*, March, 1975.

C. *Cyclical Fluctuations*
    Sales of Baxter Company
    Due to Cyclical Fluctuations

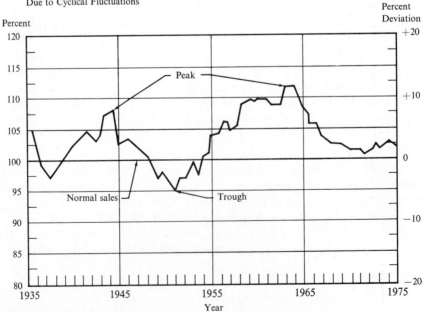

Source: Hypothetical data.

CHART 22–1

FOUR BASIC PATTERNS OF TIMES SERIES MOVEMENTS

B. *Seasonal Variation*

    Seasonal Index of Quarterly Sales
    of Abbot Company

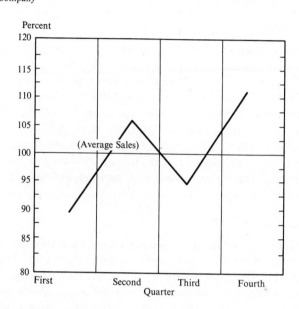

Source: Hypothetical data.

D. *Irregular Movements*

    Sales of Carthy Company
    Due to Irregular Movements

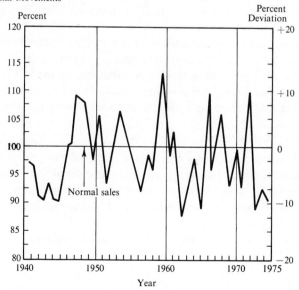

Source: Hypothetical data.

ment are the growth of population, greater capital accumulation, technique improvements, and rising standard of living. However, the farm employment for the same period as shown in the chart has a downward trend, because of the application of improved techniques that increase production and decrease the number of workers in the farm at the same time.

Seasonal Variation

Seasonal variation represents repeating periodic movement of a time series. The length of the unit period is less than a year. It may be a quarter, a month, or a day. Seasonal variation is usually expressed in index numbers. The periodic average of the index numbers is 100 percent, written as 100 on a percent scale. A sketch of the quarterly seasonal variation is shown in Chart 22–1B. The index numbers show that sales in the first quarter are generally 10 percent below the average in a year. The chief forces that cause a seasonal variation are weather (such as the winter season affecting the sale of ice cream) and traditional or habitual activities (such as Christmas bringing more trade and the week-end being a busy time for the entertainment business).

Cyclical Fluctuations

Cyclical fluctuations, also called *business cycles*, indicate the expansions (ups) and the contractions (downs) of business activities around the normal value. The length of each cycle is nonfixed and relatively short. The National Bureau of Economic Research, Inc., which in recent years has been the leader in the investigation of business cycles, indicated that there were 27 major cycles from 1854 to 1970. The durations of the 27 cycles ranged from 17 months to 117 months with an average of 52 months per cycle. The average duration is measured from previous peak to peak, including contraction for 19 months (trough from previous peak) and expansion for 33 months (trough to peak).* Cyclical fluctuations are usually expressed in percentages above or below the normal value. A sketch of the cyclical movement is shown in Chart 22–1C.

The forces which are responsible for the cyclical fluctuations are numerous and complex, but are chiefly economic factors. For example, the ups and downs of the cycles are closely connected with variations in the levels of investment, production, consumption, and government spending. Cyclical fluctuations for individual firms thus reflect the cycles for the total economy in the nation.

Irregular Movements

Irregular or *erratic* movements represent all types of movements of a time series other than secular trend, seasonal variation, and cyclical fluctuations. Forces that cause the irregularities in business activities are numerous and are of a random nature. Some of the forces are too small to be noticeable. In those cases, irregular

---

*U. S. Department of Commerce, Bureau of the Census, *Business Conditions Digest*, February 1975, page 109.

movements are usually not isolated, but are combined with cyclical fluctuations in analysis. No attempt is made to investigate the sources of the small irregular movements. However, some forces such as strikes, wars, floods, and other types of disasters have greater effect on business activities than others. When the irregular movements become very pronounced, they may be separated from the cyclical fluctuations in order to investigate the effect of the irregular forces, such as the sketch shown in Chart 22–1D.

The irregular movements may be cancelled if a longer time unit is used in classifying the date. Assume that a circus has come to a small town once a year but in different months during the past few years. When the sales in the town are classified on a monthly basis, the business activity will show an unusually high figure during the month of the circus. However, when the sales are classified on an annual basis, the irregular movement would be cancelled. The irregular movement may also be cancelled because a greater extent or area is covered in a time series. For example, a national association with more than 100,000 members changes its annual convention site each year. When we analyze the annual sales of a city, irregular movements may have occurred due to the holding of conventions in the different years. However, when the sales on a nationwide basis are analyzed, the irregular movements due to conventions during the years would be cancelled.

In summary, a value of the time series may be considered as (a) a product or (b) a sum of the above four components, or written:

$$\text{(a) } Y = T \times S \times C \times I \quad \text{or} \quad \text{(b) } Y = T + S + C + I$$

where Y = an actual value included in the original data
           T = secular trend value of Y
           S = seasonal variation for Y
           C = cyclical fluctuation for Y
           I = irregular movement for Y

Either equation is convenient in analyzing or decomposing a time series into individual components.

The remaining part of this chapter discusses the methods of obtaining trend values of a time series. The methods of measuring seasonal variation, cyclical fluctuations, and irregular movements are presented in Chapters 23 and 24.

22.2   Straight-Line Trends

A straight line drawn on a graph according to the system of rectangular coordinates (Section 3.4) may be expressed by the equation:

$$Y = a + bX$$

where Y = a point value on the straight line based on the vertical scale or Y-axis (also called a dependent variable since a Y value depends on the value of X);

$X =$ a point value on the straight line based on the horizontal scale or X-axis (also called an independent variable with respect to Y variable);

$a =$ Y intercept (the height of the ordinate from the origin to the point of intersection of the straight line and the Y-axis), which equals Y value when $X = 0$;

$b =$ the slope of the straight line, representing the average amount of change in Y variable per unit in X variable.

The relationship between a straight line and its equation is illustrated in Chart 22–2. In the chart, let

$h =$ total units of change in Y variable,

$g =$ total units of change in X variable, and the change of h corresponds to the change of g.

Then

$$b = \frac{h}{g}$$

A set of corresponding values of g and h can be obtained from any two points on the straight line. For example, the two points may be P_1 (with $X_1 = 0$ and $Y_1 = 4$) and P_2 (with $X_2 = 5$ and $Y_2 = 14$) on the line. The g value is the difference between the two X values represented by the two points, or

$$g = X_2 - X_1 = 5 - 0 = 5$$

and the h value is the difference between the two Y values, or

$$h = Y_2 - Y_1 = 14 - 4 = 10$$

The differences may also be arranged in the following manner in computing the value of b:

Point	X value	Y value
P_2	5	14
P_1	0	4
$P_2 - P_1$	$g = 5$	$h = 10$

$$b = \frac{h}{g} = \frac{Y_2 - Y_1}{X_2 - X_1} = \frac{10}{5} = 2 \quad \left(\text{or } b = \frac{Y_1 - Y_2}{X_1 - X_2} = \frac{-10}{-5} = 2 \right)$$

If the two points of a set are P_1' (with $X_1 = 1$ and $Y_1 = 6$) and P_2' (with $X_2 = 4$ and $Y_2 = 12$), the g and h values are computed:

Point	X value	Y value
P_2'	4	12
P_1'	1	6
$P_2' - P_1'$	$g = 3$	$h = 6$

$$b = \frac{h}{g} = \frac{6}{3} = 2$$

The value of b obtained from points P_1 and P_2 is the same as that obtained from points P_1' and P_2'. Thus, the equation of the straight line on the chart may be written:

$$Y = a + bX = 4 + 2X$$

Since the straight line is represented by the equation, any point on the line can be checked by the equation. For example, if we select point P on the line where $X = 3$, we may compute the Y value for the point without observing the Y scale, or

$$Y = 4 + 2(3) = 10$$

Chart 22–3 gives additional examples showing the relationships of straight lines and their respective equations. The values of a and b may be positive or negative. Observe each line on the chart. When a straight line intersects Y-axis above the origin, a is positive (Lines I and III); when Y intercept is below the origin, a is negative (Lines II and IV). When the trend of Y variable is upward (or Y values are increasing), the value of b is positive (Lines I and II); when the trend is downward (or Y values are decreasing), b is negative (Lines III and IV).

CHART 22–2

BASIC CONCEPT OF A STRAIGHT LINE AND ITS EQUATION

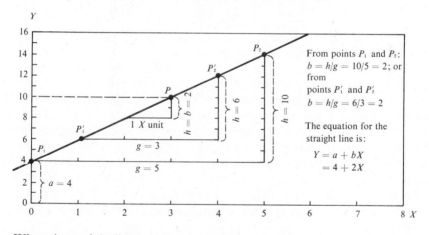

When the straight line equation is used for describing a trend movement, the equation is usually written:

$$Y_c = a + bX \qquad\qquad (22\text{--}1)$$

where Y_c = the computed trend value of a given time (Y with no subscript represents the actual value),

X = the assigned number of the given time, such as a year.

It is rather inconvenient to use the original numbers of the years in a time series as the X variable in applying the equation. In practice, a year is arbitrarily

CHART 22–3

ADDITIONAL ILLUSTRATIONS OF STRAIGHT LINES AND THEIR EQUATIONS

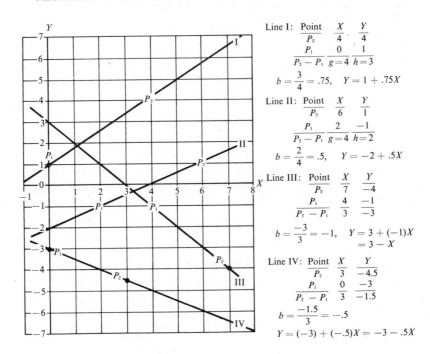

Line I:

Point	X	Y
P_2	4	4
P_1	0	1
$P_2 - P_1$	$g = 4$	$h = 3$

$$b = \frac{3}{4} = .75, \quad Y = 1 + .75X$$

Line II:

Point	X	Y
P_2	6	1
P_1	2	-1
$P_2 - P_1$	$g = 4$	$h = 2$

$$b = \frac{2}{4} = .5, \quad Y = -2 + .5X$$

Line III:

Point	X	Y
P_2	7	-4
P_1	4	-1
$P_2 - P_1$	3	-3

$$b = \frac{-3}{3} = -1, \quad Y = 3 + (-1)X$$
$$= 3 - X$$

Line IV:

Point	X	Y
P_2	3	-4.5
P_1	0	-3
$P_2 - P_1$	3	-1.5

$$b = \frac{-1.5}{3} = -.5$$
$$Y = (-3) + (-.5)X = -3 - .5X$$

assigned as zero year or the *origin* of the X scale. The numbers representing other years in the time series thus are greatly simplified. For example, if a series includes years 1970 to 1976, the earliest year, 1970, may be chosen as zero year (or July 1, 1970—the middle point of the year—as 0 on the X scale). The numbers of other years may be assigned as follows: year 1971 = 1, year 1972 = 2, and so on. If the middle year (1973) were chosen as the origin (0), the years before 1973 would be negative numbers; or year 1972 = −1, year 1971 = −2, and year 1970 = −3; the years after 1973 would be positive, or year 1974 = 1, year 1975 = 2, and year 1976 = 3. The two illustrations are diagrammed in Table 22–1.

TABLE 22–1

ASSIGNING SIMPLIFIED NUMBERS TO YEARS

Years in original numbers	1970	1971	1972	1973	1974	1975	1976
Years in assigned numbers (1970 = origin)	0	1	2	3	4	5	6
Years in assigned numbers (1973 = origin)	−3	−2	−1	0	1	2	3

A straight line and its equation used to describe secular trend may be obtained by any one of the following three methods:

1. Freehand graphic method
2. Method of semiaverages
3. Method of least squares

Freehand Graphic Method

The procedure of obtaining a straight line by the freehand graphic method for measuring secular trend is outlined below:

1. Plot the time series on a graph.
2. Examine carefully the direction of the trend based on the plotted information (dots).
3. Draw a straight line which will be the best fit to the data according to personal judgment of the drawer. The line now shows the direction of the trend.

After the line is drawn, an equation of the trend line is determined by reading two points (not too close to each other) off the line. The trend values for other years now may be read off the line or computed from the equation.

The advantage of this method is obvious, since it is very simple and easy to apply. It does not require too much time to draw a straight line based on personal judgment. Furthermore, it is usually satisfactory when the direction of the trend is clearly indicated by the plotted information. The chief disadvantage is that it is too subjective. Lines drawn by different people for the same information may have different locations on the graph, especially when the trend or direction is not obvious. Thus, it is frequently rather difficult to obtain a satisfactory result by this method.

The complete subjective method may be improved. First, the mean of the original data is computed and a point representing the mean is plotted on the graph in the middle of the period. A freehand line is then drawn through the point. By so doing, only the slope is determined subjectively by personal judgment. This method is based on the assumption that the mean of the original data is equal to the mean of the trend values. Example 1 is used to illustrate the improved method.

Example 1

Table 22–2 shows the sales of Wilson Department Store for the years 1964 to 1978 (15 years). (a) Draw a straight line to fit the data by the freehand graphic method. (b) Let the first year, 1964, be the origin, X unit be 1 year, and Y unit be $1,000. Find the straight line equation. (c) Compute the trend value for the year 1974.

Solution. (a) The original data are plotted on Chart 22–4. The data shown are in an upward trend. A straight line showing the trend is drawn through the mean, $15,400, which is located in the midde of the period of the time series, or July 1, 1971.

TABLE 22–2

SALES OF WILSON DEPARTMENT STORE, 1964 TO 1978
(DATA AND COMPUTATION FOR EXAMPLES 1 AND 2)

Year	Sales	
1964	$ 7,000	
1965	6,000	
1966	2,000	
1967	4,000	Subtotal of 7 years: $56,000
1968	8,000	Average annual sales: $56,000/7 = $8,000
1969	16,000	(based on sales of 1964 to 1970)
1970	13,000	
1971	14,000	Middle year
1972	17,000	
1973	20,000	
1974	23,000	
1975	19,000	Subtotal of 7 years: $161,000
1976	25,000	Average annual sales: $161,000/7 = $23,000
1977	28,000	(based on sales of 1972 to 1978)
1978	29,000	
Total	$231,000	Average annual sales (based on 15 years): $231,000/15 = $15,400.

(b) The equation representing the straight line is based on two points on the line: $15,400 (the mean) and $5,000 (the Y intercept, which is located on July 1, 1964, the origin). The two points indicate that the increased amount of $10,400 (= $15,400 − $5,000 as shown on the Y scale) corresponds to 7 years (or 7 unit spaces as shown on the X scale). Thus,

$$a = 5 \text{ (the } Y \text{ intercept at origin)}$$

$$b = \frac{15.4 - 5}{7} = \frac{10.4}{7} = 1.4857 \text{ (the average annual increase in}$$
$$\$1{,}000, \text{ or } \$1{,}485.70)$$

$$Y_c = 5 + 1.4857X$$

$$\text{with origin: July 1, 1964}$$

$$X \text{ unit: 1 year}$$

$$Y \text{ unit: } \$1{,}000$$

(c) The number assigned to the year 1974 is 10, since the origin (0) is the year 1964. The trend value for 1974 thus is computed as follows:

$$Y_c = 5 + 1.4857(X) = 5 + 1.4857(10)$$
$$= 19.857 \text{ or } \$19{,}857 \text{ since } Y \text{ unit is } \$1{,}000$$

Note: If a series has an even number of years, the middle of the period is on January 1 of the first year of the second half of the series. For example, if a series has 14 years, 1965 to 1978, the middle of the period is on January 1, 1972. (See Example 4.)

CHART 22–4

ILLUSTRATION OF STRAIGHT-LINE TRENDS BY FREEHAND GRAPHIC METHOD
AND METHOD OF SEMIAVERAGES (SALES OF WILSON DEPARTMENT STORE 1964–1978)

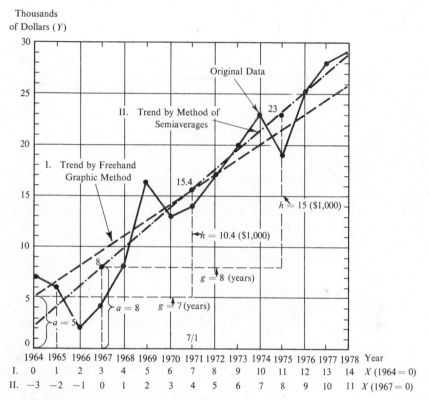

Source: Table 22–2 and Examples 1 (Line I) and 2 (Line II).

Method of Semiaverages

The method of semiaverages is the simplest method for finding the straight-line trend without involving subjective judgment in drawing. The procedure of obtaining a straight-line trend by the method of semiaverages is as follows:

1. Divide the original data into two equal groups and compute the mean of each group. If there is an odd number of years included in the original data, two methods are generally used in the division.

    a. The value of the middle year is counted twice; that is, it is included in the first group and again included in the second group.

    b. The value of the middle year is ignored. This method is used in this section because of its simplicity.

2. Plot the two means on the graph and draw a straight line through the two means. Each mean should be located in the middle of the period covered by the

group. This device is based on the assumption that the mean of the original data of a given period equals the mean of the trend values of the period.

3. Derive the trend equation based on any two points on the line. For convenience, use the two mean points and set the origin of the equation at the location of either one of the two means.

Example 2

Use the data given in Table 22–2. (a) Draw a straight line to fit the data by the semiaverage method. (b) Find the trend equation. (c) Compute the trend value of 1974 based on the equation.

Solution. (a) The original data are divided into two equal groups. The first group covers the sales from 1964 to 1970 (7 years); the second group covers the sales from 1972 to 1978 (also 7 years); and the sales of the middle year (1971) of the series is ignored.

The subtotal of the sales of the first 7 years is \$56,000 and the mean annual sales of the group is \$8,000. The subtotal of the sales of the second 7 years is \$161,000 and the mean annual sales of the group is \$23,000. The computation of the subtotals and the means are shown in Table 22–2.

The original data and the points representing the two means are also plotted on Chart 22–4. Observe that \$8,000 is plotted on the chart just above July 1, 1967, the middle of the first 7 years, and \$23,000 is just above July 1, 1975, the middle of the second 7 years. A straight line is drawn based on the two means.

(b) The trend equation for this straight line is obtained by letting 1967 be the origin. (Note: Either 1967 or 1975 can be selected as the origin.) Thus,

$a = 8$ (Y intercept on origin in \$1,000)

$b = \dfrac{23 - 8}{8} = \dfrac{15}{8} = 1.875$ (There are 8 years or spaces between years 1967 to 1975. The average annual increase is 1.875 in \$1,000, or \$1,875.)

The equation is

$$Y_c = 8 + 1.875X$$
$$\text{with origin: July 1, 1967}$$
$$X \text{ unit: 1 year}$$
$$Y \text{ unit: } \$1,000$$

(c) Trend value of 1974:

$$X = 7$$
$$Y_c = 8 + 1.875\,(7) = 21.125 \text{ (in units of } \$1,000 \text{ or } \$21,125)$$

Method of Least Squares

The method of least squares has been used in finding the arithmetic mean of a group of values as an average or a representative value for the group. The mean has two mathematical properties:

1.   The algebraic sum of the deviations of individual values from (above or below) the mean is zero.

2.   The sum of the squared deviations of individual values from the mean is the *least*.

Thus, if Y values are 1, 4, and 10, the mean of the values is 5. The properties of the mean may be illustrated as follows: (also see Section 5.2, page 126.)

TABLE 22–3

ILLUSTRATION OF LEAST SQUARES FROM THE ARITHMETIC MEAN

Values Y	Deviations from the mean $y = Y - \bar{Y}$	Squared deviations y^2
1	−4	16
4	−1	1
10	5	25
Total 15	0	42
$\bar{Y} = 15/3 = 5$	(property 1)	(property 2—least squares)

This concept may also be used in finding a straight line which is considered the best fit for the scattered points representing X and Y variables on a graph. The straight line for Y dependent variable based on the method of least squares likewise will have two mathematical properties:

1.   The algebraic sum of the deviations of individual values from (above or below) their corresponding values on the line is zero, or $\sum (Y - Y_c) = 0$.

2.   The sum of the squared deviations is the least, or $\sum (Y - Y_c)^2 = $ a minimum.

To obtain the solutions of the two unknowns, constants a and b, in the straight line equation

$$Y_c = a + bX$$

by the method of least squares, we need two normal equations as follows:

$$\text{I.} \quad na + b \sum X = \sum Y$$
$$\text{II.} \quad a \sum X + b \sum X^2 = \sum (XY) \qquad \text{(22–2)*}$$

where $n = $ the number of pairs of X and Y values (or points on the graph).

---

*The two normal equations are obtained as follows:
Let $Y_1, Y_2, \ldots Y_n$ and $X_1, X_2, \ldots X_n$ represent Y and X variables respectively, and $Y_1 = a + bX_1$, $Y_2 = a + bX_2, \ldots$

Then, multiply each of the n equations in the form $Y = a + bX$ by the coefficient of the first unknown of the equation and sum the resulting equations. The first unknown in each of the equations is a and its coefficient is 1. Thus, the equations are not changed after being multiplied by 1. The sum of the resulting equations is

$$\begin{aligned} Y_1 &= a + bX_1 \\ Y_2 &= a + bX_2 \\ .. &= .. + ... \\ Y_n &= a + bX_n \end{aligned}$$

Sum $\overline{\sum Y = na + b \sum X}$ (Normal equation I)   (footnote continued on next page)

The use of the formula is illustrated in Table 22–4. Part A of the table gives the required values for solving the unknowns a and b by the two normal equations. Part B of the table shows the mathematical properties of the straight line equation. The illustration is also shown in Chart 22–5.

TABLE 22–4

COMPUTATION OF A STRAIGHT LINE EQUATION BY THE METHOD
OF LEAST SQUARES AS COMPARED TO THE RESULT OF
AN ARBITRARILY SELECTED LINE

Points on Chart 22–5	A. Required Values for Formula (22–2)				B. From $Y_c = 0.5 + 0.75X$, Least Square Method			C. From $Y_a = 2 + 0.5X$, Arbitrarily Selected		
	X	Y	XY	X^2	Y_c	$Y-Y_c$	$(Y-Y_c)^2$	Y_a	$Y-Y_a$	$(Y-Y_a)^2$
A	4	1	4	16	3.5	−2.5	6.25	4	−3	9
B	8	4	32	64	6.5	−2.5	6.25	6	−2	4
C	6	10	60	36	5.0	5.0	25.00	5	5	25
Total (Σ)	18	15	96	116	15.0	0.0	37.50	15	0	38

Substitute the values obtained from Table 22–4A in the two normal equations of formula (22–2):

$$3a + 18b = 15 \tag{1}$$
$$18a + 116b = 96 \tag{2}$$
$$(1) \times 6 : 18a + 108b = 90 \tag{3}$$
$$(2) - (3): \quad 0 + 8b = 6$$
$$b = 6/8 = 0.75$$

Substitute the b value in (1)

$$3a + 18(0.75) = 15$$
$$3a = 15 - 13.5 = 1.5$$
$$a = 1.5/3 = 0.5$$

Now, multiply each of the n equations in the form $Y = a + bX$ by the coefficient of the second unknown of the equation and sum the resulting equations. The second unknown in each of the equations is b and its coefficients are $X_1, X_2, \ldots$. Thus,

$$X_1 Y_1 = aX_1 + bX_1^2$$
$$X_2 Y_2 = aX_2 + bX_2^2$$
$$\ldots = \ldots + \ldots$$
$$X_n Y_n = aX_n + bX_n^2$$

Sum $\overline{\Sigma(XY) = a\Sigma X + b\Sigma X^2}$ (Normal equation II)

When there are more than two unknowns, the procedure presented above is continued in a similar manner. That is, the coefficients of 3rd, 4th, ... unknowns are used in multiplying each of the n equations orderly.

CHART 22–5

STRAIGHT LINE BY THE METHOD OF LEAST SQUARES
AS COMPARED TO AN ARBITRARILY SELECTED LINE

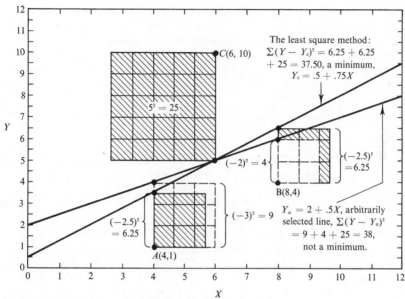

Source: Table 22–6.

The straight line equation by the method of least squares is:

$$Y_c = a + bX = 0.5 + 0.75X$$

The values of Y_c are computed from the equation:

when $X = 4$, $Y_c = 0.5 + 0.75(4) = 3.5$
when $X = 8$, $Y_c = 0.5 + 0.75(8) = 6.5$
when $X = 6$, $Y_c = 0.5 + 0.75(6) = 5.0$

Based on the equation, the mathematical properties of the straight line are:

1. The sum of the deviations is zero, or $\Sigma (Y - Y_c) = 0$.
2. The sum of the squared deviations is the least, or $\Sigma (Y - Y_c)^2 = 37.50$, a minimum as shown on Chart 22–5.

If any other line is drawn to fit points A, B, and C on the chart, though the sum of the deviations of the individual values from the line might be zero, the sum of the squared deviations must be larger than that by the least square method. For example, from the arbitrarily selected straight line represented by equation

$$Y_a = 2 + 0.5X$$

although the sum of the deviations of the individual values from the line is zero, the sum of the squared deviations is 38, which is larger than 37.50 based on the least square method. This fact is computed in Table 22–4C and also shown in Chart 22–5. The values of Y_a in the table are computed:

$$\text{when } X = 4, \quad Y_a = 2 + 0.5(4) = 4$$
$$\text{when } X = 8, \quad Y_a = 2 + 0.5(8) = 6$$
$$\text{when } X = 6, \quad Y_a = 2 + 0.5(6) = 5$$

The method of least squares is popularly used in finding a best fit trend line for a time series. In practice, formula (22–2) is usually simplified before it is used to find the unknown constants a and b. The simplification is done by letting the sum of X values, which are assigned to represent the numbers of years in a time series, be zero, or $\sum X = 0$.

When $\sum X = 0$, normal equation I becomes

$$a = \frac{\sum Y}{n} \tag{22-3a}$$

and normal equation II becomes:

$$b = \frac{\sum (XY)}{\sum X^2} \tag{22-3b}$$

In order to make $\sum X = 0$, the origin of the X variable must be located in the middle of the period included in the time series. This is illustrated in two cases: (a) the time series has an *odd number* of years, and (b) the time series has an *even number* of years.

Fitting a Straight-Line Trend for Odd Number of Years

The procedure of fitting a straight-line trend by the method of least squares for odd number of years of a time series is as follows:

1. Set up a table to compute values of $\sum Y$, $\sum X^2$, and $\sum (XY)$. Let X unit be 1 year and the mid-point (July 1) of the middle year be the origin, so that the sum of the X values $(0, -1, -2, \cdots + 1, + 2, \cdots)$ is zero, or $\sum X = 0$.

2. Obtain values of a and b of the trend equation $Y_c = a + bX$ by substituting the values computed above in formula (22–3).

3. Compute three trend values by using the derived trend equation and plot the results on the chart to draw a straight line. (Two points are required for drawing a straight line. The third point is plotted for checking the accuracy of the computation of the points.)

The procedure is shown in Example 3.

Example 3

Use the sales information for years 1964 to 1978 (15 years) as provided in Table 22–5. (a) Let Y unit be $1000. Find the straight-line trend equation by the method of least squares. (b) Compute trend values for years 1964, 1971, and 1978. (c) Plot the three trend values; then draw a straight line on a chart.

Solution. (a) Table 22–5 is set up for computing the straight-line trend equation for Example 3. Substitute the values obtained from the table in formula

TABLE 22–5

COMPUTATION OF STRAIGHT-LINE TREND EQUATION BY THE
LEAST SQUARE METHOD FOR ODD NUMBER OF YEARS

Year		Actual Sales			Trend Sales
Original Number	X (*Unit:* 1 *year*)	Y (*in* $1,000)	XY	X^2	Y_c (*in* $1,000)
1964	−7	7	−49	49	2.4
1965	−6	6	−36	36	4.3
1966	−5	2	−10	25	6.2
1967	−4	4	−16	16	8.0
1968	−3	8	−24	9	9.8
1969	−2	16	−32	4	11.7
1970	−1	13	−13	1	13.6
1971	0	14	0	0	15.4
1972	1	17	17	1	17.2
1973	2	20	40	4	19.1
1974	3	23	69	9	21.0
1975	4	19	76	16	22.8
1976	5	25	125	25	24.6
1977	6	28	168	36	26.5
1978	7	29	203	49	28.4
Total (Σ)	0	231	518	280	231.0

Source: Table 22–2 and Example 3.

(22–3) as follows:

$$a = \frac{\Sigma Y}{n} = \frac{231}{15} = 15.4 \ (Y \text{ intercept on the origin. In the present case,}$$
it equals the average annual sales, $15,400.)

$$b = \frac{\Sigma (XY)}{\Sigma X^2} = \frac{518}{280} = 1.85 \ (\text{The average annual increase } \$1,850.)$$

The trend equation is

$$Y_c = 15.4 + 1.85X$$
with origin: July 1, 1971
X unit: 1 year
Y unit: $1,000

(b) The required trend values are computed from the trend equation:

1964—$X = -7$ $Y_c = 15.4 + 1.85(-7) = 2.45$, round to 2.4
1971—$X = 0$ $Y_c = 15.4 + 1.85(0) = 15.4$
1978—$X = 7$ $Y_c = 15.4 + 1.85(7) = 28.35$, round to 28.4

(c) The three points representing the computed trend values in the above are plotted on Chart 22–6. The straight line is drawn through the three points on the chart (Line I).

CHART 22–6

STRAIGHT-LINE TREND AND SECOND-DEGREE PARABOLIC TREND FITTED BY
THE LEAST SQUARE METHOD TO SALES OF WILSON DEPARTMENT STORE, 1964 TO 1978

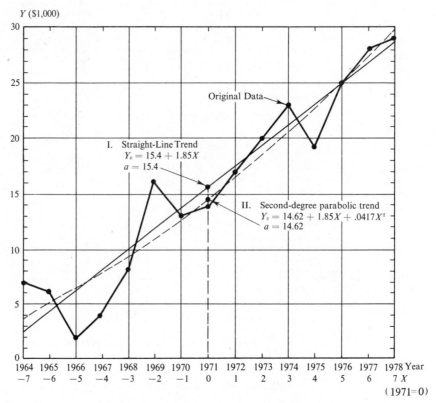

I. Straight-Line Trend
$Y_c = 15.4 + 1.85X$
$a = 15.4$

II. Second-degree parabolic trend
$Y_c = 14.62 + 1.85X + .0417X^2$
$a = 14.62$

Source: I, Table 22–5 (Example 3); II, Table 22–7 (Example 5).

Note that instead of computing the trend values by using the trend equation, it is possible to obtain the trend values by adding the b value (1.85) successively to the trend value of the first year of the series. The trend value of the first year, 1964, is 2.45 (before rounding). Thus, the trend values are:

1965: 2.45 + 1.85 = 4.30
1966: 4.30 + 1.85 = 6.15
1967: 6.15 + 1.85 = 8.00 and so on

The other trend values listed in column (6) of Table 22–5 are obtained in this manner.

Fitting a Straight-Line Trend for Even Number of Years
The procedure of computing a straight-line trend by the method of least squares for even number of years of a time series is almost identical to that for odd number of years. However, the X unit should be 1/2 year, since the origin is the

mid-point of the two middle years of the series, or January 1 of the second middle year. The X values will have a *common difference* of 2 (that is, -1, -3, $\cdots 1$, 3, $\cdots$) and the sum of the X values will be zero, or $\sum X = 0$. The procedure is illustrated in Example 4.

Example 4

Use the sales information for years 1965 to 1978 (14 years) as provided in Table 22–2. (a) Let Y unit be $1,000. Find the straight-line trend equation by the method of least squares. (b) Compute trend values for years 1965, 1971, and 1978.

Solution.   (a) Table 22–6 is set up for computing the straight-line trend equation for Example 4. Substitute the values obtained from the table in formula (22–3) as follows:

$$a = \frac{\sum Y}{n} = \frac{224}{14} = 16$$

$$b = \frac{\sum (XY)}{\sum X^2} = \frac{910}{910} = 1$$

The trend equation is:

$$Y_c = 16 + 1X$$

with origin: January 1, 1972

$$X \text{ unit: } \frac{1}{2} \text{ year}$$

$$Y \text{ unit: } \$1,000$$

TABLE 22–6

COMPUTATION OF STRAIGHT-LINE TREND EQUATION BY THE
LEAST SQUARE METHOD FOR EVEN NUMBER OF YEARS

Year		Actual Sales			Trend Sales
Original Number	X (Unit: 1/2 year)	Y (in $1,000)	XY	X²	Y_c (in $1,000)
1965	−13	6	−78	169	3
1966	−11	2	−22	121	5
1967	− 9	4	−36	81	7
1968	− 7	8	−56	49	9
1969	− 5	16	−80	25	11
1970	− 3	13	−39	9	13
1971 (7/1)	− 1	14	−14	1	15
Origin: 1/1/72	0				
1972 (7/1)	1	17	17	1	17
1973	3	20	60	9	19
1974	5	23	115	25	21
1975	7	19	133	49	23
1976	9	25	225	81	25
1977	11	28	308	121	27
1978	13	29	377	169	29
Total ($\sum$)	0	224	910	910	224

Source: Table 22–2 and Example 4.

(b) The required trend values are computed from the trend equation:

$$1965—X = -13 \quad Y_c = 16 + 1(-13) = 3$$
$$1971—X = -1 \quad Y_c = 16 + 1(-1) = 15$$
$$1978—X = 13 \quad Y_c = 16 + 1(13) = 29$$

If the three trend values were plotted on a graph, a straight line would be drawn through the three values. The X scale of the chart for Example 4 would be marked as follows:

(a)

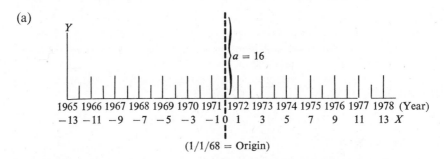

The chart for Example 4 is not constructed here since it differs very little from the chart for Example 3.

Note that since b ($=1$ or \$1,000) represents average increase per $1/2$ year (X unit), the average annual increase should be $2b = 2(1) = 2$ or \$2,000. Other trend values may be obtained by adding the average annual increase successively to the trend value of the first year of the series. The trend value of the first year, 1965, is 3. Thus, the trend values in \$1,000 are:

$$1966: \ 3 + 2 = 5$$
$$1967: \ 5 + 2 = 7$$
$$1968: \ 7 + 2 = 9, \text{ and so on.}$$

The other trend values listed in column (6) of Table 22–6 are obtained in this manner.

Also, the X unit may be a length other than $1/2$ year. However, the X variable would be used inconveniently in computation. If it were one year, for example, the X variable in Example 4 would have decimal places as in the following diagram:

(b)

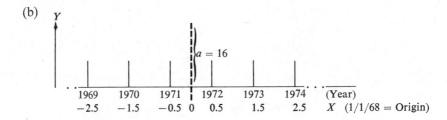

★22.3   Nonlinear Trends

A straight line on an arithmetic scale chart indicates the increase or the decrease of a time series at a constant amount. It is the simplest form for describing the secular trend movement. Frequently the description of the trend is accurate. However, in many cases, a straight line cannot fit the data adequately. For example, a time series may have a faster (or slower) increase at early stage and have a slower (or faster) increase at more recent time. In such a case, a nonlinear curve may describe the trend of the time series better than a straight line.

There are many types of nonlinear trends. A smooth curve, for example, may be drawn by the freehand graphic method. Again, however, the subjective method may not be accurate due to personal judgment. This section will present the nonlinear trends obtained by the methods as follows:

1. A parabolic trend by a second-degree polynomial equation obtained by the method of least squares.

2. A smooth-curve trend obtained by the moving average method.

Second-Degree Parabolic Trends

The general form of a polynomial equation is

$$Y_c = a + bX + cX^2 + dX^3 + eX^4 + \cdots$$

When the polynomial equation is used for describing nonlinear trend movements, it is usually written in its simplest form

$$Y_c = a + bX + cX^2 \tag{22-4}$$

which is called a *second-degree polynomial equation,* also referred to as a *parabola.* The name *second degree* indicates that the highest power of X variable in the equation is 2. Since there are three unknown constants a, b, and c in the equation, it is necessary to develop three equations for solving the unknowns. The three normal equations developed by the method of least squares are:

$$
\begin{aligned}
\text{I.} \qquad & na + b \sum (X) + c \sum (X^2) = \sum (Y) \\
\text{II.} \quad & a \sum (X) + b \sum (X^2) + c \sum (X^3) = \sum (XY) \\
\text{III.} \quad & a \sum (X^2) + b \sum (X^3) + c \sum (X^4) = \sum (X^2 Y)
\end{aligned}
\tag{22-5}*
$$

In practice, formula (22–5) is usually simplified before it is used to find the unknowns in the second-degree polynomial equation for a time series. The simplification is done by letting the sum of X values, which are assigned to represent

---

*See footnote on page 567 for the development of the three equations. For a polynomial equation of the third degree, or

$$Y_c = a + bX + cX^2 + dX^3$$

the four normal equations developed by the method of least squares are:

$$
\begin{aligned}
\text{I.} \qquad & na + b \sum (X) + c \sum (X^2) + d \sum (X^3) = \sum (Y) \\
\text{II.} \quad & a \sum (X) + b \sum (X^2) + c \sum (X^3) + d \sum (X^4) = \sum (XY) \\
\text{III.} \quad & a \sum (X^2) + b \sum (X^3) + c \sum (X^4) + d \sum (X^5) = \sum (X^2 Y) \\
\text{IV.} \quad & a \sum (X^3) + b \sum (X^4) + c \sum (X^5) + d \sum (X^6) = \sum (X^3 Y)
\end{aligned}
$$

the numbers of years in a time series, be zero. When $\sum X = 0$, $\sum (X^3) = 0$. The three normal equations become:

I. $\qquad na + c \sum (X^2) = \sum (Y)$

II. $\qquad b \sum (X^2) = \sum (XY)$

III. $a \sum (X^2) + c \sum (X^4) = \sum (X^2 Y)$

Solving equations I and III, we obtain the values of a and c. Solving equation II, we obtain the value of b, or

$$a = \frac{\sum (Y) - c \sum (X^2)}{n} \qquad\qquad (22\text{--}6a)$$

$$b = \frac{\sum (XY)}{\sum (X^2)} \qquad\qquad (22\text{--}6b)$$

$$c = \frac{n \sum (X^2 Y) - \sum (X^2) \cdot \sum (Y)}{n \sum (X^4) - (\sum X^2)^2} \qquad\qquad (22\text{--}6c)$$

In order to make $\sum X = 0$ (thus $\sum X^3 = 0$), the origin of the X variable must be located in the middle of the period included in the time series. Now, when the time series has an odd number of years, the X values of the series are $0, -1, -2, \cdots 1, 2 \cdots$ if the X unit is 1 year; when the series has an even number of years, X values are $-1, -3, -5, \cdots 1, 3, 5 \cdots$ if the X unit is $1/2$ year. Since the basic principle is the same for both cases, only the time series with an odd number years is illustrated in Example 5.

Example 5

Use the sales information for years 1964 to 1978 (15 years) as provided in Table 22-2. (a) Let Y unit be \$1,000. Find the second-degree polynomial equation by the least square method. (b) Compute the trend values for the 15 years. (c) Plot the trend values on a chart to obtain a second-degree parabola.

Solution. (a) Table 22-7 is set up for computing the second-degree polynomial equation for Example 5. Substitute the values obtained from Table 22-7 in formula (22-6) as follows:

$$c = \frac{15(4,484) - 280(231)}{15(9,352) - (280)^2} = 0.0417$$

$$a = \frac{231 - 0.0417(280)}{15} = 14.62 \text{ (Notice } a \text{ depends on the value of } c.)$$

$$b = \frac{518}{280} = 1.85 \text{ (Same as the slope } b \text{ of the straight line in Example 3.)}$$

The required equation, formula (22-4), is written:

$$Y_c = 14.62 + 1.85X + 0.0417X^2$$

with origin: July 1, 1971

X unit: 1 year

Y unit: \$1000

Note that the arrangement and the values of the first five columns in Table 22-7

TABLE 22–7

COMPUTATION OF A SECOND-DEGREE PARABOLIC TREND EQUATION
BY THE LEAST SQUARE METHOD FOR ODD NUMBER OF YEARS

Year		Actual Sales					Trend Sales
Original Number (1)	X (Unit: 1 year) (2)	Y (in $1,000) (3)	XY (4)	X^2 (5)	X^2Y (6)	X^4 (7)	Y_c (in $1,000) (8)
1964	− 7	7	−49	49	343	2,401	3.7
1965	− 6	6	−36	36	216	1,296	5.0
1966	− 5	2	−10	25	50	625	6.4
1967	− 4	4	−16	16	64	256	7.9
1968	− 3	8	−24	9	72	81	9.4
1969	− 2	16	−32	4	64	16	11.1
1970	− 1	13	−13	1	13	1	12.8
1971	0	14	0	0	0	0	14.6
1972	1	17	17	1	17	1	16.5
1973	2	20	40	4	80	16	18.5
1974	3	23	69	9	207	81	20.5
1975	4	19	76	16	304	256	22.7
1976	5	25	125	25	625	625	24.9
1977	6	28	168	36	1,008	1,296	27.2
1978	7	29	203	49	1,421	2,401	29.6
Total ($\sum$)	0	231	518	280	4,484	9,352	230.8*

Source: Table 22-2 and Example 5. (*Difference from 231 is due to rounding.)

are the same as those in Table 22–5 for the straight-line trend. However, Table 22–7 has two additional columns: X^2Y and X^4.

(b) The trend values (Y_c) listed in column (8) of Table 22–7 are computed from the above equation in the following manner:

$$1964—X = -7 \qquad Y_c = 14.62 + 1.85(-7) + 0.0417(-7)^2$$
$$= 14.62 - 12.95 + 2.0435$$
$$= 3.7135, \text{ round to } 3.7;$$

$$1965—X = -6 \qquad Y_c = 14.62 + 1.85(-6) + 0.0417(-6)^2$$
$$= 14.62 - 11.10 + 1.5012$$
$$= 5.0212, \text{ round to } 5.0; \text{ and so on}$$

(c) The trend values are plotted on Chart 22–6. The second-degree parabola is drawn through the points representing the trend values (Line II). The straight-line trend and the parabolic trend on the chart are both obtained by the method of least squares. Thus, the fact that the sum of the deviations is zero, or $\sum (Y - Y_c)$ = 0, is ture for both cases. However, the sum of the squared deviations, or $\sum (Y - Y_c)^2$, is only 96.28 (based on the values of Y_c in Table 22–7) for the parabolic trend and is 103.80 (based on the values of Y_c in Table 22–5) for the straight-line trend. The values of $\sum (Y - Y_c)^2$ are computed in a similar manner as that in Table 22–4. (The detailed computations are not shown here for the sake of simplicity.) Thus, the parabolic trend, which has a smaller sum of the squared deviations, fits to the original data better than the straight-line trend.

Moving Average Trends

Secular trend may also be measured by the method of moving averages. This method can obtain a curve which will smooth out fluctuations in a times series and thus indicate the general direction of the trend. The moving average for each year in a series is the arithmetic mean of the values of a constant number of years centered at the year. The mean *moves* one year ahead after each computation. The procedure of finding moving averages in a series is illustrated in Example 6 below.

Example 6

Use the sales information for years 1964 to 1978 as provided in Table 22–8. Find (a) the 3-year moving averages and (b) the 5-year moving averages. Draw a trend curve for each of the two sets of moving averages.

Solution. The computation of the moving averages for the example is summarized in Table 22–8. For computing each moving average, first find the *moving total*. Next, divide the moving total by the number of years included in the total. The quotient is the moving average of the middle year of the years included.

    (a) The 3-year moving averages are computed as follows: the first 3-year moving total is computed from the values of years 1964, 1965, and 1966:

$$7 + 6 + 2 = 15$$

The middle year of the first group of three years is 1965. Thus, the moving average of 1965 is the quotient of the moving total 15 divided by the number of years, 3, or

$$\frac{15}{3} = 5 \text{ (in \$1,000)}$$

The second 3-year moving total is computed from the values of years 1965, 1966, and 1967:

$$6 + 2 + 4 = 12$$

The middle year of the second group of three years is 1966. Thus, the moving average of 1966 is

$$\frac{12}{3} = 4$$

The other 3-year moving averages may also be obtained in this manner. However, there are no moving averages for the first year, 1964, and the last year, 1978.

    (b) Similarly, the 5-year moving averages are computed. Note that we may use a short method to compute moving totals when the constant number of years included in each total is large. By using the short method, each moving total,

TABLE 22–8

COMPUTATION OF TREND VALUES BY THE MOVING AVERAGE METHOD
FOR THE SALES OF WILSON DEPARTMENT STORE, 1964 TO 1978

Year	a. Based on 3-Year Periods			b. Based on 5-Year Periods		
	Sales (in $1,000)	3-Year Moving Total	3-Year Moving Averages, Col. (3) ÷ 3	Sales (in $1,000)	5-Year Moving Total	5-Year Moving Averages, Col. (6) ÷ 5
(1)	(2)	(3)	(4)	(5)	(6)	(7)
1964	7			7		
1965	6	—15—	—5	6		
1966	2	12	4	2	—27—	—5 2/5
1967	4	14	4 2/3	4	36	7 1/5
1968	8	28	9 1/3	8	43	8 3/5
1969	16	37	12 1/3	16	55	11
1970	13	43	14 1/3	13	68	13 3/5
1971	14	44	14 2/3	14	80	16
1972	17	51	17	17	87	17 2/5
1973	20	60	20	20	93	18 3/5
1974	23	62	20 2/3	23	104	20 4/5
1975	19	67	22 1/3	19	115	23
1976	25	—72—	—24	25	—124—	—24 4/5
1977	28	82	27 1/3	28		
1978	29			29		

Source: Table 22–2 and Example 6.

except the first one, is computed from its immediate preceding moving total by only one subtraction (subtracting the first value of the preceding group of years) and one addition (adding the last value of the current group of years). For example, the second and third 5-year moving totals in Table 22–8 may be obtained by either the regular method or the short method as shown in Table 22–9.

TABLE 22–9

COMPUTATION OF 5-YEAR MOVING
TOTALS BY REGULAR AND SHORT METHODS

Year	1st Moving Total (Regular Method Only)	2nd Moving Total		3rd Moving Total	
		Regular Method	Short Method	Regular Method	Short Method
			27 (1st m.t.)		36 (2nd m.t.)
1964	7		—7		
1965	6	6			—6
1966	2	2		2	
1967	4	4		4	
1968	8	8		8	
1969		16	+16	16	
1970				13	+13
Moving total	27	36	36	43	43

The 3-year moving averages (column 4), the 5-year moving averages (column 7), and the original data are plotted on Chart 22–7. The two curves based on the method of moving averages clearly show the upward trends of the series. Notice that the original data fluctuate. The 3-year moving averages smoothed the fluctuation to a smaller degree than did the 5-year moving averages. In general, the more time (or more years) included in computing the moving averages, the smoother is the moving average curve.

Note that the trend obtained by the method of moving averages cannot easily be expressed by a mathematical equation. This method will give the most effective description of the trend when the period for computing the averages is equal to, or a multiple of, the average length of the cycles in the fluctuation of a series; because when the moving averages are computed from an average length, the degree of fluctuation can be reduced to a minimum. In fact, if the cycles in the fluctuation are uniform in length (number of years) and amplitude (amounts of

CHART 22–7

MOVING AVERAGE TRENDS FITTED TO SALES OF WILSON DEPARTMENT STORE, 1964 TO 1978

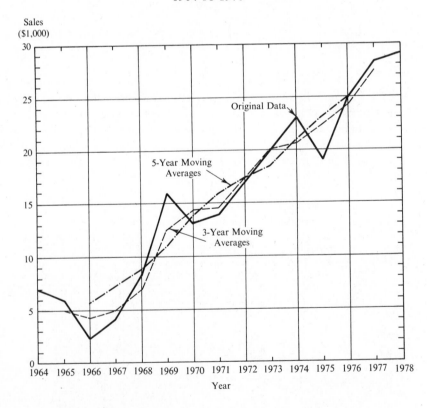

Source: Table 22–8 and Example 6.

increases and decreases), the moving averages of the period which is equal to or a multiple of the uniform length of each cycle will form a straight line. Thus the fluctuation of a series is completely eliminated. This is illustrated in Table 22–10 and Chart 22–8. The table and chart indicate that the sales of Dennis Clothing Store from 1968 to 1977 have a 3-year cycle. The increase and decrease amounts

TABLE 22–10

COMPUTATION OF TREND VALUES BY THE MOVING AVERAGE METHOD
FOR THE SALES OF DENNIS CLOTHING STORE, 1968 TO 1977
(PERIOD OF EACH MOVING TOTAL = 3 YEARS = DURATION OF EACH
CYCLE; CYCLES HAVING THE SAME AMPLITUDE, COLUMN 3)

Year	Sales (in $1,000)	Sales Increase or Decrease from Previous Year		3-Year Moving Total	3-Year Moving Average
(1)	(2)	(3)		(4)	(5)
1968	2	—		—	—
1969	3	+1		12	4
1970	7	+4	1st cycle	15	5
1971	5	−2		18	6
1972	6	+1		21	7
1973	10	+4	2nd cycle	24	8
1974	8	−2		27	9
1975	9	+1		30	10
1976	13	+4	3rd cycle	33	11
1977	11	−2		—	—

Source: Hypothetical data.

CHART 22–8

MOVING AVERAGE TRENDS FITTED TO SALES OF DENNIS CLOTHING STORE,
1968 TO 1977

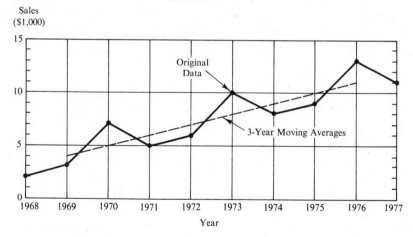

Source: Table 22–10.

in each 3-year cycle are the same: $+1, +4$, and -2 (in \$1,000). When the 3-year moving averages are plotted, a straight line can be drawn through the points on the chart. Thus, the task of selecting a proper length of period is important since the purpose of applying the method of moving averages is to reduce the fluctuation of a series to a minimum.

The method of moving averages may be extended to smooth the fluctuations for time series recorded in a unit length other than a year. In the next two chapters, we shall use the moving average method to smooth fluctuations for seasonal variations and cyclical fluctuations.

The number of years used in computing the moving averages need not be limited to odd numbers. However, if an even number of years is used, we will have to add another step in order to get the moving average for each year. The additional step will be illustrated in the next chapter.

★22.4 Measuring Trends by Logarithms

The trends discussed in the previous sections were plotted on arithmetic scales. Trends may also be plotted on a semilogarithmic (or semilog) chart in the form of a straight line or a nonlinear curve. When it is a straight line on the semilog chart, the trend shows the increase of Y values of a time series at a *constant rate*. (A straight line on an arithmetic chart indicates an increase of a *constant amount*.) When it is a nonlinear curve on the semilog chart, an upward curve shows an increase of varying rates, depending on the shapes of the slopes. The steeper the slope, the higher is the rate of increase, such as the first portion of the curve in Chart 22–9 showing greater rates of increase than the later portion of the curve.

This section discusses two types of trends which are usually computed by logarithms: (1) exponential trends and (2) growth curves. Note that in order to simplify the illustrations, the simple four-place log table (Table 4–2, page 92), is used in the examples and problems in this section. However, if the interpolation method is required on the four-place log table, the six-place log table in the Appendix (Table 2) is used.

Exponential Trends

An exponential trend is a straight line on a semilog chart, but it is a curve on an arithmetic chart. The exponential equation which is used in describing secular trend is written:

$$Y_c = ab^X \tag{22-7a}$$

Take logarithm of each side of the equation,

$$\log Y_c = \log a + (\log b)X \tag{22-7b}$$

Again let $\sum X = 0$. By applying the method of least squares, the two unknown constants $\log a$ and $\log b$ may be computed by the following formulas:

$$\log a = \frac{\Sigma \ (\log \ Y)}{n} \qquad\qquad (22\text{--}8a)$$

$$\log b = \frac{\Sigma \ (X \cdot \log \ Y)}{\Sigma \ X^2} \qquad\qquad (22\text{--}8b)^*$$

where $\log a$ = the mean of the logarithms of Y values, and

$\log b$ = the slope of the straight line on a semilog chart.

The value of b is the ratio of current year's Y_c to its immediate preceding year's Y_c, and the rate of increase or decrease (r) is the difference between b and 1 $(1 = 100\%$ or the base). When b is larger than 1, the difference is an increase rate (or $r = b - 1$); when b is smaller than 1, it is a decrease rate (or $r = 1 - b$). Thus, the exponential equation is useful for the time series that increases or decreases at a constant rate (or geometrical progression). Since formula (22–8) is obtained by the method of least squares, the sum of the squared deviations of the logarithms of Y values from the trend values, or $\Sigma \ (\log \ Y - \log \ Y_c)^2$, is a minimum. (Note: $\Sigma \ (Y - Y_c)^2$ is not a minimum in this case.)

Example 7

Use the sales information for years 1964 to 1978 as provided in Table 22–2. (a) Let Y unit be \$1,000. Find the logarithmic straight-line trend equation by the method of least squares. (b) Compute trend values for year 1964 and other years. (c) Plot the trend values on a semilog chart; then draw a straight line. Also, draw a curvilinear trend on an arithmetic chart.

Solution. (a) Table 22–11 is set up for computing the straight-line trend equation and trend values for Example 7. The middle of the period (July 1, 1971) is selected as the origin so that $\Sigma \ X = 0$.

Substitute the values obtained from Table 22–11 in formula (22–8a),

$$\log a = \frac{\Sigma \ (\log \ Y)}{n} = \frac{16.3730}{15} = 1.0915$$

in formula (22–8b),

---

*The two normal equations for a straight line on a semilogarithmic chart are:

$$\Sigma \ \log \ Y = n \cdot \log a + (\log b) \ \Sigma \ X$$
$$\Sigma \ (X \cdot \log \ Y) = (\log a) \ \Sigma \ X + (\log b)(\Sigma \ X^2)$$

They are obtained for equation $\log \ Y_c = \log a + (\log b)X$ in a similar manner to formula (22–2). When $\Sigma \ X = 0$, the normal equations become

$$\Sigma \ \log \ Y = n(\log a)$$
$$\Sigma \ (X \cdot \log \ Y) = (\log b)(\Sigma \ X^2)$$

Solving $\log a$ and $\log b$ from the two new equations, we have formulas (22–8a) and (22–8b). The equation of a second degree semilogarithmic curve is

$$\log \ Y_c = \log a + (\log b)X + (\log c)X^2$$

When $\Sigma \ X = 0$, or the origin is at the middle year, the three normal equations are:

I. $\Sigma \ \log \ Y = n(\log a) + (\log c) \ \Sigma \ X^2$

II. $\Sigma \ (X \cdot \log \ Y) = (\log b) \ \Sigma \ X^2$

III. $\Sigma \ (X^2 \cdot \log \ Y) = (\log a) \ \Sigma \ X^2 + (\log c) \ \Sigma \ X^4$

TABLE 22–11

COMPUTATION OF LOGARITHMIC STRAIGHT-LINE (EXPONENTIAL)
TREND EQUATION BY THE LEAST SQUARE METHOD

Year		Actual Sales (in $1,000) Y	$\log Y$	$X \cdot \log Y$	X^2	$\log Y_c$	Y_c
Original Number (1)	X (Unit: 1 year) (2)	(3)	(4)	(5)	(6)	(7)	(8)
1964	−7	7	0.8451	−5.9157	49	0.6365	4.3
1965	−6	6	0.7782	−4.6692	36	0.7015	5.0
1966	−5	2	0.3010	−1.5050	25	0.7665	5.8
1967	−4	4	0.6021	−2.4084	16	0.8315	6.8
1968	−3	8	0.9031	−2.7093	9	0.8965	7.9
1969	−2	16	1.2041	−2.4082	4	0.9615	9.2
1970	−1	13	1.1139	−1.1139	1	1.0265	10.6
1971	0	14	1.1461	0	0	1.0915	12.3
1972	1	17	1.2304	1.2304	1	1.1565	14.3
1973	2	20	1.3010	2.6020	4	1.2215	16.7
1974	3	23	1.3617	4.0851	9	1.2865	19.3
1975	4	19	1.2788	5.1152	16	1.3515	22.5
1976	5	25	1.3979	6.9895	25	1.4165	26.1
1977	6	28	1.4472	8.6832	36	1.4815	30.3
1978	7	29	1.4624	10.2368	49	1.5465	35.2
Total (Σ)	0	231	16.3730	18.2125	280	16.3725	226.3

Source: Table 22–2 and Example 7.

$$\log b = \frac{\Sigma \, (X \cdot \log Y)}{\Sigma \, X^2} = \frac{18.2125}{280} = 0.0650$$

The logarithmic equation of the straight-line trend is

$$\log Y_c = 1.0915 + 0.0650X$$
$$\text{with origin: July 1, 1971}$$
$$X \text{ unit: 1 year}$$
$$Y \text{ unit: } \$1,000$$

(b) The trend values are computed from the equation as follows:

Year 1964—$X = -7$, $\log Y_c = 1.0915 + 0.0650(-7) = 0.6365$

Find antilog (0.6365),

$$Y_c = 4.33, \text{ or round to } 4.3$$

Year 1971—$X = 0$, $\log Y_c = 1.0915 + 0.0650(0) = 1.0915$
$$Y_c = 12.3$$

Note: The mantissa 0915 is close to 0899, which corresponds to digits 123 in the log table.

CHART 22–9

EXPONENTIAL TREND AND GOMPERTZ CURVE FITTED TO SALES OF
WILSON DEPARTMENT STORE, 1964 TO 1978, ON LOGARITHMIC SCALE

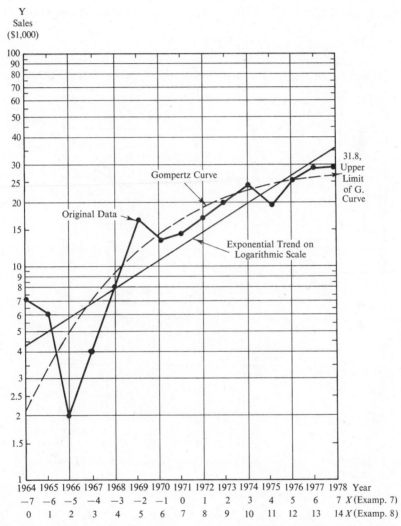

Source: Table 22–11 and Example 7 (Exponential Trend); Table 22–12 and Example
8 (Gompertz Curve).

$$\text{Year } 1978 — X = 7, \quad \log Y_c = 1.0915 + 0.0650(7) = 1.5465$$
$$Y_c = 35.2$$

Log Y_c for each succeeding year after 1964 may be obtained by adding 0.0650
($= \log b$) to log Y_c of its preceding year (log Y_c of 1964 = 0.6365), such as:

$$1965 — \log Y_c = 0.6365 + 0.0650 = 0.7015$$
$$1966 — \log Y_c = 0.7015 + 0.0650 = 0.7665, \text{ and so on}$$

(c) The trend values are listed in column (8) of Table 22–11 and are plotted on the semilog chart in the form of a straight line, Chart 22–9. When the trend values are plotted on an arithmetic chart, it is a nonlinear curve, Chart 22–10.

CHART 22–10

EXPONENTIAL TREND AND GOMPERTZ CURVE FITTED TO SALES OF
WILSON DEPARTMENT STORE, 1964 TO 1978, ON ARITHMETIC SCALE

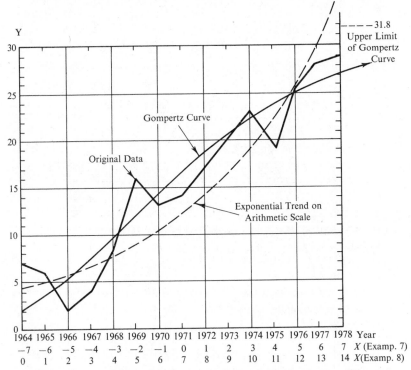

Source: Table 22–11 and Example 7 (Exponential Trend); Table 22–12 and Example 8 (Gompertz Curve).

Note that the exponential equation for the trend can be obtained as follows:

Since $\log a = 1.0915$, find antilog, $a = 12.3$
Since $\log b = 0.0650$, find antilog, $b = 1.16$

The exponential equation is

$$Y_c = 12.3(1.16)^X$$

which may be written in the form of compound interest equation:

$$Y_c = 12.3 (1 + 0.16)^X$$

Thus, the rate of increase for each year is 0.16 or 16%. The trend values Y_c are checked below:

$$1964\text{—}Y_c = \quad 4.33$$
$$+0.69 \quad (=4.33 \times 16\%)$$
$$1965\text{—}Y_c = \quad 5.02,\text{ or round to } 5.0$$
$$+0.80 \quad (=5.02 \times 16\%)$$
$$1966\text{—}Y_c = \quad 5.82,\text{ or round to } 5.8,\text{ and so on}$$

If b is a positive number less than 1, the growth is declining at a constant rate $(1 - b)$. The limit of the declining trend value is zero. For example, if $b = 0.8$, the rate of decrease is $1 - 0.8 = 0.2$ or 20%. The exponential equation can be written:

$$Y_c = ab^X = a(0.8)^X = a(1 - 0.2)^X$$

When X is very large, 0.8^X approaches zero. Thus, Y_c approaches zero also.

Growth Curves

Many types of time series concerning business and economic activities can best be described in three development stages. Assume that the activity is selling a product, such as the radio industry. At the early stage, or when the product is newly introduced to the market, the amount of growth for selling the product increases slowly; at the middle stage, or when the product is already recognized by the public, the growth becomes increasingly greater; and at the final stage, or when the product is sold on a saturated market, the growth reaches a point of stabilization. Curves that are used to describe this type of growth are generally referred to as the growth curves.

There are two well-known growth curves which have been popularly used in trend analysis for business and economic activities since the 1920's. They are the Gompertz curve and the logistic or Pearl-Reed curve. The Gompertz curve was originally used by Benjamin Gompertz in constructing mortality tables for actuarial science; the logistic curve was initially applied by Raymond Pearl and L. J. Reed in analyzing population growth. When the two types of curves are plotted on an arithmetic scale chart, such as on Chart 22–10, the curves are in the shape of an elongated S. The S shape indicates the pattern of growth in terms of actual amounts: small in the early years, increasingly greater in the middle years, and large but stabilized in the current year. When the curves are plotted on a semilogarithmic chart, such as Chart 22–9, the curves will show growth at rapidly increasing rates in the earlier period but the declining rates in the later period of the series.

Since the two types of curves are basically the same shape, only the method of constructing a Gompertz curve is illustrated in this section. The equation of Gompertz curve is

$$Y_c = ab^{(c^X)} \tag{22–9a}$$

Take logarithm of each side of the equation,

$$\log Y_c = \log a + (\log b)c^X \tag{22–9b}$$

(The letter c in Y_c is a subscript, and it denotes a computed value, whereas the letter c at the right side is an unknown constant.)

The values of the three unknown constants, $\log a$, $\log b$, and c, in formula 22–9b are not computed by the least square method but may be obtained in many different ways. The simplest objective method is to use the three subgroup sums of the logarithms of the Y values in a series.

Let $\Sigma_1 \log Y$, $\Sigma_2 \log Y$, and $\Sigma_3 \log Y =$ the sums of the logarithms of the Y values of the first, second, and third subgroups respectively, the subgroups being the equal size m. (Thus, $3m = n$, the size of the series.) Also, let

$$D_1 = \Sigma_2 \log Y - \Sigma_1 \log Y$$
$$D_2 = \Sigma_3 \log Y - \Sigma_2 \log Y$$

and the first year in the series be the origin.

Then, the three unknown constants in formula (22–9b) are:

$$\log a = \frac{1}{m}\left(\Sigma_1 \log Y - \frac{D_1}{c^m - 1}\right) \qquad \textbf{(22–10a)}$$

$$\log b = \frac{D_1(c - 1)}{(c^m - 1)^2} \qquad \textbf{(22–10b)}$$

$$c^m = \frac{D_2}{D_1} \qquad \textbf{(22–10c)*}$$

Example 8

Use the sales information for years 1964 to 1978 as provided in Table 22–2. (a) Let Y unit be \$1,000. Find the trend equation for Gompertz curve by the method of three subgroup sums of the logarithms. (b) Compute trend value for each year.

---

*Formula (22–10) may be obtained in the following manner:
Assume that there are 15 pairs of X and Y, such as the values in Example 8. Let the computed trend value Y_c be $Y_0, Y_1, Y_2, \ldots Y_{14}$ when $X = 0, 1, 2, \ldots 14$. Divide the 15 Y_c values into 3 equal subgroups, or 5 values in each subgroup. Then, based on the equation, $\log Y_c = \log a + (\log b) c^X$, we may write an equation for each of the 15 Y_c values as follows:

$$\log Y_0 = \log a + (\log b)c^0$$
$$\log Y_1 = \log a + (\log b)c^1$$
$$\ldots\ldots = \ldots\ldots$$
$$\log Y_4 = \log a + (\log b)c^4$$

Sum of first subgroup: $\quad \Sigma_1 \log Y = 5(\log a) + (\log b)c^0\,(1 + c^1 + c^2 + c^3 + c^4) \qquad (1)$
$(m = 5)$

$$\log Y_5 = \log a + (\log b)c^5$$
$$\ldots\ldots = \ldots\ldots$$
$$\log Y_9 = \log a + (\log b)c^9$$

Sum of second subgroup: $\quad \Sigma_2 \log Y = 5(\log a) + (\log b)c^5\,(1 + c^1 + c^2 + c^3 + c^4) \qquad (2)$
$(m = 5)$

$$\log Y_{10} = \log a + (\log b)c^{10}$$
$$\ldots\ldots = \ldots\ldots$$
$$\log Y_{14} = \log a + (\log b)c^{14}$$

Sum of third subgroup: $\quad \Sigma_3 \log Y = 5(\log a) + (\log b)c^{10}\,(1 + c^1 + c^2 + c^3 + c^4) \qquad (3)$
$(m = 5)$

Formula (22–10) may be obtained by solving equations (1), (2), and (3) for the unknowns.

(c) Plot the trend values on a semilog chart and on an arithmetic chart. Also, draw a smooth curve to show the trend on each chart.

Solution. (a) Table 22–12 is set up for computing the trend equation for the Gompertz curve and the trend values. The first year (1964) is selected as the origin so that formula (22–10) may be applied. Substitute the values obtained from the table in formula (22–10) as follows:

$$D_1 = 2.5660$$
$$D_2 = 0.9525$$
$$\Sigma_1 \log Y = 3.4295$$
$$m = \frac{n}{3} = \frac{15 \text{ (years)}}{3 \text{ (subgroups)}} = 5$$
$$c^5 = \frac{0.9525}{2.5660} = 0.3712$$
$$c = \sqrt[5]{0.3712} = 0.8202$$

The c value is computed by using logarithms as shown below. (Also, we may compute the c^x values in Table 22–12 by logarithms or the ordinary multiplication method.) The c value is to be used for further computation. Thus, the detailed six-place log table is used here to avoid the interpolation process from the four-place log table.

$$\log c = \frac{1}{5} (\log 0.3712) = \frac{1}{5} (-1 + 0.569608)$$
$$= -0.086078 = 9.913922 - 10$$

Find antilog $(9.913922 - 10)$, $c = 0.8202$.

$$\log b = \frac{2.5660 \, (0.8202 - 1)}{(0.3712 - 1)^2} = -1.1669$$
$$\log a = \frac{1}{5} \left(3.4295 - \frac{2.5660}{0.3712 - 1} \right) = 1.5021$$

The equation for the Gompertz curve is

$$\log Y_c = 1.5021 - 1.1669(0.8202^x)$$
$$\text{with origin: July 1, 1964}$$
$$X \text{ unit: 1 year}$$
$$Y \text{ unit: } \$1,000$$

(b) The trend value of 1964 ($X = 0$) is computed by the trend equation:

$$\log Y_c = 1.5021 - 1.1669 \, (0.8202^0) = 1.5021 - 1.1669(1) = 0.3352$$

Find antilog (0.3352),

$$Y_c = 2.16, \text{ or round to 2.2}$$

The interpolation method is not used here in finding the antilog from the log table since Y_c is not used for further computation and is rounded to only one decimal place. The mantissa 3352 is close to 3345, which corresponds to digits 216 in Table 4–2. Other Y_c values are computed in a similar manner. The computation is listed in columns (5) to (8) of Table 22–12.

(c) The trend values are plotted on Chart 22–9 (a semilog chart) and on Chart 22–10 (an arithmetic chart).

Note that the c value is usually less than 1, or D_2 is smaller than D_1, such as the values in Example 8. When X is very large, c^X (where $c < 1$) approaches zero; then $(\log b)c^X$ approaches zero. Thus, $\log Y_c = \log a + 0 = \log a$, and $Y_c = a$. The value of a, therefore, is the upper limit of the Gompertz curve. The upper limit for Example 8 is 31.8 $(=a$, since $\log a = 1.5021)$. On the other hand, if c is more than 1, or D_2 is larger than D_1, the rate of growth of the time series must be low in the early years but high in the later years. This type of growth is less common in business and economic activities.

Observe the second term (the product of constant $\log b$ and exponential c^X) on the right side of the equation for Gompertz curve. The form of the second term is the same as that of the term on the right side (the product of constant a and exponential b^x) of an exponential curve equation. The Gompertz curve is, therefore, a type of modified exponential curve, the modification being the addition of a constant $\log a$.

TABLE 22–12

COMPUTATION OF GOMPERTZ CURVE EQUATION AND TREND VALUES
FOR THE SALES OF WILSON DEPARTMENT STORE, 1964 TO 1978

Year		Actual Sales (in $1,000) Y	$\log Y$	c^X ($c = 0.8202$)	$\log b \cdot c^X =$ (−1.1669) × col. (5)	$\log Y_c = \log a + \log b \cdot c^X =$ 1.5021 + col. (6)	Y_c
Original Number	X (Unit: 1 Year)						
(1)	(2)	(3)	(4)	(5)	(6)	(7)	(8)
1964	0	7	0.8451	1.0000	−1.1669	0.3352	2.2
1965	1	6	0.7782	0.8202	−0.9571	0.5450	3.5
1966	2	2	0.3010	0.6727	−0.7850	0.7171	5.2
1967	3	4	0.6021	0.5518	−0.6439	0.8582	7.2
1968	4	8	0.9031	0.4526	−0.5281	0.9740	9.4
		$\sum_1 \log Y =$ 3.4295					
1969	5	16	1.2041	0.3712	−0.4332	1.0689	11.7
1970	6	13	1.1139	0.3045	−0.3553	1.1468	14.0
1971	7	14	1.1461	0.2497	−0.2914	1.2107	16.2
1972	8	17	1.2304	0.2048	−0.2390	1.2631	18.3
1973	9	20	1.3010	0.1680	−0.1960	1.3061	20.2
		$\sum_2 \log Y =$ 5.9955					
1974	10	23	1.3617	0.1378	−0.1608	1.3413	21.9
1975	11	19	1.2788	0.1130	−0.1319	1.3702	23.5
1976	12	25	1.3979	0.0927	−0.1082	1.3939	24.8
1977	13	28	1.4472	0.0760	−0.0887	1.4134	25.9
1978	14	29	1.4624	0.0624	−0.0728	1.4293	26.9
		$\sum_3 \log Y =$ 6.9480					

$D_1 = \sum_2 \log Y - \sum_1 \log Y = 5.9955 - 3.4295 = 2.5660$
$D_2 = \sum_3 \log Y - \sum_2 \log Y = 6.9480 - 5.9955 = 0.9525$

Source: Table 22–2 and Example 8.

★22.5   Selection of Appropriate
Method for a Trend

The previous three sections indicated the fact that a given time series, such as Table 22–2, may be fitted by different trends based on various methods. Then which one of the methods is the best method for a trend fitted to the series? Before answering this question, we must admit that none of the above methods is always better than the others. However, if we first examine the reasons for analyzing the trend movement for a series, we may select a relatively better method for trend analysis. In general, there are three important reasons for the analysis.

Historical Trend

The management may wish to know the trend of a particular activity during the past for a definite period. If management needs a quick answer and a rough estimate of the trend, the freehand graphic method for a straight line may be used, provided that the drawer of the graph is an experienced statistician. If he is not an experienced statistician, the method of semiaverages or the method of moving averages is preferred.

  If the management needs a trend which must be a good fit to data, the method of least squares for a straight-line trend, or for a second degree parabolic curve, or for a nonlinear curve by a higher degree polynomial equation may be used. As the degree of a trend equation is higher, the sum of squared deviations from the trend line gets smaller, or $\sum (Y - Y_c)^2$ gets smaller, and the trend line fits better to the data. However, if there are as many constants in the trend equation as there are Y values, the trend line will go through every point representing true Y value on a graph. In this case, the trend equation becomes meaningless since we will not eliminate any portion of the fluctuation of the series.

  If the management plans to examine the rate of change instead of the amount of change, a logarithmic chart should be used. The data may be plotted first; then a straight line or a Gompertz curve is selected for the trend.

Comparison of Trends

The management may wish to compare the trends of various groups of data. If he wants to compare the amounts of change, a businessmen should use the arithmetic straight-line trends by the method of least squares. The comparison may be made by examining the b values of the different groups of data. We recommend this type of trend because of accuracy (the sum of squared deviations is the least when compared with other types of straight-line trends) as well as simplicity (simpler in computation than nonlinear curves). For example, the straight-line trend equation by the method of least squares for Example 3 is

$$Y_c = 15.4 + 1.85X$$
with origin: July 1, 1971
X unit: 1 year
Y unit: $1,000, and
$b = 1.85$ in units of $,1000

The equation indicates that the amount of increase for each year is $1,850. If another equation obtained by the same method for the other time series indicates a higher b value, say an increase of $2,000 each year, the comparison of the two amounts of the increase will show an average difference of $150 ($= $2,000 $−$ $1,850) per year.

If the management is interested in the rates of change for comparison, logarithmic straight-line trends and the method of least squares should be used. Again a conclusion may be reached by comparing the b values of individual groups of data. For example, the logarithmic straight-line trend equation by the method of least squares for Example 7 is

$$\log Y_c = 1.0915 + 0.0650X$$
with origin: July 1, 1971
X unit: 1 year
Y unit: $1,000 and
$$\log b = 0.0650, \text{ or } b = 1.16$$

The equation indicates that the rate of increase for each year is 16 percent ($= 1.16−1$). If another equation obtained by the same method for the other time series indicates a higher b value, say $b = 1.20$ or an annual rate of increase of 20 percent, the comparison of the two rates of increase will show an average difference of 4 percent ($= 20\% − 16\%$) per year. The same procedure may be used to compare the trends of different periods for the same data.

Forecasting Future Activity

If the management wishes to forecast a given activity in the future based on the past quantitative information, the selection of an appropriate method for a trend becomes more complicated. In general a long time period should be employed so that the trend, which is affected by cyclical fluctuations to a smaller degree and therefore is more relevant for the future, may be used as a basis for predicting the future. After the original data are plotted on a graph and the plotted data are carefully examined, a method for describing the trend is then selected.

A forecast of the future activity may be made by the *extrapolation method*. The extrapolation method is used to compute the trend value on a desired future date based on a trend equation. For example, we forecast the trend values of the years 1979 and 1984 based on the trend equation

$$Y_c = 15.4 + 1.85X \text{ with origin on July 1, 1971}$$

or year $1971 = 0$, obtained from the data on Table 22–2. By the extrapolation method, the computation is as follows:

Year 1979 $X = 8$,   since $1979 − 1971 = 8$
$Y_c = 15.4 + 1.85(8) = 30.20$ or $30,200
Year 1984 $X = 13$,   since $1984 − 1971 = 13$
$Y_c = 15.4 + 1.85(13) = 39.45$ or $39,450

If the forecast is based on the trend equation,

$$\log Y_c = 1.0915 + 0.0650X \text{ with origin also on July 1, 1971}$$

or year $1971 = 0$, obtained from the same data (Table 22-2), the computation is as follows:

$$\text{Year 1979} \quad \log Y_c = 1.0915 + 0.0650(8) = 1.6115$$
$$Y_c = 40.9, \text{ or } \$40,900$$
$$\text{Year 1984} \quad \log Y_c = 1.0915 + 0.0650(13) = 1.9365$$
$$Y_c = 86.4, \text{ or } \$86,400$$

Notice that the selection of trend equation has a great effect on a forecast by the extrapolation method, especially if the forecast is for a time some distance from the last year included in the series. The arithmetic straight-line trend gives the forecast for 1979 at \$30,200, and 1984 at \$39,450, whereas the logarithmic straight-line trend gives the forecast for 1979 at \$40,900 and 1984 at \$86,400. Thus an accurate forecast cannot be reached if it is solely based on the extrapolation method. Other factors that affect the future activity must also be carefully considered in forecasting any activity.

★22.6   Changing Trend Equations

In applying a trend equation, we must define the three factors: the origin, the X unit, and the Y unit. Each of the factors may be redefined to facilitate computing trend values. However, the changes should not affect the trend values. For simplicity, only the first degree linear equations are used in the following illustrations.

Shifting Origin

Instead of placing the origin in the middle year of a series, we may shift the origin to any desired time point. To shift the origin, first compute the trend value on the desired time point based on the old trend equation. The computed trend value is the value of a. Next, use the computed a and the same b to write a new equation. The b value (the slope) is not affected by shifting the origin. For example, the equation

$$Y_c = 15.4 + 1.85X$$
$$\text{with origin: July 1, 1971}$$
$$X \text{ unit: 1 year}$$
$$Y \text{ unit: } \$1,000$$

may be changed to have its origin on July 1, 1964 as follows:

First, the trend value of 1964 is computed from the old equation.

$$\text{Year 1964, } \quad X = -7$$
$$Y_c = 15.4 + 1.85(-7) = 2.45$$

Then, let the computed value 2.45 = *a* of the new equation. The new equation with the same *b* (=1.85) is

$$Y_c = 2.45 + 1.85X$$
with origin: July 1, 1964
X unit: 1 year
Y unit: $1,000

The shift of the origin from 1971 to 1964 is diagrammed in Chart 22–11.

CHART 22–11

ILLUSTRATION OF SHIFTING TREND ORIGIN

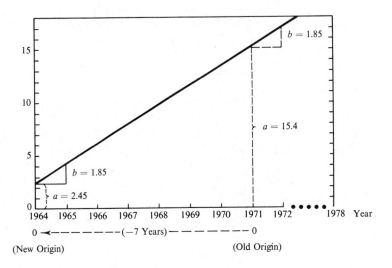

Changing X Unit

Secular trends are usually fitted to annual data. However, a trend line based on annual data may be converted to represent the trend values on a quarterly or monthly basis. To do this type of conversion, we must redefine *X* unit and change the trend equation. Thus, instead of defining *X* unit in one year, we may define the time unit in one quarter (1/4 year) or one month (1/12 year) if it is so desired. The method of converting the trend equation for annual data to that for quarterly or monthly data is illustrated under two different situations.

Actual Y Values Are Expressed in Annual Totals

First, change each value (*a*, *b*, and *X*) on the right side of the trend equation on an annual basis to the desired time unit basis. If it is changed to a quarterly basis, each value is divided by 4; if it is changed to a monthly basis, each value is divided by 12. Next, substitute the computed values in equation $Y_c = a + bX$ to obtain the new equation. The procedure of converting an old trend equation to a new one is illustrated by using the equation

$$Y_c = 15.4 + 1.85X$$
with origin: July 1, 1971
X unit: 1 year
Y unit: $1,000

of Example 3 as follows:

1. *Changing X unit from 1 year to 1 quarter.*
   First, divide each of the values $a = 15.4$, $b = 1.85$, and $X = 1$ year by 4, or

$$Y_c = \frac{15.4}{4} + \frac{1.85}{4}\left(\frac{X}{4}\right)$$

Then,

$$Y_c = 3.85 + \frac{1.85X}{16} \quad \text{or} \quad Y_c = 3.85 + 0.115625X$$
with origin: July 1, 1971
X unit: 1 quarter
Y unit: $1,000

It is rather inconvenient to compute the trend values by having the origin at the beginning of the third quarter (July 1) of 1971. Thus, we shift the origin to the middle of the quarter; that is, $1/2$ quarter away from July 1, or on August 15, 1971. The revised equation is

$$Y_c = 3.85 + 0.115625(1/2) + 0.115625X \quad \text{or}$$
$$Y_c = 3.9078125 + 0.115625X$$
with origin: August 15, 1971
X unit: 1 quarter
Y unit: $1,000

Thus, for example, the trend value on May 15, 1972 (or 3 quarters after the origin, August 15, 1971) is computed from the revised equation:

$$X = 3 \text{ (quarters)}$$
$$Y_c = 3.9078125 + 0.115625(3) = 4.2546875 \quad \text{or} \quad \$4,254.6875$$

2. *Changing X unit from 1 year to 1 month*
   First, divide each of the values a, b, and X by 12, or

$$Y_c = \frac{15.4}{12} + \frac{1.85}{12}\left(\frac{X}{12}\right)$$

Then,

$$Y_c = 1.28\tfrac{1}{3} + \frac{1.85X}{144} \quad \text{or} \quad Y_c = 1.28\tfrac{1}{3} + 0.012847\tfrac{2}{9}X$$
with origin: July 1, 1971
X unit: 1 month
Y unit: $1,000

We shift the origin to the middle of the month; that is, $1/2$ month away from

July 1, or on July 15, 1971. The revised equation is

$$Y_c = 1.28\tfrac{1}{3} + 0.012847\tfrac{2}{9}(\tfrac{1}{2}) + 0.012847\tfrac{2}{9}X \quad \text{or}$$
$$Y_c = 1.289756\tfrac{17}{18} + 0.012847\tfrac{2}{9}X$$

with origin: July 15, 1971
X unit: 1 month
Y unit: $1,000

The trend value on May 15, 1972 (or 10 months after the origin, July 15, 1971) is then computed from the revised equation:

$$X = 10 \text{ (months)}$$
$$Y_c = 1.289756\tfrac{17}{18} + 0.012847\tfrac{2}{9}(10) = 1.418229\tfrac{1}{6} \quad \text{or} \quad \$1,418.229\tfrac{1}{6}$$

The monthly trend value on May 15, 1972 may be converted to quarterly trend value on the same date by multiplying the Y_c by 3, or

$$\$1,418.229\tfrac{1}{6} \times 3 = \$4,254.6875$$

which is the same value as that obtained by the revised equation on the quarterly basis.

Also, the monthly trend value on May 15, 1972 may be converted to annual trend value on the same date by multiplying the Y_c by 12, or

$$\$1,418.229\tfrac{1}{6} \times 12 = \$17,018.75$$

which should be the same value as that obtained by the original equation on the annual basis. This statement is checked below:

$$X = \frac{10\tfrac{1}{2}}{12} = 0.875$$

(May 15, 1972 is $10\tfrac{1}{2}$ months after July 1, 1971, the origin of the equation)

$$Y_c = 15.4 + 1.85(0.875) = 17.01875, \quad \text{or} \quad \$17,018.75$$

The above illustrations for changing X unit from 1 year to 1 quarter and 1 month are diagrammed in Chart 22–12.

Actual Y Values Are Expressed in Quarterly or Monthly Averages

If the Y values are expressed in averages, only the X value is replaced by an appropriate time unit. If the averages are quarterly data, the annual X unit is divided by 4; if they are monthly data, it is divided by 12. For example, if the original equation is

$$Y_c = 4.36 + 12.72X$$

with origin: July 1, 1971
X unit: 1 year
Y unit: $1,000 (monthly averages)

the new equation with a monthly time unit is

CHART 22–12

ILLUSTRATION OF CHANGING X UNIT FROM 1 YEAR TO 1 QUARTER AND
1 MONTH FOR ACTUAL Y VALUES EXPRESSED IN ANNUAL TOTALS

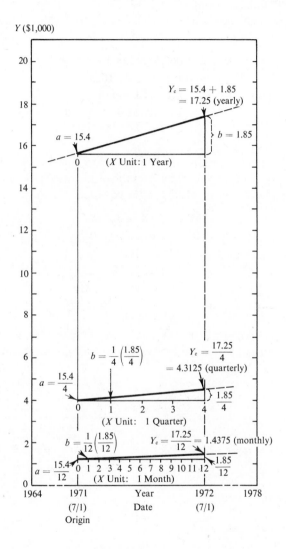

$$Y_c = 4.36 + 12.72\left(\frac{X}{12}\right) = 4.36 + 1.06X$$

with origin: July 1, 1971
X unit: 1 month
Y unit: $1.000 (monthly averages)

Note that the b value (the slope) is not affected by the change of the original equation. The slope representing 12.72 (in $1,000) increase per year in the original equation is the same slope representing 1.06 increase per month in the new equation. See the diagram in Chart 22–13.

If the origin of the new equation is shifted to the middle of the month, or on July 15, 1971, then

$$Y_c = 4.36 + 1.06(\tfrac{1}{2}) + 1.06X = 4.89 + 1.06X$$
with origin: July 15, 1971
X unit: 1 month
Y unit: $1,000 (monthly averages)

Converting Y Unit

The Y unit of a trend equation may be changed to a desired unit by multiplying the individual constants (a and b) by the ratio of the original Y unit to the desired unit. For example, Y unit of $1,000 in the equation

$$Y_c = 15.4 + 1.85X$$
with origin: July 1, 1971
X unit: 1 year
Y unit: $1,000
$a = 15.4$ and $b = 1.85,$

CHART 22–13

ILLUSTRATION OF CHANGING X UNIT FROM 1 YEAR TO 1 MONTH FOR
ACTUAL Y VALUES EXPRESSED IN MONTHLY AVERAGES

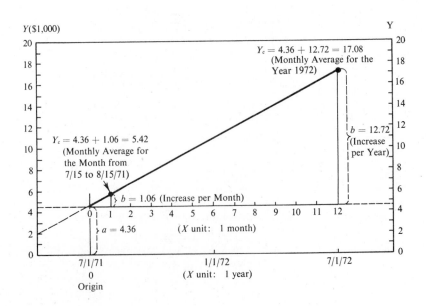

may be changed to a new Y unit of $100 as follows:

$$\text{Ratio} = \frac{Y \text{ unit (original)}}{Y \text{ unit (desired)}} = \frac{\$1,000}{\$100} = 10$$

The new equation is

$$Y_c = 15.4(10) + 1.85(10)X = 154 + 18.5X$$
$$\text{with origin: July 1, 1971}$$
$$X \text{ unit: 1 year}$$
$$Y \text{ unit: } \$100$$

22.7 Summary

A time series representing a particular business or economic activity is the result of interactions of many changing forces. The forces are usually investigated for decision-making after the movements of a time series are separated into the following four basic patterns: (1) *Secular trend* points out the direction of a time series movement over a long period of time. It may be an upward or a downward movement. (2) *Seasonal variation* represents repeating periodic movement of a time series. The length of the unit period is less than a year, such as a quarter, a month, or a day. (3) *Cyclical fluctuations* (also called business cycles) indicate the expansions (ups) and the contractions (downs) of a business activity around the normal value. The length of each cycle is nonfixed and had an average of 52 months in the United States during the period from 1954 to 1970. (4) *Irregular movements* (or erratic movements) represent all types of movements of a time series other than the above three types.

The methods of obtaining trends for a time series are presented in this chapter. Straight-line trends may be obtained by three methods: (1) Freehand graphic method requires personal judgment in drawing a straight-line trend. It is too subjective and frequently it cannot obtain a satisfactory result, especially when the artist is not an experienced statistician. (2) Method of semiaverages is the simplest method to finding the straight-line trend without involving subjective judgment in drawing the line. The original data are divided into two equal groups and the mean of each group is computed. (3) Method of least squares produces straight-line trend with the best fit for the data, because the sum of the squared deviations of individual values from the trend line is the least.

Nonlinear trends may be obtained by two methods: (1) Second-degree polynomial equation, where the unknown constants in the equation are also obtained by the method of least squares. The sum of the squared deviations from the second-degree parabola is even smaller than that from the straight-line trend by the same method. However, the computation for a parabola is more laborious than that for a straight line. (2) Method of moving averages, which will smooth out fluctuations in a time series. The smoothed curve can be used to indicate the general direction of the trend. The trend obtained by this method cannot easily be expressed by a mathematical equation. This method will give the most effective

SUMMARY OF FORMULAS

Application	Formula	Formula Number	Reference Page
Straight-line Trend			
Trend equation	$Y_c = a + bX$	22–1	561
Solution for formula (22–1) by method of least squares	I. $na + b\sum X = \sum Y$ II. $a\sum X + b\sum(X^2) = \sum(XY)$	22–2	567
Simplified formula (22–2), $\sum X = 0$	$a = \dfrac{\sum Y}{n}$	22–3a	570
	$b = \dfrac{\sum(XY)}{\sum(X^2)}$	22–3b	570
Second-Degree Parabolic Trend			
Trend equation	$Y_c = a + bX + cX^2$	22–4	575
Solution for formula (22–4) by method of least squares	I. $na + b\sum(X) + c\sum(X^2) = \sum(Y)$ II. $a\sum(X) + b\sum(X^2) + c\sum(X^3) = \sum(XY)$ III. $a\sum(X^2) + b\sum(X^3) + c\sum(X^4) = \sum(X^2 Y)$	22–5	575
Simplified formula (22–5), $\sum X = 0$ and $\sum(X^3) = 0$	$a = \dfrac{\sum(Y) - c\sum(X^2)}{n}$	22–6a	576
	$b = \dfrac{\sum(XY)}{\sum(X^2)}$	22–6b	576
	$c = \dfrac{n\sum(X^2 Y) - \sum(X^2)\cdot\sum(Y)}{n\sum(X^4) - \sum(X^2)^2}$	22–6c	576
Exponential Trend			
Trend equation General form	$Y_c = ab^X$	22–7a	582
Logarithmic form	$\log Y_c = \log a + (\log b)X$	22–7b	582
Solution for formula (22–7b) by the method of least squares	$\log a = \dfrac{\sum(\log Y)}{n}$	22–8a	583
	$\log b = \dfrac{\sum(X\cdot\log Y)}{\sum X^2}$	22–8b	583
Gompertz Curve			
Trend equation General form	$Y_c = a\cdot b^{(c^X)}$	22–9a	587
Logarithmic form	$\log Y_c = \log a + (\log b)\cdot c^X$	22–9b	587
Solution for formula (22–9b) by the three subgroup sums of the logarithms	$\log a = \dfrac{1}{m}\left(\sum_1 \log Y - \dfrac{D_1}{c^m - 1}\right)$	22–10a	588
	$\log b = \dfrac{D_1(c - 1)}{(c^m - 1)^2}$	22–10b	588
	$c^m = \dfrac{D_2}{D_1}$	22–10c	588
	where $D_1 = \sum_2 \log Y - \sum_1 \log Y$ $D_2 = \sum_3 \log Y - \sum_2 \log Y$ $m = \dfrac{n}{3}$ $n = $ number of Y values		

description of a trend when the period for computing each average is equal to, or a multiple of, the average length of the cycles (fluctuations) of a series.

The trends discussed above are solely based on arithmetic scales. Two other types of trends, which are commonly computed by logarithms and may be plotted on either an arithmetic chart or a semilog chart, are:

(1) Exponential trend, which is a straight line on a semilog chart (showing rate of increase) and is a nonlinear curve on an arithmetic chart (showing amount of increase). (2) Growth curves, which can best describe many types of time series concerning business and economic activities. The curves plotted on an arithmetic chart will show the amount of increase in a pattern: small in early years, increasingly greater in middle years, and large but stabilized in current years. The growth curves plotted on a semilog chart show the rapid increasing rates in the earlier stages but the gradual declining rates in the later stages. The two well-known growth curves since the 1920's are the Gompertz curve and the logistic or Pearl-Reed curve. The formulas used by the various methods are summarized at the end of this section.

There are three important reasons for analyzing the trend of a time series. The understanding of the reasons is important in selecting an appropriate method for a trend. The reasons are to know the historical trend of an activity, to compare the trends of various groups of data or the trends of different periods of the same data, and to forecast the activity in the future.

In applying a trend equation, we must define the three factors: the origin, the X unit, and the Y unit. Each of the factors may be redefined to facilitate the computation for trend values. However, the computed trend values must not be affected by the change of the equation due to new definitions.

Exercises 22–1

Reference: Sections 22.1 to 22.2

1.   For what purpose do we analyze time series? What are the basic patterns of time series movements? Explain briefly.

2.   Draw a straight line on an arithmetic chart for each of the following equations.
   (a) $Y = 3 + 1.25X$
   (b) $Y = -1 + 0.75X$
   (c) $Y = 2 - 0.25X$
   (d) $Y = -4 - 0.5X$

3.   The following table shows the sales of Ellis Hardware Store for the years 1970 to 1978. Plot the data on an arithmetic chart.

Year	Sales
1970	$ 2,000
1971	6,000
1972	7,000
1973	11,000
1974	12,000
1975	15,000
1976	14,000
1977	17,000
1978	16,000

(a)   Use the freehand graphic method:
   (1)   Draw a straight-line trend on the chart (Line I).
   (2)   Derive the trend equation. (Let July 1, 1970, be the origin, X unit be 1 year, and Y unit be $1,000.)
   (3)   Compute the trend values by using the equation for the years 1970, 1972, and 1976.

(b)   Use the method of semiaverages:
   (1)   Draw a straight-line trend on the chart (Line II).
   (2)   Derive the trend equation. (Let January 1, 1977, be the origin, X unit be 1/2 year, and Y unit be $1,000.)
   (3)   Compute the trend values by using the equation for the years 1970, 1972, and 1976.

(c)   Use the method of least squares:
   (1)   Derive the straight-line trend equation. (Let July 1, 1974, be the origin, X unit be 1 year, and Y unit be $1,000.
   (2)   Compute the trend values by using the equation for the years 1970, 1972, and 1976.
   (3)   Plot the trend values on the chart and draw a straight-line trend (Line III).

4.   Refer to the sales for the years 1971 to 1978 (8 years) in the table of Problem 3. Use the method of least squares:
   (a)   Derive the trend equation by letting January 1, 1975, be the origin, X unit be 1/2 year, and Y unit be $1,000.
   (b)   Compute the trend values by using the equation for years 1972 and 1976.

5.   The following table shows the expenditures on new equipment in Miller Company from the years 1968 to 1977.
   (a)   Plot the data on an arithmetic chart.
   (b)   Derive the straight-line trend equation by the method of least squares.
   (c)   Draw a straight-line trend based on the equation.

Year	Expenditures (In $1,000)	Year	Expenditures (In $1,000)
1968	34	1973	61
1969	37	1974	65
1970	39	1975	68
1971	45	1976	76
1972	52	1977	80

6. The following table gives the price indexes of commodity M in Bounty County from the years 1965 to 1976.

Year	Index (1960 = 100)	Year	Index (1960 = 100)
1965	106	1971	156
1966	109	1972	157
1967	110	1973	166
1968	118	1974	173
1969	124	1975	177
1970	143	1976	172

(a) Plot the data on an arithmetic chart.
(b) Derive the straight-line trend equation by the method of least squares.
(c) Draw a straight-line trend based on the equation.

★Exercises 22–2

Reference: Sections 22.3 to 22.6

1. The following table shows the ordinary life insurance and industrial life insurance in force in District Q for the years 1960 to 1978.
  (a) Plot the two series on the same arithmetic chart.
    (1) What shape of trend fits best to the series of ordinary life insurance? industrial life insurance?
    (2) How would you compare the two series on the chart?
  (b) Plot the two series on the same semilog chart.
    (1) What shape of the trend fits best to the series of ordinary life insurance? industrial life insurance?

(2) How would you compare the two series on the chart?

Year	Ordinary Life Insurance (billion dollars)	Industrial Life Insurance (billion dollars)
1960	149	34
1961	159	36
1962	171	38
1963	185	39
1964	198	40
1965	217	41
1966	238	41
1967	265	41
1968	288	41
1969	316	41
1970	340	40
1971	364	40
1972	389	40
1973	419	41
1974	456	41
1975	498	40
1976	539	40
1977	583	40
1978	630	39

2. Refer to the table in Problem 3, Exercises 22–1, showing the sales of Ellis Hardware Store for the years 1970 to 1978. Plot the data on an arithmetic chart.
 (a) Use the method of least squares:
   (1) Derive the second-degree ploynomial equation. (Let July 1, 1974, be the origin, X unit be 1 year, and Y unit be $1,000.)
   (2) Compute the trend value for each year by the equation.
   (3) Plot the trend values on the chart to draw a second-degree parabola.
 (b) Use the method of moving averages:
   (1) Find the 3-year moving averages.
   (2) Draw a trend curve through the averages.
   (3) indicate the direction of the trend of the series.

3. Use the same data as indicated in Problem 2 above. Plot the data on a semilog chart and on another arithmetic chart.
 (a) Use the method of least squares:
   (1) Find the logarithmic straight-line trend equation. (Let July 1, 1974, be the origin, X unit be 1 year, and Y unit be $1,000.)
   (2) Compute the trend value for each year by the equation.
   (3) Plot the trend values on each chart and draw a trend line or curve.
 (b) Use the method of three subgroup sums of the logarithms:

(1) Find the equation for the Gompertz curve. (Let July 1, 1970, be the origin, X unit be 1 year, and Y unit be $1,000.)

(2) Compute the trend value for each year by the equation.

(3) Plot the trend values on each chart and draw a trend curve.

4. Use the extrapolation method to forecast the trend values of the year 1984 from (a) the second-degree polynomial trend equation obtained in Problem 2(a), and (b) the logarithmic straight-line trend equation obtained in Problem 3(a). Should you rely completely on the extrapolation method in forecasting the sales in 1984? Explain.

5. Change the following trend equation:

$$Y_c = 27.6 + 2.88X$$

with origin: July 1, 1974

X unit: 1 year

Y unit: $5,000 (in annual totals)

to a new equation in each case when

(a) the origin is shifted to July 1, 1979

(b) the X unit is changed to:

(1) 1 quarter and the origin is shifted to the center of the third quarter of 1974.

(2) 1 month and the origin is shifted to the center of July, 1974.

(c) only the Y unit is converted to $1,000.

6. Change the following trend equation:

$$Y_c = 15.6 + 32.4X$$

with origin: July 1, 1976

X unit: 1 year

Y unit: $1,000,000 (in quarterly averages)

to a new equation in each case when:

(a) the X unit is changed to 1 quarter, and the origin is shifted to August 15, 1976.

(b) the X unit is changed to 1 month, and the origin is shifted to July 15, 1976.

23 Seasonal Variation

A time series classified into periods of less than one year such as quarters, months, or weekdays, may have repeating periodic movement. This movement is called seasonal variation. For example, a 15-year series of certain business activity classified into months may persistently show the highest amount of activity in July and the lowest amount in May in each of the 15 years. The methods of detecting the existence of seasonal variation in a series are discussed in Section 23.1.

The measures of seasonal variation are called *seasonal indexes* (percents). The methods of obtaining seasonal indexes are presented in Sections 23.2 to 23.5. In practice, quarterly or monthly data are commonly used for measuring seasonal variations. For simplicity, quarterly data are used in illustrating the basic principles for computing the seasonal indexes in these sections. Monthly data are used only for additional illustrations. The applications and uses of seasonal indexes are given in the remaining section (23.6) of this chapter.

23.1 Detecting Seasonal Variation

A time series may or may not have seasonal variation. Thus, before computing an index, a time series should be carefully examined for the variation in order to save time from unnecessary computation. A simple method to examine the variation is to compare individual values with the average of the values for each year. This can be done either with a table or a chart. Table 23–1 and Chart 23–1 show the quarterly sales of Fred Department Store for the years 1974 to 1978. The table and chart also indicate the comparisons of individual quarterly sales with the average quarterly sales for each year. The sales for the first quarters are below their respective averages, the sales in the second and the fourth quarters are close to the averages, and the sales in the third quarters are far above the averages. This clearly shows the existence of a seasonal variation in the series.

Many time series are affected by the number of calendar or working days in the time unit (such as January having 31 days and February having 28 days). In such a case, the time series should be adjusted for the calendar or working day variation before deciding whether there is a seasonal variation in the series. The adjustment may be made by classifying the series into daily averages as in Table 23–2. The table shows the production of Ellison Electrical Appliances Company for individual months of 1978.

TABLE 23–1

SALES OF FRED DEPARTMENT STORE FOR THE QUARTERS
OF THE YEARS 1974 TO 1978

Year	Quarterly Sales (Unit : $1,000), Y				Annual Total	Quarterly Average for the Year
	1st Quarter	2nd Quarter	3rd Quarter	4th Quarter		
1974	1(B)*	2(B)*	4(A)*	3(A)*	10	2.50
1975	2(B)	3(B)	5(A)	4(A)	14	3.50
1976	2(B)	4(A)	5(A)	3(B)	14	3.50
1977	3(B)	4(B)	7(A)	6(A)	20	5.00
1978	5(B)	7(A)	8(A)	7(A)	27	6.75
Total	13(5 B's)	(2 A's) 20(3 B's)	29(5 A's)	(4 A's) 23(1 B)	85	21.25
Y_a Average	2.6	4	5.8	4.6	17	4.25

    *(A) Indicates *above* quarterly average for the year.
    (B) Indicates *below* quarterly average for the year.

Source: Hypothetical data.

The number of working days varies each month because of Saturdays, Sundays, holidays, etc. The average number of working days per month in the year is 21, or

$$\frac{\text{Total number of working days in the year}}{12(\text{months})} = \frac{252}{12} = 21 \text{ days}$$

The average production per day in January, for example, is

$$\frac{520(\text{production})}{22(\text{working days})} = 23.64 \text{ (units of 1,000)}$$

If there is no working-day variation, there should be 21 working days during January. Thus, the production adjusted for working-day variation for January is

$$23.64 \times 21 = 496.44 \text{ or rounded to 496 (units of 1,000)}$$

If these computations are combined, the adjusted production may be computed as follows:

$$\frac{520}{22} \times 21 = 520 \times \frac{21}{22} = 520 \times 0.9545 = 496 \text{ (units of 1,000)}$$

CHART 23–1

ILLUSTRATION OF DETECTING SEASONAL VARIATION IN QUARTERLY SALES
OF FRED DEPARTMENT STORE FOR YEARS 1974 TO 1978

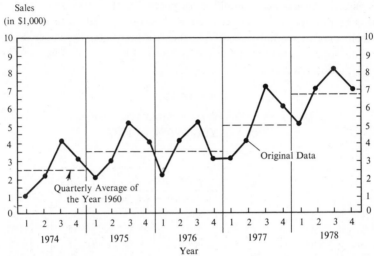

Source: Table 23–1.

The factor,

$$\frac{\text{The average number of working days per month}}{\text{The actual number of working days in January}} = \frac{21}{22} = 0.9545$$

is called the *adjusting factor* for January production, which is listed in column
(4) of Table 23–2 for simplifying the computation.

Compare the actual units produced with the units adjusted for working-day
variation for the months of January and February. The actual production in February (480,000 units) is lower than the production in January (520,000 units). The
low production in February is due to the smaller number (19) of working days
in the month. If the production were not affected by the working-day variation,
the expected production in February (531,000 units as adjusted) would be higher
than that in January (496,000 units).

Sales data are affected by calendar and working-day variations to a smaller degree than production data. Although in some cases the adjustment for the variation on sales data is desirable for seasonal analysis, the adjustment is omitted in
the following illustrations for the sake of simplicity.

The patterns of seasonal variation may be classified into two types: specific
and typical. A *specific pattern* concerns seasonal variation in a particular period
such as the variation of the quarterly sales of the year 1974 in Table 23–1;
whereas a *typical pattern* describes the *average* seasonal variation over a number
of periods, such as over the five years 1974 to 1978 in the same table. Obviously,
if the specific seasonal patterns during a number of periods are believed, after observation, to be close to each other, a single set of indexes for a typical pattern

TABLE 23–2

MONTHLY PRODUCTION OF ELLISON ELECTRICAL APPLIANCES
COMPANY ADJUSTED FOR WORKING-DAY VARIATION, 1978

(1) Month	(2) Units Produced per Month (1,000 Units)	(3) Number of Working Days	(4) Adjusting Factor 21 ÷ col. (3)	(5) Units Adjusted for Working-day Varia- tion (1,000 Units) col. (2) × col. (4)
January	520	22	0.9545	496
February	480	19	1.1053	531
March	500	21	1.0000	500
April	550	21	1.0000	550
May	500	22	0.9545	477
June	600	20	1.0500	630
July	450	21	1.0000	450
August	580	23	0.9130	530
September	490	20	1.0500	514
October	510	22	0.9545	487
November	560	21	1.0000	560
December	470	20	1.0500	494
Total		252		
Average		21		

Source: Hypothetical data.

may be conveniently used for seasonal analysis. However, if the variation among the specific seasonal patterns is large, the single set is less reliable since there is no typical seasonal movement. In such a case, it probably will be better to use the specific seasonal index for the specific time in analyzing seasonal variation.

23.2 The Method of Simple Averages of Original Data

Seasonal variation of a time series is measured *after the effects of trend, cyclical, and irregular movements on the series are eliminated.* Based on this idea, there are various methods of computing a set of seasonal indexes for a given time series. Among them, the three common methods are: (1) the method of simple averages of original data, (2) the method of simple averages adjusted for trend movement, and (3) the method of ratios to moving averages. The first method is illustrated in this section. The other two methods are presented in the next two sections individually.

The method of simple averages of original data is the simplest way to compute the index of seasonal variation. With this method, the trend movement is assumed to have little, if any, effect on the time series. It is also assumed that the upswing and downswing of the cycles in a series are fairly balanced; that is, the cycles have the same durations and amplitudes. If a sufficient number of cycles is included in the process of averaging data for each time unit (quarter or month), the cyclical fluctuations can be cancelled out by the process. Likewise, the process eliminates the effects of irregular movements by assuming that the effects are fairly balanced over the period included in the series. These assumptions are not necessarily true. However, the index obtained by this simple method usually gives a good rough estimate of the pattern of seasonal variation.

The procedure of computing the seasonal index for a time series by this method is

(1) Find the average (arithmetic mean) for each quarter (first, second, ...) from the quarterly data, or for each month (January, February, ...) from the monthly data.

(2) Compute the seasonal index from the quarterly (or monthly) averages. The seasonal index is usually expressed in percent (%) form. The total of the percents is 400% for quarterly seasonal index and is 1,200% for monthly seasonal index.

This procedure is illustrated by Example 1.

Example 1

Use the quarterly sales of the Fred Department Store for the years 1974 to 1978 given in Table 23–1. Compute the seasonal index by the method of simple averages of original data.

Solution. 1. Find the average sales for each quarter. The averages (Y_a) are computed in Table 23–1. For example, the first quarterly average is obtained by dividing the total of the sales of the first quarters 13 by 5, or $13/5 = 2.6$ (in $ 1,000). The effects of cyclical and irregular movements are thus eliminated by the averaging process. The effect of trend movement is ignored. Thus, the averages are treated as the values of a typical seasonal pattern. The typical values are now used for computing the index of the given series.

2. The computation of the index is presented in Table 23–3. The average for each quarter is listed in column (2) of the table. The index in decimal form, column (4), may be computed in either of the two ways:

TABLE 23–3

COMPUTATION OF SEASONAL INDEX BY THE METHOD OF SIMPLE
AVERAGES OF ORIGINAL DATA, FOR THE QUARTERLY
SALES OF FRED DEPARTMENT STORE, 1974 TO 1978

(1) Quarter	(2) Average Sales for the Quarter ($1,000), from Table 23–1, Y_a	(3) Ratio to Total = col.(2)÷17	(4) Index in Decimal Form=col. (3)×4 =col. (2)÷4.25	(5) Index in Percent Form(%)	(6) Index, above (+) or below (−) 100%
First	2.6	0.153	0.612	61.2	− 38.8
Second	4.0	0.235	0.940	94.0	− 6.0
Third	5.8	0.341	1.364	136.4	+ 36.4
Fourth	4.6	0.271	1.084	108.4	+ 8.4
Total	17.0	1.000	4.000	400.0	0.0
Average	4.25		1.000	100.0	

Source: Table 23–1 and Example 1.

CHART 23–2

SEASONAL INDEXES COMPUTED BY VARIOUS METHODS, FOR QUARTERLY
SALES OF FRED DEPARTMENT STORE, 1974 TO 1978

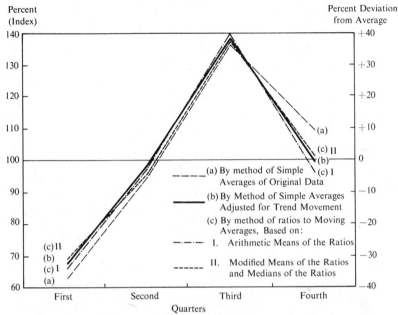

Source: (a) Example 1, Table 23–3. (b) Example 2, Table 23–4. (c) Example 3,
Table 23–10 for I and Table 23–12 for II.

a. First, express each quarterly average as a ratio to the total of the 4 quarterly averages. This is done in column (3). For example, the ratio of the first quarterly average is obtained by dividing 2.6 by 17, or $2.6 \div 17 = 0.1529$, which is
rounded to 0.153 in the table. The sum of the 4 ratios should be 1. Next, multiply
each ratio by 4 to get the index in a decimal form. Thus, $0.153 \times 4 = 0.612$ (index of the first quarter).

b. Find the average of the 4 quarterly averages, or $17 \div 4 = 4.25$. Next express each quarterly average as a ratio to the average of the 4 quarterly averages
to get the index in a decimal form. For example, the ratio of the first quarterly
average is obtained: $2.6 \div 4.25 = 0.612$.

For convenience, a decimal index is changed to a percent index (column (5))
by moving the decimal point two places to the right and annexing the percent
sign (%), such as

$.612 = 61.2\%$, or written as 61.2 in the percent (%) column.

The seasonal index in percent form may also be written as above or below the
average of the 4 quarterly averages ($= 100\%$). This is shown in column (6) of
Table 23–3 and in Chart 23–2.

23.3 The Method of Simple Averages Adjusted for Trend

The procedure of computing seasonal index by the method of simple averages adjusted for trend movement is basically the same as that by the method presented in Section 23.2. The effects of cyclical and irregular movements are again eliminated by the process of averaging data for each unit of time (as in the previous method). Thus, it is again assumed that the cyclical and the irregular fluctuations in a series are fairly balanced and can be cancelled out by including a sufficient number of years in the process of averaging. However, the effect of a trend movement is not to be ignored by this method. It is assumed that the effects of various forces on a series are based on the additive model

$$Y = T + S + C + I \qquad\qquad (23\text{--}1)$$

After the elimination of cyclical (C) and irregular (I) fluctuations by the averaging process, the model may be written

$$Y_a = T_a + S_a, \text{ or } S_a = Y_a - T_a$$

where Y_a represents a simple average of Y values for a quarter (first, second, . . .) and T_a represents a corresponding average deviation (increase or decrease) for each quarter due to trend movement. The result of the subtraction ($S_a = Y_a - T_a$) is a typical seasonal or an average adjusted for trend movement. The values of S_a are used for computing the seasonal index.

The average deviation T_a is computed from a trend equation. The trend equation may be obtained by various methods as presented in the previous chapter. In the following illustration, only the straight line trend equation by the method of least squares is used. Thus, the average deviation for each quarter is computed from the b value in the equation $Y_c = a + bX$.

The assumptions concerning the patterns of cyclical and irregular fluctuations in a series by this method are again not necessarily true. Furthermore, this method is limited to trend movement in a linear form. However, the index obtained by this method usually is very close to the index obtained by the most precise yet tedious method, the ratio to moving average method, which is discussed in the following section. The method of simple averages adjusted for trend is illustrated in detail by Example 2.

Example 2

Use the quarterly sales of the Fred Department Store for the years 1974 to 1978 given in Table 23–1. Compute the seasonal index by the method of simple averages adjusted for trend movement.

Solution. Table 23–4 is set for computing the seasonal index by this method. The computation is as follows:

1. Find the average sales (arithmetic mean) for each quarter. The averages (Y_a) are obtained from Table 23–1 and are listed in column (2) of Table 23–4. The effects of cyclical and irregular movements are thus eliminated by the aver-

TABLE 23–4

COMPUTATION OF SEASONAL INDEX BY THE METHOD OF SIMPLE
AVERAGES ADJUSTED FOR TREND MOVEMENT, FOR THE QUARTERLY
SALES OF FRED DEPARTMENT STORE, 1974 TO 1978

(1) Quarter	(2) Average Sales for the Quarter $Y_a(\$1{,}000)$	(3) Average Increase due to Trend, 0.25 (in $1,000) per quarter T_a	(4) Typical Seasonal Pattern ($1,000), $Y_a - T_a = S_a$	(5) Index in Decimals $= col.(4)$ $\div 3.875$	(6) Index in Percents (%)	(7) Index, above (+) or below (−) 100%
First	$2.6	$0.00	$2.60	0.671	67.1	− 32.9
Second	4.0	0.25	3.75	0.968	96.8	− 3.2
Third	5.8	0.50	5.30	1.367*	136.7	+ 36.7
Fourth	4.6	0.75	3.85	0.994	99.4	− 0.6
Total	17.0	1.50	15.50	4.000*	400.0	0.0
Average			3.875	1.000	100.0	

*The largest index 1.368 is adjusted to 1.367 so that the total of the column is forced to 4.000.

Source: Table 23–1 and Example 2.

aging process. The averages then form a typical seasonal pattern before adjusting for trend movement.

2. Find the average deviation for each quarter due to trend movement. First, obtain the straight-line trend equation for the quarterly data of 1974 to 1978 by the method of least squares. This is done in Table 23–5. The trend equation is

$$Y_c = 4.25 + 1\,X, \text{ with origin: July 1, 1976}$$
$$X \text{ unit: 1 year}$$
$$Y \text{ unit: } \$\,1{,}000 \text{ (for quarterly data)}$$

TABLE 23–5

COMPUTATION OF STRAIGHT–LINE TREND EQUATION BY THE METHOD
OF LEAST SQUARES, BASED ON THE QUARTERLY AVERAGE SALES IN
FRED DEPARTMENT STORE FOR THE YEARS 1974 TO 1978

(1) Year Original Number	(2) X	(3) Quarterly Average for the Year $Y(\$1{,}000)$	(4) XY	(5) X^2	(6) Y_c	
1974	− 2	$2.50	− 5.00	4	2.25	$a = \dfrac{\Sigma Y}{n} = \dfrac{21.25}{5} = 4.25$
1975	− 1	3.50	− 3.50	1	3.25	$b = \dfrac{\Sigma (XY)}{\Sigma (X^2)} = \dfrac{10}{10} = 1$
1976	0	3.50	0	0	4.25	$Y_c = 4.25 + 1X,$ with
1977	1	5.00	5.00	1	5.25	origin : July 1,1972
1978	2	6.75	13.50	4	6.25	X unit : 1 year
Total	0	21.25	10.00	10	21.25	Y unit : $1,000

Source: Table 23–1.

The b value is 1, which shows the annual increase since X unit is 1 year. Thus, the increase per quarter is \$250, or $1/4 = 0.25$ in \$1,000. The average quarterly increases due to trend movement (T_a) are listed in column (3) of Table 23–4. The average sales for the first quarter is the base, or no increase. The average sales for successive quarters are increased 0.25 per quarter, or $2(0.25) = 0.50$ for the third quarter and $3(0.25) = 0.75$ for the fourth quarter.

3. Subtract T_a from Y_a to obtain S_a. The result of the subtraction is shown in column (4). The values in the column form a typical pattern of seasonal variation after the trend, cyclical, and irregular movements are eliminated. The average of the typical seasonals is $15.50/4 = 3.875$ per quarter (in \$1,000).

4. Divide each adjusted average, S_a, by 3.875. The quotients listed in column (5) are the seasonal index in decimal form. The decimal form is expressed in percent form in column (6). The index in percent form above or below the average (100%) is presented in column (7) of the table and is also shown in Chart 23–2.

23.4   The Method of Ratios to Moving Averages

This method is based on the assumption that the result of the effects of the four forces on a time series is a product, rather than a sum. The result may be written in a multiplicative model

$$Y = T \times S \times C \times I \tag{23-2}$$

A seasonal index now may be obtained by eliminating T, C, and I in a division form as follows:

$$\frac{T \times S \times C \times I}{T \times C \times I} = S$$

where S is a relative value. The relative values of S are used in computing the seasonal index. The methods of computing seasonal indexes based on absolute values of the data as described in the preceding two sections may not be satisfactory in some cases. A few large values in a series, for example, may have a disproportionate effect upon the typical seasonal pattern obtained by the averaging process. An index computed from relatives is affected to a smaller degree by a few large values in a series and thus is considered more satisfactory for describing a typical seasonal pattern.

The elimination of $T \times C \times I$ may be done by various methods. The most satisfactory method commonly used by statisticians is the method of ratios of original data (Y values) to moving averages, or the *ratio to moving average* method.

The length of time for computing each moving average is a year; that is, 4 quarters for quarterly data and 12 months for monthly data. As stated in the previous chapter, the moving average method will smooth out the fluctuations most effectively if the period for computing moving average is equal to, or a multiple of, the average duration of the fluctuations in the series. The seasonal variation fluctuates regularly year after year with perhaps a slight varying pattern for each year. Thus, the 4-quarter or 12-month moving averages will eliminate the seasonal variation almost completely.

It was also stated in the previous chapter that irregular movements may be cancelled out, if the unit period covering the data is extended, such as extending 1 quarter or 1 month to a longer period, 1 year. Thus, most irregular movements may also be cancelled out by the one-year (4-quarter or 12-month) moving averages.

If the durations of the cycles are not too long, or are close to a one-year period, the one-year moving averages may also eliminate a major part of the cyclical fluctuations. However, business cycles are usually longer than one year. Cyclical fluctuations therefore cannot be eliminated by this method. Under normal conditions, then, a moving average can represent a close estimate of the product of the uneliminated factors, trend and cyclical effects, or

$$\text{A moving average} = T \times C = TC$$

The representation of TC by a moving average can be seen clearly in Chart 23–3. Notice that the smooth curve representing moving averages fluctuates (cyclical effect) upward (trend effect) in the chart.

After each moving average has been found, the ratio of Y (original value) to the moving average on the corresponding time point (usually on the center of a quarter or a month) can be obtained. The ratio now represents the product of seasonal variation and irregular movement, or

$$\text{Ratio of } Y \text{ to moving average} = \frac{TSCI}{TC} = SI$$

The effect of irregular movement, I, cannot be obtained at this stage. Instead, I (or at least part of I) is eliminated by the process of averaging the ratios for each quarter (or month). The averages of the ratios may be arithmetic means, modified means, or medians. The averages of the ratios are then used to obtain indexes.

Example 3

Use the quarterly sales of the Fred Department Store for the years 1974–1978 given in Table 23–1. Compute the seasonal index by the ratio to moving average method.

Solution. 1. Find the trend-cyclical value (TC) for each quarter by the method of moving averages. The 4-quarter moving averages are listed in column (5) of Table 23–6 and are plotted on Chart 23–3. The method of moving averages for odd-number time units, such as 3–year and 5–year moving averages, has been illustrated in the previous chapter. Here we have an even number of time units (4 quarters) in computing each moving average. The average of the first 4 quarterly sales in the series (based on the sales of 1974) is

$$\frac{1 + 2 + 4 + 3}{4} = \frac{10}{4} = 2.5, \text{ or } \$2,500$$

TABLE 23–6

COMPUTATION OF RATIOS OF QUARTERLY SALES TO
4-QUARTER MOVING AVERAGES FOR THE FRED
DEPARTMENT STORE, 1974 TO 1978

(1) Year Quarter		(2) Quarterly Sales (in $1,000) Y=TSCI	(3) 4-Quarter Moving Total	(4) Sum of Two 4-Quarter Moving Totals	(5) 4-Quarter Moving Average (4)÷8=TC	(6) Ratio of Y to Moving Average, (2)÷(5)=SI
1974	1st	1				
	2nd	2				
			—10—			
	3rd	4		21	2.625	1.524
			—11—			
	4th	3		23	2.875	1.043
			—12—			
1975	1st	2		25	3.125	0.640
			—13—			
	2nd	3		27	3.375	0.889
			—14—			
	3rd	5		28	3.500	1.429
			—14—			
	4th	4		29	3.625	1.103
			—15—			
1976	1st	2		30	3.750	0.533
			—15—			
	2nd	4		29	3.625	1.103
			—14—			
	3rd	5		29	3.625	1.379
			—15—			
	4th	3		30	3.750	0.800
			—15—			
1977	1st	3		32	4.000	0.750
			—17—			
	2nd	4		37	4.625	0.865
			—20—			
	3rd	7		42	5.250	1.333
			—22—			
	4th	6		47	5.875	1.021
			—25—			
1978	1st	5		51	6.375	0.784
			—26—			
	2nd	7		53	6.625	1.057
			—27—			
	3rd	8				
	4th	7				

Source: Table 23–1.

which represents the quarterly trend-cyclical value centered on July 1, 1974, or
the middle of the first 4 quarters. However, the *TC* value 2.5 neither corresponds
to the second quarterly sales 2 (centered on the middle of second quarter, or May
15, 1974) nor to the third quarterly sales 4 (centered on the middle of the third
quarter, or on August 15, 1974). Thus, a new way must be developed to obtain
the corresponding *TC* value for each quarter. An *ordinary method* requires the
total of 1/2 of the first, the entire second, third, and fourth, and 1/2 of the fifth
quarterly sales in order to get a moving average on the middle of the five quarters,
or centered on the third quarter. Thus, the centered moving average corresponds
to the third quarterly sales. This is illustrated in Table 23–7. However, a *simplified method* may yield the same result. The simplified method requires the sum
of once the first, twice the second, the third, and the fourth respectively and once

the fifth quarterly sales. This method is shown in Table 23–8. Notice that the moving average for the third quarter, 1974, by the two methods is the same— 2.625. The 4-quarter moving total 10.5 by the ordinary method is divided by 4, whereas the sum of the two 4-quarter moving totals by the simplified method 21 is divided by 8. The moving averages in Table 23–6 are obtained by the simplified method.

CHART 23–3

4–Quarter Moving Averages of the Quarterly Sales
in the Fred Department Store, 1974 to 1978

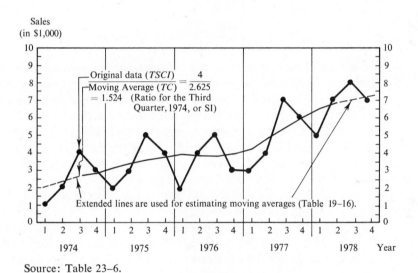

Source: Table 23–6.

TABLE 23–7

Computation of the First 4–Quarter Moving Average
in Table 23–6—By the Ordinary Method

(1) Year Quarter		(2) Quarterly Sales (in $1,000)	(3) Sales for Computing 4-Quarter Moving Total	(4) 4-Quarter Moving Total, Sum of (3)	(5) 4-Quarter Moving Average (4) ÷ 4
1974	1st	1	$(\frac{1}{2})1 = 0.5$		
	2nd	2	2		
	3rd	4	4	10.5	2.625
	4th	3	3		
1975	1st	2	$(\frac{1}{2})2 = 1.0$		

TABLE 23–8

COMPUTATION OF THE FIRST 4–QUARTER MOVING AVERAGE
IN TABLE 23–6—BY THE SIMPLIFIED METHOD

(1) Year Quarter	(2) Quarterly Sales (in $1,000)	(3) 4-Quarter Moving Total Centered	(4) Sum of Two 4-Quarter Moving Totals	(5) 4-Quarter Moving Average (4) ÷ 8
1974 1st	1			
2nd	2	10		
3rd	4	}	21	2.625
4th	3	11		
1975 1st	2			

2. Find the ratio of the quarterly sales Y to corresponding moving average, (SI). For example, the ratio of the third quarter of 1974 is obtained by dividing the Y value of the third quarter by the moving average centered on the third quarter, or $4 \div 2.625 = 1.524$. The ratios are listed in column (6) of Table 23–6.

3. Compute the seasonal index. First, average the ratios (SI). Next, compute the index from the averages. The averages may be in the forms of arithmetic means, modified means, and medians. The computations of the indexes based on the different forms of the averages are discussed below.

a. The seasonal index based on the arithmetic means is computed in Tables 23–9 and 23–10. It is also shown on Chart 23–2.

b. The seasonal index based on the modified means is computed in Tables 23–11 and 23–12. It is also shown on Chart 23–2. Each modified arithmetic mean is obtained by excluding some extremely high and low ratios. For instance,

TABLE 23–9

COMPUTATION OF ARITHMETIC MEANS OF RATIOS TO 4–QUARTER
MOVING AVERAGES, EXPRESSED IN PERCENTS (%)

Year	Quarter			
	1st	2nd	3rd	4th
1974	—	—	152.4	104.3
1975	64.0	88.9	142.9	110.3
1976	53.3	110.3	137.9	80.0
1977	75.0	86.5	133.3	102.1
1978	78.4	105.7	—	—
Total	270.7	391.4	566.5	396.7
Mean	67.7	97.8	141.6	99.2

Source: Table 23-6, column (6).

TABLE 23–10

COMPUTATION SEASONAL INDEX, BASED ON ARITHMETIC MEANS
OF RATIOS, FOR THE FRED DEPARTMENT STORE, 1974 TO 1978

Quarter (1)	Mean of Ratios (%) to 4-quarter Moving Averages (2)	Seasonal Index (%) (2) × 0.9845 (3)
1st	67.7	66.7
2nd	97.8	96.3
3rd	141.6	139.43*
4th	99.2	97.7
Total	406.3	400.0

*The largest index 139.4 is changed to 139.3 so that the total is forced to 400%.

$$\text{Adjusting factor} = \frac{400}{406.3} = 0.9845$$

if there are 7 ratios, by excluding the two highest and the two lowest ratios, according to personal judgment, the modified mean is computed from the middle three remaining ratios. In our example, the modified mean is computed by excluding the highest and the lowest ratios for each quarter. The ratios for each quarter are first arranged in an array, according to the order of values of the ratios, in Table 23–11. Next, the highest and lowest ratios are eliminated. The modified mean is then computed from the remaining ratios. The seasonal index based on the modified means is computed in Table 23–12.

c. The seasonal index based on the medians is also computed in Tables 23–11 and 23–12 and shown on Chart 23–2. The medians obtained from Table 23–11 are the same as the modified means since each modified mean is the average of

TABLE 23–11

COMPUTATION OF MODIFIED MEANS OF RATIOS TO 4–QUARTER
MOVING AVERAGES, EXPRESSED IN PERCENTS (%) AND
ARRANGED IN ORDER OF MAGNITUDE OF RATIOS

	Quarter			
	1st	*2nd*	*3rd*	*4th*
Lowest ratio	53.3	86.5	133.3	80.0
	64.0	88.9	137.9	102.1
	75.0	105.7	142.9	104.3
Highest ratio	78.4	110.3	152.4	110.3
Total of two middle ratios	139.0	194.6	280.8	206.4
Modified mean (also median)	69.5	97.3	140.4	103.2

Source: Table 23–9.

TABLE 23–12

COMPUTATION OF SEASONAL INDEX, BASED ON MODIFIED MEANS OF
RATIOS, FOR THE FRED DEPARTMENT STORE, 1974 TO 1978

Quarter (1)	Modified Mean of Ratios to 4-quarter Moving Averages, % (2)	Seasonal Index (%) (2) × 0.9747 (3)
1st	69.5	67.7
2nd	97.3	94.8
3rd	140.4	136.9
4th	103.2	100.6
Total	410.4	400.0
Average		100.0

$$\text{Adjusting factor} = \frac{400}{410.4} = .9747$$

Source: Table 23–11.

TABLE 23–13

COMPARISON OF SEASONAL INDEXES OF THE QUARTERLY SALES IN THE
FRED DEPARTMENT STORE, 1974 TO 1978, BY VARIOUS METHODS

Quarter	Indexes (%)				
	Simple Averages of Original Data Method	Simple Averages Adjusted for Trend Method	Ratios to 4-quarter Moving Averages		
			Arithmetic Means	Modified Means	Medians
1st	61.2	67.1	66.7	67.7	67.7
2nd	94.0	96.8	96.3	94.8	94.8
3rd	136.4	136.7	139.3	136.9	136.9
4th	108.4	99.4	97.7	100.6	100.6

Source: Tables 23–3, 23–4, 23–10, and 23–12.

the two middle ratios in each column. Thus, the seasonal index based on the medians is the same as that based on the modified means in the example.

Table 23–13 and Chart 23–2 show the comparison of the indexes for the same data obtained by the various methods presented in Examples 1 to 3. The indexes obtained by the method of simple averages adjusted for trend and by the ratio to moving average method are fairly close to each other. Note that if there are extreme values in actual units in a time series, the indexes obtained by the two methods may differ to a greater degree. The table and the chart also show that the indexes of the first and the fourth quarters obtained by the method of simple averages of original data are the lowest and the highest respectively among the indexes obtained by the various methods. This fact is usually true since this method ignores the effect of the upward trend movement.

For monthly data, the procedure of computing a seasonal index by the ratio to moving average method is similar to that in Example 3, except that the unit

TABLE 23–14

COMPUTATION OF RATIOS OF MONTHLY EMPLOYMENT IN GENERAL
CONSTRUCTION COMPANY TO 12–MONTH MOVING AVERAGES,
1977 TO 1978

(1) Year Month	(2) Employment (in 1,000 persons) $Y = TSCI$	(3) 12-Month Moving Total	(4) Sum of Two 12-Month Moving Totals	(5) 12-Month Moving Average $(4) \div 24 = TC$	(6) Ratio of Y to Moving Average, or Specific Seasonal $(2) \div (5) = SI$
1977 January	2.58				
February	2.47				
March	2.56				
April	2.85				
May	3.05				
June	3.23				
		36.36			
July	3.36		72.77	3.032	1.108
		36.41			
August	3.44		73.03	3.043	1.130
		36.62			
September	3.38		73.44	3.060	1.105
		36.82			
October	3.33		73.77	3.074	1.083
		36.95			
November	3.18		74.04	3.085	1.031
		37.09			
December	2.93		74.32	3.097	0.946
		37.23			
1978 January	2.63		74.59	3.108	0.846
		37.36			
February	2.68		74.83	3.118	0.860
		37.47			
March	2.76		75.01	3.125	0.883
		37.54			
April	2.98		75.18	3.132	1.019
		37.64			
May	3.19		75.38	3.141	1.073
		37.74			
June	3.37		75.58	3.149	1.108
		37.84			
July	3.49				
August	3.55				
September	3.45				
October	3.43				
November	3.28				
December	3.03				

Total 12.192

period used in computing the moving averages is 12 months. Table 23–14 shows
the computation of 12–month moving averages for the data concerning the employment in General Construction Company from January, 1977 to December,
1978. The moving total for the first 12 months (in 1977) of the series is 36.36,
which is centered at the middle of the 12 months, or on July 1, 1977. The successive moving totals centered may be computed from the total by the short

method illustrated on page 579, or

36.36 (first moving total)	36.41 (second moving total)
$-$ 2.58 (January, 1977)	$-$ 2.47 (February, 1977)
$+$ 2.63 (January, 1978)	$+$ 2.68 (February, 1978)
36.41 (second moving total)	36.62 (third moving total), and so on.

23.5 Analysis of Changing Seasonal Pattern

When a typical seasonal index is used to describe the seasonal variation of a time series, it is assumed that there has been no pronounced change in the seasonal pattern. However, a seasonal pattern may change abruptly or gradually due to the changes of business practices, customer buying habits, technological innovations, and governmental activities. For example, special promotional campaigns and changes in the time for introducing new models in television and automobile industries alter seasonal patterns of demand.

When the seasonal pattern of a time series changes abruptly, it is better to use a specific seasonal index for the specific time in analyzing seasonal variation. When the seasonal pattern changes gradually, there are usually two ways to describe the variation: (1) recompute the typical seasonal index in order to cover current specific seasonals and (2) compute the *changing seasonal index* based on a trend line instead of a single average of the specific seasonals for each time unit. The procedure of computing the changing seasonal index is illustrated in Example 4 below.

Example 4

Use the quarterly sales of the Fred Department Store for the years 1974 to 1978 given in Table 23–1. Compute the changing seasonal index for each year.

Solution. 1. Arrange the specific seasonal relatives (or the ratios to 4-quarter moving averages) for each quarter into individual series. This is done in Table 23–15 and Chart 23–4. Table 23–15 is essentially a reprint of Table 23–9 except

TABLE 23–15

ARRANGEMENT OF SPECIFIC SEASONAL RELATIVES (RATIOS TO 4-QUARTER MOVING AVERAGES) BY QUARTERS AND YEARS

Year	Quarter (%)			
	1st	*2nd*	*3rd*	*4th*
1974	50.0	87.0	152.4	104.3
1975	64.0	88.9	142.9	110.3
1976	53.3	110.3	137.9	80.0
1977	75.0	86.5	133.3	102.1
1978	78.4	105.7	118.0	97.0
Total	320.7	478.4	684.5	493.7
Mean	64.1	95.7	136.9	98.7

Source: Tables 23–9 and 23–16.

TABLE 23–16

COMPUTATION OF RATIOS TO ESTIMATED
4–QUARTER MOVING AVERAGES

Year	Quarter (1)	Quarterly Sales (in $1,000), Y (2)	4-Quarter Moving Average, estimated (3)	Ratio of Y to Moving Average (4)
1974	1st	1	2.00	0.500
	2nd	2	2.30	0.870
1978	3rd	8	6.80	1.180
	4th	7	7.20	0.970

Source: Table 23–6 for column (2); Chart 23–3 for the estimated 4-quarter moving averages in column (3).

CHART 23–4

ANALYSIS OF THE CHANGING SEASONAL PATTERN OF THE QUARTERLY
SALES OF 1974 TO 1978 IN THE FRED DEPARTMENT STORE
(UNIT: PERCENT RATIO TO 4–QUARTER MOVING AVERAGE)

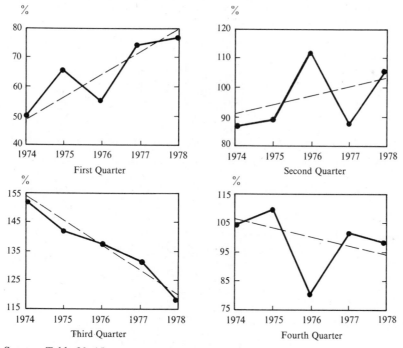

Source: Table 23–15.

the relatives of the first and second quarters of 1974 and the third and fourth quarters of 1978. The exceptions are estimated as follows and presented in Table 23–16.

First, extend the moving average line on Chart 23–3 by the dotted lines at both ends. The estimated moving averages are then read from the dotted lines.

Next, compute the ratios of the original data to the estimated 4-quarter moving averages.

2. Find a trend line to fit the specific relatives for each quarter. Any of the methods presented in the previous chapter may be used to compute the trend values for the relatives. For simplicity, the freehand graphic method is used for the trend lines on Chart 23–4.

3. Read the trend values from the trend lines. The trend values for each quarter are tabulated on Table 23–17 and are used as the changing seasonal indexes for the different years, 1974 to 1978. The total of the indexes for each year is 400%. Note that if the total for each year is not exactly equal to 400% at first, we may adjust the trend lines up or down slightly in order to make the total.

TABLE 23–17

CHANGING SEASONAL INDEXES OF THE QUARTERLY SALES OF
1974 TO 1978 IN THE FRED DEPARTMENT STORE

Quarter	Index of Each Year (%)					Annual Increase (+) or Decrease (−)
	1974	1975	1976	1977	1978	
1st	49	57	65	73	81	+ 8
2nd	91	94	97	100	103	+ 3
3rd	154	146	138	130	122	− 8
4th	106	103	100	97	94	− 3
Total	400	400	400	400	400	0
Average	100	100	100	100	100	0

Source: Chart 23–4.

Note that the "typical" indexes obtained by various methods of averages in the previous sections are fairly close to the changing index of the middle year, 1976, of the series.

23.6   Use of Seasonal Index

A seasonal index can be used in three important ways: (1) guiding current operation, (2) forecasting future seasonal activity, and (3) obtaining seasonally adjusted data.

Guiding Current Operation

Knowing the typical or changing seasonal pattern of a given business activity, management should be able to plan and control the current business operation in a rational manner. For example, if the manager of a department store knows that the typical seasonal pattern for December sales is 20 percent above average monthly sales in a year (the index being 120 (%)), he can make plans in advance for buying, selling, personnel, financial, and other problems concerning the operation during the month.

Forecasting Future Seasonal Activity

If the seasonal pattern is stable during the past and the stability is expected to continue in the future, a typical seasonal index may be used in forecasting seasonal activity. If there has been an abrupt change in seasonal pattern, the typical index cannot be used effectively in the forecast. Other factors affecting the seasonal activity must be carefully examined. In such a case, the task of forecasting becomes more difficult. If the seasonal pattern is expected to change gradually, the most recent index should be used as a forecast of the seasonal pattern of the following year. The application of a seasonal index in forecasting future seasonal activity is illustrated in Example 5.

Example 5

The Fred Department Store made an annual forecast for the sales in 1979 at $30,000. Use the changing index of 1978 given in Table 23–17 to compute the forecast on a quarterly basis.

Solution. The computation of the forecast is presented in Table 23–18. The average quarterly sales of the forecast are $7,500. The estimated quarterly sales vary according to the estimated index for the year 1979.

TABLE 23–18

COMPUTATION OF FORECAST OF QUARTERLY SALES IN THE FRED
DEPARTMENT STORE FOR 1979—ANNUAL SALES FORECAST: $30,000

(1) Quarter	(2) Estimated Seasonal Index for 1979 (based on 1978 index, Table 23–17)	(3) Forecast of Quarterly Sales $7,500 × col. (2)
1st	81 (%)	$6,075
2nd	103	7,725
3rd	122	9,150
4th	94	7,050
Total	400	$30,000
Average	100	7,500

Source: Example 5.

Obtaining Seasonally Adjusted Data

When the effect of seasonal variation is removed from a time series, the series is called *seasonally adjusted data, adjusted data for seasonal variation*, or *deseasonalized data*. A time series, which is adjusted for seasonal variation, shows what the business activity would have been if it had been affected by only the trend, cyclical, and irregular movements. Symbolically, it can be expressed as follows:

$$\text{Seasonally adjusted data} = \frac{\text{Original data}}{\text{Seasonal index}} = \frac{TSCI}{S} = TCI$$

When a time series is substantially affected by seasonal variation, by means of seasonal adjustment, the effects of current trend and cyclical movements on the series can be measured with a high degree of accuracy. The method of measuring cyclical effect on a series by using adjusted data for seasonal variation will be discussed in the next chapter. The advantage of measuring current trend based on seasonally adjusted data is illustrated in Example 6.

Example 6

The sales of the first quarter of 1978 in the Fred Department Store were $5,000 (Table 23–1). Use the quarterly sales to estimate the annual sales of 1978.

Solution.   1. Use the actual sales of the first quarter as the average quarterly sales of 1978 in the estimation. The annual rate of sales is estimated at $20,000, or

$$\$5,000 \times 4 \text{ (quarters)} = \$20,000$$

2. Use the sales of the first quarter adjusted for seasonal variation as the average quarterly sales of 1978 in the estimation. The annual rate of sales is estimated at $27,396. The estimation is computed as follows:

The changing index of the first quarter of the previous year, 1977, which is 73% or 0.73, is applied for the seasonal adjustment.

$$\text{Seasonally adjusted sales} = \frac{\$5,000}{0.73}$$

$$= \$6,849 \text{ (Quarterly average)}$$

$$\text{Estimated annual rate of sales} = \$6,849 \times 4 = \$27,396$$

which is closer to the actual annual sales than the estimate based on the actual sales of the first quarter. Note that the actual sales for the year of 1978 are $27,000 (Table 23–1),

---

Many published data are presented in annual rates based on seasonally adjusted data. The Department of Commerce, for example, reported in the *Survey of Current Business of April 1975*, that the Gross National Product in the United States in the year 1975 was 1,419.2 billion dollars. The annual rate was based on the seasonally adjusted total of the first quarter of 1975. Thus, the seasonally adjusted quarterly total of the Gross National Product was only

$$\frac{1,419.2}{4} = 354.8 \text{ billion dollars}$$

If we know first quarter index used for the adjustment, we should be able to compute the actual amount of the Gross National Product in the first quarter of 1975.

23.7 Summary

A time series classified into periods of less than one year may have repeating periodic movement. This type of movement is called seasonal variation. The measures of seasonal variation are called seasonal indexes, which are usually expressed in percentages.

A time series may or may not have seasonal variation. Before an index is computed, a time series should be carefully examined for the variation in order to save time from unnecessary computation. Many time series should be adjusted for the calendar or working day variation before the observation of whether there is a seasonal variation in the series. The adjustment may be made by classifying the series into daily averages.

A specific seasonal pattern concerns the seasonal variation in a particular period, whereas a typical seasonal pattern describes the average seasonal variation over a number of periods.

Seasonal variation of a time series is measured after the effects of trend, cyclical, and irregular movements on the series are eliminated. There are three common methods of computing seasonal indexes.

1. *The method of simple averages of original data.* This method ignores the effect of trend movement. The effects of cyclical and irregular movements are eliminated by the process of averaging data for each time unit. The index obtained by this method usually gives a good rough estimate of the seasonal pattern.

2. *The method of simple averages adjusted for trend.* The procedure of computing seasonal index by this method is basically the same as that by the previous method. However, the effect of a trend movement is not to be ignored by this method. The adjustment of trend effect is based on the additive model $Y = T + S + C + I$.

3. *The method of ratios to moving averages.* This method is based on the assumption that the result of the effects by the four forces on a time series is a product. The result may be written in a multiplicative model $Y = T \times S \times C \times I$. A seasonal index thus is obtained by eliminating trend (T), cyclical (C), and irregular (I) movements by a division form $TSCI/TCI = S$, where S is a relative value. A moving average is the product of T and C. The elimination of TC is done by finding the the ratios of the original data to moving averages, or $TSCI/TC = SI$. The effect of I is eliminated by the process of averaging ratios. The averages of the ratios may be arithmetic means, modified means, or medians. Any form of the averages of the ratios may be used to compute a set of indexes.

When a typical seasonal index is used to describe the seasonal variation of a time series, it is assumed that there has been no pronounced change in the seasonal pattern. If the seasonal pattern changes abruptly, it is better to use a specific seasonal index for the specific time in analyzing seasonal variation. If the seasonal pattern changes gradually, there are usually two ways to describe the variation: (1) recompute the typical seasonal index in order to cover current specific seasonals, and (2) compute the changing seasonal index based on trend lines.

A seasonal index can be used in three important ways: (1) guiding current operation, (2) forecasting future seasonal activity, and (3) obtaining seasonally adjusted data. When a time series is substantially affected by seasonal variation, the seasonally adjusted data may be used to measure the effects of current trend and cyclical movements with a high degree of accuracy.

SUMMARY OF FORMULAS

Application	Formula	Formula Number	Reference Page
Decomposing time series components			
Additive model	$Y = T + S + C + I$	23–1	612
Multiplicative model	$Y = T \times S \times C \times I$	23–2	614

Exercises 23

1. Refer to Table 23–2. Assume that the units produced during January and June are as follows:

> January   500,000 units
> June   520,000 units

Compute the number of units adjusted for working-day variation for each of the two months.

2. The sales of Lang's Drug Store from the first quarter, 1972, to the fourth quarter, 1978, are shown in column (2) of Table 23–19. Compute the index of seasonal variation for the sales by the method of simple averages of original data. (Compute to 1 decimal place of a percentage.)

3. Refer to Problem 2. Compute the index of seasonal variation for the sales by the method of simple averages adjusted for trend.

4. From Table 23–19 compute the seasonal index by the ratio to moving average method as follows:

(a) Complete columns (3), (4), (5) and (6).
(b) Construct a table showing the arrays of ratios to 4-quarter moving averages.
(c) Compute the seasonal index:
  (1) Use the arithmetic mean of ratios method.
  (2) Use the modified mean method—eliminate the largest and the lowest ratios, or mean of 4 central items.
  (3) Use the median of ratios method.

5. Use the index below to adjust the sales in Problem 4. (Complete column (7) of Table 23–19.)

> Typical seasonal index of the sales of Lang's Drug Store:

> First quarter   74.5(%)
> Second quarter   119.6
> Third quarter   72.1
> Fourth quarter   133.8
>
> 400.0(%)

TABLE 23–19

SALES OF LANG'S DRUG STORE, 1972–1978

Year and Quarter (1)	Sales ($1,000) Y = TSCI (2)	4-quarter Moving Total Centered (3)	Sum of Two 4-Quarter Moving Totals (4)	4-quarter Moving Average TC (5)	Ratio to Moving Average SI(%) (6)	Adjusted Sales Y/S (S = index) (7)
1972 1st	5					6.7
2nd	17	52				14.2
3rd	9	59	111	13.875	64.9	12.5
4th	21	66	125	15.625	134.4	15.7
1973 1st	12	73	139	17.375	69.1	16.1
2nd	24	84	157	19.625	122.3	20.1
3rd	16	91	175	21.875	73.1	22.2
4th	32	97	188	23.500	136.2	23.9
1974 1st	19	99	196	24.500	77.6	25.5
2nd	30	101	200	25.000	120.0	25.1
3rd	18	99	200	25.000	72.0	25.0
4th	34	96	195	24.375	139.5	25.4
1975 1st	17	92	188	23.500	72.3	22.8
2nd	27	97	189	23.625	114.3	22.6
3rd	14	100	197	24.625	56.9	19.4
4th	39				149.3	29.1
1976 1st	20				70.2	
2nd	36			30.000		
3rd	24		249	31.125		
4th	41		266	33.250	123.3	
1977 1st	27	144	282	35.250	76.6	36.2
2nd	46	151	295	36.875	124.7	38.5
3rd	30	158	309	38.625	77.7	41.6
4th	48	159	317	39.625	121.1	35.9
1978 1st	34	168	327	40.875	83.2	45.6
2nd	47	170	338	42.250	111.2	39.3
3rd	39					54.1
4th	50					37.4

6. The Lang's Drug Store made an annual forecast for the sales in 1979 at $200,000. Use the index given in Problem 5 to compute the forecast on a quarterly basis.

7. Use the quarterly sales of the Lang's Drug Store for the years 1972 to 1978 given in Problem 4. Compute the changing seasonal index for each year.

8. Table 23–20 shows the labor force employed in Benton United Company from 1974 to 1978. Compute the monthly seasonal index by the ratio to moving average method. Use the arithmetic means of ratios in the computation. (Round the decimals in the final answer.)

TABLE 23–20

LABOR FORCE EMPLOYED IN BENTON UNITED COMPANY,
1974 TO 1978 (IN THOUSANDS OF PERSONS)

Month	1974	1975	1976	1977	1978
January	4.61	4.63	4.42	4.21	3.99
February	4.62	4.71	4.58	4.05	3.93
March	4.57	4.98	4.78	4.34	4.02
April	5.39	5.00	4.96	4.67	4.43
May	5.84	5.54	5.43	5.18	5.01
June	6.86	6.67	6.29	5.95	5.85
July	6.89	6.45	6.06	5.97	5.82
August	6.45	6.33	5.77	5.50	5.40
September	6.59	5.67	5.56	5.33	5.23
October	6.25	5.96	5.48	5.35	5.13
November	5.67	5.20	4.88	4.78	4.55
December	4.95	4.42	4.07	4.04	3.78

24 Cyclical and Irregular Movements

The previous two chapters presented various methods of measuring trend and seasonal movements. This chapter will discuss the two remaining components of a time series: cyclical fluctuation and irregular movement.

The cyclical component fluctuates unevenly and is not easy to control. For this reason the study of cyclical fluctuation probably is of greater interest than the studies of other components of a time series to economists as well as businessmen. Cyclical fluctuations may be measured from annual data or from data classified into time units of less than a year. The measurements of cyclical fluctuations are presented in Sections 24.1 and 24.2. Irregular movements are usually measured from less-than-a-year data and are discussed in Section 24.3. The uses of cyclical and irregular measurements are presented in Section 24.4.

24.1 Measuring Cyclical Fluctuations from Annual Data

A seasonal variation shows repeated periodic fluctuation from year to year. When a time series is classified by years, the seasonal variation is eliminated. Also, the effect of irregular movement upon a series is eliminated mostly by the yearly classification. The multiplicative model of the time series of the annual data then becomes

$$Y = T \times C$$

The effect of cyclical fluctuation now is measured by the ratios of Y (the original data) to T (the trend values), or

$$C = \frac{Y}{T}$$

631

The ratios are also called *adjusted data for secular trend*. The trend values (T) may be obtained by various methods presented in Chapter 22. However, only the least square method for a straight-line trend is used in the following example.

Example 1

Use the information concerning the annual sales for the years 1964 to 1978 in the Wilson Department Store shown in column (2) of Table 24–1. (a) Find the trend value for each year by the least square method for a straight-line trend. (b) Draw a chart to show the effect of cyclical fluctuation for the data.

TABLE 24–1

COMPUTATION OF MEASURES OF CYCLICAL FLUCTUATION, ADJUSTED DATA FOR TREND OF THE SALES IN WILSON DEPARTMENT STORE, 1964 TO 1978

(1) Year	(2) Actual Sales Y (in \$1,000)	(3) Trend Sales T or Y_c (in \$1,000)	(4) (5) Sales Adjusted for Trend (Measures of Cycles), Y/T		(6) Percent Deviation from Trend (Trend = 100%)
			in decimals	in percents	
1964	7	2.4	2.92	292%	+192%
1965	6	4.3	1.40	140	+ 40
1966	2	6.2	0.32	32	− 68
1967	4	8.0	0.50	50	− 50
1968	8	9.8	0.82	82	− 18
1969	16	11.7	1.37	137	+ 37
1970	13	13.6	0.96	96	− 4
1971	14	15.4	0.91	91	− 9
1972	17	17.2	0.99	99	− 1
1973	20	19.1	1.05	105	+ 5
1974	23	21.0	1.10	110	+ 10
1975	19	22.8	0.83	83	− 17
1976	25	24.6	1.02	102	+ 2
1977	28	26.5	1.06	106	+ 6
1978	29	28.4	1.02	102	+ 2
Total	231	231.0			

Source: Table 22–5; Example 1.

Solution. (1) The required trend values (T or Y_c) have been computed by the method of least squares in Table 22–5, page 571. The trend values are also shown in column (3) of Table 24–1.

(2) The effect of cyclical fluctuation for the data is shown in Chart 24–1. The chart includes two parts:

a. *The effect is observed from the original data.* Consider the trend values as the normal sales for individual years. The shape of the cycles of the original data fluctuated around the normal sales or the trend line can easily be seen. However, this graph shows the fluctuation in absolute values. For ease in comparison, relative values are preferred in showing the fluctuation. This is done in part B of the chart.

CHART 24–1

EFFECT OF CYCLICAL FLUCTUATION ON THE SALES
OF THE WILSON DEPARTMENT STORE, 1964 TO 1978

(a) *Observed from Original Data*
   Y ($1,000)

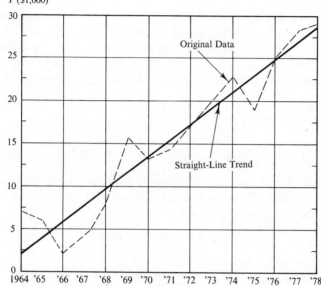

(b) *Measured by Adjusted Data for Trend*

Percent of Trend   Percent Deviation
                                              from Trend

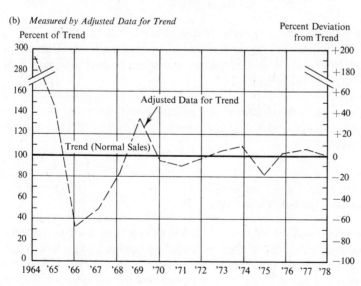

Source: Table 24–1.

b. *The effect is measured by the adjusted data for trend.* The adjusted data for trend, or the ratios of the original data to the trend values, may be expressed in decimal form or in percent form. For example, the ratio of the first year, 1964, is computed as follows:

$$C = \frac{Y}{T} = \frac{7}{2.4} = 2.916, \text{ or round to } 2.92 = 292\%$$

The ratios expressed in decimal form are shown in column (4) and in percentage form are shown in column (5) of Table 24–1. The percent ratios are shown also in part B of Chart 24–1. Observe that the upward trend line in part A is shifted to a level position in part B as the base, or 100% for each year.

The cyclical fluctuation may also be measured on the basis of the additive model,

$$Y = T + C$$

or

$$C = Y - T$$

to yield the same result. The deviations of the original data from the trend values may be expressed in absolute terms or relatives. For example, the deviation in absolute amount for the first year, 1964, is computed as follows:

$$C = Y - T = 7 - 2.4 = 4.6 \text{ (in \$1,000)}$$

The deviation in relative term based on the trend value is

$$\frac{4.6}{2.4} = 1.92 \text{ or } 192\%$$

which is the same as the difference between the adjusted data for trend ($Y/T = 292\%$) and the trend value (100% or the base). The deviations in absolute terms can be observed from part A and in percent terms can be seen from part B of Chart 24–1 and column (6) of Table 24–1.

Similarly, the adjusted data for trend of the Gross National Product from 1940 to 1975 as shown in Chart 22–1A may be computed to show the effect of cyclical fluctuation in percent form.

24.2 Measuring Cyclical Fluctuations from Less-Than-A-Year Data

When a time series is classified into periods of less than a year, such as quarters, months, or weekdays, it may be affected by trend (T), seasonal, (S), cyclical (C), and irregular (I) movements. If we assume that the trend and seasonal movements from year to year are normal activities, we may observe the effect of cyclical fluctuation above or below the normal operation. Based on the assumption, a cyclical fluctuation may be measured by three types of data as follows:

(1)　It is measured by adjusted data for seasonal variation (*TCI*).

After the effect of seasonal variation is eliminated from the original data, the effect of trend movement is considered as the normal operation. The trend movement (upward, level, or downward) can usually be observed from a graph. The effect of irregular movement is either ignored or observed from the graph. Many adjusted data for seasonal variation, either in absolute or relative values, are published for measuring cyclical fluctuations as well as trend movements. Examples of such published data are shown in Chart 24–2. A detailed illustration of measuring the cyclical fluctuation by the adjusted data for seasonal variation is presented in Example 2 (see Chart 24–3a). There is a trend line in the chart to aid a reader to observe the cyclical fluctuation, or ups and downs from the normal sales, of the series.

(2)　It is measured by adjusted data for seasonal variation and trend (*CI*). The effect of irregular movement is ignored or it may be observed from the graph of the adjusted data.

There are three methods of obtaining the adjusted data for seasonal variation and trend. The basic idea of the methods is to eliminate the effects of trend (T) and seasonal (S) from the original data ($TSCI$) in order to get the effects of cyclical (C) and irregular (I) movements.

Method A. First, obtain the adjusted data for seasonal variation, or

$$\frac{TSCI}{S} = TCI$$

Next, obtain the adjusted data for seasonal variation and trend, or

$$\frac{TCI}{T} = CI$$

This is a convenient method when the adjusted data for seasonal variation are desirable or available. This method is illustrated in Example 2 [see columns (5) and (6) of Table 24–2.]

CHART 24–2

EXAMPLES OF PUBLISHED DATA OF BUSINESS CYCLE SERIES,
CYCLES MEASURED BY ADJUSTED DATA FOR SEASONAL VARIATION

(A) *Average Workweek of Production Workers, Manufacturing, Monthly Data (A NBER Leading Indicator)*

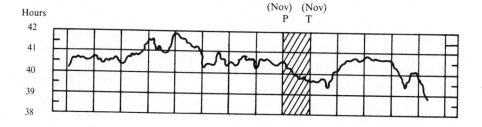

CHART 24-2 (CONTINUED)

(B) *Industrial production, Monthly Data Index: 1967 = 100% (A NBER Roughly Coincident Indicator)*

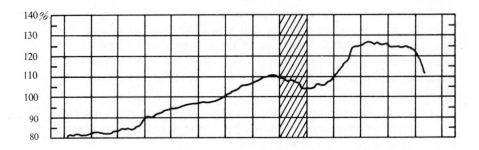

(C) *Business Expenditures, New Plant and Equipment, Quarterly Data, at Annual Rate. Unit: Billions of Dollars (A NBER Lagging Indicator)*

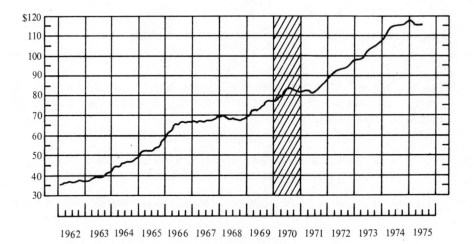

1962 1963 1964 1965 1966 1967 1968 1969 1970 1971 1972 1973 1974 1975

P—Peak of cycle indicates end of expansion and beginning of recession (shaded areas) as designated by National Bureau of Economic Research (*NBER*).

T—Trough of cycle indicates end of recession and beginning of expansion (white areas) as designated by *NBER*.

Source: U.S. Department of Commerce, Bureau of Economic Analysis, *Business Conditions Digest*, various issues.

CHART 24-3

EFFECTS OF CYCLICAL-IRREGULAR MOVEMENTS ON THE QUARTERLY SALES
OF THE FRED DEPARTMENT STORE, 1974 TO 1978

(a) Measured by Adjusted Data for Seasonal Variation (TCI)
with Trend Values (T)

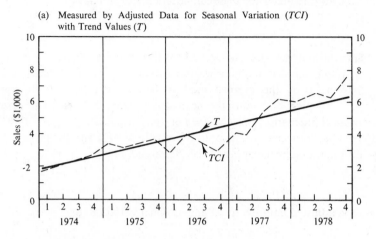

(b) Measured by Original Data ($TSCI$) with Trend-seasonal Values
(TS)

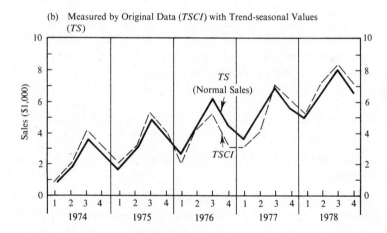

(c) Measured by Adjusted Data for Seasonal Variation and Trend
(CI) with Trend-seasonal Value $= 100\%$

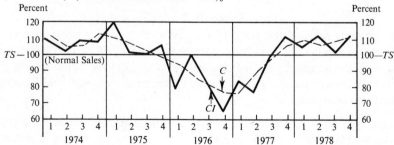

Source: Tables 24-2 and 24-3.

Method B. First, obtain the trend-seasonal value, or

$$T \times S$$

Next, obtain the adjusted data for seasonal variation and trend, or

$$\frac{TSCI}{TS} = CI$$

This is the simplest and the clearest method for measuring the effect of cyclical fluctuation. It is the simplest method because it includes one multiplication and only one division. It is the clearest method because the base of the ratio $(TSCI/TS)$ is TS, which represents the normal operation. It is easier to reason the effect of cyclical fluctuation directly based on the normal sales ($= 1$ or 100%). This method is also illustrated in Example 2 [see columns (7) and (8).]

Method C. First, obtain the adjusted data for trend, or

$$\frac{TSCI}{T} = SCI$$

Next, obtain the adjusted data for trend and seasonal variation, or

$$\frac{SCI}{S} = CI$$

This is a convenient method when the adjusted data for trend are desirable or available. A seasonal index can be obtained by averaging ratios of original data to trend values (or SCI) instead of averaging ratios to moving averages (or SI) if the effects of cyclical and irregular movements are ignored in computing the index. This method is not illustrated in Example 2 since it differs only slightly from the above two methods in procedures.

The answers obtained by the above three methods should be the same. However, there may be differences due to rounding the decimals.

(3) It is measured by the data after trend, seasonal, and irregular movements are eliminated.

This method is theoretically sound. However, there is no perfect method of eliminating the effect of irregular movement. The method of eliminating the effect of irregular movement from the data adjusted for seasonal variation and trend used in Example 2 is an approximation. (See column (5) of Table 24–3.)

Example 2

Use the quarterly sales of the years 1974 to 1978 in the Fred Department Store provided in column (2) of Table 24–2. Measure the effect of cyclical fluctuation by using: (a) the adjusted data for seasonal variation, (b) the adjusted data for seasonal variation and trend, and (c) the data after trend, seasonal, and irregular movements are eliminated.

Solution. (a) By the adjusted data for seasonal variation. This series is the same series presented in Table 23–1. The changing seasonal indexes, shown in column (3) of Table 24–2, of the series are used here to compute the adjusted data for

TABLE 24-2

COMPUTATION OF MEASURES OF CYCLICAL-IRREGULAR MOVEMENTS (*CI*), BY ADJUSTED DATA FOR SEASONAL VARIATION AND TREND OF THE QUARTERLY SALES IN THE FRED DEPARTMENT STORE, 1974 TO 1978

(UNIT: $1,000)

(1)	(2)	(3)	(4)	(5) Method A	(6) Method A	(7) Method B	(8) Method B
Year and Quarter	Original Data $Y = TSCI$	Changing Seasonal Index S	Trend Value T	Adjusted for Seasonal $\frac{TSCI}{S} = TCI$	Adjusted for Seasonal and Trend $\frac{TCI}{T} = CI$	Trend-Seasonal Value $T \times S$	Adjusted for Seasonal and Trend $\frac{TSCI}{TS} = CI$
1974–1st	$1	49%	$1.875	$2.041	109%	$0.91875	109%
2nd	2	91	2.125	2.198	103	1.93375	103
3rd	4	154	2.375	2.597	109	3.65750	109
4th	3	106	2.625	2.830	108	2.78250	108
1975–1st	2	57	2.875	3.509	122	1.63875	122
2nd	3	94	3.125	3.191	102	2.93750	102
3rd	5	146	3.375	3.425	101	4.92750	101
4th	4	103	3.625	3.883	107	3.73375	107
1976–1st	2	65	3.875	3.077	79	2.51875	79
2nd	4	97	4.125	4.124	100	4.00125	100
3rd	5	138	4.375	3.623	83	6.03750	83
4th	3	100	4.625	3.000	65	4.62500	65
1977–1st	3	73	4.875	4.110	84	3.55875	84
2nd	4	100	5.125	4.000	78	5.12500	78
3rd	7	130	5.375	5.385	100	6.98750	100
4th	6	97	5.625	6.186	110	5.45625	110
1978–1st	5	81	5.875	6.173	105	4.75875	105
2nd	7	103	6.125	6.796	111	6.30875	111
3rd	8	122	6.375	6.557	103	7.77750	103
4th	7	94	6.625	7.447	112	6.22750	112

Source: Column (2), Table 23–1; Column (3), Table 23–17; Column (4), Table 23–5.

TABLE 24–3

COMPUTATION OF CYCLICAL FLUCTUATION (C)
AND IRREGULAR MOVEMENT (I)

(1) Year and Quarter	(2) Adjusted Sales for Seasonal and Trend, CI	(3) 3-Quarter Moving Total	(4) 3-Quarter Moving Average, C	(5) Irregular Movement CI/C = I
1974 –1st	109%	—	—	—
2nd	103	321%	107%	96%
3rd	109	320	107	102
4th	108	339	113	96
1975 –1st	122	332	111	110
2nd	102	325	108	94
3rd	101	310	103	98
4th	107	287	96	111
1976 –1st	79	286	95	83
2nd	100	262	87	115
3rd	83	248	83	100
4th	65	232	77	84
1977 –1st	84	227	76	111
2nd	78	262	87	90
3rd	100	288	96	104
4th	110	315	105	105
1978 –1st	105	326	109	96
2nd	111	319	106	105
3rd	103	326	109	94
4th	112	—	—	—

Source: Table 24–2.

seasonal variation. For example, the adjusted value of the first quarter of 1974 is computed as follows:

$$TCI = \frac{TSCI}{S} = \frac{1}{49\%} = \frac{1}{0.49} = 2.041 \text{ (in \$1,000)}$$

The adjusted data for seasonal variation of other quarters are obtained similarly and are shown in column (5) of Table 24–2 and in part (A) of Chart 24–3. The trend values of the original data are shown in column (4) (see note below) and also in the part of the chart. By observing the dotted line (representing the adjusted data for seasonal variation) and the solid trend line (representing the normal sales), we can see clearly the effects of cyclical and irregular movements.

(b) By the adjusted data for seasonal variation and trend. The adjusted data are computed by the two methods below.

Method A. For example, the adjusted value of the first quarter of 1974 is computed as follows:

$$TCI = \frac{TSCI}{S} = \frac{1}{49\%} = \frac{1}{0.49} = 2.041 \text{ (in \$1,000)}$$

$$CI = \frac{TCI}{T} = \frac{2.041}{1.875} = 1.08853 \text{ or round to } 1.09 = 109\%$$

The computed data *CI* are shown in column (6) of Table 24–2 and shown in part C of Chart 24–3.

Method B. For example, the adjusted value of the first quarter of 1974 is computed as follows:

$$T \times S = 1.875 \times 49\% = 0.91875 \text{ (in } \$1,000)$$

$$CI = \frac{TSCI}{TS} = \frac{1}{0.91875} = 1.08844 \text{ or round to } 1.09 = 109\%$$

The computed values of *TS* are shown in column (7) of Table 24–2 and in part B of Chart 24–3 with the original data. Observe the cyclical fluctuation based on the original data and the normal sales (*TS*). Both are represented by the broken lines. The difference between the original value (*TSCI*) and its corresponding *TS* value represents the effect of cyclical-irregular movements.

The computed values of *CI* are shown in column (8) of the Table 24–2 and in part C of Chart 24–3 with the normal sales (*TS*) as the base or 100%. The line representing the normal sales now becomes a straight one. It is obviously more convenient to measure the cyclical-irregular fluctuation based on a single straight line than based on broken lines. Thus, the relative values shown in part C are preferred over the absolute values shown in part B in measuring the fluctuation.

(c) By the data after trend, seasonal, and irregular movements are eliminated. The adjusted data are computed by the method of 3-quarter moving averages. The computed data (*C* or 3-quarter moving averages) are shown in column (4) of Table 24–3 and in part C of Chart 24–3. The smoothed *C* curve represents the business cycles after the effect of irregular movement has been eliminated from the data adjusted for seasonal variation and trend.

Note: The trend values shown in column (4) of Table 24–2 and on Chart 24–3A are obtained by the trend equation

$$Y_c = 4.25 + 1X$$

$$\text{with origin: July 1, 1976}$$
$$X \text{ unit: 1 year}$$
$$Y \text{ unit: } \$1,000$$

(See Table 23–5, p. 613). However, there is a more convenient way to compute the trend values from the equation. We may change the *X* unit in the equation from 1 year to 1 quarter and shift the origin from the middle of the year 1976 to the middle of the quarter, July-September of 1976. When the *X* unit is changed from 1 year to 1 quarter, with the same origin and *Y* unit, the equation becomes (also see Section 22.6):

$$Y_c = 4.25 + (1/4)X = 4.25 + 0.25X$$

Further, when the origin is shifted to the middle of the quarter, it becomes:

$$Y_c = 4.25 + 0.25(1/2) + 0.25X$$
$$= 4.375 + 0.25X$$

$$\text{with origin: August 15 (middle of the third quarter), 1976}$$

X unit: 1 quarter

Y unit: $\$1,000$

Quarterly increase: $b = 0.25$ or $\$250$

The trend value of the first quarter of the first year in the series (centered on February 15, 1974) is $\$1.875$ (in $\$1,000$), or

$X = -10$ quarters, from February 15, 1974, to August 15, 1976:

Month	Day	Year
8	15	1976
(−) 2	15	1974

6 months + 0 days + 2 years = 10 quarters

$Y_c = 4.375 + 0.25(-10) = 4.375 - 2.500 = \$1.875$ (in $\$1,000$)

By adding 0.25 successively to 1.875, we may obtain the trend values for the quarters after the first quarter of 1974 as follows:

1st quarter, 1974 = 1.875

+ 0.25

2nd quarter, 1974 = 2.125

+ 0.25

3rd quarter, 1974 = 2.375, and so on.

24.3 Measuring Irregular Movements from Less-Than-A-Year Data

In order to isolate the effect of irregular movement for a time series, we must further eliminate the effect of cyclical fluctuation (C) from the computed cyclical-irregular value (CI) of the series. The CI value is computed after T and S values are determined. The determinations of T and S are rather arbitrary. There are many different methods available for computing the T and S values, such as various methods for obtaining straight-line trends and nonlinear curves and various methods for obtaining typical seasonal indexes and changing seasonal indexes. Thus, the CI value is affected by the choice of a method of computing the T and S values. Also, the length of the period included in a time series changes the answers to T and S movements. Furthermore, irregular movements are usually caused by random forces and are uncontrollable. Thus, there is no perfect method of isolating the effect of irregular movement from the CI value.

The isolation of the cyclical fluctuation in Example 2 is done by the method of 3-quarter moving averages. The short period, 3-quarter length or less than a year, is used so that the cycles, usually of a length longer than a year, cannot be smoothed out. For monthly data, 3-month or 5-month moving averages are frequently used in isolating cyclical fluctuation. The method of isolating cyclical fluctuation by the method of moving averages is not a perfect one. However, it is a logical method and is used widely by statisticians.

Once the values of *CI* and the corresponding values of *C* are isolated or determined, the values of *I* can easily be isolated or computed by the following expression:

$$\frac{CI}{C} = I$$

For example, the effect of irregular movement for the second quarter of 1974 is computed as follows:

$$I = \frac{CI}{C} = \frac{103\%}{107\%} = 0.96 = 96\%$$

The values of *I* are shown in column (5) of Table 24–3. A chart is not constructed for the effect of irregular movement since it is basically the same as the construction for the *CI* curve of Chart 24–3 C.

We now may conclude the time series analysis by using the four decomposed components based on the multiplicative model. Under the assumption of the multiplicative model, it is implied that the forces affecting the four components are interrelated. For example, the relationships among the decomposed components of the second quarter of 1974 obtained from Tables 24–2 and 24–3 are as follows:

1. The upward trend and the unfavorable seasonal variation gave the normal sales in the quarter $1,934, or $TS = \$2,125 \times 91\% = \$1,934$.

2. The upswing of cyclical fluctuation increased the sales to $2,069, or $TSC = \$1,934 \times 107\% = \$2,069$.

3. The unfavorable irregular movement, such as unusually bad weather in the quarter, decreased the final sales to $1,986, or $TSCI = \$2,069 \times 96\% = \$1,986$. The difference between the computed final sales $1,986 and the original final sales $2,000 is due to rounding the decimal places and the averaging process. In summary,

$$Y = T \times S \times C \times I = \$2,125 \times 91\% \times 107\% \times 96\%$$
$$= \$1,986$$

The decomposed relative components *S*, *C*, and *I* may also be expressed in absolute terms or dollar values as follows:

1. The upward trend gave the base sales in the quarter $2,125.

2. The usual unfavorable seasonal variation reduced the base sales by 9% ($=100\% - 91\%$), or $191($=2,125 \times 9\%$). The normal sales become $2,125 − $191 = $1,934.

3. The upswing of cyclical fluctuation increased the normal sales by 7% ($=107\% - 100\%$) or $135 ($=\$1,934 \times 7\%$). The sales then were $1,934 + $135 = $2,069.

4. The unfavorable irregular movement decreased the sales after being influenced by trend, seasonal, and cyclical movements by 4% ($=100\% - 96\%$), or $83 ($=\$2,069 \times 4\%$). The final sales were $2,069 − $83 = $1,986. The absolute amounts may be written in additive model, or

$$Y = T + S + C + I = \$2{,}125 + (-\$191) + \$135 + (-\$83)$$
$$= \$1{,}986$$

Under the assumption of additive model, it is implied that the forces affecting the four components are independent of each other. Economists generally prefer the assumption of the multiplicative model in analyzing time series since it is more logical.

It should be noted that the above four decomposed components are estimates since they are computed by the selected methods. The forces affecting economic and business time series are complex. It is unlikely that a perfect method of isolating the four components can be developed.

24.4 The Uses of Cyclical-Irregular Measurements

The cyclical-irregular components (CI) are usually used in measuring the cyclical fluctuation of a time series. The refined cyclical component (C) obtained by the time-consuming method of moving averages is not frequently used in practice for the measurement. The cyclical-irregular measurements can be used in three important ways.

Guiding Current Operation

The general business condition, either prosperity or depression, of a particular industry or a nation as a whole has direct effects upon individual business concerns. Cyclical fluctuations of representative business activities have been used extensively by industries and governmental agencies to indicate the general business conditions. The knowledge of business cycles thus is important to the management of a business concern. When the general business condition of the industry or the entire economy is reaching the peak of expansion or the beginning of recession, the manager should restrict his business from expansion. If he continues his actions of increasing inventories, acquiring large equipment or facilities, or expanding other business activities, he will suffer the loss due to the depression. On the other hand, when the condition is approaching the end of recession or the beginning of expansion, he should be ready to expand the business in order to maximize the profit.

Controlling Business Cycles

The cost of living through the period of a depression, such as during the great depression period in the 1930's, has been proved too great in terms of material losses and human suffering. Many agencies concerned have used various policies to control business cycles. For example, in periods of recession, such as 1970 and 1975, the Federal Reserve carried out a policy of monetary ease. The total bank credit then expanded rapidly to increase economic activity. In periods of high economic activity, such as 1959, 1969, and 1973, the Federal Reserve restricted the availability of bank reserves. The bank credit rose more slowly to make it difficult for further expansion.

Forecasting Business Cycles

The effect of cyclical fluctuation upon a time series is important to future operations of a firm, an industry, and a nation just as it is to current operations. To make a sound and complete forecast, we must consider the cyclical fluctuation during the past. However, based on the analysis of past record of a time series, we know that the business cycles fluctuate unevenly. That is, the ups and downs last for different time intervals.

There is no perfect method, whether quantitative or nonquantitative, that can be used to predict completely and accurately the uneven fluctuation of the cycles. Nevertheless, there are many methods which can be used to predict approximately the directions of the cycles in the future. The two methods introduced below are among the most prevailing methods of forecasting cyclical fluctuations in time series.

Lead-coincident-lag Series

The cycles of most business and economic series can be compared to the cycles of total economic activity. For many years, the National Bureau of Economic Research has provided a list of those significant series that usually lead, those that usually move with, and those that usually lag behind the cycles of total economic activity. The three different types of series are named by the NBER as indicators as follows:

a. Leading indicators—These series usually reach peaks or troughs before the measures of total economic activity. Examples of the NBER leading indicators are average workweek of production workers, manufacturing (graph A of Chart 24–2); new orders, durable goods industries; corporate profits after taxes; stock prices, 500 common stocks; industrial materials price; and so on.

b. Roughly coincident indicators—These series are direct measures of total economic activity or move roughly together with the total activity. Examples of the NBER roughly coincident indicators are industrial production (graph B of Chart 24–2), employees in nonagricultural establishments, Gross National Product, personal income, sales of retail stores, and so on.

c. Lagging indicators—These series usually reach turning points after the measures of total economic activity. Examples of the NBER lagging indicators are business expenditures, new plant and equipment (graph C of Chart 24–2); labor cost per unit of output, manufacturing; book value of manufactures' inventories; consumer installment debt; bank rates on short-term business loans; etc.

The leading indicators can be used in forecasting the turning points of the ups and downs of the cycles of total economic activity. It should be noted, however, that none of the leading indicators has been consistently successful in forecasting such points.

Diffusion Indexes

Diffusion indexes are frequently used as supplemental devices in forecasting the turning points of the cycles of total economic activity. The diffusion index is the percentage of rising industries or business enterprises, called components, in a group over a given span of time. For example, the diffusion indexes of the average workweek in the manufacturing group are computed from 21 components or industries, including electrical equipment and supplies, transportation equipment, food and kindred products, printing and publishing, chemicals and allied products, leather products, and so on. The changes of the average hours per workweek from February to March, 1975, are as follows*:

There are 4 rising components (such as the food and kindred products having 39.9 hours in February and 40.3 hours in March).

There are 15 falling components (such as the chemicals and allied products component falling from 40.5 hours in February to 40.4 hours in March).

There are 2 unchanged components (such as the electrical equipment and supplies component which has 39.0 hours in February and the same number in March).

The percentage of rising components, or the diffusion index, of March, 1975, based on a 1-month span (February to March—1-month indexes are placed on latest month), thus is 23.8%, or

$$\frac{5 \text{ (4 rising components and 2 unchanged components)}}{21 \text{ (21 industry components)}}$$
$$= 0.238 \text{ or } 23.8\%.$$

The value of 0.5 is assigned for each unchanged components. The total value of rising components is $4 + 2(0.5) = 5$.

The diffusion indexes of the selected groups of industries similar to those plotted on Chart 24–2 are shown on Chart 24–4. Diffusion indexes vary from zero (0% if all components are falling) to 1 (or 100% if all components are rising). A wide spread in the increases of the indexes is often related to rapid growth of the total economic activity, whereas a wide spread in decreases is often connected with sharp reductions in the total activity.

Chart 24–4 shows the diffusion indexes based on monthly and quarterly series. For monthly series, comparisons are made over 1-month span and either 6-month or 9-month span, depending upon the irregularity of the series. The 1-month span indexes are more current than longer span indexes but they show large irregular fluctuations. The cycles representing the longer span indexes generally are smoother and have larger amplitudes than the 1-month span indexes. The computation of longer span indexes is similar to that of 1-month span indexes. For example, the changes of the average work hours per week from June, 1974 to March, 1975, or a 9-month period, are as follows:**

There is only one rising component (the tobacco manufactures component having 37.3 hours in June, 1974, and 39.3 hours in March, 1975).

There are 20 falling components.

The percentage of rising components, or the diffusion index, of November, 1974, based on a 9-month span (9-month indexes are placed on the 6th month of

---

*U. S. Bureau of the Census, *Business Conditions Digest*, April, 1975, page 99.
**Ibid*, page 97.

CHART 24–4

SELECTED DIFFUSION INDEXES, 1962 TO 1975*

(A) *Average Workweek of Production Workers, Manufacturing—21 Industries (NBER Leading Indicator)*

21 Industries. (NBER Leading Indicator)

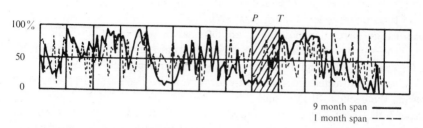

(B) *Industrial Production—24 Industries. (NBER Roughly Coincident Indicator)*

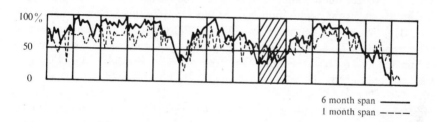

(C) *Business Expenditures for New Plant and Equipment—All Industries. (NBER Lagging Indicator)*

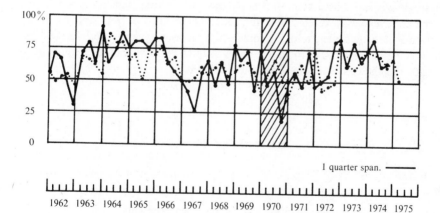

*Data are the percents of rising components and are centered within spans as follows: 1-month indexes are placed on latest month, 6-month indexes are placed on the 4th month, 9-month indexes are placed on the 6th month, and the 1-quarter indexes are placed on the 1st month of the 2nd quarter. Seasonally adjusted components are used.

**P and T (Peak and Trough)—See the footnotes of Chart 24–2.

Source: U.S. Bureau of Economic Analysis, *Business Conditions Digest*, March, 1975, pages 46–64.

the span) is thus

$$\frac{1 \text{ (rising component)}}{21 \text{ (industry components)}} = 0.048 \text{ or } 4.8\%$$

24.5 Summary

The cyclical component may be measured from annual data or from data classified into the time units of less than a year. When a time series is classified by years, the seasonal and irregular movements are eliminated. The effect of cyclical fluctuation of the annual data can thus be observed from a graph or a table showing the comparison of the original data and the corresponding trend values. However, the relative data adjusted for secular trend, or $C = Y/T$ are usually preferred in measuring the cyclical fluctuation.

When a time series is classified into periods of less than a year, it may be affected by trend, seasonal, cyclical, and irregular movements. If we assume that the trend and seasonal movements from year to year are normal activities, we may observe the effect of cyclical fluctuation above or below the normal operation. A cyclical fluctuation based on the assumption may be measured by three types of data: (1) adjusted data for seasonal variation, or TCI, (2) adjusted data for seasonal variation and trend, or CI, and (3) data after trend, seasonal, and irregular movements are eliminated, or C. Once the value of CI and the corresponding values of C are isolated or determined, the values of I can easily be computed by the division CI/I.

Note that the four components computed in this chapter are estimates. The forces affecting economic and business time series are complex. It is unlikely that a perfect method of isolating the components can be developed.

The cyclical component fluctuates unevenly and is not easy to control. The analysis of cyclical fluctuation, however, can aid businessmen, economists, and government agencies in three important ways: (1) guiding current operation, (2) controlling business cycles, and (3) forecasting business cycles. The two methods of forecasting business cycles included in this chapter are the use of lead-coincident-lag series and the construction of diffusion indexes.

Exercises 24

1. Observe Chart 24–2. Explain the cyclical fluctuation as measured by the cycles representing industrial production.

2. Discuss briefly the uses of cyclical-irregular measurements.

3. Refer to Problem 3(c) of Exercises 22–1, Chapter 22 (p. 602).
   (a) Find the trend value and the sales adjusted for trend for each year.

(b) Draw a chart to show the sales adjusted for trend for the years from 1970 to 1978.

(c) Comment on the effect of cyclical fluctuation as shown on the chart.

4. Refer to Problem 5 of Exercises 22–1.

(a) Find the trend value and the expenditures adjusted for trend for each year.

(b) Draw a chart to show the adjusted expenditures of the years from 1968 to 1977.

(c) Comment on the effect of cyclical fluctuation as shown on the chart.

5. Use the information given in Table 23–19.

(a) Measure the effect of cyclical fluctuation by using:

    (1) the adjusted data for seasonal variation,

    (2) the adjusted data for seasonal variation and trend, and

    (3) the data after trend, seasonal, and irregular movements are eliminated.

(b) Draw a chart for the adjusted data for seasonal variation and trend with trend-seasonal value = 100%.

(c) Comment on the effect of cyclical fluctuation as shown on the chart.

6. Compute the effect of irregular movement for each quarter during the year 1977. Use the findings of (a) (3) in Problem 5 above.

7. Use the information given in Table 23–20.

(a) Measure the effect of cyclical fluctuation by using:

    (1) the adjusted data for seasonal variation,

    (2) the adjusted data for seasonal variation and trend, and

    (3) the data after trend, seasonal, and irregular movements are eliminated.

(b) Draw a chart for the adjusted data for seasonal variation and trend with trend-seasonal value = 100%.

(c) Comment on the effect of cyclical fluctuation as shown on the chart.

8. Compute the effect of irregular movement for each month during the year 1977. Use the findings of (a) (3) in Problem 7 above.

part 7

Relationship

Analysis

Relationship analysis concerns the average relationship between two or more variables. The analysis of two variables based on a linear relationship is presented in Chapters 25 and 26, and that based on a nonlinear relationship is presented in Chapter 27. Chapter 28 describes methods for analyzing the linear relationship between three variables (multiple and partial correlation).

The above mentioned relationships are analyzed from the data in original units, such as dollars, miles, and grade points. The analysis may also be made by using data in ranked numbers, such as in order of first, second, etc. The relationship between two variables in ranked data is analyzed in the last part of Chapter 28.

25 Linear Regression and Correlation—General Analysis

Regression and correlation analysis concerning two variables based on a straight line will be presented in two parts: (1) general analysis, and (2) sampling analysis. The first part, which deals with general methods only for describing the average relationship between two given variables, is discussed in this chapter. In sampling analysis, the given data are used as a sample for estimating the population parameters and testing hypotheses. Sampling techniques are presented in the next chapter.

Two variables are frequently related or associated in some way or to some degree. Examples of two related variables are heights and weights of individual employees in a company, high school graduates and college freshmen enrolled in individual years during a period of time, and wages earned and amounts spent for recreation by individual workers in a nation.

Some new terms which are generally used in relationship analysis are explained in Section 25.1. The methods of obtaining the equation for the straight line and the measurement of the dispersion from the line are presented in Sections 25.2 and 25.3 respectively. The degree of closeness of the relationship based on the straight line is computed in Section 25.4. The above methods of computation are based on ungrouped data. The methods particular to grouped data are presented in Section 25.5.

25.1 Terminology in Relationship Analysis

The following new terms should be fully understood before actually analyzing relationships between variables.

Scatter Diagram

When two related variables, also called *bivariate data*, are plotted on a graph in the form of points or dots, the graph is called a *scatter diagram*, such as the diagrams shown in Chart 25–1. Each point on the diagram represents a pair of values, one based on X scale and the other based on Y scale. Making a scatter diagram usually is the first step in investigating the relationship between two variables, because the diagram shows visually the shape and degree of closeness of the relationship.

For example, diagrams A and B in Chart 25–1 indicate straight-line shaped relationships and diagram C suggests that the relationship is curvilinear. Diagrams

CHART 25–1

EXAMPLES OF SCATTER DIAGRAMS FOR BIVARIATE DATA
(A SET OF X AND Y VARIABLES)

A. *A Positive Linear Relationship*   B. *A Negative Linear Relationship*

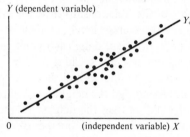

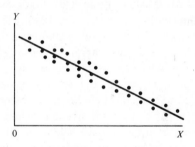

A. *A Positive Linear Relationship*   B. *A Negative Linear Relationship*

C.

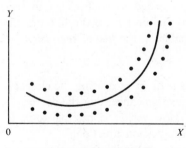

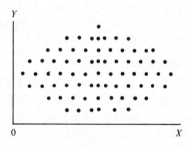

C. *A Curvilinear Relationship*   D. *No Relationship*

A and B also show the high degrees of closeness of the relationships. However, diagram C shows that the relationship is not very close. Further, diagram D indicates that there is no relationship since there is no adequate line or curve to describe the average relationship between the two variables.

The scatter diagram also indicates whether the relationship between the two variables is *positive* or *negative*. When there is a tendency for *large X* values paired with *large Y* values and for *small X* values paired with *small Y* values as shown in diagram A, the relationship between X and Y variables is said to be positive. On the other hand, when there is a tendency for *large X* values paired with *small Y* values and for *small X* values paired with *large Y* values as shown in diagram B, the relationship is said to be negative.

Regression Analysis

Regression analysis includes the techniques used in two major operations: (a) Derive an equation and a line representing the equation for describing the shape of the relationship between variables. The equation and its line, often called *regression equation* and *regression line* respectively, may be linear or curvilinear as shown in Chart 25–1. (b) Estimate one variable, called *dependent variable* (represented by Y in this chapter), from other variable (or variables), called *independent variable* (represented by X in this chapter), based on the relationship described by the regression equation.

The term *regression* was originated by Francis Galton in his work "Regression Towards Mediocrity in Hereditary Stature," published in the *Journal of the Anthropological Institute* in 1885. He analyzed the relationship between the average height of the two parents in a family and the average height of their adult children. A scatter diagram is constructed for the bivariate data, each point representing the heights of one family. The diagram resembles Chart 25–1A. As expected, in general tall parents tended to have tall children and short parents tended to have short children. However, he also found that the heights of children deviated less from the average height of all children than the heights of their parents from the average height of all parents. In other words, on the average, tall parents have tall children, but the children are not quite so tall as their parents; whereas, short parents have short children, but the children are not quite so short as their parents. The tall or short parents have children more mediocre than themselves. The heights of the children tended to go back or to *regress* toward the average height of the population. Galton termed the line describing the average relationship between the two variables as the *line of regression*.

Correlation Analysis

Correlation analysis refers to the techniques used in measuring the closeness of the relationship between variables. The computation concerning the degree of the closeness is based on the regression equation. However, it is possible to perform correlation analysis without actually having a regression equation.

Note that a high degree of correlation does not indicate a *cause and effect* relationship between variables. We may obtain a high correlation between two variables when the relationship has no real meaning. For example, when both the egg production on a small farm in Virginia and the highway accident rate in a city of Texas are rising, we can not conclude that one is the *cause* of the other. The high degree of correlation indicates only mathematical result. We should reach a conclusion based on logical reasoning and intelligent investigation on significantly related matters.

The regression and the correlation analysis may be *simple, multiple,* and *partial.* Simple analysis refers to only two variables, one dependent and the other independent. Multiple and partial analysis deals with three or more variables, one dependent and two or more independent.

25.2 Regression Equation and Line

In this chapter, only a straight line is used as a regression line in describing the shape of the average relationship between two variables. The straight line can be expressed by the linear equation

$$Y_c = a + bX$$

The methods of obtaining a regression equation and fitting a regression line are analogous to those used in time series analysis concerning secular trend. We may consider trend analysis as a special case of regression analysis. The independent variable X in trend analysis is limited to time units, whereas in regression analysis it may represent any unit desired.

As discussed in Chapter 22, a straight line can be obtained in many different ways. For example, the straight line shown in Chart 25–1A is obtained by the free-hand graphical method. The equation may then be derived from the line as in Section 22.2. However, the best way to obtain the linear equation is to use the method of least squares. The equation derived by this method will have a regression line which will give the best fit to the data. As we can see from a scatter diagram, a straight line frequently does not go through every point in the diagram. When a straight line can not fit the points perfectly, there are deviations between individual values (Y) and the values on the line (Y_c). The properties of the regression line based on the least square method concerning the deviations are similar to those of an arithmetic mean. The properties were stated on p. 567 and are restated here for reference.

1. The algebraic sum of the deviations of individual values (Y) from (above and below) the line (Y_c) is zero except the error due to rounding decimal places, or

$$\Sigma (Y - Y_c) = 0$$

2. The sum of the squared deviations from the line is the least, or

$\Sigma (Y - Y_c)^2$ is smaller than
$\Sigma (Y - $ a corresponding value on other straight line$)^2$

The two normal equations of the straight line by the method of least squares were presented in Section 22.2 and are also restated here for reference:

I. $\Sigma Y = na + b \Sigma X$
II. $\Sigma (XY) = a \Sigma X + b \Sigma X^2$

Solving the two equations simultaneously, we have constants a and b, which are also called *regression coefficients.*

$$a = \frac{\sum X^2 \cdot \sum Y - \sum X \cdot \sum (XY)}{n \sum X^2 - (\sum X)^2} \qquad \text{(25–1a)}$$

$$b = \frac{n \sum (XY) - \sum X \cdot \sum Y}{n \sum X^2 - (\sum X)^2} \qquad \text{(25–2a)}$$

Or, solving equation I only, we have constant a:

$$a = \frac{\sum Y}{n} - b\frac{\sum X}{n}$$

or $$a = \bar{Y} - b\bar{X} \qquad \text{(25–1b)}$$

The application of these formulas for obtaining a regression equation and fitting a regression line is illustrated in Example 1. The example also shows the method of estimating the dependent variable from the independent variable based on the equation.

Example 1

The first three columns of Table 25–1 show the amounts of sales (Y) made by a group of 8 salesmen in a company during a given period and the years of sales experience (X) of each salesman. (a) Plot a scatter diagram on a chart. (b) compute the linear regression equation by the least square method. (c) Draw the regression line based on the equation on the chart. (d) Estimate the amount of sales if a salesman has four years of sales experience.

TABLE 25–1

DATA AND COMPUTATION FOR EXAMPLE 1

(1) Salesman	(2) Amount of Sales (in $1,000) Y	(3) Years of Sales Experience X	(4) XY	(5) X^2	(6) Y^2
A	9	6	54	36	81
B	6	5	30	25	36
C	4	3	12	9	16
D	3	1	3	1	9
E	3	4	12	16	9
F	5	3	15	9	25
G	8	6	48	36	64
H	2	2	4	4	4
Total $(\sum)$	40	30	178	136	244

Solution. (a) The scatter diagram is plotted on Chart 25–2. Each point $(n = 8$ points) represents the amount of sales and the number of years of sales experience of a salesman.

(b) Columns (4) and (5) of Table 25–1 are used to obtain the required values $\sum (XY)$ and $\sum X^2$ for computing the linear regression equation based on the method of least squares. (Column (6) is added for use in Examples 2 and 3 and the X_c value to be presented later in the chapter.) By applying formula (25–2a),

$$b = \frac{8(178) - 30(40)}{8(136) - 30^2} = \frac{224}{188} = \frac{56}{47} = 1.19$$

By applying formula (25–1a),

$$a = \frac{136(40) - 30(178)}{8(136) - 30^2} = \frac{100}{188} = \frac{25}{47} = 0.53$$

Or, by applying formula (25–1b),

$$a = \frac{\Sigma Y}{n} - b\frac{\Sigma X}{n} = \frac{40}{8} - \frac{56}{47}\left(\frac{30}{8}\right) = \frac{25}{47} = 0.53$$

Thus, the linear regression equation $Y_c = a + bX$ can be written:

$$Y_c = 0.53 + 1.19X$$

(c) The line representing equation $Y_c = 0.53 + 1.19X$ is drawn on Chart 25–2. The line can be determined by any two points representing the coordinates

CHART 25–2

SCATTER DIAGRAM, REGRESSION LINES Y_c AND X_c BASED ON
THE METHOD OF LEAST SQUARES

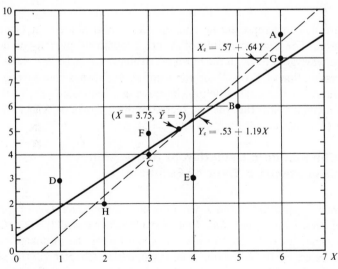

Years of Sales Experience

Source: Example 1.   Chart

of Y_c and X values. However, a third point is usually computed for checking the answers. The three points obtained must be on a straight line. For example, the three points may be obtained as follows:

$$\text{when } X = 1, \quad Y_c = 0.53 + 1.19(1) = 1.72$$
$$X = 4, \quad Y_c = 0.53 + 1.19(4) = 5.29$$
$$X = 6, \quad Y_c = 0.53 + 1.19(6) = 7.67$$

(d) The estimation of the amount of sales for a salesman who has 4 years of sales experience can be made from the regression equation by letting $X = 4$. Thus, $Y_c = 5.29$, which is computed in (c) above. The estimated amount of sales based on the average relationship is $5,290 ($=5.29 \times \$1,000$, the unit of Y value.)

Note that the regression equation of X values on Y values, or written

$$X_c = a_x + b_x Y$$

will not give the same regression line as Y_c for the same data. The values of b_x and a_x of equation X_c can be computed by formulas (25–2a) and (25–1b) respectively after interchanging X and Y in the formulas. Using the same data given in Example 1, the regression equation of X_c is obtained as follows:

$$b_x = \frac{n \sum (YX) - \sum Y \cdot \sum X}{n \sum Y^2 - (\sum Y)^2} = \frac{8(178) - 40(30)}{8(244) - 40^2} = \frac{224}{352} = \frac{7}{11} = 0.64$$

$$a_x = \frac{\sum X}{n} - b_x \frac{\sum Y}{n} = \frac{30}{8} - \frac{7}{11}\left(\frac{40}{8}\right) = \frac{25}{44} = 0.57$$

$$X_c = 0.57 + 0.64Y$$

The regression line representing equation X_c is also plotted on Chart 25–2 for comparison. The regression line of X on Y obtained by the method of least squares always goes through the mean of X values. Likewise, by the same method, the regression line of Y on X, or representing Y_c, always goes through the mean of Y values. This concept is helpful in drawing a regression line by the freehand graphic method. Note that the two regression lines shown on Chart 25–2 intersect at point ($\bar{X} = 30/8 = 3.75$, $\bar{Y} = 40/8 = 5$).

25.3 The Standard Deviation of Regression (The Standard Error of Estimate)

The standard deviation of the Y values from the regression line (Y_c) is called the *standard deviation of regression*. It is also popularly called the *standard error of estimate*, since it can be used to measure the error of the estimates of individual Y values based on the regression line. As stated in Chapter 12, the standard deviation of computed values (or a sampling distribution) is called standard error. Since Y values are original values, not computed values, the term standard error of estimate conflicts with the previous statement. We shall use the term standard deviation of regression in this text for the sake of consistency. Also, in order to avoid confusion, we shall use subscripts to indicate the type of standard deviation.

Thus,

$$s_y = \text{the standard deviation of } Y \text{ values from the mean } \bar{Y}$$
$$s_x = \text{the standard deviation of } X \text{ values from the mean } \bar{X}$$
$$s_{yx} = \text{the standard deviation of regression of } Y \text{ values from } Y_c$$
$$s_{xy} = \text{the standard deviation of regression of } X \text{ values from } X_c$$

The standard deviation of Y values from the regression line Y_c can be computed in a similar manner as the standard deviation of Y values from the arithmetic mean $\bar{Y}$. However, it is based on the points representing Y values scattered around the line. The closer the points to the line, the smaller will be the value of the standard deviation of regression. Thus, the estimates of Y values based on the line are more *reliable*. On the other hand, the wider the points scattered around the line, the larger will be the standard deviation of regression and the smaller will be the *reliability* of the estimates based on the line or the regression equation.

The general formula for the standard deviation of regression of Y values on X is

$$s_{yx} = \sqrt{\frac{\sum (Y - Y_c)^2}{n}} \tag{25-3a}$$

However, a simpler method of computing s_{yx} is to use formula:

$$s_{yx} = \sqrt{\frac{\sum Y^2 - a \sum Y - b \sum XY}{n}} \tag{25-3b}*$$

When formula (25–3b) is used, the values provided for obtaining the regression equation can be utilized. The only addition to the provided values is $\sum Y^2$, which can easily be obtained by adding a Y^2 column to Table 25–1 as shown in Example 1. Thus, the tedious work of computing Y_c and $(Y - Y_c)^2$ values can be avoided.

Example 2

Compute the standard deviation of regression of Y values for the data given in Example 1.

Solution.   By applying formula (25–3a), Table 25–2 should be arranged for the computation as follows:

---

*Proof for $\sum (Y - Y_c)^2 = \sum Y^2 - a \sum Y - b \sum XY$:

Since $\qquad (Y - Y_c)^2 = Y^2 - 2YY_c + Y_c^2$

then $\qquad \sum (Y - Y_c)^2 = \sum Y^2 - 2 \sum YY_c + \sum Y_c^2$

But, $\qquad YY_c = Y(a + bX) = aY + bXY$

and $\qquad \sum YY_c = a \sum Y + b \sum XY$

Also, $\qquad Y_c^2 = (a + bX)^2 = a^2 + 2abX + b^2 X^2$

and $\qquad \sum Y_c^2 = na^2 + 2ab \sum X + b^2 \sum X^2$
$$\qquad\qquad = a(na + b \sum X) + b(a \sum X + b \sum X^2)$$
$$\qquad\qquad = a \sum Y + b \sum XY$$

since the coefficients of a and b in the two terms are the expressions in the two normal equations.

Now, $\qquad \sum YY_c = \sum Y_c^2 = a \sum Y + b \sum XY$

Therefore, $\qquad \sum (Y - Y_c)^2 = \sum Y^2 - 2(a \sum Y + b \sum XY) + (a \sum Y + b \sum XY)$
$$\qquad\qquad\qquad\quad = \sum Y^2 - a \sum Y - b \sum XY$$

TABLE 25–2

COMPUTATION FOR EXAMPLE 2

(1) Salesman	(2) Y	(3) X	(4) $Y_c = 0.53 + 1.19X$	(5) $Y - Y_c$	(6) $(Y - Y_c)^2$
A	9	6	7.67	1.33	1.77
B	6	5	6.48	−0.48	0.23
C	4	3	4.10	−0.10	0.01
D	3	1	1.72	1.28	1.64
E	3	4	5.29	−2.29	5.24
F	5	3	4.10	0.90	0.81
G	8	6	7.67	0.33	0.11
H	2	2	2.91	−0.91	0.83
Total (Σ)	40	30	39.94	0.06	10.64*

*The two decimal places in each $(Y - Y_c)^2$ are significant digits since Y_c values are rounded to two decimal places.

Note that the sum of Y values is equal to the sum of Y_c values, or $\Sigma Y = \Sigma Y_c = 40$.** The discrepancy, $40 - 39.94 = 0.06$, is due to rounding decimal places in the regression equation.

By using the values obtained from Table 25–2, the standard deviation of regression is

$$s_{yx} = \sqrt{\frac{\Sigma (Y - Y_c)^2}{n}} = \sqrt{\frac{10.64}{8}} = \sqrt{1.33} = 1.15$$

By applying formula (25–3b), s_{yx} is computed directly from the values in Table 25–1 and the computed a and b in Example 1.

$$s_{yx} = \sqrt{\frac{\Sigma Y^2 - a \Sigma Y - b \Sigma XY}{n}}$$

$$= \sqrt{\frac{244 - \frac{25}{47}(40) - \frac{56}{47}(178)}{8}} = \sqrt{1.33}$$

$$= 1.15 \text{ (in units of \$1,000)}$$

The value of s_{yx} indicates the error range of the estimates of individual Y values. The interpretation of s_{yx} with respect to the line Y_c is similar to that of s_y with respect to the mean $\bar{Y}$. If Y values are normally distributed, 68 percent of the values will lie within a distance of one standard deviation of regression, or $1s_{yx}$, from (above and below) the line, approximately 95 percent for two

**Proof for $\Sigma Y = \Sigma Y_c$:

Since $Y_c = a + bX$

then $\Sigma Y_c = na + b \Sigma X = \Sigma Y$

(See normal equation I.)

standard deviations from the line, and approximately 99 percent for three standard deviations from the line. The 68 percent area for Example 2 is plotted on Chart 25-3 by the two dotted lines. The dotted lines representing $Y_c \pm 1s_{yx}$ can be determined by computing any two points for each line.

CHART 25-3

SCATTER DIAGRAM, REGRESSION LINE Y_c, AND STANDARD
DEVIATION OF REGRESSION s_{yx}

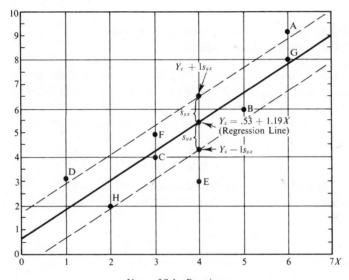

Source: Example 2.

For the line representing $Y_c + 1s_{yx}$:

when $X = 1$, $Y_c + s_{yx} = 1.72 + 1.15 = 2.87$
when $X = 4$, $Y_c + s_{yx} = 5.29 + 1.15 = 6.44$

For the line representing $Y_c - 1s_{yx}$:

when $X = 1$, $Y_c - s_{yx} = 1.72 - 1.15 = 0.57$
when $X = 4$, $Y_c - s_{yx} = 5.29 - 1.15 = 4.14$

Note that there are actually 5 (B, C, F, G, H) out of 8 points or $5/8 = 62.5\%$ within the range of $Y_c \pm 1s_{yx}$. The actual occurrence 62.5% is fairly close to the theoretical occurrence 68% in this case. If the number of points is increased, the percent of the actual occurrence can be expected to be closer to that of the theoretical occurrence for a normal distribution.

The standard deviation of regression is computed from the points representing Y values scattered around the regression line. Thus, the value of s_{yx} can be used as a measure of the degree of closeness of the relationship between two variables. For example, the higher the standard deviation of regression, the wider is the scatter of the individual points from the line and the smaller is the degree of closeness of the relationship. However, the use of the standard deviation of regression in measuring the degree of relationship is rather difficult since it is expressed in original units, such as dollars and miles. A more convenient way to measure the degree of relationship is to use a relative value as developed in the next section.

25.4   Coefficient of Determination (r^2) and Coefficient of Correlation (r)

The degree of closeness of the relationship between two variables can be measured by a relative value in either one of the two forms: (1) the coefficient of determination, denoted by r^2, or (2) the coefficient of correlation, denoted by r (which is the square root of r^2). The basic concept of the two relative measures is explained below.

CHART 25-4

DIAGRAM OF TOTAL DEVIATION $(Y - \bar{Y})$ BEING EQUAL TO
UNEXPLAINED DEVIATION $(Y - Y_c)$ + EXPLAINED DEVIATION $(Y_c - \bar{Y})$

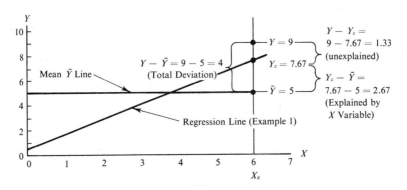

Note: $\bar{Y} = 40/8 = 5$.
Source: Based on point A, Examples 1 and 2.

The $\bar{Y}$, the arithmetic mean of Y values $= (\sum Y)/n$, is obtained without referring to X values. The Y_c, representing the regression line of Y values $= a + bX$, is obtained with the influence of X values. If Y values are related to X values to some degree, the deviations of Y values from $\bar{Y}$ must be reduced by an extent due to the introduction of X values in computing Y_c values. The extent of the reduction of the deviations is diagrammed in Chart 25-4. The diagram shows that at a given X value, denoted by X_g, the total deviation of Y from the mean $\bar{Y}$ is divided into two parts:

Total deviation = Unexplained deviation + Explained deviation, or

$$Y - \bar{Y} = (Y - Y_c) + (Y_c - \bar{Y})$$

The terms "explained" and "unexplained" are used here to indicate whether or not the part of the total deviation $(Y - \bar{Y})$ is reduced by the introduction of the X values in computing Y_c values. The explained deviation $(Y_c - \bar{Y})$ is affected or reduced by the use of the X variable. On the other hand, the unexplained deviation $(Y - Y_c)$ is retained or not reduced by the introduction of the regression line. Further, this relationship may be expressed:

Total variation = Unexplained variation + Explained variation, or

$$\sum (Y - \bar{Y})^2 = \sum (Y - Y_c)^2 + \sum (Y_c - \bar{Y})^2 \qquad \text{(25-4)}*$$

Based on the above expression, the coefficient of determination (r^2) is defined as the ratio of the explained variation to the total variation.

$$\text{Coefficient of determination} = \frac{\text{Explained variation}}{\text{Total variation}}$$

or symbolically,

$$r^2 = \frac{\sum (Y_c - \bar{Y})^2}{\sum (Y - \bar{Y})^2} \qquad \text{(25-5a)}$$

Note that when Y points all fall on the regression line, that is, $Y_c = Y$ or $\sum (Y_c - \bar{Y})^2 = \sum (Y - \bar{Y})^2$, the value of $r^2 = 1$, which indicates a *perfect correlation*. On the other hand, when the Y points are scattered far away from the regression line Y_c, then $\sum (Y - Y_c)^2$ becomes very large. Since the total variation is fixed, $\sum (Y_c - \bar{Y})^2$ now becomes very small. The value of r^2 will approach 0, which indicates that there is no correlation based on the straight regression line. The range of r^2 value is therefore from 0 to 1. Stated in a different way, when r^2 is close to 1, the Y values are very close to the regression line. Thus, the total variation of Y values is more explained by the line and the Y variable is closely related to the X variable. When r^2 is close to 0, the Y values are not close to the regression line. Thus, the total variation of Y values is mostly unexplained by the line and the Y variable is not nearly related to the X variable.

---

*Proof for equation $\sum (Y - \bar{Y})^2 = \sum (Y - Y_c)^2 + \sum (Y_c - \bar{Y})^2$:

Since
$$(Y - \bar{Y})^2 = Y^2 - 2Y\bar{Y} + \bar{Y}^2$$
the left side of the above equation can be written
$$\sum (Y - \bar{Y})^2 = \sum Y^2 - 2\bar{Y} \sum Y + n\bar{Y}^2 = \sum Y^2 - 2\bar{Y}(n\bar{Y}) + n\bar{Y}^2 = \sum Y^2 - n\bar{Y}^2$$
Likewise, extend the terms on the right side of the given equation.
$$\sum (Y - Y_c)^2 + \sum (Y_c - \bar{Y})^2 = \sum (Y^2 - 2YY_c + Y_c^2) + \sum (Y_c^2 - 2Y_c\bar{Y} + \bar{Y}^2)$$
$$= (\sum Y^2 - 2 \sum YY_c + \sum Y_c^2) + (\sum Y_c^2 - 2\bar{Y} \sum Y_c + \sum \bar{Y}^2)$$
We know that $\sum YY_c = \sum Y_c^2$ and that $\sum Y = \sum Y_c$ (pages 659–60.)
Then,
$$2\bar{Y} \sum Y_c = 2\bar{Y} \sum Y = 2\bar{Y}(n\bar{Y}) = 2n\bar{Y}^2$$
Thus,
$$\sum (Y - Y_c)^2 + \sum (Y_c - \bar{Y})^2$$
$$= (\sum Y^2 - 2 \sum Y_c^2 + \sum Y_c^2) + (\sum Y_c^2 - 2n\bar{Y}^2 + n\bar{Y}^2)$$
$$= \sum Y^2 - n\bar{Y}^2$$
which is the same as the extended left side of the given equation.

However, r^2 is always a positive number. It cannot tell whether the relationship between the two variables is positive or negative. Thus, the square root of r^2, or $\sqrt{r^2} = \pm r$, is frequently computed to indicate the direction of the relationship in addition to indicating the degree of the relationship. Since the range of r^2 is from 0 to 1, the coefficient of correlation r will vary within the range of $\sqrt{0}$ to $\sqrt{1}$, or from 0 to ± 1. The $+$ sign of r will indicate a positive correlation, whereas the $-$ sign will mean a negative correlation. The sign of r is the same as the sign of b (the slope) in the regression equation.

The coefficient of determination r^2 may also be written in terms of the variances (or standard deviations) instead of variations as follows:

$$\text{Coefficient of determination} = \frac{\text{Explained variance}}{\text{Total variance}}$$

This expression is derived from formula (25–5a). When both the numerator and the denominator of the fraction in the formula are divided by n, the explained variation and the total variation are changed to the explained variance and the total variance respectively.

$$r^2 = \frac{\dfrac{\sum (Y_c - \bar{Y})^2}{n}}{\dfrac{\sum (Y - \bar{Y})^2}{n}} = \frac{s^2_{y(c)}}{s^2_y} = \frac{\text{Explained variance}}{\text{Total variance}}$$

The symbol $s^2_{y(c)}$ is also called the variance of Y_c values. The mean of Y_c values is equal to the mean of Y values, or $\bar{Y}_c = \bar{Y}$, since $\sum Y_c = \sum Y$.

Furthermore, when each side of the equation of formula (25–4) is divided by n, the individual quotients may be written as follows:

$$s^2_y = s^2_{yx} + s^2_{y(c)}$$

which indicates that total variance = unexplained variance + explained variance.

Thus,
$$r^2 = \frac{s^2_{y(c)}}{s^2_y} = \frac{s^2_y - s^2_{yx}}{s^2_y}$$

or
$$r^2 = 1 - \frac{s^2_{yx}}{s^2_y} \tag{25-5b}*$$

The ratio of s^2_{yx} (unexplained variance) to s^2_y (total variance) in formula (25–5b) is frequently called the *coefficient of nondetermination*, denoted by k^2, or $k^2 = s^2_{yx}/s^2_y$. The square root of k^2 is called the *coefficient of alienation*, or $k = s_{yx}/s_y$. The k^2 and k values may also be used as measures of the degree of relationship between two variables. For example, the higher the unexplained variance s^2_{yx} with respect to the total variance s^2_y, the higher will be the value of k^2 and the value of k. However, r^2 and r are more convenient in interpreting the result of correlation

---

*Formula (25–5b) may also be derived from formula (25–5a) as follows:

$$r^2 = \frac{\sum (Y_c - \bar{Y})^2}{\sum (Y - \bar{Y})^2} = \frac{\sum (Y - \bar{Y})^2 - \sum (Y - Y_c)^2}{\sum (Y - \bar{Y})^2} = 1 - \frac{\sum (Y - Y_c)^2}{\sum (Y - \bar{Y})^2}$$

$$= 1 - \frac{\sum (Y - Y_c)^2/n}{\sum (Y - \bar{Y})^2/n} = 1 - \frac{s^2_{yx}}{s^2_y}$$

analysis. In the above example, we may state that the higher the unexplained variance, the smaller will be the value of r^2 and the value of r.

The application of formulas (25–5a) and (25–5b) is illustrated in Example 3. In practice, however, formula (25–5b) is preferred over formula (25–5a). The value of s_{yx}^2 is frequently available or desirable in correlation analysis. In addition, the process of computing s_y^2 may be simplified by the short methods as discussed in Chapter 7.

Example 3

Compute the coefficient of determination r^2 and the coefficient of correlation r for the data given in Example 1.

Solution.   By applying formula (25–5a), first arrange Table 25–3 for obtaining the required value of the denominator; that is, $\Sigma (Y - \bar{Y})^2 = 44$.

The numerator in formula (25–5a) is computed by using formula (25–4):

$$\Sigma (Y_c - \bar{Y})^2 = \Sigma (Y - \bar{Y})^2 - \Sigma (Y - Y_c)^2$$
$$= 44 - 10.64$$
$$= 33.36 \qquad \text{(Column [6] of Table 25–2.)}$$
$$r^2 = \frac{\Sigma (Y_c - \bar{Y})^2}{\Sigma (Y - \bar{Y})^2} = \frac{33.36}{44} = 0.7582$$
$$r = \sqrt{0.7582} = 0.87$$

TABLE 25–3

COMPUTATION FOR EXAMPLE 3

(1) Salesman	(2) Y	(3) $Y - \bar{Y}$	(4) $(Y - \bar{Y})^2$
A	9	4	16
B	6	1	1
C	4	−1	1
D	3	−2	4
E	3	−2	4
F	5	0	0
G	8	3	9
H	2	−3	9
Total (Σ)	40	0	44

$$\bar{Y} = 40/8 = 5$$

By applying formula (25–5b), the r^2 is obtained as follows: the variance of regression for Y on X has been computed in Example 2, or

$$s_{yx}^2 = 1.33$$

The variance of Y values is

$$s_y^2 = \frac{\Sigma (Y - \bar{Y})^2}{n} = \frac{44}{8} = 5.5$$

$$r^2 = 1 - \frac{1.33}{5.5} = 0.7582, \text{ or round to } 76\%$$

Thus, 76% of the variation of Y values has been reduced or explained by the regression line. Or, stated in a different way, 76% of the variation in the amounts of sales (Y) is linearly related with the variation in the years of sales experience (X) by the salesmen. Since the sign of b in the regression equation $Y_c = 0.53 + 1.19X$ is positive, or $b = +1.19$, the value of r, also the direction of correlation, must be positive.

There is a third method to compute r^2 and r. This method uses the *product-moment formula* for r:

$$r = \frac{\sum (xy)}{ns_x s_y} \tag{25-5c}*$$

where $x = X - \bar{X}$ and $y = Y - \bar{Y}$.

The mean product $[\sum (xy)]/n$, which represents the sum of the products of the paired values of x and y divided by the number of pairs n, is called the

---

*Proof—Formula (25-5c). We may shift the location of the origin from $X = 0$ and $Y = 0$ to $X = \bar{X}$ and $Y = \bar{Y}$, or the means, as shown in the diagram below. The shift does not affect the location of the regression line, r, b, s_y, s_x, and s_{yx}.

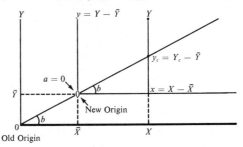

However, when the new origin is used, the equation $Y_c = a + bX$ is changed to $y_c = bx$ since the regression line goes through the new origin and $a = 0$. Thus, formula (25-5a) becomes

$$r^2 = \frac{\sum (Y_c - \bar{Y})^2}{\sum (Y - \bar{Y})^2} = \frac{\sum y_c^2}{\sum y^2} = \frac{\sum b^2 x^2}{\sum y^2} = \frac{b^2 \sum x^2}{\sum y^2} = \frac{b^2 n s_x^2}{n s_y^2}$$

and

$$r = \frac{bs_x}{s_y}$$

But, the second normal equation

$$\sum XY = a \sum X + b \sum X^2$$

now (based on New Origin) is

$$\sum xy = a \sum x + b \sum x^2$$

Since $\sum x = 0$ and $a \sum x = 0$,

$$\sum xy = b \sum x^2$$

and

$$b = \frac{\sum xy}{\sum x^2}$$

Therefore

$$r = \frac{\sum xy}{\sum x^2} \cdot \frac{s_x}{s_y} = \frac{\sum xy}{n s_x^2} \cdot \frac{s_x}{s_y} = \frac{\sum xy}{n s_x s_y}$$

covariance of the bivariate distribution of X and Y variables. The covariance is also termed the *first product-moment* of the distribution about the means $\bar{X}$ and $\bar{Y}$.

The product-moment formula can be expressed in detailed form to facilitate the computation procedure as follows:

$$r = \frac{n \sum (XY) - (\sum X)(\sum Y)}{\sqrt{[n \sum X^2 - (\sum X)^2][n \sum Y^2 - (\sum Y)^2]}} \qquad (25\text{--}5d)*$$

Formula (25–5d) avoids the computation for the deviations x and y. It gives a direct method to compute r. The sign of r computed by this formula will have the same sign as b in a regression equation. The required values are similar to those for computing the regression equation except the additional values of Y^2 for $\sum Y^2$. Note that the numerator and the first factor in the denominator of formula (25–5d) are the same as the terms of formula (25–2a) for b.

Example 3 is computed by using formula (25–5d) below. The required values are obtained from Table 25–1.

$$r = \frac{8(178) - 30(40)}{\sqrt{[8(136) - (30)^2][8(244) - (40)^2]}} = \frac{224}{\sqrt{[188][352]}}$$

$$= \frac{224}{257.25} = 0.87$$

$$r^2 = 0.87^2 = 0.7569$$

which is slightly different from the previous answer 0.7582 due to rounding in r.

From formula (25–5b), we may have a new formula for the standard deviation of regression of Y on X:

$$s_{yx} = s_y \sqrt{1 - r^2} \qquad (25\text{--}3c)$$

If we know the values of r^2 and s_y, we may compute s_{yx}. Example 2 is now computed by using formula (25–3c).

$$r^2 = 0.7582$$

and
$$s_y^2 = 5.5$$

or
$$s_y = \sqrt{5.5} = 2.35$$

$$s_{yx} = 2.35\sqrt{1 - 0.7582} = 2.35(0.49) = 1.15$$

(The same answer is on p. 671.)

---

*Proof: Formula (25–5d) is derived from formula (25–5c). (Footnote for p. 660)

$$\sum (xy) = \sum [(X - \bar{X})(Y - \bar{Y})] = \sum [XY - \bar{X}Y - X\bar{Y} + \bar{X}\bar{Y}]$$
$$= \sum XY - \bar{X} \sum Y - \bar{Y} \sum X + n\bar{X}\bar{Y} = \sum XY - \bar{X}n\bar{Y} - \bar{Y}n\bar{X} + n\bar{X}\bar{Y}$$
$$= \sum XY - n\bar{X}\bar{Y}$$

From the above expression and formula (7–7c), formula (25–5c) can be written:

$$r = \frac{\sum xy}{n s_x s_y} = \frac{\sum XY - n\bar{X}\bar{Y}}{n\sqrt{\frac{\sum X^2}{n} - \left(\frac{\sum X}{n}\right)^2} \cdot \sqrt{\frac{\sum Y^2}{n} - \left(\frac{\sum Y}{n}\right)^2}}$$

$$= \frac{n(\sum XY - n\bar{X}\bar{Y})}{n\sqrt{\frac{\sum X^2}{n} - \left(\frac{\sum X}{n}\right)^2} \cdot n\sqrt{\frac{\sum Y^2}{n} - \left(\frac{\sum Y}{n}\right)^2}}$$

$$= \frac{n \sum XY - \sum X \cdot \sum Y}{\sqrt{[n \sum X^2 - (\sum X)^2][n \sum Y^2 - (\sum Y)^2]}}$$

The formula for the standard deviation of regression of X on Y is

$$s_{xy} = s_x\sqrt{1 - r^2}$$

It is obvious that s_{yx} will be different from s_{xy} if s_y differs from s_x. In the case of linear correlation the value of r is the same regardless of whether X or Y is considered the independent variable. This fact can be seen from the product-moment formula for r. The formula clearly shows the symmetry between x and y values. Thus, the r (or r^2) can be viewed as a measure of the degree of closeness between X and Y in two ways:

1. Concerning correlation analysis—It is a measure of mutuality of relationship between two variables, which are distributed normally and vary jointly (a bivariate normal distribution). If there is only one variable and the other is a constant, there will be no correlation, or $r = 0$. For example if $(Y = 2, X = 5), (Y = 3, X = 5), (Y = 7, X = 5)$ and so on, X being a constant number 5, the value of r will be zero since $x = X - \bar{X} = 5 - 5 = 0$ and $\sum (xy) = 0$.

2. Concerning regression analysis—It is a measure of the closeness of fit of the regression line. This measure indicates the precision of the estimation of the dependent variable from the independent variable based on the regression line. We usually let Y be the dependent variable and X be the independent variable. If we changed the order, that is, Y being the independent and X being the dependent variable, the measure of precision (r or r^2) would be the same.

25.5 Grouped Data

The discussion in the previous sections dealt with ungrouped data. When the techniques used in regression and correlation analysis are extended to grouped data, care should be taken with the frequency factor f in each class interval of both X and Y variables. The formulas for computing various measures from grouped data, arranged in order for the convenience of computation, are given below.

Let ① $= n \sum fd_xd_y - (\sum f_xd_x)(\sum f_yd_y)$
② $= n \sum f_xd_x^2 - (\sum f_xd_x)^2$
③ $= n \sum f_yd_y^2 - (\sum f_yd_y)^2$

where $f = $ the frequency in a paired class of X and Y variables
$f_x = $ the frequency in a class of X variable
$f_y = $ the frequency in a class of Y variable
$n = $ the total of frequencies $= \sum f = \sum f_x = \sum f_y$
$d_x = $ the deviation of mid-point of a class of X variable from the assumed mean in class-interval units,
$d_y = $ the deviation in class-interval units for Y variable.

The formula for the constant b of the regression equation, formula (25–2a), is now written:

$$b = \frac{①}{②} \cdot \frac{i_y}{i_x} \qquad (25\text{–}2b)$$

where i_y and i_x represent the size of class interval of Y variable and X variable respectively.

The formula for the constant a of the regression equation, formula (25–1b), is now written:

$$a = \bar{Y} - b\bar{X}$$

where, based on formula (5–3c),

$$\bar{Y} = A_y + \left(\frac{\sum f_y d_y}{n}\right)i_y$$

and

$$\bar{X} = A_x + \left(\frac{\sum f_x d_x}{n}\right)i_x \tag{25–1c}$$

A being the assumed mean.

The product-moment formula for the coefficient of correlation, formula (25–5d), is now written:

$$r = \frac{①}{\sqrt{②\cdot③}} \tag{25–5e}$$

The formula for the standard deviation of regression, formula (25–3c), is now written:

$$S_{yx} = \frac{i_y\sqrt{③(1 - r^2)}}{n} \tag{25–3d}*$$

When these formulas for grouped data are applied, a *correlation table* is usually arranged to facilitate the computation. A correlation table is also called a *bivariate frequency table*. It shows the frequency distribution of two variables. The use of the table is illustrated in the example below.

Example 4

The tallies in Table 25–4 show the amounts of sales (Y) made by a group of 40 salesmen during a given period and the years of sales experience (X) corresponding to individual salesmen. The tallies in the last cell of the first column, for example, indicate that two salesmen with "0 and under 2" years of sales experience made the amounts of sales from $500 to $3,500 (the real limits 0.5 and 3.5 of class "1 — 3" in units of $1,000). Compute (a) the regression equation and draw a straight line on the table based on the equation, (b) the coefficient of correlation, and (c) the standard deviation of regression.

---

*Proof for formula (25–3d).
The standard deviation formula, formula (7–8d), can be written:

$$s_y = i_y\sqrt{\frac{\sum f_y d_y^2}{n} - \left(\frac{\sum f_y d_y}{n}\right)^2} = \frac{i_y\sqrt{n\cdot\sum f_y d_y^2 - (\sum f_y d_y)^2}}{n} = \frac{i_y\sqrt{③}}{n}$$

$$s_{yx} = s_y\sqrt{1 - r^2} = \frac{i_y\sqrt{③}}{n}\cdot\sqrt{1 - r^2} = \frac{i_y\sqrt{③(1 - r^2)}}{n}$$

TABLE 25-4

CROSS-CLASSIFICATION TABLE—EXAMPLE 4

X—Years of Sales Experience

Class Interval	0 and under 2	2 and under 4	4 and under 6	6 and under 8	8 and under 10	Total Number of Salesmen, f_y
16–18					/	1
13–15			/	/	⫪⫪ 14.1	5
10–12			⫪⫪⫪	⫪⫪ $Y_c = 3.1 + 1.1X$		7
7–9		⫪⫪⫪	⫪⫪⫪⫪⫪⫪	⫪⫪⫪ /		21
4–6		///		/		4
1–3	3.1 //					2
Total Number of Salesmen, f_x	2	8	16	10	4	40

Y—Amounts of sales (in $1,000)

Solution. 1. Table 25–5 is arranged according to the data given in Table 25–4. The number of tallies in each cell is counted and entered into the corresponding cell of Table 25–5 on the second line, denoted by the symbol f.

2. The assumed means are selected as follows:

A_x, assumed mean for X variable = 5 years, the midpoint of class "4 @ 6".
A_y, assumed mean for Y variable = \$8,000, the midpoint of class "7 – 9".

The numbers in d_x row and d_y column are thus determined— +1, +2 and so on representing the numbers of class-interval units of individual classes above (or larger than) the assumed mean class (where d_x or d_y = 0); −1, −2, and so on representing the numbers below (or smaller than) the assumed mean class.

3. The values of $f d_x d_y$ are computed for individual cells. The product of d_x and d_y is entered on the first line of each cell. The product of f, d_x, and d_y is on the third line. For example, the numbers in the last cell of the first column represent the values as follows:

1st line: $[-2$ (d_x in X class "0 @ 2")]$\cdot[-2$ (d_y in Y class "1 – 3")] = 4
or $d_x \cdot d_y = 4$

2nd line:
   2 = the frequency in the paired X class "0 @ 2" and Y class "1 – 3,"
or $f = 2$

3rd line: $8 = 4 \times 2$
or $f d_x d_y = 8$

TABLE 25–5

CORRELATION TABLE—EXAMPLE 4

X—Years of Sales Experience ($i_x = 2$ years)

Class Interval	0 @ *2	2 @ 4	4 @ 6 $A_x = 5$	6 @ 8	8 @ 10	f_y	d_y	$f_y d_y$	$f_y d_y^2$	$fd_x d_y$
16–18					$2 \cdot 3 = 6$ ×1 — 6	1	3	3	9	6
13–15			$0 \cdot 2 = 0$ ×1 — 0	$1 \cdot 2 = 2$ ×1 — 2	$2 \cdot 2 = 4$ ×3 — 12	5	2	10	20	14
10–12			$0 \cdot 1 = 0$ ×5 — 0	$1 \cdot 1 = 1$ ×2 — 2		7	1	7	7	2
7–9 $A_y = 8$		$-1 \cdot 0 = 0$ ×5 — 0	$0 \cdot 0 = 0$ ×10 — 0	$1 \cdot 0 = 0$ ×6 — 0		21	0	0	0	0
4–6		$-1 \cdot -1 = 1$ ×3 — 3		$1 \cdot -1 = -1$ ×1 — -1		4	-1	-4	4	2
1–3	$-2 \cdot -2 = 4$ ×2 $= f$ — 8					2	-2	-4	8	8
f_x	2	8	16	10	4	40	—	12	48	32
d_x	-2	-1	0	1	2	—				
$f_x d_x$	-4	-8	0	10	8	6				
$f_x d_x^2$	8	8	0	10	16	42				

Left margin label: Y — Amounts of Sales in $1,000 ($i_y = 3$)

* @ represents "and under".

Values obtained from Table 25–5:

$\sum f_x d_x = 6$, $\sum f_x d_x^2 = 42$, $n = \sum f_x = \sum f_y = \sum f = 40$, $\sum f_y d_y = 12$, $\sum f_y d_y^2 = 48$, $\sum fd_x d_y = 32$.

The sum of the products on each row is entered into the last column at right with the heading or caption $fd_x d_y$. Thus, the second number in the $fd_x d_y$ column, or $14 = 0 + 2 + 12$.

Note that the value of $fd_x d_y$ in each cell of the row or in the column of the assumed mean class is always equal to zero, since d_x and $d_y = 0$ in the classes. The computation on the row or column for the values thus is not necessary except for illustrative purposes.

4. Compute and sum the required values as shown in Table 25–5. The values obtained from Table 25–5 are summarized and placed directly under the table. The summarized values are now substituted in the formulas as follows:

$$① = n \sum fd_xd_y - (\sum f_xd_x)(\sum f_yd_y) = 40(32) - 6(12) = 1{,}208$$
$$② = n \sum f_xd_x^2 - (\sum f_xd_x)^2 = 40(42) - 6^2 = 1{,}644$$
$$③ = n \sum f_yd_y^2 - (\sum f_yd_y)^2 = 40(48) - 12^2 = 1{,}776$$

(a) The regression equation—use formula (25–2b) for b.

$$b = \frac{①}{②} \cdot \frac{i_y}{i_x} = \frac{1{,}208}{1{,}644} \cdot \frac{3}{2} = \frac{151}{137} = 1.1$$
$$a = \bar{Y} - b\bar{X} = 8.9 - 1.1(5.3) = 3.1$$

The $\bar{Y}$ and $\bar{X}$ values are computed by using formula (25–1c)

$$\bar{Y} = A_y + \left(\frac{\sum f_yd_y}{n}\right)i_y = 8 + \left(\frac{12}{40}\right)3 = 8.9$$
$$\bar{X} = A_x + \left(\frac{\sum f_xd_x}{n}\right)i_x = 5 + \left(\frac{6}{40}\right)2 = 5.3$$

The regression equation is

$$Y_c = 3.1 + 1.1X$$

The following two points are computed from the equation for drawing the regression line on Table 25–4:

When $X = 0$, the lower limit of the first X class "0 @ 2",
$$Y_c = 3.1 + 1.1(0) = 3.1$$
When $X = 10$, the upper limit of the last X class "8 @ 10",
$$Y_c = 3.1 + 1.1(10) = 14.1$$

(b) The coefficient of correlation—use formula (25–5e).

$$r = \frac{①}{\sqrt{②\cdot③}} = \frac{1{,}208}{\sqrt{(1{,}644)(1{,}776)}} = \frac{1{,}208}{\sqrt{2{,}919{,}744}} = \frac{1{,}208}{1{,}709} = 0.71$$
$$r^2 = 0.71^2 = 0.5041 \text{ or } 50.41\%$$

Thus, 50.41% of the variation of the amounts of sales is explained by the variation of the years of sales experience of the group of 40 salesmen.

(c) The standard deviation of regression—use formula (25–3d).

$$s_{yx} = \frac{i_y\sqrt{③(1-r^2)}}{n} = \frac{3\sqrt{1{,}776(1-0.5041)}}{40} = \frac{3(29.68)}{40} = 2.226 \text{ or } \$2{,}226$$

25.6   Summary

This chapter analyzes the average relationship between two variables based on a straight line. When two related variables, also called bivariate data, are plotted on a graph in the form of points or dots, the graph is called a scatter diagram. Making a scatter diagram usually is the first step in investigating the relationship between two variables.

Regression analysis includes the techniques used in two major operations: (a) Derive an equation and a line representing the equation for describing the shape of

SUMMARY OF FORMULAS

Application	Formula	Formula Number	Reference Page
Straight line			
Regression equation	$Y_c = a + bX$	22–1	561
Solution for formula (22–1) by the method of least squares from two normal equations (page 666)	$a = \dfrac{\sum X^2 \cdot \sum Y - \sum X \cdot \sum (XY)}{n \sum X^2 - (\sum X)^2}$	25–1a	656
	$a = \bar{Y} - b\bar{X} = \dfrac{\sum Y}{n} - b\dfrac{\sum X}{n}$	25–1b	656
Grouped data only	$a = \bar{Y} - b\bar{X}$, where $\bar{Y} = A_y + \dfrac{\sum f_y d_y}{n} i_y$, and $\bar{X} = A_x + \dfrac{\sum f_x d_x}{n} i_x$	25–1c	669
In general	$b = \dfrac{n \sum (XY) - \sum X \cdot \sum Y}{n \sum X^2 - (\sum X)^2}$	25–2a	656
Grouped data only	$b = \dfrac{n \sum f_x d_x d_y - (\sum f_x d_x)(\sum f_y d_y)}{n \sum f_x d_x^2 - (\sum f_x d_x)^2} \cdot \dfrac{i_y}{i_x}$	25–2b	668
Standard deviation of regression (or standard error of estimate)	$s_{yx} = \sqrt{\dfrac{\sum (Y - Y_c)^2}{n}}$	25–3a	659
For Y values on X values	$s_{yx} = \sqrt{\dfrac{\sum Y^2 - a \sum Y - b \sum XY}{n}}$	25–3b	659
	$s_{yx} = s_y \sqrt{1 - r^2}$	25–3c	667
Grouped data only	$s_{yx} = \dfrac{i_y \sqrt{[n \sum f_y d_y^2 - (\sum f_y d_y)^2](1 - r^2)}}{n}$	25–3d	669
Coefficient of determination (r^2) and coefficient of correlation (r)			
Basic relationship between variations	$\sum (Y - \bar{Y})^2 = \sum (Y - Y_c)^2 + \sum (Y_c - \bar{Y})^2$	25–4	663
Variation method	$r^2 = \dfrac{\sum (Y_c - \bar{Y})^2}{\sum (Y - \bar{Y})^2}$	25–5a	663
Variance method	$r^2 = 1 - \dfrac{s_{yx}^2}{s_y^2}$	25–5b	664
Product-moment method	$r = \dfrac{\sum (xy)}{n s_x s_y}$	25–5c	666
	$r = \dfrac{n \sum XY - (\sum X)(\sum Y)}{\sqrt{[n \sum X^2 - (\sum X)^2][n \sum Y^2 - (\sum Y)^2]}}$	25–5d	667
Grouped data only	$r = \dfrac{n \sum f_x d_x d_y - (\sum f_x d_x)(\sum f_y d_y)}{\sqrt{[n \sum f_x d_x^2 - (\sum f_x d_x)^2][n \sum f_y d_y^2 - (\sum f_y d_y)^2]}}$	25–5e	669

the relationship between variables. (b) Estimate one variable (dependent variable) from other variable or variables (independent variables) based on the relationship described by the regression equation. In addition, the standard deviation of regression (also called the standard error of estimate) is used to measure the error of the estimates of individual Y values (dependent variable) based on the equation. In a scatter diagram, the closer the points representing Y values to the line, the smaller will be the value of the standard deviation of regression. It thus indicates the reliability of the estimates.

Correlation analysis refers to the techniques used in measuring the closeness of the relationship between variables. The computation of a measure concerning the degree of the closeness is based on the regression equation. The standard deviation of regression can also be used as a measure of the degree of closeness. However, it has the weakness of being expressed in original units. The measures expressed in relative terms are called the coefficient of determination (r^2) and the coefficient of correlation (r). The coefficient of determination is a ratio of the explained variation to the total variation. It is also a ratio of the explained variance to the total variance.

When the techniques used in regression and correlation analysis are extended to grouped data, care should be taken to the frequency factor in each class interval of both X and Y variables. The computation of the required values for the analysis is usually facilitated by arranging a correlation table. A correlation table is also called a bivariate frequency table since it shows the frequency distribution of two related variables.

Exercises 25

1. Explain briefly:
   (a) Scatter diagram.
   (b) The standard deviation of regression.
   (c) The coefficient of determination.
   (d) The coefficient of correlation.
   (e) The correlation table.
   (f) Covariance.

2. State the difference:
   (a) Linear and curvilinear regression.
   (b) Regression equation and line.
   (c) Regression and correlation analysis.
   (d) Dependent and independent variable.
   (e) Simple, multiple, and partial analysis concerning the number of variables.
   (f) Explained and unexplained variance.

3. Table 25-6 shows the hours (X) studied for a test and the number of correct answers (Y) made on the test by each of the five employees in a company.
   (a) Plot a scatter diagram on a chart.
   (b) Compute the linear regression equation by the least square method. (To 2 decimal places.)
   (c) Draw the regression line based on the equation on the chart.
   (d) Estimate the number of correct answers for an employee who has studied three hours for the test.

TABLE 25–6

DATA FOR PROBLEM 3

Employee	X, Hours studied	Y, Number of correct answers
A	1	2
B	2	10
C	6	20
D	7	14
E	5	11

(e) Compute the standard deviation of regression s_{yx} by formulas (25–3a) and (25–3b).

(f) Draw two lines which are above and below the regression line by one s_{yx}. What is the interval estimate (or the range of the estimation) in (d) above if the estimation is based on 68% reliability?

(g) Compute the coefficient of determination (r^2) and coefficient of correlation (r) by formula (25–5a). Interpret the values of r^2 and r.

4. Refer to Problem 3.

(a) Compute the standard deviation of regression s_{yx} by formula (25–3b).

(b) Compute the coefficient of determination (r^2) and the coefficient of correlation (r) by formulas (25–5b) and (25–5d).

5. Table 25–7 shows the statistics grades and the economics grades of a group of nine students in a college at the end of this semester.

(a) Plot a scatter diagram on a chart.

(b) Compute the regression equations by the method of least squares for: (to two decimal places)

     1. Y_c—Consider Y as the dependent variable and X as the independent variable.

     2. X_c—Consider X as the dependent variable and Y as the independent variable.

TABLE 25–7

DATA FOR PROBLEM 5

Students	Statistics Grades Y	Economics Grades X
A	95	88
B	51	70
C	49	65
D	27	50
E	42	60
F	52	80
G	67	68
H	48	49
I	46	40
Total	477	570

$n = 9$
$\Sigma Y = 477$
$\Sigma X = 570$
$\Sigma Y^2 = 28{,}133$
$\Sigma X^2 = 37{,}994$
$\Sigma (XY) = 31{,}893$

(c) Draw the regression lines for Y_c and X_c on the same chart.

6. Use the given information and the answers obtained in Problem 5.
    (a) Compute the standard deviation of regression s_{yx} and s_{xy}.
    (b) Compute the coefficient of determination r^2 and the coefficient of correlation r.
    (c) (1) Estimate the statistics grades for students J and K who have economics grades 82 and 75 respectively in this semester.
        (2) Estimate the economics grades for students L and M who have statistics grades 50 and 90 respectively in this semester.
    (d) How reliable would you consider the estimates in (c)?

7. The relationship between the amount spent for advertising (X) and the amount of sales (Y) during the period of the past ten years in Cindy Company can be expressed in the following regression equation:

$$Y_c = 42.5 + 3.16X, \text{ with } X \text{ unit: } \$1,000$$
$$Y \text{ unit: } \$1,000$$

Estimate the amount of sales based on the given equation if the company plans to spend $10,000 for advertising this year.

8. Refer to Problem 7. Let $s_{yx} = \$500$. Find the interval estimate if it is based on the confidence level of approximately (a) 68%, and (b) 95%.

9. The cross-classification table (Table 25–8) shows the units produced in the last week by a group of 20 workers in a factory and the months of working experiences of individual workers.

TABLE 25–8

DATA FOR PROBLEM 9

X—Months of Working
Experiences of Workers

	Class Interval	0 @ 2	2 @ 4 $A_z = 3$	4 @ 6
Y—Units Produced	14 @ 18			||
	10 @ 14 $A_y = 12$	||	|||	|||
	6 @ 10	⧕|	||	
	2 @ 6	||		

@ represents "and under."

(a) Find the regression equation.
(b) Draw the regression line based on the equation on the table.
(c) Compute the coefficient of correlation and the coefficient of determination.
(d) Compute the standard deviation of regression.

10. Dale Advertising Agency collected data (Table 25–9) showing the income and the expenditures for recreation by 110 families in Lawrence City during the last year:

(a) Use the above information to complete the cross-classification table (Table 25–10).

(b) Let the assumed mean for X distribution be $7,000 and Y distribution be $450. Set up a correlation table and compute the following:

    1. The coefficient of correlation (r).

    2. The regression equation.

    3. The standard deviation of the Y variable (s_y).

    4. The standard deviation of regression for Y on X (s_{yx}).

    5. Plot the regression line on Table 25–10 [See (a)].

    6. Plot two straight lines which are above and below the regression line by the value of 3 s_{yx}.

    7. If a family has an annual income of $10,000, how much expenditures of recreation would you estimate for this family during the year based on the regression equation?

    8. How reliable would you consider the estimate in (7)?

TABLE 25–10

CROSS-CLASSIFICATION TABLE FOR PROBLEM 10

X—Family Income (Unit: $1,000)

Class Interval	0 and under 2	2 and under 4	4 and under 6	6 and under 8	8 and under 10	10 and under 12
7 and under 8						
6 and under 7						
5 and under 6						
4 and under 5						
3 and under 4						
2 under 3						
1 under 2						
0 under 1						

Y—Expenditures for Recreation (Unit: $100)

TABLE 25–9

DATA FOR PROBLEM 10

Order of Families Interviewed	Family Income	Expenditures for Recreation	Order of Families Interviewed	Family Income	Expenditures for Recreation
1	$1,500	$520	56	$6,420	$570
2	6,280	465	57	5,260	485
3	1,680	95	58	6,500	368
4	5,400	420	59	9,400	470
5	8,300	460	60	4,950	420
6	10,600	730	61	7,400	460
7	5,900	160	62	11,800	780
8	4,900	340	63	9,200	470
9	7,500	540	64	7,800	540
10	7,600	360	65	4,900	480
11	1,800	50	66	5,800	490
12	8,950	375	67	7,200	430
13	10,850	720	68	4,700	460
14	5,800	345	69	5,800	480
15	7,450	425	70	6,900	580
16	9,500	375	71	6,200	360
17	7,250	460	72	7,800	480
18	5,420	370	73	9,400	460
19	9,420	460	74	5,200	480
20	4,300	350	75	6,300	520
21	950	20	76	6,400	555
22	8,500	470	77	6,400	580
23	5,600	475	78	4,800	425
24	6,900	580	79	6,530	520
25	7,000	380	80	7,520	535
26	4,600	420	81	1,560	80
27	9,500	420	82	5,920	360
28	6,500	470	83	4,670	470
29	4,860	360	84	11,800	750
30	6,300	580	85	6,540	476
31	9,530	470	86	7,100	358
32	4,700	120	87	10,600	640
33	6,800	360	88	9,200	650
34	7,500	420	89	6,300	380
35	4,830	460	90	8,400	360
36	8,570	428	91	6,000	370
37	4,850	350	92	9,500	620
38	6,800	580	93	6,520	350
39	5,800	425	94	9,200	460
40	5,800	424	95	7,100	360
41	8,400	530	96	4,600	160
42	7,600	340	97	5,400	460
43	10,300	560	98	7,800	580
44	9,400	450	99	4,900	475
45	11,700	650	100	6,200	320
46	8,320	350	101	9,400	380
47	5,700	150	102	8,300	670
48	6,800	520	103	9,600	685
49	9,500	570	104	7,100	340
50	8,200	365	105	9,500	650
51	4,750	168	106	8,000	610
52	10,800	790	107	7,580	360
53	7,200	460	108	8,600	670
54	4,900	485	109	7,300	360
55	8,650	468	110	9,800	690

26 *Linear Regression and Correlation— Sampling Analysis

The methods introduced in this chapter for analyzing the linear relationship between two variables of a population are based on the analysis of a sample drawn from the population. The sampling methods involve the procedures of estimating a population parameter from a corresponding sample statistic and testing hypotheses as discussed in Chapters 13 to 15. In the following discussion we shall first present a population regression model for illustrating the basic concept of the population in regression analysis. The problems of estimation are presented in Sections 26.2 to 26.6, followed by the tests of hypotheses in Sections 26.7 to 26.9.

26.1 Linear Regression Model for Population

There are many types of population models for regression analysis. For simplicity, we shall limit our discussion to the population which can be expressed by a linear regression line, or linear regression model. The regression model is shown on Chart 26–1 and can be generalized under the following assumptions:

1. The independent values X are known or fixed. The dependent values Y on each X are many and are distributed normally. The distribution of Y values on each X is considered as a *subpopulation*. The population therefore consists of all subpopulations.

2. Each subpopulation of Y values on a given X has a mean, denoted by μ_{yx}. The μ_{yx} values are on a straight line, called the *population regression line*. The line can be expressed by the linear regression equation

$$\mu_{yx} = A + BX$$

where A and B are called *population regression coefficients*. μ_{yx} is the expected value of each Y.

CHART 26–1

LINEAR REGRESSION MODEL (A POPULATION MODEL)

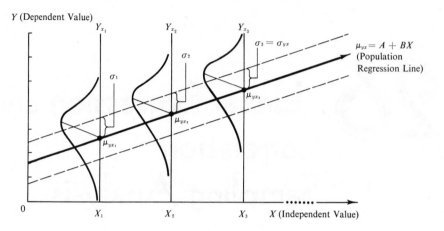

Let the deviation of each Y from μ_{yx} be ϵ (epsilon). Then,

$$Y = \mu_{yx} + \epsilon, \text{ or } Y = A + BX + \epsilon$$

We assume that the distribution of ϵ values is normal and the mean of ϵ values is 0.

3. Each subpopulation of Y values on a given X has a variance σ^2 and also a standard deviation σ. The σ values are equal for all subpopulations. The common standard deviation σ thus is the same as the standard deviation of Y values around the regression line, denoted by σ_{yx}. The σ_{yx} is also called the *population standard deviation of regression.* Let the subscripts 1, 2, 3, ... represent the subpopulations. Then,

$$\sigma_1 = \sigma_2 = \sigma_3 = \cdots = \sigma_{yx}$$

Note that these assumptions of the population model may not fit perfectly to most business and economic data. However, the model will give a satisfactory result if the relationship between X and Y values is approximately linear. The application of the above model is illustrated in Example 1. Instead of using a large number of Y values of each subpopulation, we shall use only 4 Y values on each X for simplification.

Example 1

Assume that the grade points (Y) of 20 students for a mathematics survey test are grouped according to the semester hours (X) in mathematics taken by individual students in a college as shown in Table 26–1. Consider the 20 grouped grades as a population. Find (a) the population regression equation, and (b) the population standard deviation of regression.

TABLE 26–1

DATA FOR EXAMPLE 1

	X-Semester hours				
	1	2	3	4	5
Y—Grade Points	5	6	7	8	9
	4	⑤	6	⑦	⑧*
	②	3	④	5	6
	1	2	3	4	5
Sum of Y's in each column, or $\Sigma\,Y_x$	12	16	20	24	28
Mean of Y's in each column, or μ_{yx} $=\Sigma\,Y_x \div 4$	3	4	5	6	7

*The circled Y values are selected as a sample for Example 2.

Solution.   (a) The population regression equation is based on the mean of Y values on each X, or values μ_{yx}. The scatter diagram on Chart 26–2 shows the Y values on each X as well as the means μ_{yx}. Since the points representing μ_{yx} are on a straight line, the linear regression equation can be determined by observing the points as

$$\mu_{yx} = A + BX = 2 + 1X$$

The constant A is the ordinate of Y at $X = 0$. The coefficient B is the slope, or

CHART 26–2

SCATTER DIAGRAM OF THE POPULATION DISTRIBUTION
AND THE MEANS OF SUBPOPULATIONS—EXAMPLE 1

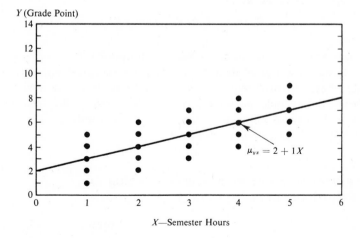

X—Semester Hours

$B = 1$ grade point/1 semester hour $= 1$. Note that the constants A and B may also be obtained by the method of least squares from all Y values.

(b) The population standard deviation of regression (or the population standard error of estimate) is the same as the common standard deviation of the Y values on each X from the mean μ_{yx}. Also, the population variance σ_{yx}^2 is the same as the common variance σ^2; that is,

$$\sigma^2 \text{ (of } Y\text{'s at } X = 1) = \sigma^2 \text{ (of } Y\text{'s at } X = 2) = \cdots = \sigma_{yx}^2 \text{ (of the population)}$$

Since the variances and the standard deviations of the five subpopulations in this example are equal, we have to compute only the values for one of the subpopulations. The variance and the standard deviation of Y values on $X = 1$, for example, are computed for the purpose of finding the population standard deviation.

	Y	$Y - \mu_{yx_1}$	$(Y - \mu_{yx_1})^2$
	5	2	4
	4	1	1
	2	-1	1
	1	-2	4
Total	12	0	10

$$\mu_{yx_1} = \frac{12}{4} = 3$$

$$\sigma_1^2 = \frac{\Sigma (Y - \mu_{yx_1})^2}{N_1} = \frac{10}{4} = 2.5$$

$$\sigma_1 = \sqrt{2.5} = 1.58$$

Thus, $\sigma_{yx}^2 = 2.5$ and $\sigma_{yx} = 1.58$ (grade points).

Note: The population variance of regression may also be computed from all the five subpopulations by the following expression:

$$\sigma_{yx}^2 = \frac{\Sigma (Y - \mu_{yx})^2}{N} = \frac{5(10)}{5(4)} = \frac{10}{4} = 2.5$$

The number of Y values in each of the five subpopulations is only four in Example 1. If we have a large number of Y values in each subpopulation and the Y's are distributed normally with $\sigma_{yx} = 1.58$, we may draw a normal curve for each subpopulation similar to those shown on Chart 26–1. We then may state that when $X = 1$, there will be 68 percent of Y values within $\mu_{yx} \pm 1\sigma_{yx} = 3 \pm 1.58 = 1.42$ to 4.58 grade points, and so on.

26.2   Estimating Population Regression Line
(Point Estimate of μ_{yx} by Y_c)

As stated in Chapter 13, an estimate of a population parameter may by expressed in two ways: a point estimate and an interval estimate. The point estimate of a population parameter is a corresponding sample statistic, such as the measures in Table 26–2.

TABLE 26–2

SYMBOLS FOR CORRESPONDING STATISTICS AND PARAMETERS

Measure	Symbol for Parameter (Population)	Symbol for Statistic (Sample)
Subpopulation Mean of Y Values on a Given X	μ_{yx}	Y_c
Population Regression Coefficients	A B	a b
Population Standard Deviation of Regression	σ_{yx}	s_{yx}
Population Coefficient of Correlation	ρ	r

The interval estimate is specified by the upper and lower confidence limits, which are computed from the sample statistic and the standard error of the statistic, or

$$\binom{\text{Confidence}}{\text{limits}} = \binom{\text{Sample}}{\text{statistic}} \pm \binom{z \ (\text{or } t \text{ for small sample}) \ times}{\text{standard error of the statistic}}$$

The estimation of population regression line is to find a sample regression line. That is, estimate the population regression equation

$$\mu_{yx} = A + BX$$

by the sample regression equation

$$Y_c = a + bX$$

where Y_c is a point estimate of μ_{yx}
     a is an unbiased estimate of A and
     b is an unbiased estimate of B

This estimation is illustrated in Example 2.

Example 2

Assume that a sample is taken from the population given in Example 1 (see circled Y values in Table 26–1) and is shown in the first two columns of Table 26–3. Estimate the population regression line based on the sample.

Solution. The sample regression coefficients a and b are computed by applying formulas (25–1b) and (25–2a) respectively as follows:

$$b = \frac{n \sum (XY) - \sum X \cdot \sum Y}{n \sum X^2 - (\sum X)^2} = \frac{5(92) - 15(26)}{5(55) - 15^2} = \frac{70}{50} = 1.4$$

$$a = \frac{\sum Y}{n} - b\frac{\sum X}{n} = \frac{26}{5} - 1.4\left(\frac{15}{5}\right) = 1$$

TABLE 26–3

DATA AND COMPUTATION FOR EXAMPLE 2

Semester hours X	Grade points Y	XY	X^2
1	2	2	1
2	5	10	4
3	4	12	9
4	7	28	16
5	8	40	25
Total 15	26	92	55

The sample regression equation is

$$Y_c = 1 + 1.4X$$

which is used as the estimated population regression equation, or Y_c is an estimate of μ_{yx}. The estimated population regression line based on the equation is plotted on Chart 26–3 along with the true population regression line for comparison purposes. Of course, in practice, the true population regression line is unknown and is to be estimated.

26.3 Estimating Population Standard Deviation of Regression (Point Estimate of σ_{yx} by $\hat{s}_{yx}$)

The unbiased estimate of the population variance of regression (σ_{yx}^2) based on a sample is

$$\hat{s}_{yx}^2 = \frac{\Sigma (Y - Y_c)^2}{n - m}$$

The estimate of the population standard deviation of regression (σ_{yx}) therefore is

$$\hat{s}_{yx} = \sqrt{\frac{\Sigma (Y - Y_c)^2}{n - m}} \qquad \text{(26–1a)}$$

where $m =$ the number of constants in the estimating equation. In the linear regression equation, there are two constants, a and b. Thus, $m = 2$. Like the unbiased estimate of population variance of X variable [formula 13–1 or $\hat{s}^2 = \Sigma x^2/(n-1)$], the denominator $(n - m = n - 2)$ is the number of degrees of freedom. Since two constants, a and b, are used in the estimation of the straight regression line, two degrees of freedom are lost.

Formula (26–1a) is similar to formula (25–3a) for s_{yx} except that n is replaced by $(n - m)$ in the denominator of formula (26–1a). Thus, formula $\hat{s}_{yx}$ can be written in terms of s_{yx} as

$$\hat{s}_{yx} = \sqrt{\frac{\Sigma (Y - Y_c)^2}{n} \cdot \frac{n}{n - m}}$$

CHART 26–3

SCATTER DIAGRAM OF THE SAMPLE DATA, THE ESTIMATED POPULATION
REGRESSION LINE (Y_c) AS COMPARED WITH THE TRUE LINE (μ_{yx}),
AND INTERVAL ESTIMATE OF μ_{yx}

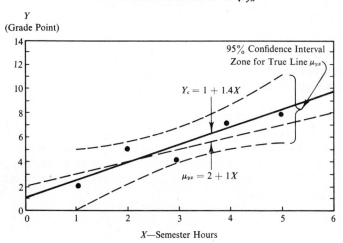

Source: Examples 2 and 4.

or

$$\hat{s}_{yx} = s_{yx}\sqrt{\frac{n}{n-m}} \qquad (26\text{–}1b)$$

The value of s_{yx} may also conveniently be computed by formula (25–3b).

Example 3

Estimate the population standard deviation of regression from the sample given in Example 2.

Solution. Table 26–4 is arranged for computing $\hat{s}_{yx}$ by formula (26–1a). The

TABLE 26–4

COMPUTATION FOR EXAMPLE 3

Semester Hours X	Grade Points Y	$Y_c = 1 + 1.4X$	$Y - Y_c$	$(Y - Y_c)^2$
1	2	2.4	−0.4	0.16
2	5	3.8	1.2	1.44
3	4	5.2	−1.2	1.44
4	7	6.6	0.4	0.16
5	8	8.0	0.0	0.00
Total 15	26	26.0	0.0	3.20

$$\hat{s}^2_{yx} = \frac{\Sigma (Y - Y_c)^2}{n-2} = \frac{3.20}{5-2} = \frac{3.20}{3} = 1.07$$

$$\hat{s}_{yx} = \sqrt{1.07} = 1.03 \text{ (grade points).}$$

values of Y_c are computed from the estimated population regression equation $Y_c = 1 + 1.4X$. For example, when

$$X = 3, Y_c = 1 + 1.4(3) = 5.2$$

The same answer may be obtained when formulas (26–1b) and (25–3b) are applied for this example.

26.4 Interval Estimate of μ_{yx}

The Y_c value is a point estimate of μ_{yx} on a given X. It is a sample statistic since there is a Y_c on a given X for each sample. The standard error of statistic Y_c, denoted by σ_{y_c} or called the standard deviation of a sampling distribution of statistic Y_c on a given X, depends on the value of population standard deviation of regression σ_{yx}. When σ_{yx} is estimated by $\hat{s}_{yx}$, the estimated standard error of Y_c, represented by s_{y_c}, is written

$$s_{y_c} = \hat{s}_{yx}\sqrt{\frac{1}{n} + \frac{(X - \bar{X})^2}{\sum (X - \bar{X})^2}} \qquad (26\text{–}2a)^*$$

Note that the estimated standard error of Y_c, or s_{y_c}, also depends on the values of sample size n, given X [for $(X - \bar{X})^2$], and the sum of variation $\sum (X - \bar{X})^2$. There is a s_{y_c} for each given X.

This distribution of the Y_c values based on straight line equations is a t-distribution with $(n - 2)$ degrees of freedom. The mean or the expected value of the

---

*The variance of the distribution of Y_c's (respresented by $\sigma_{y_c}^2$):

Since $\qquad\qquad\qquad\qquad\qquad a = \bar{Y} - b\bar{X}$

then $\qquad\qquad Y_c = a + bX = (\bar{Y} - b\bar{X}) + bX = \bar{Y} + b(X - \bar{X})$

Table 15-2 shows that the variance of the *difference* between two sample means is the *sum* of variances of the two sample means. It is also true that the variance of the *sum* of two sample means is the *sum* of variances of the two sample means. In a similar manner,

$$\text{Variance of } Y_c = \text{Variance of } [\bar{Y} + b(X - \bar{X})] = \text{Var. } \bar{Y} + \text{Var. } b(X - \bar{X})$$

The factor $(X - \bar{X})$ is a fixed number. When each value (b) of a distribution is multiplied by a fixed number, the variance of the distribution is multiplied by the fixed number squared. Thus, the above equation may be written:

$$\sigma_{y_c}^2 = \sigma_{\bar{y}}^2 + (X - \bar{X})^2 \cdot \sigma_b^2$$

Like the variance of the sample mean $\bar{X}$, or $\sigma_{\bar{x}}^2 = \sigma^2/n$, the variance of the sample mean $\bar{Y}$ related to X's is

$$\sigma_{\bar{y}}^2 = \frac{\sigma_{yx}^2}{n}$$

because Variance of $\bar{Y}$ = Variance of $(\sum Y/n)$, where n is a fixed number, or sample size. Variance of $\bar{Y} = (1/n^2)$ (Variance of $\sum Y$), or

$$\sigma_{\bar{y}}^2 = \frac{1}{n^2}(\sigma_{y(x=1)}^2 + \sigma_{y(x=2)}^2 + \cdots + \sigma_{y(x=n)}^2) = \frac{n\sigma_{yx}^2}{n^2} = \frac{\sigma_{yx}^2}{n}$$

The variance of the slope b is:

$$\sigma_b^2 = \frac{\sigma_{yx}^2}{\sum (X - \bar{X})^2}$$

Now, $\qquad \sigma_{y_c}^2 = \frac{\sigma_{yx}^2}{n} + (X - \bar{X})^2 \cdot \frac{\sigma_{yx}^2}{\sum (X - \bar{X})^2} = \sigma_{yx}^2\left(\frac{1}{n} + \frac{(X - \bar{X})^2}{\sum (X - \bar{X})^2}\right)$

sampling distribution of Y_c is μ_{yx}.* Based on the concept illustrated in Chapter 13, we may state the interval estimate of μ_{yx} from Y_c according to the t-distribution:

$$\text{Confidence limits} = Y_c \pm t \cdot s_{y_c}$$

Example 4

Refer to the sample given in Example 2. Estimate the mean of the grade points for each given number of semester hours at a 95% confidence interval.

Solution.   (1) First, find the estimated standard error of Y_c, or s_{y_c}, by applying formula (26–2a). The computation is shown in Table 26–5. Each s_{y_c} is obtained by substituing $\hat{s}_{yx} = 1.03$ (Example 3), $n = 5$, $(X - \bar{X})^2 = 4, 1, 0, 1, 4$ respectively as shown in column (3), and $\sum (X - \bar{X})^2 = 10$ in the formula.

TABLE 26–5

COMPUTATION FOR EXAMPLE 4

X	$X - \bar{X}$	$(X - \bar{X})^2$	$\hat{s}_{yx}\sqrt{\dfrac{1}{n} + \dfrac{(X-\bar{X})^2}{\sum(X-\bar{X})^2}} = s_{y_c}$
1	-2	4	$1.03\sqrt{\dfrac{1}{5} + \dfrac{4}{10}} = 1.03(0.7746) = 0.80$
2	-1	1	$1.03\sqrt{\dfrac{1}{5} + \dfrac{1}{10}} = 1.03(0.5477) = 0.56$
3	0	0	$1.03\sqrt{\dfrac{1}{5} + \dfrac{0}{10}} = 1.03(0.4472) = 0.46$
4	1	1	$1.03\sqrt{\dfrac{1}{5} + \dfrac{1}{10}} = 1.03(0.5477) = 0.56$
5	2	4	$1.03\sqrt{\dfrac{1}{5} + \dfrac{4}{10}} = 1.03(0.7746) = 0.80$
Total 15	0	10	

$$\bar{X} = 15/5 = 3$$

(2) Next, find the upper and lower confidence limits for Y_c. The limits depend upon the confidence coefficient t. At 95% confidence interval, the value of t with $3 (= n - 2 = 5 - 2)$ degrees of freedom is 3.182 (5% or 0.05 area in both tails, or 0.025 area in each tail). The limits are computed in Table 26–6 and are plotted on Chart 26–3.

The limits are symmetric around $X = \bar{X} = 3$. When the points representing the limits are connected on Chart 26–3, the two lines, which indicate the confidence intervals, are not straight. Note that the width of the confidence interval on $X = \bar{X} = 3$ is the smallest. The farther X is from $\bar{X} = 3$, the larger will be the width of the confidence interval on that X.

---

*The mean or the expected value of the sampling distribution of Y_c:

$$\bar{Y}_c = \frac{\sum Y_c}{N} = \frac{\sum (a + bX)}{N} = \frac{\sum a}{N} + \frac{\sum bX}{N} = A + BX = \mu_{yx}$$

where N represents the number of possible samples.

TABLE 26-6

COMPUTATION FOR EXAMPLE 4, CONTINUED

X	Y_c^*	$t \cdot s_{y_c}$	Lower Limit $Y_c - t \cdot s_{y_c}$	Upper Limit $Y_c + t \cdot s_{y_c}$
1	2.4	$3.182(0.80) = 2.55$	-0.15	4.95
2	3.8	$3.182(0.56) = 1.75$	2.05	5.55
3	5.2	$3.182(0.46) = 1.46$	3.74	6.66
4	6.6	$3.182(0.56) = 1.75$	4.85	8.35
5	8.0	$3.182(0.80) = 2.55$	5.45	10.55

*Table 26-4.

The confidence interval zone shown by the dotted lines on Chart 26-3 is constructed from a single sample. If we take many samples from a large population, we may construct a confidence interval zone for each of the samples. The interpretation of the confidence interval zone is the same as that for the confidence interval stated in Chapter 13. There are 95 out of 100 chances that the true population regression line (or μ_{yx}) will fall within the zone constructed for a sample. In other words, if we constructed 100 sample confidence interval zones, we will expect that 95 of the zones will contain the true population regression line.

When the sample size is large, say $n = 100$ or more, the ratio of the single squared deviation $(X - \bar{X})^2$ to the sum of n squared deviations $\sum (X - \bar{X})^2$ becomes small even for the X values not close to $\bar{X}$. At the same time, when n is large, $\hat{s}_{yx}$ approaches s_{yx}. (See formula 26-1b.) If the ratio is ignored and $\hat{s}_{yx}$ is replaced by s_{yx}, formula (26-2a) may be written:

$$s_{y_c} = s_{yx}\sqrt{\frac{1}{n}} = \frac{s_{yx}}{\sqrt{n}} \qquad (26\text{-}2b)$$

Also, when the sample size is large, the value of t approaches the normal deviate z, such as at 95% confidence interval $z = 1.96$, which is the limit of t with ∞ degrees of freedom. Thus, the interval estimate of μ_{yx} may be expressed for large sample size as follows:

$$\text{Confidence limits} = Y_c \pm z \cdot s_{y_c} = Y_c \pm z \cdot \frac{s_{yx}}{\sqrt{n}}$$

The computation of the above expression is based on the table of the areas under the normal curve.

Note that formula (26-2b) resembles the formula for the standard error of the sampling distribution of the mean, formula (13-3),

$$s_{\bar{x}} = \frac{s}{\sqrt{n}}$$

where s is the estimated population standard deviation based on a large sample.

26.5 Interval Estimate of Individual Y

Sometimes we may wish to predict an individual Y value on a given X, such as predicting the grade points of mathematics survey test for a student who has 4 (or $X = 4$) semester hours in mathematics. We are not asking the average grade μ_{yx} where $X = 4$.

The Y values on a given X are normally distributed. If we know the true mean μ_{yx}, the interval estimate of the Y values on a given X will be $\mu_{yx} \pm$ a number of standard deviations of Y values specified by the level of confidence interval. However, we usually do not know the μ_{yx} value, but estimate it by a Y_c value from a sample. The distribution of the sample Y_c values on a given X is normal. When we wish to estimate an individual Y value, therefore, we should consider both the variance (or standard deviation) of Y values as well as the variance (or standard error) of Y_c values. In fact, the sum of the two variances is equal to the variance of the difference between Y and Y_c, or

$$\text{Variance of } (Y - Y_c) = \text{Variance of } Y + \text{Variance of } Y_c$$

or

$$\sigma^2_{(y-y_c)} = \sigma^2_{yx} + \sigma^2_{y_c}$$

This expression can be proved intuitively in a similar manner to those shown in Table 15–2 concerning the variance of the difference between two sample means being equal to the sum of variances of the two sample means.

When the variances of Y and Y_c are estimated from a sample, the above equation is written:

$$s^2_{(y-y_c)} = \hat{s}^2_{yx} + s^2_{y_c} \tag{26-3a}$$

Thus, the variance of the difference $(Y - Y_c)$ is obtained by combining the variance of the regression ($\hat{s}^2_{yx}$) and the variance of the estimated mean of Y's on a given X ($s^2_{y_c}$). Substitute the value of $s^2_{y_c}$ as indicated by formula (26–2a) in formula (26–3a), we have

$$s^2_{(y-y_c)} = \hat{s}^2_{yx} + \hat{s}^2_{yx}\left(\frac{1}{n} + \frac{(X - \bar{X})^2}{\Sigma(X - \bar{X})^2}\right)$$

or

$$s_{(y-y_c)} = \hat{s}_{yx}\sqrt{1 + \frac{1}{n} + \frac{(X - \bar{X})^2}{\Sigma(X - \bar{X})^2}} \tag{26-3b}$$

The distribution of the difference $(Y - Y_c)$ for small samples is a t distribution with $n - 2$ degrees of freedom.

Example 5

Refer to the sample given in Example 2. Estimate the individual grades for each given number of semester hours at a 95% confidence interval.

Solution. (1) Find the estimated standard error of $(Y - Y_c)$, or $s_{(y-y_c)}$. Since the values of $s^2_{y_c}$ are already computed in Example 4–1, formula (26–3a) is applied in this case. The results are shown in Table 26–7.

TABLE 26–7

COMPUTATION FOR EXAMPLE 5

(1) X	(2) $\hat{s}^2_{yx}$ (Example 3)	(3) $s^2_{y_c}$ (Example 4–1)	(4) $s^2_{(y-y_c)}$ (2) + (3)	(5) $s_{(y-y_c)}$ Square Root of (4)
1	1.07	$0.80^2 = 0.64$	1.71	1.31
2	1.07	$0.56^2 = 0.31$	1.38	1.17
3	1.07	$0.46^2 = 0.21$	1.28	1.13
4	1.07	$0.56^2 = 0.31$	1.38	1.17
5	1.07	$0.80^2 = 0.64$	1.71	1.31

(2) Find the upper and lower confidence limits for Y's. Again, the limits depend upon the confidence coefficient t. At 95% confidence interval, the value of t with 3 ($= n - 2 = 5 - 2$) degrees of freedom is 3.182. The limits are computed in Table 26–8 and are plotted on Chart 26–4. Note that the confidence interval for individual Y values is wider than the confidence interval for the mean of the Y values on a given X. This is always true since the interval estimate of a specific Y value must consider the standard deviation of Y distribution ($\hat{s}_{yx}$) in addition to the standard error of Y_c distribution (s_{y_c}).

TABLE 26–8

COMPUTATION FOR EXAMPLE 5, CONTINUED

X	Y_c^*	$t \cdot s_{(y-y_c)}$	Lower Limit $Y_c - t \cdot s_{(y-y_c)}$	Upper Limit $Y_c + t \cdot s_{(y-y_c)}$
1	2.4	$3.182(1.31) = 4.17$	−1.77	6.57
2	3.8	$3.182(1.17) = 3.72$	0.08	7.52
3	5.2	$3.182(1.13) = 3.60$	1.60	8.80
4	6.6	$3.182(1.17) = 3.72$	2.88	10.32
5	8.0	$3.182(1.31) = 4.17$	3.83	12.17

*Table 26–4.

Thus, the predicted grades for the students who have 3 semester hours are from 1.60 to 8.80 points. Again, if we take many samples from a large population, we may construct a confidence interval zone of the Y's for each of the samples. If we constructed 100 sample confidence interval zones, we will except that 95 of the zones will contain the Y value on a given X.

Note that when the sample size n is *very large*, the variance of Y_c, or $s^2_{y_c} = s^2_{yx}/n$ (formula 26–2b), becomes very small. At the same time, when n is very large, $\hat{s}^2_{yx}$ approaches s^2_{yx} (formula 26–1b). If $s^2_{y_c}$ is ignored and $\hat{s}^2_{yx}$ is replaced by s^2_{yx}, formula (26–3a) may be written:

$$s^2_{(y-y_c)} = s^2_{yx}$$

CHART 26–4

Interval Estimate of Individual Y Values on a Given X as
Compared with the True Y Values of the Population

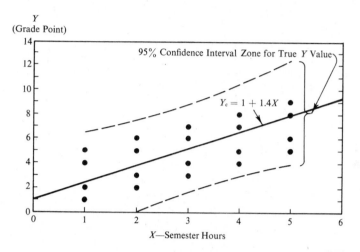

X—Semester Hours

Also, when the sample size is very large, the value of t approaches the normal deviate z. Thus, the interval estimate of Y may be approximated for large sample size as follows:

$$\text{Confidence limits} = Y_c \pm z \cdot s_{yx}$$

The computation of the above expression should be based on the table of the areas under the normal curve. If the upper and lower limits at a 68% confidence interval are plotted on a chart, the points representing the limits should be on two straight lines similar to the two dotted lines on Chart 25–3.

26.6   Estimating Coefficient of Determination ρ^2 by $\hat{r}^2$

The population correlation coefficient of a bivariate normal distribution is represented by the Greek letter rho ρ. The population coefficient of determination is the letter squared, or ρ^2, and can be derived in a similar manner to formula (25–5b) for r^2. It is usually written:

$$\rho^2 = 1 - \frac{\sigma_{yx}^2}{\sigma_y^2} \tag{26–4}$$

When a sample is used to estimate ρ^2, the formula becomes

$$\hat{r}^2 = 1 - \frac{\hat{s}_{yx}^2}{\hat{s}_y^2} \tag{26–5a}$$

or

$$\hat{r}^2 = 1 - (1 - r^2)\left(\frac{n-1}{n-m}\right) \qquad (26\text{–}5b)^*$$

where r^2 can be obtained from formula (25–5), and $m = 2$ for a simple linear regression equation.

Example 6

Estimate the population coefficient of determination and the population coefficient of correlation based on the sample given in Example 2.

Solution. By formula (26–5a),

$$\hat{s}_{yx}^2 = \frac{3.20}{5-2} = 1.07 \qquad \text{(Example 3, p. 685)}$$

$$\hat{s}_y^2 = \frac{22.80}{5-1} = 5.7 \qquad \text{(See Table 26–9)}$$

$$\hat{r}^2 = 1 - \frac{1.07}{5.7} = 0.81, \text{ or } 81\%$$

$$\hat{r} = \sqrt{0.81} = 0.90$$

TABLE 26–9

COMPUTATION FOR EXAMPLE 6

Y	$Y - \bar{Y}$	$(Y - \bar{Y})^2$
2	−3.2	10.24
5	−0.2	0.04
4	−1.2	1.44
7	1.8	3.24
8	2.8	7.84
26	0.0	22.80
$\bar{Y} = 26/5 = 5.2$		

By formula (26–5b),

$$r^2 = 1 - \frac{s_{yx}^2}{s_y^2} = 1 - \frac{3.20/5}{22.80/5} = 1 - \frac{3.20}{22.80} = \frac{19.60}{22.80} = 0.8596$$

$$r = \sqrt{0.8596} = 0.927, \text{ or round to } 0.93$$

$$\hat{r}^2 = 1 - (1 - 0.8596)\left(\frac{5-1}{5-2}\right) = 0.81 \qquad \text{(Same as above.)}$$

For small samples the factor $(n-1)/(n-2)$ in formula (26–5b) is important since it will give a downward adjustment for r^2 and for r, such as $r^2 = 0.86$

---

*Proof of formula (26–5b):

$$\hat{r}^2 = 1 - \frac{s_{yx}^2}{s_y^2} = 1 - \frac{s_{yx}^2[n/(n-m)]}{s_y^2[n/(n-1)]} = 1 - (1 - r^2)\left(\frac{n}{n-m} \cdot \frac{n-1}{n}\right)$$

$$= 1 - (1 - r^2)\left(\frac{n-1}{n-m}\right)$$

versus $\hat{r}^2 = 0.81$ and $r = 0.93$ versus $\hat{r} = 0.90$ in Example 6 when $n = 5$. For large samples, however, the factor is usually ignored since it approaches 1. When the factor is ignored, $\hat{r}^2$ equals r^2, since $\hat{r}^2 = 1 - (1 - r^2) = r^2$ by formula (26–5b).

Example 7

Estimate the population coefficient of determination if the sample coefficient of determination is 0.64 and the sample size is 101.

Solution. By applying formula (26–5b), $r^2 = 0.64$, $n = 101$, and

$$\hat{r}^2 = 1 - (1 - 0.64)\left(\frac{101 - 1}{101 - 2}\right) = 1 - 0.36(0.99) = 0.6436, \text{ or round to } 0.64$$

Thus,

$$r^2 = {}^2\hat{r} = 0.64$$

and

$$r = \hat{r} = \sqrt{0.64} = 0.80$$

The standard error of the correlation coefficient r can be estimated from a large sample based on the formula below:

$$s_r = \frac{1 - r^2}{\sqrt{n - 1}} \qquad (26\text{–}6)*$$

In the above formula, r^2, instead of $\hat{r}^2$ since it is a large sample, is the estimated population coefficient of determination.

Note that the distribution of sample r's is only approximately normal for large samples. The formula is therefore inadequate for small samples. Even with large samples, the distribution of r's is quite skewed when the true value of correlation coefficient approaches ± 1. This is obvious since the range of ρ is from -1 to $+1$. If ρ is close to $+1$, such as $+0.99$, the sample r's will have a maximum range of only 0.01 in the direction to upper limit, and a possible range of 1.99 in the direction to lower limit.

Examples 6 and 7 illustrated the method of finding the point estimates of ρ^2 and ρ. Example 8 will illustrate the method in finding an interval estimate of ρ.

Example 8

Find the interval estimate of the population coefficient of correlation based on the sample given in Example 7 at a 95% confidence interval.

Solution. By applying formula (26–6),

$$s_r = \frac{1 - r^2}{\sqrt{n - 1}} = \frac{1 - 0.64}{\sqrt{101 - 1}} = 0.036$$

Upper confidence limit $= r + z \cdot s_r = 0.80 + 1.96(0.036)$
$$= 0.80 + 0.07 = 0.87$$

---

*The population standard error of r is

$$\sigma_r = \frac{1 - \rho^2}{\sqrt{n - 1}}$$

Lower confidence limit $= r - z \cdot s_r = 0.80 - 1.96(0.036)$
$$= 0.73$$

Thus, there are 95 out of 100 chances that the true population coefficient of correlation ρ will fall within the range from 0.73 to 0.87.

26.7   Testing Hypothesis—the Population Coefficient of Correlation $\rho = \rho_0$ by the Z-Transformation

Since the distribution of r is only approximately normal for large samples and not normal for small samples, R. A. Fisher transformed r into Z by using logarithms as follows:

$$Z = 1.1513 \log_{10} \left(\frac{1+r}{1-r} \right) \tag{26-7}*$$

The distribution of Z values is close to normal except for very small samples. The standard error of Z is

$$\sigma_z = \frac{1}{\sqrt{n-3}} \tag{26-8}$$

The procedure of testing hypothesis by using Z is similar to those presented in Chapter 15 for sample mean $\bar{X}$ and sample proportion p. The null hypothesis, or H_0 in the present case, is that there is no difference between a given population coefficient of correlation ρ and the assumed population coefficient of correlation ρ_0, or $\rho = \rho_0$. The procedure is illustrated in the following example.

Example 9

The correlation coefficient based on a sample of 101 items was computed in Example 7, $r = 0.80$. Can we accept the hypothesis that the population correlation coefficient is as large as 0.85? Let the level of significance be 5%.

Solution.   1. The null hypothesis, or H_0: There is no difference between the assumed population correlation coefficient $\rho_0 = 0.85$ and the sample $r = 0.80$. The sample r is the estimated ρ, the true population correlation coefficient.

2. Find z, the standard normal deviate:
Transform r to Z by formula (26-7),

---

*Formula (26-7) is frequently written in the form of

$$Z = \frac{1}{2} \log_e \left(\frac{1+r}{1-r} \right), \quad \text{where } e = 2.71828$$

Common logarithms on the base 10 may be shifted to natural logarithms on the base e by multiplying the factor 2.3026. For example, $2.3026 \cdot \log_{10} X = \log_e X$, where X represents a positive number. Since $(1/2)(2.3026) = 1.1513$, the above form is written as formula (26-7) for the convenience in computation. This formula can also be used for a partial, but not for a multiple, correlation coefficient. However, the standard error formula for partial correlation coefficient is different; that is,

$$\sigma_{z12.3} = \sqrt{\frac{3}{3n-11}}$$

$$Z_r = 1.1513 \log \left(\frac{1 + 0.80}{1 - 0.80}\right) = 1.1513 \,(\log 9) = 1.1513(0.954243) = 1.0986$$

Transform p_0 also to Z by formula (26–7),

$$Z_{p_0} = 1.1513 \log \left(\frac{1 + 0.85}{1 - 0.85}\right) = 1.1513 \,(\log 12.3)$$
$$= 1.1513(1.089905) = 1.2548$$

The standard error of Z is

$$\sigma_z = \frac{1}{\sqrt{101 - 3}} = 0.1010 \qquad \text{(Formula 26–8)}$$

Now,

$$z = \frac{Z_r - Z_{p_0}}{\sigma_z} = \frac{1.0986 - 1.2548}{0.1010} = -1.55$$

3. Make a decision. At a 5% level of significance for a one-tail test, the critical value of z is -1.64. We would reject the hypothesis only if z were smaller than -1.64. The computed $z = -1.55$ is larger than -1.64. We thus accept the hypothesis that the population correlation coefficient is as large as 0.85. In other words, the difference between p_0 and r is not significant. This example is diagrammed in Chart 26–5 for a clearer understanding of the above decision.

CHART 26–5

DIAGRAM FOR EXAMPLE 9

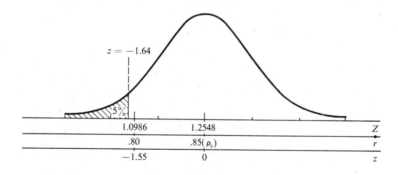

The conclusion of this example is consistent with the interval estimate of p computed in Example 8 from the same data ($r = 0.73$ to 0.87).

26.8 Testing Hypothesis—The Population Coefficient of Correlation $p = 0$ by the Analysis of Variance

The population coefficient of correlation p indicates the degree of relationship between X and Y variables of a bivariate population. When we assume $p = 0$, we are making the hypothesis that there is no relationship between the two variables. This hypothesis may be accepted or rejected by the F or t test.

We know that a sample drawn from a bivariate population has

$$\underset{\Sigma (Y - \bar{Y})^2}{\text{total variation}} = \underset{\Sigma (Y - Y_c)^2}{\text{unexplained variation}} + \underset{\Sigma (Y_c - \bar{Y})^2}{\text{explained variation}}$$

If we express the unexplained and explained variations as two independent unbiased estimates of the population variance, we may have the F ratio as presented in Chapter 17. That is,

$$F = \frac{\text{Explained variance}}{\text{Unexplained variance}} = \frac{\dfrac{\Sigma (Y_c - \bar{Y})^2}{m - 1}}{\dfrac{\Sigma (Y - Y_c)^2}{n - m}}$$

or simply

$$F = \frac{\hat{s}_{y(c)}^2}{\hat{s}_{yx}^2} \qquad\qquad (26\text{--}9)^*$$

where
$\quad n =$ number of Y values in the sample,

$\quad m =$ number of constants in the sample regression equation,

$\quad m - 1 = D_1 =$ degrees of freedom for the numerator of F ratio, and

$\quad n - m = D_2 =$ degrees of freedom for the denominator.

Note: The symbol $s_{y(c)}$ in formula (26–9) (also see the symbol on page 664) and the symbol s_{yc} in formula (26–2a) represent different expressions.

The sample variances are not additive, but the degrees of freedom are:

$$\text{total } (n - 1) = \text{unexplained } (n - m) + \text{explained } (m - 1)$$

If the explained variance does not differ significantly from the unexplained variance, the two independent estimates of the population variance should tend to be approximately the same. Then, F approaches 1 or fluctuates within the extent attributed to chance. This happens when the degree of correlation is moderate.

If the explained variance does differ significantly from the unexplained variance, the value of F will occur in either one of the following two ways:

1. The F ratio will be less than 1 or close to zero—the explained variance (the numerator) is smaller than the unexplained variance (the denominator).

This happens when the hypothesis is true, that is, when there is no correlation between the two variables. When there is no correlation, Y_c is close to or equal to $\bar{Y}$. The explained variation, $\Sigma (Y_c - \bar{Y})^2$, will be close to or equal to zero. This implies that the unexplained variation $\Sigma (Y - Y_c)^2$ will be relatively large since the total variation $\Sigma (Y - \bar{Y})^2$ is fixed. The variances are computed from the variations. Thus, the F ratio will be close to or equal to zero when there is a low or no correlation.

---

*The techniques of the analysis of variance by F ratio used in simple correlation for testing significance may also be used in multiple and partial correlation and the coefficients in a regression equation.

2. The F ratio will be larger than 1 and will exceed the extent due to chance as shown in the F distribution table—the explained variance is larger than the unexplained variance.

This happens when the hypothesis is not true, that is, when there is a high correlation between the two variables. When the correlation is high, the explained variation will be large since Y_c will differ greatly from $\bar{Y}$. This implies that the unexplained variation will be relatively small. Thus, F is expected to be large when there is a high correlation.

In testing the hypothesis, we are interested in knowing whether or not the value of F is significantly large according to the F table at a given level of significance. If F is significantly large, we reject the hypothesis that there is no correlation. Alternatively, we may state that there is a high correlation between two variables. If F is not significantly large, we accept the hypothesis, although it is possible that there is a moderate correlation. If F is very small or close to zero, we can accept the hypothesis with no hesitation.

The number of constants in a linear regression equation is 2, or $m = 2$ (constants a and b). Thus, $D_1 = m - 1 = 2 - 1$, and $D_2 = n - m = n - 2$. When $D_1 = 1$, at a given level of significance,

$$t = \sqrt{F} \qquad\qquad (26\text{–}10a)*$$

where t is based on two tails with $n - 2$ (or D_2) degrees of freedom.

The above formula may also be written in the following form which permits to compute t directly from r and r^2:

$$t = r\sqrt{\frac{n-2}{1-r^2}} \qquad\qquad (26\text{–}10b)$$

We now may use the t table with $n - 2$ degrees of freedom to obtain the same result as the F table in the analysis of variance. The t table has the advantage of having more entries and levels of significance than the F table.

Example 10

Use the sample given in Example 2. (a) Compute the F ratio. (b) Determine whether or not Y variable is related to X variable in the population based on the F ratio at the 5% level of significance.

Solution.   (a) First, the unbiased estimates of the population variance are obtained in Table 26–10.

---

*Also see the F properties (item 5) concerning the relationship between t and F on p. 455. The t test cannot be used for a multiple correlation coefficient since the number of constants (m) in a multiple regression equation is 3 or more and $D_1 = m - 1$ is not equal to 1. However, the test can be used for a coefficient of partial correlation. The formula for the coefficient of partial correlation between X_1 and X_3, keeping X_2 constant, is

$$t = r_{13.2}\sqrt{\frac{n-3}{1-r_{13.2}^2}}$$

The t-distribution has $n - 3$ degrees of freedom. See the footnote on page 737 for the application of the above formula.

TABLE 26–10

COMPUTATION FOR EXAMPLE 10

Type of variation	Variation	Computation of variation	Degrees of freedom	Estimate of population variance
Total	$\sum (Y - \bar{Y})^2 = 22.80$	Example 6, page 704	$n-1=5-1=4$	$\hat{s}_y^2 = \dfrac{22.80}{4} = 5.70$
Unexplained	$\sum (Y - Y_c)^2 = 3.20$	Example 3, page 697	$D_2 = n - m$ $= 5 - 2 = 3$	$\hat{s}_{yx}^2 = \dfrac{3.20}{3} = 1.07$
Explained	$\sum (Y_c - \bar{Y})^2 = 19.60$	Total minus unexplained	$D_1 = m - 1$ $= 2 - 1 = 1$	$\hat{s}_{y(c)}^2 = \dfrac{19.60}{1} = 19.60$

$$F = \frac{\hat{s}_{y(c)}^2}{\hat{s}_{yx}^2} = \frac{19.60}{1.07} = 18.32 \qquad \text{(Formula 26–9)}$$

(b) At the 5% level, with $D_1 = 1$ and $D_2 = 3$, the F is significantly large if it is 10.13 or larger. The computed F value, 18.32, is far above 10.13. We may state that the explained variance is significantly greater than the unexplained variance. We then reject the hypothesis that there is no correlation in the population. In other words, there is a relationship between X and Y variables. The relationship between the two variables stated by a coefficient of correlation r, therefore, is significant at a 5% level.

However, if we based on the 1% table, the F is significantly large if it is 34.12 or larger. We then would accept the hypothesis. A better conclusion for Example 10, therefore, is to state that the correlation is *probably* significant.

Example 10 may also be solved by the t test as follows:

$$t = \sqrt{F} = \sqrt{18.32} = 4.28 \qquad \text{(Formula 26–10a)}$$

or

$$t = r\sqrt{\frac{n-2}{1-r^2}} = 0.927\sqrt{\frac{5-2}{1-0.8596}} = 0.927(4.622) = 4.28$$

(page 698 and Formula 26–10b)

Use the t table with $D = D_2 = 3$: at 5%, $t = 3.182$; at 1%, $t = 5.841$. The computed $t = 4.28$ is between the two tabulated values. Thus, we may draw the same conclusion as in the F test.

26.9 Testing Hypothesis—the Population Regression Coefficient $B = 0$

The population regression coefficient B, or the slope of the population regression line, also indicates the relationship between X and Y variables. For instance, $B = 1$ in the population regression equation of Example 1,

$$\mu_{yx} = A + BX = 2 + 1X$$

indicates the increase in the average grade points μ_{yx} when there is a unit increase in the number of semester hours (X) in mathematical courses taken by the students. If $B = 0$, the population regression line becomes horizontal since

$$\mu_{yx} = A + 0X = A$$

The horizontal line indicates that there is no relationship between X and Y variables. In other words, the increase or decrease in X variable has nothing to do with that in Y variable. Since we are interested in the relationship between the two variables, we may wish to check whether or not $B = 0$. This can be done by testing the hypothesis that $B = 0$. The standard error of b, or the standard deviation of the sampling distribution of b, denoted by σ_b, can be estimated from a sample by formula

$$s_b = \sqrt{\frac{\hat{s}_{yx}^2}{\sum (X - \bar{X})^2}} \tag{26–11}*$$

The distribution of b is a t-distribution with $n - 2$ degrees of freedom. The method of testing the hypothesis based on the standard error of b is illustrated in Example 11.

Example 11
Based on the sample regression coefficient $b = 1.4$ computed in Example 2, find whether or not Y variable is related to X values. Let the level of significance be 5%.

Solution.   The method of testing the hypothesis for Example 11 is presented also according to the procedure outlined on p. 366.

1. The null hypothesis, or H_0: $B = 0$, or there is no relationship between X and Y variables.

2. Express the difference between the sample statistic b and the population parameter B in units of the standard error of the statistic, or

$$t = \frac{b - B}{s_b}$$

The standard error of b is computed by formula (26–11),

$$s_b = \sqrt{\frac{\hat{s}_{yx}^2}{\sum (X - \bar{X})^2}} = \sqrt{\frac{1.07}{10}} = \sqrt{0.107} = 0.327 \qquad \text{(pages 685 and 686)}$$

$$t = \frac{1.4 - 0}{0.327} = 4.281$$

The t value is the same as that obtained in Example 10.**

3. Make a decision. At 0.05 level of significance, the critical value of t with $n - 2 = 5 - 2 = 3$ degress of freedom is 3.182 (both tails). The computed t value $= 4.281$ is greater than the critical value 3.182. Thus, we reject the hy-

---

*For the variance of the sampling distribution of b, see the footnote on page 686.
**It can be proved that when $D_1 = m - 1 = 2 - 1 = 1$, t obtained from $\sqrt{F}$ (formulas 26–9 and 26–10a) is the same as that obtained from $(b - 0)/s_b$; that is, $t = \sqrt{F} = b/s_b$.

pothesis that $B = 0$. We then conclude that since $B \neq 0$, there is a relationship between X and Y variables.

Note that if the sample size is large, we may use the normal distribution instead of the t-distribution in step 2 of Example 11, or t is approximated by z as

$$z = \frac{b - B}{s_b}$$

Note also that the same conclusion may be obtained by finding the interval estimate of B based on the sample b. At a 95% confidence interval, which corresponds to a 5% level of significance, the interval estimate of B for Example 11 is as follows:

Confidence limits $= b \pm t \cdot s_b = 1.40 \pm 3.182(0.327) = 1.40 \pm 1.04$
$$= 0.36 \text{ to } 2.44$$

Since the estimated value of B is not zero, but is within the range from 0.36 to 2.44, we may conclude that there is a relationship between X and Y variables. The actual value of B is equal to 1 (Example 1). Of course, in practice we do not have the actual value of a population regression coefficient B for comparison.

26.10 Summary

This chapter discusses the methods in analyzing the linear relationship between two variables of a population by a sample drawn from the population. Those methods involve the procedures of estimating a population parameter from a corresponding sample statistic and testing hypotheses. The basic concept of the population in the regression analysis is illustrated by presenting a linear regression model for the population. The population model can be generalized under the three assumptions: (1) The independent values X are known or fixed. The dependent values Y on each X are many and are distributed normally as a subpopulation. (2) The means of the subpopulations of the population are on a straight line. (3) The variances (also the standard deviations) of the subpopulations are equal. Also, the common subpopulation variance is equal to the population variance of regression.

An estimate of a population parameter may be expressed in two ways: a point estimate and an interval estimate. The point estimate of a population parameter is a corresponding sample statistic. The interval estimate is specified by the upper and lower confidence limits, which are computed from the sample statistic and the standard error of the statistic. Estimates made in this chapter are:

  1. Population regression line μ_{yx}:
     a. Point estimate of μ_{yx} by Y_c. (Section 26.2)
     b. Interval estimate of μ_{yx} by the confidence limits of Y_c computed from s_{yc}. (Section 26.4)
  2. Population variance (or standard deviation) of regression σ_{yx}^2 (or σ_{yx}): Point estimate of σ_{yx}^2 by $\hat{s}_{yx}^2$, or point estimate of σ_{yx} by $\hat{s}_{yx}$. (Section 26.3)

3. Individual Y value of a subpopulation: Interval estimate of Y by the confidence limits of Y_c computed from $s_{(y-y_c)}$. (Section 26.5)

4. Population coefficient of determination ρ^2 (or population coefficient of correlation ρ):

   a. Point estimate of ρ^2 by $\hat{r}^2$ or ρ by $\hat{r}$. (Section 26.6)

   b. Interval estimate of ρ by the confidence limits of r computed from s_r for large sample size. (Section 26.6)

SUMMARY OF FORMULAS

Application	Formula	Formula number	Reference page
Estimation			
Estimated population standard deviation of regression	$\hat{s}_{yx} = \sqrt{\dfrac{\sum (Y - Y_c)^2}{n - m}}$	26-1a	684
	$\hat{s}_{yx} = s_{yx}\sqrt{\dfrac{n}{n - m}}$	26-1b	685
Estimated standard error of Y_c—general case	$s_{y_c} = \hat{s}_{yx}\sqrt{\dfrac{1}{n} + \dfrac{(X - \bar{X})^2}{\sum (X - \bar{X})^2}}$	26-2a	686
Large samples only	$s_{y_c} = \dfrac{s_{yx}}{\sqrt{n}}$	26-2b	688
Estimated variance and standard error of the difference between Y and Y_c	$s^2_{(y-y_c)} = \hat{s}^2_{yx} + s^2_{y_c}$	26-3a	689
	$s_{(y-y_c)} = \hat{s}_{yx}\sqrt{1 + \dfrac{1}{n} + \dfrac{(X - \bar{X})^2}{\sum (X - \bar{X})^2}}$	26-3b	689
Population coefficient of determination	$\rho^2 = 1 - \dfrac{\sigma^2_{yx}}{\sigma^2_y}$	26-4	691
Estimated population coefficient of determination ($\hat{r}^2$) and estimated population coefficient of correlation ($\hat{r}$, or square root of $\hat{r}^2$)	$\hat{r}^2 = 1 - \dfrac{\hat{s}^2_{yx}}{\hat{s}^2_y}$	26-5a	691
	$\hat{r}^2 = 1 - (1 - r^2)\left(\dfrac{n - 1}{n - m}\right)$	26-5b	692
Estimated standard error of r for large samples	$s_r = \dfrac{1 - r^2}{\sqrt{n - 1}}$	26-6	693
Tests of hypotheses			
Population coefficient of correlation $\rho = \rho_0$			
Transform r into Z	$Z = 1.1513 \log_{10}\left(\dfrac{1 + r}{1 - r}\right)$	26-7	694
Standard error of Z	$\sigma_Z = \dfrac{1}{\sqrt{n - 3}}$	26-8	694
Population coefficient of correlation $\rho = 0$			
Analysis of variance by F ratio	$F = \dfrac{\hat{s}^2_{y(c)}}{\hat{s}^2_{yx}}$	26-9	696
Analysis of variance by t distribution	$t = \sqrt{F}$	26-10a	697
	$t = r\sqrt{\dfrac{n - 2}{1 - r^2}}$	26-10b	697
Population regression coefficient $B = 0$			
Estimated standard error of b, the slope of regression line	$s_b = \sqrt{\dfrac{\hat{s}^2_{yx}}{\sum (X - \bar{X})^2}}$	26-11	699

5. Population regression coefficient B:

    a. Point estimate of B by b. (Section 26.2)

    b. Interval estimate of B by the confidence limits of b computed from s_b. (Section 26.9)

Tests of hypotheses made in this chapter are:

1. There is no difference between a given population coefficient of correlation ρ (usually estimated by r) and the assumed population coefficient of correlation ρ_0, or $\rho = \rho_0$. This test is made by the Z-transformation method. (Section 26.7)

2. There is no relationship between X and Y variables, or the population coefficient of correlation is zero, $\rho = 0$. This test is done by the analysis of variance (F ratio test or t test). (Section 26.8)

3. The population regression coefficient B is zero, or $B = 0$. This test is based on the standard error of b. The distribution of b is a t-distribution with $n - 2$ degrees of freedom (Section 26.9).

Exercises 26

1. What is a population regression model? What are the assumptions concerning the linear regression model used in this text?

2. Which one of the two forms of standard deviation of regression (or standard error of estimate), s_{yx} and $\hat{s}_{yx}$, is a better device in estimating the population standard deviation of regression? Why?

3. Assume that the following information represents a sample taken from the population given in Example 1, p. 680. Based on the sample, find:

Semester Hours	Grade Points
X	Y
1	1
2	3
3	4
4	8
5	9

(a) The estimated population regression equation.

(b) The estimated population standard deviation of regression.

(c) The estimated mean of the grade points for each given number of semester hours at a 95% confidence interval.

(d) The estimated individual grades for each given number of semester hours at a 95% confidence interval. Check the actual Y values of the given population with the computed interval estimates. Do all Y values fall within the 95% confidence interval zone? If not, would you think that the accuracy of the estimation could be improved, such as at a 99% confidence interval? Why?

4. Refer to Problem 3.
   (a) Find the estimated population coefficient of determination and the estimated population coefficient of correlation.
   (b) Which one of the two samples—the sample referred to in this problem or
   the sample in Example 2, page 683, gives a better description of the population? Why?

Note: The assumptions for the linear population regression model stated in Section
26.1 are applicable to the following problems.

5. Assume that Table 25–6 (p. 675), Problem 3 of the Exercises for Chapter 25,
represents a sample drawn from the population of all employees in the company.
Find the following based on the sample:
   (a) The estimated population regression equation.
   (b) The estimated population standard deviation of regression.
   (c) The estimated mean of the numbers of correct answers of the employees
   who studied two hours for the test at a 90% confidence interval.
   (d) The interval estimate of the number of correct answers for an employee
   who studied two hours at a 90% confidence interval.
6. Refer to Problem 5.
   (a) Find the estimated population coefficient of determination and the
   estimated population coefficient of correlation.
   (b) Compare r (obtained in Problem 3 of the Exercises for Chapter 25) and
   $\hat{r}$ (obtained in (a) above). Explain the difference between them.

7. Assume that Table 25–7 (p. 675), Problem 5 of the Exercises for Chapter 25,
represents a sample drawn from the population of all students in the college.
Consider the statistics grades (Y) as the dependent variable and the economics
grades (X) as the independent variable. Find the following based on the sample:
   (a) The estimated population regression equation.
   (b) The estimated population standard deviation of regression.
   (c) The estimated mean of the statistics grades of the students who had 70
   points for economics at a 99% confidence interval.
$$(\bar{X} = 63.33 \text{ and } \Sigma (X - \bar{X})^2 = 1{,}894)$$
   (d) The estimated statistics grade of a student who had 70 points for economics at a 99% confidence interval.
   (e) The estimated population coefficient of determination and the estimated
   population coefficient of correlation.

8. The estimated population coefficient of determination based on a sample of
145 items is 0.76. What is the interval estimate of the population coefficient of
correlation at a 99% confidence interval?

9. The sample regression coefficient b obtained from Problem 3 (a) above is 2.1.
Determine whether or not Y variable (grade points) is related to X variable (semester hours) by the following two methods:
   (a) Test the hypothesis that $B = 0$ at the level of significance (1) 5%, and (2)
   1%.

(b) Find the interval estimate of B at the confidence interval of (1) 95% and (2) 99%.

Check the actual B value of the given population with the computed interval estimates. Does the true B value fall within (1) 95%, and (2) 99% interval estimate?

10. The estimated population regression coefficient b based on the sample of Problem 5 (a) above is 2.07. Find whether or not there is a relationship between Y variable (number of correct answers) and X variable (hours studied) by

(a) the method of testing hypothesis that $B = 0$ at the level of significance of 5%, and

(b) the method of finding the interval estimate of B at the 95% confidence interval.

27

*Nonlinear Regression and Correlation

The analysis presented in Chapters 25 and 26 was based on the assumption that the average relationship between two variables could be described by a straight line. However, the assumption may become inadequate for some data. Very frequently the relationships concerning business and economic activities can be described better by curves than by straight lines.

The first three sections of this chapter will present the nonlinear regression and correlation analysis based on (1) a parabolic curve determined by a second-degree polynomial equation, (2) a smooth curve drawn by the freehand graphic method, and (3) a broken line computed for grouped data by a correlation table.

The application of matrix algebra to the area of statistics has greatly increased recently. Matrix algebra is especially useful in studying more advanced topics of correlation analysis. Section 27.4 will first introduce the basic terminology and operations of the matrix. Next, it will present selected methods of solving linear equations by matrix algebra.

27.1 Second-Degree Parabolic Curve

This regression curve is based on the second-degree polynomial equation presented in Chapter 22, p. 575.

$$Y_c = a + bX + cX^2 \qquad \text{(Formula 22-4)}$$

Since there are three unknown constants a, b, and c in the equation, it is necessary to developed three equations for solving the unknowns. The three normal equations based on the method of least squares are restated here for the convenience of reference.

$$\text{I.} \qquad \Sigma\,(Y) = na + b\,\Sigma\,(X) + c\,\Sigma\,(X^2)$$
$$\text{II.} \quad \Sigma\,(XY) = a\,\Sigma\,(X) + b\,\Sigma\,(X^2) + c\,\Sigma\,(X^3)$$
$$\text{III.} \quad \Sigma\,(X^2Y) = a\,\Sigma\,(X^2) + b\,\Sigma\,(X^3) + c\,\Sigma\,(X^4)$$

$$\text{(Formula 22–5)}$$

The three normal equations may be simplified if X is assigned so that $\Sigma\,X = 0$, such as formulas (22–6a), (22–6b), and (22–6c), in the time series analysis. However, the sum of X values is frequently not equal to zero in regression analysis. In such a case, that is $\Sigma\,X \neq 0$, the constants, a, b, and c are obtained by simultaneously solving the three normal equations.

The methods of computing the standard deviation of regression and the coefficient of determination based on a nonlinear regression equation are the same as those based on a straight-line regression equation. The coefficient of correlation, which is the square root of the coefficient of determination for nonlinear correlation, is also called the *index of correlation*. The coefficients for nonlinear correlation are indicated by subscripts, arranged in order of dependent variable and independent variable. For example, the coefficient of correlation of Y variable (dependent) on X variable (independent) is indicated by subscripts as r_{yx}. The subscripts were omitted in the previous chapters for linear correlation since the value of r was the same either for Y or X as the dependent variable. The subscripts are necessary for the measure of the degree of nonlinear correlation since r_{yx} is not the same as r_{xy} (now X as the dependent and Y as the independent variable). We may consider r for linear correlation as a special case of nonlinear correlation.

The value of r^2_{yx} for nonlinear correlation can be interpreted in the same way as r^2 for linear correlation. However, the computed value of r_{yx} or r_{xy} is not indicated by a $+$ or $-$ sign. *The slope of a nonlinear regression line, or a curve, may be positive at some parts and negative at other parts of the curve.* This fact can easily be seen on a scatter diagram with a curve fitted to the data.

The topics usually included in the regression and correlation analysis based on the second-degree parabolic curve are as follows:

(a) Plot a scatter diagram on a chart.

(b) Compute the regression equation $Y_c = a + bX + cX^2$.

(c) Draw the regression curve based on the above equation on the chart.

(d) Computed the standard deviation of regression by the formula presented in Chapter 25 (page 659),

$$s_{yx} = \sqrt{\frac{\Sigma\,(Y - Y_c)^2}{n}} \qquad \text{(Formula 25–3a)}^*$$

(e) Compute the coefficient of determination r^2_{yx} and the coefficient of correlation r_{yx} either by formula (25–5a), which is now written:

$$r^2_{yx} = \frac{\text{Explained variation}}{\text{Total variation}} = \frac{\Sigma\,(Y_c - \bar{Y})^2}{\Sigma\,(Y - \bar{Y})^2} \qquad \textbf{(27–1a)}$$

---

*It can also be obtained by modifying formula (25–3b) as follows:

$$s_{yx} = \sqrt{\frac{\Sigma\,Y^2 - a\,\Sigma\,Y - b\,\Sigma\,XY - c\,\Sigma\,X^2Y}{n}}$$

or by formula (25–5b), which is now written:

$$r_{yx}^2 = 1 - \frac{s_{yx}^2}{s_y^2} \qquad\qquad (27\text{–}1b)$$

Example 1

The first three columns of Table 27–1 show the amounts of sales (Y) made by a group of 8 salesmen during a given period and the years of sales experience (X) of each salesmen. Perform the usual regression and correlation analysis. (Note: This example uses the same data as Examples 1, 2, and 3 in Chapter 25.)

TABLE 27–1

CALCULATION FOR THE SECOND-DEGREE POLYNOMIAL EQUATION
BY THE METHOD OF LEAST SQUARES—EXAMPLE 1 (b)

(1) Salesman	(2) Amount of Sales (in $1,000) Y	(3) Years of Sales Experience X	(4) X^2	(5) X^3	(6) X^4	(7) XY	(8) X^2Y
A	9	6	36	216	1,296	54	324
B	6	5	25	125	625	30	150
C	4	3	9	27	81	12	36
D	3	1	1	1	1	3	3
E	3	4	16	64	256	12	48
F	5	3	9	27	81	15	45
G	8	6	36	216	1,296	48	288
H	2	2	4	8	16	4	8
$n = 8$ Total	40	30	136	684	3,652	178	902

Solution. (a) A scatter diagram is plotted on Chart 27–1.

(b) The parabolic regression equation is computed by the method of least squares. The required values for applying the three normal equations are obtained from columns (2) to (8) of Table 27–1. Substitute the column sums in the normal equations as follows:

$$\text{I.} \qquad 8a + 30b + 136c = 40$$
$$\text{II.} \qquad 30a + 136b + 684c = 178$$
$$\text{III.} \qquad 136a + 684b + 3{,}652c = 902$$

These three equations may be solved simultaneously by the method of elimination as discussed in Section 4.2 or by using matrix algebra (to be presented in Section 27.4, Example 10). The computation can be simplified if each equation is reduced to have its lowest terms. For example, divide each side of equation I by 2, the greatest common divisor of numbers 8, 30, 136, and 40 in the equation, we have a simplified equation I:

$$4a + 15b + 68c = 20$$

CHART 27–1

SCATTER DIAGRAM AND PARABOLIC REGRESSION LINE
COMPARED WITH STRAIGHT REGRESSION LINE

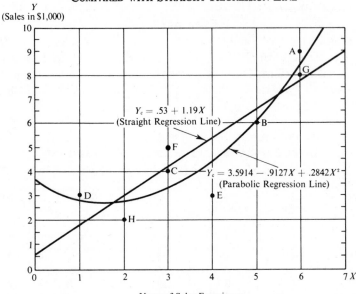

Years of Sales Experience

Source: Example 1 and Chart 25–3.

The solutions are:*

$$a = 3.5914, \qquad b = -0.9127, \qquad c = 0.2842$$

The required equation is:

$$Y_c = 3.5914 - 0.9127X + 0.2842X^2$$

---

*The answers may also be obtained by solving the following two equations simultaneously for b and c:

$$\sum xy = b \sum x^2 + c \sum xt \dots (1), \quad \text{where } x = X - \bar{X}, y = Y - \bar{Y}$$
$$\sum ty = b \sum xt + c \sum t^2 \dots (2), \qquad t = T - \bar{T}, \text{and } T = X^2$$

The required values for the two equations are expressed below and are obtained from Table 27-1.

$$\sum x^2 = \sum X^2 - \frac{(\sum X)^2}{n} = 136 - \frac{30^2}{8} = 23.5$$

$$\sum xt = \sum X^3 - \frac{\sum X \cdot \sum X^2}{n} = 684 - \frac{30(136)}{8} = 174$$

$$\sum xy = \sum XY - \frac{\sum X \cdot \sum Y}{n} = 178 - \frac{30(40)}{8} = 28$$

$$\sum t^2 = \sum X^4 - \frac{(\sum X^2)^2}{n} = 3{,}652 - \frac{136^2}{8} = 1{,}340$$

$$\sum ty = \sum X^2Y - \frac{\sum X^2 \cdot \sum Y}{n} = 902 - \frac{136(40)}{8} = 222$$

Substitute the above 5 values in equations (1) and (2).

(c) The regression line based on the computed equation is also drawn on Chart 27–1. The points on the line are determined by the Y_c values obtained as follows:

$$\text{For salesman A,} \quad X = 6$$
$$Y_c = 3.5914 - 0.9127(6) + 0.2842(6^2)$$
$$= 3.5914 - 5.4762 - 10.2312$$
$$= 8.3464, \text{ or round to } 8.35$$

In a similar manner, the Y_c values for other salesmen are computed and listed in column (4) of Table 27–2.

TABLE 27–2

CALCULATION FOR Y_c AND THE STANDARD DEVIATION OF
REGRESSION s_{yx}—EXAMPLE 1 (c) AND (d)

(1) Salesman	(2) Y	(3) X	(4) $Y_c = 3.5914 -$ $0.9127X + 0.2842X^2$	(5) $Y - Y_c$	(6) $(Y - Y_c)^2$
A	9	6	8.35	0.65	0.42
B	6	5	6.13	−0.13	0.02
C	4	3	3.41	0.59	0.35
D	3	1	2.96	0.04	0.00
E	3	4	4.49	−1.49	2.22
F	5	3	3.41	1.59	2.53
G	8	6	8.35	−0.35	0.12
H	2	2	2.90	−0.90	0.81
Total	40	30	40.00	0.00	6.47

$$23.5b + 174c = 28. \ldots (1)$$
$$174b + 1{,}340c = 222 \ldots (2)$$
$$\text{The solutions are:} \quad c = 0.284185$$
$$b = -0.912689$$

Divide normal equation I by n,

$$a = \frac{\sum Y}{n} - b\frac{\sum X}{n} - c\frac{\sum X^2}{n} = \frac{40}{8} - (-0.912689)\left(\frac{30}{8}\right) - 0.284185\left(\frac{136}{8}\right)$$
$$= 3.591439$$

The five expressions can be proved as follows:

$$\sum x^2 = \sum (X - \bar{X})^2 = \sum (X^2 - 2X\cdot\bar{X} + \bar{X}^2) = \sum X^2 - 2\bar{X}\cdot\sum X + n\bar{X}^2$$
$$= \sum X^2 - 2\bar{X}\cdot n\bar{X} + n\bar{X}^2 = \sum X^2 - n\bar{X}^2 = \sum X^2 - \frac{(\sum X)^2}{n}$$

$$\sum xt = \sum (X - \bar{X})(T - \bar{T}) = \sum (XT - \bar{X}T - X\bar{T} + \bar{X}\bar{T})$$
$$= \sum XT - \bar{X}\cdot\sum T - \bar{T}\cdot\sum X + n\bar{X}\bar{T} = \sum XT - \bar{X}\cdot n\bar{T} - \bar{T}\cdot n\bar{X} + n\bar{X}\bar{T}$$
$$= \sum XT - n\bar{X}\bar{T}$$

Since $T = X^2$,

$$\sum xt = \sum XX^2 - n\bar{X}\cdot\bar{X}^2 = \sum X^3 - \frac{\sum X\cdot\sum X^2}{n}$$

In a similar manner we can prove the extened expressions for $\sum xy$, $\sum t^2$, and $\sum ty$.

The Y_c values also indicate the estimated amounts of sales. For example, the estimated amount of sales for a salesman having 4 years ($X = 4$) of sales experience is \$4,490 ($= 4.49 \times \$1,000$, the Y unit). Note that the straight regression line gives the estimate for the salesman having 4 years of sales experience as \$5,290 (page 658). The straight regression line computed in Example 1 of Chapter 25 is also drawn on the chart for comparison purposes.

(d) The standard deviation of regression is computed by using formula (25–3a), or

$$S_{yx} = \sqrt{\frac{\sum (Y - Y_c)^2}{n}} = \sqrt{\frac{6.47}{8}} = \sqrt{0.80875} = 0.90 \qquad \text{(in units of \$1,000)}$$

The measure of dispersion s_{yx} based on the parabolic curve ($= 0.90$ above) is smaller than that based on the straight line ($= 1.15$, page 660).

(e) The coefficient of determination is computed by using formula (27–1b), or

$$r_{yx}^2 = 1 - \frac{s_{yx}^2}{s_y^2} = 1 - \frac{0.80875}{5.5} = 1 - 0.14705 = 0.85295, \text{ or round to } 85\%$$

($s_y^2 = 44/8 = 5.5$, page 665.) The coefficient of correlation is

$$r_{yx} = \sqrt{0.85295} = 0.92$$

The values of r_{yx}^2 and r_{yx} can be interpreted in the same manner as r^2 and r respectively. Thus, 85% of the variation of the amounts of sales (Y) is related to, or explained by, the variation of the years of sales experience (X) of the salesmen based on the parabolic regression line.

Compare the two measures of the degree of correlation based on the parabolic equation with those based on the straight line for the same data given in Example 1. $r_{yx}^2 = 0.85$ versus $r^2 = 0.76$ and $r_{yx} = 0.92$ versus $r = 0.87$ (page 665). The higher values of r_{yx}^2 and r_{yx} indicate that the parabolic curve gives a better fit to the data than the straight line. This is based on the fact that the smaller the dispersion s_{yx}, the higher is the coefficient of determination r_{yx}^2 (also the coefficient of correlation r_{yx}).

When the given data are considered as a sample for estimating population parameters, the unbiased estimates of the parameters should be used. The general formulas for the unbiased estimates can be obtained in the same way as those presented in Chapter 26.*

---

*For example, the formula (based on formula 26–1) and the computation (based on the data given in Example 1 as a sample) for the estimate of the population standard deviation of regression are:

$$\hat{s}_{yx} = \sqrt{\frac{\sum (Y - Y_c)^2}{n - m}} = \sqrt{\frac{6.47}{8 - 3}} = \sqrt{1.2940} = 1.14$$

or,

$$\hat{s}_{yx} = s_{yx}\sqrt{\frac{n}{n - m}} = 0.90\sqrt{\frac{8}{8 - 3}} = 0.90\sqrt{1.6} = 1.14$$

$$m = 3$$

since there are three constants, a, b, and c, in the parabolic equation. The formula (based on formula 26–5) and the computation (based on the same data as above) for the estimate of the

27.2 Smoothed Curve by Freehand Drawing

Since the computation of a second- or a higher-degree polynomial equation based on the least square method is very burdensome for a large set of data, a smooth curve drawn by the freehand graphic method is frequently employed in practice for the regression and correlation analysis. The advantages and disadvantages of a freehand regression curve are similar to those discussed in Chapter 22 for a freehand trend line representing a time series. However, when a scatter diagram shows a high degree of correlation between two variables, a freehand curve should be very close to a mathematical curve since it is less subjective in locating the curve.

The procedure of finding the important measures for the regression and correlation analysis by the freehand graphic method is outlined below.

1. Draw a smooth curve on the scatter diagram to fit the given data. This curve may resemble the parabolic curve on Chart 27–1 and is the desired regression line.

2. Read the Y_c value from the smooth curve for each given X value.

3. Compute each $(Y - Y_c)$, the difference between each Y and its corresponding Y_c on the given X. When the differences are squared and added, or $\sum (Y - Y_c)^2$, the standard deviation of regression s_{yx}, the coefficient of determination r_{yx}^2, and the coefficient of correlation r_{yx} may be computed from the sum in the usual manner as in the previous section.

27.3 Grouped Data

The procedure of fitting a parabolic curve by the method of least squares to ungrouped data can be extended for grouped data. However, the added complexity reduces its value for practical uses. The detailed procedure for grouped data thus is not discussed in this section. The graphic method for ungrouped data as discussed in the previous section has its weakness being too subjective in drawing an appropriate curve. This weakness is even more serious for grouped data. Furthermore, when an appropriate smooth curve is obtained, the computation for the differences, $Y - Y_c$, and the squared differences, $(Y - Y_c)^2$ is also burdensome.

---

population coefficient of determination are:

$$\hat{s}_y^2 = \frac{\sum (Y - \bar{Y})^2}{n - 1} = \frac{44}{8 - 1} = 6.2857 \qquad \text{(page 665)}$$

$$\hat{r}_{yx}^2 = 1 - \frac{\hat{s}_{yx}^2}{\hat{s}_y^2} = 1 - \frac{1.2940}{6.2857} = 1 - 0.2059 = 0.7941 \text{ or round to } 79\%$$

or, $\qquad \hat{r}_{yx}^2 = 1 - (1 - r_{yx}^2)\left(\frac{n-1}{n-m}\right) = 1 - (1 - 0.85295)\left(\frac{8-1}{8-3}\right) = 0.7941$

The estimate of the population coefficient of correlation is:

$$\hat{r}_{yx} = \sqrt{0.7941} = 0.89$$

A practical way to measure nonlinear correlation for a large set of data is to use a correlation table, which presents bivariate grouped data. A broken line instead of a smooth curve is used to represent the average relationship between two variables. There is no need for a regression equation for the broken line. The broken line is determined by the points centered in the columns heading X class intervals. There is one point for each column as shown in the correlation table, Table 27–3. Each point in a column represents the mean of the Y values in the X column, denoted by $\bar{Y}_X$. The number of points thus is equal to the number of columns having Y values.

The coefficient of determination for a correlation table showing the grouped data, Y being dependent variable and X being independent variable, is denoted by η^2. The Greek letter η (eta) is called the *correlation ratio* and is interpreted in the same manner as r, the correlation coefficient based on a straight-line relationship.

The simplest way to compute η^2 is based on the basic relationship as stated in formula (25–5a),

$$\text{coefficient of determination} = \frac{\text{explained variation}}{\text{total variation}}$$

Since the regression line in the form of a broken line goes through every point representing $\bar{Y}_x$ value in the correlation table, we rewrite formula (25–5a) for the grouped data as follows:

$$\eta^2 = \frac{\Sigma f_x(\bar{Y}_x - \bar{Y})^2}{\Sigma f_y(Y - \bar{Y})^2} \tag{27–2a}$$

where f_x = the frequency of Y values in each X column of the correlation table,
f_y = the frequency of Y values in each Y row,
$\bar{Y}_x$ = the mean of Y values in each X column or simply
  = Y_c, the value on the regression line,
Y = the midpoint of each Y class interval, and
$\bar{Y}$ = the grand mean of the entire Y values of the table.

For convenience in computation, formula (27–2a) may be written:

$$\eta^2 = \frac{\Sigma f_x(\bar{Y}_x - \bar{Y})^2}{ns_y^2} \tag{27–2b}$$

since $\Sigma f_y(Y - \bar{Y})^2 = ns_y^2$. The variance of Y values, s_y^2, can be computed by formula (7–8d) written for grouped data as follows:

$$s_y^2 = i_y^2 \left[\frac{\Sigma f_y d_y^2}{n} - \left(\frac{\Sigma f_y d_y}{n} \right)^2 \right] \tag{27–3}$$

Example 2

Table 27–3, a correlation table, shows the amounts of sales (Y) made by a group of 40 salesmen during a given period and the years of sales experience (X) corre-

TABLE 27–3

CORRELATION TABLE FOR EXAMPLE 2

X—Years of Sales Experience ($i_x = 2$ years)

Y—Amounts of Sales in $1,000 ($i_y = 3$)

Y Class Interval	Midpoint	0 @*2	2 @ 4	4 @ 6	6 @ 8	8 @ 10	f_v
16–18	Y 17					(17) 1	1
13–15	14			(14) 1	(14) 1	(42) 3	5
10–12	11			(55) 5	(22) 2		7
7–9	8		(40) 5	(80) 10	(48) 6		21
4–6	5	$\bar{Y}_x$	(15) 3		(5) 1		4
1–3	2	(4) 2					2
Sum of Y values		(4)	(55)	(149)	(89)	(59)	(356)=ΣY
f_x		2	8	16	10	4	40 = n
$\bar{Y}_x$		2	6.875	9.3125	8.9	14.75	8.9 = $\bar{Y}$

* @ represents "and under".

sponding to individual salesmen. (Same as in Example 4, page 669.) Compute the coefficient of determination and the correlation ratio.

Solution. (1) Compute the $\bar{Y}_x$, the mean of Y values in each X column, from the correlation table. The total of Y values of each cell is written in the parenthesis. Each total is obtained by multiplying the midpoint of the Y class interval on the row by the frequency in the cell. The totals of Y values in the cells in each column are then added. The sum is now divided by the total frequency f_x in the column to obtain the $\bar{Y}_x$. For example, the $\bar{Y}_x$ in the X column of class "2 @ 4" is computed as follows:

Y in the cell with stub "7 — 9" = 8 (midpoint) × 5 (frequency) = 40
Y in the cell with stub "4 — 6" = 5 (midpoint) × 3 (frequency) = 15

Sum of Y values in the column = 55

f_x (the total frequency in the X column) = 5 + 3 = 8

$$\bar{Y}_x = \frac{55}{8} = 6.875, \text{ or round to } 6.9$$

(2) Compute the explained variation. The explained variation, $\Sigma f_x(\bar{Y}_x - \bar{Y})^2 = 269.02$, is computed from Table 27–4. This table is based on the correlation table (Table 27–3).

(3) Compute the total variation. The total variation is 399.60, which can be computed in two ways.

(a) By expression $\Sigma f_y(Y - \bar{Y})^2$ as stated in formula (27–2a). This ex-

TABLE 27–4

COMPUTATION FOR THE EXPLAINED VARIATION—EXAMPLE 2 (2)

(1) X Class Interval (Years)	(2) Frequency of X Column (Salesmen) f_x	(3) Mean of Y Values in X Column ($1,000) $\bar{Y}_x$	(4) Deviation of Column Mean from Grand Mean $(\bar{Y} = 8.9)*$ $\bar{Y}_x - \bar{Y}$	(5) $(\bar{Y}_x - \bar{Y})^2$	(6) $f_x(\bar{Y}_x - \bar{Y})^2$
0 @ 2	2	2	−6.9	47.61	95.22
2 @ 4	8	6.9	−2.0	4.00	32.00
4 @ 6	16	9.3	0.4	0.16	2.56
6 @ 8	10	8.9	0.0	0.00	0.00
8 @ 10	4	14.8	5.9	34.81	139.24
Total	40	—	—	—	269.02

Source: Table 27–3.

*The grand mean of all Y values $= 356/40 = 8.9 = \bar{Y}$.

pression is computed from Table 27–5. The Y value in the table represents the midpoint of each Y class interval.

TABLE 27–5

COMPUTATION FOR THE TOTAL VARIATION—EXAMPLE 2 (3a)

(1) Y Class Interval	(2) Midpoint (in $1,000) Y	(3) Frequency of Y Row f_y	(4) Deviation of Y from Grand Mean $(\bar{Y} = 8.9)$ $Y - \bar{Y}$	(5) $(Y - \bar{Y})^2$	(6) $f_y(Y - \bar{Y})^2$
1— 3	2	2	−6.9	47.61	95.22
4— 6	5	4	−3.9	15.21	60.84
7— 9	8	21	−0.9	0.81	17.01
10—12	11	7	2.1	4.41	30.87
13—15	14	5	5.1	26.01	130.05
16—18	17	1	8.1	65.61	65.61
Total		40			399.60

Source: Table 27–3.

(b) By expression ns_y^2 as stated in formula (27–2b). Using formula (27–3), the variance s_y^2 may be computed from the value already provided in Table 25–5, page 671.

$$s_y^2 = i_y^2 \left[\frac{\sum f_y d_y^2}{n} - \left(\frac{\sum f_y d_y}{n} \right)^2 \right] = 3^2 \left[\frac{48}{40} - \left(\frac{12}{40} \right)^2 \right] = 9(1.20 - 0.09) = 9.99$$
$$ns_y^2 = 40(9.99) = 399.60$$

(4) Compute the coefficient of determination from the variations obtained in (2) and (3) above.

$$\eta^2 = \frac{\text{Explained variation}}{\text{Total variation}} = \frac{269.02}{399.60} = 0.6732, \text{ or round to } 67\%$$

(formula 27–2a or 27–2b)

The correlation ratio is

$$\eta = \sqrt{0.6732} = 0.82*$$

Thus, the computed coefficient of determination η^2 indicates that 67% variation of the amounts of sales (Y) are explained by the variation of years of sales experience (X) of the salesmen in the company based on the broken line shown in the correlation table (Table 27–3).

At this point, it is possible to compute the standard deviation of regression s_{yx} for Example 2 in the usual manner.** However, the use of a standard deviation based on a broken line on a correlation table is limited.

Comments on the coefficient of determination η^2 and the correlation ratio η

(1) The value of η^2 indicates the maximum possible correlation for the bivariate data presented in a correlation table. In other words, η^2 sets an upper limit to other types of coefficient of determination (r^2 for a straight line and r_{yx}^2 for any curve, such as a parabolic curve). We know that

$$\begin{pmatrix} \text{total} \\ \text{variation} \end{pmatrix} = \begin{pmatrix} \text{unexplained} \\ \text{variation} \end{pmatrix} + \begin{pmatrix} \text{explained} \\ \text{variation} \end{pmatrix}$$

or

$$\sum f_y (Y - \bar{Y})^2 = \sum f(Y - \bar{Y}_x)^2 + \sum f_x(\bar{Y}_x - \bar{Y})^2$$

Since $\bar{Y}_x$ is an arithmetic mean of the Y values in each column, the unexplained variation, or the sum of the squared deviations $\sum f(Y - \bar{Y}_x)^2$, is the least. The f is the frequency of Y values in each cell of the correlation table. The total variation is fixed, or is not affected by the X variable. Thus, the explained variation,

---

*If the data of Example 2 are considered as a sample to be used for obtaining the estimate of the population coefficient of determination, the computation for the estimate, based on formula (26–5b), is

$$\hat{\eta}^2 = 1 - (1 - \eta^2)\left(\frac{n-1}{n-m}\right) = 1 - (1 - 0.6732)\left(\frac{40-1}{40-5}\right) = 0.6359$$

or round to 64%, where m = number of X columns in the correlation table = 5. The estimate of population correlation ratio is

$$\hat{\eta} = \sqrt{0.6359} = 0.80$$

The estimates $\hat{\eta}^2$ and $\hat{\eta}$ are smaller than the unadjusted values of η^2 and η respectively.
**Since the regression line in the form of a broken line represents column means, $\bar{Y}_x$ must replace Y_c in formula (25–3a), or

$$s_{yx} = \sqrt{\frac{\text{Unexplained variation}}{n}} = \sqrt{\frac{\sum f(Y - \bar{Y}_x)^2}{n}}$$

where f = the frequency of Y values in each cell of the correlation table. The unexplained variation can be obtained by arranging a table showing $\sum f(Y - \bar{Y}_x)^2$. However, a simple way to compute the unexplained variation is to use the computed variations shown on Tables 27–4 and 27–5:

Unexplained variation = Total variation − Explained variation
$$= 399.60 - 269.02 = 130.58$$

Thus,
$$s_{yx} = \sqrt{130.58/40} = \sqrt{3.2645} = 1.81$$

explained or affected by X variable, is the largest and the η^2, the ratio of the explained variation to the total variation, is also the largest.

(2) If the column means $\bar{Y}_x$ fall on a straight line, η^2 will be equal to r^2. As the relationship between the X and Y variable departs from linear, η^2 will differ from r^2, but greater than r^2. Compare η^2 $(= 67\%)$ to r^2 $(= 50\%$, page 672) for the same data given in Example 2. It is obvious that the coefficient of determination based on the broken line, η^2, is far greater than that based on the straight line, r^2. Likewise, if the column means $\bar{Y}_x$ fall on a curve, η^2 will be equal to r^2_{yx}. As the relationship between the two variables departs from curvilinear, η^2 will be greater than r^2_{yx}.

(3) If the Y values in each X column are concentrated in only one cell, $Y = \bar{Y}_x$. The broken line then goes through all Y values on the correlation table. Since there is no scatter about the broken line, the correlation is perfect, or $\eta^2 = 1$. However, this perfect correlation will not increase our knowledge or understanding of the relationship between two variables. Our experience shows that we always can draw a broken line which will go through every point on a scatter diagram. We should therefore select an appropriate straight line or curve for the regression and correlation analysis in such a case.

27.4 Solving Linear Equations by Matrix Algebra

A system of n linear equations in n unknowns may be solved by the ordinary algebraic methods (as discussed in Section 4.2). However, the work of solving a system of three or more linear equations becomes increasingly difficult with the ordinary methods. Matrix algebra offers simplified and systematic methods that are convenient in using an electronic computer to solve these equations.

Terminology and Basic Operations

A *matrix* is a rectangular array of numbers, or *elements*, enclosed in brackets (or in bold-faced parentheses.) Capital letters are usually used to represent the matrices, such as

$$A = [1 \quad 2], \quad B = \begin{bmatrix} 1 \\ 0 \\ 5 \end{bmatrix}, \quad C = \begin{bmatrix} 2 & 1 \\ 3 & -2 \end{bmatrix},$$

$$D = \begin{bmatrix} 1 & 3 & 0 \\ 2 & 4 & 1 \end{bmatrix}, \quad E = \begin{bmatrix} 1 & 3 \\ 0 & -4 \\ 2 & 1 \end{bmatrix}, \quad I = \begin{bmatrix} 1 & 0 & 0 \\ 0 & 1 & 0 \\ 0 & 0 & 1 \end{bmatrix}$$

A matrix may be specified by using the *order* of the matrix. The order is first based on the number of rows and then based on the number of columns. Thus, the order of matrix D is 2×3 (read 2 by 3) and that of E is 3×2.

A matrix which has only one row (or column) is also called a *vector*. Thus, A, which is a 1×2 matrix, is a *row vector* and B, which is a 3×1 matrix, is a *column vector*.

When the number of rows and the number of columns in a matrix are equal, it is a *square matrix*, such as matrices C and I. Matrix I is also called an *identity* (or *unit*) *matrix* of order 3. An identity matrix, usually denoted by the letter I, is a square matrix with all elements on its principal diagonal (the line from upper left corner to the lower right corner) being 1's and all other elements being 0's.

The basic operations with matrices are addition, subtraction, and multiplication. These operations are illustrated below.

Addition

In addition, the two matrices must have the same number of rows and the same number of columns. The corresponding elements of the two matrices are added to obtain the elements of the sum.

Example 3

Add two 2×2 matrices.

Solution.

$$\begin{bmatrix} 2 & 1 \\ 3 & -2 \end{bmatrix} + \begin{bmatrix} 1 & 3 \\ 2 & 4 \end{bmatrix} = \begin{bmatrix} 2+1 & 1+3 \\ 3+2 & (-2)+4 \end{bmatrix} = \begin{bmatrix} 3 & 4 \\ 5 & 2 \end{bmatrix}$$

Subtraction

In subtraction, the two matrices also must have the same number of rows and the same number of columns. The subtraction is performed on the corresponding elements of the two matrices in order to obtain the elements of the remainder.

Example 4

Subtract a 2×2 matrix from another 2×2 matrix.

Solution.

$$\begin{bmatrix} 2 & 1 \\ 3 & -2 \end{bmatrix} - \begin{bmatrix} 1 & 3 \\ 2 & 4 \end{bmatrix} = \begin{bmatrix} 2-1 & 1-3 \\ 3-2 & (-2)-4 \end{bmatrix} = \begin{bmatrix} 1 & -2 \\ 1 & -6 \end{bmatrix}$$

Multiplication

When a matrix is multiplied by a number (scalar), every element of the matrix must be multiplied by the number.

Example 5

Multiply a 2×3 matrix by 2.

Solution.

$$2\begin{bmatrix} 1 & 3 & 0 \\ 2 & 4 & 1 \end{bmatrix} = \begin{bmatrix} 2 & 6 & 0 \\ 4 & 8 & 2 \end{bmatrix}$$

Two vectors with the same number of elements can be multiplied if the first (multiplicand) is a row vector and the second (multiplier) is a column vecter. Each element of the row vector is multiplied by the corresponding element of the column vector to obtain the partial product. The sum of all partial products is the product of the multiplication; the product is a number, not a vector.

Example 6

Multiply a row vector (3 elements) by a column vector (also 3 elements).

Solution.

$$[2 \quad 1 \quad -3] \cdot \begin{bmatrix} 3 \\ 4 \\ 1 \end{bmatrix} = (2 \times 3) + (1 \times 4) + [(-3) \times 1] = 7$$

In multiplying two matrices, the number of columns in the first matrix must be equal to the number of rows in the second matrix. Each row vector of the first matrix is multiplied by each column vector of the second matrix to obtained the corresponding element of the product. The product is a matrix in which the number of rows is the same as that of the first matrix and the number of columns is the same as that of the second matrix.

Example 7

Multiply a 2×2 matrix by a 2×3 matrix.

Solution.

$$\begin{bmatrix} 2 & 1 \\ 3 & -2 \end{bmatrix}_{2\times2} \cdot \begin{bmatrix} 1 & 3 & 0 \\ 2 & 4 & 1 \end{bmatrix}_{2\times3} = \begin{bmatrix} [2 \ 1]\begin{bmatrix}1\\2\end{bmatrix} & [2 \ 1]\begin{bmatrix}3\\4\end{bmatrix} & [2 \ 1]\begin{bmatrix}0\\1\end{bmatrix} \\ [3 \ -2]\begin{bmatrix}1\\2\end{bmatrix} & [3 \ -2]\begin{bmatrix}3\\4\end{bmatrix} & [3 \ -2]\begin{bmatrix}0\\1\end{bmatrix} \end{bmatrix}$$

$$= \begin{bmatrix} 2+2 & 6+4 & 0+1 \\ 3-4 & 9-8 & 0-2 \end{bmatrix} = \begin{bmatrix} 4 & 10 & 1 \\ -1 & 1 & -2 \end{bmatrix}_{2\times3}$$

The order of each matrix in the multiplication is written directly under the lower right corner of the matrix. Observe that the product of the 2×2 matrix and the 2×3 matrix is a 2×3 matrix. This is diagrammed below.

Multiplication of Two Matrices

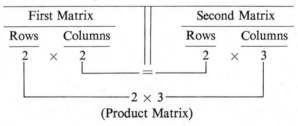

(Product Matrix)

Example 8

Let matrix $A = \begin{bmatrix} 2 & 1 \\ 3 & -2 \end{bmatrix}$ and matrix $A^{-1} = \begin{bmatrix} 2/7 & 1/7 \\ 3/7 & -2/7 \end{bmatrix}$. Find the product of $A \cdot A^{-1}$

Solution.

$$A \cdot A^{-1} = \begin{bmatrix} 2 & 1 \\ 3 & -2 \end{bmatrix} \cdot \begin{bmatrix} 2/7 & 1/7 \\ 3/7 & -2/7 \end{bmatrix} = \frac{1}{7} \begin{bmatrix} 2 & 1 \\ 3 & -2 \end{bmatrix} \begin{bmatrix} 2 & 1 \\ 3 & -2 \end{bmatrix}$$

$$= \frac{1}{7} \begin{bmatrix} 7 & 0 \\ 0 & 7 \end{bmatrix} = \begin{bmatrix} 1 & 0 \\ 0 & 1 \end{bmatrix}$$

The product is a 2×2 identity matrix, which has the same order as A and A^{-1}. A^{-1} is called the *inverse* of the square matrix A since the product of the two matrices is an identity matrix.

Solving Linear Equations

There are various methods of solving linear equations by matrix algebra. Basically, however, each method is involved with finding the inverse of a square matrix. In the following discussion, the basic concept is first introduced; then, a systematic method, called *Gaussian Method*, is presented to solve a system of linear equations.

A system of linear equations can be written in a matrix form, such as the system of two given equations

$$\begin{cases} 2x + y = 5 \\ 3x - 2y = 11 \end{cases}$$

being written

$$\begin{bmatrix} 2 & 1 \\ 3 & -2 \end{bmatrix} \cdot \begin{bmatrix} x \\ y \end{bmatrix} = \begin{bmatrix} 5 \\ 11 \end{bmatrix}$$

This is true because by multiplying the two matrices on the left side of the equation, the product is the same as the expressions on the left sides of the two given equations. Notice that the first matrix on the left side of the above equation is formed by the coefficients of unknowns x and y, the second matrix by the unknowns, and the matrix on the right side by the constants.

Let A = the coefficients matrix, or $A = \begin{bmatrix} 2 & 1 \\ 3 & -2 \end{bmatrix}$, and

A^{-1} = the inverse of the square matrix A

Substitute A in the above equation and multiply both sides by A^{-1}. Then,

$$A^{-1} \cdot A \cdot \begin{bmatrix} x \\ y \end{bmatrix} = A^{-1} \cdot \begin{bmatrix} 5 \\ 11 \end{bmatrix}$$

It can be proved that $A^{-1} \cdot A = \begin{bmatrix} 1 & 0 \\ 0 & 1 \end{bmatrix}$ (see Example 8), and $\begin{bmatrix} 1 & 0 \\ 0 & 1 \end{bmatrix} \cdot \begin{bmatrix} x \\ y \end{bmatrix} = \begin{bmatrix} x \\ y \end{bmatrix}$.

Thus,

$$\begin{bmatrix} x \\ y \end{bmatrix} = A^{-1} \cdot \begin{bmatrix} 5 \\ 11 \end{bmatrix}$$

Using the information given in Example 8,

$$A^{-1} \cdot \begin{bmatrix} 5 \\ 11 \end{bmatrix} = \begin{bmatrix} 2/7 & 1/7 \\ 3/7 & -2/7 \end{bmatrix} \cdot \begin{bmatrix} 5 \\ 11 \end{bmatrix} = \begin{bmatrix} 10/7 + 11/7 \\ 15/7 - 22/7 \end{bmatrix} = \begin{bmatrix} 3 \\ -1 \end{bmatrix}$$

or

$$\begin{bmatrix} x \\ y \end{bmatrix} = \begin{bmatrix} 3 \\ -1 \end{bmatrix}$$

The elements in the corresponding positions of the two matrices are equal. Thus,

$$x = 3 \text{ and } y = -1$$

In general, let X = the matrix formed by the unknowns, and C = the matrix formed by the constants. The solution to a system of equations can be written:

$$X = A^{-1} \cdot C$$

Therefore, the problem of solving a system of linear equations becomes a problem of finding the product of the inverse of the coefficients matrix, A^{-1}, and the matrix of constants, C.

The Gaussian Method

The Gaussian Method is used to find the inverse and the solution of a system of equations simultaneously. The steps of the method are as follows:

(1) Write an initial tableau formed by the coefficients matrix A, an identity matrix I of the same order as A, and the constants matrix C, or

$$[A|I|C] = \begin{bmatrix} 2 & 1 & 1 & 0 & 5 \\ 3 & -2 & 0 & 1 & 11 \end{bmatrix}$$

(if the equations in the preceding illustration are used).

(2) Perform *elementary row operations* (see below for details) on the three matrices in the tableau until matrix A becomes an identity matrix.

The old matrices I and C are now replaced respectively by the inverse of matrix A, or A^{-1}, and the solution matrix, denoted by X. Thus, we are transforming the initial tableau

$$[A|I|C] \text{ into a final tableau } [I|A^{-1}|X]$$

or, we are transforming the initial tableau

$$\begin{bmatrix} 2 & 1 & 1 & 0 & 5 \\ 3 & -2 & 0 & 1 & 11 \end{bmatrix}$$

into the final tableau

$$\begin{bmatrix} 1 & 0 & 2/7 & 1/7 & 3 \\ 0 & 1 & 3/7 & -2/7 & -1 \end{bmatrix}$$

The *elementary row operations* can be performed in two ways:

(a) A row in a tableau may be multiplied or divided by a real number k (other than zero). Thus, the first row in the initial tableau above

$$2, \quad 1, \quad 1, \quad 0, \quad 5$$

may be converted to a new row by dividing each element by 2. The new first row now becomes

$$2/2 = 1, \quad 1/2, \quad 1/2, \quad 0/2 = 0, \quad 5/2 = 2 \ 1/2$$

(b) A row may be added or subtracted by the product of a real number k multiplied by another row. Thus, the second row in the initial tableau

$$3, \quad -2, \quad 0, \quad 1, \quad 11$$

may be converted to a new row by subtracting the product of the real number 3 multiplied by the new first row (or the old first row if so desired), or

Old row 2	3	-2	0	1	11 $\cdots\cdots$ (1)
$3 \times$ new row 1	$3(1)$	$3(1/2)$	$3(1/2)$	$3(0)$	$3(5/2)$ $\cdots\cdots$ (2)
$(1) - (2)$	0	$-7/2$	$-3/2$	1	$7/2$ $\cdots$ *New row 2*

The work of transforming the square matrix A into the identity matrix I may be done systematically. The system is to convert the elements of matrix A column by column, from left to right. Within each column,

First convert the appropriate element to 1 for obtaining the 1's on the principal diagonal of the identity matrix—use row operation (a).

Next convert all other elements to 0—use row operation (b).

Example 9

Given the system of two equations:

$$\begin{cases} 2x + y = 5 \\ 3x - 2y = 11 \end{cases}$$

Let the coefficients matrix $A = \begin{bmatrix} 2 & 1 \\ 3 & -2 \end{bmatrix}$. Find the inverse of A and the solution to the equations.

Solution. (1) Write the initial tableau,

$$[A|I|C] = \begin{bmatrix} 2 & 1 & | & 1 & 0 & | & 5 \\ 3 & -2 & | & 0 & 1 & | & 11 \end{bmatrix} \begin{matrix} \cdots\cdots & (1) \\ \cdots\cdots & (2) \end{matrix}$$

(2) Perform row operations.

1. Convert the elements in the first column of A from $\frac{2}{3}$ to $\frac{1}{0}$.

  a. Convert the first element 2 in row (1) to 1. See the illustration for operation (a) above.

$$\text{Row } (1) \div 2 : \quad 1 \quad 1/2 \quad 1/2 \quad 0 \quad 5/2 \quad \cdots\cdots \ (1)'$$

  b. Convert the first element 3 in row (2) to 0. See the illustration for operation (b) above.

Row (2) − [3 × row (1)′]: 0   −7/2   −3/2   1   7/2 · · · · · (2)′

The initial tableau thus is converted into the second tableau as follows:

$$\begin{bmatrix} 1 & 1/2 & 1/2 & 0 & 5/2 \\ 0 & -7/2 & -3/2 & 1 & 7/2 \end{bmatrix} \begin{matrix} \cdots\cdots (1)' \\ \cdots\cdots (2)' \end{matrix}$$

2. Convert the elements in the second column of the second tableau from
$\dfrac{1/2}{-7/2}$ to $\dfrac{0}{1}$

a. Convert the second element (−7/2) in row (2)′ to 1. Use operation (a):

Row (2)′ ÷ (−7/2), or row (2)′ × (−2/7): 0   1   3/7   −2/7   −1 · · · · (2)″

b. Convert the second element 1/2 in row (1)′ to 0. Use operation (b).

Row (1)′	1	1/2	1/2	0	5/2
Row (2)″ × (1/2)	0	1/2	3/14	−1/7	−1/2
Subtraction	1	0	4/14	1/7	6/2 · · · · · (1)″
			=2/7		=3

The final tableau is

$$[I|A^{-1}|X] = \begin{bmatrix} 1 & 0 & 2/7 & 1/7 & 3 \\ 0 & 1 & 3/7 & -2/7 & -1 \end{bmatrix} \begin{matrix} \cdots\cdots (1)'' \\ \cdots\cdots (2)'' \end{matrix}$$

Thus,

$$A^{-1} = \begin{bmatrix} 2/7 & 1/7 \\ 3/7 & -2/7 \end{bmatrix} \qquad \text{Check: } A \cdot A^{-1} = I \text{ (See Example 8.)}$$

$$X = \begin{bmatrix} x \\ y \end{bmatrix} = \begin{bmatrix} 3 \\ -1 \end{bmatrix} \text{ or } x = 3 \text{ and } y = -1 \qquad \text{(Also see Example 4, Section 4.2)}$$

Example 9 illustrated the procedure of finding the inverse of the coefficients matrix and the solution of a given system of linear equations simultaneously. When the inverse is not required, the procedure of finding the solution of a system of equations can be simplified, or simply transforming the tableau

$$[A|C] \text{ into a new tableau } [I|X]$$

This simplified procedure is illustrated in Example 10.

Example 10

Solve the system of three equations: (same as the three normal equations in Example 1, page 707.)

$$8a + 30b + 136c = 40$$
$$30a + 136b + 684c = 178$$
$$136a + 684b + 3652c = 902$$

Solution. (1) Write the initial tableau

$$[A|C] = \begin{bmatrix} 8 & 30 & 136 & 40 \\ 30 & 136 & 684 & 178 \\ 136 & 684 & 3652 & 902 \end{bmatrix} \begin{matrix} \cdots\cdots & (1) \\ \cdots\cdots & (2) \\ \cdots\cdots & (3) \end{matrix}$$

(2) Perform row operations. Let R represent "row."

1. Convert the elements in column 1 of A from $\begin{matrix} 8 & 1 \\ 30 & \text{to } 0. \\ 136 & 0 \end{matrix}$ The order of

the three computations for the conversion is: $R(1)'$, $R(2)'$, and $R(3)'$. The results of the computations are shown in the second tableau below.

$$\begin{matrix} R(1) \div 8 \\ R(2) - 30 \times R(1)' \\ R(3) - 136 \times R(1)' \end{matrix} \begin{bmatrix} 1 & 3.75 & 17 & 5 \\ 0 & 23.50 & 174 & 28 \\ 0 & 174.00 & 1340 & 222 \end{bmatrix} \begin{matrix} \cdots\cdots & R(1)' \\ \cdots\cdots & R(2)' \\ \cdots\cdots & R(3)' \end{matrix}$$

Computation for $R(2)'$:				
$R(2)$	30	136.00	684	178
$30 \times R(1)'$	30	112.50	510	150
Subtract	0	23.50	174	28

Computation for $R(3)'$:				
$R(3)$	136	684	3652	902
$136 \times R(1)'$	136	510	2312	680
Subtract	0	174	1340	222

2. Convert the elements in column 2 of the second tableau above from $\begin{matrix} 3.75 & 0 \\ 23.50 \text{ to } 1. \\ 174.00 & 0 \end{matrix}$ The order the three computations for the conversion is: $R(2)''$, $R(1)''$, and $R(3)''$. The results of the computations are shown in the third tableau below.

$$\begin{matrix} R(1)' - 3.75 \times R(2)'' \\ R(2)' \div 23.5 \\ R(3)' - 174 \times R(2)'' \end{matrix} \begin{bmatrix} 1 & 0 & -10.765956 & 0.531916 \\ 0 & 1 & 7.404255 & 1.191489 \\ 0 & 0 & 51.659630 & 14.680914 \end{bmatrix} \begin{matrix} \cdots\cdots & R(1)'' \\ \cdots\cdots & R(2)'' \\ \cdots\cdots & R(3)'' \end{matrix}$$

Computation for $R(1)''$:				
$R(1)'$	1	3.75	17	5
$3.75 \times R(2)''$	0	3.75	27.765956	4.468084
Subtract	1	0	-10.765956	0.531916

Computation for $R(3)''$:				
$R(3)'$	0	174	1340	222
$174 \times R(2)''$	0	174	1288.340370	207.319086
Subtract	0	0	51.659630	14.680914

3. Convert the elements in column 3 of the third tableau above from

$-10.765956 \quad 0$

7.404255 to 0. The order of the three computations for the conversion

$51.659630 \quad 1$

is: $R(3)'''$, $R(1)'''$, and $R(2)'''$. The results of the computations are shown in the from of $[I|X]$, which is the final tableau.

$$
\begin{array}{l}
R(1)'' - (-10.765956)R(3)''' \\
R(2)'' - (7.404255)R(3)''' \\
R(3)'' \div 51.659630
\end{array}
\begin{bmatrix}
1 & 0 & 0 \\
0 & 1 & 0 \\
0 & 0 & 1
\end{bmatrix}
\begin{bmatrix}
3.591439 \\
-0.912689 \\
0.284185
\end{bmatrix}
\begin{array}{l}
\cdots\cdots R(1)'' \\
\cdots\cdots R(2)'' \\
\cdots\cdots R(3)''
\end{array}
$$

Computation for $R(1)'''$:				
$R(1)''$	1	0	-10.765956	0.531916
$(-10.765956)R(3)'''$	0	0	-10.765956	-3.059523
Subtract	1	0	0	3.591439

Computation for $R(2)'''$:				
$R(2)''$	0	1	7.404255	1.191489
$(7.404255)R(3)'''$	0	0	7.404255	2.104178
Subtract	0	1	0	-0.912689

The final tableau gives: $X = \begin{bmatrix} a \\ b \\ c \end{bmatrix} = \begin{bmatrix} 3.591439 \\ -0.912689 \\ 0.284185 \end{bmatrix}$. Thus, $a = 3.591439$, $b = -0.912689$, and $c = 0.284185$.

27.5 Summary

The relationship between two variables concerning business and economic activities can frequently be described better by a curve than by a straight line. The coefficient of determination obtained from a nonlinear regression line is denoted by r^2 with subscripts y and x. The value of r_{yx}^2 (Y being the dependent and X the independent variable) is not the same as the value of r_{xy}^2 (X being the dependent and Y the independent variable), except that the curve becomes a straight line. The coefficient of correlation, r_{yx} or r_{xy}, is also called the index of correlation and is not prefixed by a $+$ or $-$ sign. The slope of a curve may be positive in some parts and negative in other parts of the curve.

A regression curve can be obtained by a mathematical equation or by the freehand graphic method. The mathematical equation presented in this chapter is a second-degree polynomial equation. The steps usually included in regression and correlation analysis based on the equation are: (1) plot a scatter diagram, (2) compute the regression equation, (3) draw the regression curve, (4) compute the standard deviation of regression, and (5) compute the coefficient of determination and the coefficient. When the freehand graphic method is used, the computation of regression equation may be omitted. The Y_c values may be read directly from the smooth curve drawn by the investigator.

A practical way in measuring nonlinear correlation for a large set of data is to use a correlation table, which presents bivariate grouped data. A broken line is used to represent the average relationship between two variables. The broken line is determined by the points representing $\bar{Y}_x$, the mean of the Y values in each X column. The coefficient of determination for a correlation table, Y being the dependent variable and X the independent variable on the table, is denoted by η^2. The Greek letter η is called the correlation ratio. The η^2 is expressed as a ratio of the explained variation to the total variation for the convenience in computation.

A matrix is a rectangular array of numbers. The basic operations with matrices are addition, subtraction, and multiplication. Matrix algebra can offer simplified and systematic methods that are convenient in using an electronic computer to solve a system of linear equations. Basically, the methods are involved with the process of finding the inverse of a square matrix. This chapter introduced the Gaussian method of finding the inverse and the solution of a system of linear equations simultaneously (Example 9). When the inverse is not required, this method can be simplified in finding the solution of the equations (Example 10.)

SUMMARY OF FORMULAS

Application	Formula	Formula Number	Reference Page
Parabolic curve Second-degree polynomial equation	$Y_c = a + bX + cX^2$	22–4	675
Normal equations for formula (22-4)	I. $\sum (Y) = na + b \sum (Y) + c \sum (X^2)$ II. $\sum (XY) = a \sum (X) + b \sum (X^2) + c \sum (X^3)$ III. $\sum (X^2 Y) = a \sum (X^2) + b \sum (X^3) + c \sum (X^4)$	22–5	675
Standard deviation of regression	$s_{yx} = \sqrt{\dfrac{\sum (Y - Y_c)^2}{n}}$	25–3a	659
Coefficient of determination	$r_{yx}^2 = \dfrac{\sum (Y_c - \bar{Y})^2}{\sum (Y - \bar{Y})^2}$	27–1a	706
	$r_{yx}^2 = 1 - \dfrac{s_{yx}^2}{s_y^2}$	27–1b	707
Correlation table for grouped data Coefficient of determination	$\eta^2 = \dfrac{\sum f_x(\bar{Y}_x - \bar{Y})^2}{\sum f_y(Y - \bar{Y})^2}$	27–2a	712
	$\eta^2 = \dfrac{\sum f_x(\bar{Y}_x - \bar{Y})^2}{ns_y^2}$	27–2b	712
Variance for formula (27-2b)	$s_y^2 = i_y^2 \left[\dfrac{\sum f_y d_y^2}{n} - \left(\dfrac{\sum f_y d_y}{n} \right)^2 \right]$	27–3	712

Exercises 27

1. Explain briefly:
    (a) Second-degree regression curve.
    (b) Index of correlation.
    (c) Correlation ratio.
    (d) Identity matrix.

2. How can you determine whether the relationship between two variables is linear or nonlinear?

3. Is it true that a curve based on the method of least squares will generally be better fitted to data concerning business and economic activities than a straight line based on the same method?

4. Based on the method of least squares, why is the value of correlation ratio (η) usually larger than the value of coefficient of correlation of nonlinear relationship (r_{yx}), which in turn is usually larger than the value of correlation coefficient of linear relationship (r) for the same data? Under what condition will the three values be the same?

5. How can you tell whether the nonlinear relationship between two varables is positive or negative?

6. Is it true that a regression line having a higher value of coefficient of determination, such as r_{yx}^2 generally being higher than r^2, is always a better line to be used in describing the relationship between two given variables for the correlation analysis? If so, why do we not use a broken line which goes through every point on a scatter diagram for the analysis?

7. Use the information given in Problem 3, in the Exercises for Chapter 25, page 674, to do the following:
    (a) Plot a scatter diagram on a chart.
    (b) Compute the regression equation $Y_c = a + bX + cX^2$ by the method of least squares.
    (c) Draw the regression curve based on the above equation on the chart.

8. Refer to Problem 7 above.
    (a) Compute the standard deviation of regression s_{yx}.
    (b) Compute the coefficient of determination r_{yx}^2 and the coefficient of correlation r_{yx}.
    (c) Compare the value of r_{yx}^2 with the value of r^2 ($= 0.67$). What type of regression line, straight line or parabolic curve, is a better device for describing the average relationship between the hours (X) studied and the number of correct answers (Y) made on the test by each of the five employees in the company?

9. Use the information given in Problem 5 of the Exercises for Chapter 25, to do the following (consider Y as the dependent variable and X as the independent variable):
    (a) Plot a scatter diagram on a chart.

(b) Draw a smooth curve to fit the data by the freehand graphic method. This curve should fit the data better than the straight line obtained by the least square method in the original problem.

(c) Read the Y_c value from the smooth curve for each given X value.

(d) Compute each $(Y - Y_c)$, the difference between each Y and its corresponding Y_c on the given X.

(e) Compute the standard deviation of regression s_{yx}.

(f) Compute the coefficient of determination r_{yx}^2 and the coefficient of correlation r_{yx}.

(g) Estimate the statistics grades for students J and K who have economics grades 82 and 75 respectively in this semester. Do you think the estimates based on the regression curve are better than those based on the straight-line regression ($r^2 = 0.52$ and $r = 0.72$)?

10. Use Table 25–8 given in Problem 9 in the Exercises for Chapter 25. Compute the coefficient of determination (η^2) and the correlation ratio (η).

11. How would you determine approximately (without computation) the values of r_{yx}^2 and r_{yx} based on the parabolic curve fitted to the data referred to in Problem 10 above by the method of least squares? (Hint: $r^2 = 0.50$ and $r = 0.71$ based on the straight line for the same data.)

12. Use Table 25–9 given in Problem 10, of the Exercises for Chapter 25 to compute the coefficient of determination (η^2) and the correlation ratio (η).

13. From each of the following systems of linear equations, find (a) the inverse of coefficients matrix, and (b) the solution to the equations by the Gaussian method:

(1) $\begin{cases} 5x + 7y = 11 \\ 2x - 4y = -16 \end{cases}$

(2) $\begin{cases} x + 3y = 1 \\ 3x - 2y = 14 \end{cases}$

(3) $\begin{cases} x + y + 2z = -3 \\ 6x - y + z = 1 \\ 2x + 3y - 3z = 17 \end{cases}$

(4) $\begin{cases} x - y + z = 15 \\ 3x - 2y - 3z = -2 \\ 2x + y - 2z = -13 \end{cases}$

14. Solve the system of three equations (same as the three normal equations in Example 1, Chapter 28, page 732.) by the Gaussion method:

$$\begin{cases} 8a + 30b_2 + 16b_3 = 40 \\ 30a + 136b_2 + 68b_3 = 178 \\ 16a + 68b_2 + 38b_3 = 94 \end{cases}$$

28 *Multiple, Partial, and Rank Correlation

This chapter continues the discussion of regression and correlation analysis. Frequently, one dependent variable may be related or associated not only with one independent variable, but also with two or more independent variables. The analysis of the relationship between three or more variables concerning multiple and partial correlation is presented in Sections 28.1 and 28.2. In addition, two related variables may be expressed in ranked numbers. The measurement of rank correlation is discussed in Section 28.3.

28.1 Multiple Linear Regression and Correlation

This section will discuss methods for measuring the linear relationship between three or more variables. When the dependent variable is related to two or more independent variables, a more realistic result can be achieved by analyzing the variables together at the same time.

In simple regression and correlation analysis, only two variables were included; the dependent variable was represented by Y and the independent variable by X. In multiple analysis, however, it is more convenient to use X with subscripts to represent all the variables included in the analysis. There is still only one dependent variable, which is represented by X_1. The independent variables are represented by X_2, X_3, X_4, ... and so on. Based on the new symbols, the simple linear regression equation

$$Y_c = a + bX$$

may be written:

$$X_{1c} = a + bX_2$$

Examples of three or more related variables are as follows:

Dependent Variable (X_1)	Related Independent Variables $(X_2, X_3, \ldots)$
Amount of sales by each salesman in a company	Years of sales experience (X_2), and intelligence test score (X_3) of each salesman
Weight of each boy in a group	Height (X_2) and age (X_3)
Corn yield per acre in each year during a period of years	Amount of rainfall (X_2), amount of fertilizer (X_3), and average temperature (X_4)
Highway accidents in each state during a given period	Cars registered (X_2), miles traveled (X_3), population size (X_4), and highway maintenance expenditure (X_5)

For simplicity in illustration, only the relationship between three variables is analyzed in the following discussion. The regression equation of X_1 on only two independent variables X_2 and X_3 is

$$X_{1c} = a + b_2 X_2 + b_3 X_3 \qquad (28\text{-}1)^*$$

Formula (28-1) may be written in a more detailed form:

$$X_{1(23)c} = a_{1(23)} + b_{12.3} X_2 + b_{13.2} X_3$$

The subscripts of the dependent variable X and the constant a are the same, except the letter c. The letter c indicates the computed value, the same indication as for Y_c. The numbers in the parenthesis (23) represent the independent variables X_2 and X_3 that are included in the multiple analysis.

---

*This regression equation can easily be extended for more than two independent variables. For example, the multiple regression equation of X_1 on three independent variables X_2, X_3, and X_4 is written:

$$X_{1c} = a + b_2 X_2 + b_3 X_3 + b_4 X_4$$

The four normal equations for solving the four unknown constants, a, b_2, b_3, and b_4 are obtained by multiplying the regression equation successively by 1, X_2, X_3, and X_4, and summing on both sides:

I. $\sum (X_1) = na + b_2 \sum (X_2) + b_3 \sum (X_3) + b_4 \sum (X_4)$
II. $\sum (X_1 X_2) = a \sum (X_2) + b_2 \sum (X_2^2) + b_3 \sum (X_2 X_3) + b_4 \sum (X_2 X_4)$
III. $\sum (X_1 X_3) = a \sum (X_3) + b_2 \sum (X_2 X_3) + b_3 \sum (X_3^2) + b_4 \sum (X_3 X_4)$
IV. $\sum (X_1 X_4) = a \sum (X_4) + b_2 \sum (X_2 X_4) + b_3 \sum (X_3 X_4) + b_4 \sum (X_4^2)$

Or, let $x_1 = X_1 - \bar{X}_1$, $x_2 = X_2 - \bar{X}_2$, $x_3 = X_3 - \bar{X}_3$,   and $x_4 = X_4 - \bar{X}_4$

Then, solve the following three equations simultaneously for the constants b_2, b_3, and b_4: (Normal equation I now becomes $0 = 0$.)

From II— $\sum (x_1 x_2) = b_2 \sum (x_2^2) + b_3 \sum (x_2 x_3) + b_4 \sum (x_2 x_4) \ldots (1)$
From III— $\sum (x_1 x_3) = b_2 \sum (x_2 x_3) + b_3 \sum (x_3^2) + b_4 \sum (x_3 x_4) \ldots (2)$
From IV— $\sum (x_1 x_4) = b_2 \sum (x_2 x_4) + b_3 \sum (x_3 x_4) + b_4 \sum (x_4^2) \ldots (3)$

The summations in the above three equations may be computed from the following two general equations: (See the footnote on page 709 for proofs.)

$$\sum (x_i^2) = \sum (X_i^2) - \frac{(\sum X_i)^2}{n}, \text{ and } \sum (x_i x_j) = \sum (X_i X_j) - \frac{\sum X_i \cdot \sum X_j}{n}$$

The constant a can be obtained by dividing the original normal equation I by n. See the footnote on page 732 for application.

The subscripts of b coefficients are separated into two groups in a detailed notation. The two numbers to the left of the point are called the *primary subscripts* and are used respectively to indicate the dependent variable and the independent variable of which this b is the coefficient. The numbers, if any, to the right of the point are called the *secondary subscripts* and are used to indicate other independent variables which are included in the multiple regression equation.

Thus, $b_{12.3}$ represents the coefficient of the independent variable X_2 when the dependent variable is X_1 and the other independent variable in the equation is X_3. If there were two other independent variables, X_3 and X_4, in the equation, the coefficient of X_2 would be written $b_{12.34}$. Likewise, $b_{13.2}$ represents the coefficient of the independent variable X_3, the other independent variable in the equation being X_2. We shall use the simplified notation unless it becomes ambiguous. That is, $b_{12.3} = b_2$ and $b_{13.2} = b_3$.

Since there are three unknown constants a, b_2, and b_3 in the equation of formula (28–1), it is necessary to develop three equations for solving the unknowns. The three normal equations based on the method of least squares can be developed in the same manner as those for the second-degree parabolic curve, formula (22–5).

Let $X_1 = Y$, $X_2 = X$, $X_3 = X^2$, $a = $ the same a, $b_2 = b$, and $b_3 = c$

Formula (22–5) may be written for the multiple regression equation as follows:

$$\begin{array}{ll} \text{I.} & \Sigma\,(X_1) = na + b_2\,\Sigma\,(X_2) + b_3\,\Sigma\,(X_3) \\ \text{II.} & \Sigma\,(X_1 X_2) = a\,\Sigma\,(X_2) + b_2\,\Sigma\,(X_2^2) + b_3\,\Sigma\,(X_2 X_3) \\ \text{III.} & \Sigma\,(X_1\,X_3) = a\,\Sigma\,(X_3) + b_2\,\Sigma\,(X_2 X_3) + b_3\,\Sigma\,(X_3^2) \end{array} \qquad \text{(28–2)}$$

The standard deviation of the X_1 values from the computed X_{1c} values is denoted by the symbol $s_{1.23}$, where the primary subscript indicates the dependent variable, and the secondary subscripts indicate the independent variables. It is computed in a similar manner to the standard deviation of simple regression, or

$$s_{1.23} = \sqrt{\frac{\Sigma\,(X_1 - X_{1c})^2}{n}} \qquad \text{(28–3a)}$$

Also, there is a simpler method to compute the standard deviation without computing $(X_1 - X_{1c})^2$ values. This method is an extension of formula (25–3b) and can be stated as follows:

$$s_{1.23} = \sqrt{\frac{\Sigma\,X_1^2 - a\,\Sigma\,X_1 - b_2\,\Sigma\,(X_1 X_2) - b_3\,\Sigma\,(X_1 X_3)}{n}} \qquad \text{(28–3b)}^*$$

The coefficient of determination for multiple correlation may be computed from the variance $s_{1.23}^2$ in the same manner as that for simple correlation. However, it is usually represented by the capital letter R with the same subscripts as the

---

*If the extension includes one more independent variable, X_4, the standard deviation is

$$s_{1.234} = \sqrt{\frac{\Sigma\,X_1^2 - a\,\Sigma\,X_1 - b_2\,\Sigma\,(X_1 X_2) - b_3\,\Sigma\,(X_1 X_3) - b_4\,\Sigma\,(X_1 X_4)}{n}}$$

Also see the footnote on page 706 for a similar extension of formula (25–3b).

variance. The standard deviation of X_1 values from the mean $\bar{X}_1$ is represented by the symbol s_1. Thus,

$$R^2_{1.23} = 1 - \frac{s^2_{1.23}}{s^2_1} \tag{28-4}$$

The standard deviation, or variance, of the multiple regression is also used as a measure of the closeness of the estimates based on the regression equation. The value $\sum (X_1 - X_{1c})^2$ represents the variation that is not explained or not reduced by the introduction of the independent variables. The coefficient of determination of the multiple correlation, as in the case of simple correlation, is the ratio of the explained variation $\sum (X_{1c} - \bar{X}_1)^2$ to the total variation $\sum (X_1 - \bar{X}_1)^2$. It is written in the form of formula (28-4) for the convenience of computation. Thus, the smaller the deviation of the actual values X_1 from the computed values X_{1c}, the smaller will be the standard deviation $s_{1.23}$, but the larger will be the coefficient of determination $R^2_{1.23}$ or the coefficient of correlation $R_{1.23}$.

Example 1

The first four columns of Table 28-1 show the amounts of sales (X_1) made by a group of 8 salesmen during a given period, the years of sales experience (X_2), and the intelligence test scores (X_3) of each salesman. Compute (a) the multiple regression equation, (b) the standard deviation of regression, and (c) the coefficient of determination and the coefficient of correlation. (Note: The X_1 and X_2 variables of this example are the same data as the Y and X variables of Example 1, Chapter 27.)

TABLE 28-1

CALCULATION FOR THE MULTIPLE REGRESSION EQUATION
BY THE METHOD OF LEAST SQUARES—EXAMPLE 1

(1) Sales- man	(2) Amount of Sales (in $1,000)	(3) Years of Sales Ex- perience	(4) Intelli- gence Test Score	(5)	(6)	(7)	(8)	(9)	(10)
	X_1	X_2	X_3	X_1^2	X_2^2	X_3^2	X_1X_2	X_1X_3	X_2X_3
A	9	6	3	81	36	9	54	27	18
B	6	5	2	36	25	4	30	12	10
C	4	3	2	16	9	4	12	8	6
D	3	1	1	9	1	1	3	3	1
E	3	4	1	9	16	1	12	3	4
F	5	3	3	25	9	9	15	15	9
G	8	6	3	64	36	9	48	24	18
H	2	2	1	4	4	1	4	2	2
Total	40	30	16	244	136	38	178	94	68

Solution.   (a) Substitute the totals obtained from Table 28-1 in the three normal equations of formula (28-2) as follows:

$$\text{I.} \quad 8a + 30b_2 + 16b_3 = 40$$
$$\text{II.} \quad 30a + 136b_2 + 68b_3 = 178$$
$$\text{III.} \quad 16a + 68b_2 + 38b_3 = 94$$

The above three equations may be solved simultaneously by the method of elimination as discussed in Section 4.2 or by matrix algebra. The solutions are:*

$$a = -0.4545, \quad b_2 = 0.7273, \quad b_3 = 1.3636$$

The required equation is:

$$X_{1c} = -0.4545 + 0.7273X_2 + 1.3636X_3$$

(b) If the standard deviation of regression is computed by formula (28–3a), the values of X_{1c} must be computed first by the multiple regression equation:

$$\text{For salesman } A: X_2 = 6, \text{ and } X_3 = 3$$
$$X_{1c} = -0.4545 + 0.7273(6) + 1.3636(3)$$
$$= -0.4545 + 4.3638 + 4.0908$$
$$= 8.0001, \text{ or round to } 8.00$$

In a similar manner, the X_{1c} values for other salesmen are computed and listed in column (5) of Table 28–2.

The standard deviation of regression is computed from the variation shown in column (7) of Table 28–2.

---

*The answers may also be obtained by solving the following two equations simultaneously for b_2 and b_3:

$$\sum (x_1 x_2) = b_2 \sum (x_2^2) + b_3 \sum (x_2 x_3) \ldots (1)$$
$$\sum (x_1 x_3) = b_2 \sum (x_2 x_3) + b_3 \sum (x_3^2) \ldots (2)$$

where $x_1 = X_1 - \bar{X}_1, x_2 = X_2 - \bar{X}_2$, and $x_3 = X_3 - \bar{X}_3$. The required values for the two equations are expressed below and can also be obtained from Table 28–1.

$$\sum x_2^2 = \sum X_2^2 - \frac{(\sum X_2)^2}{n} = 136 - \frac{30^2}{8} = 23.5$$

$$\sum x_3^2 = \sum X_3^2 - \frac{(\sum X_3)^2}{n} = 38 - \frac{16^2}{8} = 6$$

$$\sum (x_1 x_2) = \sum (X_1 X_2) - \frac{\sum X_1 \cdot \sum X_2}{n} = 178 - \frac{40(30)}{8} = 28$$

$$\sum (x_1 x_3) = \sum (X_1 X_3) - \frac{\sum X_1 \cdot \sum X_3}{n} = 94 - \frac{40(16)}{8} = 14$$

$$\sum (x_2 x_3) = \sum (X_2 X_3) - \frac{\sum X_2 \cdot \sum X_3}{n} = 68 - \frac{30(16)}{8} = 8$$

Substitute the 5 values in equations (1) and (2) and solve the equations.

$$28 = 23.5b_2 + 8b_3 \ldots (1)$$
$$14 = 8b_2 + 6b_3 \ldots (2)$$

The solutions are

$$b_2 = \frac{8}{11} = 0.7273, \quad b_3 = 1\frac{4}{11} = 1.3636$$

Divide normal equation I by n,

$$a = \frac{\sum (X_1)}{n} - b_2 \frac{\sum (X_2)}{n} - b_3 \frac{\sum (X_3)}{n} = \frac{40}{8} - \frac{8}{11}\left(\frac{30}{8}\right) - \frac{15}{11}\left(\frac{16}{8}\right) = -\frac{5}{11} = -0.4545$$

The five expressions can be proved in the same manner as the proofs in the footnote on page 708.

TABLE 28–2

CALCULATION FOR X_{1c} AND THE STANDARD DEVIATION OF
REGRESSION $s_{1.23}$—EXAMPLE 1 (b)

(1) Salesman	(2) X_1	(3) X_2	(4) X_3	(5) $X_{1c} = -0.4545 + 0.7273X_2 + 1.3636X_3$	(6) $X_1 - X_{1c}$	(7) $(X_1 - X_{1c})^2$
A	9	6	3	8.00	1.00	1.00
B	6	5	2	5.91	0.09	0.01
C	4	3	2	4.45	−0.45	0.20
D	3	1	1	1.64	1.36	1.85
E	3	4	1	3.82	−0.82	0.67
F	5	3	3	5.82	−0.82	0.67
G	8	6	3	8.00	0.00	0.00
H	2	2	1	2.36	−0.36	0.13
Total	40	30	16	40.00	0.00	4.53

$$s_{1.23} = \sqrt{\frac{4.53}{8}} = \sqrt{0.56625} = 0.75 \qquad \text{(in units of \$1,000)}$$

If the standard deviation of regression is computed by formula (28–3b), the values shown in Table 28–1 are used in the computation:

$$s_{1.23} = \sqrt{\frac{244 - (-0.4545)(40) - 0.7273(178) - 1.3636(94)}{8}}$$

$$= \sqrt{\frac{4.5422}{8}} = \sqrt{0.56778} = 0.75$$

This measure of dispersion is based on the linear relationship between the sales and the two independent variables, years of sales experience, and intelligence test scores. Its value ($s_{1.23} = 0.75$) is smaller than that based on the linear relationship between the sales and only one variable, years of sales experience ($s_{yx} = 1.15$, page 660). In general, therefore, the multiple regression equation gives better estimates than the simple regression equation.

Further, compare individually the squared deviation $(X_1 - X_{1c})^2$ based on the multiple regression in Table 28–2 with the squared deviation $(Y - Y_c)^2$ based on the simple regression in Table 25–2 on page 660. The estimates by Y_c for salesmen C and D are closer to the actual sales (Y or X_1) than the estimates by X_{1c}. However, the estimates by X_{1c} for the other six salesmen are closer to the actual sales than those by Y_c. For example, the estimated amount of sales for salesman E who has 4 years of sales experience ($X_2 = 4$) and the intelligence test score 1 point ($X_3 = 1$) is \$3,820 ($X_{1c} = 3.82 \times \$1,000$, the X_{1c} unit). This estimate is closer to the actual sales of salesman E ($X_1 = \$3,000$) than the estimate made in Chapter 25 ($Y_c = 5.29 \times \$1,000 = \$5,290$) without taking the intelligence test score into consideration.

(c) The coefficient of determination $R^2_{1.23}$ and the coefficient of correlation $R_{1.23}$ are computed below by formula (28–4).

$$R_{1.23}^2 = 1 - \frac{s_{1.23}^2}{s_1^2} = 1 - \frac{0.56625}{5.5} = 0.8970 \text{ or round to } 90\%.$$

$$(s_1^2 = s_y^2 = 44/8 = 5.5, \text{ page } 665.)$$

$$R_{1.23} = \sqrt{0.8970} = 0.95$$

The values of $R_{1.23}^2$ and $R_{1.23}$ can be interpreted in the same manner as r^2 and r for simple correlation. Thus, 90% of the variation of the amounts of sales (X_1) is related to, or explained by, the variation of the years of sales experience (X_2) and the intelligence test scores (X_3) of the salesmen based on the multiple regression equation.

The linear relationship between X_1 and the two independent variables X_2 and X_3 for Example 1 is diagrammed in Chart 28–1. Part I of the chart shows a portion of X_{1c} variable—the sum of the first two terms of the regression equation, or

$$- 0.4545 + 0.7273X_2 \, (= X_{1c} - 1.3636X_3)$$

where X_2 variable represents 0, 1, 2, 3, 4, 5, and 6. For example,

$$\text{when } X_2 = 0, \, - 0.4545 + 0.7273(0) = - 0.4545$$
$$X_2 = 1, \, - 0.4545 + 0.7273(1) = 0.2728$$
$$X_2 = 6, \, - 0.4545 + 0.7273(6) = 3.9093$$

Part II of the chart includes six graphs. Each graph represents the values of X_{1c} on a given X_2 and the variable $X_3 \, (= 0, 1, 2, \text{ and } 3)$. The actual values of X_1 are also plotted on the graphs for comparison with corresponding X_{1c} values.

For example, graph II–6 shows the values of X_{1c} on $X_2 = 6$ and the variable X_3.

$$\text{When } X_3 = 3, X_{1c} = 3.9093 + 1.3636X_3$$
$$= 3.9093 + 1.3636(3)$$
$$= 3.9093 + 4.0908$$
$$= 8.0001 \text{ or round to } 8$$

The corresponding actual value of $X_{1c} = 8$ is the sales made by salesman A, or $X_1 = 9$, which is also plotted on the same graph.

When the given data are considered a sample for estimating population parameters, the unbiased estimates of the parameters should be used. Again, the general formulas for the unbiased estimates can be obtained in a similar manner to those presented in Chapter 26.*

---

*For example, the formula (based on formula 26–1) and the computation (based on the data given in Example 1 as a sample) for the estimate of the population standard deviation of multiple regression are

$$\hat{s}_{1.23} = \sqrt{\frac{\sum (X_1 - X_{1c})^2}{n - m}} = \sqrt{\frac{4.53}{8 - 3}} = \sqrt{0.906} = 0.95$$

or

$$\hat{s}_{1.23} = s_{1.23}\sqrt{\frac{n}{n - m}} = 0.75\sqrt{\frac{8}{8 - 3}} = 0.75\sqrt{1.60} = 0.95$$

$$m = 3$$

footnote continued page 736

CHART 28–1

ILLUSTRATION OF MULTIPLE LINEAR RELATIONSHIP—EXAMPLE 1
$$X_{1c} = -.4545 + .7273X_2 + 1.3636X_3$$

Part I—*Showing a Portion of Each X_{1c} Value (The Effect of X_3 is Excluded, or a Portion of $X_{1c} = X_{1c} - 1.3636X_3 = -.4545 + .7273X_2$)*

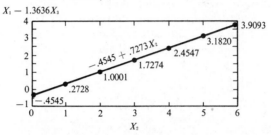

Part II—*Showing Entire X_{1c} Values (The Effect of X_3 is Included, or $X_{1c} = (-.4545 + .7273X_2) + 1.3636X_3 = Part I + 1.3636 X_3$)*

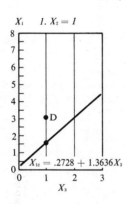

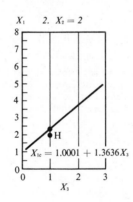

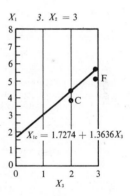

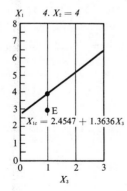

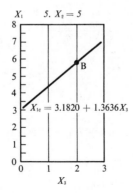

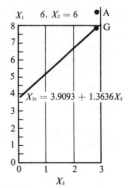

28.2   Partial Correlation

When three or more variables are involved in correlation analysis, the correlation between the dependent variable and only one particular independent variable is called *partial correlation* or *net correlation*. The influence of other independent variables is kept constant in the partial correlation analysis.

The *coefficient of partial determination* for measuring the correlation between X_1 and X_3, keeping X_2 constant, is denoted by the symbol $r_{13.2}^2$. Its square root is called the *coefficient of partial correlation*, or $r_{13.2}$. The primary subscripts represent the variables for which the partial correlation is being measured, and the secondary subscript represents the variable that is to be kept constant. The formula is

$$r_{13.2}^2 = 1 - \frac{s_{1.23}^2}{s_{1.2}^2} \qquad (28\text{-}5)**$$

where $s_{1.2}^2$ is the same as $s_{y.x}^2$ in the simple correlation analysis.

This formula may be written in the form of

$$r_{13.2}^2 = \frac{s_{1.2}^2 - s_{1.23}^2}{s_{1.2}^2}$$

where $s_{1.2}^2 =$ the variance of variable X_1 unexplained by X_2 in the simple correlation analysis, and

---

the number of constants in the regression equation (a, b_2, and b_3). The formula (based on formula 26-5) and the computation (based on the same data as above) for the estimate of the population coefficient of determination for multiple correlation are

$$\hat{R}_{1.23}^2 = 1 - \frac{s_{1.23}^2}{s_1^2} = 1 - \frac{0.906}{6.2857} = 0.85586, \text{ or round to } 86\%$$

$$s_{1.23}^2 = 0.906 \qquad\qquad\qquad \text{(see above computation)}$$

$$s_1^2 = \frac{\sum (X_1 - \bar{X}_1)^2}{n - 1} = \frac{44}{8 - 1} = 6.2857 \qquad \text{(page 665 by letting } X_1 = Y)$$

$$\hat{R}_{1.23} = \sqrt{0.85586} = 0.93$$

Or, from the computed value of $R_{1.23}^2 = 0.8970$ above,

$$\hat{R}_{1.23}^2 = 1 - (1 - R_{1.23}^2)\left(\frac{n-1}{n-m}\right) = 1 - (1 - 0.8970)\left(\frac{8-1}{8-3}\right) = 0.86$$

For large samples, the estimated standard error of the multiple correlation coefficient is (also see formula 26-6)

$$s_R = \frac{1 - R^2}{\sqrt{n - m}}$$

where R may represent $R_{1.23}$ and $m = 3$.

  **The coefficient of partial determination for measuring the correlation between X_1 and X_2, keeping X_3 constant, is

$$r_{12.3}^2 = 1 - \frac{s_{1.23}^2}{s_{1.3}^2}$$

Further, if X_3 and X_4 are kept constant, it is written:

$$r_{12.34}^2 = 1 - \frac{s_{1.234}^2}{s_{1.34}^2}$$

$s_{1.23}^2$ = the variance of variable X_1 unexplained by X_2 and X_3 in the multiple correlation analysis.

The difference, $s_{1.2}^2 - s_{1.23}^2$ in the numerator of the fraction, therefore represents the reduced portion of the variance unexplained by X_2. The reduction is due to the added variable X_3 in the multiple analysis.

If the difference is zero, $r_{13.2}^2 = 0$, which indicates that the unexplained variance remains the same. The added variable X_3 contributed nothing in explaining the variance of variable X_1 in the multiple analysis. Thus, there is no partial correlation between X_1 and X_3, keeping X_2 constant. However, if the difference is equal to $s_{1.2}^2$, the denominator of the fraction, $r_{13.2}^2 = 1$. The value 1 indicates that the variance unexplained by X_2 in the simple correlation analysis is now completely explained by the added X_3 in the multiple analysis. In this case, there is a perfect partial correlation between X_1 and X_3, keeping X_2 constant. If the unexplained variance $s_{1.2}^2$ is large, the inclusion of X_3 is highly desirable since it can contribute a great amount in explaining the variance of variable X_1 in the multiple correlation analysis.

Example 2

Refer to the data given in Example 1. Compute the partial correlation between X_1 (amounts of sales) and X_3 (scores of intelligence test), keeping X_2 (years of sales experience) constant.

Solution. $s_{1.23}^2 = 0.56625$ (page 733), and $s_{1.2}^2 = 1.33$ (page 660, s_{yx}^2). Substitute the above values in formula (28–5),

$$r_{13.2}^2 = 1 - \frac{0.56625}{1.33} = 1 - 0.4258 = 0.5742, \text{ or round to } 57\%$$

$$r_{13.2} = \sqrt{0.5742} = +0.76*$$

The signs of the partial correlation coefficients r are taken from the signs the corresponding net regression coefficients b. The sign of $r_{13.2}$ is the same as that

---

*To test the significance of the partial correlation between X_1 and X_3, with X_2 being held constant, for the data given in Example 2 as a sample, use the formula on page 697 as follows:

$$t = r_{13.2}\sqrt{\frac{n-3}{1 - r_{13.2}^2}} = 0.76\sqrt{\frac{8-3}{1 - 0.5742}} = 0.76\sqrt{11.7426} = 0.76(3.427) = 2.605$$

At 5% level of significance, $t = 2.571$ when $D = n - 3 = 5$. The computed t value 2.605 is larger than the tabulated value. Thus, the coefficient of partial correlation $r_{13.2} = 0.76$ is significant. We thus conclude that X_3 variable, or the intelligence test scores, should be included for estimating the sales amount made by individual salesmen.

The standard error of the coefficient of partial correlation, based on formula (26–6), may be written:

$$s_{r13.2} = \frac{1 - r_{13.2}^2}{\sqrt{n-2}}$$

where $n - 2 = (n - 1) - 1$ (the number of secondary subscript, or number of variable that is held constant).

The standard error can be used for the interval estimate of the population coefficient of partial correlation. However, the above formula is appropriate only for large samples. Like formula (26–6), it is not accurate for small samples.

of $b_{13.2}$ or b_3 and is positive in this example ($b_3 = +1.3636$). Likewise, the sign of $r_{12.3}$ should be the same as that of $b_{12.3}$ or b_2.

The coefficient of partial determination may be computed in a different form,

$$r_{13.2}^2 = \frac{1.33 - 0.56625}{1.33} = \frac{0.76375}{1.33} = 0.5742, \text{ or } 57\%$$

The difference, 0.76375, is the reduced amount of unexplained variance due to the added variable X_3 in the multiple regression. In other words, it is the part of the variance of X_1, which was not explained by X_2 alone in the simple analysis, which is now explained by X_3 while X_2 is kept constant or unchanged in the multiple analysis. Thus, the interpretation of the value of $r_{13.2}^2$ is that 57% of the sales amount variance, which is not explained by years of sales experience, is explained by the intelligence test scores of the eight salesmen.

28.3 Rank Correlation

The relationship between two variables may be analyzed according to the ranks of the values of each variable. The values may be ranked on the basis of qualities, quantities, or other standards established by a person. The advantages of using the ranked data in the relationship analysis are many, such as:

(1) The computation of the measure of the relationship for the ranked data, called the *coefficient of rank correlation* and denoted by r_k, is simple. This fact is true since the ranked data are a set of simple ordered numbers.

(2) Limitations on samples drawn from the populations, which are not normally distributed or which may have no distributions, may be avoided. There are no assumptions about the parameters of those types of populations. The method employed in the rank correlation analysis, called a *nonparametric method* (see Chapter 16), thus provides another way in making statistical inferences.

(3) The ranked data may be obtained from the items which are difficult to express by exact measurement. For example, it is rather a difficult task to measure exactly the conduct of each salesman in a group. However, it could be a simple job for the sales manager to rank the conduct of each salesman.

A widely used coefficient of rank correlation is Spearman's coefficient. It is written:

$$r_k = 1 - \frac{6 \sum d^2}{n(n^2 - 1)} \qquad (28\text{-}6)^*$$

where d = the difference between two paired ranks, and
$\quad\;\; n$ = the number of pairs included in the data or the size of sample.

---

*The sampling distribution of r_k is symmetrical around 0 and is between -1 and $+1$. For small samples, the distribution is not normal. It approaches the normal curve when sample size n becomes large. The standard error of r_k is

$$\sigma_{rk} = \frac{1}{\sqrt{n - 1}}$$

which may be used in the sampling analysis when n is large, say 20 or more.

Example 3

The grades of first and second tests of seven students in a mathematics class are given in columns (2) and (3) of Table 28–3. The ranks of the grades for each of the two tests, in columns (4) and (5), are assigned according to the grade points, from the highest to the lowest. Compute the coefficient of rank correlation.

TABLE 28–3

CALCULATION FOR THE COEFFICIENT OF RANK CORRELATION–EXAMPLE 3

Student	Grade Points		Rank of Grade Points		$d = Y - X$ $= (4) - (5)$	d^2
	1st Test Y	2nd Test X	1st Test Y	2nd Test X		
(1)	(2)	(3)	(4)	(5)	(6)	(7)
A	90	94	3	1	2	4
B	81	66	4	6	-2	4
C	65	69	6	5	1	1
D	69	87	5	3	2	4
E	94	72	2	4	-2	4
F	97	89	1	2	-1	1
G	60	64	7	7	0	0
Total					0	18

Solution. The required values for the computation are obtained from Table 28–3. Substitute $\sum d^2 = 18$ and $n = 7$ in formula (28–6).

$$r_k = 1 - \frac{6(18)}{7(7^2 - 1)} = 1 - 0.32 = 0.68$$

The range of r_k values is between $+1$ and -1. If the ranks of the first and the second test grades are identical for each student, or $d = Y - X = 0$, $\sum d^2$ will be zero and $r_k = +1$. Then, r_k indicates a perfect positive correlation between the two groups of ranks. If the ranks are exactly inverse, that is, if $n = 7$ pairs, $d = Y - X = 1 - 7 = -6$, $d = 2 - 6 = -4$, $d = 3 - 5 = -2$, and so on, $\sum d^2$ will have the maximum value and $r_k = -1$. The r_k therefore indicates a perfect negative correlation. If $r_k = 0$, there is no correlation between the two groups of ranks.

When two or more items are tied in rank, each item is given the mean of the ranks. Thus, if two items are tied for sixth and seventh places, each item is given a rank of 6.5, such as the ranks of the sales amounts for salesmen D and E in Example 4 below, since $(6 + 7)/2 = 6.5$. If three items are tied for second, third, and fourth places, each item is given a rank of 3 ($= (2 + 3 + 4)/3$).

Example 4

The Y and X variables in columns (2) and (3) of the following table are the same data given in Example 1, Chapter 27. Compute the coefficient of rank correlation.

TABLE 28–4

CALCULATION FOR THE COEFFICIENT OF RANK CORRELATION—
EXAMPLE 4 (WHEN TWO ITEMS ARE TIED IN RANK)

Sales-man	Original Data		Ranked Data		$d = Y - X$ $= (4) - (5)$	d^2
	Amount of Sales Y	Years of Sales Experience X	Amount of Sales Y	Years of Sales Experience X		
(1)	(2)	(3)	(4)	(5)	(6)	(7)
A	$9,000	6	1.0	1.5**	−0.5	0.25
B	6,000	5	3.0	3.0	0.0	0.00
C	4,000	3	5.0	5.5***	−0.5	0.25
D	3,000	1	6.5*	8.0	−1.5	2.25
E	3,000	4	6.5*	4.0	2.5	6.25
F	5,000	3	4.0	5.5***	−1.5	2.25
G	8,000	6	2.0	1.5**	0.5	0.25
H	2,000	2	8.0	7.0	1.0	1.00
Total			36.0	36.0	0.0	12.50

*The mean of sixth and seventh ranks $= (6 + 7)/2 = 6.5$.
**The mean of first and second ranks $= (1 + 2)/2 = 1.5$.
***The mean of fifth and sixth ranks $= (5 + 6)/2 = 5.5$.

Solution. Substitute $\sum d^2 = 12.5$ and $n = 8$ in formula (28–6),

$$r_k = 1 - \frac{6(12.50)}{8(8^2 - 1)} = 1 - 0.1488 = 0.8512 \text{ or round to } 0.85$$

This coefficient of rank correlation is also positive and is relatively higher than that of Example 3 ($r_k = 0.85$ versus $r_k = 0.68$).

28.4 Summary

Multiple linear regression and correlation analysis concerns the linear relationship between the dependent variable and two or more independent variables. All the variables included in the analysis are represented by the letter X with different subscripts. The standard deviation of the X_1 values is denoted by $s_{1.23}$, where the primary subscript indicates the dependent variable X_1 and the secondary subscripts indicate the independent variables X_2 and X_3. The coefficients of determination and correlation are represented by R^2 and R with the same subscripts as the standard deviation. The values of $R_{1.23}^2$ and $R_{1.23}$ can be computed and interpreted in the same manner as r^2 and r for simple correlation.

When three or more variables are involved in correlation analysis, the correlation between the dependent variable and only one particular independent variable is called partial correlation or net correlation. The coefficient of partial determination for measuring the correlation between X_1 and X_3, keeping X_2 constant, is denoted by $r_{13.2}^2$. Its square root is called the coefficient of partial correlation, or $r_{13.2}$. The signs of the coefficients of partial correlation are taken from the signs of the corresponding net regression coefficients b.

The relationship between two variables may be analyzed according to the ranks of the values of each variable. The values may be ranked on the basis of qualities, quantities, or other standards established by a person. The range of values of r_k (the coefficient of rank correlation) is between $+1$ (indicating a perfect positive correlation) and -1 (indicating a perfect negative correlation).

SUMMARY OF FORMULAS

Application	Formula	Formula Number	Reference Page
Multiple linear regression and correlation			
Regression equation	$X_{1c} = a + b_2 X_2 + b_3 X_3$	28–1	729
Normal equations for formula (28–1)	I. $\sum (X_1) = na + b_2 \sum (X_2) + b_3 \sum (X_3)$ II. $\sum (X_1 X_2) = a \sum (X_2) + b_2 \sum (X_2^2) + b_3 \sum (X_2 X_3)$ III. $\sum (X_1 X_3) = a \sum (X_3) + b_2 \sum (X_2 X_3) + b_3 \sum (X_3^2)$	28–2	730
Standard deviation of regression	$s_{1.23} = \sqrt{\dfrac{\sum (X_1 - X_{1c})^2}{n}}$	28–3a	730
	$s_{1.23} = \sqrt{\dfrac{\sum X_1^2 - a \sum X_1 - b_2 \sum (X_1 X_2) - b_3 \sum (X_1 X_3)}{n}}$	28–3b	730
Coefficient of determination	$R_{1.23}^2 = 1 - \dfrac{s_{1.23}^2}{s_1^2}$	28–4	731
Partial correlation			
Coefficient of partial determination	$r_{13.2}^2 = 1 - \dfrac{s_{1.23}^2}{s_{1.2}^2}$	28–5	736
Rank correlation			
Coefficient of rank correlation	$r_k = 1 - \dfrac{6 \sum d^2}{n(n^2 - 1)}$	28–6	738

Exercises 28

1. Using the detailed form of subscript notation, write the regression equations of (a) X_2 (dependent variable) on X_1 and X_3 (independent variables), and (b) X_3 (dependent variable) on X_1 and X_2 (independent variables).

2. Indicate the difference, if any:

   (a) Between r_{12}, $R_{1.23}$, and r.

   (b) Between $s_{1.2}$ and $s_{1.23}$.

3. Table 28–5 shows the number of correct answers (X_1) made on a test, the hours (X_2) studied for the test, and the years (X_3) of college education for each of the five employees in a company. (Note: The X_1 and X_2 variables of this problem are the same data as the Y and X variables of Problem 3 of the Exercises for Chapter 25.) Compute:

TABLE 28–5

DATA FOR PROBLEM 3

Employee	X_1, Number of Correct Answers	X_2, Hours Studied	X_3, Years of College Education
A	2	1	1
B	10	2	1
C	20	6	4
D	14	7	2
E	11	5	2

(a) The regression equation of X_1 on X_2 and X_3 by the least square method. (Round the final answers to two decimal places.)
(b) The standard deviation of regression $s_{1.23}$ by formula (28–3a).
(c) The coefficient of determination $R_{1.23}^2$ and the coefficient of correlation $R_{1.23}$.

4. Refer to Problem 3. Estimate the number of correct answers to be made by employee G, who studied 4 hours for the test and had 3 years of college education. How reliable would you consider this estimate?

5. Using the data given in Problem 3, compute (a) the coefficient of partial determination $r_{13.2}^2$, and (b) the coefficient of partial correlation $r_{13.2}$.

6. Table 28–6 shows the grades of the three courses—statistics (X_1), economics (X_2), and mathematics (X_3)—of a group of 9 students in a college at the end of this semester. (Note: The X_1 and X_2 variables of this problem are the same data as the Y and X variables of Problem 5 in the Exercises for Chapter 25.) The following

TABLE 28–6

DATA FOR PROBLEM 6

Students	Statistics Grades X_1	Economics Grades X_2	Mathematics Grades X_3
A	95	88	92
B	51	70	57
C	49	65	54
D	27	50	30
E	42	60	46
F	52	80	56
G	67	68	70
H	48	49	50
I	46	40	52
Total	477	570	507

sums are computed from Table 28–6:

$$\Sigma X_1^2 = 28{,}133 \qquad \Sigma X_2^2 = 37{,}994 \qquad \Sigma X_3^2 = 30{,}885$$
$$\Sigma (X_1 X_2) = 31{,}893 \qquad \Sigma (X_1 X_3) = 29{,}429 \qquad \Sigma (X_2 X_3) = 33{,}626$$

Compute:
(a) The regression equation of X_1 on X_2 and X_3 by the least square method. (Round the final answers to three decimal places.)
(b) The standard deviation of regression $s_{1.23}$ by formula (28–3b).
(c) The coefficient of determination $R_{1.23}^2$ and the coefficient of correlation $R_{1.23}$.

7. Refer to Problem 6. Estimate the statistics grade for student J who has economics grade 82 and mathematics grade 90 in this semester. How reliable would you consider this estimate?

8. Using the data given in Problem 6, compute (a) the coefficient of partial determination $r_{13.2}^2$, and (b) the coefficient of partial correlation $r_{13.2}$.

9. Compute the coefficient of rank correlation for the data given in Problem 5 of the Exercises for Chapter 25. Compare the r_k with the correlation coefficient r obtained in that problem. ($r = 0.72$) Discuss your finding from the comparison.

10. The points listed in Table 28–7 are given by two judges for each of the six candidates in a contest. Compute the coefficient of rank correlation. Do you think that the two judges are in close agreement?

TABLE 28–7

DATA FOR PROBLEM 10

Candidate Number	Points Given by the First Judge	Points Given by the Second Judge
1	90	94
2	55	62
3	85	93
4	88	86
5	79	86
6	79	60

Appendix

Tables
Selected References
Answers to
Odd-Numbered Problems
Index

TABLE 1

Squares, Square Roots, and Radicals

n	n^2	$\sqrt{n}$	$\sqrt{10n}$	$1/n$	n	n^2	$\sqrt{n}$	$\sqrt{10n}$	$1/n$
1.00	1.0000	1.00000	3.16228	1.000000	**1.40**	1.9600	1.18322	3.74166	.714286
1.01	1.0201	1.00499	3.17805	.990099	1.41	1.9881	1.18743	3.75500	.709220
1.02	1.0404	1.00995	3.19374	.980392	1.42	2.0164	1.19164	3.76829	.704225
1.03	1.0609	1.01489	3.20936	.970874	1.43	2.0449	1.19583	3.78153	.699301
1.04	1.0816	1.01980	3.22490	.961538	1.44	2.0736	1.20000	3.79473	.694444
1.05	1.1025	1.02470	3.24037	.952381	1.45	2.1025	1.20416	3.80789	.689655
1.06	1.1236	1.02956	3.25576	.943396	1.46	2.1316	1.20830	3.82099	.684932
1.07	1.1449	1.03441	3.27109	.934579	1.47	2.1609	1.21244	3.83406	.680272
1.08	1.1664	1.03923	3.28634	.925926	1.48	2.1904	1.21655	3.84708	.675676
1.09	1.1881	1.04403	3.30151	.917431	1.49	2.2201	1.22066	3.86005	.671141
1.10	1.2100	1.04881	3.31662	.909091	**1.50**	2.2500	1.22474	3.87298	.666667
1.11	1.2321	1.05357	3.33167	.900901	1.51	2.2801	1.22882	3.88587	.662252
1.12	1.2544	1.05830	3.34664	.892857	1.52	2.3104	1.23288	3.89872	.657895
1.13	1.2769	1.06301	3.36155	.884956	1.53	2.3409	1.23693	3.91152	.653595
1.14	1.2996	1.06771	3.37639	.877193	1.54	2.3716	1.24097	3.92428	.649351
*1.15	1.3225	1.07238	3.39116	.869565	1.55	2.4025	1.24499	3.93700	.645161
1.16	1.3456	1.07703	3.40588	.862069	1.56	2.4336	1.24900	3.94968	.641026
1.17	1.3689	1.08167	3.42053	.854701	1.57	2.4649	1.25300	3.96232	.636943
1.18	1.3924	1.08628	3.43511	.847458	1.58	2.4964	1.25698	3.97492	.632911
1.19	1.4161	1.09087	3.44964	.840336	1.59	2.5281	1.26095	3.98748	.628931
1.20	1.4400	1.09545	3.46410	.833333	**1.60**	2.5600	1.26491	4.00000	.625000
1.21	1.4641	1.10000	3.47851	.826446	1.61	2.5921	1.26886	4.01248	.621118
1.22	1.4884	1.10454	3.49285	.819672	1.62	2.6244	1.27279	4.02492	.617284
1.23	1.5129	1.10905	3.50714	.813008	1.63	2.6569	1.27671	4.03733	.613497
1.24	1.5376	1.11355	3.52136	.806452	1.64	2.6896	1.28062	4.04969	.609756
1.25	1.5625	1.11803	3.53553	.800000	1.65	2.7225	1.28452	4.06202	.606061
1.26	1.5876	1.12250	3.54965	.793651	1.66	2.7556	1.28841	4.07431	.602410
1.27	1.6129	1.12694	3.56371	.787402	1.67	2.7889	1.29228	4.08656	.598802
1.28	1.6384	1.13137	3.57771	.781250	1.68	2.8224	1.29615	4.09878	.595238
1.29	1.6641	1.13578	3.59166	.775194	1.69	2.8561	1.30000	4.11096	.591716
1.30	1.6900	1.14018	3.60555	.769231	**1.70**	2.8900	1.30384	4.12311	.588235
1.31	1.7161	1.14455	3.61939	.763359	1.71	2.9241	1.30767	4.13521	.584795
1.32	1.7424	1.14891	3.63318	.757576	1.72	2.9584	1.31149	4.14729	.581395
1.33	1.7689	1.15326	3.64692	.751880	1.73	2.9929	1.31529	4.15933	.578035
1.34	1.7956	1.15758	3.66060	.746269	1.74	3.0276	1.31909	4.17133	.574713
1.35	1.8225	1.16190	3.67423	.740741	1.75	3.0625	1.32288	4.18330	.571429
1.36	1.8496	1.16619	3.68782	.735294	1.76	3.0976	1.32665	4.19524	.568182
1.37	1.8769	1.17047	3.70135	.729927	1.77	3.1329	1.33041	4.20714	.564972
1.38	1.9044	1.17473	3.71484	.724638	1.78	3.1684	1.33417	4.21900	.561798
1.39	1.9321	1.17898	3.72827	.719424	1.79	3.2041	1.33791	4.23084	.558659
1.40	1.9600	1.18322	3.74166	.714286	**1.80**	3.2400	1.34164	4.24264	.555556

*Examples :

When $n = 1.15$,
$$n^2 = 1.15^2 = 1.3225,$$
$$\sqrt{n} = \sqrt{1.15} = 1.07238,$$
$$\sqrt{10n} = \sqrt{11.50} = 3.39116,$$
$$1/n = 1/1.15 = 0.869565$$

When $n = 115$,
$$n^2 = 115^2 = 13{,}225,$$
$$\sqrt{n} = \sqrt{115} = 10.7238,$$
$$\sqrt{10n} = \sqrt{1{,}150} = 33.9116,$$
$$1/n = 1/115 = 0.00869565$$

For detailed illustrations of finding square roots, see Section 4.5, p. 89.

TABLE 1 (Continued)

n	n^2	$\sqrt{n}$	$\sqrt{10n}$	$1/n$	n	n^2	$\sqrt{n}$	$\sqrt{10n}$	$1/n$
1.80	3.2400	1.34164	4.24264	.555556	**2.30**	5.2900	1.51658	4.79583	.434783
1.81	3.2761	1.34536	4.25441	.552486	2.31	5.3361	1.51987	4.80625	.432900
1.82	3.3124	1.34907	4.26615	.549451	2.32	5.3824	1.52315	4.81664	.431034
1.83	3.3489	1.35277	4.27785	.546448	2.33	5.4289	1.52643	4.82701	.429185
1.84	3.3856	1.35647	4.28952	.543478	2.34	5.4756	1.52971	4.83735	.427350
1.85	3.4225	1.36015	4.30116	.540541	2.35	5.5225	1.53297	4.84768	.425532
1.86	3.4596	1.36382	4.31277	.537634	2.36	5.5696	1.53623	4.85798	.423729
1.87	3.4969	1.36748	4.32435	.534759	2.37	5.6169	1.53948	4.86826	.421941
1.88	3.5344	1.37113	4.33590	.531915	2.38	5.6644	1.54272	4.87852	.420168
1.89	3.5721	1.37477	4.34741	.529101	2.39	5.7121	1.54596	4.88876	.418410
1.90	3.6100	1.37840	4.35890	.526316	**2.40**	5.7600	1.54919	4.89898	.416667
1.91	3.6481	1.38203	4.37035	.523560	2.41	5.8081	1.55242	4.90918	.414938
1.92	3.6864	1.38564	4.38178	.520833	2.42	5.8564	1.55563	4.91935	.413223
1.93	3.7249	1.38924	4.39318	.518135	2.43	5.9049	1.55885	4.92950	.411523
1.94	3.7636	1.39284	4.40454	.515464	2.44	5.9536	1.56205	4.93964	.409836
1.95	3.8025	1.39642	4.41588	.512821	2.45	6.0025	1.56525	4.94975	.408163
1.96	3.8416	1.40000	4.42719	.510204	2.46	6.0516	1.56844	4.95984	.406504
1.97	3.8809	1.40357	4.43847	.507614	2.47	6.1009	1.57162	4.96991	.404858
1.98	3.9204	1.40712	4.44972	.505051	2.48	6.1504	1.57480	4.97996	.403226
1.99	3.9601	1.41067	4.46094	.502513	2.49	6.2001	1.57797	4.98999	.401606
2.00	4.0000	1.41421	4.47214	.500000	**2.50**	6.2500	1.58114	5.00000	.400000
2.01	4.0401	1.41774	4.48330	.497512	2.51	6.3001	1.58430	5.00999	.398406
2.02	4.0804	1.42127	4.49444	.495050	2.52	6.3504	1.58745	5.01996	.396825
2.03	4.1209	1.42478	4.50555	.492611	2.53	6.4009	1.59060	5.02991	.395257
2.04	4.1616	1.42829	4.51664	.490196	2.54	6.4516	1.59374	5.03984	.393701
2.05	4.2025	1.43178	4.52769	.487805	2.55	6.5025	1.59687	5.04975	.392157
2.06	4.2436	1.43527	4.53872	.485437	2.56	6.5536	1.60000	5.05964	.390625
2.07	4.2849	1.43875	4.54973	.483092	2.57	6.6049	1.60312	5.06952	.389105
2.08	4.3264	1.44222	4.56070	.480769	2.58	6.6564	1.60624	5.07937	.387597
2.09	4.3681	1.44568	4.57165	.478469	2.59	6.7081	1.60935	5.08920	.386100
2.10	4.4100	1.44914	4.58258	.476190	**2.60**	6.7600	1.61245	5.09902	.384615
2.11	4.4521	1.45258	4.59347	.473934	2.61	6.8121	1.61555	5.10882	.383142
2.12	4.4944	1.45602	4.60435	.471698	2.62	6.8644	1.61864	5.11859	.381679
2.13	4.5369	1.45945	4.61519	.469484	2.63	6.9169	1.62173	5.12835	.380228
.14	4.5796	1.46287	4.62601	.467290	2.64	6.9696	1.62481	5.13809	.378788
2.15	4.6225	1.46629	4.63681	.465116	2.65	7.0225	1.62788	5.14782	.377358
2.16	4.6656	1.46969	4.64758	.462963	2.66	7.0756	1.63095	5.15752	.375940
2.17	4.7089	1.47309	4.65833	.460829	2.67	7.1289	1.63401	5.16720	.374532
2.18	4.7524	1.47648	4.66905	.458716	2.68	7.1824	1.63707	5.17687	.373134
2.19	4.7961	1.47986	4.67974	.456621	2.69	7.2361	1.64012	5.18652	.371747
2.20	4.8400	1.48324	4.69042	.454545	**2.70**	7.2900	1.64317	5.19615	.370370
2.21	4.8841	1.48661	4.70106	.452489	2.71	7.3441	1.64621	5.20577	.369004
2.22	4.9284	1.48997	4.71169	.450450	2.72	7.3984	1.64924	5.21536	.367647
2.23	4.9729	1.49332	4.72229	.448430	2.73	7.4529	1.65227	5.22494	.366300
2.24	5.0176	1.49666	4.73286	.446429	2.74	7.5076	1.65529	5.23450	.364964
2.25	5.0625	1.50000	4.74342	.444444	2.75	7.5625	1.65831	5.24404	.363636
2.26	5.1076	1.50333	4.75395	.442478	2.76	7.6176	1.66132	5.25357	.362319
2.27	5.1529	1.50665	4.76445	.440529	2.77	7.6729	1.66433	5.26308	.361011
2.28	5.1984	1.50997	4.77493	.438596	2.78	7.7284	1.66733	5.27257	.359712
2.29	5.2441	1.51327	4.78539	.436681	2.79	7.7841	1.67033	5.28205	.358423
2.30	5.2900	1.51658	4.79583	.434783	**2.80**	7.8400	1.67332	5.29150	.357143

TABLE 1   (Continued)

n	n^2	$\sqrt{n}$	$\sqrt{10n}$	$1/n$	n	n^2	$\sqrt{n}$	$\sqrt{10n}$	$1/n$
2.80	7.8400	1.67332	5.29150	.357143	**3.30**	10.8900	1.81659	5.74456	.303030
2.81	7.8961	1.67631	5.30094	.355872	3.31	10.9561	1.81934	5.75326	.302115
2.82	7.9524	1.67929	5.31037	.354610	3.32	11.0224	1.82209	5.76194	.301205
2.83	8.0089	1.68226	5.31977	.353357	3.33	11.0889	1.82483	5.77062	.300300
2.84	8.0656	1.68523	5.32917	.352113	3.34	11.1556	1.82757	5.77927	.299401
2.85	8.1225	1.68819	5.33854	.350877	3.35	11.2225	1.83030	5.78792	.298507
2.86	8.1796	1.69115	5.34790	.349650	3.36	11.2896	1.83303	5.79655	.297619
2.87	8.2369	1.69411	5.35724	.348432	3.37	11.3569	1.83576	5.80517	.296736
2.88	8.2944	1.69706	5.36656	.347222	3.38	11.4244	1.83848	5.81378	.295858
2.89	8.3521	1.70000	5.37587	.346021	3.39	11.4921	1.84120	5.82237	.294985
2.90	8.4100	1.70294	5.38516	.344828	**3.40**	11.5600	1.84391	5.83095	.294118
2.91	8.4681	1.70587	5.39444	.343643	3.41	11.6281	1.84662	5.83952	.293255
2.92	8.5264	1.70880	5.40370	.342466	3.42	11.6964	1.84932	5.84808	.292398
2.93	8.5849	1.71172	5.41295	.341297	3.43	11.7649	1.85203	5.85662	.291545
2.94	8.6436	1.71464	5.42218	.340136	3.44	11.8336	1.85472	5.86515	.290698
2.95	8.7025	1.71756	5.43139	.338983	3.45	11.9025	1.85742	5.87367	.289855
2.96	8.7616	1.72047	5.44059	.337838	3.46	11.9716	1.86011	5.88218	.289017
2.97	8.8209	1.72337	5.44977	.336700	3.47	12.0409	1.86279	5.89067	.288184
2.98	8.8804	1.72627	5.45894	.335570	3.48	12.1104	1.86548	5.89915	.287356
2.99	8.9401	1.72916	5.46809	.334448	3.49	12.1801	1.86815	5.90762	.286533
3.00	9.0000	1.73205	5.47723	.333333	**3.50**	12.2500	1.87083	5.91608	.285714
3.01	9.0601	1.73494	5.48635	.332226	3.51	12.3201	1.87350	5.92453	.284900
3.02	9.1204	1.73781	5.49545	.331126	3.52	12.3904	1.87617	5.93296	.284091
3.03	9.1809	1.74069	5.50454	.330033	3.53	12.4609	1.87883	5.94138	.283286
3.04	9.2416	1.74356	5.51362	.328947	3.54	12.5316	1.88149	5.94979	.282486
3.05	9.3025	1.74642	5.52268	.327869	3.55	12.6025	1.88414	5.95819	.281690
3.06	9.3636	1.74929	5.53173	.326797	3.56	12.6736	1.88680	5.96657	.280899
3.07	9.4249	1.75214	5.54076	.325733	3.57	12.7449	1.88944	5.97495	.280112
3.08	9.4864	1.75499	5.54977	.324675	3.58	12.8164	1.89209	5.98331	.279330
3.09	9.5481	1.75784	5.55878	.323625	3.59	12.8881	1.89473	5.99166	.278552
3.10	9.6100	1.76068	5.56776	.322581	**3.60**	12.9600	1.89737	6.00000	.277778
3.11	9.6721	1.76352	5.57674	.321543	3.61	13.0321	1.90000	6.00833	.277008
3.12	9.7344	1.76635	5.58570	.320513	3.62	13.1044	1.90263	6.01664	.276243
3.13	9.7969	1.76918	5.59464	.319489	3.63	13.1769	1.90526	6.02495	.275482
3.14	9.8596	1.77200	5.60357	.318471	3.64	13.2496	1.90788	6.03324	.274725
3.15	9.9225	1.77482	5.61249	.317460	3.65	13.3225	1.91050	6.04152	.273973
3.16	9.9856	1.77764	5.62139	.316456	3.66	13.3956	1.91311	6.04979	.273224
3.17	10.0489	1.78045	5.63028	.315457	3.67	13.4689	1.91572	6.05805	.272480
3.18	10.1124	1.78326	5.63915	.314465	3.68	13.5424	1.91833	6.06630	.271739
3.19	10.1761	1.78606	5.64801	.313480	3.69	13.6161	1.92094	6.07454	.271003
3.20	10.2400	1.78885	5.65685	.312500	**3.70**	13.6900	1.92354	6.08276	.270270
3.21	10.3041	1.79165	5.66569	.311526	3.71	13.7641	1.92614	6.09098	.269542
3.22	10.3684	1.79444	5.67450	.310559	3.72	13.8384	1.92873	6.09918	.268817
3.23	10.4329	1.79722	5.68331	.309598	3.73	13.9129	1.93132	6.10737	.268097
3.24	10.4976	1.80000	5.69210	.308642	3.74	13.9876	1.93391	6.11555	.267380
3.25	10.5625	1.80278	5.70088	.307692	3.75	14.0625	1.93649	6.12372	.266667
3.26	10.6276	1.80555	5.70964	.306748	3.76	14.1376	1.93907	6.13188	.265957
3.27	10.6929	1.80831	5.71839	.305810	3.77	14.2129	1.94165	6.14003	.265252
3.28	10.7584	1.81108	5.72713	.304878	3.78	14.2884	1.94422	6.14817	.264550
3.29	10.8241	1.81384	5.73585	.303951	3.79	14.3641	1.94679	6.15630	.263852
3.30	10.8900	1.81659	5.74456	.303030	**3.80**	14.4400	1.94936	6.16441	.263158

TABLE 1 (Continued)

n	n^2	$\sqrt{n}$	$\sqrt{10n}$	$1/n$	n	n^2	$\sqrt{n}$	$\sqrt{10n}$	$1/n$
3.80	14.4400	1.94936	6.16441	.263158	**4.30**	18.4900	2.07364	6.55744	.232558
3.81	14.5161	1.95192	6.17252	.262467	4.31	18.5761	2.07605	6.56506	.232019
3.82	14.5924	1.95448	6.18061	.261780	4.32	18.6624	2.07846	6.57267	.231481
3.83	14.6689	1.95704	6.18870	.261097	4.33	18.7489	2.08087	6.58027	.230947
3.84	14.7456	1.95959	6.19677	.260417	4.34	18.8356	2.08327	6.58787	.230415
3.85	14.8225	1.96214	6.20484	.259740	4.35	18.9225	2.08567	6.59545	.229885
3.86	14.8996	1.96469	6.21289	.259067	4.36	19.0096	2.08806	6.60303	.229358
3.87	14.9769	1.96723	6.22093	.258398	4.37	19.0969	2.09045	6.61060	.228833
3.88	15.0544	1.96977	6.22896	.257732	4.38	19.1844	2.09284	6.61816	.228311
3.89	15.1321	1.97231	6.23699	.257069	4.39	19.2721	2.09523	6.62571	.227790
3.90	15.2100	1.97484	6.24500	.256410	**4.40**	19.3600	2.09762	6.63325	.227273
3.91	15.2881	1.97737	6.25300	.255754	4.41	19.4481	2.10000	6.64078	.226757
3.92	15.3664	1.97990	6.26099	.255102	4.42	19.5364	2.10238	6.64831	.226244
3.93	15.4449	1.98242	6.26897	.254453	4.43	19.6249	2.10476	6.65582	.225734
3.94	15.5236	1.98494	6.27694	.253807	4.44	19.7136	2.10713	6.66333	.225225
3.95	15.6025	1.98746	6.28490	.253165	4.45	19.8025	2.10950	6.67083	.224719
3.96	15.6816	1.98997	6.29285	.252525	4.46	19.8916	2.11187	6.67832	.224215
3.97	15.7609	1.99249	6.30079	.251889	4.47	19.9809	2.11424	6.68581	.223714
3.98	15.8404	1.99499	6.30872	.251256	4.48	20.0704	2.11660	6.69328	.223214
3.99	15.9201	1.99750	6.31664	.250627	4.49	20.1601	2.11896	6.70075	.222717
4.00	16.0000	2.00000	6.32456	.250000	**4.50**	20.2500	2.12132	6.70820	.222222
4.01	16.0801	2.00250	6.33246	.249377	4.51	20.3401	2.12368	6.71565	.221729
4.02	16.1604	2.00499	6.34035	.248756	4.52	20.4304	2.12603	6.72309	.221239
4.03	16.2409	2.00749	6.34823	.248139	4.53	20.5209	2.12838	6.73053	.220751
4.04	16.3216	2.00998	6.35610	.247525	4.54	20.6116	2.13073	6.73795	.220264
4.05	16.4025	2.01246	6.36396	.246914	4.55	20.7025	2.13307	6.74537	.219780
4.06	16.4836	2.01494	6.37181	.246305	4.56	20.7936	2.13542	6.75278	.219298
4.07	16.5649	2.01742	6.37966	.245700	4.57	20.8849	2.13776	6.76018	.218818
4.08	16.6464	2.01990	6.38749	.245098	4.58	20.9764	2.14009	6.76757	.218341
4.09	16.7281	2.02237	6.39531	.244499	4.59	21.0681	2.14243	6.77495	.217865
4.10	16.8100	2.02485	6.40312	.243902	**4.60**	21.1600	2.14476	6.78233	.217391
4.11	16.8921	2.02731	6.41093	.243309	4.61	21.2521	2.14709	6.78970	.216920
4.12	16.9744	2.02978	6.41872	.242718	4.62	21.3444	2.14942	6.79706	.216450
4.13	17.0569	2.03224	6.42651	.242131	4.63	21.4369	2.15174	6.80441	.215983
4.14	17.1396	2.03470	6.43428	.241546	4.64	21.5296	2.15407	6.81175	.215517
4.15	17.2225	2.03715	6.44205	.240964	4.65	21.6225	2.15639	6.81909	.215054
4.16	17.3056	2.03961	6.44981	.240385	4.66	21.7156	2.15870	6.82642	.214592
4.17	17.3889	2.04206	6.45755	.239808	4.67	21.8089	2.16102	6.83374	.214133
4.18	17.4724	2.04450	6.46529	.239234	4.68	21.9024	2.16333	6.84105	.213675
4.19	17.5561	2.04695	6.47302	.238663	4.69	21.9961	2.16564	6.84836	.213220
4.20	17.6400	2.04939	6.48074	.238095	**4.70**	22.0900	2.16795	6.85565	.212766
4.21	17.7241	2.05183	6.48845	.237530	4.71	22.1841	2.17025	6.86294	.212314
4.22	17.8084	2.05426	6.49615	.236967	4.72	22.2784	2.17256	6.87023	.211864
4.23	17.8929	2.05670	6.50384	.236407	4.73	22.3729	2.17486	6.87750	.211416
4.24	17.9776	2.05913	6.51153	.235849	4.74	22.4676	2.17715	6.88477	.210970
4.25	18.0625	2.06155	6.51920	.235294	4.75	22.5625	2.17945	6.89202	.210526
4.26	18.1476	2.06398	6.52687	.234742	4.76	22.6576	2.18174	6.89928	.210084
4.27	18.2329	2.06640	6.53452	.234192	4.77	22.7529	2.18403	6.90652	.209644
4.28	18.3184	2.06882	6.54217	.233645	4.78	22.8484	2.18632	6.91375	.209205
4.29	18.4041	2.07123	6.54981	.233100	4.79	22.9441	2.18861	6.92098	.208768
4.30	18.4900	2.07364	6.55744	.232558	**4.80**	23.0400	2.19089	6.92820	.208333

TABLE 1   (Continued)

n	n^2	$\sqrt{n}$	$\sqrt{10n}$	$1/n$	n	n^2	$\sqrt{n}$	$\sqrt{10n}$	$1/n$
4.80	23.0400	2.19089	6.92820	.208333	**5.30**	28.0900	2.30217	7.28011	.188679
4.81	23.1361	2.19317	6.93542	.207900	5.31	28.1961	2.30434	7.28697	.188324
4.82	23.2324	2.19545	6.94262	.207469	5.32	28.3024	2.30651	7.29383	.187970
4.83	23.3289	2.19773	6.94982	.207039	5.33	28.4089	2.30868	7.30068	.187617
4.84	23.4256	2.20000	6.95701	.206612	5.34	28.5156	2.31084	7.30753	.187266
4.85	23.5225	2.20227	6.96419	.206186	5.35	28.6225	2.31301	7.31437	.186916
4.86	23.6196	2.20454	6.97137	.205761	5.36	28.7296	2.31517	7.32120	.186567
4.87	23.7169	2.20681	6.97854	.205339	5.37	28.8369	2.31733	7.32803	.186220
4.88	23.8144	2.20907	6.98570	.204918	5.38	28.9444	2.31948	7.33485	.185874
4.89	23.9121	2.21133	6.99285	.204499	5.39	29.0521	2.32164	7.34166	.185529
4.90	24.0100	2.21359	7.00000	.204082	**5.40**	29.1600	2.32379	7.34847	.185185
4.91	24.1081	2.21585	7.00714	.203666	5.41	29.2681	2.32594	7.35527	.184843
4.92	24.2064	2.21811	7.01427	.203252	5.42	29.3764	2.32809	7.36206	.184502
4.93	24.3049	2.22036	7.02140	.202840	5.43	29.4849	2.33024	7.36885	.184162
4.94	24.4036	2.22261	7.02851	.202429	5.44	29.5936	2.33238	7.37564	.183824
4.95	24.5025	2.22486	7.03562	.202020	5.45	29.7025	2.33452	7.38241	.183486
4.96	24.6016	2.22711	7.04273	.201613	5.46	29.8116	2.33666	7.38918	.183150
4.97	24.7009	2.22935	7.04982	.201207	5.47	29.9209	2.33880	7.39594	.182815
4.98	24.8004	2.23159	7.05691	.200803	5.48	30.0304	2.34094	7.40270	.182482
4.99	24.9001	2.23383	7.06399	.200401	5.49	30.1401	2.34307	7.40945	.182149
5.00	25.0000	2.23607	7.07107	.200000	**5.50**	30.2500	2.34521	7.41620	.181818
5.01	25.1001	2.23830	7.07814	.199601	5.51	30.3601	2.34734	7.42294	.181488
5.02	25.2004	2.24054	7.08520	.199203	5.52	30.4704	2.34947	7.42967	.181159
5.03	25.3009	2.24277	7.09225	.198807	5.53	30.5809	2.35160	7.43640	.180832
5.04	25.4016	2.24499	7.09930	.198413	5.54	30.6916	2.35372	7.44312	.180505
5.05	25.5025	2.24722	7.10634	.198020	5.55	30.8025	2.35584	7.44983	.180180
5.06	25.6036	2.24944	7.11337	.197628	5.56	30.9136	2.35797	7.45654	.179856
5.07	25.7049	2.25167	7.12039	.197239	5.57	31.0249	2.36008	7.46324	.179533
5.08	25.8064	2.25389	7.12741	.196850	5.58	31.1364	2.36220	7.46994	.179211
5.09	25.9081	2.25610	7.13442	.196464	5.59	31.2481	2.36432	7.47663	.178891
5.10	26.0100	2.25832	7.14143	.196078	**5.60**	31.3600	2.36643	7.48331	.178571
5.11	26.1121	2.26053	7.14843	.195695	5.61	31.4721	2.36854	7.48999	.178253
5.12	26.2144	2.26274	7.15542	.195312	5.62	31.5844	2.37065	7.49667	.177936
5.13	26.3169	2.26495	7.16240	.194932	5.63	31.6969	2.37276	7.50333	.177620
5.14	26.4196	2.26716	7.16938	.194553	5.64	31.8096	2.37487	7.50999	.177305
5.15	26.5225	2.26936	7.17635	.194175	5.65	31.9225	2.37697	7.51665	.176991
5.16	26.6256	2.27156	7.18331	.193798	5.66	32.0356	2.37908	7.52330	.176678
5.17	26.7289	2.27376	7.19027	.193424	5.67	32.1489	2.38118	7.52994	.176367
5.18	26.8324	2.27596	7.19722	.193050	5.68	32.2624	2.38328	7.53658	.176056
5.19	26.9361	2.27816	7.20417	.192678	5.69	32.3761	2.38537	7.54321	.175747
5.20	27.0400	2.28035	7.21110	.192308	**5.70**	32.4900	2.38747	7.54983	.175439
5.21	27.1441	2.28254	7.21803	.191939	5.71	32.6041	2.38956	7.55645	.175131
5.22	27.2484	2.28473	7.22496	.191571	5.72	32.7184	2.39165	7.56307	.174825
5.23	27.3529	2.28692	7.23187	.191205	5.73	32.8329	2.39374	7.56968	.174520
5.24	27.4576	2.28910	7.23878	.190840	5.74	32.9476	2.39583	7.57628	.174216
5.25	27.5625	2.29129	7.24569	.190476	5.75	33.0625	2.39792	7.58288	.173913
5.26	27.6676	2.29347	7.25259	.190114	5.76	33.1776	2.40000	7.58947	.173611
5.27	27.7729	2.29565	7.25948	.189753	5.77	33.2929	2.40208	7.59605	.173310
5.28	27.8784	2.29783	7.26636	.189394	5.78	33.4084	2.40416	7.60263	.173010
5.29	27.9841	2.30000	7.27324	.189036	5.79	33.5241	2.40624	7.60920	.172712
5.30	28.0900	2.30217	7.28011	.188679	**5.80**	33.6400	2.40832	7.61577	.172414

TABLE 1 (Continued)

n	n^2	$\sqrt{n}$	$\sqrt{10n}$	$1/n$	n	n^2	$\sqrt{n}$	$\sqrt{10n}$	$1/n$
5.80	33.6400	2.40832	7.61577	.172414	**6.30**	39.6900	2.50998	7.93725	.158730
5.81	33.7561	2.41039	7.62234	.172117	6.31	39.8161	2.51197	7.94355	.158479
5.82	33.8724	2.41247	7.62889	.171821	6.32	39.9424	2.51396	7.94984	.158228
5.83	33.9889	2.41454	7.63544	.171527	6.33	40.0689	2.51595	7.95613	.157978
5.84	34.1056	2.41661	7.64199	.171233	6.34	40.1956	2.51794	7.96241	.157729
5.85	34.2225	2.41868	7.64853	.170940	6.35	40.3225	2.51992	7.96869	.157480
5.86	34.3396	2.42074	7.65506	.170649	6.36	40.4496	2.52190	7.97496	.157233
5.87	34.4569	2.42281	7.66159	.170358	6.37	40.5769	2.52389	7.98123	.156986
5.88	34.5744	2.42487	7.66812	.170068	6.38	40.7044	2.52587	7.98749	.156740
5.89	34.6921	2.42693	7.67463	.169779	6.39	40.8321	2.52784	7.99375	.156495
5.90	34.8100	2.42899	7.68115	.169492	**6.40**	40.9600	2.52982	8.00000	.156250
5.91	34.9281	2.43105	7.68765	.169205	6.41	41.0881	2.53180	8.00625	.156006
5.92	35.0464	2.43311	7.69415	.168919	6.42	41.2164	2.53377	8.01249	.155763
5.93	35.1649	2.43516	7.70065	.168634	6.43	41.3449	2.53574	8.01873	.155521
5.94	35.2836	2.43721	7.70714	.168350	6.44	41.4736	2.53772	8.02496	.155280
5.95	35.4025	2.43926	7.71362	.168067	6.45	41.6025	2.53969	8.03119	.155039
5.96	35.5216	2.44131	7.72010	.167785	6.46	41.7316	2.54165	8.03741	.154799
5.97	35.6409	2.44336	7.72658	.167504	6.47	41.8609	2.54362	8.04363	.154560
5.98	35.7604	2.44540	7.73305	.167224	6.48	41.9904	2.54558	8.04984	.154321
5.99	35.8801	2.44745	7.73951	.166945	6.49	42.1201	2.54755	8.05605	.154083
6.00	36.0000	2.44949	7.74597	.166667	**6.50**	42.2500	2.54951	8.06226	.153846
6.01	36.1201	2.45153	7.75242	.166389	6.51	42.3801	2.55147	8.06846	.153610
6.02	36.2404	2.45357	7.75887	.166113	6.52	42.5104	2.55343	8.07465	.153374
6.03	36.3609	2.45561	7.76531	.165837	6.53	42.6409	2.55539	8.08084	.153139
6.04	36.4816	2.45764	7.77174	.165563	6.54	42.7716	2.55734	8.08703	.152905
6.05	36.6025	2.45967	7.77817	.165289	6.55	42.9025	2.55930	8.09321	.152672
6.06	36.7236	2.46171	7.78460	.165017	6.56	43.0336	2.56125	8.09938	.152439
6.07	36.8449	2.46374	7.79102	.164745	6.57	43.1649	2.56320	8.10555	.152207
6.08	36.9664	2.46577	7.79744	.164474	6.58	43.2964	2.56515	8.11172	.151976
6.09	37.0881	2.46779	7.80385	.164204	6.59	43.4281	2.56710	8.11788	.151745
6.10	37.2100	2.46982	7.81025	.163934	**6.60**	43.5600	2.56905	8.12404	.151515
6.11	37.3321	2.47184	7.81665	.163666	6.61	43.6921	2.57099	8.13019	.151286
6.12	37.4544	2.47386	7.82304	.163399	6.62	43.8244	2.57294	8.13634	.151057
6.13	37.5769	2.47588	7.82943	.163132	6.63	43.9569	2.57488	8.14248	.150830
6.14	37.6996	2.47790	7.83582	.162866	6.64	44.0896	2.57682	8.14862	.150602
6.15	37.8225	2.47992	7.84219	.162602	6.65	44.2225	2.57876	8.15475	.150376
6.16	37.9456	2.48193	7.84857	.162338	6.66	44.3556	2.58070	8.16088	.150150
6.17	38.0689	2.48395	7.85493	.162075	6.67	44.4889	2.58263	8.16701	.149925
6.18	38.1924	2.48596	7.86130	.161812	6.68	44.6224	2.58457	8.17313	.149701
6.19	38.3161	2.48797	7.86766	.161551	6.69	44.7561	2.58650	8.17924	.149477
6.20	38.4400	2.48998	7.87401	.161290	**6.70**	44.8900	2.58844	8.18535	.149254
6.21	38.5641	2.49199	7.88036	.161031	6.71	45.0241	2.59037	8.19146	.149031
6.22	38.6884	2.49399	7.88670	.160772	6.72	45.1584	2.59230	8.19756	.148810
6.23	38.8129	2.49600	7.89303	.160514	6.73	45.2929	2.59422	8.20366	.148588
6.24	38.9376	2.49800	7.89937	.160256	6.74	45.4276	2.59615	8.20975	.148368
6.25	39.0625	2.50000	7.90569	.160000	6.75	45.5625	2.59808	8.21584	.148148
6.26	39.1876	2.50200	7.91202	.159744	6.76	45.6976	2.60000	8.22192	.147929
6.27	39.3129	2.50400	7.91833	.159490	6.77	45.8329	2.60192	8.22800	.147710
6.28	39.4384	2.50599	7.92465	.159236	6.78	45.9684	2.60384	8.23408	.147493
6.29	39.5641	2.50799	7.93095	.158983	6.79	46.1041	2.60576	8.24015	.147275
6.30	39.6900	2.50998	7.93725	.158730	**6.80**	46.2400	2.60768	8.24621	.147059

TABLE 1   (Continued)

n	n^2	$\sqrt{n}$	$\sqrt{10n}$	$1/n$	n	n^2	$\sqrt{n}$	$\sqrt{10n}$	$1/n$
6.80	46.2400	2.60768	8.24621	.147059	**7.30**	53.2900	2.70185	8.54400	.136986
6.81	46.3761	2.60960	8.25227	.146843	7.31	53.4361	2.70370	8.54985	.136799
6.82	46.5124	2.61151	8.25833	.146628	7.32	53.5824	2.70555	8.55570	.136612
6.83	46.6489	2.61343	8.26438	.146413	7.33	53.7289	2.70740	8.56154	.136426
6.84	46.7856	2.61534	8.27043	.146199	7.34	53.8756	2.70924	8.56738	.136240
6.85	46.9225	2.61725	8.27647	.145985	7.35	54.0225	2.71109	8.57321	.136054
6.86	47.0596	2.61916	8.28251	.145773	7.36	54.1696	2.71293	8.57904	.135870
6.87	47.1969	2.62107	8.28855	.145560	7.37	54.3169	2.71477	8.58487	.135685
6.88	47.3344	2.62298	8.29458	.145349	7.38	54.4644	2.71662	8.59069	.135501
6.89	47.4721	2.62488	8.30060	.145138	7.39	54.6121	2.71846	8.59651	.135318
6.90	47.6100	2.62679	8.30662	.144928	**7.40**	54.7600	2.72029	8.60233	.135135
6.91	47.7481	2.62869	8.31264	.144718	7.41	54.9081	2.72213	8.60814	.134953
6.92	47.8864	2.63059	8.31865	.144509	7.42	55.0564	2.72397	8.61394	.134771
6.93	48.0249	2.63249	8.32466	.144300	7.43	55.2049	2.72580	8.61974	.134590
6.94	48.1636	2.63439	8.33067	.144092	7.44	55.3536	2.72764	8.62554	.134409
6.95	48.3025	2.63629	8.33667	.143885	7.45	55.5025	2.72947	8.63134	.134228
6.96	48.4416	2.63818	8.34266	.143678	7.46	55.6516	2.73130	8.63713	.134048
6.97	48.5809	2.64008	8.34865	.143472	7.47	55.8009	2.73313	8.64292	.133869
6.98	48.7204	2.64197	8.35464	.143266	7.48	55.9504	2.73496	8.64870	.133690
6.99	48.8601	2.64386	8.36062	.143062	7.49	56.1001	2.73679	8.65448	.133511
7.00	49.0000	2.64575	8.36660	.142857	**7.50**	56.2500	2.73861	8.66025	.133333
7.01	49.1401	2.64764	8.37257	.142653	7.51	56.4001	2.74044	8.66603	.133156
7.02	49.2804	2.64953	8.37854	.142450	7.52	56.5504	2.74226	8.67179	.132979
7.03	49.4209	2.65141	8.38451	.142248	7.53	56.7009	2.74408	8.67756	.132802
7.04	49.5616	2.65330	8.39047	.142045	7.54	56.8516	2.74591	8.68332	.132626
7.05	49.7025	2.65518	8.39643	.141844	7.55	57.0025	2.74773	8.68907	.132450
7.06	49.8436	2.65707	8.40238	.141643	7.56	57.1536	2.74955	8.69483	.132275
7.07	49.9849	2.65895	8.40833	.141443	7.57	57.3049	2.75136	8.70057	.132100
7.08	50.1264	2.66083	8.41427	.141243	7.58	57.4564	2.75318	8.70632	.131926
7.09	50.2681	2.66271	8.42021	.141044	7.59	57.6081	2.75500	8.71206	.131752
7.10	50.4100	2.66458	8.42615	.140845	**7.60**	57.7600	2.75681	8.71780	.131579
7.11	50.5521	2.66646	8.43208	.140647	7.61	57.9121	2.75862	8.72353	.131406
7.12	50.6944	2.66833	8.43801	.140449	7.62	58.0644	2.76043	8.72926	.131234
7.13	50.8369	2.67021	8.44393	.140252	7.63	58.2169	2.76225	8.73499	.131062
7.14	50.9796	2.67208	8.44985	.140056	7.64	58.3696	2.76405	8.74071	.130890
7.15	51.1225	2.67395	8.45577	.139860	7.65	58.5225	2.76586	8.74643	.130719
7.16	51.2656	2.67582	8.46168	.139665	7.66	58.6756	2.76767	8.75214	.130548
7.17	51.4089	2.67769	8.46759	.139470	7.67	58.8289	2.76948	8.75785	.130378
7.18	51.5524	2.67955	8.47349	.139276	7.68	58.9824	2.77128	8.76356	.130208
7.19	51.6961	2.68142	8.47939	.139082	7.69	59.1361	2.77308	8.76926	.130039
7.20	51.8400	2.68328	8.48528	.138889	**7.70**	59.2900	2.77489	8.77496	.129870
7.21	51.9841	2.68514	8.49117	.138696	7.71	59.4441	2.77669	8.78066	.129702
7.22	52.1284	2.68701	8.49706	.138504	7.72	59.5984	2.77849	8.78635	.129534
7.23	52.2729	2.68887	8.50294	.138313	7.73	59.7529	2.78029	8.79204	.129366
7.24	52.4176	2.69072	8.50882	.138122	7.74	59.9076	2.78209	8.79773	.129199
7.25	52.5625	2.69258	8.51469	.137931	7.75	60.0625	2.78388	8.80341	.129032
7.26	52.7076	2.69444	8.52056	.137741	7.76	60.2176	2.78568	8.80909	.128866
7.27	52.8529	2.69629	8.52643	.137552	7.77	60.3729	2.78747	8.81476	.128700
7.28	52.9984	2.69815	8.53229	.137363	7.78	60.5284	2.78927	8.82043	.128535
7.29	53.1441	2.70000	8.53815	.137174	7.79	60.6841	2.79106	8.82610	.128370
7.30	53.2900	2.70185	8.54400	.136986	**7.80**	60.8400	2.79285	8.83176	.128205

TABLE 1   (Continued)

n	n^2	$\sqrt{n}$	$\sqrt{10n}$	$1/n$	n	n^2	$\sqrt{n}$	$\sqrt{10n}$	$1/n$
7.80	60.8400	2.79285	8.83176	.128205	**8.30**	68.8900	2.88097	9.11043	.120482
7.81	60.9961	2.79464	8.83742	.128041	8.31	69.0561	2.88271	9.11592	.120337
7.82	61.1524	2.79643	8.84308	.127877	8.32	69.2224	2.88444	9.12140	.120192
7.83	61.3089	2.79821	8.84873	.127714	8.33	69.3889	2.88617	9.12688	.120048
7.84	61.4656	2.80000	8.85438	.127551	8.34	69.5556	2.88791	9.13236	.119904
7.85	61.6225	2.80179	8.86002	.127389	8.35	69.7225	2.88964	9.13783	.119760
7.86	61.7796	2.80357	8.86566	.127226	8.36	69.8896	2.89137	9.14330	.119617
7.87	61.9369	2.80535	8.87130	.127065	8.37	70.0569	2.89310	9.14877	.119474
7.88	62.0944	2.80713	8.87694	.126904	8.38	70.2244	2.89482	9.15423	.119332
7.89	62.2521	2.80891	8.88257	.126743	8.39	70.3921	2.89655	9.15969	.119190
7.90	62.4100	2.81069	8.88819	.126582	**8.40**	70.5600	2.89828	9.16515	.119048
7.91	62.5681	2.81247	8.89382	.126422	8.41	70.7281	2.90000	9.17061	.118906
7.92	62.7264	2.81425	8.89944	.126263	8.42	70.8964	2.90172	9.17606	.118765
7.93	62.8849	2.81603	8.90505	.126103	8.43	71.0649	2.90345	9.18150	.118624
7.94	63.0436	2.81780	8.91067	.125945	8.44	71.2336	2.90517	9.18695	.118483
7.95	63.2025	2.81957	8.91628	.125786	8.45	71.4025	2.90689	9.19239	.118343
7.96	63.3616	2.82135	8.92188	.125628	8.46	71.5716	2.90861	9.19783	.118203
7.97	63.5209	2.82312	8.92749	.125471	8.47	71.7409	2.91033	9.20326	.118064
7.98	63.6804	2.82489	8.93308	.125313	8.48	71.9104	2.91204	9.20869	.117925
7.99	63.8401	2.82666	8.93868	.125156	8.49	72.0801	2.91376	9.21412	.117786
8.00	64.0000	2.82843	8.94427	.125000	**8.50**	72.2500	2.91548	9.21954	.117647
8.01	64.1601	2.83019	8.94986	.124844	8.51	72.4201	2.91719	9.22497	.117509
8.02	64.3204	2.83196	8.95545	.124688	8.52	72.5904	2.91890	9.23038	.117371
8.03	64.4809	2.83373	8.96103	.124533	8.53	72.7609	2.92062	9.23580	.117233
8.04	64.6416	2.83549	8.96660	.124378	8.54	72.9316	2.92233	9.24121	.117096
8.05	64.8025	2.83725	8.97218	.124224	8.55	73.1025	2.92404	9.24662	.116959
8.06	64.9636	2.83901	8.97775	.124069	8.56	73.2736	2.92575	9.25203	.116822
8.07	65.1249	2.84077	8.98332	.123916	8.57	73.4449	2.92746	9.25743	.116686
8.08	65.2864	2.84253	8.98888	.123762	8.58	73.6164	2.92916	9.26283	.116550
8.09	65.4481	2.84429	8.99444	.123609	8.59	73.7881	2.93087	9.26823	.116414
8.10	65.6100	2.84605	9.00000	.123457	**8.60**	73.9600	2.93258	9.27362	.116279
8.11	65.7721	2.84781	9.00555	.123305	8.61	74.1321	2.93428	9.27901	.116144
8.12	65.9344	2.84956	9.01110	.123153	8.62	74.3044	2.93598	9.28440	.116009
8.13	66.0969	2.85132	9.01665	.123001	8.63	74.4769	2.93769	9.28978	.115875
8.14	66.2596	2.85307	9.02219	.122850	8.64	74.6496	2.93939	9.29516	.115741
8.15	66.4225	2.85482	9.02774	.122699	8.65	74.8225	2.94109	9.30054	.115607
8.16	66.5856	2.85657	9.03327	.122549	8.66	74.9956	2.94279	9.30591	.115473
8.17	66.7489	2.85832	9.03881	.122399	8.67	75.1689	2.94449	9.31128	.115340
8.18	66.9124	2.86007	9.04434	.122249	8.68	75.3424	2.94618	9.31665	.115207
8.19	67.0761	2.86182	9.04986	.122100	8.69	75.5161	2.94788	9.32202	.115075
8.20	67.2400	2.86356	9.05539	.121951	**8.70**	75.6900	2.94958	9.32738	.114943
8.21	67.4041	2.86531	9.06091	.121803	8.71	75.8641	2.95127	9.33274	.114811
8.22	67.5684	2.86705	9.06642	.121655	8.72	76.0384	2.95296	9.33809	.114679
8.23	67.7329	2.86880	9.07193	.121507	8.73	76.2129	2.95466	9.34345	.114548
8.24	67.8976	2.87054	9.07744	.121359	8.74	76.3876	2.95635	9.34880	.114416
8.25	68.0625	2.87228	9.08295	.121212	8.75	76.5625	2.95804	9.35414	.114286
8.26	68.2276	2.87402	9.08845	.121065	8.76	76.7376	2.95973	9.35949	.114155
8.27	68.3929	2.87576	9.09395	.120919	8.77	76.9129	2.96142	9.36483	.114025
8.28	68.5584	2.87750	9.09945	.120773	8.78	77.0884	2.96311	9.37017	.113895
8.29	68.7241	2.87924	9.10494	.120627	8.79	77.2641	2.96479	9.37550	.113766
8.30	68.8900	2.88097	9.11043	.120482	**8.80**	77.4400	2.96648	9.38083	.113636

TABLE 1 (Continued)

n	n^2	$\sqrt{n}$	$\sqrt{10n}$	$1/n$	n	n^2	$\sqrt{n}$	$\sqrt{10n}$	$1/n$
8.80	77.4400	2.96648	9.38083	.113636	**9.30**	86.4900	3.04959	9.64365	.107527
8.81	77.6161	2.96816	9.38616	.113507	9.31	86.6761	3.05123	9.64883	.107411
8.82	77.7924	2.96985	9.39149	.113379	9.32	86.8624	3.05287	9.65401	.107296
8.83	77.9689	2.97153	9.39681	.113250	9.33	87.0489	3.05450	9.65919	.107181
8.84	78.1456	2.97321	9.40213	.113122	9.34	87.2356	3.05614	9.66437	.107066
8.85	78.3225	2.97489	9.40744	.112994	9.35	87.4225	3.05778	9.66954	.106952
8.86	78.4996	2.97658	9.41276	.112867	9.36	87.6096	3.05941	9.67471	.106838
8.87	78.6769	2.97825	9.41807	.112740	9.37	87.7969	3.06105	9.67988	.106724
8.88	78.8544	2.97993	9.42338	.112613	9.38	87.9844	3.06268	9.68504	.106610
8.89	79.0321	2.98161	9.42868	.112486	9.39	88.1721	3.06431	9.69020	.106496
8.90	79.2100	2.98329	9.43398	.112360	**9.40**	88.3600	3.06594	9.69536	.106383
8.91	79.3881	2.98496	9.43928	.112233	9.41	88.5481	3.06757	9.70052	.106270
8.92	79.5664	2.98664	9.44458	.112108	9.42	88.7364	3.06920	9.70567	.106157
8.93	79.7449	2.98831	9.44987	.111982	9.43	88.9249	3.07083	9.71082	.106045
8.94	79.9236	2.98998	9.45516	.111857	9.44	89.1136	3.07246	9.71597	.105932
8.95	80.1025	2.99166	9.46044	.111732	9.45	89.3025	3.07409	9.72111	.105820
8.96	80.2816	2.99333	9.46573	.111607	9.46	89.4916	3.07571	9.72625	.105708
8.97	80.4609	2.99500	9.47101	.111483	9.47	89.6809	3.07734	9.73139	.105597
8.98	80.6404	2.99666	9.47629	.111359	9.48	89.8704	3.07896	9.73653	.105485
8.99	80.8201	2.99833	9.48156	.111235	9.49	90.0601	3.08058	9.74166	.105374
9.00	81.0000	3.00000	9.48683	.111111	**9.50**	90.2500	3.08221	9.74679	.105263
9.01	81.1801	3.00167	9.49210	.110988	9.51	90.4401	3.08383	9.75192	.105152
9.02	81.3604	3.00333	9.49737	.110865	9.52	90.6304	3.08545	9.75705	.105042
9.03	81.5409	3.00500	9.50263	.110742	9.53	90.8209	3.08707	9.76217	.104932
9.04	81.7216	3.00666	9.50789	.110619	9.54	91.0116	3.08869	9.76729	.104822
9.05	81.9025	3.00832	9.51315	.110497	9.55	91.2025	3.09031	9.77241	.104712
9.06	82.0836	3.00998	9.51840	.110375	9.56	91.3936	3.09192	9.77753	.104603
9.07	82.2649	3.01164	9.52365	.110254	9.57	91.5849	3.09354	9.78264	.104493
9.08	82.4464	3.01330	9.52890	.110132	9.58	91.7764	3.09516	9.78775	.104384
9.09	82.6281	3.01496	9.53415	.110011	9.59	91.9681	3.09677	9.79285	.104275
9.10	82.8100	3.01662	9.53939	.109890	**9.60**	92.1600	3.09839	9.79796	.104167
9.11	82.9921	3.01828	9.54463	.109769	9.61	92.3521	3.10000	9.80306	.104058
9.12	83.1744	3.01993	9.54987	.109649	9.62	92.5444	3.10161	9.80816	.103950
9.13	83.3569	3.02159	9.55510	.109529	9.63	92.7369	3.10322	9.81326	.103842
9.14	83.5396	3.02324	9.56033	.109409	9.64	92.9296	3.10483	9.81835	.103734
9.15	83.7225	3.02490	9.56556	.109290	9.65	93.1225	3.10644	9.82344	.103627
9.16	83.9056	3.02655	9.57079	.109170	9.66	93.3156	3.10805	9.82853	.103520
9.17	84.0889	3.02820	9.57601	.109051	9.67	93.5089	3.10966	9.83362	.103413
9.18	84.2724	3.02985	9.58123	.108932	9.68	93.7024	3.11127	9.83870	.103306
9.19	84.4561	3.03150	9.58645	.108814	9.69	93.8961	3.11288	9.84378	.103199
9.20	84.6400	3.03315	9.59166	.108696	**9.70**	94.0900	3.11448	9.84886	.103093
9.21	84.8241	3.03480	9.59687	.108578	9.71	94.2841	3.11609	9.85393	.102987
9.22	85.0084	3.03645	9.60208	.108460	9.72	94.4784	3.11769	9.85901	.102881
9.23	85.1929	3.03809	9.60729	.108342	9.73	94.6729	3.11929	9.86408	.102775
9.24	85.3776	3.03974	9.61249	.108225	9.74	94.8676	3.12090	9.86914	.102669
9.25	85.5625	3.04138	9.61769	.108108	9.75	95.0625	3.12250	9.87421	.102564
9.26	85.7476	3.04302	9.62289	.107991	9.76	95.2576	3.12410	9.87927	.102459
9.27	85.9329	3.04467	9.62808	.107875	9.77	95.4529	3.12570	9.88433	.102354
9.28	86.1184	3.04631	9.63328	.107759	9.78	95.6484	3.12730	9.88939	.102249
9.29	86.3041	3.04795	9.63846	.107643	9.79	95.8441	3.12890	9.89444	.102145
9.30	86.4900	3.04959	9.64365	.107527	**9.80**	96.0400	3.13050	9.89949	.102041

TABLE 1 (Continued)

n	n^2	$\sqrt{n}$	$\sqrt{10n}$	$1/n$	n	n^2	$\sqrt{n}$	$\sqrt{10n}$	$1/n$
9.80	96.0400	3.13050	9.89949	.102041	**9.90**	98.0100	3.14643	9.94987	.101010
9.81	96.2361	3.13209	9.90454	.101937	9.91	98.2081	3.14802	9.95490	.100908
9.82	96.4324	3.13369	9.90959	.101833	9.92	98.4064	3.14960	9.95992	.100806
9.83	96.6289	3.13528	9.91464	.101729	9.93	98.6049	3.15119	9.96494	.100705
9.84	96.8256	3.13688	9.91968	.101626	9.94	98.8036	3.15278	9.96995	.100604
9.85	97.0225	3.13847	9.92472	.101523	9.95	99.0025	3.15436	9.97497	.100503
9.86	97.2196	3.14006	9.92975	.101420	9.96	99.2016	3.15595	9.97998	.100402
9.87	97.4169	3.14166	9.93479	.101317	9.97	99.4009	3.15753	9.98499	.100301
9.88	97.6144	3.14325	9.93982	.101215	9.98	99.6004	3.15911	9.98999	.100200
9.89	97.8121	3.14484	9.94485	.101112	9.99	99.8001	3.16070	9.99500	.100100
9.90	98.0100	3.14643	9.94987	.101010	**10.00**	100.000	3.16228	10.0000	.100000

TABLE 2

Logarithms of Number (N) 1,000 to 10,499
Six- and Seven-Place Mantissas
(With the highest difference [D] between adjacent mantissas on each line)

N	0	1	2	3	4	5	6	7	8	9	D#
100	00 0000	0434	0868	1301	1734	2166	2598	3029	3461	3891	434
01	4321	4751	5181	5609	6038	6466	6894	7321	7748	8174	430
02	00 8600	9026	9451	9876	*0300	*0724	*1147	*1570	*1993	*2415	426
03	01 2837	3259	3680	4100	4521	4940	5360	5779	6197	6616	422
04	01 7033	7451	7868	8284	8700	9116	9532	9947	*0361	*0775	418
05	02 1189	1603	2016	2428	2841	3252	3664	4075	4486	4896	414
06	5306	5715	6125	6533	6942	7350	7757	8164	8571	8978	410
07	02 9384	9789	*0195	*0600	*1004	*1408	*1812	*2216	*2619	*3021	406
08	03 3424	3826	4227	4628	5029	5430	5830	6230	6629	7028	402
09	03 7426	7825	8223	8620	9017	9414	9811	*0207	*0602	*0998	399
110	04 1393	1787	2182	2576	2969	3362	3755	4148	4540	4932	395
11	5323	5714	6105	6495	6885	7275	7664	8053	8442	8830	391
12	04 9218	9606	9993	*0380	*0766	*1153	*1538	*1924	*2309	*2694	388
13	05 3078	3463	3846	4230	4613	4996	5378	5760	6142	6524	385
14	05 6905	7286	7666	8046	8426	8805	9185	9563	9942	*0320	381
15	06 0698	1075	1452	1829	2206	2582	2958	3333	3709	4083	377
16	4458	4832	5206	5580	5953	6326	6699	7071	7443	7815	374
17	06 8186	8557	8928	9298	9668	*0038	*0407	*0776	*1145	*1514	371
18	07 1882	2250	2617	2985	3352	3718	4085	4451	4816	5182	368
19	5547	5912	6276	6640	7004	7368	7731	8094	8457	8819	365
120	07 9181	9543	9904	*0266	*0626	*0987	1347	*1707	*2067	*2426	362
21	08 2785	3144	3503	3861	4219	4576	4934	5291	5647	6004	359
22	6360	6716	7071	7426	7781	8136	8490	8845	9198	9552	356
23	08 9905	*0258	*0611	*0963	*1315	*1667	*2018	*2370	*2721	*3071	353
24	09 3422	3772	4122	4471	4820	5169	5518	5866	6215	6562	350
25	09 6910	7257	7604	7951	8298	8644	8990	9335	9681	*0026	347
26	10 0371	0715	1059	1403	1747	2091	2434	2777	3119	3462	344
27	3804	4146	4487	4828	5169	5510	5851	6191	6531	6871	342
28	10 7210	7549	7888	8227	8565	8903	9241	9579	9916	*0253	339
29	11 0590	0926	1263	1599	1934	2270	2605	2940	3275	3609	337
130	3943	4277	4611	4944	5278	5611	5943	6276	6608	6940	334
31	11 7271	7603	7934	8265	8595	8926	9256	9586	9915	*0245	332
32	12 0574	0903	1231	1560	1888	2216	2544	2871	3198	3525	329
33	3852	4178	4504	4830	5156	5481	5806	6131	6456	6781	326
34	12 7105	7429	7753	8076	8399	8722	9045	9368	9690	*0012	324
35	13 0334	0655	0977	1298	1619	1939	2260	2580	2900	3219	322
36	3539	3858	4177	4496	4814	5133	5451	5769	6086	6403	319
37	6721	7037	7354	7671	7987	8303	8618	8934	9249	9564	317
38	13 9879	*0194	*0508	*0822	*1136	*1450	*1763	*2076	*2389	*2702	315
39	14 3015	3327	3639	3951	4263	4574	4885	5196	5507	5818	312
140	6128	6438	6748	7058	7367	7676	7985	8294	8603	8911	310
41	14 9219	9527	9835	*0142	*0449	*0756	*1063	*1370	*1676	*1982	308
42	15 2288	2594	2900	3205	3510	3815	4120	4424	4728	5032	306
43	5336	5640	5943	6246	6549	6852	7154	7457	7759	8061	304
44	15 8362	8664	8965	9266	9567	9868	*0168	*0469	*0769	*1068	302
45	16 1368	1667	1967	2266	2564	2863	3161	3460	3758	4055	300
46	4353	4650	4947	5244	5541	5838	6134	6430	6726	7022	297
47	16 7317	7613	7908	8203	8497	8792	9086	9380	9674	9968	296
48	17 0262	0555	0848	1141	1434	1726	2019	2311	2603	2895	293
49	3186	3478	3769	4060	4351	4641	4932	5222	5512	5802	292
N	0	1	2	3	4	5	6	7	8	9	D

*Prefix first two places on next line.

Example: The mantissa for number (N) 1072 is 03 0195.

#The *bighest difference* between adjacent mantissas on the *individual line*. It is also the *lowest difference* between adjacent mantissas on the *preceding line* in many cases.

From Stephen P. Shao, *Mathematics for Management and Finance*, 1969, South-Western Publishing Company, by permission of the publisher.

TABLE 2 (Continued)

N	0	1	2	3	4	5	6	7	8	9	D
150	17 6091	6381	6670	6959	7248	7536	7825	8113	8401	8689	290
51	17 8977	9264	9552	9839	*0126	*0413	*0699	*0986	*1272	*1558	288
52	18 1844	2129	2415	2700	2985	3270	3555	3839	4123	4407	286
53	4691	4975	5259	5542	5825	6108	6391	6674	6956	7239	284
54	18 7521	7803	8084	8366	8647	8928	9209	9490	9771	*0051	282
55	19 0332	0612	0892	1171	1451	1730	2010	2289	2567	2846	280
56	3125	3403	3681	3959	4237	4514	4792	5069	5346	5623	278
57	5900	6176	6453	6729	7005	7281	7556	7832	8107	8382	277
58	19 8657	8932	9206	9481	9755	*0029	*0303	*0577	*0850	*1124	275
59	20 1397	1670	1943	2216	2488	2761	3033	3305	3577	3848	273
160	4120	4391	4663	4934	5204	5475	5746	6016	6286	6556	272
61	6826	7096	7365	7634	7904	8173	8441	8710	8979	9247	270
62	20 9515	9783	*0051	*0319	*0586	*0853	*1121	*1388	*1654	*1921	268
63	21 2188	2454	2720	2986	3252	3518	3783	4049	4314	4579	266
64	4844	5109	5373	5638	5902	6166	6430	6694	6957	7221	265
65	21 7484	7747	8010	8273	8536	8798	9060	9323	9585	9846	263
66	22 0108	0370	0631	0892	1153	1414	1675	1936	2196	2456	262
67	2716	2976	3236	3496	3755	4015	4274	4533	4792	5051	260
68	5309	5568	5826	6084	6342	6600	6858	7115	7372	7630	259
69	22 7887	8144	8400	8657	8913	9170	9426	9682	9938	*0193	257
170	23 0449	0704	0960	1215	1470	1724	1979	2234	2488	2742	256
71	2996	3250	3504	3757	4011	4264	4517	4770	5023	5276	254
72	5528	5781	6033	6285	6537	6789	7041	7292	7544	7795	253
73	23 8046	8297	8548	8799	9049	9299	9550	9800	*0050	*0300	251
74	24 0549	0799	1048	1297	1546	1795	2044	2293	2541	2790	250
75	3038	3286	3534	3782	4030	4277	4525	4772	5019	5266	248
76	5513	5759	6006	6252	6499	6745	6991	7237	7482	7728	247
77	24 7973	8219	8464	8709	8954	9198	9443	9687	9932	*0176	246
78	25 0420	0664	0908	1151	1395	1638	1881	2125	2368	2610	244
79	2853	3096	3338	3580	3822	4064	4306	4548	4790	5031	243
180	5273	5514	5755	5996	6237	6477	6718	6958	7198	7439	241
81	25 7679	7918	8158	8398	8637	8877	9116	9355	9594	9833	240
82	26 0071	0310	0548	0787	1025	1263	1501	1739	1976	2214	239
83	2451	2688	2925	3162	3399	3636	3873	4109	4346	4582	237
84	4818	5054	5290	5525	5761	5996	6232	6467	6702	6937	236
85	7172	7406	7641	7875	8110	8344	8578	8812	9046	9279	235
86	26 9513	9746	9980	*0213	*0446	*0679	*0912	*1144	*1377	*1609	234
87	27 1842	2074	2306	2538	2770	3001	3233	3464	3696	3927	232
88	4158	4389	4620	4850	5081	5311	5542	5772	6002	6232	231
89	6462	6692	6921	7151	7380	7609	7838	8067	8296	8525	230
190	27 8754	8982	9211	9439	9667	9895	*0123	*0351	*0578	*0806	229
91	28 1033	1261	1488	1715	1942	2169	2396	2622	2849	3075	228
92	3301	3527	3753	3979	4205	4431	4656	4882	5107	5332	226
93	5557	5782	6007	6232	6456	6681	6905	7130	7354	7578	225
94	28 7802	8026	8249	8473	8696	8920	9143	9366	9589	9812	224
95	29 0035	0257	0480	0702	0925	1147	1369	1591	1813	2034	223
96	2256	2478	2699	2920	3141	3363	3584	3804	4025	4246	222
97	4466	4687	4907	5127	5347	5567	5787	6007	6226	6446	221
98	6665	6884	7104	7323	7542	7761	7979	8198	8416	8635	220
99	29 8853	9071	9289	9507	9725	9943	*0161	*0378	*0595	*0813	218
N	0	1	2	3	4	5	6	7	8	9	D

TABLE 2 (Continued)

N	0	1	2	3	4	5	6	7	8	9	D
200	30 1030	1247	1464	1681	1898	2114	2331	2547	2764	2980	217
01	3196	3412	3628	3844	4059	4275	4491	4706	4921	5136	216
02	5351	5566	5781	5996	6211	6425	6639	6854	7068	7282	215
03	7496	7710	7924	8137	8351	8564	8778	8991	9204	9417	214
04	30 9630	9843	*0056	*0268	*0481	*0693	*0906	*1118	*1330	*1542	213
05	31 1754	1966	2177	2389	2600	2812	3023	3234	3445	3656	212
06	3867	4078	4289	4499	4710	4920	5130	5340	5551	5760	211
07	5970	6180	6390	6599	6809	7018	7227	7436	7646	7854	210
08	31 8063	8272	8481	8689	8898	9106	9314	9522	9730	9938	209
09	32 0146	0354	0562	0769	0977	1184	1391	1598	1805	2012	208
210	2219	2426	2633	2839	3046	3252	3458	3665	3871	4077	207
11	4282	4488	4694	4899	5105	5310	5516	5721	5926	6131	206
12	6336	6541	6745	6950	7155	7359	7563	7767	7972	8176	205
13	32 8380	8583	8787	8991	9194	9398	9601	9805	*0008	*0211	204
14	33 0414	0617	0819	1022	1225	1427	1630	1832	2034	2236	203
15	2438	2640	2842	3044	3246	3447	3649	3850	4051	4253	202
16	4454	4655	4856	5057	5257	5458	5658	5859	6059	6260	201
17	6460	6660	6860	7060	7260	7459	7659	7858	8058	8257	200
18	33 8456	8656	8855	9054	9253	9451	9650	9849	*0047	*0246	200
19	34 0444	0642	0841	1039	1237	1435	1632	1830	2028	2225	199
220	2423	2620	2817	3014	3212	3409	3606	3802	3999	4196	198
21	4392	4589	4785	4981	5178	5374	5570	5766	5962	6157	197
22	6353	6549	6744	6939	7135	7330	7525	7720	7915	8110	196
23	34 8305	8500	8694	8889	9083	9278	9472	9666	9860	*0054	195
24	35 0248	0442	0636	0829	1023	1216	1410	1603	1796	1989	194
25	2183	2375	2568	2761	2954	3147	3339	3532	3724	3916	193
26	4108	4301	4493	4685	4876	5068	5260	5452	5643	5834	192
27	6026	6217	6408	6599	6790	6981	7172	7363	7554	7744	191
28	7935	8125	8316	8506	8696	8886	9076	9266	9456	9646	191
29	35 9835	*0025	*0215	*0404	*0593	*0783	*0972	*1161	*1350	*1539	190
230	36 1728	1917	2105	2294	2482	2671	2859	3048	3236	3424	189
31	3612	3800	3988	4176	4363	4551	4739	4926	5113	5301	188
32	5488	5675	5862	6049	6236	6423	6610	6796	6983	7169	187
33	7356	7542	7729	7915	8101	8287	8473	8659	8845	9030	187
34	36 9216	9401	9587	9772	9958	*0143	*0328	*0513	*0698	*0883	186
35	37 1068	1253	1437	1622	1806	1991	2175	2360	2544	2728	185
36	2912	3096	3280	3464	3647	3831	4015	4198	4382	4565	184
37	4748	4932	5115	5298	5481	5664	5846	6029	6212	6394	184
38	6577	6759	6942	7124	7306	7488	7670	7852	8034	8216	183
39	37 8398	8580	8761	8943	9124	9306	9487	9668	9849	*0030	182
240	38 0211	0392	0573	0754	0934	1115	1296	1476	1656	1837	181
41	2017	2197	2377	2557	2737	2917	3097	3277	3456	3636	180
42	3815	3995	4174	4353	4533	4712	4891	5070	5249	5428	180
43	5606	5785	5964	6142	6321	6499	6677	6856	7034	7212	179
44	7390	7568	7746	7923	8101	8279	8456	8634	8811	8989	178
45	38 9166	9343	9520	9698	9875	*0051	*0228	*0405	*0582	*0759	178
46	39 0935	1112	1288	1464	1641	1817	1993	2169	2345	2521	177
47	2697	2873	3048	3224	3400	3575	3751	3926	4101	4277	.76
48	4452	4627	4802	4977	5152	5326	5501	5676	5850	6025	175
49	6199	6374	6548	6722	6896	7071	7245	7419	7592	7766	175
N	0	1	2	3	4	5	6	7	8	9	D

TABLE 2 (Continued)

N	0	1	2	3	4	5	6	7	8	9	D
250	39 7940	8114	8287	8461	8634	8808	8981	9154	9328	9501	174
51	39 9674	9847	*0020	*0192	*0365	*0538	*0711	*0883	*1056	*1228	173
52	40 1401	1573	1745	1917	2089	2261	2433	2605	2777	2949	172
53	3121	3292	3464	3635	3807	3978	4149	4320	4492	4663	172
54	4834	5005	5176	5346	5517	5688	5858	6029	6199	6370	171
55	6540	6710	6881	7051	7221	7391	7561	7731	7901	8070	171
56	8240	8410	8579	8749	8918	9087	9257	9426	9595	9764	170
57	40 9933	*0102	*0271	*0440	*0609	*0777	*0946	*1114	*1283	*1451	169
58	41 1620	1788	1956	2124	2293	2461	2629	2796	2964	3132	169
59	3300	3467	3635	3803	3970	4137	4305	4472	4639	4806	168
260	4973	5140	5307	5474	5641	5808	5974	6141	6308	6474	167
61	6641	6807	6973	7139	7306	7472	7638	7804	7970	8135	167
62	8301	8467	8633	8798	8964	9129	9295	9460	9625	9791	166
63	41 9956	*0121	*0286	*0451	*0616	*0781	*0945	*1110	*1275	*1439	165
64	42 1604	1768	1933	2097	2261	2426	2590	2754	2918	3082	165
65	3246	3410	3574	3737	3901	4065	4228	4392	4555	4718	164
66	4882	5045	5208	5371	5534	5697	5860	6023	6186	6349	163
67	6511	6674	6836	6999	7161	7324	7486	7648	7811	7973	163
68	8135	8297	8459	8621	8783	8944	9106	9268	9429	9591	162
69	42 9752	9914	*0075	*0236	*0398	*0559	*0720	*0881	*1042	*1203	162
270	43 1364	1525	1685	1846	2007	2167	2328	2488	2649	2809	161
71	2969	3130	3290	3450	3610	3770	3930	4090	4249	4409	161
72	4569	4729	4888	5048	5207	5367	5526	5685	5844	6004	160
73	6163	6322	6481	6640	6799	6957	7116	7275	7433	7592	159
74	7751	7909	8067	8226	8384	8542	8701	8859	9017	9175	159
75	43 9333	9491	9648	9806	9964	*0122	*0279	*0437	*0594	*0752	158
76	44 0909	1066	1224	1381	1538	1695	1852	2009	2166	2323	158
77	2480	2637	2793	2950	3106	3263	3419	3576	3732	3889	157
78	4045	4201	4357	4513	4669	4825	4981	5137	5293	5449	156
79	5604	5760	5915	6071	6226	6382	6537	6692	6848	7003	156
280	7158	7313	7468	7623	7778	7933	8088	8242	8397	8552	155
81	44 8706	8861	9015	9170	9324	9478	9633	9787	9941	*0095	155
82	45 0249	0403	0557	0711	0865	1018	1172	1326	1479	1633	154
83	1786	1940	2093	2247	2400	2553	2706	2859	3012	3165	154
84	3318	3471	3624	3777	3930	4082	4235	4387	4540	4692	153
85	4845	4997	5150	5302	5454	5606	5758	5910	6062	6214	153
86	6366	6518	6670	6821	6973	7125	7276	7428	7579	7731	152
87	7882	8033	8184	8336	8487	8638	8789	8940	9091	9242	152
88	45 9392	9543	9694	9845	9995	*0146	*0296	*0447	*0597	*0748	151
89	46 0898	1048	1198	1348	1499	1649	1799	1948	2098	2248	151
290	2398	2548	2697	2847	2997	3146	3296	3445	3594	3744	150
91	3893	4042	4191	4340	4490	4639	4788	4936	5085	5234	149
92	5383	5532	5680	5829	5977	6126	6274	6423	6571	6719	149
93	6868	7016	7164	7312	7460	7608	7756	7904	8052	8200	148
94	8347	8495	8643	8790	8938	9085	9233	9380	9527	9675	148
95	46 9822	9969	*0116	*0263	*0410	*0557	*0704	*0851	*0998	*1145	147
96	47 1292	1438	1585	1732	1878	2025	2171	2318	2464	2610	147
97	2756	2903	3049	3195	3341	3487	3633	3779	3925	4071	147
98	4216	4362	4508	4653	4799	4944	5090	5235	5381	5526	146
99	5671	5816	5962	6107	6252	6397	6542	6687	6832	6976	146
N	0	1	2	3	4	5	6	7	8	9	D

TABLE 2 (Continued)

N	0	1	2	3	4	5	6	7	8	9	D
300	47 7121	7266	7411	7555	7700	7844	7989	8133	8278	8422	145
01	47 8566	8711	8855	8999	9143	9287	9431	9575	9719	9863	145
02	48 0007	0151	0294	0438	0582	0725	0869	1012	1156	1299	144
03	1443	1586	1729	1872	2016	2159	2302	2445	2588	2731	144
04	2874	3016	3159	3302	3445	3587	3730	3872	4015	4157	143
05	4300	4442	4585	4727	4869	5011	5153	5295	5437	5579	143
06	5721	5863	6005	6147	6289	6430	6572	6714	6855	6997	142
07	7138	7280	7421	7563	7704	7845	7986	8127	8269	8410	142
08	8551	8692	8833	8974	9114	9255	9396	9537	9677	9818	141
09	48 9958	*0099	*0239	*0380	*0520	*0661	*0801	*0941	*1081	*1222	141
310	49 1362	1502	1642	1782	1922	2062	2201	2341	2481	2621	140
11	2760	2900	3040	3179	3319	3458	3597	3737	3876	4015	140
12	4155	4294	4433	4572	4711	4850	4989	5128	5267	5406	139
13	5544	5683	5822	5960	6099	6238	6376	6515	6653	6791	139
14	6930	7068	7206	7344	7483	7621	7759	7897	8035	8173	139
15	8311	8448	8586	8724	8862	8999	9137	9275	9412	9550	138
16	49 9687	9824	9962	*0099	*0236	*0374	*0511	*0648	*0785	*0922	138
17	50 1059	1196	1333	1470	1607	1744	1880	2017	2154	2291	137
18	2427	2564	2700	2837	2973	3109	3246	3382	3518	3655	137
19	3791	3927	4063	4199	4335	4471	4607	4743	4878	5014	136
320	5150	5286	5421	5557	5693	5828	5964	6099	6234	6370	136
21	6505	6640	6776	6911	7046	7181	7316	7451	7586	7721	136
22	7856	7991	8126	8260	8395	8530	8664	8799	8934	9068	135
23	50 9203	9337	9471	9606	9740	9874	*0009	*0143	*0277	*0411	135
24	51 0545	0679	0813	0947	1081	1215	1349	1482	1616	1750	134
25	1883	2017	2151	2284	2418	2551	2684	2818	2951	3084	134
26	3218	3351	3484	3617	3750	3883	4016	4149	4282	4415	133
27	4548	4681	4813	4946	5079	5211	5344	5476	5609	5741	133
28	5874	6006	6139	6271	6403	6535	6668	6800	6932	7064	133
29	7196	7328	7460	7592	7724	7855	7987	8119	8251	8382	132
330	8514	8646	8777	8909	9040	9171	9303	9434	9566	9697	132
31	51 9828	9959	*0090	*0221	*0353	*0484	*0615	*0745	*0876	*1007	132
32	52 1138	1269	1400	1530	1661	1792	1922	2053	2183	2314	131
33	2444	2575	2705	2835	2966	3096	3226	3356	3486	3616	131
34	3746	3876	4006	4136	4266	4396	4526	4656	4785	4915	130
35	5045	5174	5304	5434	5563	5693	5822	5951	6081	6210	130
36	6339	6469	6598	6727	6856	6985	7114	7243	7372	7501	130
37	7630	7759	7888	8016	8145	8274	8402	8531	8660	8788	129
38	52 8917	9045	9174	9302	9430	9559	9687	9815	9943	*0072	129
39	53 0200	0328	0456	0584	0712	0840	0968	1096	1223	1351	128
340	1479	1607	1734	1862	1990	2117	2245	2372	2500	2627	128
41	2754	2882	3009	3136	3264	3391	3518	3645	3772	3899	128
42	4026	4153	4280	4407	4534	4661	4787	4914	5041	5167	127
43	5294	5421	5547	5674	5800	5927	6053	6180	6306	6432	127
44	6558	6685	6811	6937	7063	7189	7315	7441	7567	7693	127
45	7819	7945	8071	8197	8322	8448	8574	8699	8825	8951	126
46	53 9076	9202	9327	9452	9578	9703	9829	9954	*0079	*0204	126
47	54 0329	0455	0580	0705	0830	0955	1080	1205	1330	1454	126
48	1579	1704	1829	1953	2078	2203	2327	2452	2576	2701	125
49	2825	2950	3074	3199	3323	3447	3571	3696	3820	3944	125
N	0	1	2	3	4	5	6	7	8	9	D

TABLE 2 (Continued)

N	0	1	2	3	4	5	6	7	8	9	D
350	54 4068	4192	4316	4440	4564	4688	4812	4936	5060	5183	124
51	5307	5431	5555	5678	5802	5925	6049	6172	6296	6419	124
52	6543	6666	6789	6913	7036	7159	7282	7405	7529	7652	124
53	7775	7898	8021	8144	8267	8389	8512	8635	8758	8881	123
54	54 9003	9126	9249	9371	9494	9616	9739	9861	9984	*0106	123
55	55 0228	0351	0473	0595	0717	0840	0962	1084	1206	1328	123
56	1450	1572	1694	1816	1938	2060	2181	2303	2425	2547	122
57	2668	2790	2911	3033	3155	3276	3398	3519	3640	3762	122
58	3883	4004	4126	4247	4368	4489	4610	4731	4852	4973	122
59	5094	5215	5336	5457	5578	5699	5820	5940	6061	6182	121
360	6303	6423	6544	6664	6785	6905	7026	7146	7267	7387	121
61	7507	7627	7748	7868	7988	8108	8228	8349	8469	8589	121
62	8709	8829	8948	9068	9188	9308	9428	9548	9667	9787	120
63	55 9907	*0026	*0146	*0265	*0385	*0504	*0624	*0743	*0863	*0982	120
64	56 1101	1221	1340	1459	1578	1698	1817	1936	2055	2174	120
65	2293	2412	2531	2650	2769	2887	3006	3125	3244	3362	119
66	3481	3600	3718	3837	3955	4074	4192	4311	4429	4548	119
67	4666	4784	4903	5021	5139	5257	5376	5494	5612	5730	119
68	5848	5966	6084	6202	6320	6437	6555	6673	6791	6909	118
69	7026	7144	7262	7379	7497	7614	7732	7849	7967	8084	118
370	8202	8319	8436	8554	8671	8788	8905	9023	9140	9257	118
71	56 9374	9491	9608	9725	9842	9959	*0076	*0193	*0309	*0426	117
72	57 0543	0660	0776	0893	1010	1126	1243	1359	1476	1592	117
73	1709	1825	1942	2058	2174	2291	2407	2523	2639	2755	117
74	2872	2988	3104	3220	3336	3452	3568	3684	3800	3915	116
75	4031	4147	4263	4379	4494	4610	4726	4841	4957	5072	116
76	5188	5303	5419	5534	5650	5765	5880	5996	6111	6226	116
77	6341	6457	6572	6687	6802	6917	7032	7147	7262	7377	116
78	7492	7607	7722	7836	7951	8066	8181	8295	8410	8525	115
79	8639	8754	8868	8983	9097	9212	9326	9441	9555	9669	115
380	57 9784	9898	*0012	*0126	*0241	*0355	*0469	*0583	*0697	*0811	115
81	58 0925	1039	1153	1267	1381	1495	1608	1722	1836	1950	114
82	2063	2177	2291	2404	2518	2631	2745	2858	2972	3085	114
83	3199	3312	3426	3539	3652	3765	3879	3992	4105	4218	114
84	4331	4444	4557	4670	4783	4896	5009	5122	5235	5348	113
85	5461	5574	5686	5799	5912	6024	6137	6250	6362	6475	113
86	6587	6700	6812	6925	7037	7149	7262	7374	7486	7599	113
87	7711	7823	7935	8047	8160	8272	8384	8496	8608	8720	113
88	8832	8944	9056	9167	9279	9391	9503	9615	9726	9838	112
89	58 9950	*0061	*0173	*0284	*0396	*0507	*0619	*0730	*0842	*0953	112
390	59 1065	1176	1287	1399	1510	1621	1732	1843	1955	2066	112
91	2177	2288	2399	2510	2621	2732	2843	2954	3064	3175	111
92	3286	3397	3508	3618	3729	3840	3950	4061	4171	4282	111
93	4393	4503	4614	4724	4834	4945	5055	5165	5276	5386	111
94	5496	5606	5717	5827	5937	6047	6157	6267	6377	6487	111
95	6597	6707	6817	6927	7037	7146	7256	7366	7476	7586	110
96	7695	7805	7914	8024	8134	8243	8353	8462	8572	8681	110
97	8791	8900	9009	9119	9228	9337	9446	9556	9665	9774	110
98	9883	9992	*0101	*0210	*0319	*0428	*0537	*0646	*0755	*0864	109
99	60 0973	1082	1191	1299	1408	1517	1625	1734	1843	1951	109
N	0	1	2	3	4	5	6	7	8	9	D

TABLE 2 (Continued)

N	0	1	2	3	4	5	6	7	8	9	D
400	60 2060	2169	2277	2386	2494	2603	2711	2819	2928	3036	109
01	3144	3253	3361	3469	3577	3686	3794	3902	4010	4118	109
02	4226	4334	4442	4550	4658	4766	4874	4982	5089	5197	108
03	5305	5413	5521	5628	5736	5844	5951	6059	6166	6274	108
04	6381	6489	6596	6704	6811	6919	7026	7133	7241	7348	108
05	7455	7562	7669	7777	7884	7991	8098	8205	8312	8419	108
06	8526	8633	8740	8847	8954	9061	9167	9274	9381	9488	107
07	60 9594	9701	9808	9914	*0021	*0128	*0234	*0341	*0447	*0554	107
08	61 0660	0767	0873	0979	1086	1192	1298	1405	1511	1617	107
09	1723	1829	1936	2042	2148	2254	2360	2466	2572	2678	107
410	2784	2890	2996	3102	3207	3313	3419	3525	3630	3736	106
11	3842	3947	4053	4159	4264	4370	4475	4581	4686	4792	106
12	4897	5003	5108	5213	5319	5424	5529	5634	5740	5845	106
13	5950	6055	6160	6265	6370	6476	6581	6686	6790	6895	106
14	7000	7105	7210	7315	7420	7525	7629	7734	7839	7943	105
15	8048	8153	8257	8362	8466	8571	8676	8780	8884	8989	105
16	61 9093	9198	9302	9406	9511	9615	9719	9824	9928	*0032	105
17	62 0136	0240	0344	0448	0552	0656	0760	0864	0968	1072	104
18	1176	1280	1384	1488	1592	1695	1799	1903	2007	2110	104
19	2214	2318	2421	2525	2628	2732	2835	2939	3042	3146	104
420	3249	3353	3456	3559	3663	3766	3869	3973	4076	4179	104
21	4282	4385	4488	4591	4695	4798	4901	5004	5107	5210	104
22	5312	5415	5518	5621	5724	5827	5929	6032	6135	6238	103
23	6340	6443	6546	6648	6751	6853	6956	7058	7161	7263	103
24	7366	7468	7571	7673	7775	7878	7980	8082	8185	8287	103
25	8389	8491	8593	8695	8797	8900	9002	9104	9206	9308	103
26	62 9410	9512	9613	9715	9817	9919	*0021	*0123	*0224	*0326	102
27	63 0428	0530	0631	0733	0835	0936	1038	1139	1241	1342	102
28	1444	1545	1647	1748	1849	1951	2052	2153	2255	2356	102
29	2457	2559	2660	2761	2862	2963	3064	3165	3266	3367	102
430	3468	3569	3670	3771	3872	3973	4074	4175	4276	4376	101
31	4477	4578	4679	4779	4880	4981	5081	5182	5283	5383	101
32	5484	5584	5685	5785	5886	5986	6087	6187	6287	6388	101
33	6488	6588	6688	6789	6889	6989	7089	7189	7290	7390	101
34	7490	7590	7690	7790	7890	7990	8090	8190	8290	8389	100
35	8489	8589	8689	8789	8888	8988	9088	9188	9287	9387	100
36	63 9486	9586	9686	9785	9885	9984	*0084	*0183	*0283	*0382	100
37	64 0481	0581	0680	0779	0879	0978	1077	1177	1276	1375	100
38	1474	1573	1672	1771	1871	1970	2069	2168	2267	2366	100
39	2465	2563	2662	2761	2860	2959	3058	3156	3255	3354	99
440	3453	3551	3650	3749	3847	3946	4044	4143	4242	4340	99
41	4439	4537	4636	4734	4832	4931	5029	5127	5226	5324	99
42	5422	5521	5619	5717	5815	5913	6011	6110	6208	6306	99
43	6404	6502	6600	6698	6796	6894	6992	7089	7187	7285	98
44	7383	7481	7579	7676	7774	7872	7969	8067	8165	8262	98
45	8360	8458	8555	8653	8750	8848	8945	9043	9140	9237	98
46	64 9335	9432	9530	9627	9724	9821	9919	*0016	*0113	*0210	98
47	65 0308	0405	0502	0599	0696	0793	0890	0987	1084	1181	97
48	1278	1375	1472	1569	1666	1762	1859	1956	2053	2150	97
49	2246	2343	2440	2536	2633	2730	2826	2923	3019	3116	97
N	0	1	2	3	4	5	6	7	8	9	D

TABLE 2 (Continued)

N	0	1	2	3	4	5	6	7	8	9	D
450	65 3213	3309	3405	3502	3598	3695	3791	3888	3984	4080	97
51	4177	4273	4369	4465	4562	4658	4754	4850	4946	5042	97
52	5138	5235	5331	5427	5523	5619	5715	5810	5906	6002	97
53	6098	6194	6290	6386	6482	6577	6673	6769	6864	6960	96
54	7056	7152	7247	7343	7438	7534	7629	7725	7820	7916	96
55	8011	8107	8202	8298	8393	8488	8584	8679	8774	8870	96
56	8965	9060	9155	9250	9346	9441	9536	9631	9726	9821	96
57	65 9916	*0011	*0106	*0201	*0296	*0391	*0486	*0581	*0676	*0771	95
58	66 0865	0960	1055	1150	1245	1339	1434	1529	1623	1718	95
59	1813	1907	2002	2096	2191	2286	2380	2475	2569	2663	95
460	2758	2852	2947	3041	3135	3230	3324	3418	3512	3607	95
61	3701	3795	3889	3983	4078	4172	4266	4360	4454	4548	95
62	4642	4736	4830	4924	5018	5112	5206	5299	5393	5487	94
63	5581	5675	5769	5862	5956	6050	6143	6237	6331	6424	94
64	6518	6612	6705	6799	6892	6986	7079	7173	7266	7360	94
65	7453	7546	7640	7733	7826	7920	8013	8106	8199	8293	94
66	8386	8479	8572	8665	8759	8852	8945	9038	9131	9224	94
67	66 9317	9410	9503	9596	9689	9782	9875	9967	*0060	*0153	93
68	67 0246	0339	0431	0524	0617	0710	0802	0895	0988	1080	93
69	1173	1265	1358	1451	1543	1636	1728	1821	1913	2005	93
470	2098	2190	2283	2375	2467	2560	2652	2744	2836	2929	93
71	3021	3113	3205	3297	3390	3482	3574	3666	3758	3850	93
72	3942	4034	4126	4218	4310	4402	4494	4586	4677	4769	92
73	4861	4953	5045	5137	5228	5320	5412	5503	5595	5687	92
74	5778	5870	5962	6053	6145	6236	6328	6419	6511	6602	92
75	6694	6785	6876	6968	7059	7151	7242	7333	7424	7516	92
76	7607	7698	7789	7881	7972	8063	8154	8245	8336	8427	92
77	8518	8609	8700	8791	8882	8973	9064	9155	9246	9337	91
78	67 9428	9519	9610	9700	9791	9882	9973	*0063	*0154	*0245	91
79	68 0336	0426	0517	0607	0698	0789	0879	0970	1060	1151	91
480	1241	1332	1422	1513	1603	1693	1784	1874	1964	2055	91
81	2145	2235	2326	2416	2506	2596	2686	2777	2867	2957	91
82	3047	3137	3227	3317	3407	3497	3587	3677	3767	3857	90
83	3947	4037	4127	4217	4307	4396	4486	4576	4666	4756	90
84	4845	4935	5025	5114	5204	5294	5383	5473	5563	5652	90
85	5742	5831	5921	6010	6100	6189	6279	6368	6458	6547	90
86	6636	6726	6815	6904	6994	7083	7172	7261	7351	7440	90
87	7529	7618	7707	7796	7886	7975	8064	8153	8242	8331	90
88	8420	8509	8598	8687	8776	8865	8953	9042	9131	9220	89
89	68 9309	9398	9486	9575	9664	9753	9841	9930	*0019	*0107	89
490	69 0196	0285	0373	0462	0550	0639	0728	0816	0905	0993	89
91	1081	1170	1258	1347	1435	1524	1612	1700	1789	1877	89
92	1965	2053	2142	2230	2318	2406	2494	2583	2671	2759	89
93	2847	2935	3023	3111	3199	3287	3375	3463	3551	3639	88
94	3727	3815	3903	3991	4078	4166	4254	4342	4430	4517	88
95	4605	4693	4781	4868	4956	5044	5131	5219	5307	5394	88
96	5482	5569	5657	5744	5832	5919	6007	6094	6182	6269	88
97	6356	6444	6531	6618	6706	6793	6880	6968	7055	7142	88
98	7229	7317	7404	7491	7578	7665	7752	7839	7926	8014	88
99	8101	8188	8275	8362	8449	8535	8622	8709	8796	8883	87
N	0	1	2	3	4	5	6	7	8	9	D

TABLE 2 (Continued)

N	0	1	2	3	4	5	6	7	8	9	D
500	69 8970	9057	9144	9231	9317	9404	9491	9578	9664	9751	87
01	69 9838	9924	*0011	*0098	*0184	*0271	*0358	*0444	*0531	*0617	87
02	70 0704	0790	0877	0963	1050	1136	1222	1309	1395	1482	87
03	1568	1654	1741	1827	1913	1999	2086	2172	2258	2344	87
04	2431	2517	2603	2689	2775	2861	2947	3033	3119	3205	86
05	3291	3377	3463	3549	3635	3721	3807	3893	3979	4065	86
06	4151	4236	4322	4408	4494	4579	4665	4751	4837	4922	86
07	5008	5094	5179	5265	5350	5436	5522	5607	5693	5778	86
08	5864	5949	6035	6120	6206	6291	6376	6462	6547	6632	86
09	6718	6803	6888	6974	7059	7144	7229	7315	7400	7485	86
510	7570	7655	7740	7826	7911	7996	8081	8166	8251	8336	86
11	8421	8506	8591	8676	8761	8846	8931	9015	9100	9185	85
12	70 9270	9355	9440	9524	9609	9694	9779	9863	9948	*0033	85
13	71 0117	0202	0287	0371	0456	0540	0625	0710	0794	0879	85
14	0963	1048	1132	1217	1301	1385	1470	1554	1639	1723	85
15	1807	1892	1976	2060	2144	2229	2313	2397	2481	2566	85
16	2650	2734	2818	2902	2986	3070	3154	3238	3323	3407	85
17	3491	3575	3659	3742	3826	3910	3994	4078	4162	4246	84
18	4330	4414	4497	4581	4665	4749	4833	4916	5000	5084	84
19	5167	5251	5335	5418	5502	5586	5669	5753	5836	5920	84
520	6003	6087	6170	6254	6337	6421	6504	6588	6671	6754	84
21	6838	6921	7004	7088	7171	7254	7338	7421	7504	7587	84
22	7671	7754	7837	7920	8003	8086	8169	8253	8336	8419	84
23	8502	8585	8668	8751	8834	8917	9000	9083	9165	9248	83
24	71 9331	9414	9497	9580	9663	9745	9828	9911	9994	*0077	83
25	72 0159	0242	0325	0407	0490	0573	0655	0738	0821	0903	83
26	0986	1068	1151	1233	1316	1398	1481	1563	1646	1728	83
27	1811	1893	1975	2058	2140	2222	2305	2387	2469	2552	83
28	2634	2716	2798	2881	2963	3045	3127	3209	3291	3374	83
29	3456	3538	3620	3702	3784	3866	3948	4030	4112	4194	82
530	4276	4358	4440	4522	4604	4685	4767	4849	4931	5013	82
31	5095	5176	5258	5340	5422	5503	5585	5667	5748	5830	82
32	5912	5993	6075	6156	6238	6320	6401	6483	6564	6646	82
33	6727	6809	6890	6972	7053	7134	7216	7297	7379	7460	82
34	7541	7623	7704	7785	7866	7948	8029	8110	8191	8273	82
35	8354	8435	8516	8597	8678	8759	8841	8922	9003	9084	82
36	9165	9246	9327	9408	9489	9570	9651	9732	9813	9893	81
37	72 9974	*0055	*0136	*0217	*0298	*0378	*0459	*0540	*0621	*0702	81
38	73 0782	0863	0944	1024	1105	1186	1266	1347	1428	1508	81
39	1589	1669	1750	1830	1911	1991	2072	2152	2233	2313	81
540	2394	2474	2555	2635	2715	2796	2876	2956	3037	3117	81
41	3197	3278	3358	3438	3518	3598	3679	3759	3839	3919	81
42	3999	4079	4160	4240	4320	4400	4480	4560	4640	4720	81
43	4800	4880	4960	5040	5120	5200	5279	5359	5439	5519	80
44	5599	5679	5759	5838	5918	5998	6078	6157	6237	6317	80
45	6397	6476	6556	6635	6715	6795	6874	6954	7034	7113	80
46	7193	7272	7352	7431	7511	7590	7670	7749	7829	7908	80
47	7987	8067	8146	8225	8305	8384	8463	8543	8622	8701	80
48	8781	8860	8939	9018	9097	9177	9256	9335	9414	9493	80
49	73 9572	9651	9731	9810	9889	9968	*0047	*0126	*0205	*0284	80
N	0	1	2	3	4	5	6	7	8	9	D

765

TABLE 2 (Continued)

N	0	1	2	3	4	5	6	7	8	9	D
550	74 0363	0442	0521	0600	0678	0757	0836	0915	0994	1073	79
51	1152	1230	1309	1388	1467	1546	1624	1703	1782	1860	79
52	1939	2018	2096	2175	2254	2332	2411	2489	2568	2647	79
53	2725	2804	2882	2961	3039	3118	3196	3275	3353	3431	79
54	3510	3588	3667	3745	3823	3902	3980	4058	4136	4215	79
55	4293	4371	4449	4528	4606	4684	4762	4840	4919	4997	79
56	5075	5153	5231	5309	5387	5465	5543	5621	5699	5777	78
57	5855	5933	6011	6089	6167	6245	6323	6401	6479	6556	78
58	6634	6712	6790	6868	6945	7023	7101	7179	7256	7334	78
59	7412	7489	7567	7645	7722	7800	7878	7955	8033	8110	78
560	8188	8266	8343	8421	8498	8576	8653	8731	8808	8885	78
61	8963	9040	9118	9195	9272	9350	9427	9504	9582	9659	78
62	74 9736	9814	9891	9968	*0045	*0123	*0200	*0277	*0354	*0431	78
63	75 0508	0586	0663	0740	0817	0894	0971	1048	1125	1202	78
64	1279	1356	1433	1510	1587	1664	1741	1818	1895	1972	77
65	2048	2125	2202	2279	2356	2433	2509	2586	2663	2740	77
66	2816	2893	2970	3047	3123	3200	3277	3353	3430	3506	77
67	3583	3660	3736	3813	3889	3966	4042	4119	4195	4272	77
68	4348	4425	4501	4578	4654	4730	4807	4883	4960	5036	77
69	5112	5189	5265	5341	5417	5494	5570	5646	5722	5799	77
570	5875	5951	6027	6103	6180	6256	6332	6408	6484	6560	77
71	6636	6712	6788	6864	6940	7016	7092	7168	7244	7320	76
72	7396	7472	7548	7624	7700	7775	7851	7927	8003	8079	76
73	8155	8230	8306	8382	8458	8533	8609	8685	8761	8836	76
74	8912	8988	9063	9139	9214	9290	9366	9441	9517	9592	76
75	75 9668	9743	9819	9894	9970	*0045	*0121	*0196	*0272	*0347	76
76	76 0422	0498	0573	0649	0724	0799	0875	0950	1025	1101	76
77	1176	1251	1326	1402	1477	1552	1627	1702	1778	1853	76
78	1928	2003	2078	2153	2228	2303	2378	2453	2529	2604	76
79	2679	2754	2829	2904	2978	3053	3128	3203	3278	3353	75
580	3428	3503	3578	3653	3727	3802	3877	3952	4027	4101	75
81	4176	4251	4326	4400	4475	4550	4624	4699	4774	4848	75
82	4923	4998	5072	5147	5221	5296	5370	5445	5520	5594	75
83	5669	5743	5818	5892	5966	6041	6115	6190	6264	6338	75
84	6413	6487	6562	6636	6710	6785	6859	6933	7007	7082	75
85	7156	7230	7304	7379	7453	7527	7601	7675	7749	7823	75
86	7898	7972	8046	8120	8194	8268	8342	8416	8490	8564	74
87	8638	8712	8786	8860	8934	9008	9082	9156	9230	9303	74
88	76 9377	9451	9525	9599	9673	9746	9820	9894	9968	*0042	74
89	77 0115	0189	0263	0336	0410	0484	0557	0631	0705	0778	74
590	0852	0926	0999	1073	1146	1220	1293	1367	1440	1514	74
91	1587	1661	1734	1808	1881	1955	2028	2102	2175	2248	74
92	2322	2395	2468	2542	2615	2688	2762	2835	2908	2981	74
93	3055	3128	3201	3274	3348	3421	3494	3567	3640	3713	74
94	3786	3860	3933	4006	4079	4152	4225	4298	4371	4444	74
95	4517	4590	4663	4736	4809	4882	4955	5028	5100	5173	73
96	5246	5319	5392	5465	5538	5610	5683	5756	5829	5902	73
97	5974	6047	6120	6193	6265	6338	6411	6483	6556	6629	73
98	6701	6774	6846	6919	6992	7064	7137	7209	7282	7354	73
99	7427	7499	7572	7644	7717	7789	7862	7934	8006	8079	73
N	0	1	2	3	4	5	6	7	8	9	D

TABLE 2 (Continued)

N	0	1	2	3	4	5	6	7	8	9	D
600	77 8151	8224	8296	8368	8441	8513	8585	8658	8730	8802	73
01	8874	8947	9019	9091	9163	9236	9308	9380	9452	9524	73
02	77 9596	9669	9741	9813	9885	9957	*0029	*0101	*0173	*0245	73
03	78 0317	0389	0461	0533	0605	0677	0749	0821	0893	0965	72
04	1037	1109	1181	1253	1324	1396	1468	1540	1612	1684	72
05	1755	1827	1899	1971	2042	2114	2186	2258	2329	2401	72
06	2473	2544	2616	2688	2759	2831	2902	2974	3046	3117	72
07	3189	3260	3332	3403	3475	3546	3618	3689	3761	3832	72
08	3904	3975	4046	4118	4189	4261	4332	4403	4475	4546	72
09	4617	4689	4760	4831	4902	4974	5045	5116	5187	5259	72
610	5330	5401	5472	5543	5615	5686	5757	5828	5899	5970	72
11	6041	6112	6183	6254	6325	6396	6467	6538	6609	6680	71
12	6751	6822	6893	6964	7035	7106	7177	7248	7319	7390	71
13	7460	7531	7602	7673	7744	7815	7885	7956	8027	8098	71
14	8168	8239	8310	8381	8451	8522	8593	8663	8734	8804	71
15	8875	8946	9016	9087	9157	9228	9299	9369	9440	9510	71
16	78 9581	9651	9722	9792	9863	9933	*0004	*0074	*0144	*0215	71
17	79 0285	0356	0426	0496	0567	0637	0707	0778	0848	0918	71
18	0988	1059	1129	1199	1269	1340	1410	1480	1550	1620	71
19	1691	1761	1831	1901	1971	2041	2111	2181	2252	2322	71
620	2392	2462	2532	2602	2672	2742	2812	2882	2952	3022	70
21	3092	3162	3231	3301	3371	3441	3511	3581	3651	3721	70
22	3790	3860	3930	4000	4070	4139	4209	4279	4349	4418	70
23	4488	4558	4627	4697	4767	4836	4906	4976	5045	5115	70
24	5185	5254	5324	5393	5463	5532	5602	5672	5741	5811	70
25	5880	5949	6019	6088	6158	6227	6297	6366	6436	6505	70
26	6574	6644	6713	6782	6852	6921	6990	7060	7129	7198	70
27	7268	7337	7406	7475	7545	7614	7683	7752	7821	7890	70
28	7960	8029	8098	8167	8236	8305	8374	8443	8513	8582	70
29	8651	8720	8789	8858	8927	8996	9065	9134	9203	9272	69
630	79 9341	9409	9478	9547	9616	9685	9754	9823	9892	9961	69
31	80 0029	0098	0167	0236	0305	0373	0442	0511	0580	0648	69
32	0717	0786	0854	0923	0992	1061	1129	1198	1266	1335	69
33	1404	1472	1541	1609	1678	1747	1815	1884	1952	2021	69
34	2089	2158	2226	2295	2363	2432	2500	2568	2637	2705	69
35	2774	2842	2910	2979	3047	3116	3184	3252	3321	3389	69
36	3457	3525	3594	3662	3730	3798	3867	3935	4003	4071	69
37	4139	4208	4276	4344	4412	4480	4548	4616	4685	4753	69
38	4821	4889	4957	5025	5093	5161	5229	5297	5365	5433	68
39	5501	5569	5637	5705	5773	5841	5908	5976	6044	6112	68
640	6180	6248	6316	6384	6451	6519	6587	6655	6723	6790	68
41	6858	6926	6994	7061	7129	7197	7264	7332	7400	7467	68
42	7535	7603	7670	7738	7806	7873	7941	8008	8076	8143	68
43	8211	8279	8346	8414	8481	8549	8616	8684	8751	8818	68
44	8886	8953	9021	9088	9156	9223	9290	9358	9425	9492	68
45	80 9560	9627	9694	9762	9829	9896	9964	*0031	*0098	*0165	68
46	81 0233	0300	0367	0434	0501	0569	0636	0703	0770	0837	68
47	0904	0971	1039	1106	1173	1240	1307	1374	1441	1508	68
48	1575	1642	1709	1776	1843	1910	1977	2044	2111	2178	67
49	2245	2312	2379	2445	2512	2579	2646	2713	2780	2847	67
N	0	1	2	3	4	5	6	7	8	9	D

TABLE 2 (Continued)

N	0	1	2	3	4	5	6	7	8	9	D
650	81 2913	2980	3047	3114	3181	3247	3314	3381	3448	3514	67
51	3581	3648	3714	3781	3848	3914	3981	4048	4114	4181	67
52	4248	4314	4381	4447	4514	4581	4647	4714	4780	4847	67
53	4913	4980	5046	5113	5179	5246	5312	5378	5445	5511	67
54	5578	5644	5711	5777	5843	5910	5976	6042	6109	6175	67
55	6241	6308	6374	6440	6506	6573	6639	6705	6771	6838	67
56	6904	6970	7036	7102	7169	7235	7301	7367	7433	7499	67
57	7565	7631	7698	7764	7830	7896	7962	8028	8094	8160	67
58	8226	8292	8358	8424	8490	8556	8622	8688	8754	8820	66
59	8885	8951	9017	9083	9149	9215	9281	9346	9412	9478	66
660	81 9544	9610	9676	9741	9807	9873	9939	*0004	*0070	*0136	66
61	82 0201	0267	0333	0399	0464	0530	0595	0661	0727	0792	66
62	0858	0924	0989	1055	1120	1186	1251	1317	1382	1448	66
63	1514	1579	1645	1710	1775	1841	1906	1972	2037	2103	66
64	2168	2233	2299	2364	2430	2495	2560	2626	2691	2756	66
65	2822	2887	2952	3018	3083	3148	3213	3279	3344	3409	66
66	3474	3539	3605	3670	3735	3800	3865	3930	3996	4061	66
67	4126	4191	4256	4321	4386	4451	4516	4581	4646	4711	65
68	4776	4841	4906	4971	5036	5101	5166	5231	5296	5361	65
69	5426	5491	5556	5621	5686	5751	5815	5880	5945	6010	65
670	6075	6140	6204	6269	6334	6399	6464	6528	6593	6658	65
71	6723	6787	6852	6917	6981	7046	7111	7175	7240	7305	65
72	7369	7434	7499	7563	7628	7692	7757	7821	7886	7951	65
73	8015	8080	8144	8209	8273	8338	8402	8467	8531	8595	65
74	8660	8724	8789	8853	8918	8982	9046	9111	9175	9239	65
75	9304	9368	9432	9497	9561	9625	9690	9754	9818	9882.	65
76	82 9947	*0011	*0075	*0139	*0204	*0268	*0332	*0396	*0460	*0525	65
77	83 0589	0653	0717	0781	0845	0909	0973	1037	1102	1166	65
78	1230	1294	1358	1422	1486	1550	1614	1678	1742	1806	64
79	1870	1934	1998	2062	2126	2189	2253	2317	2381	2445	64
680	2509	2573	2637	2700	2764	2828	2892	2956	3020	3083	64
81	3147	3211	3275	3338	3402	3466	3530	3593	3657	3721	64
82	3784	3848	3912	3975	4039	4103	4166	4230	4294	4357	64
83	4421	4484	4548	4611	4675	4739	4802	4866	4929	4993	64
84	5056	5120	5183	5247	5310	5373	5437	5500	5564	5627	64
85	5691	5754	5817	5881	5944	6007	6071	6134	6197	6261	64
86	6324	6387	6451	6514	6577	6641	6704	6767	6830	6894	64
87	6957	7020	7083	7146	7210	7273	7336	7399	7462	7525	64
88	7588	7652	7715	7778	7841	7904	7967	8030	8093	8156	64
89	8219	8282	8345	8408	8471	8534	8597	8660	8723	8786	63
690	8849	8912	8975	9038	9101	9164	9227	9289	9352	9415	63
91	83 9478	9541	9604	9667	9729	9792	9855	9918	9981	*0043	63
92	84 0106	0169	0232	0294	0357	0420	0482	0545	0608	0671	63
93	0733	0796	0859	0921	0984	1046	1109	1172	1234	1297	63
94	1359	1422	1485	1547	1610	1672	1735	1797	1860	1922	63
95	1985	2047	2110	2172	2235	2297	2360	2422	2484	2547	63
96	2609	2672	2734	2796	2859	2921	2983	3046	3108	3170	63
97	3233	3295	3357	3420	3482	3544	3606	3669	3731	3793	63
98	3855	3918	3980	4042	4104	4166	4229	4291	4353	4415	63
99	4477	4539	4601	4664	4726	4788	4850	4912	4974	5036	63
N	0	1	2	3	4	5	6	7	8	9	D

TABLE 2 (Continued)

N	0	1	2	3	4	5	6	7	8	9	D
700	84 5098	5160	5222	5284	5346	5408	5470	5532	5594	5656	62
01	5718	5780	5842	5904	5966	6028	6090	6151	6213	6275	62
02	6337	6399	6461	6523	6585	6646	6708	6770	6832	6894	62
03	6955	7017	7079	7141	7202	7264	7326	7388	7449	7511	62
04	7573	7634	7696	7758	7819	7881	7943	8004	8066	8128	62
05	8189	8251	8312	8374	8435	8497	8559	8620	8682	8743	62
06	8805	8866	8928	8989	9051	9112	9174	9235	9297	9358	62
07	84 9419	9481	9542	9604	9665	9726	9788	9849	9911	9972	62
08	85 0033	0095	0156	0217	0279	0340	0401	0462	0524	0585	62
09	0646	0707	0769	0830	0891	0952	1014	1075	1136	1197	62
710	1258	1320	1381	1442	1503	1564	1625	1686	1747	1809	62
11	1870	1931	1992	2053	2114	2175	2236	2297	2358	2419	61
12	2480	2541	2602	2663	2724	2785	2846	2907	2968	3029	61
13	3090	3150	3211	3272	3333	3394	3455	3516	3577	3637	61
14	3698	3759	3820	3881	3941	4002	4063	4124	4185	4245	61
15	4306	4367	4428	4488	4549	4610	4670	4731	4792	4852	61
16	4913	4974	5034	5095	5156	5216	5277	5337	5398	5459	61
17	5519	5580	5640	5701	5761	5822	5882	5943	6003	6064	61
18	6124	6185	6245	6306	6366	6427	6487	6548	6608	6668	61
19	6729	6789	6850	6910	6970	7031	7091	7152	7212	7272	61
720	7332	7393	7453	7513	7574	7634	7694	7755	7815	7875	61
21	7935	7995	8056	8116	8176	8236	8297	8357	8417	8477	61
22	8537	8597	8657	8718	8778	8838	8898	8958	9018	9078	61
23	9138	9198	9258	9318	9379	9439	9499	9559	9619	9679	61
24	85 9739	9799	9859	9918	9978	*0038	*0098	*0158	*0218	*0278	60
25	86 0338	0398	0458	0518	0578	0637	0697	0757	0817	0877	60
26	0937	0996	1056	1116	1176	1236	1295	1355	1415	1475	60
27	1534	1594	1654	1714	1773	1833	1893	1952	2012	2072	60
28	2131	2191	2251	2310	2370	2430	2489	2549	2608	2668	60
29	2728	2787	2847	2906	2966	3025	3085	3144	3204	3263	60
730	3323	3382	3442	3501	3561	3620	3680	3739	3799	3858	60
31	3917	3977	4036	4096	4155	4214	4274	4333	4392	4452	60
32	4511	4570	4630	4689	4748	4808	4867	4926	4985	5045	60
33	5104	5163	5222	5282	5341	5400	5459	5519	5578	5637	60
34	5696	5755	5814	5874	5933	5992	6051	6110	6169	6228	60
35	6287	6346	6405	6465	6524	6583	6642	6701	6760	6819	60
36	6878	6937	6996	7055	7114	7173	7232	7291	7350	7409	59
37	7467	7526	7585	7644	7703	7762	7821	7880	7939	7998	59
38	8056	8115	8174	8233	8292	8350	8409	8468	8527	8586	59
39	8644	8703	8762	8821	8879	8938	8997	9056	9114	9173	59
740	9232	9290	9349	9408	9466	9525	9584	9642	9701	9760	59
41	86 9818	9877	9935	9994	*0053	*0111	*0170	*0228	*0287	*0345	59
42	87 0404	0462	0521	0579	0638	0696	0755	0813	0872	0930	59
43	0989	1047	1106	1164	1223	1281	1339	1398	1456	1515	59
44	1573	1631	1690	1748	1806	1865	1923	1981	2040	2098	59
45	2156	2215	2273	2331	2389	2448	2506	2564	2622	2681	59
46	2739	2797	2855	2913	2972	3030	3088	3146	3204	3262	59
47	3321	3379	3437	3495	3553	3611	3669	3727	3785	3844	59
48	3902	3960	4018	4076	4134	4192	4250	4308	4366	4424	58
49	4482	4540	4598	4656	4714	4772	4830	4888	4945	5003	58
N	0	1	2	3	4	5	6	7	8	9	D

TABLE 2 (Continued)

N	0	1	2	3	4	5	6	7	8	9	D
750	87 5061	5119	5177	5235	5293	5351	5409	5466	5524	5582	58
51	5640	5698	5756	5813	5871	5929	5987	6045	6102	6160	58
52	6218	6276	6333	6391	6449	6507	6564	6622	6680	6737	58
53	6795	6853	6910	6968	7026	7083	7141	7199	7256	7314	58
54	7371	7429	7487	7544	7602	7659	7717	7774	7832	7889	58
55	7947	8004	8062	8119	8177	8234	8292	8349	8407	8464	58
56	8522	8579	8637	8694	8752	8809	8866	8924	8981	9039	58
57	9096	9153	9211	9268	9325	9383	9440	9497	9555	9612	58
58	87 9669	9726	9784	9841	9898	9956	*0013	*0070	*0127	*0185	58
59	88 0242	0299	0356	0413	0471	0528	0585	0642	0699	0756	58
760	0814	0871	0928	0985	1042	1099	1156	1213	1271	1328	58
61	1385	1442	1499	1556	1613	1670	1727	1784	1841	1898	57
62	1955	2012	2069	2126	2183	2240	2297	2354	2411	2468	57
63	2525	2581	2638	2695	2752	2809	2866	2923	2980	3037	57
64	3093	3150	3207	3264	3321	3377	3434	3491	3548	3605	57
65	3661	3718	3775	3832	3888	3945	4002	4059	4115	4172	57
66	4229	4285	4342	4399	4455	4512	4569	4625	4682	4739	57
67	4795	4852	4909	4965	5022	5078	5135	5192	5248	5305	57
68	5361	5418	5474	5531	5587	5644	5700	5757	5813	5870	57
69	5926	5983	6039	6096	6152	6209	6265	6321	6378	6434	57
770	6491	6547	6604	6660	6716	6773	6829	6885	6942	6998	57
71	7054	7111	7167	7223	7280	7336	7392	7449	7505	7561	57
72	7617	7674	7730	7786	7842	7898	7955	8011	8067	8123	57
73	8179	8236	8292	8348	8404	8460	8516	8573	8629	8685	57
74	8741	8797	8853	8909	8965	9021	9077	9134	9190	9246	57
75	9302	9358	9414	9470	9526	9582	9638	9694	9750	9806	56
76	88 9862	9918	9974	*0030	*0086	*0141	*0197	*0253	*0309	*0365	56
77	89 0421	0477	0533	0589	0645	0700	0756	0812	0868	0924	56
78	0980	1035	1091	1147	1203	1259	1314	1370	1426	1482	56
79	1537	1593	1649	1705	1760	1816	1872	1928	1983	2039	56
780	2095	2150	2206	2262	2317	2373	2429	2484	2540	2595	56
81	2651	2707	2762	2818	2873	2929	2985	3040	3096	3151	56
82	3207	3262	3318	3373	3429	3484	3540	3595	3651	3706	56
83	3762	3817	3873	3928	3984	4039	4094	4150	4205	4261	56
84	4316	4371	4427	4482	4538	4593	4648	4704	4759	4814	56
85	4870	4925	4980	5036	5091	5146	5201	5257	5312	5367	56
86	5423	5478	5533	5588	5644	5699	5754	5809	5864	5920	56
87	5975	6030	6085	6140	6195	6251	6306	6361	6416	6471	56
88	6526	6581	6636	6692	6747	6802	6857	6912	6967	7022	56
89	7077	7132	7187	7242	7297	7352	7407	7462	7517	7572	55
790	7627	7682	7737	7792	7847	7902	7957	8012	8067	8122	55
91	8176	8231	8286	8341	8396	8451	8506	8561	8615	8670	55
92	8725	8780	8835	8890	8944	8999	9054	9109	9164	9218	55
93	9273	9328	9383	9437	9492	9547	9602	9656	9711	9766	55
94	89 9821	9875	9930	9985	*0039	*0094	*0149	*0203	*0258	*0312	55
95	90 0367	0422	0476	0531	0586	0640	0695	0749	0804	0859	55
96	0913	0968	1022	1077	1131	1186	1240	1295	1349	1404	55
97	1458	1513	1567	1622	1676	1731	1785	1840	1894	1948	55
98	2003	2057	2112	2166	2221	2275	2329	2384	2438	2492	55
99	2547	2601	2655	2710	2764	2818	2873	2927	2981	3036	55
N	0	1	2	3	4	5	6	7	8	9	D

TABLE 2 (Continued)

N	0	1	2	3	4	5	6	7	8	9	D
800	90 3090	3144	3199	3253	3307	3361	3416	3470	3524	3578	55
01	3633	3687	3741	3795	3849	3904	3958	4012	4066	4120	55
02	4174	4229	4283	4337	4391	4445	4499	4553	4607	4661	55
03	4716	4770	4824	4878	4932	4986	5040	5094	5148	5202	54
04	5256	5310	5364	5418	5472	5526	5580	5634	5688	5742	54
05	5796	5850	5904	5958	6012	6066	6119	6173	6227	6281	54
06	6335	6389	6443	6497	6551	6604	6658	6712	6766	6820	54
07	6874	6927	6981	7035	7089	7143	7196	7250	7304	7358	54
08	7411	7465	7519	7573	7626	7680	7734	7787	7841	7895	54
09	7949	8002	8056	8110	8163	8217	8270	8324	8378	8431	54
810	8485	8539	8592	8646	8699	8753	8807	8860	8914	8967	54
11	9021	9074	9128	9181	9235	9289	9342	9396	9449	9503	54
12	90 9556	9610	9663	9716	9770	9823	9877	9930	9984	*0037	54
13	91 0091	0144	0197	0251	0304	0358	0411	0464	0518	0571	54
14	0624	0678	0731	0784	0838	0891	0944	0998	1051	1104	54
15	1158	1211	1264	1317	1371	1424	1477	1530	1584	1637	54
16	1690	1743	1797	1850	1903	1956	2009	2063	2116	2169	54
17	2222	2275	2328	2381	2435	2488	2541	2594	2647	2700	54
18	2753	2806	2859	2913	2966	3019	3072	3125	3178	3231	54
19	3284	3337	3390	3443	3496	3549	3602	3655	3708	3761	53
820	3814	3867	3920	3973	4026	4079	4132	4184	4237	4290	53
21	4343	4396	4449	4502	4555	4608	4660	4713	4766	4819	53
22	4872	4925	4977	5030	5083	5136	5189	5241	5294	5347	53
23	5400	5453	5505	5558	5611	5664	5716	5769	5822	5875	53
24	5927	5980	6033	6085	6138	6191	6243	6296	6349	6401	53
25	6454	6507	6559	6612	6664	6717	6770	6822	6875	6927	53
26	6980	7033	7085	7138	7190	7243	7295	7348	7400	7453	53
27	7506	7558	7611	7663	7716	7768	7820	7873	7925	7978	53
28	8030	8083	8135	8188	8240	8293	8345	8397	8450	8502	53
29	8555	8607	8659	8712	8764	8816	8869	8921	8973	9026	53
830	9078	9130	9183	9235	9287	9340	9392	9444	9496	9549	53
31	91 9601	9653	9706	9758	9810	9862	9914	9967	*0019	*0071	53
32	92 0123	0176	0228	0280	0332	0384	0436	0489	0541	0593	53
33	0645	0697	0749	0801	0853	0906	0958	1010	1062	1114	53
34	1166	1218	1270	1322	1374	1426	1478	1530	1582	1634	52
35	1686	1738	1790	1842	1894	1946	1998	2050	2102	2154	52
36	2206	2258	2310	2362	2414	2466	2518	2570	2622	2674	52
37	2725	2777	2829	2881	2933	2985	3037	3089	3140	3192	52
38	3244	3296	3348	3399	3451	3503	3555	3607	3658	3710	52
39	3762	3814	3865	3917	3969	4021	4072	4124	4176	4228	52
840	4279	4331	4383	4434	4486	4538	4589	4641	4693	4744	52
41	4796	4848	4899	4951	5003	5054	5106	5157	5209	5261	52
42	5312	5364	5415	5467	5518	5570	5621	5673	5725	5776	52
43	5828	5879	5931	5982	6034	6085	6137	6188	6240	6291	52
44	6342	6394	6445	6497	6548	6600	6651	6702	6754	6805	52
45	6857	6908	6959	7011	7062	7114	7165	7216	7268	7319	52
46	7370	7422	7473	7524	7576	7627	7678	7730	7781	7832	52
47	7883	7935	7986	8037	8088	8140	8191	8242	8293	8345	52
48	8396	8447	8498	8549	8601	8652	8703	8754	8805	8857	52
49	8908	8959	9010	9061	9112	9163	9215	9266	9317	9368	52
N	0	1	2	3	4	5	6	7	8	9	D

TABLE 2 (Continued)

N	0	1	2	3	4	5	6	7	8	9	D
850	92 9419	9470	9521	9572	9623	9674	9725	9776	9827	9879	52
51	92 9930	9981	*0032	*0083	*0134	*0185	*0236	*0287	*0338	*0389	51
52	93 0440	0491	0542	0592	0643	0694	0745	0796	0847	0898	51
53	0949	1000	1051	1102	1153	1204	1254	1305	1356	1407	51
54	1458	1509	1560	1610	1661	1712	1763	1814	1865	1915	51
55	1966	2017	2068	2118	2169	2220	2271	2322	2372	2423	51
56	2474	2524	2575	2626	2677	2727	2778	2829	2879	2930	51
57	2981	3031	3082	3133	3183	3234	3285	3335	3386	3437	51
58	3487	3538	3589	3639	3690	3740	3791	3841	3892	3943	51
59	3993	4044	4094	4145	4195	4246	4296	4347	4397	4448	51
860	4498	4549	4599	4650	4700	4751	4801	4852	4902	4953	51
61	5003	5054	5104	5154	5205	5255	5306	5356	5406	5457	51
62	5507	5558	5608	5658	5709	5759	5809	5860	5910	5960	51
63	6011	6061	6111	6162	6212	6262	6313	6363	6413	6463	51
64	6514	6564	6614	6665	6715	6765	6815	6865	6916	6966	51
65	7016	7066	7117	7167	7217	7267	7317	7367	7418	7468	51
66	7518	7568	7618	7668	7718	7769	7819	7869	7919	7969	51
67	8019	8069	8119	8169	8219	8269	8320	8370	8420	8470	51
68	8520	8570	8620	8670	8720	8770	8820	8870	8920	8970	50
69	9020	9070	9120	9170	9220	9270	9320	9369	9419	9469	50
870	93 9519	9569	9619	9669	9719	9769	9819	9869	9918	9968	50
71	94 0018	0068	0118	0168	0218	0267	0317	0367	0417	0467	50
72	0516	0566	0616	0666	0716	0765	0815	0865	0915	0964	50
73	1014	1064	1114	1163	1213	1263	1313	1362	1412	1462	50
74	1511	1561	1611	1660	1710	1760	1809	1859	1909	1958	50
75	2008	2058	2107	2157	2207	2256	2306	2355	2405	2455	50
76	2504	2554	2603	2653	2702	2752	2801	2851	2901	2950	50
77	3000	3049	3099	3148	3198	3247	3297	3346	3396	3445	50
78	3495	3544	3593	3643	3692	3742	3791	3841	3890	3939	50
79	3989	4038	4088	4137	4186	4236	4285	4335	4384	4433	50
880	4483	4532	4581	4631	4680	4729	4779	4828	4877	4927	50
81	4976	5025	5074	5124	5173	5222	5272	5321	5370	5419	50
82	5469	5518	5567	5616	5665	5715	5764	5813	5862	5912	50
83	5961	6010	6059	6108	6157	6207	6256	6305	6354	6403	50
84	6452	6501	6551	6600	6649	6698	6747	6796	6845	6894	50
85	6943	6992	7041	7090	7140	7189	7238	7287	7336	7385	50
86	7434	7483	7532	7581	7630	7679	7728	7777	7826	7875	49
87	7924	7973	8022	8070	8119	8168	8217	8266	8315	8364	49
88	8413	8462	8511	8560	8609	8657	8706	8755	8804	8853	49
89	8902	8951	8999	9048	9097	9146	9195	9244	9292	9341	49
890	9390	9439	9488	9536	9585	9634	9683	9731	9780	9829	49
91	94 9878	9926	9975	*0024	*0073	*0121	*0170	*0219	*0267	*0316	49
92	95 0365	0414	0462	0511	0560	0608	0657	0706	0754	0803	49
93	0851	0900	0949	0997	1046	1095	1143	1192	1240	1289	49
94	1338	1386	1435	1483	1532	1580	1629	1677	1726	1775	49
95	1823	1872	1920	1969	2017	2066	2114	2163	2211	2260	49
96	2308	2356	2405	2453	2502	2550	2599	2647	2696	2744	49
97	2792	2841	2889	2938	2986	3034	3083	3131	3180	3228	49
98	3276	3325	3373	3421	3470	3518	3566	3615	3663	3711	49
99	3760	3808	3856	3905	3953	4001	4049	4098	4146	4194	49
N	0	1	2	3	4	5	6	7	8	9	D

TABLE 2 (Continued)

N	0	1	2	3	4	5	6	7	8	9	D
900	95 4243	4291	4339	4387	4435	4484	4532	4580	4628	4677	49
01	4725	4773	4821	4869	4918	4966	5014	5062	5110	5158	49
02	5207	5255	5303	5351	5399	5447	5495	5543	5592	5640	49
03	5688	5736	5784	5832	5880	5928	5976	6024	6072	6120	48
04	6168	6216	6265	6313	6361	6409	6457	6505	6553	6601	48
05	6649	6697	6745	6793	6840	6888	6936	6984	7032	7080	48
06	7128	7176	7224	7272	7320	7368	7416	7464	7512	7559	48
07	7607	7655	7703	7751	7799	7847	7894	7942	7990	8038	48
08	8086	8134	8181	8229	8277	8325	8373	8421	8468	8516	48
09	8564	8612	8659	8707	8755	8803	8850	8898	8946	8994	48
910	9041	9089	9137	9185	9232	9280	9328	9375	9423	9471	48
11	9518	9566	9614	9661	9709	9757	9804	9852	9900	9947	48
12	95 9995	*0042	*0090	*0138	*0185	*0233	*0280	*0328	*0376	*0423	48
13	96 0471	0518	0566	0613	0661	0709	0756	0804	0851	0899	48
14	0946	0994	1041	1089	1136	1184	1231	1279	1326	1374	48
15	1421	1469	1516	1563	1611	1658	1706	1753	1801	1848	48
16	1895	1943	1990	2038	2085	2132	2180	2227	2275	2322	48
17	2369	2417	2464	2511	2559	2606	2653	2701	2748	2795	48
18	2843	2890	2937	2985	3032	3079	3126	3174	3221	3268	48
19	3316	3363	3410	3457	3504	3552	3599	3646	3693	3741	48
920	3788	3835	3882	3929	3977	4024	4071	4118	4165	4212	48
21	4260	4307	4354	4401	4448	4495	4542	4590	4637	4684	48
22	4731	4778	4825	4872	4919	4966	5013	5061	5108	5155	48
23	5202	5249	5296	5343	5390	5437	5484	5531	5578	5625	47
24	5672	5719	5766	5813	5860	5907	5954	6001	6048	6095	47
25	6142	6189	6236	6283	6329	6376	6423	6470	6517	6564	47
26	6611	6658	6705	6752	6799	6845	6892	6939	6986	7033	47
27	7080	7127	7173	7220	7267	7314	7361	7408	7454	7501	47
28	7548	7595	7642	7688	7735	7782	7829	7875	7922	7969	47
29	8016	8062	8109	8156	8203	8249	8296	8343	8390	8436	47
930	8483	8530	8576	8623	8670	8716	8763	8810	8856	8903	47
31	8950	8996	9043	9090	9136	9183	9229	9276	9323	9369	47
32	9416	9463	9509	9556	9602	9649	9695	9742	9789	9835	47
33	96 9882	9928	9975	*0021	*0068	*0114	*0161	*0207	*0254	*0300	47
34	97 0347	0393	0440	0486	0533	0579	0626	0672	0719	0765	47
35	0812	0858	0904	0951	0997	1044	1090	1137	1183	1229	47
36	1276	1322	1369	1415	1461	1508	1554	1601	1647	1693	47
37	1740	1786	1832	1879	1925	1971	2018	2064	2110	2157	47
38	2203	2249	2295	2342	2388	2434	2481	2527	2573	2619	47
39	2666	2712	2758	2804	2851	2897	2943	2989	3035	3082	47
940	3128	3174	3220	3266	3313	3359	3405	3451	3497	3543	47
41	3590	3636	3682	3728	3774	3820	3866	3913	3959	4005	47
42	4051	4097	4143	4189	4235	4281	4327	4374	4420	4466	47
43	4512	4558	4604	4650	4696	4742	4788	4834	4880	4926	46
44	4972	5018	5064	5110	5156	5202	5248	5294	5340	5386	46
45	5432	5478	5524	5570	5616	5662	5707	5753	5799	5845	46
46	5891	5937	5983	6029	6075	6121	6167	6212	6258	6304	46
47	6350	6396	6442	6488	6533	6579	6625	6671	6717	6763	46
48	6808	6854	6900	6946	6992	7037	7083	7129	7175	7220	46
49	7266	7312	7358	7403	7449	7495	7541	7586	7632	7678	46
N	0	1	2	3	4	5	6	7	8	9	D

TABLE 2 (Continued)

N	0	1	2	3	4	5	6	7	8	9	D
950	97 7724	7769	7815	7861	7906	7952	7998	8043	8089	8135	46
51	8181	8226	8272	8317	8363	8409	8454	8500	8546	8591	46
52	8637	8683	8728	8774	8819	8865	8911	8956	9002	9047	46
53	9093	9138	9184	9230	9275	9321	9366	9412	9457	9503	46
54	97 9548	9594	9639	9685	9730	9776	9821	9867	9912	9958	46
55	98 0003	0049	0094	0140	0185	0231	0276	0322	0367	0412	46
56	0458	0503	0549	0594	0640	0685	0730	0776	0821	0867	46
57	0912	0957	1003	1048	1093	1139	1184	1229	1275	1320	46
58	1366	1411	1456	1501	1547	1592	1637	1683	1728	1773	46
59	1819	1864	1909	1954	2000	2045	2090	2135	2181	2226	46
960	2271	2316	2362	2407	2452	2497	2543	2588	2633	2678	46
61	2723	2769	2814	2859	2904	2949	2994	3040	3085	3130	46
62	3175	3220	3265	3310	3356	3401	3446	3491	3536	3581	46
63	3626	3671	3716	3762	3807	3852	3897	3942	3987	4032	46
64	4077	4122	4167	4212	4257	4302	4347	4392	4437	4482	45
65	4527	4572	4617	4662	4707	4752	4797	4842	4887	4932	45
66	4977	5022	5067	5112	5157	5202	5247	5292	5337	5382	45
67	5426	5471	5516	5561	5606	5651	5696	5741	5786	5830	45
68	5875	5920	5965	6010	6055	6100	6144	6189	6234	6279	45
69	6324	6369	6413	6458	6503	6548	6593	6637	6682	6727	45
970	6772	6817	6861	6906	6951	6996	7040	7085	7130	7175	45
71	7219	7264	7309	7353	7398	7443	7488	7532	7577	7622	45
72	7666	7711	7756	7800	7845	7890	7934	7979	8024	8068	45
73	8113	8157	8202	8247	8291	8336	8381	8425	8470	8514	45
74	8559	8604	8648	8693	8737	8782	8826	8871	8916	8960	45
75	9005	9049	9094	9138	9183	9227	9272	9316	9361	9405	45
76	9450	9494	9539	9583	9628	9672	9717	9761	9806	9850	45
77	98 9895	9939	9983	*0028	*0072	*0117	*0161	*0206	*0250	*0294	45
78	99 0339	0383	0428	0472	0516	0561	0605	0650	0694	0738	45
79	0783	0827	0871	0916	0960	1004	1049	1093	1137	1182	45
980	1226	1270	1315	1359	1403	1448	1492	1536	1580	1625	45
81	1669	1713	1758	1802	1846	1890	1935	1979	2023	2067	45
82	2111	2156	2200	2244	2288	2333	2377	2421	2465	2509	45
83	2554	2598	2642	2686	2730	2774	2819	2863	2907	2951	45
84	2995	3039	3083	3127	3172	3216	3260	3304	3348	3392	45
85	3436	3480	3524	3568	3613	3657	3701	3745	3789	3833	45
86	3877	3921	3965	4009	4053	4097	4141	4185	4229	4273	44
87	4317	4361	4405	4449	4493	4537	4581	4625	4669	4713	44
88	4757	4801	4845	4889	4933	4977	5021	5065	5108	5152	44
89	5196	5240	5284	5328	5372	5416	5460	5504	5547	5591	44
990	5635	5679	5723	5767	5811	5854	5898	5942	5986	6030	44
91	6074	6117	6161	6205	6249	6293	6337	6380	6424	6468	44
92	6512	6555	6599	6643	6687	6731	6774	6818	6862	6906	44
93	6949	6993	7037	7080	7124	7168	7212	7255	7299	7343	44
94	7386	7430	7474	7517	7561	7605	7648	7692	7736	7779	44
95	7823	7867	7910	7954	7998	8041	8085	8129	8172	8216	44
96	8259	8303	8347	8390	8434	8477	8521	8564	8608	8652	44
97	8695	8739	8782	8826	8869	8913	8956	9000	9043	9087	44
98	9131	9174	9218	9261	9305	9348	9392	9435	9479	9522	44
99	99 9565	9609	9652	9696	9739	9783	9826	9870	9913	9957	44
N	0	1	2	3	4	5	6	7	8	9	D

TABLE 2 (Continued)

N	0	1	2	3	4	5	6	7	8	9	D#
1000	000 0000	0434	0869	1303	1737	2171	2605	3039	3473	3907	435
1001	4341	4775	5208	5642	6076	6510	6943	7377	7810	8244	434
1002	8677	9111	9544	9977	*0411	*0844	*1277	*1710	*2143	*2576	434
1003	001 3009	3442	3875	4308	4741	5174	5607	6039	6472	6905	433
1004	7337	7770	8202	8635	9067	9499	9932	*0364	*0796	*1228	433
1005	002 1661	2093	2525	2957	3389	3821	4253	4685	5116	5548	432
1006	5980	6411	6843	7275	7706	8138	8569	9001	9432	9863	432
1007	003 0295	0726	1157	1588	2019	2451	2882	3313	3744	4174	432
1008	4605	5036	5467	5898	6328	6759	7190	7620	8051	8481	431
1009	8912	9342	9772	*0203	*0633	*1063	*1493	*1924	*2354	*2784	431
1010	004 3214	3644	4074	4504	4933	5363	5793	6223	6652	7082	430
1011	7512	7941	8371	8800	9229	9659	*0088	*0517	*0947	*1376	430
1012	005 1805	2234	2663	3092	3521	3950	4379	4808	5237	5666	429
1013	6094	6523	6952	7380	7809	8238	8666	9094	9523	9951	429
1014	006 0380	0808	1236	1664	2092	2521	2949	3377	3805	4233	429
1015	4660	5088	5516	5944	6372	6799	7227	7655	8082	8510	428
1016	8937	9365	9792	*0219	*0647	*1074	*1501	*1928	*2355	*2782	428
1017	007 3210	3637	4064	4490	4917	5344	5771	6198	6624	7051	427
1018	7478	7904	8331	8757	9184	9610	*0037	*0463	*0889	*1316	427
1019	008 1742	2168	2594	3020	3446	3872	4298	4724	5150	5576	426
1020	6002	6427	6853	7279	7704	8130	8556	8981	9407	9832	426
1021	009 0257	0683	1108	1533	1959	2384	2809	3234	3659	4084	426
1022	4509	4934	5359	5784	6208	6633	7058	7483	7907	8332	425
1023	8756	9181	9605	*0030	*0454	*0878	*1303	*1727	*2151	*2575	425
1024	010 3000	3424	3848	4272	4696	5120	5544	5967	6391	6815	424
1025	7239	7662	8086	8510	8933	9357	9780	*0204	*0627	*1050	424
1026	011 1474	1897	2320	2743	3166	3590	4013	4436	4859	5282	424
1027	5704	6127	6550	6973	7396	7818	8241	8664	9086	9509	423
1028	9931	*0354	*0776	*1198	*1621	*2043	*2465	*2887	*3310	*3732	423
1029	012 4154	4576	4998	5420	5842	6264	6685	7107	7529	7951	422
1030	8372	8794	9215	9637	*0059	*0480	*0901	*1323	*1744	*2165	422
1031	013 2587	3008	3429	3850	4271	4692	5113	5534	5955	6376	421
1032	6797	7218	7639	8059	8480	8901	9321	9742	*0162	*0583	421
1033	014 1003	1424	1844	2264	2685	3105	3525	3945	4365	4785	421
1034	5205	5625	6045	6465	6885	7305	7725	8144	8564	8984	420
1035	9403	9823	*0243	*0662	*1082	*1501	*1920	*2340	*2759	*3178	420
1036	015 3598	4017	4436	4855	5274	5693	6112	6531	6950	7369	419
1037	7788	8206	8625	9044	9462	9881	*0300	*0718	*1137	*1555	419
1038	016 1974	2392	2810	3229	3647	4065	4483	4901	5319	5737	419
1039	6155	6573	6991	7409	7827	8245	8663	9080	9498	9916	418
1040	017 0333	0751	1168	1586	2003	2421	2838	3256	3673	4090	418
1041	4507	4924	5342	5759	6176	6593	7010	7427	7844	8260	418
1042	8677	9094	9511	9927	*0344	*0761	*1177	*1594	*2010	*2427	417
1043	018 2843	3259	3676	4092	4508	4925	5341	5757	6173	6589	417
1044	7005	7421	7837	8253	8669	9084	9500	9916	*0332	*0747	416
1045	019 1163	1578	1994	2410	2825	3240	3656	4071	4486	4902	416
1046	5317	5732	6147	6562	6977	7392	7807	8222	8637	9052	415
1047	9467	9882	*0296	*0711	*1126	*1540	*1955	*2369	*2784	*3198	415
1048	020 3613	4027	4442	4856	5270	5684	6099	6513	6927	7341	415
1049	7755	8169	8583	8997	9411	9824	*0238	*0652	*1066	*1479	414
N	0	1	2	3	4	5	6	7	8	9	D

TABLE 3

BINOMIAL PROBABILITY DISTRIBUTION—VALUES OF PROBABILITY

$$P(X; n, P) = {}_nC_X \cdot P^X \cdot Q^{n-X}$$

Probability $P(X; n, P)$

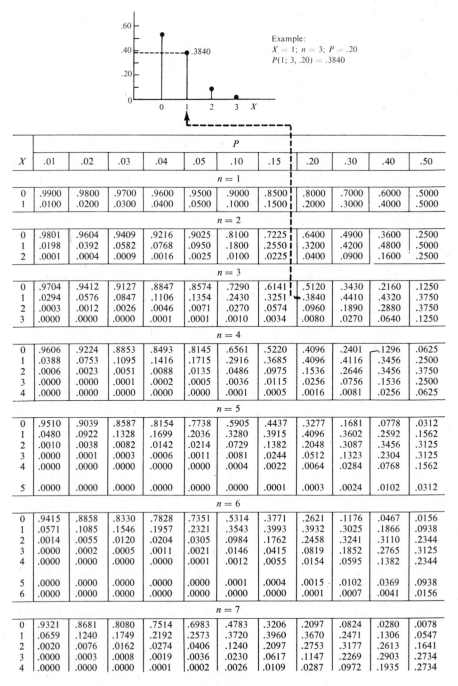

Example:
$X = 1; n = 3; P = .20$
$P(1; 3, .20) = .3840$

X	.01	.02	.03	.04	.05	.10	.15	.20	.30	.40	.50
							P				
						$n = 1$					
0	.9900	.9800	.9700	.9600	.9500	.9000	.8500	.8000	.7000	.6000	.5000
1	.0100	.0200	.0300	.0400	.0500	.1000	.1500	.2000	.3000	.4000	.5000
						$n = 2$					
0	.9801	.9604	.9409	.9216	.9025	.8100	.7225	.6400	.4900	.3600	.2500
1	.0198	.0392	.0582	.0768	.0950	.1800	.2550	.3200	.4200	.4800	.5000
2	.0001	.0004	.0009	.0016	.0025	.0100	.0225	.0400	.0900	.1600	.2500
						$n = 3$					
0	.9704	.9412	.9127	.8847	.8574	.7290	.6141	.5120	.3430	.2160	.1250
1	.0294	.0576	.0847	.1106	.1354	.2430	.3251	.3840	.4410	.4320	.3750
2	.0003	.0012	.0026	.0046	.0071	.0270	.0574	.0960	.1890	.2880	.3750
3	.0000	.0000	.0000	.0001	.0001	.0010	.0034	.0080	.0270	.0640	.1250
						$n = 4$					
0	.9606	.9224	.8853	.8493	.8145	.6561	.5220	.4096	.2401	.1296	.0625
1	.0388	.0753	.1095	.1416	.1715	.2916	.3685	.4096	.4116	.3456	.2500
2	.0006	.0023	.0051	.0088	.0135	.0486	.0975	.1536	.2646	.3456	.3750
3	.0000	.0000	.0001	.0002	.0005	.0036	.0115	.0256	.0756	.1536	.2500
4	.0000	.0000	.0000	.0000	.0000	.0001	.0005	.0016	.0081	.0256	.0625
						$n = 5$					
0	.9510	.9039	.8587	.8154	.7738	.5905	.4437	.3277	.1681	.0778	.0312
1	.0480	.0922	.1328	.1699	.2036	.3280	.3915	.4096	.3602	.2592	.1562
2	.0010	.0038	.0082	.0142	.0214	.0729	.1382	.2048	.3087	.3456	.3125
3	.0000	.0001	.0003	.0006	.0011	.0081	.0244	.0512	.1323	.2304	.3125
4	.0000	.0000	.0000	.0000	.0000	.0004	.0022	.0064	.0284	.0768	.1562
5	.0000	.0000	.0000	.0000	.0000	.0000	.0001	.0003	.0024	.0102	.0312
						$n = 6$					
0	.9415	.8858	.8330	.7828	.7351	.5314	.3771	.2621	.1176	.0467	.0156
1	.0571	.1085	.1546	.1957	.2321	.3543	.3993	.3932	.3025	.1866	.0938
2	.0014	.0055	.0120	.0204	.0305	.0984	.1762	.2458	.3241	.3110	.2344
3	.0000	.0002	.0005	.0011	.0021	.0146	.0415	.0819	.1852	.2765	.3125
4	.0000	.0000	.0000	.0000	.0001	.0012	.0055	.0154	.0595	.1382	.2344
5	.0000	.0000	.0000	.0000	.0000	.0001	.0004	.0015	.0102	.0369	.0938
6	.0000	.0000	.0000	.0000	.0000	.0000	.0000	.0001	.0007	.0041	.0156
						$n = 7$					
0	.9321	.8681	.8080	.7514	.6983	.4783	.3206	.2097	.0824	.0280	.0078
1	.0659	.1240	.1749	.2192	.2573	.3720	.3960	.3670	.2471	.1306	.0547
2	.0020	.0076	.0162	.0274	.0406	.1240	.2097	.2753	.3177	.2613	.1641
3	.0000	.0003	.0008	.0019	.0036	.0230	.0617	.1147	.2269	.2903	.2734
4	.0000	.0000	.0000	.0001	.0002	.0026	.0109	.0287	.0972	.1935	.2734

TABLE 3 (Continued)

					P						
X	.01	.02	.03	.04	.05	.10	.15	.20	.30	.40	.50
5	.0000	.0000	.0000	.0000	.0009	.0002	.0012	.0043	.0250	.0774	.1641
6	.0000	.0000	.0000	.0000	.0000	.0000	.0001	.0004	.0036	.0172	.0547
7	.0000	.0000	.0000	.0000	.0000	.0000	.0000	.0000	.0002	.0016	.0078

					$n = 8$						
0	.9227	.8508	.7837	.7214	.6634	.4305	.2725	.1678	.0576	.0168	.0039
1	.0746	.1389	.1939	.2405	.2793	.3826	.3847	.3355	.1977	.0896	.0312
2	.0026	.0099	.0210	.0351	.0515	.1488	.2376	.2936	.2965	.2090	.1094
3	.0001	.0004	.0013	.0029	.0054	.0331	.0839	.1468	.2541	.2787	.2188
4	.0000	.0000	.0001	.0002	.0004	.0046	.0185	.0459	.1361	.2322	.2734
5	.0000	.0000	.0000	.0000	.0000	.0004	.0026	.0092	.0467	.1239	.2188
6	.0000	.0000	.0000	.0000	.0000	.0000	.0002	.0011	.0100	.0413	.1094
7	.0000	.0000	.0000	.0000	.0000	.0000	.0000	.0001	.0012	.0079	.0312
8	.0000	.0000	.0000	.0000	.0000	.0000	.0000	.0000	.0001	.0007	.0039

					$n = 9$						
0	.9135	.8337	.7602	.6925	.6302	.3874	.2316	.1342	.0404	.0101	.0020
1	.0830	.1531	.2116	.2597	.2985	.3874	.3679	.3020	.1556	.0605	.0176
2	.0034	.0125	.0262	.0433	.0629	.1722	.2597	.3020	.2668	.1612	.0703
3	.0001	.0006	.0019	.0042	.0077	.0446	.1069	.1762	.2668	.2508	.1641
4	.0000	.0000	.0001	.0003	.0006	.0074	.0283	.0661	.1715	.2508	.2461
5	.0000	.0000	.0000	.0000	.0000	.0008	.0050	.0165	.0735	.1672	.2461
6	.0000	.0000	.0000	.0000	.0000	.0001	.0006	.0028	.0210	.0743	.1641
7	.0000	.0000	.0000	.0000	.0000	.0000	.0000	.0003	.0039	.0212	.0703
8	.0000	.0000	.0000	.0000	.0000	.0000	.0000	.0000	.0004	.0035	.0176
9	.0000	.0000	.0000	.0000	.0000	.0000	.0000	.0000	.0000	.0003	.0020

					$n = 10$						
0	.9044	.8171	.7374	.6648	.5987	.3487	.1969	.1074	.0282	.0060	.0010
1	.0914	.1667	.2281	.2770	.3151	.3874	.3474	.2684	.1211	.0403	.0098
2	.0042	.0153	.0317	.0519	.0746	.1937	.2759	.3020	.2335	.1209	.0439
3	.0001	.0008	.0026	.0058	.0105	.0574	.1298	.2013	.2668	.2150	.1172
4	.0000	.0000	.0001	.0004	.0010	.0112	.0401	.0881	.2001	.2508	.2051
5	.0000	.0000	.0000	.0000	.0001	.0015	.0085	.0264	.1029	.2007	.2461
6	.0000	.0000	.0000	.0000	.0000	.0001	.0012	.0055	.0368	.1115	.2051
7	.0000	.0000	.0000	.0000	.0000	.0000	.0001	.0008	.0090	.0425	.1172
8	.0000	.0000	.0000	.0000	.0000	.0000	.0000	.0001	.0014	.0106	.0439
9	.0000	.0000	.0000	.0000	.0000	.0000	.0000	.0000	.0001	.0016	.0098
10	.0000	.0000	.0000	.0000	.0000	.0000	.0000	.0000	.0000	.0001	.0010

					$n = 11$						
0	.8953	.8007	.7153	.6382	.5688	.3138	.1673	.0859	.0198	.0036	.0005
1	.0995	.1798	.2433	.2925	.3293	.3835	.3248	.2362	.0932	.0266	.0054
2	.0050	.0183	.0376	.0609	.0867	.2131	.2866	.2953	.1998	.0887	.0269
3	.0002	.0011	.0035	.0076	.0137	.0710	.1517	.2215	.2568	.1774	.0806
4	.0000	.0000	.0002	.0006	.0014	.0158	.0536	.1107	.2201	.2365	.1611
5	.0000	.0000	.0000	.0000	.0001	.0025	.0132	.0388	.1321	.2207	.2256
6	.0000	.0000	.0000	.0000	.0000	.0003	.0023	.0097	.0566	.1471	.2256
7	.0000	.0000	.0000	.0000	.0000	.0000	.0003	.0017	.0173	.0701	.1611
8	.0000	.0000	.0000	.0000	.0000	.0000	.0000	.0002	.0037	.0234	.0806
9	.0000	.0000	.0000	.0000	.0000	.0000	.0000	.0000	.0005	.0052	.0269
10	.0000	.0000	.0000	.0000	.0000	.0000	.0000	.0000	.0000	.0007	.0054
11	.0000	.0000	.0000	.0000	.0000	.0000	.0000	.0000	.0000	.0000	.0005

					$n = 12$						
0	.8864	.7847	.6938	.6127	.5404	.2824	.1422	.0687	.0138	.0022	.0002
1	.1074	.1922	.2575	.3064	.3413	.3766	.3012	.2062	.0712	.0174	.0029
2	.0060	.0216	.0438	.0702	.0988	.2301	.2924	.2835	.1678	.0639	.0161
3	.0002	.0015	.0045	.0098	.0173	.0852	.1720	.2362	.2397	.1419	.0537
4	.0000	.0001	.0003	.0009	.0021	.0213	.0683	.1329	.2311	.2128	.1208

TABLE 3 (Continued)

X	.01	.02	.03	.04	.05	.10	.15	.20	.30	.40	.50

P

n = 12

X	.01	.02	.03	.04	.05	.10	.15	.20	.30	.40	.50
5	.0000	.0000	.0000	.0001	.0002	.0038	.0193	.0532	.1585	.2270	.1934
6	.0000	.0000	.0000	.0000	.0000	.0005	.0040	.0155	.0792	.1766	.2256
7	.0000	.0000	.0000	.0000	.0000	.0000	.0006	.0033	.0291	.1009	.1934
8	.0000	.0000	.0000	.0000	.0000	.0000	.0001	.0005	.0078	.0420	.1208
9	.0000	.0000	.0000	.0000	.0000	.0000	.0000	.0001	.0015	.0125	.0537
10	.0000	.0000	.0000	.0000	.0000	.0000	.0000	.0000	.0002	.0025	.0161
11	.0000	.0000	.0000	.0000	.0000	.0000	.0000	.0000	.0000	.0003	.0029
12	.0000	.0000	.0000	.0000	.0000	.0000	.0000	.0000	.0000	.0000	.0002

n = 13

X	.01	.02	.03	.04	.05	.10	.15	.20	.30	.40	.50
0	.8775	.7690	.6730	.5882	.5133	.2542	.1209	.0550	.0097	.0013	.0001
1	.1152	.2040	.2706	.3186	.3512	.3672	.2774	.1787	.0540	.0113	.0016
2	.0070	.0250	.0502	.0797	.1109	.2448	.2937	.2680	.1388	.0453	.0095
3	.0003	.0019	.0057	.0122	.0214	.0997	.1900	.2457	.2181	.1107	.0349
4	.0000	.0001	.0004	.0013	.0028	.0277	.0838	.1535	.2337	.1845	.0873
5	.0000	.0000	.0000	.0001	.0003	.0055	.0266	.0691	.1803	.2214	.1571
6	.0000	.0000	.0000	.0000	.0000	.0008	.0063	.0230	.1030	.1968	.2095
7	.0000	.0000	.0000	.0000	.0000	.0001	.0011	.0058	.0442	.1312	.2095
8	.0000	.0000	.0000	.0000	.0000	.0000	.0001	.0011	.0142	.0656	.1571
9	.0000	.0000	.0000	.0000	.0000	.0000	.0000	.0001	.0034	.0243	.0873
10	.0000	.0000	.0000	.0000	.0000	.0000	.0000	.0000	.0006	.0065	.0349
11	.0000	.0000	.0000	.0000	.0000	.0000	.0000	.0000	.0001	.0012	.0095
12	.0000	.0000	.0000	.0000	.0000	.0000	.0000	.0000	.0000	.0001	.0016
13	.0000	.0000	.0000	.0000	.0000	.0000	.0000	.0000	.0000	.0000	.0001

n = 14

X	.01	.02	.03	.04	.05	.10	.15	.20	.30	.40	.50
0	.8687	.7536	.6528	.5647	.4877	.2288	.1028	.0440	.0068	.0008	.0001
1	.1229	.2153	.2827	.3294	.3593	.3559	.2539	.1539	.0407	.0073	.0009
2	.0081	.0286	.0568	.0892	.1229	.2570	.2912	.2501	.1134	.0317	.0056
3	.0003	.0023	.0070	.0149	.0259	.1142	.2056	.2501	.1943	.0845	.0222
4	.0000	.0001	.0006	.0017	.0037	.0349	.0998	.1720	.2290	.1549	.0611
5	.0000	.0000	.0000	.0001	.0004	.0078	.0352	.0860	.1963	.2066	.1222
6	.0000	.0000	.0000	.0000	.0000	.0013	.0093	.0322	.1262	.2066	.1833
7	.0000	.0000	.0000	.0000	.0000	.0002	.0019	.0092	.0618	.1574	.2095
8	.0000	.0000	.0000	.0000	.0000	.0000	.0003	.0020	.0232	.0918	.1833
9	.0000	.0000	.0000	.0000	.0000	.0000	.0000	.0003	.0066	.0408	.1222
10	.0000	.0000	.0000	.0000	.0000	.0000	.0000	.0000	.0014	.0136	.0611
11	.0000	.0000	.0000	.0000	.0000	.0000	.0000	.0000	.0002	.0033	.0222
12	.0000	.0000	.0000	.0000	.0000	.0000	.0000	.0000	.0000	.0005	.0056
13	.0000	.0000	.0000	.0000	.0000	.0000	.0000	.0000	.0000	.0001	.0009
14	.0000	.0000	.0000	.0000	.0000	.0000	.0000	.0000	.0000	.0000	.0001

n = 15

X	.01	.02	.03	.04	.05	.10	.15	.20	.30	.40	.50
0	.8601	.7386	.6333	.5421	.4633	.2059	.0874	.0352	.0047	.0005	.0000
1	.1303	.2261	.2938	.3388	.3658	.3432	.2312	.1319	.0305	.0047	.0005
2	.0092	.0323	.0636	.0988	.1348	.2669	.2856	.2309	.0916	.0219	.0032
3	.0004	.0029	.0085	.0178	.0307	.1285	.2184	.2501	.1700	.0634	.0139
4	.0000	.0002	.0008	.0022	.0049	.0428	.1156	.1876	.2186	.1268	.0417
5	.0000	.0000	.0001	.0002	.0006	.0105	.0449	.1032	.2061	.1859	.0916
6	.0000	.0000	.0000	.0000	.0000	.0019	.0132	.0430	.1472	.2066	.1527
7	.0000	.0000	.0000	.0000	.0000	.0003	.0030	.0138	.0811	.1771	.1964
8	.0000	.0000	.0000	.0000	.0000	.0000	.0005	.0035	.0348	.1181	.1964
9	.0000	.0000	.0000	.0000	.0000	.0000	.0001	.0007	.0116	.0612	.1527

TABLE 3 (Continued)

X	.01	.02	.03	.04	.05	.10	.15	.20	.30	.40	.50
							P				

$n = 15$

X	.01	.02	.03	.04	.05	.10	.15	.20	.30	.40	.50
10	.0000	.0000	.0000	.0000	.0000	.0000	.0000	.0001	.0030	.0245	.0916
11	.0000	.0000	.0000	.0000	.0000	.0000	.0000	.0000	.0006	.0074	.0417
12	.0000	.0000	.0000	.0000	.0000	.0000	.0000	.0000	.0001	.0016	.0139
13	.0000	.0000	.0000	.0000	.0000	.0000	.0000	.0000	.0000	.0003	.0032
14	.0000	.0000	.0000	.0000	.0000	.0000	.0000	.0000	.0000	.0000	.0005

$n = 16$

X	.01	.02	.03	.04	.05	.10	.15	.20	.30	.40	.50
0	.8515	.7238	.6143	.5204	.4401	.1853	.0743	.0281	.0033	.0003	.0000
1	.1376	.2363	.3040	.3469	.3706	.3294	.2097	.1126	.0228	.0030	.0002
2	.0104	.0362	.0705	.1084	.1463	.2745	.2775	.2111	.0732	.0150	.0018
3	.0005	.0034	.0102	.0211	.0359	.1423	.2285	.2463	.1465	.0468	.0085
4	.0000	.0002	.0010	.0029	.0061	.0514	.1311	.2001	.2040	.1014	.0278
5	.0000	.0000	.0001	.0003	.0008	.0137	.0555	.1201	.2099	.1623	.0667
6	.0000	.0000	.0000	.0000	.0001	.0028	.0180	.0550	.1649	.1983	.1222
7	.0000	.0000	.0000	.0000	.0000	.0004	.0045	.0197	.1010	.1889	.1746
8	.0000	.0000	.0000	.0000	.0000	.0001	.0009	.0055	.0487	.1417	.1964
9	.0000	.0000	.0000	.0000	.0000	.0000	.0001	.0012	.0185	.0840	.1746
10	.0000	.0000	.0000	.0000	.0000	.0000	.0000	.0002	.0056	.0392	.1222
11	.0000	.0000	.0000	.0000	.0000	.0000	.0000	.0000	.0013	.0142	.0667
12	.0000	.0000	.0000	.0000	.0000	.0000	.0000	.0000	.0002	.0040	.0278
13	.0000	.0000	.0000	.0000	.0000	.0000	.0000	.0000	.0000	.0008	.0085
14	.0000	.0000	.0000	.0000	.0000	.0000	.0000	.0000	.0000	.0001	.0018
15	.0000	.0000	.0000	.0000	.0000	.0000	.0000	.0000	.0000	.0000	.0002

$n = 17$

X	.01	.02	.03	.04	.05	.10	.15	.20	.30	.40	.50
0	.8429	.7093	.5958	.4996	.4181	.1668	.0631	.0225	.0023	.0002	.0000
1	.1447	.2461	.3133	.3539	.3741	.3150	.1893	.0957	.0169	.0019	.0001
2	.0117	.0402	.0775	.1180	.1575	.2800	.2673	.1914	.0581	.0102	.0010
3	.0006	.0041	.0120	.0246	.0415	.1556	.2359	.2393	.1245	.0341	.0052
4	.0000	.0003	.0013	.0036	.0076	.0605	.1457	.2093	.1868	.0796	.0182
5	.0000	.0000	.0001	.0004	.0010	.0175	.0668	.1361	.2081	.1379	.0472
6	.0000	.0000	.0000	.0000	.0001	.0039	.0236	.0680	.1784	.1839	.0944
7	.0000	.0000	.0000	.0000	.0000	.0007	.0065	.0267	.1201	.1927	.1484
8	.0000	.0000	.0000	.0000	.0000	.0001	.0014	.0084	.0644	.1606	.1855
9	.0000	.0000	.0000	.0000	.0000	.0000	.0003	.0021	.0276	.1070	.1855
10	.0000	.0000	.0000	.0000	.0000	.0000	.0000	.0004	.0095	.0571	.1484
11	.0000	.0000	.0000	.0000	.0000	.0000	.0000	.0001	.0026	.0242	.0944
12	.0000	.0000	.0000	.0000	.0000	.0000	.0000	.0000	.0006	.0081	.0472
13	.0000	.0000	.0000	.0000	.0000	.0000	.0000	.0000	.0001	.0021	.0182
14	.0000	.0000	.0000	.0000	.0000	.0000	.0000	.0000	.0000	.0004	.0052
15	.0000	.0000	.0000	.0000	.0000	.0000	.0000	.0000	.0000	.0001	.0010
16	.0000	.0000	.0000	.0000	.0000	.0000	.0000	.0000	.0000	.0000	.0001

$n = 18$

X	.01	.02	.03	.04	.05	.10	.15	.20	.30	.40	.50
0	.8345	.6951	.5780	.4796	.3972	.1501	.0536	.0180	.0016	.0001	.0000
1	.1517	.2554	.3217	.3597	.3763	.3002	.1704	.0811	.0126	.0012	.0001
2	.0130	.0443	.0846	.1274	.1683	.2835	.2556	.1723	.0458	.0069	.0006
3	.0007	.0048	.0140	.0283	.0473	.1680	.2406	.2297	.1046	.0246	.0031
4	.0000	.0004	.0016	.0044	.0093	.0700	.1592	.2153	.1681	.0614	.0117
5	.0000	.0000	.0001	.0005	.0014	.0218	.0787	.1507	.2017	.1146	.0327
6	.0000	.0000	.0000	.0000	.0002	.0052	.0301	.0816	.1873	.1655	.0708
7	.0000	.0000	.0000	.0000	.0000	.0010	.0091	.0350	.1376	.1892	.1214
8	.0000	.0000	.0000	.0000	.0000	.0002	.0022	.0120	.0811	.1734	.1669
9	.0000	.0000	.0000	.0000	.0000	.0000	.0004	.0033	.0386	.1284	.1855

TABLE 3 (Continued)

X	.01	.02	.03	.04	.05	.10	.15	.20	.30	.40	.50
						P					

$n = 18$

X	.01	.02	.03	.04	.05	.10	.15	.20	.30	.40	.50
10	.0000	.0000	.0000	.0000	.0000	.0000	.0001	.0008	.0149	.0771	.1669
11	.0000	.0000	.0000	.0000	.0000	.0000	.0000	.0001	.0046	.0374	.1214
12	.0000	.0000	.0000	.0000	.0000	.0000	.0000	.0000	.0012	.0145	.0708
13	.0000	.0000	.0000	.0000	.0000	.0000	.0000	.0000	.0002	.0045	.0327
14	.0000	.0000	.0000	.0000	.0000	.0000	.0000	.0000	.0000	.0011	.0117
15	.0000	.0000	.0000	.0000	.0000	.0000	.0000	.0000	.0000	.0002	.0031
16	.0000	.0000	.0000	.0000	.0000	.0000	.0000	.0000	.0000	.0000	.0006
17	.0000	.0000	.0000	.0000	.0000	.0000	.0000	.0000	.0000	.0000	.0001

$n = 19$

X	.01	.02	.03	.04	.05	.10	.15	.20	.30	.40	.50
0	.8262	.6812	.5606	.4604	.3774	.1351	.0456	.0144	.0011	.0001	.0000
1	.1586	.2642	.3294	.3645	.3774	.2852	.1529	.0685	.0093	.0008	.0000
2	.0144	.0485	.0917	.1367	.1787	.2852	.2428	.1540	.0358	.0046	.0003
3	.0008	.0056	.0161	.0323	.0533	.1796	.2428	.2182	.0869	.0175	.0018
4	.0000	.0005	.0020	.0054	.0112	.0798	.1714	.2182	.1491	.0467	.0074
5	.0000	.0000	.0002	.0007	.0018	.0266	.0907	.1636	.1916	.0933	.0222
6	.0000	.0000	.0000	.0001	.0002	.0069	.0374	.0955	.1916	.1451	.0518
7	.0000	.0000	.0000	.0000	.0000	.0014	.0122	.0443	.1525	.1797	.0961
8	.0000	.0000	.0000	.0000	.0000	.0002	.0032	.0166	.0981	.1797	.1442
9	.0000	.0000	.0000	.0000	.0000	.0000	.0007	.0051	.0514	.1464	.1762
10	.0000	.0000	.0000	.0000	.0000	.0000	.0001	.0013	.0220	.0976	.1762
11	.0000	.0000	.0000	.0000	.0000	.0000	.0000	.0003	.0077	.0532	.1442
12	.0000	.0000	.0000	.0000	.0000	.0000	.0000	.0000	.0022	.0237	.0961
13	.0000	.0000	.0000	.0000	.0000	.0000	.0000	.0000	.0005	.0085	.0518
14	.0000	.0000	.0000	.0000	.0000	.0000	.0000	.0000	.0001	.0024	.0222
15	.0000	.0000	.0000	.0000	.0000	.0000	.0000	.0000	.0000	.0005	.0074
16	.0000	.0000	.0000	.0000	.0000	.0000	.0000	.0000	.0000	.0001	.0018
17	.0000	.0000	.0000	.0000	.0000	.0000	.0000	.0000	.0000	.0000	.0003

$n = 20$

X	.01	.02	.03	.04	.05	.10	.15	.20	.30	.40	.50
0	.8179	.6676	.5438	.4420	.3585	.1216	.0388	.0115	.0008	.0000	.0000
1	.1652	.2725	.3364	.3683	.3774	.2702	.1368	.0576	.0068	.0005	.0000
2	.0159	.0528	.0988	.1458	.1887	.2852	.2293	.1369	.0278	.0031	.0002
3	.0010	.0065	.0183	.0364	.0596	.1901	.2428	.2054	.0716	.0123	.0011
4	.0000	.0006	.0024	.0065	.0133	.0898	.1821	.2182	.1304	.0350	.0046
5	.0000	.0000	.0002	.0009	.0022	.0319	.1028	.1746	.1789	.0746	.0148
6	.0000	.0000	.0000	.0001	.0003	.0089	.0454	.1091	.1916	.1244	.0370
7	.0000	.0000	.0000	.0000	.0000	.0020	.0160	.0545	.1643	.1659	.0739
8	.0000	.0000	.0000	.0000	.0000	.0004	.0046	.0222	.1144	.1797	.1201
9	.0000	.0000	.0000	.0000	.0000	.0001	.0011	.0074	.0654	.1597	.1602
10	.0000	.0000	.0000	.0000	.0000	.0000	.0002	.0020	.0308	.1171	.1762
11	.0000	.0000	.0000	.0000	.0000	.0000	.0000	.0005	.0120	.0710	.1602
12	.0000	.0000	.0000	.0000	.0000	.0000	.0000	.0001	.0039	.0355	.1201
13	.0000	.0000	.0000	.0000	.0000	.0000	.0000	.0000	.0010	.0146	.0739
14	.0000	.0000	.0000	.0000	.0000	.0000	.0000	.0000	.0002	.0049	.0370
15	.0000	.0000	.0000	.0000	.0000	.0000	.0000	.0000	.0000	.0013	.0148
16	.0000	.0000	.0000	.0000	.0000	.0000	.0000	.0000	.0000	.0003	.0046
17	.0000	.0000	.0000	.0000	.0000	.0000	.0000	.0000	.0000	.0000	.0011
18	.0000	.0000	.0000	.0000	.0000	.0000	.0000	.0000	.0000	.0000	.0002

TABLE 3 (Continued)

X	P										
	.01	.02	.03	.04	.05	.10	.15	.20	.30	.40	.50

					$n = 25$						
0	.7778	.6035	.4670	.3604	.2774	.0718	.0172	.0038	.0001	.0000	.0000
1	.1964	.3079	.3611	.3754	.3650	.1994	.0759	.0236	.0014	.0000	.0000
2	.0238	.0754	.1340	.1877	.2305	.2659	.1607	.0708	.0074	.0004	.0000
3	.0018	.0118	.0318	.0600	.0930	.2265	.2174	.1358	.0243	.0019	.0001
4	.0001	.0013	.0054	.0137	.0269	.1384	.2110	.1867	.0572	.0071	.0004
5	.0000	.0001	.0007	.0024	.0060	.0646	.1564	.1960	.1030	.0199	.0016
6	.0000	.0000	.0001	.0003	.0010	.0239	.0920	.1633	.1472	.0442	.0053
7	.0000	.0000	.0000	.0000	.0001	.0072	.0441	.1108	.1712	.0800	.0143
8	.0000	.0000	.0000	.0000	.0000	.0018	.0175	.0623	.1651	.1200	.0322
9	.0000	.0000	.0000	.0000	.0000	.0004	.0058	.0294	.1336	.1511	.0609
10	.0000	.0000	.0000	.0000	.0000	.0001	.0016	.0118	.0916	.1612	.0974
11	.0000	.0000	.0000	.0000	.0000	.0000	.0004	.0040	.0536	.1465	.1328
12	.0000	.0000	.0000	.0000	.0000	.0000	.0001	.0012	.0268	.1140	.1550
13	.0000	.0000	.0000	.0000	.0000	.0000	.0000	.0003	.0115	.0760	.1550
14	.0000	.0000	.0000	.0000	.0000	.0000	.0000	.0001	.0042	.0434	.1328
15	.0000	.0000	.0000	.0000	.0000	.0000	.0000	.0000	.0013	.0212	.0974
16	.0000	.0000	.0000	.0000	.0000	.0000	.0000	.0000	.0004	.0088	.0609
17	.0000	.0000	.0000	.0000	.0000	.0000	.0000	.0000	.0001	.0031	.0322
18	.0000	.0000	.0000	.0000	.0000	.0000	.0000	.0000	.0000	.0009	.0143
19	.0000	.0000	.0000	.0000	.0000	.0000	.0000	.0000	.0000	.0002	.0053
20	.0000	.0000	.0000	.0000	.0000	.0000	.0000	.0000	.0000	.0000	.0016
21	.0000	.0000	.0000	.0000	.0000	.0000	.0000	.0000	.0000	.0000	.0004
22	.0000	.0000	.0000	.0000	.0000	.0000	.0000	.0000	.0000	.0000	.0001

					$n = 30$						
0	.7397	.5455	.4010	.2939	.2146	.0424	.0076	.0012	.0000	.0000	.0000
1	.2242	.3340	.3721	.3673	.3389	.1413	.0404	.0093	.0003	.0000	.0000
2	.0328	.0988	.1669	.2219	.2586	.2277	.1034	.0337	.0018	.0000	.0000
3	.0031	.0188	.0482	.0863	.1270	.2361	.1703	.0785	.0072	.0003	.0000
4	.0002	.0026	.0101	.0243	.0451	.1771	.2028	.1325	.0208	.0012	.0000
5	.0000	.0003	.0016	.0053	.0124	.1023	.1861	.1723	.0464	.0041	.0001
6	.0000	.0000	.0002	.0009	.0027	.0474	.1368	.1795	.0829	.0115	.0006
7	.0000	.0000	.0000	.0001	.0005	.0180	.0828	.1538	.1219	.0263	.0019
8	.0000	.0000	.0000	.0000	.0001	.0058	.0420	.1106	.1501	.0505	.0055
9	.0000	.0000	.0000	.0000	.0000	.0016	.0181	.0676	.1573	.0823	.0133
10	.0000	.0000	.0000	.0000	.0000	.0004	.0067	.0355	.1416	.1152	.0280
11	.0000	.0000	.0000	.0000	.0000	.0001	.0022	.0161	.1103	.1396	.0509
12	.0000	.0000	.0000	.0000	.0000	.0000	.0006	.0064	.0749	.1474	.0806
13	.0000	.0000	.0000	.0000	.0000	.0000	.0001	.0022	.0444	.1360	.1115
14	.0000	.0000	.0000	.0000	.0000	.0000	·0000	.0007	.0231	.1101	.1354
15	.0000	.0000	.0000	.0000	.0000	.0000	.0000	.0002	.0106	.0783	.1445
16	.0000	.0000	.0000	.0000	.0000	.0000	.0000	.0000	.0042	.0489	.1354
17	.0000	.0000	.0000	.0000	.0000	.0000	.0000	.0000	.0015	.0269	.1115
18	.0000	.0000	.0000	.0000	.0000	.0000	.0000	.0000	.0005	.0129	.0806
19	.0000	.0000	.0000	.0000	.0000	.0000	.0000	.0000	.0001	.0054	.0509
20	.0000	.0000	.0000	.0000	.0000	.0000	.0000	.0000	.0000	.0020	.0280
21	.0000	.0000	.0000	.0000	.0000	.0000	.0000	.0000	.0000	.0006	.0133
22	.0000	.0000	.0000	.0000	.0000	.0000	.0000	.0000	.0000	.0002	.0055
23	.0000	.0000	.0000	.0000	.0000	.0000	.0000	.0000	.0000	.0000	.0019
24	.0000	.0000	.0000	.0000	.0000	.0000	.0000	.0000	.0000	.0000	.0006
25	.0000	.0000	.0000	.0000	.0000	.0000	.0000	.0000	.0000	.0000	.0001

TABLE 4

POISSON PROBABILITY DISTRIBUTION—VALUES OF $P(X) = \dfrac{\mu^X \cdot e^{-\mu}}{X!}$

Probability $P(X)$

.60

.40

.3293

.20

0

0   1   2   3   4   5 X

Example:
$\mu = .60,\ X = 1$
$P(1) = .3293$

X	.005	.01	.02	.03	.04	.05	.06	.07	.08	.09
						μ				
0	.9950	.9900	.9802	.9704	.9608	.9512	.9418	.9324	.9231	.9139
1	.0050	.0099	.0192	.0291	.0384	.0476	.0565	.0653	.0738	.0823
2	.0000	.0000	.0002	.0004	.0008	.0012	.0017	.0023	.0030	.0037
3	.0000	.0000	.0000	.0000	.0000	.0000	.0000	.0001	.0001	.0001

X	0.10	0.20	0.30	0.40	0.50	0.60	0.70	0.80	0.90	1.00
						μ				
0	.9048	.8187	.7408	.6703	.6065	.5488	.4966	.4493	.4066	.3679
1	.0905	.1637	.2222	.2681	.3033	.3293	.3476	.3595	.3659	.3679
2	.0045	.0164	.0333	.0536	.0758	.0988	.1217	.1438	.1647	.1839
3	.0002	.0011	.0033	.0072	.0126	.0198	.0284	.0383	.0494	.0613
4	.0000	.0001	.0002	.0007	.0016	.0030	.0050	.0077	.0111	.0153
5	.0000	.0000	.0000	.0001	.0002	.0004	.0007	.0012	.0020	.0031
6	.0000	.0000	.0000	.0000	.0000	.0000	.0001	.0002	.0003	.0005
7	.0000	.0000	.0000	.0000	.0000	.0000	.0000	.0000	.0000	.0001

X	1.10	1.20	1.30	1.40	1.50	1.60	1.70	1.80	1.90	2.00
						μ				
0	.3329	.3012	.2725	.2466	.2231	.2019	.1827	.1653	.1496	.1353
1	.3662	.3614	.3543	.3452	.3347	.3230	.3106	.2975	.2842	.2707
2	.2014	.2169	.2303	.2417	.2510	.2584	.2640	.2678	.2700	.2707
3	.0738	.0867	.0998	.1128	.1255	.1378	.1496	.1607	.1710	.1804
4	.0203	.0260	.0324	.0395	.0471	.0551	.0636	.0723	.0812	.0902
5	.0045	.0062	.0084	.0111	.0141	.0176	.0216	.0260	.0309	.0361
6	.0008	.0012	.0018	.0026	.0035	.0047	.0061	.0078	.0098	.0120
7	.0001	.0002	.0003	.0005	.0008	.0011	.0015	.0020	.0027	.0034
8	.0000	.0000	.0001	.0001	.0001	.0002	.0003	.0005	.0006	.0009
9	.0000	.0000	.0000	.0000	.0000	.0000	.0001	.0001	.0001	.0002

X	2.10	2.20	2.30	2.40	2.50	2.60	2.70	2.80	2.90	3.00
						μ				
0	.1225	.1108	.1003	.0907	.0821	.0743	.0672	.0608	.0550	.0498
1	.2572	.2438	.2306	.2177	.2052	.1931	.1815	.1703	.1596	.1494
2	.2700	.2681	.2652	.2613	.2565	.2510	.2450	.2384	.2314	.2240
3	.1890	.1966	.2033	.2090	.2138	.2176	.2205	.2225	.2237	.2240
4	.0992	.1082	.1169	.1254	.1336	.1414	.1488	.1557	.1622	.1680
5	.0417	.0476	.0538	.0602	.0668	.0735	.0804	.0872	.0940	.1008
6	.0146	.0174	.0206	.0241	.0278	.0319	.0362	.0407	.0455	.0504
7	.0044	.0055	.0068	.0083	.0099	.0118	.0139	.0163	.0188	.0216
8	.0011	.0015	.0019	.0025	.0031	.0038	.0047	.0057	.0068	.0081
9	.0003	.0004	.0005	.0007	.0009	.0011	.0014	.0018	.0022	.0027
10	.0001	.0001	.0001	.0002	.0002	.0003	.0004	.0005	.0006	.0008
11	.0000	.0000	.0000	.0000	.0000	.0001	.0001	.0001	.0002	.0002
12	.0000	.0000	.0000	.0000	.0000	.0000	.0000	.0000	.0000	.0001

TABLE 4 (Continued)

X	\multicolumn{10}{c}{μ}
	3.10	3.20	3.30	3.40	3.50	3.60	3.70	3.80	3.90	4.00
0	.0450	.0408	.0369	.0334	.0302	.0273	.0247	.0224	.0202	.0183
1	.1397	.1304	.1217	.1135	.1057	.0984	.0915	.0850	.0789	.0733
2	.2165	.2087	.2008	.1929	.1850	.1771	.1692	.1615	.1539	.1465
3	.2237	.2226	.2209	.2186	.2158	.2125	.2087	.2046	.2001	.1954
4	.1734	.1781	.1823	.1858	.1888	.1912	.1931	.1944	.1951	.1954
5	.1075	.1140	.1203	.1264	.1322	.1377	.1429	.1477	.1522	.1563
6	.0555	.0608	.0662	.0716	.0771	.0826	.0881	.0936	.0989	.1042
7	.0246	.0278	.0312	.0348	.0385	.0425	.0466	.0508	.0551	.0595
8	.0095	.0111	.0129	.0148	.0169	.0191	.0215	.0241	.0269	.0298
9	.0033	.0040	.0047	.0056	.0066	.0076	.0089	.0102	.0116	.0132
10	.0010	.0013	.0016	.0019	.0023	.0028	.0033	.0039	.0045	.0053
11	.0003	.0004	.0005	.0006	.0007	.0009	.0011	.0013	.0016	.0019
12	.0001	.0001	.0001	.0002	.0002	.0003	.0003	.0004	.0005	.0006
13	.0000	.0000	.0000	.0000	.0001	.0001	.0001	.0001	.0002	.0002
14	.0000	.0000	.0000	.0000	.0000	.0000	.0000	.0000	.0000	.0001

X	\multicolumn{10}{c}{μ}
	4.10	4.20	4.30	4.40	4.50	4.60	4.70	4.80	4.90	5.00
0	.0166	.0150	.0136	.0123	.0111	.0101	.0091	.0082	.0074	.0067
1	.0679	.0630	.0583	.0540	.0500	.0462	.0427	.0395	.0365	.0337
2	.1393	.1323	.1254	.1188	.1125	.1063	.1005	.0948	.0894	.0842
3	.1904	.1852	.1798	.1743	.1687	.1631	.1574	.1517	.1460	.1404
4	.1951	.1944	.1933	.1917	.1898	.1875	.1849	.1820	.1789	.1755
5	.1600	.1633	.1662	.1687	.1708	.1725	.1738	.1747	.1753	.1755
6	.1093	.1143	.1191	.1237	.1281	.1323	.1362	.1398	.1432	.1462
7	.0640	.0686	.0732	.0778	.0824	.0869	.0914	.0959	.1002	.1044
8	.0328	.0360	.0393	.0428	.0463	.0500	.0537	.0575	.0614	.0653
9	.0150	.0168	.0188	.0209	.0232	.0255	.0280	.0307	.0334	.0363
10	.0061	.0071	.0081	.0092	.0104	.0118	.0132	.0147	.0164	.0181
11	.0023	.0027	.0032	.0037	.0043	.0049	.0056	.0064	.0073	.0082
12	.0008	.0009	.0011	.0014	.0016	.0019	.0022	.0026	.0030	.0034
13	.0002	.0003	.0004	.0005	.0006	.0007	.0008	.0009	.0011	.0013
14	.0001	.0001	.0001	.0001	.0002	.0002	.0003	.0003	.0004	.0005
15	.0000	.0000	.0000	.0000	.0001	.0001	.0001	.0001	.0001	.0002

X	\multicolumn{10}{c}{μ}
	5.10	5.20	5.30	5.40	5.50	5.60	5.70	5.80	5.90	6.00
0	.0061	.0055	.0050	.0045	.0041	.0037	.0033	.0030	.0027	.0025
1	.0311	.0287	.0265	.0244	.0225	.0207	.0191	.0176	.0162	.0149
2	.0793	.0746	.0701	.0659	.0618	.0580	.0544	.0509	.0477	.0446
3	.1348	.1293	.1239	.1185	.1133	.1082	.1033	.0985	.0938	.0892
4	.1719	.1681	.1641	.1600	.1558	.1515	.1472	.1428	.1383	.1339
5	.1753	.1748	.1740	.1728	.1714	.1697	.1678	.1656	.1632	.1606
6	.1490	.1515	.1537	.1555	.1571	.1584	.1594	.1601	.1605	.1606
7	.1086	.1125	.1163	.1200	.1234	.1267	.1298	.1326	.1353	.1377
8	.0692	.0731	.0771	.0810	.0849	.0887	.0925	.0962	.0998	.1033
9	.0392	.0423	.0454	.0486	.0519	.0552	.0586	.0620	.0654	.0688
10	.0200	.0220	.0241	.0262	.0285	.0309	.0334	.0359	.0386	.0413
11	.0093	.0104	.0116	.0129	.0143	.0157	.0173	.0190	.0207	.0225
12	.0039	.0045	.0051	.0058	.0065	.0073	.0082	.0092	.0102	.0113
13	.0015	.0018	.0021	.0024	.0028	.0032	.0036	.0041	.0046	.0052
14	.0006	.0007	.0008	.0009	.0011	.0013	.0015	.0017	.0019	.0022
15	.0002	.0002	.0003	.0003	.0004	.0005	.0006	.0007	.0008	.0009
16	.0001	.0001	.0001	.0001	.0001	.0002	.0002	.0002	.0003	.0003
17	.0000	.0000	.0000	.0000	.0000	.0001	.0001	.0001	.0001	.0001

TABLE 4 (Continued)

					μ					
X	6.10	6.20	6.30	6.40	6.50	6.60	6.70	6.80	6.90	7.00
0	.0022	.0020	.0018	.0017	.0015	.0014	.0012	.0011	.0010	.0009
1	.0137	.0126	.0116	.0106	.0098	.0090	.0082	.0076	.0070	.0064
2	.0417	.0390	.0364	.0340	.0318	.0296	.0276	.0258	.0240	.0223
3	.0848	.0806	.0765	.0726	.0688	.0652	.0617	.0584	.0552	.0521
4	.1294	.1249	.1205	.1162	.1118	.1076	.1034	.0992	.0952	.0912
5	.1579	.1549	.1519	.1487	.1454	.1420	.1385	.1349	.1314	.1277
6	.1605	.1601	.1595	.1586	.1575	.1562	.1546	.1529	.1511	.1490
7	.1399	.1418	.1435	.1450	.1462	.1472	.1480	.1486	.1489	.1490
8	.1066	.1099	.1130	.1160	.1188	.1215	.1240	.1263	.1284	.1304
9	.0723	.0757	.0791	.0825	.0858	.0891	.0923	.0954	.0985	.1014
10	.0441	.0469	.0498	.0528	.0558	.0588	.0618	.0649	.0679	.0710
11	.0245	.0265	.0285	.0307	.0330	.0353	.0377	.0401	.0426	.0452
12	.0124	.0137	.0150	.0164	.0179	.0194	.0210	.0227	.0245	.0264
13	.0058	.0065	.0073	.0081	.0089	.0098	.0108	.0119	.0130	.0142
14	.0025	.0029	.0033	.0037	.0041	.0046	.0052	.0058	.0064	.0071
15	.0010	.0012	.0014	.0016	.0018	.0020	.0023	.0026	.0029	.0033
16	.0004	.0005	.0005	.0006	.0007	.0008	.0010	.0011	.0013	.0014
17	.0001	.0002	.0002	.0002	.0003	.0003	.0004	.0004	.0005	.0006
18	.0000	.0001	.0001	.0001	.0001	.0001	.0001	.0002	.0002	.0002
19	.0000	.0000	.0000	.0000	.0000	.0000	.0000	.0001	.0001	.0001

					μ					
	7.10	7.20	7.30	7.40	7.50	7.60	7.70	7.80	7.90	8.00
0	.0008	.0007	.0007	.0006	.0006	.0005	.0005	.0004	.0004	.0003
1	.0059	.0054	.0049	.0045	.0041	.0038	.0035	.0032	.0029	.0027
2	.0208	.0194	.0180	.0167	.0156	.0145	.0134	.0125	.0116	.0107
3	.0492	.0464	.0438	.0413	.0389	.0366	.0345	.0324	.0305	.0286
4	.0874	.0836	.0799	.0764	.0729	.0696	.0663	.0632	.0602	.0573
5	.1241	.1204	.1167	.1130	.1094	.1057	.1021	.0986	.0951	.0916
6	.1468	.1445	.1420	.1394	.1367	.1339	.1311	.1282	.1252	.1221
7	.1489	.1486	.1481	.1474	.1465	.1454	.1442	.1428	.1413	.1396
8	.1321	.1337	.1351	.1363	.1373	.1382	.1388	.1392	.1395	.1396
9	.1042	.1070	.1096	.1121	.1144	.1167	.1187	.1207	.1224	.1241
10	.0740	.0770	.0800	.0829	.0858	.0887	.0914	.0941	.0967	.0993
11	.0478	.0504	.0531	.0558	.0585	.0613	.0640	.0667	.0695	.0722
12	.0283	.0303	.0323	.0344	.0366	.0388	.0411	.0434	.0457	.0481
13	.0154	.0168	.0181	.0196	.0211	.0227	.0243	.0260	.0278	.0296
14	.0078	.0086	.0095	.0104	.0113	.0123	.0134	.0145	.0157	.0169
15	.0037	.0041	.0046	.0051	.0057	.0062	.0069	.0075	.0083	.0090
16	.0016	.0019	.0021	.0024	.0026	.0030	.0033	.0037	.0041	.0045
17	.0007	.0008	.0009	.0010	.0012	.0013	.0015	.0017	.0019	.0021
18	.0003	.0003	.0004	.0004	.0005	.0006	.0006	.0007	.0008	.0009
19	.0001	.0001	.0001	.0002	.0002	.0002	.0003	.0003	.0003	.0004
20	.0000	.0000	.0001	.0001	.0001	.0001	.0001	.0001	.0001	.0002
21	.0000	.0000	.0000	.0000	.0000	.0000	.0000	.0000	.0001	.0001

					μ					
	8.10	8.20	8.30	8.40	8.50	8.60	8.70	8.80	8.90	9.00
0	.0003	.0003	.0002	.0002	.0002	.0002	.0002	.0002	.0001	.0001
1	.0025	.0023	.0021	.0019	.0017	.0016	.0014	.0013	.0012	.0011
2	.0100	.0092	.0086	.0079	.0074	.0068	.0063	.0058	.0054	.0050
3	.0269	.0252	.0237	.0222	.0208	.0195	.0183	.0171	.0160	.0150
4	.0544	.0517	.0491	.0466	.0443	.0420	.0398	.0377	.0357	.0337

TABLE 4 (Continued)

					μ				
X	8.10	8.20	8.30	8.40	8.50	8.60	8.70	8.80	8.90	9.00
5	.0882	.0849	.0816	.0784	.0752	.0722	.0692	.0663	.0635	.0607
6	.1191	.1160	.1128	.1097	.1066	.1034	.1003	.0972	.0941	.0911
7	.1378	.1358	.1338	.1317	.1294	.1271	.1247	.1222	.1197	.1171
8	.1395	.1392	.1388	.1382	.1375	.1366	.1356	.1344	.1332	.1318
9	.1256	.1269	.1280	.1290	.1299	.1306	.1311	.1315	.1317	.1318
10	.1017	.1040	.1063	.1084	.1104	.1123	.1140	.1157	.1172	.1186
11	.0749	.0776	.0802	.0828	.0853	.0878	.0902	.0925	.0948	.0970
12	.0505	.0530	.0555	.0579	.0604	.0629	.0654	.0679	.0703	.0728
13	.0315	.0334	.0354	.0374	.0395	.0416	.0438	.0459	.0481	.0504
14	.0182	.0196	.0210	.0225	.0240	.0256	.0272	.0289	.0306	.0324
15	.0098	.0107	.0116	.0126	.0136	.0147	.0158	.0169	.0182	.0194
16	.0050	.0055	.0060	.0066	.0072	.0079	.0086	.0093	.0101	.0109
17	.0024	.0026	.0029	.0033	.0036	.0040	.0044	.0048	.0053	.0058
18	.0011	.0012	.0014	.0015	.0017	.0019	.0021	.0024	.0026	.0029
19	.0005	.0005	.0006	.0007	.0008	.0009	.0010	.0011	.0012	.0014
20	.0002	.0002	.0002	.0003	.0003	.0004	.0004	.0005	.0005	.0006
21	.0001	.0001	.0001	.0001	.0001	.0002	.0002	.0002	.0002	.0003
22	.0000	.0000	.0000	.0000	.0001	.0001	.0001	.0001	.0001	.0001

					μ				
	9.10	9.20	9.30	9.40	9.50	9.60	9.70	9.80	9.90	10.00
0	.0001	.0001	.0001	.0001	.0001	.0001	.0001	.0001	.0001	.0000
1	.0010	.0009	.0009	.0008	.0007	.0007	.0006	.0005	.0005	.0005
2	.0046	.0043	.0040	.0037	.0034	.0031	.0029	.0027	.0025	.0023
3	.0140	.0131	.0123	.0115	.0107	.0100	.0093	.0087	.0081	.0076
4	.0319	.0302	.0285	.0269	.0254	.0240	.0226	.0213	.0201	.0189
5	.0581	.0555	.0530	.0506	.0483	.0460	.0439	.0418	.0398	.0378
6	.0881	.0851	.0822	.0793	.0764	.0736	.0709	.0682	.0656	.0631
7	.1145	.1118	.1091	.1064	.1037	.1010	.0982	.0955	.0928	.0901
8	.1302	.1286	.1269	.1251	.1232	.1212	.1191	.1170	.1148	.1126
9	.1317	.1315	.1311	.1306	.1300	.1293	.1284	.1274	.1263	.1251
10	.1198	.1210	.1219	.1228	.1235	.1241	.1245	.1249	.1250	.1251
11	.0991	.1012	.1031	.1049	.1067	.1083	.1098	.1112	.1125	.1137
12	.0752	.0776	.0799	.0822	.0844	.0866	.0888	.0908	.0928	.0948
13	.0526	.0549	.0572	.0594	.0617	.0640	.0662	.0685	.0707	.0729
14	.0342	.0361	.0380	.0399	.0419	.0439	.0459	.0479	.0500	.0521
15	.0208	.0221	.0235	.0250	.0265	.0281	.0297	.0313	.0330	.0347
16	.0118	.0127	.0137	.0147	.0157	.0168	.0180	.0192	.0204	.0217
17	.0063	.0069	.0075	.0081	.0088	.0095	.0103	.0111	.0119	.0128
18	.0032	.0035	.0039	.0042	.0046	.0051	.0055	.0060	.0065	.0071
19	.0015	.0017	.0019	.0021	.0023	.0026	.0028	.0031	.0034	.0037
20	.0007	.0008	.0009	.0010	.0011	.0012	.0014	.0015	.0017	.0019
21	.0003	.0003	.0004	.0004	.0005	.0006	.0006	.0007	.0008	.0009
22	.0001	.0001	.0002	.0002	.0002	.0002	.0003	.0003	.0004	.0004
23	.0000	.0001	.0001	.0001	.0001	.0001	.0001	.0001	.0002	.0002
24	.0000	.0000	.0000	.0000	.0000	.0000	.0000	.0001	.0001	.0001

TABLE 4 (Continued)

X	11.	12.	13.	14.	15.	16.	17.	18.	19.	20.
					μ					
0	.0000	.0000	.0000	.0000	.0000	.0000	.0000	.0000	.0000	.0000
1	.0002	.0001	.0000	.0000	.0000	.0000	.0000	.0000	.0000	.0000
2	.0010	.0004	.0002	.0001	.0000	.0000	.0000	.0000	.0000	.0000
3	.0037	.0018	.0008	.0004	.0002	.0001	.0000	.0000	.0000	.0000
4	.0102	.0053	.0027	.0013	.0006	.0003	.0001	.0001	.0000	.0000
5	.0224	.0127	.0070	.0037	.0019	.0010	.0005	.0002	.0001	.0001
6	.0411	.0255	.0152	.0087	.0048	.0026	.0014	.0007	.0004	.0002
7	.0646	.0437	.0281	.0174	.0104	.0060	.0034	.0019	.0010	.0005
8	.0888	.0655	.0457	.0304	.0194	.0120	.0072	.0042	.0024	.0013
9	.1085	.0874	.0661	.0473	.0324	.0213	.0135	.0083	.0050	.0029
10	.1194	.1048	.0859	.0663	.0486	.0341	.0230	.0150	.0095	.0058
11	.1194	.1144	.1015	.0844	.0663	.0496	.0355	.0245	.0164	.0106
12	.1094	.1144	.1099	.0984	.0829	.0661	.0504	.0368	.0259	.0176
13	.0926	.1056	.1099	.1060	.0956	.0814	.0658	.0509	.0378	.0271
14	.0728	.0905	.1021	.1060	.1024	.0930	.0800	.0655	.0514	.0387
15	.0534	.0724	.0885	.0989	.1024	.0992	.0906	.0786	.0650	.0516
16	.0367	.0543	.0719	.0866	.0960	.0992	.0963	.0884	.0772	.0646
17	.0237	.0383	.0550	.0713	.0847	.0934	.0963	.0936	.0863	.0760
18	.0145	.0256	.0397	.0554	.0706	.0830	.0909	.0936	.0911	.0844
19	.0084	.0161	.0272	.0409	.0557	.0699	.0814	.0887	.0911	.0888
20	.0046	.0097	.0177	.0286	.0418	.0559	.0692	.0798	.0866	.0888
21	.0024	.0055	.0109	.0191	.0299	.0426	.0560	.0684	.0783	.0846
22	.0012	.0030	.0065	.0121	.0204	.0310	.0433	.0560	.0676	.0769
23	.0006	.0016	.0037	.0074	.0133	.0216	.0320	.0438	.0559	.0669
24	.0003	.0008	.0020	.0043	.0083	.0144	.0226	.0329	.0442	.0557
25	.0001	.0004	.0010	.0024	.0050	.0092	.0154	.0237	.0336	.0446
26	.0000	.0002	·0005	.0013	.0029	.0057	.0101	.0164	.0246	.0343
27	.0000	.0001	.0002	.0007	.0016	.0034	.0063	.0109	.0173	.0254
28	.0000	.0000	.0001	.0003	.0009	.0019	.0038	.0070	.0117	.0181
29	.0000	.0000	.0001	.0002	.0004	.0011	.0023	.0044	.0077	.0125
30	.0000	.0000	.0000	.0001	.0002	.0006	.0013	.0026	.0049	.0083
31	.0000	.0000	.0000	.0000	.0001	.0003	.0007	.0015	.0030	.0054
32	.0000	.0000	.0000	.0000	.0001	.0001	.0004	.0009	.0018	.0034
33	.0000	.0000	.0000	.0000	.0000	.0001	.0002	.0005	.0010	.0020
34	.0000	.0000	.0000	.0000	.0000	.0000	.0001	.0002	.0006	.0012
35	.0000	.0000	.0000	.0000	.0000	.0000	.0000	.0001	.0003	.0007
36	.0000	.0000	.0000	.0000	.0000	.0000	.0000	.0001	.0002	.0004
37	.0000	.0000	.0000	.0000	.0000	.0000	.0000	.0000	.0001	.0002
38	.0000	.0000	.0000	.0000	.0000	.0000	.0000	.0000	.0000	.0001
39	.0000	.0000	.0000	.0000	.0000	.0000	.0000	.0000	.0000	.0001

786

TABLE 5
O?DINATES OF THE NORMAL CURVE-VALUES OF $f(z)$

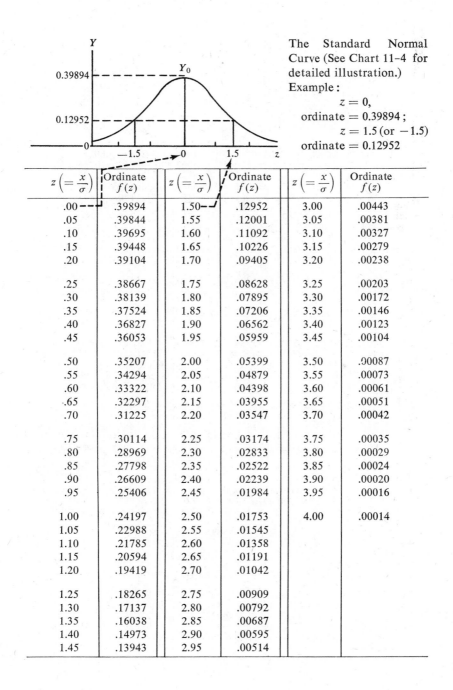

The Standard Normal Curve (See Chart 11–4 for detailed illustration.)
Example:
$$z = 0,$$
ordinate $= 0.39894$;
$$z = 1.5 \,(\text{or} -1.5)$$
ordinate $= 0.12952$

$z\left(=\dfrac{x}{\sigma}\right)$	Ordinate $f(z)$	$z\left(=\dfrac{x}{\sigma}\right)$	Ordinate $f(z)$	$z\left(=\dfrac{x}{\sigma}\right)$	Ordinate $f(z)$
.00	.39894	1.50	.12952	3.00	.00443
.05	.39844	1.55	.12001	3.05	.00381
.10	.39695	1.60	.11092	3.10	.00327
.15	.39448	1.65	.10226	3.15	.00279
.20	.39104	1.70	.09405	3.20	.00238
.25	.38667	1.75	.08628	3.25	.00203
.30	.38139	1.80	.07895	3.30	.00172
.35	.37524	1.85	.07206	3.35	.00146
.40	.36827	1.90	.06562	3.40	.00123
.45	.36053	1.95	.05959	3.45	.00104
.50	.35207	2.00	.05399	3.50	.90087
.55	.34294	2.05	.04879	3.55	.00073
.60	.33322	2.10	.04398	3.60	.00061
.65	.32297	2.15	.03955	3.65	.00051
.70	.31225	2.20	.03547	3.70	.00042
.75	.30114	2.25	.03174	3.75	.00035
.80	.28969	2.30	.02833	3.80	.00029
.85	.27798	2.35	.02522	3.85	.00024
.90	.26609	2.40	.02239	3.90	.00020
.95	.25406	2.45	.01984	3.95	.00016
1.00	.24197	2.50	.01753	4.00	.00014
1.05	.22988	2.55	.01545		
1.10	.21785	2.60	.01358		
1.15	.20594	2.65	.01191		
1.20	.19419	2.70	.01042		
1.25	.18265	2.75	.00909		
1.30	.17137	2.80	.00792		
1.35	.16038	2.85	.00687		
1.40	.14973	2.90	.00595		
1.45	.13943	2.95	.00514		

TABLE 6

AREAS UNDER THE NORMAL CURVE

VALUES OF $A(z)$ BETWEEN ORDINATE AT MEAN (Y_o) AND ORDINATE AT z

Example:
$z = 0.52$ (or -0.52),
$A(z) = 0.19847$ or 19.847%

$z\left(=\dfrac{x}{\sigma}\right)$	.00	.01	.02	.03	.04	.05	.06	.07	.08	.09
0.0	.00000	.00399	.00798	.01197	.01595	.01994	.02392	.02790	.03188	.03586
0.1	.03983	.04380	.04776	.05172	.05567	.05962	.06356	.06749	.07142	.07535
0.2	.07926	.08317	.08706	.09095	.09483	.09871	.10257	.10642	.11026	.11409
0.3	.11791	.12172	.12552	.12930	.13307	.13683	.14058	.14431	.14803	.15173
0.4	.15542	.15910	.16276	.16640	.17003	.17364	.17724	.18082	.18439	.18793
0.5	.19146	.19497	.19847	.20194	.20540	.20884	.21226	.21566	.21904	.22240
0.6	.22575	.22907	.23237	.23565	.23891	.24215	.24537	.24857	.25175	.25490
0.7	.25804	.26115	.26424	.26730	.27035	.27337	.27637	.27935	.28230	.28524
0.8	.28814	.29103	.29389	.29673	.29955	.30234	.30511	.30785	.31057	.31327
0.9	.31594	.31859	.32121	.32381	.32639	.32894	.33147	.33398	.33646	.33891
1.0	.34134	.34375	.34614	.34850	.35083	.35314	.35543	.35769	.35993	.36214
1.1	.36433	.36650	.36864	.37076	.37286	.37493	.37698	.37900	.38100	.38298
1.2	.38493	.38686	.38877	.39065	.39251	.39435	.39617	.39796	.39973	.40147
1.3	.40320	.40490	.40658	.40824	.40988	.41149	.41309	.41466	.41621	.41774
1.4	.41924	.42073	.42220	.42364	.42507	.42647	.42786	.42922	.43056	.43189
1.5	.43319	.43448	.43574	.43699	.43822	.43943	.44062	.44179	.44295	.44408
1.6	.44520	.44630	.44738	.44845	.44950	.45053	.45154	.45254	.45352	.45449
1.7	.45543	.45637	.45728	.45818	.45907	.45994	.46080	.46164	.46246	.46327
1.8	.46407	.46485	.46562	.46638	.46712	.46784	.46856	.46926	.46995	.47062
1.9	.47128	.47193	.47257	.47320	.47381	.47441	.47500	.47558	.47615	.47670
2.0	.47725	.47778	.47831	.47882	.47932	.47982	.48030	.48077	.48124	.48169
2.1	.48214	.48257	.48300	.48341	.48382	.48422	.48461	.48500	.48537	.48574
2.2	.48610	.48645	.48679	.48713	.48745	.48778	.48809	.48840	.48870	.48899
2.3	.48928	.48956	.48983	.49010	.49036	.49061	.49086	.49111	.49134	.49158
2.4	.49180	.49202	.49224	.49245	.49266	.49286	.49305	.49324	.49343	.49361
2.5	.49379	.49396	.49413	.49430	.49446	.49461	.49477	.49492	.49506	.49520
2.6	.49534	.49547	.49560	.49573	.49585	.49598	.49609	.49621	.49632	.49643
2.7	.49653	.49664	.49674	.49683	.49693	.49702	.49711	.49720	.49728	.49736
2.8	.49744	.49752	.49760	.49767	.49774	.49781	.49788	.49795	.49801	.49807
2.9	.49813	.49819	.49825	.49831	.49386	.49841	.49846	.49851	.49856	.49861
3.0	.49865	.49869	.49874	.49878	.49882	.49886	.49889	.49893	.49897	.49900
3.1	.49903	.49906	.49910	.49913	.49916	.49918	.49921	.49924	.49926	.49929
3.2	.49931	.49934	.49936	.49938	.49940	.49942	.49944	.49946	.49948	.49950
3.3	.49952	.49953	.49955	.49957	.49958	.49960	.49961	.49962	.49964	.49965
3.4	.49966	.49968	.49969	.49970	.49971	.49972	.49973	.49974	.49975	.49976
3.5	.49977	.49978	.49978	.49979	.49980	.49981	.49981	.49982	.49983	.49983
3.6	.49984	.49985	.49985	.49986	.49986	.49987	.49987	.49988	.49988	.49989
3.7	.49989	.49990	.49990	.49990	.49991	.49991	.49992	.49992	.49992	.49992
3.8	.49993	.49993	.49993	.49994	.49994	.49994	.49994	.49995	.49995	.49995
3.9	.49995	.49995	.49996	.49996	.49996	.49996	.49996	.49996	.49997	.49997
4.0	.49997									

TABLE 7

VALUES OF *t* FOR SELECTED PROBABILITIES

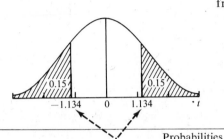

Example.
D (Number of degrees of freedom) = 6:
One tail above $t = 1.134$ *or* below $t = -1.134$ represents 0.15 or 15% of the area under the curve.
Two tails above $t = 1.134$ *and* below $t = -1.134$ represent 0.30 or 30%.

Probabilities
(or Areas Under *t* Distribution Curve)

One tail	.45	.35	.25	.15	.10	.05	.025	.01	.005
Two tails	.90	.70	.50	.30	.20	.10	.05	.02	.01
D						Values of *t*			
1	.158	.510	1.000	1.963	3.078	6.314	12.706	31.821	63.657
2	.142	.445	.816	1.386	1.886	2.920	4.303	6.965	9.925
3	.137	.424	.765	1.250	1.638	2.353	3.182	4.541	5.841
4	.134	.414	.741	1.190	1.533	2.132	2.776	3.747	4.604
5	.132	.408	.727	1.156	1.476	2.015	2.571	3.365	4.032
6	.131	.404	.718	1.134	1.440	1.943	2.447	3.143	3.707
7	.130	.402	.711	1.119	1.415	1.895	2.365	2.998	3.499
8	.130	.399	.706	1.108	1.397	1.860	2.306	2.896	3.355
9	.129	.398	.703	1.100	1.383	1.833	2.262	2.821	3.250
10	.129	.397	.700	1.093	1.372	1.812	2.228	2.764	3.169
11	.129	.396	.697	1.088	1.363	1.796	2.201	2.718	3.106
12	.128	.395	.695	1.083	1.356	1.782	2.179	2.681	3.055
13	.128	.394	.694	1.079	1.350	1.771	2.160	2.650	3.012
14	.128	.393	.692	1.076	1.345	1.761	2.145	2.624	2.977
15	.128	.393	.691	1.074	1.341	1.753	2.131	2.602	2.947
16	.128	.392	.690	1.071	1.337	1.746	2.120	2.583	2.921
17	.128	.392	.689	1.069	1.333	1.740	2.110	2.567	2.898
18	.127	.392	.688	1.067	1.330	1.734	2.101	2.552	2.878
19	.127	.391	.688	1.066	1.328	1.729	2.093	2.539	2.861
20	.127	.391	.687	1.064	1.325	1.725	2.086	2.528	2.845
21	.127	.391	.686	1.063	1.323	1.721	2.080	2.518	2.831
22	.127	.390	.686	1.061	1.321	1.717	2.074	2.508	2.819
23	.127	.390	.685	1.060	1.319	1.714	2.069	2.500	2.807
24	.127	.390	.685	1.059	1.318	1.711	2.064	2.492	2.797
25	.127	.390	.684	1.058	1.316	1.708	2.060	2.485	2.787
26	.127	.390	.684	1.058	1.315	1.706	2.056	2.479	2.779
27	.127	.389	.684	1.057	1.314	1.703	2.052	2.473	2.771
28	.127	.389	.683	1.056	1.313	1.701	2.048	2.467	2.763
29	.127	.389	.683	1.055	1.311	1.699	2.045	2.462	2.756
30	.127	.389	.683	1.055	1.310	1.697	2.042	2.457	2.750
40	.126	.388	.681	1.050	1.303	1.684	2.021	2.423	2.704
60	.126	.387	.679	1.046	1.296	1.671	2.000	2.390	2.660
120	.126	.386	.677	1.041	1.289	1.658	1.980	2.358	2.617
∞	.126	.385	.674	1.036	1.282	1.645	1.960	2.326	2.576

TABLE 8

VALUES OF x^2 FOR SELECTED PROBABILITIES

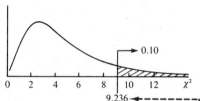

Example.
D (Number of degrees
of freedom) = 5,
the tail above $\chi^2 = 9.236$
represents 0.10 or 10%
of the area under the
curve.

Probabilities
(or Areas Under χ^2 Distribution Curve Above Given χ^2 Values)

D	.90	.70	.50	.30	.20	.10	.05	.02	.01
					Values of χ^2				
1	.016	.148	.455	1.074	1.642	2.706	3.841	5.412	6.635
2	.211	.713	1.386	2.408	3.219	4.605	5.991	7.824	9.210
3	.584	1.424	2.366	3.665	4.642	6.251	7.815	9.837	11.345
4	1.064	2.195	3.357	4.878	5.989	7.779	9.488	11.668	13.277
5	1.610	3.000	4.351	6.064	7.289	9.236	11.070	13.388	15.086
6	2.204	3.828	5.348	7.231	8.558	10.645	12.592	15.033	16.812
7	2.833	4.671	6.346	8.383	9.803	12.017	14.067	16.622	18.475
8	3.490	5.527	7.344	9.524	11.030	13.362	15.507	18.168	20.090
9	4.168	6.393	8.343	10.656	12.242	14.684	16.919	19.679	21.666
10	4.865	7.267	9.342	11.781	13.442	15.987	18.307	21.161	23.209
11	5.578	8.148	10.341	12.899	14.631	17.275	19.675	22.618	24.725
12	6.304	9.034	11.340	14.011	15.812	18.549	21.026	24.054	26.217
13	7.042	9.926	12.340	15.119	16.985	19.812	22.362	25.472	27.688
14	7.790	10.821	13.339	16.222	18.151	21.064	23.685	26.873	29.141
15	8.547	11.721	14.339	17.322	19.311	22.307	24.996	28.259	30.578
16	9.312	12.624	15.338	18.418	20.465	23.542	26.296	29.633	32.000
17	10.085	13.531	16.338	19.511	21.615	24.769	27.587	30.995	33.409
18	10.865	14.440	17.338	20.601	22.760	25.989	28.869	33.346	34.805
19	11.651	15.352	18.338	21.689	23.900	27.204	30.144	33.687	36.191
20	12.443	16.266	19.337	22.775	25.038	28.412	31.410	35.020	37.566
21	13.240	17.182	20.337	23.858	26.171	29.615	32.671	36.343	38.932
22	14.041	18.101	21.337	24.939	27.301	30.813	33.924	37.659	40.289
23	14.848	19.021	22.337	26.018	28.429	32.007	35.172	38.968	41.638
24	15.659	19.943	23.337	27.096	29.553	33.196	36.415	40.270	42.980
25	16.473	20.867	24.337	28.172	30.675	34.382	37.652	41.566	44.314
26	17.292	21.792	25.336	29.246	31.795	35.563	38.885	42.856	45.642
27	18.114	22.719	26.336	30.319	32.912	36.741	40.113	44.140	46.963
28	18.939	23.647	27.336	31.391	34.027	37.916	41.337	45.419	48.278
29	19.768	24.577	28.336	32.461	35.139	39.087	42.557	46.693	49.588
30	20.599	25.508	29.336	33.530	36.250	40.256	43.773	47.962	50.892

TABLE 9A

VALUES OF F FOR UPPER 5% PROBABILITY
(OR 5% AREA UNDER F DISTRIBUTION CURVE)

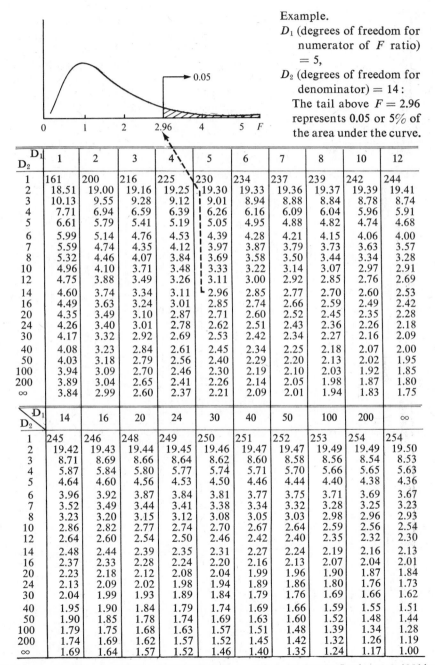

Example.
D_1 (degrees of freedom for numerator of F ratio) = 5,

D_2 (degrees of freedom for denominator) = 14: The tail above $F = 2.96$ represents 0.05 or 5% of the area under the curve.

D_1 / D_2	1	2	3	4	5	6	7	8	10	12
1	161	200	216	225	230	234	237	239	242	244
2	18.51	19.00	19.16	19.25	19.30	19.33	19.36	19.37	19.39	19.41
3	10.13	9.55	9.28	9.12	9.01	8.94	8.88	8.84	8.78	8.74
4	7.71	6.94	6.59	6.39	6.26	6.16	6.09	6.04	5.96	5.91
5	6.61	5.79	5.41	5.19	5.05	4.95	4.88	4.82	4.74	4.68
6	5.99	5.14	4.76	4.53	4.39	4.28	4.21	4.15	4.06	4.00
7	5.59	4.74	4.35	4.12	3.97	3.87	3.79	3.73	3.63	3.57
8	5.32	4.46	4.07	3.84	3.69	3.58	3.50	3.44	3.34	3.28
10	4.96	4.10	3.71	3.48	3.33	3.22	3.14	3.07	2.97	2.91
12	4.75	3.88	3.49	3.26	3.11	3.00	2.92	2.85	2.76	2.69
14	4.60	3.74	3.34	3.11	2.96	2.85	2.77	2.70	2.60	2.53
16	4.49	3.63	3.24	3.01	2.85	2.74	2.66	2.59	2.49	2.42
20	4.35	3.49	3.10	2.87	2.71	2.60	2.52	2.45	2.35	2.28
24	4.26	3.40	3.01	2.78	2.62	2.51	2.43	2.36	2.26	2.18
30	4.17	3.32	2.92	2.69	2.53	2.42	2.34	2.27	2.16	2.09
40	4.08	3.23	2.84	2.61	2.45	2.34	2.25	2.18	2.07	2.00
50	4.03	3.18	2.79	2.56	2.40	2.29	2.20	2.13	2.02	1.95
100	3.94	3.09	2.70	2.46	2.30	2.19	2.10	2.03	1.92	1.85
200	3.89	3.04	2.65	2.41	2.26	2.14	2.05	1.98	1.87	1.80
∞	3.84	2.99	2.60	2.37	2.21	2.09	2.01	1.94	1.83	1.75

D_1 / D_2	14	16	20	24	30	40	50	100	200	∞
1	245	246	248	249	250	251	252	253	254	254
2	19.42	19.43	19.44	19.45	19.46	19.47	19.47	19.49	19.49	19.50
3	8.71	8.69	8.66	8.64	8.62	8.60	8.58	8.56	8.54	8.53
4	5.87	5.84	5.80	5.77	5.74	5.71	5.70	5.66	5.65	5.63
5	4.64	4.60	4.56	4.53	4.50	4.46	4.44	4.40	4.38	4.36
6	3.96	3.92	3.87	3.84	3.81	3.77	3.75	3.71	3.69	3.67
7	3.52	3.49	3.44	3.41	3.38	3.34	3.32	3.28	3.25	3.23
8	3.23	3.20	3.15	3.12	3.08	3.05	3.03	2.98	2.96	2.93
10	2.86	2.82	2.77	2.74	2.70	2.67	2.64	2.59	2.56	2.54
12	2.64	2.60	2.54	2.50	2.46	2.42	2.40	2.35	2.32	2.30
14	2.48	2.44	2.39	2.35	2.31	2.27	2.24	2.19	2.16	2.13
16	2.37	2.33	2.28	2.24	2.20	2.16	2.13	2.07	2.04	2.01
20	2.23	2.18	2.12	2.08	2.04	1.99	1.96	1.90	1.87	1.84
24	2.13	2.09	2.02	1.98	1.94	1.89	1.86	1.80	1.76	1.73
30	2.04	1.99	1.93	1.89	1.84	1.79	1.76	1.69	1.66	1.62
40	1.95	1.90	1.84	1.79	1.74	1.69	1.66	1.59	1.55	1.51
50	1.90	1.85	1.78	1.74	1.69	1.63	1.60	1.52	1.48	1.44
100	1.79	1.75	1.68	1.63	1.57	1.51	1.48	1.39	1.34	1.28
200	1.74	1.69	1.62	1.57	1.52	1.45	1.42	1.32	1.26	1.19
∞	1.69	1.64	1.57	1.52	1.46	1.40	1.35	1.24	1.17	1.00

Reproduced by permission from *Statistical Methods*, 5th ed., by George Snedecor, © 1956 by the Iowa State University Press.

VALUES OF F FOR UPPER 1% PROBABILITY

(OR 1% AREA UNDER F DISTRIBUTION CURVE)

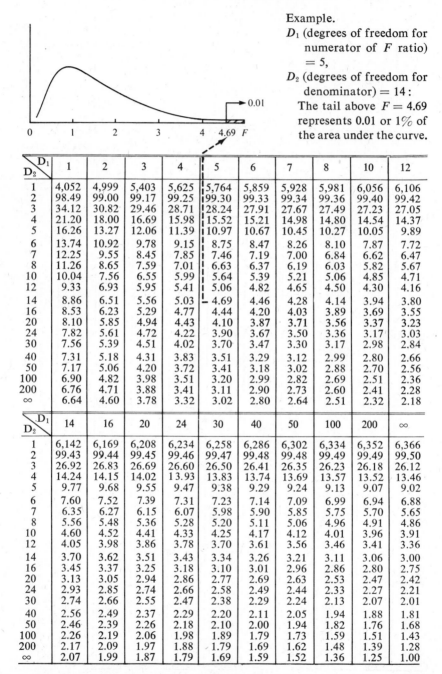

Example.
D_1 (degrees of freedom for numerator of F ratio) = 5,
D_2 (degrees of freedom for denominator) = 14:
The tail above $F = 4.69$ represents 0.01 or 1% of the area under the curve.

D_1 / D_2	1	2	3	4	5	6	7	8	10	12
1	4,052	4,999	5,403	5,625	5,764	5,859	5,928	5,981	6,056	6,106
2	98.49	99.00	99.17	99.25	99.30	99.33	99.34	99.36	99.40	99.42
3	34.12	30.82	29.46	28.71	28.24	27.91	27.67	27.49	27.23	27.05
4	21.20	18.00	16.69	15.98	15.52	15.21	14.98	14.80	14.54	14.37
5	16.26	13.27	12.06	11.39	10.97	10.67	10.45	10.27	10.05	9.89
6	13.74	10.92	9.78	9.15	8.75	8.47	8.26	8.10	7.87	7.72
7	12.25	9.55	8.45	7.85	7.46	7.19	7.00	6.84	6.62	6.47
8	11.26	8.65	7.59	7.01	6.63	6.37	6.19	6.03	5.82	5.67
10	10.04	7.56	6.55	5.99	5.64	5.39	5.21	5.06	4.85	4.71
12	9.33	6.93	5.95	5.41	5.06	4.82	4.65	4.50	4.30	4.16
14	8.86	6.51	5.56	5.03	4.69	4.46	4.28	4.14	3.94	3.80
16	8.53	6.23	5.29	4.77	4.44	4.20	4.03	3.89	3.69	3.55
20	8.10	5.85	4.94	4.43	4.10	3.87	3.71	3.56	3.37	3.23
24	7.82	5.61	4.72	4.22	3.90	3.67	3.50	3.36	3.17	3.03
30	7.56	5.39	4.51	4.02	3.70	3.47	3.30	3.17	2.98	2.84
40	7.31	5.18	4.31	3.83	3.51	3.29	3.12	2.99	2.80	2.66
50	7.17	5.06	4.20	3.72	3.41	3.18	3.02	2.88	2.70	2.56
100	6.90	4.82	3.98	3.51	3.20	2.99	2.82	2.69	2.51	2.36
200	6.76	4.71	3.88	3.41	3.11	2.90	2.73	2.60	2.41	2.28
∞	6.64	4.60	3.78	3.32	3.02	2.80	2.64	2.51	2.32	2.18

D_1 / D_2	14	16	20	24	30	40	50	100	200	∞
1	6,142	6,169	6,208	6,234	6,258	6,286	6,302	6,334	6,352	6,366
2	99.43	99.44	99.45	99.46	99.47	99.48	99.48	99.49	99.49	99.50
3	26.92	26.83	26.69	26.60	26.50	26.41	26.35	26.23	26.18	26.12
4	14.24	14.15	14.02	13.93	13.83	13.74	13.69	13.57	13.52	13.46
5	9.77	9.68	9.55	9.47	9.38	9.29	9.24	9.13	9.07	9.02
6	7.60	7.52	7.39	7.31	7.23	7.14	7.09	6.99	6.94	6.88
7	6.35	6.27	6.15	6.07	5.98	5.90	5.85	5.75	5.70	5.65
8	5.56	5.48	5.36	5.28	5.20	5.11	5.06	4.96	4.91	4.86
10	4.60	4.52	4.41	4.33	4.25	4.17	4.12	4.01	3.96	3.91
12	4.05	3.98	3.86	3.78	3.70	3.61	3.56	3.46	3.41	3.36
14	3.70	3.62	3.51	3.43	3.34	3.26	3.21	3.11	3.06	3.00
16	3.45	3.37	3.25	3.18	3.10	3.01	2.96	2.86	2.80	2.75
20	3.13	3.05	2.94	2.86	2.77	2.69	2.63	2.53	2.47	2.42
24	2.93	2.85	2.74	2.66	2.58	2.49	2.44	2.33	2.27	2.21
30	2.74	2.66	2.55	2.47	2.38	2.29	2.24	2.13	2.07	2.01
40	2.56	2.49	2.37	2.29	2.20	2.11	2.05	1.94	1.88	1.81
50	2.46	2.39	2.26	2.18	2.10	2.00	1.94	1.82	1.76	1.68
100	2.26	2.19	2.06	1.98	1.89	1.79	1.73	1.59	1.51	1.43
200	2.17	2.09	1.97	1.88	1.79	1.69	1.62	1.48	1.39	1.28
∞	2.07	1.99	1.87	1.79	1.69	1.59	1.52	1.36	1.25	1.00

Answers to Odd-Numbered Problems

Exercise 2 (p. 36)

7. (a) $1 + 11 = 100$ (b) $10 + 100 + 101 = 1011$

(c) $110 + 111 + 1000 = 10101$ (d) $10 \times 11 = 110$

(e) $100 \times 101 = 10100$ (f) $110 \times 111 = 101010$

Exercise 3 (p. 62)

5. (a) 96; 44; 3; 622; 218.

(b) 53,000; 67,000; 318,000; 2,674,000; 7,986,000.

15. (a) 1 sq. in. (1 inch each side); 2.38 sq. in. (1.5 inch each side).

(b) .785 sq. in. (r = 0.5 inch); 1.767 sq. in. (r = 0.75 inch).

17. 0.125 cu. in. (0.50 inch each side); 0.5 cu. in. (0.79 inch each side).

Exercise 4-1 (p. 103)

1. 4   **3.** -5 **5.** $\dfrac{y - a}{b}$ **7.** 20

9. $\dfrac{5y - 7z}{4d}$ **11.** $x = 3$, $y = 2$.   **13.** $x = -6$, $y = 5$.   **15.** $x = 2$, $y = 9$.

17. $x = -2$, $y = 3$.   **19.** $x = 1$, $y = 2$, $z = -3$.

Exercise 4-2 (p. 104)

1. $10 : 40 : 35 : 15$ **3.** $\dfrac{72}{13} = 5\dfrac{7}{13}$ **5.** $\dfrac{21}{22}$

7. 0.0052 **9.** 0.43 **11.** 86.00

13. $\dfrac{3}{50}$ **15.** $\dfrac{3,186}{100} = 31\dfrac{43}{50}$ **17.** $\dfrac{28}{100,000} = \dfrac{7}{25,000}$

19. 45% **21.** $.32\%$ **23.** 518%

25. 12% **27.** 475% **29.** 12.71428, or $1,271\%$

31. $\$147.20$ **33.** $\$5,625$; $\$3,750$; $\$3,125$.   **35.** 70%

37. 75 **39.** $\$2,500$

Exercise 4-3 (p. 105)

1. $3^6 = 729$ **3.** $4^{-1} = \dfrac{1}{4}$ **5.** $a^0 = 1$

7. $15^2 = 225$ **9.** $(ab)^4$ **11.** $c^{2/5}$

13. $3^3 = 27$ **15.** a^3 **17.** $-20x^3y^2z^3$

19. $\dfrac{-9yz^2}{w^2}$ **21.** $\sqrt{a}$ **23.** $\sqrt[7]{x^4}$

25. $\sqrt[5]{185^2}$ **27.** $b^{2/3}$ **29.** $y^{4/5}$

31. $258^{1/3}$ **33.** $\sqrt{a^3b}$ **35.** $\sqrt[4]{2k^4m}$

37. $\sqrt[4]{162}$ **39.** $6\sqrt{2a}$ **41.** $5\sqrt{3}$

43. $12\sqrt{10}$ **45.** $-90\sqrt{55}$ **47.** $30\sqrt{210}$

49. 2 **51.** $\dfrac{\sqrt{21}}{3}$ **53.** 0.24

55. 15 **57.** 137 **59.** 13.4

Exercise 4-4 (p. 106)

1. 2.5328 **3.** 0.4942 **5.** 3.8820 **7.** $9.9138 - 10$
9. 7.42 **11.** 352 **13.** 0.452 **15.** 0.00291
17. 3.1623 **19.** 802.6 **21.** $1,710$ **23.** 13.3
25. $181\ (10^{27})$ **27.** 12.4

Exercise 5 (p. 148)

3. (a) 7, (b) 4.5, (c) 2 and 11.
5. 58   **7.** 82
9. (a) $\$6,000$, (b) $\$6,500$, (c) no   **11.** 18
13. (a) 65, (b) $65\dfrac{1}{7}$, (c) (1) 65, (2) and (3) $65\dfrac{1}{3}$, (4) $65\dfrac{3}{7}$.

Exercise 6 (p. 169)

3. (a) 7.2, (b) 5.83, (c) 4.61
5. (a) 0.2, 5, 0.5, 2; (b) 1; (c) 1.7; (d) 1.925; (e) 1.7.
7. (a) 68% (or 0.683) increase; (b) 7.04% (or 0.07037) increase.
9. (a) 11.25¢; (b) $10\frac{2}{3}$¢

Exercise 7 (p. 191)

5. (a) (1) 19, (2) 4.5, (3) (I) 4.75, (II) 5.25,
     (4) 6.12.
6. (Used in Problem 11).
     Granby: (1) 90, (2) 18.75, (3) 21.82 ($\overline{X} = 200$),
         (4) 26.97.
     Norview: (1) 90, (2) 30, (3) 27.27 ($\overline{X} = 200$),
         (4) 31.04.
7. 31; 5.57.   **9.** (a) 1.47, (b) 1.56, (c) 2.11.
11. (a) 13.5% (Granby); 15.5% (Norview).

Exercise 8 (p. 208)

3. (a) (1) $-.16$; (2) -0.006.
5. (a) 0.99; (b) 2.8192.   **7.** (a) -0.28; (b) 2.525

Exercise 9-1 (p. 233)

1. (1) Finite; (2) Finite; (3) Infinite; (4) Empty set;
     (5) Finite.
3. (a) f, h, i, j, k, l (b) f, h
     (c) g, j (d) g, i, k, l
5. 16   **7.** 12
9. 126   **11.** 3,024

Exercise 9-2 (p. 234)

1. 4/7   **3.** (a) 5/9; (b) 4/9; (c) 0
5. (a) 1/10; (b) 9/10.   **7.** 1/10.

Exercise 10 (p. 270)

3. $x^5 + 5x^4y + 10x^3y^2 + 10x^2y^3 + 5xy^4 + y^5$
5. (a) 0.216, 0.432; 0.288; 0.064; (b) 1.2; (c) 0.85
7. (a) 0.1296; 0.3456; 0.3456; 0.1536; 0.0256; (b) 0.4 (c) 0.24
9. (a) 0; 0.4; 0.6; 0; 0 (b) 1.6; (c) 0.49
11. In Problem 5: (a) 0.4 (b) 0.49
      In Problem 8: (a) 0.4 (b) 0.49   **15.** (a) 0.0006048
13. 34,650   (b) 0.00036288
17. (a) .27068 (b) 0.09023
19. $0.216 + 0.064 = 0.280$ (smaller than $1/1^2 = 1$)

Exercise 11 (p. 293)

7. (a) 764 (b) 273 (c) 620 (d) 229
9. (a): (1) 0.58581 (2) 0.10565 (3) 0.30854
   (b): (1) $65.2 (2) $54.8

Exercise 12 (p. 319)

5. 64
7. (a) 1.5; 2.0; 2.5(2); 3.0(2); 3.5(2); 4.0; 4.5
   (b) 3 (c) 0.87 (d) (1) 19.215% (2) 20%
   (e) (1) 19.215% (2) 20%
9. (a) 40% (b) 0.19 (c) 29.806%
11. 15.866%

Exercise 13 (p. 343)

1. (a) $74.2 to $125.80
   (b) $74.2 and $125.80
   (c) 99%

3. (a) 35.13% to 54.87%
   (b) 35.13% and 54.87%
   (c) 90%

5. 7 = 7

7. $\dfrac{1}{4} = \dfrac{1}{4}$

9. 19

11. 95.59 to 100.41

13. 256

15. (1) 2,305 (2) 2,401

Exercise 14 (p. 365)

7. (a) $0.33147 + 0.49846 = 0.82993$; $1 - 0.82993 = 0.17007$
   (b) $0.50000 - 0.20540 = 0.29460$; $1 - 0.29460 = 0.70540.$

Exercise 15 (p. 384)

3. $z = 1.8$; yes.
5. $z = -3$; (a) and (b): The company's claim is overstated.
7. $z = 2.2$. (a) There is no significant difference.
             (b) There is a significant difference.
9. $z = -1.69$. The claim is not valid.
11. $z = 1.06$. Accept the hypothesis.

Exercise 16 (p. 414)

11. UCL = 64 hours; LCL = 40 hours.
13. (a): (1) UCL = 0.13948, LCL = 0.02852;
         (2) UCL = 0.12561, LCL = 0.04239;
         (3) UCL = 0.16722. LCL = 0.00078.

15. (a) $\bar{\bar{X}} = 35$, $\bar{R} = 38$.   UCL = 56.926 (or 0.556926 inches)
                                   LCL = 13.074 (or 0.513074 inches)
   (b) UCL = 80.37 (or 0.58037 inches)
   LCL = 0

17. (a) $\bar{p} = 0.05$,   (b) $\overline{np} = 25$,
    UCL = 0.07924,   UCL = 39.62,
    LCL = 0.02076.   LCL = 10.38.

19. The first sample size is 80 items. He should take another sample of 80 items.

Exercise 17-1 (p. 447)

1. Coin X is fair: $\chi^2 = 1.96$.   Coin Y is not fair: $\chi^2 = 4$.
3. $\chi^2 = 18.656$. The difference is significant.
5. $\chi^2 = 2.842$. The claim is not valid.
7. $\chi^2 = 4.1999$. Accept the hypothesis.

Exercise 17-2 (p. 449)

1. $z = -1.25$. (a) and (b): There is no significant difference.
3. $z = -1.44778$. (a) and (b): There is no significant difference.
5. $z = -0.88$. There is no significant difference.
7. $H = 0.0364$. The means are not significantly different.
9. $z = -1.46$. Yes, this sample was taken at random.

Exercise 18 (p. 479)

5. $F = 8$. (a) Reject the hypothesis.
            (b) Accept the hypothesis.
7. $F = 1.65$. (a) Accept the hypothesis.
            (b) Accept the hypothesis.
9. $F = 22.79$. Reject the hypothesis.
10. (Used in Problem 11.)
     $S_t = 108; S_r = 10; S_w = 98; F = 0.61$; Accept the hypothesis.
11. (1) $F_r = 1.2$; Accept the hypothesis that the means are equal.
     (2) $F_c = 3.88$; Reject the hypothesis.

Exercise 19-1 (p. 449)

Problems 4 to 10 are continuations of Problem 3.

3. Payoff Table

Event	21	22	Act 23	24	25
		Possible Profit			
21	3.15	2.90	2.65	2.40	2.15
22	3.15	3.30	3.05	2.80	2.55
23	3.15	3.30	3.45	3.20	2.95
24	3.15	3.30	3.45	3.60	3.35
25	3.15	3.30	3.45	3.60	3.75

4.

Weekly Sales, Copies	No. of Weeks	(a) Prob.	(b) Exp. Sales
21	5	0.1	2.1
22	10	0.2	4.4
23	15	0.3	6.9
24	15	0.3	7.2
25	5	0.1	2.5
Total	50	1.0	23.1

5. Expected monetary profits

| | | \multicolumn{10}{c}{Act} | | | | | | | | |
| | | 21 | | 22 | | 23 | | 24 | | 25 | |
Event	Prob.	Con.	Exp.	Con.	Exp.	Con.	Exp.	Con.	Exp.	Con.	Exp.
21	0.1	3.15	0.315	2.90	0.290	2.65	0.265	2.40	0.240	2.15	0.215
22	0.2	3.15	0.630	3.30	0.660	3.05	0.610	2.80	0.560	2.55	0.510
23	0.3	3.15	0.945	3.30	0.990	3.45	1.035	3.20	0.960	2.95	0.885
24	0.3	3.15	0.945	3.30	0.990	3.45	1.035	3.60	1.080	3.35	1.005
25	0.1	3.15	0.315	3.30	0.330	3.45	0.345	3.60	0.360	3.75	0.375
Total	1.0		3.150		3.260		3.290		3.200		2.990

Optimum Act

6. (a) Expected profit under certainty

Event	Probability	Profit	Expected
21	0.1	$3.15	0.315
22	0.2	3.30	0.660
23	0.3	3.45	1.035
24	0.3	3.60	1.080
25	0.1	3.75	0.375
Total	1.0		$3.465

(b) Expected value of perfect information:

$$\begin{array}{r} 3.465 \\ -\ 3.290 \\ \hline \$\ .175 \end{array}$$

7. (a) Conditional loss table

| | | \multicolumn{5}{c}{Act} |
Event	21	22	23	24	25
		\multicolumn{5}{c}{Conditional Loss}
21	0	0.25	0.50	0.75	1.00
22	0.15	0	0.25	0.50	0.75
23	0.30	0.15	0	0.25	0.50
24	0.45	0.30	0.15	0	0.25
25	0.60	0.45	0.30	0.15	0

Cost losses:

$0.25 × copies overstocked

Profit loss per copy:

$0.40 − $0.25 = $0.15

(b) Expected monetary losses

| | | \multicolumn{10}{c}{Act} | | | | | | | | |
| | | 21 | | 22 | | 23 | | 24 | | 25 | |
Event	Prob.	Con.	Exp.	Con.	Exp.	Con.	Exp.	Con.	Exp.	Con.	Exp.
21	0.1	0	0	0.25	0.025	0.50	0.050	0.75	0.075	1.00	0.100
22	0.2	0.15	0.030	0	0	0.25	0.050	0.50	0.100	0.75	0.150
23	0.3	0.30	0.090	0.15	0.045	0	0	0.25	0.075	0.50	0.150
24	0.3	0.45	0.135	0.30	0.090	0.15	0.045	0	0	0.25	0.075
25	0.1	0.60	0.060	0.45	0.045	0.30	0.030	0.15	0.015	0	0
Total	1.0		0.315		0.205		0.175		0.265		0.475

Optimum Act
& EVPI
(c)

8. Payoff Table

Event	21	22	23	24	25
			Act		
			Possible Profit		
21	3.15	2.93	2.71	2.49	2.27
22	3.15	3.30	3.08	2.86	2.64
23	3.15	3.30	3.45	3.23	3.01
24	3.15	3.30	3.45	3.60	3.38
25	3.15	3.30	3.45	3.60	3.75

9. (a) Expected monetary profits

		21		22		23		24		25	
					Act						
Event	Prob.	Con.	Exp.	Con.	Exp.	Con.	Exp.	Con.	Exp.	Con.	Exp.
21	0.1	3.15	0.315	2.93	0.293	2.71	0.271	2.49	0.249	2.27	0.227
22	0.2	3.15	0.630	3.30	0.660	3.08	0.616	2.86	0.572	2.64	0.528
23	0.3	3.15	0.945	3.30	0.990	3.45	1.035	3.23	0.969	3.01	0.903
24	0.3	3.15	0.945	3.30	0.990	3.45	1.035	3.60	1.080	3.38	1.014
25	0.1	3.15	0.315	3.30	0.330	3.45	0.345	3.60	0.360	3.75	0.375
Total	1.0		3.150		3.263		3.302		3.230		3.047

Optimum Act

(b) Expected
Value of perfect information:
$$\begin{array}{r} 3.465 \\ -\ 3.302 \\ \hline \$0.163 \end{array}$$

10. (a) Conditional loss table

Event	21	22	23	24	25
		Act			
		Conditional Loss			
21	0	0.22	0.44	0.66	0.88
22	0.15	0	0.22	0.44	0.66
23	0.30	0.15	0	0.22	0.44
24	0.45	0.30	0.15	0	0.22
25	0.60	0.45	0.30	0.15	0

Cost losses:

($0.25 − $0.03) ×
copies overstocked = $0.22 × copies ov.

Profit loss = $0.15
(Same as Problem 7.)

(b) Expected monetary losses

		21		22		23		24		25	
						Act					
Event	Prob.	Con.	Exp.	Con.	Exp.	Con.	Exp.	Con.	Exp.	Con.	Exp.
21	0.1	0	0	0.22	0.022	0.44	0.044	0.66	0.066	0.88	0.088
22	0.2	0.15	0.030	0	0	0.22	0.044	0.44	0.088	0.66	0.132
23	0.3	0.30	0.090	0.15	0.045	0	0	0.22	0.066	0.44	0.132
24	0.3	0.45	0.135	0.30	0.090	0.15	0.045	0	0	0.22	0.066
25	0.1	0.60	0.060	0.45	0.045	0.30	0.030	0.15	0.015	0	0
Total	1.0		0.135		0.202		0.163		0.235		0.418

Optimum Act
& EVPI
(c)

Exercise 19-2 (p. 500)

1. 40   **3.** 44   **5.** $86\frac{2}{3}$ **7.** -140

9. 37.5%; 62.5%.

10. (a) $0, not buy   (b) $0, not buy   (c) $-$$100, buy

    (d) $0, not buy

Exercise 20 (p. 530)

1. (a) P (red) = 18/30 = 3/5 P (blue) = 12/30 = 2/5

    P (yellow) = 15/30 = 1/2 P (green) = 15/30 = 1/2

    P (red and yellow) = 12/30 = 2/5

    P (red and green) = 6/30 = 1/5

    P (blue and yellow) = 3/30 = 1/10

    P (blue and green) = 9/30 = 3/10

    (b) P (red | yellow) = 12/15 = 4/5

    P (blue | yellow) = 3/15 = 1/5

3. 0.2727; 0.3636; 0.2273; 0.1364

5. X: 0.333   Y: 0.667

7. X: 0.5   Y: 0.5

8. (Used in Problem 9.)

    0.493; 0.076; 0.408; 0.023.

9. Expected, prior distribution = 0.067

    Expected, posterior distribution = 0.11947

    s = 0.034

11.

Midpoint	Probability	
0.070	0.0001	9495
0.075	0.0072	1318
0.080	0.0797	3487
0.085	0.2921	3374
0.090	0.3929	2329

Exercise 21 (p. 552)

1. (a) 1.20 or 120%

    (b) 0.80 or 80%

    (c) 0.96 or 96%

3.

Year	Relatives
1973	45%
1974	68%
1975	91%
1976	109%
1977	127%

5. (a) 106.0%

    (b) 71.5%

    (c) 75.8%

9. (a) 106.0%

    (b) 71.48%

7. (a) 94.31%

    (b) 0.999686

11. 1972: 0.83963

    1973: 0.74239

    1974: 0.58309

Exercise 22-1 (p. 601)

3. (a) 1. Freehand drawing a straight line.

    2. $Y_c = 4 + 1.7778X$ (approximately)

    3. 1970: $4,000.

      1972: $7,555.60.

      1976: $14,666.80.

(b) 1. Point between 1971 and 1972, or January 1, 1972:   6.5.
         Point between 1976 and 1977, or January 1, 1977: 15.5
    2. $Y_c = 15.5 + 0.9X$
    3. 1970: $3,800.
       1972: $7,400.
       1976: $14,600.
3. (c) 1. $Y_c = 11.1111 + 1.7833X$
    2. 1970: $3,977.9
       1972: $7,544.50
       1976: $14,677.70
    3. A straight line
5. (b) $Y_c = 55.7 + 2.7303X$

Exercise 22-2 (p. 603)

3. (a) 1. $\log Y_c = 0.97796 + 0.09510X$
    2. 1970: 3.96; 1978: 22.80
   (b) 1. $\log Y_c = 1.2182 + (-0.8913)(0.5915^X)$
    2. 1970: 2.12; 1978: 16.03
5. (a) $Y_c = 42 + 2.88X$
   (b) 1. $Y_c = 6.99 + 0.18X$
    2. $Y_c = 2.31 + 0.02X$
   (c) $Y_c = 138 + 14.4X$

Exercise 23 (p. 628)

1. 477,250; 546,000
3. 73.5%, 120.3%, 73.6%, 132.6%
5. 1976: 26.8, 30.1, 33.3, 30.6
7. Possible solution:

| Quarter | Index of Each Year | | | | | | | Annual change |
	1972	1973	1974	1975	1976	1977	1978	
1st	64	67	70	73	76	79	82	+3
2nd	133	129.5	126	122.5	119	115.5	112	−3.5
3rd	60	64.5	69	73.5	78	82.5	87	+4.5
4th	143	139	135	131	127	123	119	−4
Total	400	400.0	400	400.0	400	400.0	400	0
Average	100	100.0	100	100.0	100	100.0	100	0

Exercise 24 (p. 648)

3. (a)

Year	Trend Value ($1,000)	Sales Adjusted for Trend %
1970	4.0	50
1971	5.8	103
1972	7.5	93
1973	9.3	118
1974	11.1	108
1975	12.9	116
1976	14.7	95
1977	16.5	103
1978	18.2	88

5. (a) (1) See column (7), Y/S, in Table 23-19 and the
solution of Exercise 23, Problem 5.
(2) 1972-first quarter: 0.5403 or 54%.
1978-fourth quarter: 0.8697 or 87%.
(3) 1972-second quarter: 81%
1978-third quarter: 104%
7. (a) (1) 1974-January: 5.588
1978-December: 4.549
(2) 1974-January: 96%
1978-December: 98%
(3) 1974-February: 94%
1978-November: 100%

Exercise 25 (p. 674)

3. (b) $Y_c = 2.69 + 2.07X$ (d) 9
(e) 3.34   (f) 5.56 to 12.24
(g) $r^2 = 0.673773$, $r = 0.82$
5. (b) 1. $Y_c = -3.28 + 0.89X$ 2. $X_c = 32.06 + 0.59Y$
7. $74,100.
9. (a) $Y_c = 6.55 + 1.38X$
(d) 2.27   (c) $r = 0.7106$; $r^2 = 0.5049$

Exercise 26 (p. 702)

3. (a) $Y_c = -1.3 + 2.1X$ (b) 0.8

(c)

X	Lower limit	Upper limit
1	-1.17	2.77
2	1.50	4.30
3	3.85	6.15
4	5.70	8.50
5	7.23	11.17

(d)

X	Lower limit	Upper limit
1	-2.38	3.98
2	0.00	5.80
3	2.23	7.77
4	4.20	10.00
5	6.02	12.38

4. (Continued from Problem 3.)   (a) 0.95; 0.97
5. (a) $Y_c = 2.69 + 2.07X$ (b) 4.31
(c) 0.57 to 13.09   (d) -5.10 to 18.76
6. (Continued from Problem 5.)   (a) 0.56; 0.75
7. (a) $Y_c = -3.28 + 0.89X$ (b) 13.92
(c) 41 to 77.04   (d) 6.95 to 101.09
(e) 0.456; 0.68
9. (a) $H_o: B = 0$.   Reject H_o for both (1) and (2).
(b) (1) 1.30 to 2.90
(2) 0.63 to 3.57

Exercise 27 (p. 726)

7. (b) $Y_c = -2.1529 + 5.9009X - 0.4883X^2$
8. (Continued from Problem 7.)
   (a) 2.62   (b) 0.80; 0.89
9. A possible solution:

Student	(c) Y_c	(d) $Y - Y_c$
A	95	0
B	58	-7
C	52	-3
D	39	-12
E	47	-5
F	72	-20
G	55	12
H	38	10
I	33	13
Total	489	-12

(e) 10.75

(f) 0.635; 0.80

(g) J: 74
    K: 65

10. (Used in Problem 11.)
    $\eta^2 = 0.5077$; $\eta = 0.71$

13. (1) (a) $A^{-1} = \begin{bmatrix} 2/17 & 7/34 \\ 1/17 & -5/34 \end{bmatrix}$ (b) $x = -2$
                                                                                       $y = 3$

(2) (a)
    $A^{-1} = \begin{bmatrix} 2/11 & 3/11 \\ 3/11 & -1/11 \end{bmatrix}$ (b) $x = 4$
                                                                                 $y = -1$

(3) (a)
    $A^{-1} = \begin{bmatrix} 0 & 3/20 & 1/20 \\ 1/3 & -7/60 & 11/60 \\ 1/3 & -1/60 & -7/60 \end{bmatrix} = \dfrac{1}{60}\begin{bmatrix} 0 & 9 & 3 \\ 20 & -7 & 11 \\ 20 & -1 & -7 \end{bmatrix}$

    (b) $x = 1; y = 2; z = -3$

(4) (a)
    $A^{-1} = \begin{bmatrix} 1/2 & -1/14 & 5/14 \\ 0 & -2/7 & 3/7 \\ 1/2 & -3/14 & 1/14 \end{bmatrix} = \dfrac{1}{14}\begin{bmatrix} 7 & -1 & 5 \\ 0 & -4 & 6 \\ 7 & -3 & 1 \end{bmatrix}$

    (b) $x = 3; y = -5; z = 7$

Exercise 28 (p. 741)

1. (a) $X_2(13)c = a_2(13) + b_{21.3}X_1 + b_{23.1}X_3$
   (b) $X_3(12)c = a_3(12) + b_{31.2}X_1 + b_{32.1}X_2$
3. (a) $X_{1c} = 0.8396 + 1.0226X_2 + 3.1328X_3$
   (b) 2.31   (c) 0.8444; 0.92
5. (a) 0.523   (b) 0.72
6. (Used in Problem 7.)
   (a) $X_{1c} = -9.4271 + 0.0159X_2 + 1.0903X_3$
   (b) 2.01   (c) 0.9873; 0.99
7. Estimated grade: 90
   Reliability: 90 $\pm$ 2.01 at 68%
                  90 $\pm$ 4.02 at 95%
                  90 $\pm$ 6.03 at 99%
9. $r_k = 0.82$

Selected References

Current Statistical Data in Business and Economics

See Section 1.5 Collecting Published Data of Chapter 1 under the following two headings:
(1) Guides for Locating Published Data, pages 11-12.
(2) Sources of Published Data, pages 12-15.

General References

Croxton, Frederick E., Dudley J. Cowden, and Sidney Klein. *Applied General Statistics*, 3rd ed. Englewood Cliffs, N.J.: Prentice-Hall, Inc., 1967.

Dixon, Wilfrid J. and Frank J. Massey. *Introduction to Statistical Analysis*, 3rd ed. New York: McGraw-Hill Book Company, 1969.

Hernett, Donald L. *Introduction to Statistical Methods*. Reading, Mass.: Addison-Wesley Publishing Co., Inc., 1970.

Hoel, Paul G. *Statistics*. New York: John Wiley & Sons, Inc., 1966.

Hogg, Robert V. and Allen T. Craig. *Introduction to Mathematical Statistics*, 2nd ed. New York: The Macmillan Company, 1965.

Huff, Darrell. *How to Lie with Statistics*. New York: W. W. Norton & Company, Inc., 1954.

RAND Corporation. *A Million Random Digits with 100,000 Normal Deviates*. New York: Free Press of Clencoe, 1955.

Schmid, Calvin F. *Handbook of Graphic Presentation*. New York: The Ronald Press Company, 1954.

Yamane, Taro. *Statistics: An Introductory Analysis*, 3rd ed. New York: Harper & Row, Publishers, 1973.

Statistical Induction

Alder, Henry L. and Edward B. Roessler. *Introduction to Probability and Statistics*, 3rd. ed. San Francisco: W. H. Freeman & Co., Publishers, 1964.

Cochran, William G. *Sampling Techniques,* 2nd. ed. New York: John Wiley & Sons, Inc., 1963.

Deming, William E. *Sample Design in Business Research.* New York: John Wiley & Sons, Inc., 1960.

_____ *Some Theories of Sampling.* New York: John Wiley & Sons, Inc., 1950.

Ewart, Park J., James S. Ford, and Chi-Yuan Lin. *Probability for Statistical Decision Making.* Englewood Cliffs, N. J.: Prentice-Hall, 1974.

Fisher, Ronald A. *Statistical Methods and Scientific Inference.* London: Oliver & Boyd, Ltd., 1956.

_____ *Statistical Methods for Research Workers*, 13th ed. New York: Hafner Publishing Co., Inc., 1958.

Goldberg, Samuel. *Probability: An Introduction.* Englewood Cliffs, N. J.: Prentice-Hall, Inc., 1960.

Guenther, William C. *Concepts of Statistical Inference.* New York: McGraw-Hill Book Company, 1965.

Hansen, Morris H., William N. Hurwitz, and William G. Madow. *Sampling Survey Methods and Theory.* New York: John Wiley & Sons, Inc., 1953.

Keeping, E. S. *Introduction to Statistical Inference.* Princeton, N. J.: D. Van Nostrand Co., Inc., 1962.

Kish, Leslie. *Survey Sampling.* New York: John Wiley & Sons, Inc., 1965.

Larson, H. J. *Introduction to Probability Theory and Statistical Inference.* New York: John Wiley & Sons, Inc., 1969.

Lindgren, B. W. *Statistical Theory.* New York: The Macmillan Company, 1962.

Mendenhall, William. *Introduction to Probability and Statistics.* Belmont, Calif.: Wadsworth Publishing Co., Inc., 1967.

Mosteller, Frederick, Robert E. K. Rourke, and George B. Thomas, Jr. *Probability and Statistical Applications,* 2nd ed. Reading, Mass.: Addison-Wesley Publishing Co., Inc., 1970.

Scheffe, Henry. *The Analysis of Variance.* New York: John Wiley & Sons, Inc., 1959.

Schlaifer, Robert. *Probability and Statistics for Business Decisions.* New York: McGraw-Hill Book Company, 1959.

Shook, Robert C. and Harold J. Highland. *Probability Models: With Business Applications.* Homewood, Ill.: Richard D. Irwin, Inc., 1969.

Snedecor, George W. *Statistical Methods.* 6th ed. Ames, Iowa: Iowa State University Press, 1967.

Statistical Quality Control

Cowden, Dudley J. *Statistical Methods in Quality Control.* Englewood Cliffs, N. J.: Prentice-Hall, Inc., 1957.

Duncan, Acheson J. *Quality Control and Industrial Statistics,* 3rd ed. Homewood, Ill.: Richard D. Irwin, Inc., 1965.

Grant, Eugene L. *Statistical Quality Control,* 3rd ed. New York: McGraw-Hill Book Company, 1964.

Shewhart, W. A. *Economic Control of Quality of Manufactured Product.* Princeton, N. J.: D. Van Nostrand Co., Inc., 1931.

Statistical Research Group, Columbia University. *Sampling Inspection.* New York: McGraw-Hill Book Company, 1948.

United States Department of Defense. *Sampling Procedures and Tables for Inspection by Attributes* (MIL-STD-105D). Washington, D. C., 1963.

Decision Theory

Alwan, A. J., and D. G. Parisi. *Quantitative Methods for Decision Making.* Morristown, N. J.: General Learning Press, 1974.

Bierman, Harold, Jr., Charles P. Bonini, and Warren H. Hausman. *Quantitative Analysis for Business Decisions,* 4th ed. Homewood, Ill.: Richard D. Irwin, Inc., 1973.

Bradley, James V. *Distribution-Free Statistical Tests.* Englewood Cliffs, N. J.: Prentice-Hall, 1968.

Fishburn, Peter C. *Utility Theory for Decision Making.* New York: John Wiley & Sons, Inc., 1970.

Levin, Richard I., and C. A. Kirkpatrick. *Quantitative Approaches to Management.* New York: McGraw-Hill Book Company, 1965.

Luce, R. Duncan, and Howard Raiffa. *Games and Decisions.* New York: John Wiley & Sons, Inc., 1957.

Martin, J. J. *Bayesian Decision Problems and Markov Chains.* New York: John Wiley & Sons, Inc., 1967.

Morgan, Bruce W. *An Introduction to Bayesian Statistical Decision Processes.* Englewood Cliffs, N. J.: Prentice-Hall, Inc., 1968.

Raiffa, Howard. *Decision Analysis.* Reading, Mass.: Addison-Wesley Publishing Co., Inc., 1968.

Richmond, Samuel B. *Operations Research for Management Decisions.* New York: The Ronald Press Company, 1968.

Schlaifer, Robert O. *Analysis of Decisions Under Uncertainty.* New York: McGraw-Hill Book Company, 1969.

Siegel, Sidney. *Non-Parametric Statistics for the Behavioral Sciences.* New York: McGraw-Hill, 1956.

Thierauf, Robert J., and Robert C. Klekamp. *Decision Making Through Operations Research.* New York: John Wiley & Sons, Inc., 1975.

Trueman, Richard E. *An Introduction to Quantitative Methods for Decision Making.* New York: Holt, Rinehart and Winston, Inc., 1974.

Time Series Analysis

Abramson, Adolph G., and Russell H. Mack. *Business Forecasting in Practice: Principles and Cases.* New York: John Wiley & Sons, Inc., 1956.

Bratt, Elmer C. *Business Cycles and Forecasting.* Homewood, Ill.: Richard D. Irwin, Inc., 1961.

Brown, Robert G. *Smoothing, Forecasting and Prediction of Discrete Time Series.* Englewood Cliffs, N. J.: Prentice-Hall, Inc., 1963.

Dauten, Carl A. *Business Cycles and Forecasting,* 3rd ed. Cincinnati, Ohio: South-Western Publishing Co., 1968.

Fisher, Irving. *The Making of Index Numbers.* Boston: Houghton Mifflin Company, 1927.

Gordon, Robert A. *Business Fluctuations,* 2nd ed. New York: Harper & Row, Publishers, 1961.

Kendrick, John W. *Productivity Trends in the United States.* New York: National Bureau of Economic Research, Inc., 1961.

Kuznets, Simon. *Seasonal Variations in Industry and Trade*. New York: National Bureau of Economic Research, 1933.

Macaulay, Frederick R. *The Smoothing of Time Series*. New York: National Bureau of Economic Research, 1931.

Mitchell, Wesley C. *Business Cycles: The Problem and Its Setting*. New York: National Bureau of Economic Research, Inc., 1927.

_____ *The Making and Using of Index Numbers*. Washington, D. C.: U. S. Government Printing Office, 1938.

_____ *What Happens During Business Cycles: A Progress Report*. New York: National Bureau of Economic Research, 1951.

Moore, Geoffrey H. *Business Cycle Indicators*. New York: National Bureau of Economic Research, 1961.

Mudgett, Bruce D. *Index Numbers*. New York: John Wiley & Sons, 1951.

Nelson, Charles R. *Applied Time Series Analysis for Managerial Forecasting*. San Francisco: Holden-Day, Inc., 1973.

Persons, Warren M. *The Construction of Index Numbers*. Boston: Houghton Mifflin Company, 1928.

United States Department of Commerce, Bureau of the Census. *The X-11 Variant of the Census Method II, Seasonal Adjustment Program* (Technical Paper No. 15). Washington, D. C., 1965.

Wolfe, Harry Deane. *Business Forecasting Methods*. New York: Holt, Rinehart and Winston, Inc., 1966.

Regression and Correlation Analysis

Acton, Forman S. *Analysis of Straight-Line Data*. New York: John Wiley & Sons, Inc., 1959.

Baggaley, Andrew R. *Intermediate Correlational Methods*. New York: John Wiley & Sons., Inc., 1964.

Ezekiel, Mordecai, and Karl A. Fox. *Methods of Correlation and Regression Analysis*. New York: John Wiley & Sons, Inc., 1959.

Neter, John and William Wasserman. *Applied Linear Statistical Models*. Homewood, Ill.: Richard D. Irwin, Inc., 1974.

Williams, E. J. *Regression Analysis*. New York: John Wiley & Sons, Inc., 1959.

Index

Abscissa, 43
Acceptable quality level, 410
Acceptance region, defined, 351
Acceptance sampling, 409
Accounting machine, 28
Acts, 482
Additive model, 559, 612, 644
Adjusted data
   for calender or working day
     variation, 607
   for seasonal variation, 625, 639
   for seasonal variation and trend, 639
   for trend, 634
Adjusting factor, 608
Aggregates, methods of, 541
Algorithmic Language (ALGOL)
American Society for Quality Control,
   386
Analytical table, 37
Arithmetic mean (*see also* mean)
   grouped data, 120
   ungrouped data, 117
   weighted, 120
Array, 111
   frequency, 111
Assumed mean, 119
Attribute, 386
Average
   deviation, 178

of relatives, 544
quality level, 390
seasonal variation, 608
Averages, 110, 152

Bar chart, 49
Bayes' theorem, 231, 502, 506
Bayesian decision rule, 497
Beginner's All-purpose Symbolic
   Instruction Code (BASIC), 35
Bernoulli, Jacob, 237
Bernoulli Process, 236
Biased estimate, 327
Bimodal, 136
Binary indications, 30
Binomial distribution, 240
   characteristics of, 241
Binomial probability density function,
   240
Binomial probability distribution table
   of values, 776
Binomial theorem, 240
Bivariate
   data, 653, 667, 669
   distribution, 667
   frequency table, 669
Block diagram, 34
Bowley, Arthur, 197
Bureau of the Census Catalog, 11

808